Praise for *Throes of Democracy*

"His approach is intelligent, incisive, often original, and commendably thought-provoking. His work belongs on the bookshelves of all serious aficionados of history."   —Jay Winik, *National Review*

"McDougall, a professor of history at the University of Pennsylvania, covers all the major events and social forces of the era with great energy and wit. . . . Because he writes exceedingly well, this makes for a history that is bracing to read. . . . It's useful to have a rebuttal of the progressive views most contemporary historians hold, and McDougall does so with verve."
    —*New York Times Book Review*

"*Throes of Democracy*, covering the years from 1829 to 1877, is the sequel to *Freedom Just Around the Corner*, Mr. McDougall's chronicle of the American experience from 1585 to 1828. Both volumes are unconventional histories brimming with idiosyncratic detail and offering impressionistic and surprisingly powerful sketches of their eras. . . . Among much else, *Throes of Democracy* is a chronicle where myths come to die and where hard historical facts are given an especially bitter twist. There's a lot in here."
    —*Wall Street Journal*

"A broad-ranging portrait of America in a time of torment. . . . McDougall ventures that in the Civil War era, something of the nation's essential nature came through: progressive yet conservative, pious yet sanguinary. . . . Provocative and richly detailed—a welcome contribution to popular history."   —*Kirkus Reviews* (starred review)

"History buffs will definitely gravitate to this thick book. . . . It proves as boisterous as the busy, mid-nineteenth-century Americans whose expanding, industrializing, and warring McDougall chronicles. . . . A provocative survey from a premier historian."   —*Booklist* (starred review)

"An exciting new multifaceted synthesis of American history between 1829 and 1877. . . . Those of us who find history endlessly fascinating usually like to argue about it. . . . The scope and skill of his book make him a worthy interlocutor."   —*New York Sun*

Jerry Bauer

About the Author

WALTER A. MCDOUGALL is a professor of history at the University of Pennsylvania and the author of many books, including the Pulitzer Prize–winning *The Heavens and the Earth: A Political History of the Space Age* and *Let the Sea Make a Noise . . . : A History of the North Pacific from Magellan to MacArthur.* A native of Illinois, he lives in Bryn Mawr, Pennsylvania, with his wife and two teenage children.

ALSO BY WALTER A. McDOUGALL

Freedom Just Around the Corner:
A New American History 1585–1828

Promised Land, Crusader State:
The American Encounter with the World since 1776

Let the Sea Make a Noise . . .
A History of the North Pacific from Magellan to MacArthur

. . . the Heavens and the Earth: A Political History of the Space Age

The Grenada Papers
(coeditor, with Paul Seabury)

France's Rhineland Diplomacy, 1914–1924:
The Last Bid for a Balance of Power in Europe

THROES OF DEMOCRACY

The American Civil War Era 1829–1877

WALTER A. McDOUGALL

HARPER ● PERENNIAL

NEW YORK ● LONDON ● TORONTO ● SYDNEY ● NEW DELHI ● AUCKLAND

To all my former teachers and professors
thanks to whom I became a historian

HARPER ● PERENNIAL

Maps on pages 244, 278, 287, 419, 427, 467, 479, and the Civil War maps key are from the *Encyclopedia of American History*, by Richard B. Morris, Editor; and Jeffrey B. Morris, Assistant Editor, copyright © 1953, 1961, 1965, 1970, 1976 by Harper & Row, Publishers, Inc.; reprinted by permission of HarperCollins Publishers. Maps on pages 18, 53, 81, 100, 101, 210, 333, 336, 393, 440, 530, 587 are from *Harper's Atlas of American History with Map Studies*, by Dixon Ryan Fox, Ph.D.; copyright 1920 by Harper & Brothers.

A hardcover edition of this book was published in 2008 by HarperCollins Publishers.

FIRST HARPER PERENNIAL EDITION PUBLISHED 2009.

The Library of Congress has catalogued the hardcover edition as follows:

McDougall, Walter A., 1946–
  Throes of democracy : the American Civil War era, 1829–1877 / Walter A. McDougall.—1st ed.
      p. cm.
  Includes bibliographical references and index.
  ISBN 978-0-06-056751-4
  1. United States—History—1815–1861. 2. United States—History—Civil War, 1861–1865. 3. United States—History—1865–1898. I. Title.
E338.M38 2008
973.7—dc22   2007026499

ISBN 978-0-06-056753-8 (pbk.)

09 10 11 12 13 WBC/RRD 10 9 8 7 6 5 4 3 2 1

By blue Ontario's shore,

*As I mused of these warlike days and of peace return'd, and the
dead that return no more,*

A Phantom gigantic superb, with stern visage accosted me,

*Chant me the poem, it said, that comes from the soul of America,
chant me the carol of victory,*

And strike up the marches of Libertad, marches more powerful yet,

And sing me before you go the song of the throes of Democracy.

*(Democracy, the destin'd conqueror, yet treacherous lip-smiles
everywhere,*

And death and infidelity at every step.)

—WALT WHITMAN, "By Blue Ontario's Shore," *Leaves of Grass*
(1881–1882 edition)

CONTENTS

MAPS

The maps are from the Encyclopedia of American History *(New York: Harper, 1953), ed. Richard B. Morris; and* Harper's Atlas of American History *(New York: Harper, 1920).*

PREFACE

*Y*ears ago, while stuck in a railroad station cursing a tardy train, I noticed a man in a jacket bedecked with insignia from Vietnam. Experience had taught me that veterans still advertising their service decades after the fact were probably bitter and down-at-the-heels. But to relieve the tedium I mentioned that I, too, was a Vietnam veteran and struck up a conversation. We exchanged the usual questions—When were you there? What was your outfit? Where were you based?—but then the dialogue flagged. His speech, demeanor, and clothing suggested that he had a blue-collar background and no success in civilian life. We had little in common. But then he surprised me by saying, "I'm glad I went, 'cause I learned the lesson of Vietnam." What was that, I inquired. "Say no to bullsh–t," he replied. I feigned a cynical chuckle, then quickly waved good-bye to him because I suddenly found myself choking back tears.

Thank you for sampling this book. If you love history, or just love the United States of America, you may relish this volume. That is not to say you will find all its judgments congenial. Indeed, I hope you do not, because we truly engage the past only when confronted with evidence and arguments that challenge our preconceptions. I found my own challenged repeatedly while researching this book, with the result that I was led into subjects and down paths of argument I never could have anticipated. That is the joy of empiricism. But it harbors a danger as well. The history written by those of us who imagine ourselves the "clay" and the evidence the "sculptor" (instead of the other way around) is often misunderstood by fans and critics alike. That is a risk I am willing to run, because the only way a historian can avoid misunderstanding is to distort the past in pursuit of some unmistakable present agenda. Such a procedure leaves no doubt as to "where the author is coming from," but it prostitutes history. So where am

I coming from? If obliged to put it all in a nutshell I guess I would venture the following. I believe the United States (so far) is the greatest success story in history. I believe Americans (on balance) are experts at self-deception. And I believe the "creative corruption" born of their pretense goes far to explain their success. The upshot is that American history is chock-full of cruelty and love, hypocrisy and faith, cowardice and courage, plus no small measure of tongue-in-cheek humor. American history is a tale of human nature set free. So how you, the reader, respond to this book will depend in good part on how you yourself (all pretense aside!) regard human nature.

As for me, I have just tried to teach myself American history with two precepts in mind. The first one—"Write as if you are already dead"—I learned from Nadine Gordimer via the *Los Angeles Times*'s book editor Steve Wasserman. The second precept I learned in a railroad station from an anonymous veteran to whom I feel obligated in ways I cannot explain.

Throes of Democracy: The American Civil War Era 1829–1877 is the sequel to *Freedom Just Around the Corner: A New American History 1585–1828* (Harper-Collins, 2004). That earlier volume traced the colonial and early national history of the American people until the election of Andrew Jackson to the presidency. The present volume carries the story to the end of Reconstruction. You are free to tackle *Throes* without first having experienced *Freedom*. Each book is meant to stand alone. But in that case I recommend you peruse the synopsis of the first volume that follows this preface. It will alert you to various themes that inform the present volume as well.

Two admonitions are in order. First, the endnotes are meant to be read, or at least glanced at, because many of them contain substantive, sometimes humorous details I could not bear to leave out. However, to keep the sheer volume of endnotes under control I often insert them after every second or third paragraph rather than every one. So please don't be frustrated when you encounter quotations lacking immediate citations. Just refer to the next endnote and you will find what you seek. Second, you will notice that periodically I interrupt the narrative with sidebars celebrating the entry of new states into the Union. I began this project determined to pay due attention to all regions and states because most general American histories do not. They proceed as if Kansas, for instance, did not exist except when it was "bleeding." My solution was to craft thumbnail biographies describing the origins, distinctive characteristics, and local color (again, often humorous) of every new state. The sidebars also illustrate larger themes in the book. But if you are eager to press on with the main narrative, you may skip these sidebars or else return to them at your leisure.

—*Philadelphia, Pennsylvania*
December 2006

SYNOPSIS OF *FREEDOM JUST AROUND THE CORNER*

A New American History

1585–1828

*T*he creation of the United States of America is the central historical event of the past 400 years. If some ghostly ship, some *Flying Dutchman*, were transported from the year 1600 into the present the crew would be amazed by our technology and the sheer numbers of people on the globe, but the array of civilizations would be recognizable. The crew would find today, as in 1600: a huge Chinese empire run by an authoritarian but beleaguered bureaucracy; a homogeneous, anxious, suspicious Japan; a teeming mosaic of Hindus and Muslims trying to make a great nation of India; an amorphous Russian empire pulsing outward or inward in proportion to Muscovy's projection of force; a vast Islamic crescent hostile to infidels but beset by rival centers of power; a dynamic, more or less Christian civilization in Europe aspiring to unity but vexed by its dense congeries of nations and tongues; a sub-Saharan Africa plagued by poverty, disease, and tribal feuds; and a Latin America both blessed with a rich Iberian-Amerindian culture and cursed with strategic impotence. The only continent that would astound the Renaissance time travelers would be North America, which was primitive and nearly vacant as late as 1607, but which today hosts the mightiest, richest, most creative civilization on earth—a civilization that perturbs the trajectories of all other civilizations just by existing. Not only was the United States born of revolution; it is one.

When I accepted the offer from HarperCollins to take a fresh look at American history I imagined a whole list of themes worthy of emphasis, including the influence of geography, technology, demography, ethnicity, mythology, and political culture. Not least, I meant to examine America's experiment in religious liberty, which is to say the kaleidoscopic effects of competing sectarian and civic faiths on law, society, culture, and politics. But no sooner did I begin to research the very roots of the thirteen colonies than it dawned on me that one of my major themes must be none of the above. That theme is the American people's penchant for *hustling* in both the positive and negative senses. That is what emboldened me to call the book candid.

Americans as hustlers? Yes, because the colonists who settled the coast from New England to Georgia embodied the four original spirits that drove English expansion in the Tudor-Stuart era.

The first spirit was economic and social, because something happened in England around 1450 that happened nowhere else on the globe: a whole society began spontaneously to organize itself on the basis of capitalism. We usually associate capitalism with merchants such as the Italian and Dutch who invented the joint-stock company, marine insurance, double-entry bookkeeping, floating debt, and mercantilist theory. But long-distance trade in luxury goods existed in every culture and era. For a whole society to become capitalist the 95 percent of the people who earned their daily bread on the land had to dissolve the bonds of slavery or serfdom, feudal dues, land tenures, and village commons, and replace them with more or less free markets in land, labor, and food. In short, they had to turn the necessities of life into commodities.

In England, that began to happen around 1485, when the Wars of the Roses came to an end. Ever since the bubonic plague arrived in 1348, feudal landlords had faced a persistent labor shortage and huge swaths of Europe fell back into wasteland. Elsewhere various means were applied to bind peasants to the soil, but English lords cleverly learned to exploit feudal privileges to dismantle feudalism. They maximized revenue by enclosing common pastures to expand cultivation and sheepherding, then farmed out land to "farmers" in exchange for rent. England's common-law tradition and independent judiciary blessed the transition by sustaining contracts and property rights. The new printing industry boosted rural capitalism through books and pamphlets teaching husbandry and preaching an "improvement ethic." The process was gradual and uneven, but over time landlords and farmers responded to market incentives with more efficient methods, cost-cutting, and specialization. Successful ones expanded their acreage by purchase or reclamation, while brokers, merchants, and lawyers moved in to manage the real estate and commodities markets and export the surpluses England began to produce.

By the late 1500s lords and merchants invested their profits in other enterprises ranging from mining and textiles to overseas trade and finally colonization. Capitalism, frowned upon by the nobility elsewhere, obsessed the English: and the market economy forced people of all social classes to adjust to a new notion of freedom that meant in effect bowing to the "dictates of the market." The losers included crofters and day laborers of both sexes so that England's port cities swelled with rootless young people. Dreaming of new land to improve and thereby improve themselves, they would provide the indentured servants for overseas colonization.

The second spirit of English expansion was religious and political. Enter Henry VIII, who broke with the papacy in 1534 and declared England an "empire" subordinate to no power on earth. Henry seized and sold off the Catholic church's vast properties in England and Wales, giving another big boost to rural capitalism and incidentally financing England's first seagoing navy. Henry also vastly increased the power and popularity of the monarchy. But he was careful to make it all legal by asking Parliament to approve his measures, a precedent that would haunt the Stuart kings. Finally, Henry stirred up support for the crown through propaganda that called the pope the Antichrist and called all who served him slaves of the devil. But since many English and Welsh, not to mention Scots and Irish, people remained Catholic, the crown grew paranoid about popish plots. Paranoia increased during the subsequent reign of the Catholic queen Bloody Mary. But when Mary died childless, Protestant clergymen returned from exile exulting, "God is English!"—meaning, of course, "God is Protestant, and the English are his chosen people."

Queen Elizabeth I tried to conciliate her subjects by making the Church of England broad and tolerant. She failed to reconcile either Catholics or Presbyterians, especially the radical Puritans. But all English

Protestants could agree on their archenemy when the Spanish armada sailed in 1588, sparking fifteen years of total war. England was God's earthly sword, slaying the Spanish dragon like Saint George. Hence, the second spirit in which Englishmen settled America was fierce, at times paranoid, anti-Catholicism.

The third was geopolitical. If Spain and then France were religious enemies, so, too were they England's chief competitors for the wealth of the New World. As yet, no English captain had sailed beyond the sight of land. Then an English spy pilfered a copy of Spain's secret textbook for sailors. Translated in 1561, this *Arte of Navigation* launched England's career as the world's greatest sea power. Soon Oxford and Cambridge dons were teaching celestial navigation and drawing accurate maps, while craftsmen designed excellent astrolabes and sextants. Privateers such as Francis Drake raided the Spanish Main and Sir Walter Raleigh dreamed of outflanking the Spanish empire by settling Virginia. When Spain protested, Queen Elizabeth replied, "Prescription without possession is not valid. Moreover, all are at liberty to navigate . . . since use of the sea and air are free to all." Thus did she invoke common law and the improvement ethic to argue that only settlement confers ownership. Proprietors, indentured servants, and squatters galore sailed to America with those principles close to their hearts.

What of lands already occupied, for instance by Native Americans? That question inspired the fourth spirit of English colonization, the racial and legal one. It held simply that the rights of savage or indigent peoples were forfeit. The English first applied that principle in Ireland, the colonial laboratory where they learned contempt and fear for uncivilized neighbors and asserted eminent domain over unimproved lands. John Locke justified that right in his *Second Treatise of Government* in 1689. He began by asserting, "In the beginning all the world was *America*," by which he meant undeveloped, and ended by asserting that American Indians "exercise very little Dominion, and have but a very moderate sovereignty." In short, the English believed the displacement of indolent people by dynamic civilizers was not only inevitable but just.

The people who settled the thirteen colonies were imbued with those spirits. They displaced the Indians, damned their Catholic neighbors in French Quebec and Spanish Florida, and hustled to seize the opportunities afforded by a rich, forested, well-watered continent. The colonists were almost all loyal to the British Empire (as it came to be known after the merger of England, Scotland, and Wales in 1707). But they were prepared from the start to resist British authority if it violated the spirits of empire as the colonists understood them. As early as 1650 the assemblies of New England and Virginia repudiated Parliament's first Navigation Acts. Later all the colonies resisted the Stuart kings' efforts to impose centralized rule. The colonies were saved from that by the Glorious Revolution, which deposed James II in 1688 and established Protestant, parliamentary supremacy. In the decades that followed, Britain's Whig governments were pleased to exhibit "salutary neglect" of the colonies. Americans took advantage, not only by shrugging off the authority of the chartered companies and proprietors that gave birth to their colonies, but by turning smuggling, tax evasion, cooking the books, sailing under false colors, and jury nullification into fine arts. Meanwhile, the thirteen colonies grew rapidly in population, wealth, and self-government: hustling in every good sense.

Ironically, what finally provoked an imperial crisis was a glorious event that, for more than seventy-five years, all British subjects had prayed and fought for: the conquest of Canada during the Seven Years' War (or French and Indian War) of 1756–1763. That meant Parliament was responsible for governing all the Catholics and Indians of the former French realm, as well as for paying off the debt incurred during the war. But the policies Parliament and the crown adopted to meet those responsibilities seemed to Americans a double cross, a repudiation of all four spirits of English expansion. First, in violation of the capitalist improvement ethic, Parliament imposed new taxes and restrictions on colonial manufactures and commerce. Second, Parliament legalized the Catholic church in Canada. Third, Britain cleared the French out of the Mississippi basin (Louisiana), only to hand it over to Catholic Spain. Fourth, George III sheltered his Native American subjects with the Proclamation Line of 1763 prohibiting white settlements west of the Alleghenies. It seemed to colonists as if the British had become heretics in their own church, strangling American growth for the benefit of a corrupt aristocratic oligarchy. So the colonists trained their rhetorical, then literal cannon, so recently used on the French, against their own sovereign.

What about liberty? Did not many colonists sail for America in the 1600s and 1700s in search of religious and civil liberty? Some did, to be sure, but as Oscar Handlin shrewdly put it, "liberty was something Americans discovered, gradually, while in pursuit of something else." In any event, the four main British cultures that placed their stamps on the colonies understood liberty differently. Hence, when the Continental Congress and later the Constitutional Convention met in Philadelphia to make a nation, their delegates had to negotiate a hybrid definition of liberty that goes far to explain the rest of American history.

Consider the Puritans who swarmed out of East Anglia and London after 1630 to settle New England. They were devout Calvinists harried out of England by the Stuart kings James I and Charles I. Under John Winthrop they seized control of the Massachusetts Bay Company's board of directors, voted to sail for America, and spirited the company's charter away in their baggage. The Puritans claimed three missions: to live in godly harmony as a "shining city on a hill"; to convert the Indians before the French Jesuits got to them; and to return to England when it was ready to be "purified." They accomplished none of those goals. To be sure, New England was a theocracy of sorts, but the Puritans pursued material goals at least as assiduously as spiritual ones, converted few Indians, and lost their mission to England when a fellow Puritan, Oliver Cromwell, rose against King Charles I in the English Civil War.

How did New Englanders define liberty? Knowing there is no freedom in moral anarchy wherein all men are slaves to sin, but also no freedom to be found beneath bishops or kings, Puritans preached "ordered liberty" rooted in the rule of law, private virtue, the family, and communal self-government of towns and churches. They hated hierarchy and privilege, placed public service above private wants, and believed men free in all things except willful sin. Such was the liberty loved by John Hancock, and Sam and John Adams.

Virginia, by contrast, was from the start a rural capitalist land speculation turned tobacco plantation. Its neighbor Maryland followed as a haven for English Catholics, but Catholics were soon a resented minority even there. What really placed a stamp on the Chesapeake was the influx of distressed Cavaliers, or King's Men, who lost their lands fighting Cromwell and accepted an offer by Governor Berkeley of Virginia granting land in America. They stemmed mostly from England's southwest and spoke in a languid rhythm (Guv'nah Baaahkley). Planters in Virginia cared little for religious disputes, but made a religion of horsemanship, hospitality, and hierarchy. Their pride was expressed by William Byrd II, who likened himself to a biblical patriarch endowed with herds, flocks, manservants, maidservants, and independence from all authority save that of Providence. Thus was born the American South, and a notion of liberty contrasting sharply with that of New England. Planters took human inequality for granted; hence to be free meant not to be dominated *by* others—such domination is slavery—and thus to exert authority *over* others, be it wife, children, or servants. Such "hegemonic liberty" was the direct descendent of the aristocratic ethic that inspired the Magna Carta, English Parliament, and Bill of Rights of 1690. To a highborn Virginian the salient question was not who should govern, but whether the natural aristocrats governed with honor, justice, and wisdom. George Mason, George Washington, Thomas Jefferson, and James Madison cherished such freedom.

The Quaker political culture dominating the Delaware Valley abominated the Puritans for their belligerence and intolerance, and the Cavaliers for their hierarchy and slavery. The Society of Friends dispensed with all clergy, churches, ceremony, symbols, and dogmatic theology in the belief that all men and women possess a divine "inner light." They admonished each other to live in simplicity and equality, albeit their shrewd merchants elevated countinghouse values over those of the meetinghouse. William Penn's religious toleration, for instance, doubled as a shrewd promotional gimmick to attract settlers, and his far-flung advertising agents helped make Pennsylvania a stunning commercial success. But the Quakers, of course, were pacifists with a "live and let live" idea of liberty. A simple enough proposition, but its corollaries were in fact complex, including resistance to taxes for waging war and any taxes imposed by a nonrepresentative body. Hence, Pennsylvanians such as John Dickinson would shout for liberty as loudly as other colonists, but drive the others batty by refusing to fight.

The fourth great migration followed the Act of Union creating Great Britain. Many Scots hated England, the Anglican church, or Parliament's assault on their woolen trade. After the Highland rebellion of 1746, tens of thousands lost their land as well. Some 200,000 Scots and Ulstermen sailed for the New World. They included men of high education and Enlightenment philosophy, but most were poor and lawless Borderers whose ancestors had feuded, rustled cattle, and fought the English for centuries. They swarmed inland on the Great Philadelphia Wagon Road and fanned out from western Pennsylvania through the Shenandoah Valley and into Virginia and the Carolinas. They planted the original "country and western," or hillbilly, culture in America, and they proved ungovernable.

Scots-Irish Borderers were pure libertarians for whom freedom meant the absence of outside authority, political or religious. They reckoned all who imposed or collected taxes and rents to be glorified thieves. They reckoned all who drew or enforced boundaries to be glorified jailers. As a German observer wrote, they rejected all external constraint, not because they were wicked, but simply because they were wild. That was the liberty precious to Patrick Henry, Francis Marion, the Jacksons, and the Calhouns.

Colonists representing all four major cultures were primed to resist Parliament's impositions after 1763. But few Americans were prepared to conclude that resistance required a declaration of independence and treasonous war until Benjamin Franklin encouraged a ne'er-do-well drifter to seek his fortune in Philadelphia. After just eighteen months in the colonies Tom Paine wrote *Common Sense*, the pamphlet that accused any American who shrank from independence of lacking manhood itself. With brilliant intuition Paine tapped both vocabularies that resonated with colonists groping to know their destiny. The first was Whig "country-party" ideology informed by the commonsense philosophy of the Scottish Enlightenment. That ideology extolled human rights and equality, self-government, free markets, and the pursuit of happiness. The other was the evangelical Protestantism stamped on the colonies by the revivals known as the Great Awakening and led by Jonathan Edwards and George Whitefield. The evangelical message identified Americans as a new chosen people in a new promised land with a mission to usher in a new order for the ages, perhaps even the kingdom of God.

Yet *Common Sense* was ambiguous, as if Paine sensed the danger of liberating human nature. He urged independence. But he was not sure Americans would prove up to it. He imagined the new nation becoming the greatest power on earth. But he warned that its unity might be as fragile as glass. He damned repression. But he urged patriots to repress all Tories, traitors, and slackers. He extolled equality. But he warned against the fickle mind of the multitude. He saw America becoming the richest country in history. But he warned that materialism bored and corrupted. He praised Americans' rebelliousness. But he chided them for their lawlessness. He told Americans they were free to begin the world over again. But he feared Americans lacked the virtue to do so. Paine extolled reason. But he played on a keyboard of emotions ranging from hatred, anger, and vengeance to fear, self-love, and self-doubt. Indeed, *Common Sense* was so demagogic that sober John Adams, who also favored independence, wrote a rebuttal. But Paine had his finger on the colonial pulse. He conflated biblical and secular principles so deftly that Puritans, Presbyterians, Baptists, Quakers, and deists all nodded in agreement. He derived a politics of liberty from religious principle in order to make a religion of politics. He pronounced the colonies' rebellion a moral crusade. In sum, he made America itself a sort of religion, which in turn made its fight for independence a holy war, which in turn made opposition to the American dream a sacrilege. That is why Paine repeated the not-so-veiled threat that "we the people" will get anyone who won't rally to the cause.

After 1776 only a minority of the colonists displayed the virtue Paine sought. Adams figured that a third of the colonists were Tory in sentiment and another third just lay low. Most of the new state governments snubbed Congress's pleas for recruits and supplies. Farmers dodged military service and hid their crops, or else sold them for hard money to the redcoats. As has been the case throughout much of American history, the honor and sacrifice of a few, with George Washington at their head, secured the freedoms of the many.

But the United States also survived and prospered because, as Edmund Burke tried to warn Parliament, Americans were "full of chicane and take whatever they want." What is more, their chicanery and apparent corruption were almost always creative. The supposedly unanimous passage of the Declaration of Independence was in fact a product of desperate arm-twisting, logrolling, and possibly back-alley threats or bribes. Robert Morris's financial sleight of hand saw the Continental Congress through the War of Independence, but he also lined his own pockets. Benjamin Franklin engaged in magnificent diplomatic duplicity, first to win the French alliance that made victory possible, then to betray that alliance by making a separate peace with the British. When the Articles of Confederation proved inadequate in the 1780s, James Madison, Alexander Hamilton, and the other Federalists used delightful deceptions to convoke what none dared call a Constitutional Convention. They then plotted and politicked to ram the Constitution through ratifying conventions in state after state.

Moreover, the framers of that document took human nature as they found it. Inspired by the skeptical Scottish Enlightenment, they purposely designed a government that did *not* rely on republican virtue—a commodity always in short supply—but instead harnessed private ambitions for the public good, while checking the potential for harmful corruption by balancing the powers of factions, regions, and branches of government. To be sure, the founders condemned political parties, and none more so than Thomas Jefferson. But when he decided in the 1790s that Hamilton and the Federalists were conspiring to replicate British-style institutions in the United States, Jefferson organized the first political party, spun the news like a modern press agent, and won the presidency in 1800 in league with the first boss of an urban machine, Aaron Burr. Out on the frontier, Americans' freewheeling manners were even more on display. Westward expansion, the clearing of Indians, and admission of new states to the Union were invariably steeped in the politics of land speculation.

So Paine's vision of empire came true, but in ways neither he, Jefferson, nor Adams expected. The United States became, with surpassing speed, the largest, richest, most dynamic republic in history because Americans were uniquely free to exploit their ingenuity, take risks, and reap the fruits of their labor. They multiplied at a frenetic pace, thanks to immigration and native fertility, and set out to seize and develop their continent by hook, crook, or pocketbook. They hustled in the sense of scoffing at or finding a way around any law or authority that stood in the way of their pursuits of happiness. They conveniently ignored grotesque contradictions such as chattel slavery, the cult of republican womanhood that dictated "separate spheres" for the sexes, and the abject failure of federal Indian policies. But Americans also proved consummate hustlers in the positive sense of builders, doers, go-getters, dreamers, inventors, organizers, engineers, team players, and philanthropists, thanks in good part to their technical ingenuity and national market crafted by John Marshall's Supreme Court decisions.

Perhaps most amazing of all, Americans embraced a civic religion all the more powerful for being unspoken, unwritten, and for the most part unacknowledged. It amounted to a covenant binding the federal union together despite the diversity of its regions and peoples and promising the nation a millenarian destiny so long as the Union survived. Why did delegates to the Constitutional Convention not even mention religion or slavery? Because they knew each issue was a potential deal-breaker. The

refusal to confront slavery, of course, would condemn the nation to decades of masquerade and potential disaster. But free exercise of religion,
sanctified in the First Amendment, proved to be a half-accidental stroke
of genius. It made the market in spirituality as free as the markets in
power and wealth, and thus invited Americans to select whatever belief
system allowed them to *feel good about doing well*. Many chose evangelical-
ism, which exploded anew in revivals on the frontier, then in towns beginning in Jefferson's term. Others shed Christian orthodoxy for Unitarianism
or lost interest in religion altogether. It did not matter to the body politic
so long as no sectarian faith encouraged disloyalty to the civic faith that
ensured free exercise of religion in the first place. Indeed, the American
civic religion owed as much to the influence of Freemasonry as to various
forms of Christianity. A striking proportion of the founding fathers embraced that male fraternity and its devotion to toleration, reason, republican virtue, unity, and material progress. Masons were master builders, and
American Freemasons imagined their new nation an unfinished pyramid
rising toward heaven under the watchful, benevolent all-seeing eye of the
God whose name is Geometry. The Great Seal of the United States, the
American flag, and the eventual one-dollar bill were eloquent expressions
of such Masonic symbolism.

What was the *purpose* of the United States? What were American
politics all about? In the early national period statesmen and philosophers
imagined the purpose of their national experiment was to foster republican virtues, empires of liberty, New Jerusalems, more perfect unions, and
other lofty ideals. But the elegant political machine designed in Philadelphia and running like clockwork did not run for any preordained purpose
except *to keep running*. In the 1790s Federalists and Democratic Republicans nearly came to blows over their imagined differences, only to have
Jefferson declare upon his inauguration in 1801, "We are all Federalists.
We are all Republicans." During the War of 1812 the parties flirted again
with coercion, rebellion, even secession, only to have the Federalist Party
dissolve and the Republicans embrace much of the Federalist platform. By
1820 one-party politics seemed to prevail, only to spawn anything but the
"era of good feelings" Americans pretended it was. A National Republican
faction split from the Democratic Republicans and hotly debated the purpose and destiny of the United States. Yet in the decade that followed,
citizens learned (albeit rarely admitted) that the principal purpose of their

politics was simply to win elections! Almost no discussion of policy or principle graced the presidential campaign in 1828. It was simply a contest in mudslinging, the outcome of which was a function of party organization, propaganda, and patronage. But it was hailed as a triumph for the people, democracy, equality, and liberty because Andy Jackson, the rowdy Scots-Irish chieftain from Tennessee, buried the sober, pious, but allegedly "aristocratic" New Englander, John Quincy Adams.

At Jackson's inauguration Washington City teemed with Democratic "Hurra Boys" waving hickory sticks, federal job-seekers, panhandlers, laborers, and immigrants. Justice Joseph Story named the brawling, boozing Democracy "King Mob." What had happened? Elite scions of Puritans, New Yorkers, Quakers, and southern planters along the east coast had always assumed their settled notions of liberty, hierarchy, law, and order would gradually spread west and tame the frontier. Instead, it seemed the rollicking frontier had spread east and conquered the nation! To be sure, the founding fathers had imagined the American experiment coming to all sorts of bad ends. But they never imagined the Federal City overrun by frontiersmen who loved only cheap land and credit, whiskey, tobacco, guns, fast women, fast horses, and Jesus—though not necessarily in that order.

THROES OF DEMOCRACY

Chapter 1

PRETENDERS?

The Melees and Masks of White
Man's Democracy Post-1830

*A*nd then there came a day of fire, to New York City, in 1835. The catastrophe and its aftermath displayed in sharp relief the glories and flaws of a city fast becoming the symbol of a nation drunk on democracy.

"How shall I record the events of last night, or how attempt to describe the most awful calamity which has ever visited the United States? The greatest loss by fire that has ever been known, with the exception perhaps of the conflagration of Moscow, and that was an incidental concomitant of war. I am fatigued in body, disturbed in mind, and my fancy filled with images of horror which my pen is inadequate to describe. Nearly one half of the first ward is in ashes; 500 to 700 stores, which with their contents are valued at $20,000,000 to $40,000,000, are now lying in an indistinguishable mass of ruins. There is not perhaps in the world the same space of ground covered by so great an amount of real and personal property as the scene of this dreadful conflagration. The fire broke out at nine o'clock last evening. I was waiting in the library when the alarm was given and went immediately down. The night was intensely cold, which was one cause of the unprecedented progress of the flames, for the water froze in the hydrants, and the engines and their hose could not be worked without great difficulty. The firemen, too, had been on duty all last night, and were almost incapable of performing their usual services.

"The fire originated in the store of Comstock & Adams in Merchant Street, a narrow crooked street, filled with high stores lately erected and occupied by dry goods and hardware merchants. . . . When I arrived at the spot the scene exceeded all description; the progress of the flames, like flashes of lightning, communicated in every direction, and a few minutes sufficed to level the lofty edifices on every side. . . . At this period the flames were unmanageable, and the crowd, including the firemen, appeared to look on with the apathy of despair, and the destruction continued until it reached Coenties Slip, in that direction, and Wall Street down to the river. . . . The Merchants' Exchange, one of the ornaments of the city, took fire in the rear, and is now a heap of ruins. The façade and magnificent marble columns fronting on Wall Street are all that remains of this noble building, and resemble the ruins of an ancient temple. . . . When the dome of the edifice fell in, the sight was awfully grand. In its fall it demolished the statue of [Alexander] Hamilton executed by Ball Hughes, which was erected in the rotunda only eight months ago by the public spirit of the merchants."

Philip Hone wrote that account in his diary. He was born to penurious German immigrants in 1780. New York, at that time, lay under British occupation, its commerce wrecked by the thirteen colonies' war for independence. By 1797, when Philip and his brother teamed up to auction off cargoes down by the docks, New York City was already emerging as North America's premier port of entry for the goods of the world. By 1821, the handsome, blond Hone was already rich enough to retire. Over the next thirty years he traveled abroad, collected books and artwork, and won a reputation as a man-about-town second only, perhaps, to the millionaire John Jacob Astor. Hone served a term in City Hall, the last Federalist to do so, and later advised such giants of the Whig Party as Daniel Webster, Henry Clay, and William Seward. He also befriended Martin Van Buren, kingpin of New York's Jacksonians. But Hone had no love for "King Andrew" Jackson himself. "That such a man should have governed this great country with a rule more absolute than that of any hereditary monarch of Europe and that the people should not only have submitted to it, but upheld and supported him . . . will equally occasion the surprise and indignation of future generations."

An even more dangerous threat to the fair city beloved by Hone was the influx of Irish whose violence and vice undermined law and order, and whose votes tightened the Tammany Hall Democrats' grip on power. He concluded his account of the fire by damning "the miserable wretches who prowled about the ruins, and became beastly drunk on the champagne and other wines and liquors with which the streets and wharves were lined." They "seemed to exult in the misfortune, and such expressions were heard as 'Ah! They'll make no more five percent dividends!' and 'This will make the aristocracy haul in their horns!' Poor deluded wretches, little do they know that their own horns 'live and move and have their being' in these very horns of the aristocracy, as their instigators teach them to call it. This cant is the very text from which their leaders teach their deluded followers. It forms part of the warfare of the poor against the rich. . . . This class of men are the most ignorant, and consequently the most obstinate white men in the world. . . . These Irishmen, strangers among us, without a feeling of patriotism or affection in common with American citizens, decide the elections in the city of New York."[1]

That city was fast becoming the locus and metaphor of the American dream when suddenly, a week before Christmas in 1835, lower Manhattan ceased to exist. On the night of December 16 a watchman breasted arctic northwester gales that measured eighteen degrees below zero. At the corner of Pearl and Exchange streets the odor of smoke reached his nostrils. He whistled up comrades, broke open the door to a five-story warehouse, and gasped to find the building a furnace. Stove coals had ignited gas from a leaking pipe. The flames exploded the roof, caught the stiff breeze, and kindled the whole dry-goods district in fifteen minutes. "Imagine," wrote an eyewitness, "the loud and incessant ringing out of church bells at the commencement of the fire; the rattling of all the engines of the city and suburbs; the shouts of the firemen, and the glare of their torches; the terrible roaring of the volumes of flame that poured from the windows of the burning houses, while the shivered glass came jingling down in showers . . . ; the horror depicted in the upturned faces of thousands of spectators, looking white in the red glare around them, many of whom saw the whole of their property being consumed before their eyes with a rapidity which nothing seemed to check." This eyewitness saw hot kegs of nails fuse into blobs and copper run like melted butter off roofs.

Nothing could check the inferno. New York's volunteer fire companies were undermanned because of cholera and drained from extinguishing two large fires during the day. They lacked the latest steam-powered firefighting equipment and dragged their pump engines over icy streets because the men were too proud to use horses. The rival companies' "plug-uglies" fought over access to hydrants. In any event, the cisterns were inadequate and the pipes frozen solid. Firemen furiously chopped through the East River's ice only to curse when their hoses froze or weep when the wind blew their pitiful streams back in their faces.

Everyone gasped at the nocturnal hell. To observers in Brooklyn the masts of the merchant fleet were limned against the curtain of flame as sailors frantically cast off before sparks reached their sails. Jersey City and Newark dispatched fire companies. People as far afield as New Haven and Philadelphia wondered at the glow on the horizon. Manhattan's own men and women streamed down Broadway and the Bowery in hopes of salvaging silks, satins, wines, feathers, and flowers—all the finery of their Vanity Fair. In Hanover Square citizens hurled merchandise out of windows into a pile thirty feet high. Drayers, charging extortionist rates, carted some of the booty to safety. Most of it turned into oily black smoke, scenting the air "with the various articles that were burning, particularly with large quantities of coffee." Delmonico's opened its doors to serve hot drinks to the fagged and the frigid until the restaurant itself caught fire. When the old Dutch church proved another false refuge, a man mounted the organ loft to play Mozart's Requiem for the city.[2]

The Tontine Coffee House, a cradle of American capitalism, went up about four in the morning. With that the fire threatened to cross Wall Street, explode a nearby warehouse filled with brandy, and march uptown. Mayor Cornelius Lawrence made the desperate decision to blow up Wall Street. Marines from the Navy Yard arrived in the nick of time with just enough powder kegs to blast out a fire wall. But everything southeast of William and Wall burned for thirty-six hours until 700 buildings covering seventeen blocks were reduced to embers.

Amazingly, only two people perished, because the fire started at night in a district devoid of homes and hotels. But the monetary cost exceeded that of three Erie Canals. The city's twenty-six insurance companies (in which the unfortunate mayor had a big stake) went bankrupt. Hundreds of merchants lost their capital and thousands of laborers their employment. Except for looters, the only people who profited from the fire were printers. The *New York Sun* set a circulation record with its morning-after edition. The *New York Herald* pioneered the use of illustrations and maps in its coverage of the calamity. A twenty-two-year old Yankee, who had recently hung the shingle "N. Currier, Lithographer" at Nassau and Wall, made his reputation with dramatic prints of the fire. Seventeen years later Currier made his fortune as well, when he teamed up with Ives. Otherwise, the city seemed ruined, humbled, visited with a "flagrant providential warning."[3]

Back in 1792 the traders who first gathered under the buttonwood (sycamore) tree on Wall Street dealt in the bonds issued by Hamilton's First Bank of the United States. President Jefferson, who cursed government bonds as "federal filth," allowed the bank's charter to lapse. Moreover, since most American businesses were still family affairs, just 100 shares per day might change hands on the exchange. The real action on Wall Street was found in its gambling parlors and on the East River auction blocks where Hone got his start. Save for the spires of Trinity Church, the financial district was a tangle of wooden and brick shops and warehouses. But over those same years the great semicircle of piers around the tip of Manhattan's "thumb" bustled like no other spot in America. Between 1792 and 1807 (the year Washington Irving dubbed New York "Gotham") imports increased five and a half times, netting the city a third of the nation's foreign trade and a fourth of its coastal shipping.[4]

New York's progress abruptly ceased, thanks to Jefferson's embargo of 1808 on foreign trade and the ensuing War of 1812. But when news of the peace treaty reached the city in 1815, New Yorkers paraded by torchlight all through the night. The seven lean years were over. The port of New York was free to resume the competition with Philadelphia and Boston that Knickerbockers were sure they would win. Why such confidence, given the most popular explanation for New York's preeminence—the Erie Canal—was not yet under construction? First, you can't argue with geography. Manhattan boasted not one but two excellent seaward approaches. The channel between the sandbars off Brooklyn and New Jersey's Sandy Hook accommodated ships with a draft up to twenty-one feet at low tide, and the Main Ship Channel north to the Hudson was a commodious twenty-three feet. Long Island Sound, the back door to New York, offered a second well-sheltered route for mariners willing to brave the rocks and tides of Hell Gate on the tidal strait called the East River. Philadelphia and Baltimore, by contrast, had channels just eighteen feet deep and lay, respectively, far up the Delaware River and Chesapeake Bay. Boston enjoyed direct access to the sea, but not to the interior, and was troubled by ice and fog far more than New York. Second, New York harbor could be expanded at will as demand required. Counting both shores of Manhattan, Brooklyn, Staten Island, and the Jersey towns, the Upper Bay contained over 700 miles of potential wharf frontage.

British shrewdness was a second cause of the city's sudden prosperity. Following the end of the Napoleonic and American wars, merchants in Liverpool decided that New York was the optimal port of entry for manufactures they meant to "dump" on the American market. This allowed New York middlemen to capture the lion's share of the market for redistribution of imports to other U.S. ports and the interior. In 1818 English Quakers led by Jeremiah Thompson founded the famous Black Ball Line, the first regularly scheduled Atlantic packets. They made New York their American hub. With what did Black Ball supercargoes fill their holds after unloading their exports on the East River docks? Mostly cotton, a commodity harvested by slaves in the American South. Those were the years when plantations spread like kudzu across the Gulf coast. Thompson himself swallowed his scruples to become the world's biggest player in the market for cotton. That gave every incentive to his New York–based suppliers to set up subsidiary offices in Charleston (South Carolina), Savannah (Georgin), Mobile (Alabama), and New Orleans. By 1825 they captured three-quarters of the coastal commerce in cotton, carried in their outbound vessels most of the domestic manufactures purchased by southern planters, and habitually advanced money against next year's crop so that many planters were permanently in their debt. So tight was New York's grip on the throat of King Cotton that 5 to 10 percent of white citizens in the cotton ports were of Yankee origin. Overseas, New York merchants bested their domestic competitors in Latin American markets, garnering two-thirds of American sugar imports and half the coffee imports. The Erie Canal, which opened in 1825, was an additional stimulus, mostly because of its timing. Two decades later railroad trunk lines afforded other ports access to trans-Allegheny markets. But for the time being, New York City enjoyed a near monopoly in the Great Lakes Northwest. By 1835 half of all American exports and a third of all imports passed through the city.[5]

The concentration of commercial wealth in turn allowed New York's wheelers and dealers to profit from enterprise elsewhere in the country. The national frenzy for canals, railroads, and land speculation in the 1830s turned Wall Street into a magnet for corporations hungry for capital. Daily volume on the floor of the Stock Exchange soared to 5,000 or 6,000 shares, enough to turn occasional traders into full-time professional brokers, touts, underwriters, loan sharks, and confidence men exploiting inside information, dispensing false information (especially to foreign investors), floating rumors to incite booms and panics, buying on margin, selling short, or attempting to corner a market. Jacob Little, the first "bear" who specialized in "selling the bear's skin before one has caught the bear," mastered the art of turning a profit at others' expense. Fellow traders hated and feared him; they soon adopted or countered his ploys. The Street was born, and Americans noticed.[6]

"Dress strictly respectable; hat well down on forehead; face thin, dry, close-shaven; mouth with a grip like a vice; eye sharp and quick; brows bent; forehead scowling; step jerky and bustling": thus did Walt Whitman describe (and instruct) youngsters seeking to make their way on Wall Street. The young poet's passion for New York City was almost erotic, yet he acknowledged that the merchants and money traders had "their brains full and throbbing with greedy hopes or bare fears about the almighty dollar, the only real god of their i-dollar-try." Highbrow New Englanders rued the spirit of speculation moving over the continent from New York. Preachers such as Henry Ward Beecher, Horace Bushnell, and William Ellery Channing described stock markets as a sinful temptation to "feverish, insatiable cupidity." Yankee intellectuals such as Henry David Thoreau and Ralph Waldo Emerson denounced both speculation in capital and the industrial market it fueled (albeit Emerson clipped 6 percent coupons on his own portfolio). Spokesmen for the southern plantation economy despised the ethic and monopoly power of northern financiers, whose "code of laws is that of the gambler, the sharper, the imposter, the cheat, and the swindler." Westerners deemed their dependence on moneyed men back East to be a species of servitude. Their leader was Andrew Jackson himself, who roared against "brokers and stock speculators," hoped they would all "break" (go broke), feared Americans "shall yet be punished for their idolatry," and in his farewell address endorsed "sober pursuits of honest industry." Charles Dickens's Martin Chuzzlewit spoke for foreign visitors when he described Americans' conversation as "barren of interest, to say the truth; and the greatest part of it may be summed up in one word: Dollars. All their cares, hopes, joys, affections, virtues, and associations, seemed to be melted down into dollars. Whatever the chance contributions that fell into the slow cauldron of their talk, they made the gruel thick and slab with dollars. Men were weighed by their dollars, measures gauged by their dollars; life was auctioneered, appraised, put up, and knocked down for its dollars."[7]

New Yorkers shrugged off such preachments, condemnations, gripes, accusations, and complaints. On December 19, 1835, while lower Manhattan's cinders still glowed, Mayor Lawrence assembled 125 prominent citizens to draft plans for the city's resurgence. The next day Wall Street traders were back at their posts doing deals. Lawrence and Albert Gallatin, President Jefferson's former treasurer, traveled to Washington City (as the nation's capital was known) in search of assistance. They argued that the federal government should imagine *itself* "a large capitalist, with an overflowing treasury . . . whose duty it is to promote the welfare and prosperity of the people." Capitol Hill's reply to New York was "Drop dead," a sentiment shared by President Jackson, who did not believe in federal largesse to states and localities. All New York got was remittance of duties paid on goods destroyed in the fire. The state government in Albany, however, favored Philip Hone's petition with $6 million in loans, $5 million of which City Hall in turn lent to developers and stricken merchants. That was all New Yorkers needed to rout the bears and confound the moralists. More than 600 firms relocated uptown, causing rents and property values to double all along Broadway from Saint Paul's Chapel past City Hall and the Masonic Lodge. The Swiss brothers John and Peter Delmonico broke ground for a new and more lavish culinary cathedral that would be fronted by marble columns shipped from the excavations of Pompeii. John Jacob Astor, who in his old age became a forerunner of Donald Trump, was at work on a grand hotel sure to transform the neighborhood around City Hall. Completed in 1836, Astor House boasted over 300 gaslit rooms, baths and toilets on every corridor, and dining rooms offering thirty entrees per day. It helped make Broadway a magnet for tourists, a fact not lost on Phineas T. Barnum, who would open his American Museum two blocks away. Hone, too, was a beneficiary of the fire. He sold his town house at 235 Broadway, for which he had paid $25,000, to the neighboring American Hotel. "I have turned myself out of doors," he wrote, "but $60,000 is a lot of money."

Even more spectacular were the rocket-propelled real estate prices in the burned-out district. Public officials and private developers realized that the immolation of antiquated structures on narrow, haphazard streets was no catastrophe, but a golden opportunity to carve out broader boulevards, streets, and plazas; erect modern buildings; install gas lighting; and make the district "more eligible for the transaction of business than it has heretofore been." Workers who briefly feared that their jobs had gone up in smoke were hired to clear the rubble and lay the foundations for ornate, fire-resistant, brick-and-marble office buildings. The crowning glory was the rebuilt Merchants' Exchange, designed in the Corinthian style by the Greek Revivalist Isaiah Rogers. Merchants and banks subscribed $1 million in six weeks for this temple, which was intended to house the stock exchange, the chamber of commerce, and a commercial library with up-to-date business news from all over the United States, Europe, and the rest of the world. Everybody with access to cash, credit, or friends in high places jostled for a piece of the action until lots south of Wall Street sold for seven or eight times more than they had when warehouses sat on them. By spring 1836, a correspondent exulted, New Yorkers were buoyant again. "Smiling faces and cheerful countenances meet us at every corner, and demonstrate that there is an elasticity in the character of our people which always enables them to rise above the most overwhelming evils."[8]

Finally, the fire silenced opponents of a plan to construct an aqueduct from the Croton River, in upper Westchester County, and funnel an "inexhaustible supply" of water to New York City's factories, businesses, utilities, and fire companies. This triumph of will on the part of the city fathers in league with municipal and state government was second only to that represented by the Erie Canal. First, the town corporation had to buy out the Manhattan waterworks monopoly established in 1799 by the slippery Aaron Burr. Then legal challenges from landowners in Westchester had to be beaten back. Some $2.5 million in start-up capital, and $12 million in all, had to be raised. Thousands of graves were dug up and moved because a cemetery lay in the valley flooded by the Croton dam. Scores of bickering rival contractors had to be managed. The laborers, exceeding 3,000, repeatedly went on strike and also brawled with each other over which Irish communities should be favored with jobs. John B. Jervis, already the dean of American civil engineers, oversaw the design of the forty-one-mile aqueduct, the 1,450-foot High Bridge over the Harlem River, the 180-million-gallon reservoir in the midst of what is now

Central Park, the distributing reservoir on Forty-Second Street, and the subterranean pipes to the existing tank on Thirteenth Street. In October 1842 the water flowed on time and on budget to the hysterical cheers of a parade five miles long and a 100-gun salute. "Water! water! is the universal note which is sounded through every part of the city, and infuses joy and exultation in the masses," Hone recorded. "It was a day for a New Yorker to be proud of."[9]

Fountains of water irrigated New York, but the city mushroomed so fast that nothing could wash it clean. Numbering 100,000 people in 1815—the first American city to reach that mark—Manhattan surpassed 200,000 by 1830; 300,000 by 1840; and 500,000 by 1850, by which time a fifth of the population consisted of immigrants. Development grew up the island like Jack's beanstalk. But it never kept pace with the influx. Every neighborhood's density doubled or tripled; the density of Greenwich Village quadrupled. Even in poor neighborhoods rents rose almost every year, making May 1, "moving day" when all city leases expired, especially traumatic. Poorest of all was the Sixth Ward's so-called Five Points, named for the crossing where Anthony Street met the oblique intersection of Orange and Cross streets. It lay midway between Broadway and the Bowery, in what is now Chinatown. Early in the nineteenth century Five Points became home to much of the city's free black community, and to both Christian and Jewish German immigrants. But the first ripples in a tide that would wash 1.6 million Irish Catholics into New York by 1860 utterly changed the sleepy neighborhood. In the mid-1830s (a decade before the famous potato blight) Irish were already debarking at the rate of 30,000 per year. Landlords met the demand for housing by dividing their two- and three-story buildings into smaller and smaller apartments. These buildings became teeming tenements, made all the more ramshackle because Five Points was built on a landfill and structures often buckled or settled. Most notorious was the Old Brewery, converted into a rookery for hundreds of indigent (but rent-paying) Irish in 1837. They were by and large "bog-trotters" from desperately poor rural villages. They had few skills, bathed seldom or never, and took solace from hard labor and unemployment alike with strong drink. After all, a worker's daily wage might only amount to a dollar, but that sufficed to buy four gallons of liquor. Irishwomen sought jobs as domestics, only to learn polite Protestant ladies wanted no part of them. Instead, they worked for pennies as seamstresses and "hucksters" peddling vegetables and rags, or else became prostitutes in Five Points' flourishing brothels. Often besotted, often beaten by their husbands, the women ignored their own children, forcing them to fend for themselves. Many waifs knew nothing but disease, crime, drunkenness, debauchery, and sewage from sunup to sundown, birth until death.

Most Irish arrivals, however, were single, unskilled males hoping for work in construction. Failing that, they turned to rackets or else fought the city's 14,000 free blacks for jobs, self-esteem, and survival. The first "gangs of New York" were organized in the years around 1830, taking names like the Roach Guards, Bowery B'hoys, Hookers, and Swamp Angels. Alcohol, poverty, racial tensions, and gangs made murder and manslaughter a daily routine. Bosses of English and Dutch descent were pleased to hire the Irish and profit from them in other ways. The tobacco tycoon George Lorillard, Robert Livingston's brother John, and John Jacob Astor owned chains of brothels or theaters where clients rendezvoused with prostitutes in the upper tiers. Otherwise, polite Protestant society turned up its nose at the noisome "paddies," while Irish muscle, in turn, drove Negroes from the docks, stables, laundries, groceries, barbershops, oyster stands, and other humble places where they were accustomed to make their living. Hence, the place of African-Americans in the northern economy became even more marginal than it had been under slavery (not that Irish enjoyed having to "work like a nigger," as the new saying went).[10]

Exploitation of the Irish and persecution of Negroes less than half a mile northeast of City Hall Park disturbed respectable Gotham's pursuit of happiness; such harsh realities made it difficult to feel good about doing well. In the case of the Irish, Americans' old, original contempt for Catholicism served as a balm to their conscience. Surely the Irishman's stubborn adherence to the autocratic, corrupt popish religion made him naturally superstitious, ignorant, filthy, immoral, and poor. The cholera epidemic of 1832, New Yorkers duly noted, arrived in North America on a ship carrying Irish passengers to Catholic Quebec. This epidemic claimed more than 4,000 lives: 2 percent of the population, and a much higher percentage in the Sixth Ward. A doctor from Kentucky who ministered to the sick in Five Points that summer found its denizens "more filthy, degraded, and wretched than any *slave* I have ever beheld, under the most cruel and tyrannical master." Knickerbockers were quick to conclude this was because Catholics were slaves of a sort. In 1836 Harper Brothers, the publishing house, went to the bank on Maria Monk's *Awful Disclosures of the Hotel Dieu Nunnery of Montreal*. A purported memoir, it told of lupine, licentious priests forcing themselves upon nuns and burying the resulting babies in the convent's cellar. What is more, the Irish had to be kept out of the American mainstream lest the pope and his allies among Europe's monarchs employ them as shock troops in a conspiracy to overthrow America's republican government. Samuel F. B. Morse took time out from inventing the telegraph to write two best-selling nativist tracts "exposing" this popish plot. When the New York Protestant Association met in 1835 to debate "Is Popery Compatible with Civil Liberty?" an Irish crowd burst into the hall to smash furniture. The organizers beat an undignified retreat out the back, insisting that the incident just proved their point.[11]

In the case of Negroes, an itinerant entertainer born in the city provided a balm for Americans' conscience. Thomas Dartmouth Rice had deserted his carpenter's bench to join theater troupes barnstorming in the West. In 1828, so legend has it, the twenty-year-old Rice observed an elderly slave performing a song-and-dance shuffle. His name was Jim Crow. Rice bought the clothes off Crow's back; blackened his own face; mimicked the slave's voice and movements; added lyrics and patter confirming the image of Negroes as shiftless, funny, and dumb (read: *harmless*); and then took his show on the road. In November 1832 "T. D." or "Daddy" Rice made a triumphant return home to "jump Jim Crow" before 3,000 spectators at the grand, gaslit Bowery Theater. Street life in Five Points inspired him to create for his slave an urban sidekick: Zip Coon, the frilly-shirted, fast-talking northern black on the make. Whites of all classes rolled in the aisles with laughter; blackface minstrelsy began its long run in American theater.[12]

More than 200 brothels, 3,000 grog shops, dozens of bawdy theaters, and countless gambling dens made New York a byword for vice. City fathers approved. Tourism generated by vice profited retailers and hoteliers, who hired "drummers" to ensure wealthy clients a safe night on the town. Tammany Hall recruited the Irish, found them jobs, sheltered them from prosecution, and hastened their naturalization by providing false papers. The Irish reciprocated by voting en bloc for the Democrats. The police rarely raided immoral establishments. Judges gave standing in court to brothels whose unruly customers harmed women or property. The only official crackdown on vice targeted the lotteries that according to *Niles' Weekly Register* sucked more than $50 million each year from the pockets of American dupes. But New York's prohibition on lotteries in 1834 only cleared the way for a more democratic racket: the policy, or "numbers game," that allowed people with just a few pennies to gamble twice every day. Its kingpin, John Frink, ran more than 300 policy "agencies" in New York, plus a battalion of wandering hawkers, by 1839.[13]

Still, Manhattan lay too near pious New England to escape the zeal of reformers. Evangelicals believed that only God could redeem mankind's sinful nature and sanctify the American republic in advance of the millennium, the Second Coming of Christ. (Lydia Maria Child sighed in Five Points, "What a place to ask oneself, 'Will the millennium ever come!'") The Puritans' devolved Unitarian cousins, by contrast, believed human nature innately good and capable of self-improvement. But both believed Americans were called to perfect society and create heaven on earth. The Tappan brothers of Northampton, Massachusetts, combined both strains. Arthur, an orthodox Calvinist, moved to New York when the mother of his eight children died. He went into business selling silk to the boomtowns along the Erie Canal. Arthur's appeal was to charge customers a modest, fixed markup rather than haggle for whatever the market would bear. His ploy was to demand payment in cash: no trust. He also dismissed out of hand any employee suspected of drinking, smoking, or theatergoing. Arthur soon ranked among Manhattan's wealthiest men. Lewis Tappan, by contrast, turned Unitarian and set up his shop in Boston—until, in 1827, he fell on hard times. Arthur bailed Lewis out with $50,000, but only on condition that the prodigal sibling move to New York and join him in a moral crusade.[14]

The Tappan brothers, aged forty-one and thirty-nine, failed at first. Their *New York Journal of Commerce*, launched in 1828 to teach workers temperance, abolitionism, and self-reliance (no labor unions!), had few subscribers. From that they drew a lesson. Reform could not be a top-down pursuit urged on common folk by an elite: that was the mistake of Episcopalians, Freemasons, and their women. Reform must well up from below, motivated by mass evangelical religious conversions. So the Tappans bankrolled the American Bible Society, Tract Society, Home Missionary Society, Temperance Union, Sunday School Union, Education Society, and Magdalen Society.[15] Not least, they and other men with roots in New England invited the spellbinding preacher Charles Grandison Finney to redeem "our Stupid, Poluted, and Perishing City." Finney's successes in Rochester and elsewhere had proved that the ecstatic "tent meeting" revivals characteristic of the Second Great Awakening could be just as effective in cities as on the frontier. Indeed, so many New Yorkers of all social classes crowded the "anxious bench" in Finney's chapel on Chatham Street that Tappan's gentlemen booked the Bowery Theater for him and then, in 1835, the cavernous Broadway Tabernacle. As early as 1832 Finney's backers and converts founded New York University as a Presbyterian rival to the Episcopalians' Columbia, and in 1839 they founded the Union Theological Seminary.[16]

Above all, Arthur Tappan insisted that New Yorkers confront the evil of slavery in the American South and the need to assist free blacks in the North. Heretofore, the most popular "solution" to slavery was the American Colonization Society's plan to ship emancipated blacks back to Africa. The Reverend Samuel Cornish, an eloquent graduate of Princeton and a leader of New York's black community, argued that this plan was absurd and unjust: colored people were native-born Americans deserving of full citizenship, yet were treated worse than foreigners like the Irish. Cornish convinced Boston's William Lloyd Garrison, who transformed the abolitionist movement in 1831 with his journal *The Liberator*. Tappan was likewise convinced, and began publishing *The Emancipator* in 1833. The newspaper editor James Watson Webb locked horns with Tappan at once. If emancipated slaves were left on American soil, then tens of thousands, perhaps hundreds of thousands, of Negroes would migrate to northern cities, exacerbating poverty, crime, and racial antipathy. "Are we tamely to look on, and see this most dangerous species of fanaticism extending itself through society?" He vowed to crush the "many-headed hydra in the bud." White workers had no trouble siding with Webb; nor did the city's business elite, given its stake in the cotton trade. "We are not such fools as not to know that slavery is a great evil," said one businessman, but we "cannot afford, sir, to let your associates succeed in your endeavor to overthrow slavery. It is not a matter of principle with us. It is a matter of business necessity."

Arthur Tappan received death threats. A committee in New Orleans put a $20,000 bounty on him. When Tappan declared his intention to form an antislavery society, Webb's paper called on patriotic New Yorkers (Irish included, for this purpose) to storm the proceedings at Clinton Hall. The abolitionists cleverly repaired to Finney's chapel and elected Tappan their president. In December 1833 he merged his society with Garrison's to form the biracial American Anti-Slavery Society, with headquarters on Nassau Street in New York. Ugly rumors flew around town on the wings of the penny press. Tappan meant to install "nigger pews" in churches and encourage mixed marriages. During a heat wave in July 1834 the situation exploded. A white club tried to deny African-American worshippers the use of Finney's chapel. Webb's paper blamed the ensuing fisticuffs on Tappan's nefarious influence. Two nights later mobs numbering in the thousands stormed the chapel, burst into the Bowery Theater to protest against an abolitionist English actor who was appearing there, and then converged on Arthur Tappan's town house, which they reduced to rubble. Well-lubricated, their "Irish up," the rioters careened to Five Points, bent on destroying all houses, churches, and shops owned by Negroes. At length the militia and special constables deputized by Mayor Lawrence restored a semblance of order ("rather tardily," thought Hone). But the violence triggered copycat riots. In the otherwise placid city of Boston enraged Protestants sacked and burned Charlestown's Ursuline Convent, terrorizing the nuns. In Philadelphia anti-abolitionists went on rampages in black neighborhoods.[17]

Tappan left his gutted house standing, as a silent sermon; moved to Long Island; and went back to work. In hopes of raising the consciousness of both northern and southern Americans he paid for thousands of tracts to be mailed throughout the country. All he achieved was to help make 1835 "the crest of rioting in the United States." The year witnessed 147 riots nationwide, including forty-six directed against abolitionists and fourteen against enslaved and free Africans. New York itself remained relatively calm, perhaps because Tappan took the advice of those who said that "the time has not come to mix with people of color in public." Instead, Gothamites spent the summer marveling over a serial in the *New York Sun* reporting that the astronomer Sir John Herschel had discovered forests, oceans, pyramids, and winged human beings on the moon. When professors at Yale gave credence to the articles, everyone's morning refrain became, "Have you read the *Sun*?" Its editors finally admitted the possibility of a hoax, but not before boosting the daily circulation to 20,000, making the *Sun* for a time the biggest newspaper in the world.[18]

That was the city visited, in December, by fire—the city whose people displayed "remarkable elasticity" and "cheerful countenances" in a matter of months. But just as disaster bred unity, recovery bred discord. By mid-1836 inflation was out of control. President Jackson's war of attrition against the Second Bank of the United States put financial markets at risk by encouraging local banks to print dubious paper money. The fire's destruction and the subsequent real estate boom drove prices up faster than wages. Then a poor grain harvest tripled the price of bread and set off more riots of "pillaging canaille, the colored people, thieves, and Irish."[19]

The following year the bottom fell out for rich and poor alike when disaster struck Wall Street again. It began on Saint Patrick's Day when the mighty firm of Joseph and Joseph went bust just two days after its new marbled headquarters collapsed into the street! This bankruptcy touched off so many business failures that Philip Hone quit counting. Stocks and real estate plummeted; assets evaporated. Hone initially blamed all this on the "straits to which men have been driven by the wicked interference of the government with the currency of the country." In so doing, he displayed Americans' penchant for wanting to believe that they are in control of their destiny. Reality proved otherwise. The real captain of global markets was the Old Lady of Threadneedle Street—the Bank of England. Its directors fretted that British investors were overexposed in American real estate, canal, and railroad bubbles. So they pushed up the discount rate on pounds sterling. That curtailed demand in English mills for American cotton, driving down prices already pressured by oversupply. That forced cotton factors in New Orleans to renege on their debts, leaving hundreds of northern firms high and dry. In the contraction that followed, the firms of Mayor Lawrence, Hone's son, and even the Tappans shut their doors. (The young George Templeton Strong wisecracked: "Tappan has failed! Help him all ye niggers!") For the first time anti-Semitism entered the mix, because the Josephs, who had started it all, were Jewish. New York banks suspended payment, causing thousands of people to beat on their doors crying, "Pay!" But on Wall Street, Hone wrote, "All is still as death. No business is transacted, no bargains made. . . . The fever has broken, but the patient lies in a sort of syncope, exhausted by the violence of the disease and the severity of the remedies."[20]

That was the Panic of 1837. Work on railroads and canals all but ceased. State governments and corporations in Ohio, Indiana, and Illinois defaulted, enraging foreign bondholders and spoiling America's credit.

Farm prices collapsed; urban jobs disappeared. From the Atlantic seaboard west to the Mississippi, from the Great Lakes south to the Gulf of Mexico, the "great nation of futurity" toasted by New York's ebullient John O'Sullivan collapsed in depression. The ensuing distress divided Americans: poor against rich, white against black, Protestant against Catholic, native against immigrant, tippler against teetotaler, Whig against Democrat, abolitionist against nearly everyone, and North against South against West. Riots, reform movements, and religious revivals curdled cities and towns, mocking the "ordered liberty" and equality the republic claimed to embody. Above all, the hard times exposed the private anxieties of a people whose public professions were all about progress, pride, and providence. The United States on display in the 1830s and 1840s was a barroom brawl of contesting cultures, each trumpeting its own vision of what the founding fathers had meant to achieve while blaming others for wrecking that vision. It was enough to prompt visitors from abroad to suppose that democracy in America was built, not on rock or on sand, but on *pretense*.

In the decades following 1815 no fewer than 200 Europeans toured the United States and published impressions of what, to them, seemed a wildly improbable experiment in popular government.[21] Alas, the most celebrated of them was among the least reliable. As late as 1997 a historian with some pretensions to veracity wrote (albeit tongue in cheek) that "complete objectivity about America is a characteristic only of God and Alexis de Tocqueville."[22] In truth, the young Frenchman's methods were highly subjective. He was an aristocrat whose parents narrowly escaped the guillotine during the Reign of Terror in the French Revolution. So he came to America inclined to believe that government in the hands of the envious masses was far more dangerous than rule by disinterested aristocrats. Tocqueville was raised a Catholic, but exposure to Enlightenment philosophy hobbled his faith: "I believe, but I cannot practice." So he came to America with little appreciation of what made religious people tick, especially Protestants of British stock. His classical education and training for a French legal career biased his mind toward deduction rather than empirical, historical thought. So he came to America with little sense of the profound experience that inspired the thirteen colonies to found the United States.

Most tellingly, Tocqueville's motive was didactic. He meant to argue that democracy either did not work in America or else could not be made to work back in France. What prompted his journey in the first place was the French Revolution of 1830 that overthrew the last Bourbon king and installed the "bourgeois monarchy" of Louis-Philippe. Its motto was *juste milieu*, meaning government prudently balanced between royal and popular sovereignty. But it was born of mob violence and its empowerment of the French middle class was bound to provoke demands for representation from the lower classes. Tocqueville feared persecution even as he hoped to forge a career in the ministry of justice. So he conjured an official excuse to flee France for a time: a sabbatical to study prison reform in the United States. In April 1831, the twenty-five-year-old Tocqueville and his friend Gustave Beaumont weighed anchor at Le Havre.[23]

During the thirty-eight-day voyage Tocqueville revealed another damaging trait, which was to turn the opinions of others or his own first impressions into postulates. He took the New York chandler Peter Schermerhorn at his word when Schermerhorn insisted that there was

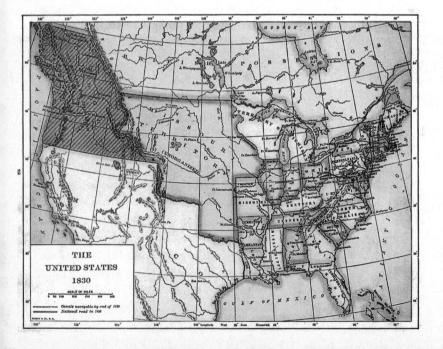

no rancorous party spirit in America, that the Union was in no danger of dissolution, and that the "greatest blot on the national character was the avidity to get rich and to do it by any means whatever." Tocqueville's diary and letters reveal a plethora of instant judgments. Upon meeting the governor of New York in a humble boardinghouse, he concluded, "The greatest equality seems to reign, even among those who occupy very different positions in society." Upon hearing Americans boast over their "very copious" suppers he decided, "These people seem to me stinking with national conceit; it pierces through all their courtesy." After one Sunday in Manhattan he wrote, "Taken together," the Americans "seem a religious people." After one stroll down Broadway he agreed with Beaumont that fine arts in America "are still in their infancy." After one conversation he decided, "Political passions here are only on the surface. The profound passion . . . is the acquisition of riches; and there are a thousand ways of acquiring them," many of which involved "cupidity, fraud, and bad faith." All those judgments were rendered during Tocqueville's first week in the country even though, he confessed in a letter, his English was still laughably poor.[24]

Tocqueville and Beaumont spent only forty-one weeks on American soil. Their inspections of Sing Sing and other prisons consumed eight weeks. Beaumont's desire to study Indians around the Great Lakes cost five more weeks. A sentimental journey to French Quebec lasted ten days. Nor did they invest the remaining six months wisely: they devoted to the entire South just one month, during which Tocqueville was frequently ill. Snap judgments multiplied. He decided that Philadelphians "know only arithmetic" because of their grid of numbered streets, and that their women "practice an unrestrained coquetry" even though "*tout le monde* agrees in acknowledging they stop there." A Quaker merchant who had lived in New Orleans told him that northerners were active, intelligent, cold, and calculating, whereas southerners were open and lively, but inclined to a certain hauteur and laziness. This primed Tocqueville to admire New England and to sense something "disorderly, revolutionary, passionate" about the South. On the shores of Lake Huron an old Catholic priest dismissed America's 450-odd Protestant sects as *rienists* ("nothingarians"), wherefore Tocqueville deduced that all Americans must someday gravitate to natural religion or Catholicism! In Cincinnati he was told Ohio thrived while Kentucky languished on account of slavery in the latter. Accordingly, Tocqueville failed to notice the fantastic wealth of planters and merchants as he raced through the South on steamboats and stagecoaches. In Washington City, the former president John Quincy Adams explained that whereas New England was founded "by a race of very enlightened and profoundly religious men," the new western states were "populated by all the adventurers to be found in the Union, people who for the most part were without principles or morality" or "who knew only the passion to get rich." Had Tocqueville visited Saint Louis and talked with Senator Thomas Hart Benton, perhaps he would have realized how quickly the West was replicating the civilization beloved by Adams. He even squandered his entrée in the federal city by engaging in small talk with President Jackson and evening frivolities with his cabinet. In any event, Tocqueville thought the rustic capital a sorry excuse for America's Paris, and couldn't wait to board ship for home. He admitted having spent less than half the time needed to get a true picture of the United States. "Yet I hope I have not wasted my time." [25]

Insofar as *De la Démocratie en Amérique* made him famous he had not wasted his time. But the volumes were neither systematic nor thorough. Tocqueville failed to cover the country in terms of geography; never set foot in a college or factory; and met no authors, artists, or scientists. He studied no canals and failed to grasp how important government-private partnerships were in constructing them. He inspected no military or naval facilities, and witnessed no revivals or camp meetings. He heard complaints about Jackson, but missed the new two-party system then in gestation. He suspected the potential of railroads, but did not imagine the transport, energy, and industrial revolutions they would spark. It could even be said the society he tried to describe in *Democracy in America* ceased to exist before the decade was out. [26]

Tocqueville deserves renown. He helped to pioneer sociology and posed questions of timeless relevance. What institutions, laws, and customs make a democracy? Are democratic ideals universal, and if so must democratic forms vary across nations and cultures? What social conventions and popular habits buttress or threaten democracy? Moreover, he got some things right. Although irony was not in his nature, he suspected democracy in America was a great truth maintained by a delicate balance of fictions. A brilliant young lawyer in Cincinnati gave him the clue. "We have carried 'Democracy' to its last limits," the man said. "The right of voting is universal. Thence result, especially in our towns, some very bad elections." The unworthy candidates won "by mingling with the populace, by base flattery, by drinking with it," whereas distinguished men simply "can't struggle against the flood of public opinion." Thus did the future chief justice Salmon P. Chase encourage Tocqueville to conclude that "the ablest men in the United States are rarely placed at the head of affairs."

Yet democracy thrived, thanks to a "thousand circumstances." The uniquely blessed location "in which Providence has placed the Americans" freed them from threatening neighbors, and hence from the need for dangerous, costly armies. The size and abundant resources of the United States provided so much opportunity that even the lower orders of people demanded laws to shelter, not challenge, private property. The federal Constitution dispersed political power into the hands of the people while superimposing a judiciary to serve as a check on the people's own passions. In any event, the hand of government was so light that people could pursue most of their goals without it. Frequent elections and popular suspicion of power seekers made democracy self-correcting. Last but not least, democracy in America was sustained by a breathtaking *conformity* imposed by public opinion shaped in turn by religion.[27]

Tocqueville learned that last point from a doctor in Baltimore. Don't be misled, the doctor said, by separation of church and state and the seeming indifference to doctrine prevailing among Protestants. The great majority of Americans were fervent believers and quick to ostracize unbelief. Such social pressure must create many hypocrites, Tocqueville replied. Yes, said the doctor, but "it prevents people's *speaking* of it. Public opinion accomplishes with us what the Inquisition was never able to do." If American men (not to say women) desired success in society, matrimony, business, or politics, they learned "to keep their mouths shut" about religious doubts. Tocqueville did not grimace; he marveled. So their secret was this. Americans were a people "seeking with almost equal eagerness material wealth and moral satisfaction; heaven in the world beyond, and well-being and liberty in this one." They established religious tolerance to which even Catholics conformed. But a dominant Protestant ethic regulated domestic life, especially through the influence of women, forging a public consensus that in turn regulated the state. To be sure, sects proliferated. But all conflated civil and religious liberty and all preached the same moral code.

"Thus, while the law permits the Americans to do what they please, religion prevents them from conceiving, and forbids them to commit, what is rash or unjust."

Now Tocqueville knew why missionaries from New England were so anxious to proselytize the West. Since the liberty and equality Americans cherished were derived from religion, the future of their democracy hinged on saving the frontier from heathenism. "Despotism may govern without faith," he wrote, "but liberty cannot. . . . How is it possible that society should escape destruction if the *moral* tie is not strengthened in proportion as the *political* tie is relaxed? And what can be done with a people who are their own masters if they are not submissive to the Deity?" Evidently, the American answer was to help them believe, and if they still could not believe make them fake it. Thus were the spirits of religion and liberty, so often at war in Europe, "intimately united" in America. The separation of church and state empowered Protestant public opinion to make civil society a sort of church.[28]

"The American," Tocqueville quipped, "is the Englishman left to himself."[29] Every American travel book, sniffed John Stuart Mill, was a party pamphlet. Nothing demonstrated those points better than Frances Trollope's *Domestic Manners of the Americans*, published in 1832. That was the year Parliament passed its first great Reform Bill to ensure fair representation and enfranchise Britain's broad middle classes. Where might such democracy lead? To perdition, Conservatives feared, and they found their proof in Fanny Trollope's vivid sketch of the United States. An obscure woman of radical, even utopian views, she said she "awoke one morning and found myself famous" thanks to a torrent of Tory publicity.[30]

Fanny Trollope, the daughter of an Anglican clergyman, made what seemed a good marriage to a country lawyer with ties to the petty nobility. She did her part, bearing five children. But he suffered from migraines, had a nasty temper, and failed at business and farming. She took solace in literary society, eventually falling under the sway of the "angelic" Fanny Wright, a wealthy Scots orphan who apotheosized America, knew James Fenimore Cooper and Washington Irving, and founded a commune in Tennessee to educate slaves for freedom. In November 1827 Fanny Trollope, aged forty-eight, went off on an adventure to visit Wright's commune, take a break from her husband, and help her son Henry get rich in America. In just five days she was boarding a ship with three of the children and their handsome French tutor (a subject of much innuendo). She could not conceive that in just five years everyone from London to Natchez would ask, "Have you read Mrs. Trollope?"[31]

Fanny Trollope's first glimpse of the United States was the "utterly desolate" mouth of the Mississippi River. Uprooted trees of "enormous length" floated past. Pelicans patrolled overhead. Alligators (she called them crocodiles) peered from monstrous bulrushes sprouting on mudflats. Her first exposure to New World society was at New Orleans, where minute racial distinctions and "incessant, remorseless spitting" seemed the principal American traits. She fell in love with the landscapes, fauna, and flora, noting the graceful palmetto, noble ilex, gay-plumaged birds, and towering pines draped with vines. America was an outsize Eden. On her way upriver to Tennessee, Fanny admired the majestic steamboats and was prepared to admire American men. They were considerably taller than Englishmen and ruggedly handsome. But their manners were loathsome! Foul-mouthed, rarely sober, they left off gambling and arguing only to wolf down whole platters of viands, then pick their teeth with their knives. Even more distressing was the Nashoba commune that inspired her pilgrimage. Wright had gathered some thirty Negroes in a rude collection of shacks, but planted no school, farm, or anything to indicate that she was "raising the African to the level of European intellect." Deciding the "angel" was a fanatic, Trollope hurried her children aboard a boat bound for Cincinnati, the Queen City of the West. Her account of the passage later won plaudits from no less an authority than Mark Twain.[32]

The 1830 census counted 20,000 residents in Cincinnati's riparian blocks of low wooden buildings. When the Trollopes arrived in 1828 only Main Street was fully paved and no streets had sewers. People just tossed sewage into the gutters for meandering pigs to devour. Without a church steeple in sight, the town's most imposing structure was the brick abattoir that made Cincinnati hog butcher to the Ohio Valley and bestowed on it a less glorious nickname: Porkopolis. The Trollopes rented a house and spent two years searching for the American dream. No luck. Fanny enrolled Henry in the utopian farm school New Harmony, but he fled from its twelve-hour shifts in the fields. She ran advertisements offering Henry as a tutor of Latin. No customers called. He even worked for a time as an oracular, multilingual "Invisible Girl" in a carnival act promoted by a frontier peddler of bunkum. Finally, on receiving a bequest from her deceased father, Mrs. Trollope constructed a garish bazaar with Gothic and Greek motifs, expecting it to become Ohio's leading emporium. The contractor cheated her. The sheriff foreclosed. She quit Cincinnati in March 1830, then quit America in July 1831 after failing to interest publishers in New York in her journal. Once her book proved a sensation, of course, American publishers made a killing with pirated editions. Given all this, one would expect a cultured Englishwoman, trapped and repeatedly bilked on a foreign frontier, to give a jaundiced account. But she also exposed unwelcome truths that irked American critics and pleased English Tories.

In all her time on the Ohio River, wrote Fanny Trollope, "I neither saw a beggar, nor a man of sufficient fortune to permit his ceasing his efforts to increase it; thus every bee in the hive is actively employed in search of that honey of Hybla, vulgarly called money." Like bees, Americans buzzed about on seemingly random, individual flights, competing for pollen. Yet, as with bees, the sum of their exertions was a bounteous hive. Cincinnati's markets were a year-round cornucopia of the finest beef, veal, pork, poultry, freshwater fish, eggs, butter, cheese, vegetables, and beans, and that American delicacy, the tomato, all for sale at prices even the relatively poor could afford. As for the rich, they offered up capital for any project promising future returns, but never squandered their fortunes in vain quests for glory of the sort that "make man forget he is a thing of clay." Indeed, Ohioans were so bent on business that they seemed not to have any fun. They held few balls, concerts, or dinners; rarely attended their theater; and outlawed billiards and playing cards. (Was she too proper to note the "wolf traps" by the steamboat landing, where river sharps were making poker America's favorite game?) Yet it was not materialism, prudery, or lack of refinement that soured Fanny on the United States. What she damned was the pretense born of "that phrase of mischievous sophistry, 'all men are born free and equal.'" [33]

Hiring housemaids must be called "getting help," she learned, because it was "petty treason to the republic to call a free citizen a *servant*." It was hard to find "help" because American girls preferred mill work to domestic service. The girls Trollope did engage were insolent—for instance, demanding silk finery, the better to catch a suitor's eye. Yet Americans' egalitarianism was partly a pose. She marveled how the whole town mingled at a Washington's Birthday fete until she inquired after the absent Miss C—. "You do not yet understand our aristocracy," it was explained.

"The family of Miss C—are mechanics." Democratic equality was rarely extended to women because almost all social activities were segregated by sex. Of course, that allowed men to drink, smoke, chew, spit, gamble, and fight as they pleased. A favorite haunt was the whiskey shop, where men got their news. "You spend a good deal of time reading the newspapers," she observed to a denizen. "And I'd like you to tell me how we can spend it better," he said. "How should freemen spend their time, but looking after their government." Was it a sense of duty, then, that drove American men to drink? "To be sure it is, and he'd be no true-born American as didn't. . . . I'd rather have my son drunk three times in a week, than not look after the affairs of his country."[34]

The weapon wielded by women in their war to civilize frontiersmen was religion. But that only reflected the power male preachers exercised over American women, which Trollope likened to that of the priesthood in Catholic countries. Everyone told her the constitutional proscription on established churches was a great blessing. But Trollope, like Tocqueville, concluded that "religious tyranny may be exerted very effectually without the aid of the government, in a way much more oppressive than the paying of tithe." Americans might not care what you believe, but "you are said to be *not a Christian*, unless you attach yourself to a particular congregation."[35]

The most obvious blight on the egalitarian creed was slavery. It recalled to Trollope the stinging stanza: "Oh! Freedom, Freedom, how I hate thy cant! . . . Where motley laws admitting no degree betwixt the vilely slaved, and madly free, alike the bondage and the license suit, the brute made ruler, and the man made brute!" But at least southerners made no pretense that all men were created equal. That is why she thought slavery's influence "far less injurious to the manners and morals of the people than the fallacious ideas of equality, which are so fondly cherished by the working classes of the white population in America." Another case in point was the treatment of Indians. Americans despised Europe's monarchs because "they favour the powerful and oppress the weak," yet they dealt treacherously with their own weak. Was there no probity in the American soul?[36]

Fanny Trollope's love for America's natural stage never flagged. She found Maryland a garden of plump strawberries and cherries, wild roses, sweetbriar, acacia, locust and dogwood trees, azaleas of all hues, rhododendrons, and sassafras. Butterflies and hummingbirds like "flowers on the wing" abounded. That American original, the "fire fly," made evenings a visit to fairyland. During a total eclipse of the sun in February 1831, Trollope used words such as *mystery* and *awe* to describe the uncanny glow of the snowy Potomac. No words sufficed for the falls of Niagara. But she had only coarse words for the masked players on the American stage, as summed up in this pithy genesis:

> *The wealth, the learning, the glory of Britain, was to them nothing; having their own way in every thing. . . . Their elders drew together, and said: "Let us make a government that shall suit us all; let it be rude, and rough, and noisy; let it not affect either dignity, glory, or splendour; let it interfere with no man's will, nor meddle with any man's business; . . . let every man have a hand in making the laws, and no man be troubled about keeping them; . . . let every man take care of himself, and if England should come to bother us again, why then we will fight all together."*[37]

Trollope was disillusioned by democracy. Harriet Martineau was not, because she never entertained illusions. "I went with a mind, I believe, as nearly as possible unprejudiced about America, with a strong disposition to admire democratic institutions, but an entire ignorance how far the people of the United States lived up to, or fell below, their own theory." That disposition derived from her family's pure English Liberalism of the sort that pushed through Parliament the Reform Bill of 1832 and the abolition of slavery in the British Empire in 1833. Her father owned a textile plant and believed in free trade. Both parents were Unitarians, sotto voce republicans, and advocates of women's rights. Never marrying, Harriet devoted herself to journalism and wrote a popular text on political economy while she was still young. In 1834, she hit on an even more precocious project. She would dissect American society with her own scientific methods for "how to observe manners and morals." She docked in New York in September 1834.[38]

Martineau meant to do the thing right. She sought out the wealthy, middling, and poor. She met merchants, mechanics, farmers, and fishermen, politicians, professors, and preachers, reformers and antireformers, women and men. Over two years she traversed 10,000 miles by steamboat, barge, stagecoach, horseback, and railroad. She went by the Erie Canal to

Buffalo, turned south to Pittsburgh, and crossed Pennsylvania. She lingered for six weeks in Philadelphia, three in Baltimore, and five in Washington City. Then she invaded Virginia to learn American history and law from James Madison and Chief Justice John Marshall. Not content to make a brief foray into the slaveholding states, as so many visitors did, she visited Jefferson's university at Charlottesville and crossed the tobacco country. After two weeks' rest at Charleston, she meandered through the Deep South to New Orleans, then up the Mississippi and Ohio rivers, and back to New York in July. She wisely spent the autumn—or "fall" as the Americans called it—in upper New England and wintered in Boston (thereby missing the fire in New York). The last months of her trip included a tour of the Great Lakes and a ride on the new railroad crossing the Alleghenies. In August 1836 she sailed for England. By 1837 her book was in print.

Martineau meant to measure American society according to the tenets of its own civic religion: that all men are created equal and endowed with unalienable rights, and that only government by the consent of the governed is legitimate. She hypothesized what Tocqueville discovered: that in American eyes politics and morals were the same, politics was every citizen's duty, and in politics the majority is always right. She discerned, as Tocqueville did not, that two parties existed in the United States and must always exist because those who have gained wealth or otherwise realized their hopes will always be challenged by those who have not. She reasoned that foreigners misunderstood American politics because the issues seemed trivial and government weak. But every accretion or diminution of political power stoked fears that America's liberty and destiny were endangered either by an entrenched aristocracy or by an ignorant, leveling democracy. Martineau imagined the American people "a great embryo poet: now moody, now wild, but bringing out results of absolute good sense . . . exulting that he has caught the true aspect of things past, and at the depth of futurity which lies before him, wherein to create something so magnificent as the world has scarcely begun to dream of."[39]

Heady stuff, but Martineau's keen observations obliged her to conclude that Americans, despite or because of their liberty, were cynics. "To hunger and thirst after righteousness has been naturally, as it were, supposed a disqualification for affairs" and any man proposing to advance truth and love through public life "would hear it reported on every hand that he had a demon." Her interlocutors assured her the American people did not want the truth. Accordingly, those who flatter and lie were elected to office, while conscientious men either shunned public service or were wrecked by it. A gentleman told her of three upright men who entered politics hoping to transcend faction and interest. "Look at them," said he, "A is a slave, B is a slave, and C is a worm in the dust." Harriet witnessed a candidate, reputedly a scholar and gentleman, running in Massachusetts, reputedly the most scrupulous state. He mounted a platform beneath a shabby walnut tree, "grimacing like a mountebank before the assemblage whose votes he desired to have. . . . He spoke of the 'stately tree' (the poor walnut) and the 'mighty assemblage' (a little flock . . .) and offered them shreds of tawdry sentiment, without the intermixture of one sound thought." Because newspapers were partisan, people learned only one side of a question—and that a dissimulation. "It is hard to tell which is worse: the wide diffusion of things that are not true, or the suppression of things that are true." Thus was opinion formed, and "the worship of Opinion is, at this day, the established religion of the United States."[40]

Martineau bristled at Americans' ethnic prejudice. Could they not see how mixed ancestry uniquely blessed their nation? "Let the United States then cherish their industrious Germans and Dutch; their hardy Irish; their intelligent Scotch; their kindly Africans, as well as the intellectual Yankee, the insouciant Southerner, and the complacent Westerner. All are good in their way; and augment the moral value of their country, as diversities of soil, climate, and productions, do its material wealth." Such stereotypes are no longer in vogue, but her point may be taken. Slavery was another debilitating character flaw. The "tumult of opinions and prophecies" on the subject amused Martineau. But she judged them all pretense. Planters never even used the word *slaves*, speaking instead of their force, hands, Negroes, or people. They protested too much the affection between masters and slaves, and did so in language one might use for a horse. She suspected such masks hid a deep fear and hatred of any slave who displayed intellect or emotion. Morals aside, slavery held back the South by making labor disgraceful in the eyes of wealthy and "mean whites" alike. In the North children attended school and were obliged to work hard. But northerners had their own sins, such as speculation, bankruptcy laws that allowed men to exploit others' trust without penalty, and exploitation of slavery by cotton merchants and mill owners. "The day will come when their eyes will be cleansed from the gold-dust which blinds them."[41]

One might imagine an egalitarian republic free of caste, but Martineau devoted a whole chapter to it. Precisely because Americans had to compete for status in a market whose currencies were wealth, pedigree, ostentation, reputation, and deference, they proved to be the most status-conscious of all. Boston's "first people" freely admitted their vanity. The phrase "first families of Virginia" spoke for itself. In Philadelphia, a matron who lived on Chestnut Street explained that she and her friends never mingled with the ladies of Arch Street, owing to "the *fathers* of the Arch Street ladies having made their fortunes, while the Chesnut [sic] Street ladies owed theirs to their *grandfathers*." Like Trollope, Martineau thought the status of American women was generally lower than that of women in the Old World. The republican ethos turned women into ornaments and helpmates for their striving husbands and vessels of virtues for the children. Women received enough education to make agreeable conversation, but not enough to realize their personalities. "All American ladies are more or less literary. . . . Readers are plentiful; thinkers are rare." They made "some pretension to mental and moral philosophy; but the less that is said on that head the better."[42]

Martineau reserved her final chapters for the fundamental force in American society. It was the same one chosen by Tocqueville—religion— and its faults were the same as those noted by Trollope: conformity and hypocrisy. Belief in the biblical basis of self-government required Americans to profess religion of sorts, but respect for others' pursuit of happiness required that religion do "little molestation to their vices, little rectification of their errors." The upshot was a "monstrous superstition" that consigned the honest unbeliever to the torment of living a lie, and the honest believer to the frustration of living with lies. Martineau was disappointed to encounter so little candor among the clergy. Even her own Unitarians were "self-exiled from the great moral questions of the time." They wasted their Sundays making parishioners feel good, damning Catholics, extolling God's glorious work in nature, or at most addressing "third and fourth-rate objects of human exertion and amelioration." In short, they left it up to the laity to decide whether and how to confront the pressing crises of the day. And no one petted these kept, cotton-mouthed ministers more than the women.

What was Harriet Martineau's interim judgment on the United States? Hopeful: for so long as Americans clung to their great founding principles they must sooner or later "live down all contempt." But to do so they must overcome "their deficiency of moral independence."[43]

By 1840 American quarterly journals began to voice suspicions that the British government was sponsoring these slanderous caricatures, perhaps to discourage emigration. The United States devoid of intellectual life? The faculties and graduates of Harvard, Yale, Princeton, Bowdoin, Dartmouth, Pennsylvania, Virginia, Georgetown, Transylvania, and a host of younger colleges had cause to dispute that canard. American Christianity merely a pose? Tens of thousands of fiery Methodists, Baptists, and Presbyterians were surprised to learn that. American politics a species of mob rule orchestrated by demagogues and mudslingers? No Oxford-trained orator in the House of Commons could outdo senators Clay, Webster, Calhoun, and Benton in eloquence and devotion to principle. Anyway, what was so wrong about a country where men of all ranks shook hands, addressed each other as "mister" or "sir," and gave up their seats to women? Southern writers called the attacks on slavery malicious exaggerations. Northern ones bade the English chew on their own corruption, their "dark Satanic mills," and their imperial rule over lesser breeds.

There was reason to hope this "war of the quarterlies" might end in an honorable peace when Mr. and Mrs. Charles Dickens docked in Boston in January 1842. Though Dickens was not yet thirty, his *Sketches by Boz* and *The Pickwick Papers* had made him beloved and trusted. Americans lionized him like no European since Lafayette.[44] Moreover, the trip began swimmingly. Dickens thought Boston a beautiful city of high refinement. The people were mercantile, but the "golden calf" they worshipped was subsumed in a "whole Pantheon of better gods." Hence the "public institutions and charities of the capital of Massachusetts are as nearly perfect, as the most considerate wisdom, benevolence, and humanity, can make them." On the railroad Dickens praised the easy equality of American men and their deference to women, albeit they talked of little but politics and the price of cotton. Lowell's textile factories impressed Dickens as engines of social uplift as well as production. Their young female employees were models of deportment and hygiene, and resided in neat, ventilated boardinghouses. They shared a piano, published a literary review, and saved enough from their wages to float a bank. The Connecticut valley was lovely, and New Haven and Yale proved Boston was not unique. So Dickens's mood could not have been higher when he sailed into "a noble bay, whose waters sparkled in the now cloudless sunshine," and caught his first glimpse of New York.[45]

What happened next? Did a satirical imp swipe Dickens's notebook, or did the author suffer "topographic disorientation"? Whatever the cause, the tone of his *American Notes* changed abruptly. Manhattan's "excellent" hospitals, schools, and library; its "singularly beautiful" women; and the busy docks served by "the finest Packet System in the world" evidently counted for nothing. For it was in New York that Dickens decided America was best described in terms of swine, spit, and squalor. His supercilious vignette about the portly sows and gentlemen-hogs promenading on Broadway seemed to equate the people with the pigs who rooted amid their sewage. He singled out one who "is in every respect a republican pig, going wherever he pleases, and mingling with the best of society, on an equal, if not superior footing." He was a philosopher, too. Upon spying carcasses at a butcher's shop, he said that "the pig seemed to grunt, 'Such is life; all flesh is pork.'" When evening descended and the streets grew quiet, Dickens asked, "Are there no amusements?" His bitter reply was that New York's amusements were found in the countinghouse, the noisome pubs, and the fifty-odd newspapers hawked by bawling urchins. The screeds' business was "abuse . . . pimping and pandering for all degrees of vicious taste, and gorging with coined lies the most voracious maw; imputing to every man in public life the coarsest and the vilest motives." In search of more amusements Dickens "plunged" into Five Points, which appeared to be home to the pigs, and to human filth, disease, and debauchery, albeit every barroom displayed prints of George Washington and the American eagle. Here buxom mulatto madams presided; flash Negro pimps dressed like white gentlemen, and white entertainers put on blackface; gin mills fronted alleys knee-deep in mud; and the police dragged Irish drunks to the Tombs, a stinking dungeon that would "disgrace the most despotic empire in the world." Here was republican virtue stripped naked.[46]

Washington City, "head-quarters of tobacco-tinctured saliva," was a different manner of pigsty. What did Dickens witness in that temple of equality known as Congress? Slave drivers whose pursuit of happiness claimed an inalienable right to praise liberty "to the music of clanking chains and bloody stripes"; political machinery that turned on the wheels of electoral tricks, official bribes, cowardly newspaper attacks, sharp dealing, and artful suppression of truth: "Dishonest Faction in its most depraved and most unblushing form." His mood grew darker still when he shared a carriage to Richmond with a slave monger whose "biped beasts of burden" traipsed alongside in manacles.[47]

Dickens's charity did not desert him completely. Out West he chuckled at the myriad ways Americans used the word "fix" and wondered what it meant to be "darned." In Cincinnati he was delighted to see even some Irish go "on the wagon" in a temperance parade. Niagara Falls enraptured him. But his closing indictment was in the voice of an angry prophet. Beneath Americans' democratic cosmetics lurked the blemish of universal distrust symptomatic of their crooked politics, business swindles, fickle public opinion, and poisonous press. "You will strain at a gnat in the way of trustfulness and confidence, however fairly won and well deserved; but you will swallow a whole caravan of camels, if they be laden with unworthy doubts and mean suspicions." He damned the pretense with which "Liberty in America doth hew and hack her slaves." He lamented that so many people "cannot bear the truth in any form."[48]

Dickens also scolded Americans for their brazen violation of foreign copyrights. Sure enough, within months of its publication, 100,000 pirated copies of *American Notes* were flying off bookstalls in the United States. Bostonians loved it because they emerged from its pages as America's only civilized people. Longfellow wrote in Dickens's defense, and the *North American Review* published a thirty-page rave. But South Carolinians censored the book. Westerners revived the idiom "give him the dickens." New Yorkers felt betrayed. Philip Hone had taken pains to organize the Boz Ball, a grand reception in a hall festooned with Pickwickian decor: "Everybody was there." Dickens's response was to smear the city and fail even to mention the ball! Hone wrote, "[We] made fools of ourselves to do him honor."[49]

After the Panic of 1837 the Knickerbocker elite acted swiftly to rekindle hope in the American dream. Hard-currency loans from the stolid Erie Canal Fund and the chastened Bank of England permitted bank tellers to reopen their cages by spring 1838. Wall Street lobbied hard for a lenient federal bankruptcy law and got its way in 1841. The Croton Aqueduct provided much-needed jobs. Religious revivals and temperance movements helped to comfort and discipline workers. Perhaps most important for the peace of the streets, Irish Catholics were given a leader who could play the game called democracy.

In 1814 a poor farmer from County Tyrone sailed the Atlantic in steerage to began life anew in Chambersburg, Pennsylvania. Three years later his younger son, John Joseph Hughes, scraped together the shillings to join him. Hard labor and self-instruction won the younger Hughes admittance to Mount Saint Mary's Seminary, and ordination in 1826. The hawk-nosed, well-muscled priest was dispatched to Philadelphia, where he guided several parishes, founded an orphanage, and earned the sobriquet "Dagger John" for his piercing defense of Catholics' civil rights. In 1837 the church selected the feisty forty-year-old for the toughest post in America: bishop coadjutor in the diocese of New York. Tough indeed, for New York's septuagenarian bishop John Du Bois was too old and too French to shepherd the city's Irish flock or protect them from Protestant wolves. Hughes, by contrast, was vigorous, Irish, and trained in the rough-and-tumble school of American urban politics. When Du Bois died in 1842, Hughes was consecrated in old Saint Patrick's Cathedral and began a twenty-two-year reign over the see of New York.[50]

Hughes decided at once that what Irish New Yorkers needed was schooling. He sailed for Europe in search of money and returned with the pledges that launched Fordham College. Next he preached a veritable crusade against the Public School Society that ran New York's eighty-six elementary schools. The society ignored Five Points entirely, while mandating curricula elsewhere that equated Protestant and American values. Hughes crafted a lawyer's brief: if public funds went for schools with a Protestant bias, then public funds should also be granted to Catholic schools. Metropolitan leaders hotly refused to do this, but at the state level winds of change were blowing. The Democratic "Albany regency" forged long before by Van Buren had relinquished its grip to the new anti-Jacksonian party known as the Whigs. Their governor, William Henry Seward, from the upstate town of Auburn, considered free immigration an ornament to the nation and a spur to its growth. He thought that immigrants from all countries could be assimilated through education. His wily political handler Thurlow Weed also saw a chance to compete for the Irish vote. So Seward proposed that publicly funded schools should hire teachers of the same faith and language as the pupils. Bishop Hughes, supported this time by a synagogue and a Scottish church, renewed his appeal to the Public School Society. Again the council turned him away, arguing that in a diverse city schools needed to be more uniform and centralized, not less.

Hughes would not give up. Instead, he urged parishioners to defend their rights at the ballot box, endorsed a slate of Catholic candidates, and made education the "wedge issue" in the elections of 1842. Samuel Morse's American Protestant Union countered with a nativist slate, which was certain to win if Hughes's Sixth Ward Catholics split the Democrat vote. That was enough to persuade the legislature in Albany to meet Seward halfway. It passed a law establishing citywide boards of education, placing supervision of schools in the hands of local wards, and banning any sectarian education in public schools.

In the short run, Hughes's only reward was violence. The Spartans, a notorious Democratic gang that specialized in thuggery at the polls, recruited an outraged army styling itself "the Americans" and invaded the Sixth Ward. Its members clubbed Catholics indiscriminately all the way to Five Points. The police, when they arrived, hauled away to the Tombs only the Irish, many "so beaten about the head that they could not be recognized as human beings." When the police departed the mobs resumed their rampage, then capped off the night by vandalizing Bishop Hughes's residence. Walt Whitman thought that this comeuppance was well-deserved, lamenting only that the rocks aimed at the bishop's house were not aimed at his head.

In the long run, Hughes's only reward was frustration. Protestant pastors and politicians soon mobilized their citywide majority to pack the new board of education. It judged nondenominational Protestant instruction acceptable while ruling out Catholic content and denying public funds to parochial schools. There was nothing for it, the bishop told his impoverished people, but to "build your own schools. Raise arguments in the shape of the best educated and most moral citizens of the republic, and the day will come when you will enforce recognition." Mother Madeleine-Sophie Barat answered his call, founding the first academy to train teachers for what would become the largest Catholic school system in America.[51]

One climactic battle remained before Catholics could feel secure as American citizens. A nativist party, the American Republicans, arose in New York under the staunch Methodist James Harper, one of the publishing brothers. When his party swept City Hall and the council in 1844 Mayor Harper enforced Sunday blue laws, swept Irish hucksters from their curbside booths, and purged the employment roster of Tammany

Irish. Nativist ranks swelled with new numbers and gumption until, in May 1844, the moment to save America from the pope seemed at hand. Philadelphia had risen! Protestant crowds visited purifying fire upon three Catholic churches and the homes of 200 families in Irish "Nanny Goat Square." Thirty people died in the brawling. Within days, nativists in New York and Brooklyn were busy arranging a premeditated riot of their own. Bishop Hughes made ready. He took personal command of 1,000 Irish worthies, deployed them in ramparts around his churches, and then paid a call on the mayor. "Are you afraid for your churches?" asked Harper. "No, sir," the bishop replied, "but I am afraid that some of *yours* will be burned." He had sternly cautioned against violence, but warned that if arson was done to a Catholic church his people might get "out of control" and burn down the city. It was pure and simple deterrence, and the mayor understood. At his behest Protestant leaders reluctantly canceled their rally, and nativist choler went into remission. The next year city voters took Governor Seward's advice and ousted Harper, even though this meant putting Tammany Hall back in power. "As for Bishop Hughes," wrote Philip Hone acidly, "he deserves a cardinal's hat at least for what he has done in placing Irish Catholics upon the necks of native New Yorkers."[52]

"Few public men of his day possessed a more statesmanlike grasp of the genius of the American Republic," *The Catholic Encyclopaedia* wrote of John Hughes.[53] One of those few who shared Hughes's grasp lived a remarkably similar life. He, too, began life in foreign poverty, but rose to become the leader of a religious minority, an educational pioneer, a student of American ways, and a lover of his adopted homeland. He even began his career in south-central Pennsylvania less than twenty miles as the crow flies from Hughes's Chambersburg. Of all the foreign visitors who tried to explain the United States to Europeans, he probably did the job best. And he wasn't even English or Irish or French. Rather, as Philip Schaff liked to boast, "I am Swiss by birth, German by education, and American by choice."

Obscure parish records suggest that Schaff was born in 1819, the illegitimate son of a carpenter and a farmer's daughter. His father promptly died and his banished mother gave him up to an orphanage. In school the boy showed promise but was expelled at age fifteen for "the secret sin" (masturbation?). A local pastor placed Philip in a pietist seminary in

Württemberg, where he experienced a mighty conversion. Studies at Tübingen, Halle, and Berlin introduced him to a "historical" approach to theology, inspired by Hegel, which in turn drew him into the "evangelical catholic" movement. Simply put, the movement sought to bridge the chasm dug by the Reformation, explain how Protestantism emerged from Catholicism, and combine the best in both traditions. Its Anglican parallel was E. B. Pusey's and John Henry Newman's Tractarian movement, which Schaff studied firsthand during six weeks at Oxford.

Following his ordination in 1844 Schaff was called to save a struggling seminary in Mercersburg, Pennsylvania. It was a humorous echo of the misunderstanding that had brought the Scotsman John Witherspoon to Princeton in the previous century. The trustees thought Schaff was a traditional Lutheran ready to damn Catholics and Protestant dissenters alike. When they heard him preach the historical method, they twice tried him for heresy (the last such trials in the German-American church). Like Witherspoon, Schaff not only overcame the assaults but he turned his Mercersburg catechism, hymnal, and liturgy into models for imitation. He also married, had eight children, and wrote extensively on church history. Back in Germany, famous professors noticed the bastard orphan's stunning success. In 1854 he was invited to Berlin to explain what democracy in America was all about.[54]

Upon his arrival Schaff was asked how he could live in a "barbarian country where the mob reigned supreme, and where neither person nor property is safe." Such notions were absurd, he replied. But the charges Europeans leveled at slavery, materialism, radicalism, and sectarianism required rebuttal. So without hesitation he addressed the ugliest features of American life.[55]

"Slavery is, without question, the political and social canker, the *tendo-Achilles*, in the otherwise vigorous system of the United States, and contradicts alike its own republican symbol and the spirit of Christianity and philanthropy." Schaff thought the institution must die out over time, but might yet cause disunion, for the reason that Americans did not know how to abolish it. Moreover, the issue was complicated by an "almost irreconcilable conflict of races" sure to persist even after abolition. He was more sanguine about materialism. Such were the inexhaustible resources of the great continent that the temptation to "race for earthly gain and pleasure" was irresistible. But foreigners failed to appreciate how equally zealous most Americans were for education, philanthropy, and religion, which turned "millions among them away from the vain glories of time to the imperishable riches of heaven." The best advice he could give aspiring emigrants was the adage "Pray and work." For just as genuine Americans despised idleness, so did they consider monetary gain a chance "to do good."[56]

The dangerous radicalism European observers discerned in American democracy was simply a myth. To be sure, the most elevated communities in the Union, such as Philadelphia, New York, and Boston, had lately been scenes of despicable riots. But such incidents violated the true sentiments of the American people. Far from tending toward revolutions of the sort that periodically convulsed continental Europe, Americans shared the "strong conservatism and a deeply-rooted reverence for the divine law and order" characteristic of Anglo-Saxons. Moreover, the moderate constitutional liberalism prevailing in the United States was the best safeguard against the despotism that provokes revolution. Radicalism in America "continually breaks on the free institutions of the country and the sound sense of order in the people." Finally, foreigners indicted American society for its wild sectarianism. Schaff, an orthodox, ecumenical pastor, could only confess that this was an evil consequence of Protestantism's centrifugal tendencies and America's experiment in freedom of worship. But the benefits far outweighed the drawbacks. He quoted the observation of a friend ("an Israelite, hence an impartial observer") to the effect that "the United States are by far the most religious and Christian country in the world; and that, just because religion is there most free."[57]

What then of America's political mayhem? Schaff asked his Prussian audience to imagine a world characterized not by remnants of feudalism, but rather by civil and religious liberty, equality, free speech, popular sovereignty, and election of almost all officials. If Europeans could imagine such a thing, they would surely think it a dangerous surfeit of freedom. But it proved not to be so, because the vast majority of the American people displayed reverence for the law, Christian values, and conservative government. That was because the revolution of 1776 was "entirely different in principle, character, and tendency from all the revolutions of the European continent since 1789." Except for Tom Paine, who was quickly repudiated, the American founders were men of "sound practical judgment" to whom freedom was not licentious indulgence. "True national freedom, in the American view, rests upon a moral groundwork, upon the virtue of self-possession and self-control in individual citizens. He alone is worthy of this great blessing and capable of enjoying it, who *holds his passions in check*."

Were not U.S. politics incorrigibly naughty? Without question, said Schaff. Elections were circuses in which all manner of "passions, falsehood, calumny, bribery, and wickedness" were unleashed. The conduct of business in legislatures was such that honorable citizens, "disgusted with the wire-pulling and mean selfishness of self-styled friends of the people," shunned public life. But the end product of corruption and chaos was, ironically, stable, conservative government precisely because it *belonged to the people*. Hence the "imposing spectacle, when immediately after the election of a president or a governor, a universal calm at once succeeds the furious storm of party strife." Schaff, a historian, attributed the American people's success at self-government ("this is the secret of their national greatness") to their belief that "individual liberty and communal order rest on the same moral foundation."

Finally, Schaff returned to the cult of the almighty dollar, since Europeans were convinced that the accumulation of wealth must create aristocracy even if birth and politics did not. The United States indeed spawned "fops and quack-aristocrats" who were devoid of true nobility and "have no sense for any thing but outward show." The most princely mansions in New York and Philadelphia, he noted, belonged to bogus doctors who peddled sarsaparilla as a cure-all. But excess and pretense did not erode Americans' egalitarian ethic, because fortunes derived from business, rather than entailed estates or privilege, rarely lasted through three generations. "The middle classes are there, more than in any other country, the proper bone and sinew of society, and always restore the equilibrium."[58]

Schaff dared not predict what role the United States was meant to play in God's providential script. He knew only that it would be tremendous, because "America, favored by the most extensive emigration from all other countries, will become more and more the receptacle of all the elements of the old world's good and evil, which will wildly ferment together, and from the most fertile soil bring forth fruit for the weal or woe of generations to come."[59]

* * *

The anxiety and pretension of the Jacksonian era were unforeseen consequences of the achievements wrought by the Continental Congress, the Constitutional Convention, and the jurists and lawmakers of the early national period. Those founders were inspired inter alia by the Scottish Enlightenment, English common law and representative government, Protestant understandings of Christian liberty, and the Masonic blueprint for a tolerant scientific nation of builders. The founders hoped that a free people would display sufficient republican virtue to preserve their new order and bequeath it to the ages. But taking human nature as they found it, they wisely designed a "mixed" government whose circumscribed powers and balanced branches and levels checked the potential for tyranny and corruption. The Constitution fashioned a central government sufficient to defend the Union and assert its interests abroad without threat to the various notions of liberty planted by Puritans in New England, Quakers in the Delaware Valley, Cavalier planters in the Chesapeake, Scots-Irish Borderers on the frontier, and ethnically mixed commercial elites from Newport to Savannah. The founders meant to give the American people maximum opportunity to pursue material and spiritual happiness in a free and unified national market. The founders also knew that there were two issues they dared not address if the Constitution was to win universal respect. Those issues were slavery and religion.

So the founders invented something new under the sun: a republic determined to realize continental ambitions *without* a strong executive, a large army, an established church, or a commercial caste; and *without* succumbing to the tyranny that wrecked the ancient Greek and Roman republics. They designed, in short, an empire shorn of overarching authorities. Democratic institutions and nearly universal white male suffrage by 1830 made the United States a virtually free market in power. The power of Congress to regulate commerce and the federalist rulings of Marshall's Supreme Court made the United States a virtually free market in goods and services. The First Amendment made the United States a virtually free market in religion, ideas, and association. The Second Amendment, upholding the citizens' right to bear arms, made the United States a virtually free market in violence. Not least, decisions by Congress to eschew restrictions on naturalization made all those markets available to virtually the whole human race.

What the founders could not do was bestow on the boisterous, disparate nation a unanimous or even dominant ethic or faith that imbued it with purpose. Hence paranoia and party strife erupted as early as the 1790s. But the triumph of "white man's democracy" by 1830 rendered even more vexing such awesome questions as these: What was the United States all about? What was the purpose of liberty? By what transcendent glue might a teeming democracy adhere, especially as new western migrations, new industries, new immigrants, and new religious denominations rendered society vastly more complicated than it had been in George Washington's day? Did the very freedom of citizens (except the enslaved) to seek their own meaning and purpose ensure that the nation at large could have no meaning and purpose? This was one source of pervasive anxiety in American life. Or might some faction or coalition "corner the market" in political, economic, or spiritual power, and seek to impose its own version of the national purpose on everyone else? This was a second source of pervasive anxiety. In the Jacksonian era alone it erupted in panics over conspiratorial threats posed by Freemasonry, Wall Street, the Second United States Bank, a "Christian Party in Politics," abolitionism, slavocracy, immigration, popery, and Mormonism. No wonder foreigners sensed that Americans were excessively but defensively boastful about their democracy. No wonder foreigners suspected that the glue holding the Union together was pretense.

Chapter 2

OLD HICKORY, INDIANS, BANKERS, AND WHIGS

The Politics of Anger Breeds a New Party System, 1829–1840

*T*here was no "antebellum era" in nineteenth-century American history. Granted the phrase is useful—too useful—as a catchall for the decades prior to 1860. Granted that numerous warnings and threats of civil war were voiced in the era. But inasmuch as secession had not yet happened and was desired by very few, it is unhistorical to speak of an antebellum. Alternative tags such as the Age of Jackson, the era of the common man, white man's democracy, the second party system, the Second Great Awakening, the sectional crisis, Manifest Destiny, industrialism, laissez-faire capitalism, and Romanticism have the virtue of being rooted in the era itself. Taken together, they also convey a keen sense of the many creative, destructive, centripetal, and centrifugal movements roiling American society. By 1815 Americans knew that the drama of their Union's founding had reached a happy ending. By 1829 the drama defining their Union's character moved out of rehearsal. Would it unfold as tragedy or comedy? Wise patriots hoped the answer was neither, for one ends in ruin, the other in farce. Alas, the first act opened with a series of scenes decidedly tragicomic.

* * *

Andrew Jackson entered the White House in a foul mood. Except for occasional sweet moments of vengeance he stayed angry for eight years. By contrast, the gun-toting, cider-swilling supporters who poured into the Federal City for his inauguration were ecstatic. So were thousands of farmers, planters, mechanics, day workers, and immigrants, and even some merchants in every region of the United States save parts of New England. To them Jackson's landslide victory over John Quincy Adams marked the triumph of democracy over elites, reform over corruption, and Jeffersonian liberty over power brokers and moneylenders who claimed to be good republicans but raised statues to Alexander Hamilton. Jackson was the first Scots-Irish president, first westerner, first born in a log cabin, and first "self-made man," a sobriquet coined for him. Orphaned during the American War of Independence, self-schooled in Tennessee, Jackson willed himself to become a sharp lawyer and politician, deadly marksman, wealthy planter, land speculator, fierce Indian fighter, war hero, architect of a national political machine, and at last chief executive. He shrewdly refused to elaborate what he stood for, thereby inviting voters of all stations in life to project their hopes and fears onto him. Jackson was an "against-er," and voters relished that.

On the outside, the gaunt frontier chieftain who mounted his silver steed to parade through Pennsylvania Avenue's mud looked the part more than any president since Washington. On the inside, Jackson hurt in body and soul. He was nearly sixty-two and in terrible health. He carried one bullet in his left arm and another that pinched a lung: relics of long-ago duels. Every day he coughed up "great quantities of slime," and sometimes blood. He suffered the symptoms of malaria and dysentery caught while fighting Creeks in the Alabama swamps in 1813. He was poisoned by a lead-based potion his doctor prescribed for his chest and stomach maladies. His rotten, aching teeth had to be pulled shortly before the election. His joints ached with rheumatism. He never knew a day without pain or a night of untroubled sleep. Jackson's adopted children and wards likewise vexed his soul. One was a prodigal forever in debt; another was a bully who tormented slaves at the Hermitage; and the third, an Indian boy he saved on the battlefield, rebelled against white men's ways. What sustained Jackson was his beloved wife, with whom he had eloped in 1791. But Rachel's health broke during the campaign of 1828 when mudslingers called her an adulteress for having "married" Jackson before her divorce was final. Nashville had planned a grand victory party for the Jacksons on December 23. Instead, 10,000 people mourned Rachel at her funeral on Christmas Eve. Old Hickory prayed for the grace to forgive the slanderers, knowing he never would. His wife's Presbyterianism had kindled in him a faith of sorts. But if he believed in a heaven for souls such as Rachel's, he took at least as much solace knowing that a hell awaited his enemies.[1]

During the deathwatch Senator John Henry Eaton of Tennessee paid a call at the Hermitage. A loyal factotum in banking and politics, he wrote a campaign biography of Jackson and was treated like one of the family. Now Eaton announced his intention to marry a Washington widow notorious for alleged affairs. Jackson knew the woman in question and thought her as much a victim of smears as his Rachel. He urged Eaton to salvage the woman's honor prior to his own inauguration. Others in Jackson's brain trust feared he was incurring a needless scandal, but Old Hickory would hear none of it. Eaton got married on New Year's Day 1829, with consequences that plagued Jackson for over two years.

The president-elect boarded a steamboat on the Cumberland River on Monday, January 19 (a day late, in deference to Sabbatarians), bound for the capital. He arrived, still in deep mourning, on February 11. That left him just three weeks to pick his cabinet, a chore complicated by the vagueness of his platform. "Jackson and Reform"—the slogan coined by the journalist Amos Kendall of Kentucky—gave no hint of where his administration would stand on tariffs, internal improvements, banks, and other issues, about which Jackson's lieutenants held disparate views. Accordingly, every name mentioned for a cabinet post provoked warm objections from one faction or another. Jackson made matters worse by refusing to pull distinguished members out of the Congress, and worse still by announcing that he meant to serve just one term. This ensured that presidential aspirants would take a partisan interest in his cabinet appointments. Chief among them were Vice President John C. Calhoun, the formidable South Carolinian, and New York's "little magician," Martin Van Buren, whose key role in the campaign earned him the State Department. Jackson's other selections were awful. He named North Carolina's John Branch to the Navy Department for the wealth and social prestige he would impart. He chose Georgia's John Berrian as attorney general for his hostility to Native Americans. He picked Pennsylvania's Samuel D. Ingham for the

Treasury because he waffled on tariffs. He appointed Adams's postmaster general, the Ohioan John McLean, to the Supreme Court in reward for his surreptitious use of the mails against Adams, then gave the Post Office to the negligent William T. Barry of Kentucky. Finally, to ensure the presence of at least one confidant in his cabinet, Jackson made the newly-wed Eaton secretary of war.

What a fiasco! Every man on Capitol Hill "knew" Peggy Eaton by reputation, if not in the biblical sense. Born Margaret O'Neale in 1799 (or was it 1796?), she grew up in Franklin House, a popular boardinghouse owned by her Irish father, a veteran of the war of independence. Everyone drank there and many members of Congress resided there, sans spouses, while Congress was in session. The tavern paid Peggy's way to finishing schools where she acquired graces as fetching as her physical charms. During the years when Dolley Madison presided over the White House, Peggy—"astonishingly pretty, lively, impudent, and full of blarney"—presided over Congress. She later boasted of causing duels, a suicide, an abortive elopement, and dozens of broken hearts. In 1816 she wed a ship's purser named John Bowie Timberlake and had three children. Timberlake's career prospered, thanks to the patronage of Senator Eaton, who arranged for him a series of postings including the USS *Constitution* ("Old Ironsides"). It was the tale of David and Bathsheba, with poor Timberlake playing the role of Uriah the Hittite. During Timberlake's lengthy voyages Eaton squired Peggy about town and handled her business. In 1828 news arrived of Timberlake's death, possibly by his own hand. Eaton promptly proposed and Peggy, now twenty-nine, accepted before her year of mourning was up. Senator Louis McLane expressed the sense of the Congress: "Eaton has just married his mistress, and the mistress of eleven doz. others!" [2]

Jackson's cabinet, sneered William Wirt, was "the millennium of minnows." Britain's minister reported a "general expression of disappointment." The chronicler Margaret Bayard Smith noted that the roster "grieved" friends of the president: "Even Van Buren, altho' a profound politician is not supposed to be an able statesman. . . . Yet on him, all rests." Van Buren feared that this was so. He was a short, balding man with unruly red whiskers who spent his entire career wondering if he was worthy to climb the next step on the ladder. He had been born in Kinderhook on the Hudson, the very Dutch village after which Washington

Irving modeled Rip Van Winkle's hometown. His father owned a few slaves, but was otherwise poor. The farmhouse doubled as a tavern. Martin had little schooling and few prospects until 1796, when, at age thirteen, he was apprenticed to a prominent Federalist lawyer. In this lawyer's office Martin learned how to dress, converse, persuade, collude, publicize, and ingratiate himself with useful "aristocrats." But he stubbornly rejected his benefactor's politics. After campaigning for Jefferson and Burr in 1800, Van Buren joined a Republican firm in New York City with close ties to Tammany Hall. He married, moved to Albany, and over two decades mastered the arts of deftly managing grassroots campaigns, patronage, a partisan press, and shifting alliances.[3]

Van Buren might have lost heart in 1819 when his parents and wife died, leaving him alone with four children; his rival DeWitt Clinton came to power; and he lost his post as attorney general. But the dogged Dutchman merely confessed to a certain "anxiety" and then threw himself back into politics. Within a year Van Buren's Bucktail coalition took over the legislature and elected him to the Senate. At once the Little Magician scouted Capitol Hill in search of potential patrons, especially among southerners. He found Calhoun "a fascinating man," John Randolph "most extraordinary," and Jackson "uniformly kind." He played cards and dined with such men while exploiting his status as a widower to befriend their wives. After February 1825, when the House of Representatives elected John Quincy Adams president despite Jackson's plurality in the popular vote, Van Buren took up his next challenge. He would reinvent the Democratic Republican Party as a union of "the planters of the South and the plain Republicans of the North," and obliterate Adams's National Republicans as thoroughly as Jefferson had effaced the old Federalists. Sure enough, in 1828 Van Buren's Democratic machine carried twenty of New York's thirty-eight electoral votes for Jackson. An apparent draw, that result in fact dashed Adams's hopes of breaking out of his New England bastion and helped to swing pivotal Pennsylvania behind Jackson.[4]

When Jackson, as president-elect, arrived at Gadsby's Hotel to await his inauguration, Van Buren's lobbyists were already on the ground. Churchill C. Cambreleng gathered intelligence, including the spicy details about Peggy Eaton. Alexander Hamilton's son James, now a loyal Democrat, communicated Jackson's plea that Van Buren devote his "intelligence and sound judgment" to the State Department. Van Buren had just been elected governor, but decided once and for all to elevate national above state ambitions. His friends thought it a fatal mistake, given Jackson's debilities, scheming brain trust, and laughable cabinet. So Van Buren was already apprehensive when he was met in Washington City by a mob of toadies begging for diplomatic assignments. The trying day ended long after dark when he finally dropped in at the White House and spied, by the light of a lone candle, the president hunched over in coughs and despair. Jackson croaked an affectionate greeting and then sent Van Buren to bed.[5]

That the administration was sure to get off to a terrible start was evident from the day in March 1829 when John and Peggy Eaton called on Floride Bonneau Calhoun, the vice president's petite, raven-haired Charlestonian wife. She did not slam the door on her unwelcome company, but made it clear she would not be returning the courtesy. The wives of Eaton's own colleagues in the cabinet followed Floride's example. They ostentatiously shunned Peggy at receptions and dinners. Clergymen denounced Peggy's adulteries. Even Andrew Jackson Donelson's wife Emily, the official White House hostess, urged Jackson to rid himself of the Eatons. Predictably, Old Hickory went to war on the other side. He insisted that Peggy was "as chaste as those who attempt to slander her," and trumped the clergy's insinuations with his Freemason card. "Every person who is acquainted with the obligations of masons must know that Mr. Eaton, as a mason, could not have criminal intercourse with another mason's wife, without being one of the most abandoned of men." Jackson ordered the husbands to order their wives to behave, because "I did not come here to make a cabinet for the Ladies of this place."[6]

That was another mistake. Women were the nation's repository of Christian and republican virtue. How could they countenance, at the pinnacle of government, such effrontery as the Eatons displayed, especially when Peggy invited resentment by acting haughty (i.e., as their equal)? The wives' refusal turned the affair into a virility contest. Why, asked Eaton (and Jackson) did this nonsense continue? Were the husbands all henpecked weaklings? Or were they prolonging the split in the Cabinet for political motives? Needless to say, the husbands preferred people to draw the second conclusion, even if false. Jackson assumed a conspiracy. Sick again, this time with dropsy, he wasted months examining Peggy's past, even debating a pastor over the alleged date of an alleged miscarriage eight years before. The comedy climaxed in September when Jackson screamed, "She is chaste as a virgin!" before the stunned leadership of the executive branch of the United States government. When Congress convened two months later, however, Peggy was still shunned.

Whose stock tumbled as a result of this "Eaton malaria"? John C. Calhoun had not even been in the capital since March, but Jackson assumed that he and Floride encouraged the ostracism to wreck the administration and propel their own ambitions. Whose stock soared? Martin Van Buren, who had no wife to worry about, smoothly entertained the Eatons, and ingratiated himself with the president. They rode horses, hiked, and discovered much in common, including their humble origins, Jeffersonian roots, political savvy, common opponents, and status as widowers. By the end of the year Jackson was praising Van Buren's "frank, open, candid, and manly" character. So the upshot of the "petticoat war," as John Quincy Adams described it, was that "Calhoun leads the moral party, Van Buren that of the frail sisterhood; and he is notoriously engaged in canvassing for the Presidency by paying his court to Mrs. Eaton."[7]

The only achievement of Jackson's first year was a minor purge of officials. He had pandered to voters by declaring any honest citizen capable of government service, and his vow to clean the Augean stable caused bureaucrats to expect a "reign of terror." In fact, Jackson removed just 9 percent of the federal register in 1829 and just 20 percent over eight years. It was Senator William L. Marcy from the patronage-laden state of New York who boasted (in 1832), "To the victor belong the spoils of the enemy." Of more moment was the auditor Amos Kendall's discovery that "rats" in the Treasury had embezzled $250,000 and those in the Navy Department more than $1 million. That persuaded the president to emphasize economy and probity in his first annual message to Congress.

Given Jackson's lackluster inaugural speech, his message of December 8, 1829, on the state of the Union was the first indication of what his vaunted Democracy was about. The speech (read by a clerk, not the president) greeted "the Federal Legislature of twenty-four sovereign States and 12,000,000 happy people," gave thanks to a benign Providence, then droned through a 10,000-word list of initiatives. It called for rotation of federal employees, a judicious tariff (whatever that meant), retirement of the national debt, respect for states' rights, liberal pensions for war veterans, swift action to determine "the ulterior destiny of the Indian tribes," reconsideration of the Second United States Bank inasmuch as "it has failed in the great end of establishing a uniform and sound currency," and an amendment to abolish the electoral college (a slap at the outcome of 1824). Save for veterans' benefits, these were all hot-button issues. But the only one susceptible to immediate action was the "Indian question." Jackson meant to expel all "redskins" east of the Mississippi or see them go the way of "the Mohegans and other dead tribes" if they resisted.[8]

Why? Was Andrew Jackson simply a racist who had won fame and fortune by killing and despoiling Native Americans? Was there something psychotic about his pose as "Great Father" to Indian "children"? Was Jackson sincere when he said the Indians would surely perish unless the federal government separated and sheltered them? Was "ethnic cleansing" a logical corollary of Jackson's egalitarian program to expand opportunity for the white majority? Or was Jackson just forced to act boldly when a noxious witches' brew boiled over on his watch?[9]

Whatever the merits of the first four theories, the last is certain. After thirty years the government's original Indian policy of gradual assimilation was almost universally judged a failure. Teachers and missionaries exhorted Native Americans to embrace agriculture, literacy, and Christianity. Most Indians resisted, but even those trying to imitate the white man were frustrated by a lack of time and resources. In the years after Mad Anthony Wayne's victory at Fallen Timbers in 1791, the northern tribes were forced to cede most of the land between the Ohio River and the Great Lakes. In the years after Jackson's victory at Horseshoe Bend in 1813, the southeastern tribes were forced to cede most of the land between the Appalachians and the Mississippi. By 1820 a quarter of all U.S. citizens already lived beyond the Appalachians. Violence, distrust, and hatred prevailed wherever the races were in proximity. To pioneers it was ludicrous to suppose that the march of civilization must halt in deference to aborigines. As Jackson put the matter to Congress: "What good man would prefer a country covered with forest and ranged by a few thousand savages to our extensive Republic, studded with cities, towns, and prosperous farms, embellished with all the improvements which art can devise or industry execute . . . and filled with all the blessings of liberty, civilization, and religion?" Indian Commissioner Thomas L. McKenney concurred. "Two centuries have gone round," he reported, "and the remnants of these people, that remain, are more wretched than when they were a great and numerous people. . . . They catch fish—and plant patches of corn; dance, paint, hunt, fight, get drunk when they can get liquor and often starve. There is no sketching that can convey a clear perception of [their] misery and degradation."[10]

The last effort to foster assimilation was the Indian Civilization Act of 1819, but its appropriation of just $10,000 per year made it a pretense. By 1824 President James Monroe came around to the alternative first broached by Jefferson in 1803: separation of the races. Of course, forcible removal of Indians beyond the Mississippi would be "revolting to humanity, and utterly unjustifiable." So Monroe hoped to persuade Indians that voluntary exile was the only way they might preserve their customs or else buy time to civilize. During the 1820s several thousand Native Americans straggled west on their own. But the so-called Five Civilized Tribes clung to the treaties that granted the Choctaw and Chickasaw perpetual title to central and northern Mississippi, the Creeks eastern Alabama, the Seminoles the heart of the Florida territory, and the Cherokee northwestern Georgia. They refused to cede any more land, much less abandon the woods of their ancestors. The patience of white cotton planters, settlers, land speculators, and—in the case of Georgia—gold prospectors wore thin. If the federal government was not willing or able to condemn Indian lands, then state governments would act on their own. In December 1828 Georgia's legislature in effect nullified federal treaties with the Cherokee nation. The state refused to countenance within its borders an independent "nation" keeping 6,000 square miles of real estate off the market. Henceforth, the state would enforce its laws uniformly in all parts of Georgia, take title to Indian lands, and use force to ensure compliance. An incipient constitutional crisis thus greeted the incoming president.[11]

Before urging swift legislation in December 1829, Jackson dispatched generals to persuade southern Indians—and McKenney to persuade northern whites—that voluntary removal was the "just and humane" solution. Vocal church leaders and members of Congress choked on that. Senator Theodore Frelinghuysen, New Jersey's "Christian statesman," damned a "solution" that trampled Indians' civil rights and exposed the Indians to greedy, bloodthirsty state governments. So much for Jackson's commitment to liberty and the little man! Senator Henry R. Storrs of New York warned southerners that endowing the executive branch with the power to condemn private property and remove human beings had ominous implications for their doctrine of states' rights. A Yale graduate, Jeremiah Evarts of the American Board of Commissioners for Foreign Missions, aroused religious sentiments. Removal was not only immoral, he cried, but ill-founded, given the progress that missionaries were making among Indians. Evarts's articles (written under the pseudonym "William Penn") inspired a flood of petitions that gave northern congressmen pause. But southerners (with the notable exception of Davy Crockett) hurled back the charge of hypocrisy. It was easy for easterners to weep for Indians: the tribes in their neck of the woods had long since been exterminated or subjected to state law.

What tipped the balance was probably an authoritative opinion from the West: that of Lewis Cass, whose reputation as an Indian expert won him election to the American Philosophical Society. A native of New Hampshire and a graduate of Exeter, he served for eighteen years as governor of the Michigan Territory (and grand master of its Masonic lodge). Cass wrote long articles in the influential *North American Review* arguing that Indians had regressed since contact with the white man, owing to disease, demoralization, and whiskey. Moreover, Indians were incapable of civilization because their languages precluded concrete, rational thought. As late as 1827 Cass had opposed Indian removal as "controversial and intolerably difficult and expensive." But in 1830 he rallied to Jackson. The Great Father of Detroit assured wavering consciences that no race of mankind was less provident, industrious, peaceful, governable, or intelligent than the American Indian: prudence and charity dictated separation. The Indian removal bill passed the Senate handily and squeaked by, 102 to ninety-seven, in the House. Jackson signed it into law on May 28, 1830.[12]

Evarts suspected Cass of trimming his sails "in such a manner as to catch the breeze of government favor and patronage." But the worst feature of Cass's testimony was his pretense of knowing anything about the Five Civilized Tribes 1,000 miles to the south. They were called civilized precisely because of the strides they had made to conform with American institutions and values. Beginning in 1808 the Cherokee, Chickasaw, and Choctaw replaced government by consensus under chiefs and councils with representative assemblies, the rule of law, militias, courts, and (by the 1820s) written constitutions. Their governments supplemented federal and church contributions with $20,000 per year to support, by 1825, eleven English-language schools in the Choctaw nation, six in the Cherokee, two in the Chickasaw, and one in the Creek. Gifted Indians won scholarships to the missionary board's academy in Connecticut. The brilliant Sequoya devised an eighty-five-character Cherokee syllabary, enabling his people to read and write; translate the Bible; and start their own newspaper, *The Cherokee Phoenix*, in 1828. A census conducted in 1824 revealed that the Cherokee population was growing, not dying out. Missionaries reported a rapid spread of agriculture. Roads linked all corners of the reserve to the new capital of New Echota. Far from wandering in the woods, most Cherokee resided in alluvial lowlands on 1,700 farms; raised 269,000 bushels of corn per year; tended 80,000 head of livestock and 63,000 peach trees; and even owned 776 slaves. Good Americans, in every respect! They even knew about politics and class conflict. When "common Indians" threatened rebellion against the wealthy mixed-breed planters in 1827, Chief John Ross (who was seven-eighths white) mollified them with an American-style constitution that ensured private ownership of improvements made on the land and government loans to help Indians improve their lot.[13]

One can imagine how this Cherokee renaissance terrified Georgians. However much they wanted the rest of America to think of Indians as savages, their impatience was born of the knowledge that a growing number were not. So the governor cleverly informed Jackson that the seeming progress of the Cherokee was really the work of an "oppressive system of Government which the Cherokee chiefs (principally the educated sons of white men) are now enforcing upon the body of the Indians." Of course, this line of reasoning implied the worst Indians were those with the most white blood. But it allowed Jackson to imagine that removal would "liberate" common Indians and thus dovetail with his program to save white commoners from privileged elites. Secretary of War Eaton assured the president that the common Indians would "soon burst their bonds of slavery, and compel their chiefs to propose terms for their removal." So Jackson expected the Five Civilized Tribes to welcome his offers of fair payment, transport west "under the care of discreet and careful persons," plentiful food and supplies, and rich lands that would be theirs "as long as grass grows and water runs . . . and I never speak with forked tongue."[14]

The truth was otherwise. Most Indians would not depart voluntarily, and the administration was mad to think it could transplant tens of thousands of people over hundreds of miles of wilderness on the cheap.

General John Coffee resorted to outright bribes to win Choctaw chiefs over to the Treaty of Dancing Rabbit Creek in September 1830. The first 1,000 Choctaw migrants trekked all winter across rugged Arkansas accompanied only by missionaries forced to spend their own money to stave off starvation. Less than half made it alive to their arid, windswept reserve in Oklahoma. The Indian Bureau's thrifty response was to hire private contractors to ship the remaining Choctaw by steamboat up the Arkansas River. Predictably, these contractors defrauded the government, gave the Indians rotten meat or no food at all, and crammed them like cattle on boats stricken with cholera. In the end, just 9,000 Choctaw made it west; 5,000 died; and 7,000 melted into the swamps. By 1832 the Chickasaw and the Creeks accepted Coffee's treaties, but only because authorities in Mississippi and Alabama refused to police the crooked speculators and gun-toting squatters who cheated Indians of land in anticipation of their removal. The War Department dispatched federal troops, then withdrew them to prevent clashes with white state militias. Instead, the Army waged a Second Creek War (the first being Jackson's in 1813) against young braves defying removal. Captured and manacled, the "hostiles" marched west in the vanguard of 3,000 Creeks, 700 of whom perished.

The Cherokee appealed over the heads of Congress and the executive to the Supreme Court, where wizened John Marshall, the "last Federalist," still presided. In *Cherokee Nation v. Georgia* (March 1831) the court ruled that Indians were "domestic dependent nations" beyond state jurisdiction. In *Worcester v. Georgia* (March 1832) the court overruled a Georgia law that prohibited "un-licensed" whites (i.e., missionaries) from entering Cherokee territory. Those judgments were famously impotent. Jackson never said, "Marshall has made his decision; now let him enforce it" (Horace Greeley put those words in his mouth), but Georgians ignored the Supreme Court with the president's imprimatur. The state simply asserted title to Cherokee land and sold it off in lotteries. Whites poured in. Chief John Ross himself was evicted and later arrested. Still, most of the chiefs rejected a treaty offering them $4.5 million. So Jackson's envoy, the Reverend John Schermerhorn, rigged the process. His Indian commission ruled that any Cherokee who failed to show up for a council in December 1835 would be assumed to favor removal. The Ross faction boycotted it. A minority party did not; hence Schermerhorn left with an "agreement" whereby the Cherokee traded their empire for $15 million and 7 million acres of Red River country. That bogus Treaty of New Echota sparked another debate, but the Senate ratified it, with one vote to spare, in May 1836.

Jackson's term had expired when the federal government at last pushed the Cherokee on to the *Nunna daul Isunyi* ("Trail Where We Cried" or "Trail of Tears"). General Winfield Scott made earnest efforts to minimize suffering during the Army's roundup of Indians in the summer of 1838 and their march to Oklahoma over the fall and winter. But inevitably soldiers got out of hand. An eyewitness described men dragged from their fields, children from play, and mothers from nursing infants; herds driven off; aged and lame people prodded by bayonets; and Indian graves looted for silver. Years later a veteran of bloody Civil War battles recalled the removal of the Cherokee as "the cruelest work I ever knew." Scott also did his best to spare the exiles hunger and cold. But since most Indians refused to prepare for a journey they did not want to make, the crabbed federal commissaries were grossly inadequate. A missionary who witnessed the march of the 18,000 said he felt "in the midst of death itself." The mortality, usually estimated at 4,000, might well have been twice that.[15]

Native Americans got a measure of vengeance, thanks to the Seminoles. In 1832 their chiefs agreed to send a party to Oklahoma just to reconnoiter

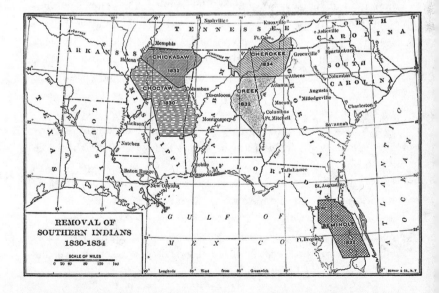

REMOVAL OF
SOUTHERN INDIANS
1830-1834

SCALE OF MILES
0 20 40   80   120   160

the lands offered them. When the scouts arrived at Fort Gibson, Schermer-horn presented them with a creatively edited draft of the treaty endowing the scouts with power to approve removal itself. Bribes closed the deal. When the news reached Florida, however, angry warriors led by the charis-matic Osceola assaulted the Indian agency, killing five whites. The Second Seminole War (the first being Jackson's in 1818) broke out in earnest in 1836 when Indians and fugitive Negroes killed 107 U.S. soldiers in an ambush. General Scott was dispatched, but the rebels eluded him in their swamps. Zachary Taylor made his reputation in the bloody victory at Okeechobee, but the rebels fought on. In 1838 General Thomas Sidney Jesup arrived with thousands of reinforcements. All he managed to do was burn villages, re-enslave some captured blacks, and seize Osceola by treachery. This was hardly edifying, yet it was Jesup who told the truth to the War Department. "In regard to the Seminoles we have committed the error of attempting to remove them when their lands were not required for agricultural purposes; when they were not in the way of the white inhabitants; and when the greater part of their country was an unexplored wilderness, of the interior of which we were as ignorant of as the interior of China." By 1842, when the Army declared victory and left, the war had cost 1,500 American lives and $20 million. So much for Indian removal on the cheap.[16]

North of the Ohio River, removal of the Shawnee, Seneca, Ottawas, Potawatomis, Chippewa, Wyandots, Menominees, and Winnebagos oc-casioned less resistance, corruption, and violence. Their only vengeance was wreaked by a Winnebago prophet's disciple, the Sauk warrior Black Hawk. In 1832 his war party reclaimed their ancestral lands near Rock Island, Illinois, and routed the local militia. Federal and state troops (including Abraham Lincoln) supported by armed steamboats soon trapped the renegades on the banks of the Mississippi. Black Hawk survived to write an autobiography proclaiming "I am a man!" Senti-mental whites wept, enabling themselves to feel better about the whole business.[17]

By setting in train the coercive resettlement of 46,000 Native Ameri-cans Jackson gambled that most whites would either approve or not care. He won his bet. Jackson was hailed for solving one of the most intractable problems on the national agenda. In fact he solved nothing. He merely established a cruel precedent for future federal officials when white civili-zation burst into the Great Plains.

* * *

The United States were (at this time, the name of the nation was construed as plural) a republic of words. The Declaration of Independence gave it life, the Constitution its form, the *corpus juris* its vestments, and words both gilded and debased the coinage of the nation's transactions. Foreigners mocked the speech of Americans, failing to grasp how they exploited the English language's pliable syntax and rich vocabulary (thanks to its Germanic, Latin, Norse, and Celtic roots) in their efforts to master a continent and themselves. Americans did not care if words borrowed from Indians and immigrants or coined by farmers, sailors, and gamblers offended the King's English so long as they greased the business of daily life, preferably while eliciting a guffaw. In his encyclopedic survey of the American language H. L. Mencken observed that drinking was one spouting font of new words; hence the multitude of names for "cock-tails" and drunkenness. Another font was politics. Already by Jackson's time, current words and phrases included "run" for office, endorse, candidacy, favorite son, omnibus bill, gag rule, gerrymander, filibuster, executive session, spoils system, mass meeting, steering committee, straight or split ticket, pork barrel, landslide, machine, dark horse, lame duck, on the stump, on the fence, mugwump, horse-trade, logroll, eat crow, buncombe, and lobby. That patterns of speech needed no regulation was "especially true in a country where scholarship is strange, cloistered and timorous, and the overwhelming majority of the people are engaged upon new and highly exhilarating tasks, far away from schools and with a gigantic cockiness in their hearts." The election of Andy Jackson, thought Mencken, shook the linguistic hegemony of the New England and Tidewater gentry, so that "an urbane habit of mind and utterance began to be confined to the narrowing feudal areas of the South and the still narrower refuge of the Boston Brahmins."[18]

Yet those same decades during which politicians and peddlers learned to make stump speeches and sales pitches in the vernacular were also the golden age of American oratory. Even hayseeds were ready to hoot down candidates, preachers, and actors who failed to soar on rhetorical wings. The American, observed Tocqueville, "speaks to you as if he were addressing a society." That should not surprise us, since legislatures, courts, tent meetings, clubs, societies, and public debates were often the only entertainment available in rural communities. People rode miles just to hear others talk. Pupils learned elocution by memorizing speeches from copybooks. Grandiloquence seasoned with poetry, biblical verses, and classical allusions flattered the masses by dignifying even mundane causes and those who embraced them. Aristotle had laid it all out in his *Rhetoric*. Republican government was a battle of words. The essence of demagogy was to seduce listeners into becoming coauthors of incomplete syllogisms. That is, one baited the voters, jury, or lynch mob with a seemingly relevant principle, analogy, or emotion, inviting them to leap to the intended conclusion themselves. Aristotle wrote his treatise to alert ancient Athenians to orators' tricks. All he did was give future tricksters a manual.[19]

How did orators operate when obliged to address each other, for instance in the halls of the Capitol? Since celebrated speeches were often published around the country, senators had to consider public opinion even as they worked to persuade colleagues to vote aye or nay on a bill or else win allies on some other issue unrelated to the matter at hand. Multiple audiences required that rhetoric sometimes cloak motives different from or even opposite to those implied by the literal words. That made rivals suspicious and turned even revered notions like liberty, union, compact, and sovereignty into invidious code words. Senators launched them, like billiard balls, in one direction, but the ricochets were what counted. Liberty for whom? Union at what price? Why compact rather than contract? Sovereignty resting where? Even silence could be ambiguous. For instance, Jackson's ponderous compendium of pressing national issues never once mentioned slavery. Ever since the Constitutional Convention Americans finessed slavery through a conspiracy of silence broken only by the traumatic debate of 1819 over Missouri's statehood. But Congress failed to honor that conspiracy when even permissible code words got out of control.

It began innocently enough. Senator Samuel Foot of Connecticut suggested that a committee inquire into the wisdom of limiting sales of western land to sections already surveyed. For fifteen years the federal government had done a brisk and corrupt "land-office business." Perhaps it was time for a breather. The suggestion alarmed Missouri's brawler turned statesman Thomas Hart Benton because he sniffed two ulterior motives: Foot meant to slow the growth, and hence the political clout, of western states, and ipso facto maintain the cheap labor supply needed by New England's mills. Robert Y. Hayne of South Carolina smelled an additional rat. The loss of receipts from land sales would make the government more dependent on customs receipts from protective tariffs, which Yankee manufacturers loved and southern consumers hated. Moreover, since South Carolinians deemed discriminatory federal laws unconstitutional, Foot's "innocent" suggestion could be construed as an assault on states' rights. Fantastic or not, that logical leap suited Hayne's own ulterior motive, which was to trade southern votes over land policy for western votes over tariff policy. On January 19, 1830, he stated the opinion that westerners "have had some cause for complaint." Such were the demands of florid speech that it took him an hour to say it.[20]

Hayne studied law in Charleston, commanded a regiment in the War of 1812, and entered politics as a disciple of John C. Calhoun, the brilliant graduate of Yale and Tapping Reeve's law school in Connecticut. When Calhoun supported the charter of the Second Bank of the United States and internal improvements such as roads and canals, Hayne did likewise. When Calhoun broke with John Quincy Adams and swung behind Jackson in 1825, so did Hayne. When the Tariff of Abominations of 1828 caused Calhoun to draft an "Exposition" arguing a state's right to nullify unjust federal laws, Hayne also jettisoned nationalism. But Hayne was no lackey. His father owned the lush rice plantation Pon Pon; his legal mentor was Speaker of the House Langdon Cheves; his wife was a daughter of Governor Charles Pinckney. Hayne was less polished than Calhoun, but just as prominent socially and nearly as thrilling rhetorically. He was also gorgeous and just thirty-eight. So many women turned out for his speeches that spectators in the Senate gallery spilled onto the floor.

By contrast, the craggy forty-seven-year-old Daniel Webster had a face that scared children. His protruding brow, jowls, and pursed lips hardened his visage into a perpetual scowl. But once he began to speak, auditors almost stopped breathing for fear of missing a word, cadence, or gesture. A graduate of Dartmouth, Webster represented New Hampshire in Congress during the years when the War of 1812 wrecked New England's economy. His support for the secessionist Hartford Convention of 1814 called his patriotism into question. So Webster started over in Boston and soon was renowned as a spellbinder. He argued dramatic cases before the Supreme Court and delivered famous orations at the Bunker Hill monument and Faneuil Hall. The latter event was a tour de force in which Webster used rhetoric to belittle rhetoric and feed his own fame with a pretense of humility. "True eloquence, indeed, does not consist in speech. . . . Words and phrases may be marshaled in every way, but they cannot compass it. It must exist in the man. . . . [T]his, this, is eloquence, or rather it is something greater and higher than all eloquence, it is action, noble, sublime, *godlike* action." The speech was about John Adams, but the public came away lauding the "godlike Daniel." Webster also craved money to support his formidable appetites and his five children. So he supplemented the rewards of a thriving law practice with a generous retainer from the Bank of the United States. That became a notorious conflict of interest when Massachusetts elected him to Congress and then to the Senate. What really made Webster's fortune, however, was the death of his pious, sweet wife in 1828. Within a year he saved from spinsterhood the thirty-one-year-old daughter of the mogul Herman Le Roy of New York, purchased a sprawling estate with her dowry, and reveled in high society. A man of strong Yankee principles, he was never religious except insofar as he worshipped himself.[21]

In January 1830 Webster rose in the Senate to rebut Robert Hayne. Perhaps he hoped that the suspension of western land sales would boost the value of his own western investments. Perhaps he spied a chance to play patriot and bury the memory of Hartford. Perhaps he meant to maneuver the West into alliance with the Northeast instead of the South. Most likely his motives were murky and manifold. Webster began with the disingenuous claim that he had no intention of speaking until the gentleman from South Carolina impugned the "obnoxious, the rebuked, the always reproached East. . . . Sir, I rise to defend the East!" He recited the history of New England's support for roads and canals in the West, defended the tariff, then administered a gratuitous slap by crediting a New Englander with the clause in the Northwest Ordinance that banned slavery north of the Ohio River, thereby ensuring the West's prosperity.[22]

Hayne countered with a verbal barrage lasting three hours. As for this New Englander, all that southerners knew of him was his presence at the treasonous Hartford Convention (i.e., *your* treason, Mr. Webster). As for slavery, it was not southern sailors and merchants who profited from that "accursed traffic." Indeed, the "States North of the Potomac actually derive greater profits from the labor of our slaves, than we do ourselves," thanks to the wicked workings of markets and tariffs that skimmed plantation proceeds into the pockets of northern brokers and manufacturers. Nor was it southerners who traded in "false philanthropy" and shed "weak tears" over people in bondage. If the fanatical "friends of humanity" wished to observe Negroes' suffering they need only attend to "the unfortunate blacks of Philadelphia, and New York, and Boston," whose poverty and wretchedness were unmatched on earth. But northern do-gooders had no interest in truth. "Their first principle of action is to leave their own affairs, and neglect their own duties, to regulate the affairs and the duties of *others*." The gentleman from Massachusetts claimed "consolidation" of central government was the object of the Constitution. But the gentleman was mistaken. The object of the framers "was not the consolidation of the Government, but the consolidation of the *Union*," which purported friends of the tariff endangered by assaulting the welfare of the South. By his words, the gentleman had "invaded the State of South Carolina, is making war upon her citizens. . . . I will drive back the invader discomfited. Nor shall I stop there. If the gentleman provokes the war, he shall have war." [23]

How did a bland discussion about public lands escalate into a fight for the soul of the Constitution? Webster slyly provoked it when he shifted the issue from western lands to the tariff, knowing the proud South Carolinian was bound to take a radical states' rights position. This in turn allowed Webster to champion the Union in a speech that enthralled the packed chamber for two days. He started by cooing that Mr. Hayne ought not to feel threatened. "The domestic slavery of the South I leave where I find it." But the Union was not strengthened by those who threatened rebellion over the tariff or asked, "What interest has South Carolina in a canal in Ohio?" Rather, the Union was strengthened by the "selfish men of New England" who applauded projects of benefit to the nation at large. "*The great question is, whose prerogative is it to decide on the constitutionality or unconstitutionality of the laws?*" Did four-and-twenty state legislatures have the authority to annul federal law? That was a formula for anarchy. Only the Supreme Court acting in the name of the sovereign people had the power to strike laws. After winding the stem, Webster concluded: "When my eyes shall be turned to behold, for the last time, the sun in Heaven, may I not see him shining on the broken and dishonored fragments of a once glorious Union . . . drenched, it may be, in fraternal blood! Let their last feeble and lingering glance, rather behold the gorgeous Ensign of the

Republic . . . bearing for its motto, no such miserable interrogatory as, *What is all this worth?* Nor those other words of delusion and folly, *Liberty first, and Union afterward*, but . . . that other sentiment, dear to every true American heart—Liberty *and* Union, now and forever, one and inseparable!"[24]

The Webster-Hayne debate revealed how hard it was for Americans to hold truth a safe distance away when even innocent talk of liberty, union, tariffs, and land sales could conjure the Banquo's ghost of slavery. Where did President Jackson stand? Calhoun meant to find out when Democrats gathered at the Indian Queen Hotel on April 13 to salute Jefferson's birthday. Unlike their party's timid, conflicted founder, Jackson felt no guilt over slavery and loved a fight. Surely he must side with the South, especially after seventeen toasts to states' rights and liberty. Instead the president raised his glass to "Our Union: *It must be preserved.*" Everyone gasped. Hayne begged the president to amend his toast to "Our *Federal* Union." Calhoun cried, "The Union: next to our liberty, the most dear." Van Buren purred, "Mutual forbearance and reciprocal concessions."[25] If words bound the republic together, words could tear it apart. But the actors seemed trapped in their roles.

Jackson imagined his role as that of a Roman tribune or dictator, summoned to executive power for a season to defend the plebeians against corrupt patricians. That meant, among other things, slashing federal expenses and retiring the national debt. Yet even as they debated lofty constitutional principles, members of Congress persisted in cutting deals to fund internal improvements benefiting states and localities. The president itched to send them a message, and got an opportunity when Van Buren briefed him on the Maysville Road bill. It appropriated Treasury funds to build a road entirely within the borders of one state, which Jackson considered unjust and probably unconstitutional. The favored state was that of Jackson's old rival Henry Clay, and the bill was favored by his new rival Calhoun. So Jackson waited until Indian Removal was passed, then vetoed the Maysville Road the next day. Congress, thinking this an isolated case born of pique, passed more spending bills on the eve of adjournment. The apoplectic Jackson vetoed them all in what was then an unprecedented assertion of presidential power. His justification was bolder still. He, the president, was the only official elected by all the people. He, not the Congress, was vox populi.

* * *

In 1831, the year of the solar eclipse, shadows fell on the land.[26] To put it another way, much that was hidden became painfully clear after the moon passed on. Just five days after the eclipse of February 12, Duff Green's *United States Telegraph* killed hopes of reconciliation between Jackson and Calhoun, plus any chance the latter, a two-term vice president, had of winning the White House. Incredibly, Calhoun himself was responsible. The origins of his blunder dated from 1818, when Jackson was accused of malfeasance for his invasion of Spanish Florida. He assumed that Calhoun, Monroe's secretary of war at the time, had defended him. But soon after the Jefferson Day dinner William Crawford leaked the twelve-year-old secret that Calhoun had pressed for Jackson's punishment. Old Hickory spent eight months trying to dig up the truth while Calhoun tried to muddle the truth. He called it a conspiracy between Crawford and Van Buren. He even pestered old Monroe, who was dying in poverty. At length Calhoun might have reasoned the only way to salvage his presidential ambitions was to appeal to public opinion. But whatever his motive (none makes sense), Calhoun gave his correspondence on the Florida imbroglio to the *Telegraph*, which eagerly printed a fifty-two-page pamphlet. To say it backfired is an understatement. From the White House the word went forth to every Democratic organ: "Calhoun is a dead man." Jackson expelled him from the party and for good measure decided to run himself for a second term. Calhoun, his prospects ruined, set his teeth against Jackson and embarked on a new career as extortionist for the South.[27]

Calhoun's fall clarified Van Buren's next move, which was to get out of town. That was the only way to avoid stigmatization by Calhoun's supporters and Jackson's enemies, pose as a statesman above the battle, and do the president a great favor at the same time. When Van Buren first floated the idea on the bridle path, Jackson did not understand. After sleeping on it he did. The paralysis over Peggy Eaton had to end, but without Eaton's having to make a humiliating resignation that would appear a defeat for Jackson. The solution was for Van Buren, the most respected officer, to resign in tandem with Eaton; afterward, the president could purge the whole cabinet. Yes, replied Jackson: "a House divided cannot stand." In spring 1831 everybody resigned save the postmaster, whereupon Jackson appointed a loyal, regionally balanced "reelection" cabinet. Michigan's Cass got the War Department as his reward for dooming the Indians, Levi Woodbury of New Hampshire the Navy, Louisiana's Edward Livingston the State Department, and Delaware's Louis McLane the Treasury. For attorney general, Jackson chose Roger Taney, the squinting, bookish flack who chaired his Maryland campaign. Taney's Catholicism appealed to the Irish, his relation by marriage to Francis Scott Key cheered nationalists, and his contentment with slavery reassured southerners. The appointment launched Taney on a trajectory leading to the post of chief justice. Jackson appointed Eaton governor of the Florida territory, then U.S. minister to Spain (where Peggy thrived). Van Buren became minister to Great Britain, where he spent eleven delightful months hobnobbing with royalty. When Congress convened, however, Calhoun lobbied to deny confirmation to Van Buren, then arranged for the Senate vote to end in a tie so that he as vice president could cast the decisive nay. Thomas Hart Benton thought this a blunder that was certain to make Van Buren Jackson's next choice for vice president. But the vindictive Calhoun crowed, "It will kill him, sir, kill him dead."[28]

Various dates may be proposed for the functional end of Democratic Republican one-party rule. But its symbolic end surely was the Fourth of July 1831, when James Monroe gave up the ghost exactly five years after the deaths of John Adams and Jefferson. He was the last of the Virginia dynasty and the last to preside over a unified party. Everyone seemed to know there was no going back, or forward, to the times Monroe loved to call "happy." In Old South Church, Boston, John Quincy Adams was the featured eulogist. He worked on his remarks for eight weeks, but found little of comfort to say. The atmosphere was steamy, the crowd restless, the ambience glum. Ralph Waldo Emerson came away thinking there had been "nothing heroic in the subject, & not much in the feelings of the orator." In Charleston on July 4, Robert Hayne's address in another packed church dripped with the heroic. The invidious tariff imposed by the North was like a thief who "comes not upon us like the strong-man armed, at noon-day, but steals on the wings of the wind, and, like the 'pestilence walking in darkness,' infuses the fatal poison in our veins." Three-eighths of Charleston's exports, he judged, went to pay federal duties on imports. He ended with an echo of Tom Paine's cry for independence in 1776. If the only deliverance for a people "held in bonds of iron" was nullification, "who is there that would not be a *nullifier*?"[29]

Three weeks later Calhoun buttressed the case with a lengthy essay of political theory known as the Fort Hill address. He avoided the word "nullification" but argued the Constitution was a contract which sovereign states had a right to suspend if federal law trampled on "the peculiar and local interests of the country." Rejecting majority rule through Congress, Calhoun proposed that federal law be subject to "concurrent majorities" representing all sections and interests. Webster thought it dangerously plausible. John McLean lamented, "Our friend Calhoun is gone, I fear, forever."[30]

Southerners talked of tariffs at the top of their lungs, but failed to silence talk of the most dangerous source of sectional strife. In the wee hours of August 22, 1831, seven enslaved Negroes led by thirty-one-year-old Nat Turner crept into the big house on Joseph Travis's Tidewater plantation near Cross Keys, Virginia. Armed with hatchets and axes, they beheaded Travis, his wife, two other adults, and a baby. That began forty-eight hours of terror in Southampton County, during which Turner's eighty or so disciples murdered at least fifty-seven white people, including ten pupils whose severed limbs were found "in a bleeding heap" on their schoolroom floor. Militiamen, soldiers from Fort Monroe, and disorganized mobs of furious whites counterattacked, often without discrimination or mercy. More than 100 blacks, marauders and innocent alike, were killed or brutalized during the nine weeks Turner remained at large. But even his capture, trial, and swift execution did not stop violence from spreading through Virginia and North Carolina, where an eerie replay of the Salem witch trials ensued. Suspected black agitators who implicated others were pardoned while those who insisted on their innocence were tortured and killed. Farther afield, from Norfolk to New Orleans, white mobs rioted against free blacks, abolitionists, and teachers or preachers who ministered to the enslaved.

What caused Nat Turner's revolt? It was scary enough to assume that fugitive slaves had burst forth from their hideouts in the Dismal Swamp in search of plunder and food. When it turned out the "blood-stained monster" and his cohort were "the property of kind and indulgent masters" from local plantations, Virginians recoiled. They dared not entertain the thought that slavery itself might foster such savage hatred; hence they assumed Turner must be deranged. Thomas R. Gray, a lawyer who represented the rebels, assured everyone that this was the case in his sensational pamphlet, *The Confessions of Nat Turner*. His motive was to cash in on his exclusive interviews with the ringleader. But his revelations met a desperate public need. According to Gray, Turner was a "gloomy fanatic" who believed himself born a prophet. Beginning in 1825 he had a series of apocalyptic visions in which white and black spirits battled in rivers of blood beneath a sun that went dark. Baptized by a white preacher, Turner at first preached Christian forbearance, but the "negroes found fault, and murmured against me." Then lo! the sun did go dark in 1831, and Turner took this as a sign. He felt called to bring "terror and devastation" on the white race so "the blood of Christ" might return to earth and "the first should be last and the last should be first."[31] The obvious lesson was that slaves could not be permitted to read or hear the Bible without supervision by whites, lest they be driven mad by millenarian images beyond their comprehension. But insofar as slaves did get ideas into their heads, southern whites shuddered to think what might lurk beneath the calm surface of daily life. A niece of George Washington called it "a smothered volcano—we know not when, or where, the flame will burst forth, but we know that death in the most repulsive forms awaits us."[32]

Many Virginians, perhaps a majority, insisted on lifting the veil of silence over the future of slavery in their state. Ever since 1787, when the planter George Mason hoped that the Constitutional Convention would mandate abolition, Virginians fretted over the scourge their ancestors first introduced to Anglo-America. The original tobacco culture had long ago lost its luster, owing to exhausted soil and low prices. Breeding slaves for the purpose of exporting them to the cotton states was contemptible. Once the largest state, Virginia had been surpassed in population by dynamic New York and Pennsylvania. Inside Virginia, growth was concentrated in the western counties, where plantation slavery was rare. Even in the Tidewater and Piedmont voices began to call for the extinction of slavery. Those voices were female. Moved by tender sentiment and candid fear, women such as the Presbyterian Louisa Cocke and the Episcopalians Ann Page and Mary Blackford tried to rejuvenate the American Colonization Society, devoted to shipping slaves back to Africa. Their tenderness showed itself in a commitment to the education and conversion of slaves, and a faith that their destiny was to carry the "glorious beams of christian revelation" to the dark continent. Their fear was expressed by the Female Citizens of Fredericksburg, who implored Virginia's assembly to consider their "defenseless state in the absence of our Lords, in times of apparent peace." In late 1831 more than 1,000 women signed petitions urging the "speedy extirpation of slavery in the Commonwealth."[33]

Breaking with precedent, the legislature entertained an antislavery proposal filed by a grandson of Patrick Henry and supported by a grandson of Thomas Jefferson. Suddenly, in January 1832, distinguished white men could stand up in Richmond and speak words once taboo: human bondage was evil, a curse, a calamity; abolition was a matter of when and how, not whether. Alas, it became readily apparent that no one knew *how* to get rid of slavery, and even the question of *when* was ominous. Tidewater burgesses shouted that the state had no power to confiscate private property or raise enough money to compensate planters. They asked what would become of the manumitted. It was ludicrous to talk of transporting 450,000 slaves to Africa when the government was powerless to deal with the 50,000 free blacks in their midst. Mere mention of abolition at some future time was bound to enrage blacks who would not live to see it. Indeed, any precipitate action would be perceived as rewarding the late insurrectionists and encourage more revolts. They claimed Virginia's slaves were "as happy a laboring class as exists upon the habitable globe," but might not remain so if infused with false hopes.

The great debate climaxed with a vote over whether it was expedient even to inquire about the prospect of some future plan. If the decision had been favorable, Virginia might have inaugurated a frank, ongoing discussion of ways to liquidate slavery, with Maryland, Delaware, and Kentucky probably following suit. But the assembly judged an inquiry "inexpedient" by seventy-three to fifty-eight, and instead passed laws to hasten the removal of free blacks and outlaw unsupervised worship among slaves.[34] Pretense triumphed, the "republic of words" moved on, and many a southern woman learned how to handle a gun.

"I am determined," wrote Andy Jackson, to "lay the perfidy, meanness, and wickedness, of Clay, naked before the american people. I have lately got an intimation of some of his secrete movements, which, if I can reach with possitive and responsible proof, I will wield to his political, and perhaps, his actual destruction. he is certainly the bases[t], meanest, scoundrel that ever disgraced the image of his god." Once upon a time Jackson, Calhoun, Adams, Crawford, Webster, and Clay had been comrades in arms. That time was long gone by 1831 when a fully self-conscious National Republican party rallied behind Henry Clay in an effort to oust the "wicked, passionate, and corrupt" Democrats and their "tyrant" Jackson. For Clay, the campaign was one part frontier feud, one part ambition, and one part high patriotism.

Clay was a native Virginian who received little schooling, but learned law from John Marshall's mentor George Wythe and oratory from Patrick Henry. In 1797 he joined his mother in Kentucky and married into the bluegrass gentry. So swiftly did Clay make a name for himself that the governor appointed him to fill out a Senate term while he was still under the age limit of thirty (nobody noticed). Clay, known as "Harry of the West," was homely—his head was too small for his frame and too thin for his large nose and mouth. But his eloquence, pluck, drinking, and gambling charmed frontiersmen, while his brains, ambition, and Masonry ingratiated him with the elite. Clay loved Ashland, his tasteful estate on a prosperous, slave-worked farm near Lexington. But for most of his life he spent part of each year in Washington City. Clay made Speaker of the House a powerful post, promoted the American System for economic growth, and won renown as the "great pacificator" while brokering the Missouri Compromise. Of course, the alleged "corrupt bargain" that made Adams president and Clay secretary of state earned him Jackson's enmity, a sentiment he returned after Jackson's landslide four years later. Accordingly, when Kentucky elected Clay to the Senate in the year of the eclipse, he made up his mind to outfox Old Hickory, save the country, and win the White House.[35]

Save the country indeed, for even though Clay's tactics were as slippery as an otter, he clung to his goals like a bulldog. The components of his American System were to him sacred tenets of the American civic religion for which he would "defy the South, the President, and the devil." They included preservation of the Union, because the nation had no future without that; protective tariffs to stimulate American industry, diversify the economy, and make the United States self-sufficient in the sinews of war; federal support for roads, canals, harbors, levees, and other internal improvements; and a sound currency regulated by the Second Bank of the United States (BUS). Jackson shared Clay's devotion to national unity, but otherwise used executive power to quash managed expansion of growth and opportunity. Clay stood for one sort of capitalism, which was associated with the North but which the South needed, too, in his judgment. Jackson stood for another sort of capitalism, which was associated with southern planters and homesteaders but which had little to offer the Northeast or the emerging Northwest. Clay ached for a vibrant future; Jackson clung to a dead Jeffersonian past.[36]

Clay returned to Capitol Hill deeply depressed—his son had just been committed to an asylum. Still, he made clear at once by word and posture that he considered himself the leader of the opposition. This meant opposing Indian removal, which he had previously cheered; supporting a mild tariff reduction, which he had previously opposed; and opposing hasty renewal of the BUS charter, which he had previously urged. Clay also fudged his Masonic tie in deference to the Anti-Masonic Party, the first genuine tertium quid or third force in American politics. A generation before, Freemasons had been instrumental in the design of a republican civic religion devoted to tolerance and material progress. They were still a ubiquitous presence in the legal, military, and engineering professions. But the scandalous disappearance in 1828 of a turncoat Mason in upstate New York aroused paranoid fear and loathing. Secret societies had no place in a democracy! Heretical cults had no place in a Christian nation! When the Anti-Masonic Party won surprising victories in New England and New York, it emerged as a serious threat to political Masons such as Jackson and Clay. The latter issued a tortured statement to the effect he had "never attached much importance to Masonry" and believed it did little good or harm, but neither would he renounce his craft in a bid to become president. William Wirt did renounce it. In September 1831 the Anti-Masonic convention in Baltimore's Atheneum Hall made Wirt its candidate for the White House.[37]

Two months later the National Republicans met in the same hall to nominate a ticket of Clay and Pennsylvania's John Sergeant, counsel and close friend to Nicholas Biddle, president of the BUS. Clay put out feelers to Wirt in hopes of an anti-Jackson alliance, but his main electoral ploy was to lever Congress into a decisive act that would permanently discredit Jackson. No sooner had he returned from Baltimore than Clay wrote Biddle a stunning letter. He had changed his mind. The moment to seek a recharter of the BUS was *now*.

It cannot be said often enough: few topics in history are more important—or boring—than public finance. How a government farms and consumes its revenue determines which oxen are gored, which geese grow fat, and whether the barnyard at large prospers. Limber fiscal policies can turn the Netherlands or England into a world power. Calcified fiscal systems can wreck great monarchies such as France and Russia. The first American political parties emerged from debate over the First BUS. That debate appeared settled when even Jeffersonians learned from the War of 1812 the value of a central bank to shelter the Treasury's assets, finance government, regulate loans and currency issued by local banks, and ensure the integrity of the dollar. In 1816, Congress granted a twenty-year charter for the Second BUS. It got off to a terrible start when corrupt, incompetent directors inflated a land bubble that burst in the Panic of 1819. But Langdon Cheves righted the ship, then relinquished the helm to Biddle in 1824. Thanks to his prudential care, Americans never knew, before or since, a sounder currency than in the very years when Jackson claimed that the BUS "failed." The charge was all the more scurrilous, given the indispensable help Biddle provided for Jackson's program to retire the national debt.[38]

Yet Biddle was an irresistible target for a president posing as the common man's champion against privileged elites. He had been born at the acme of Philadelphia society, and everything came easily to him. At age fifteen he graduated as valedictorian from Princeton. At eighteen he was in Paris assisting the Louisiana Purchase and attending Napoléon's coronation. Back home he won the hand of a merchant princess and admiration as an aficionado of literature. But Wall Streeters judged Biddle a "clumsy" investor, and Pennsylvanians found his politics haughty. A Philadelphia blue blood saying to frontier legislators "I mean to tell you plainly and simply that you are wrong" was going to turn their heads nowhere but to their spittoons. His tragic flaws were really naïveté and presumption, but his management style seemed that of an imperious snob. When the BUS kept state banks honest by reminding them "gently" (Biddle's favorite word) that their bonds must be payable in specie, friends of easy credit railed against the "monster bank." When the BUS reined in irrational exuberance to prevent speculative land or railroad bubbles, Jacksonians called it a conspiracy to crush opportunity. In fact, the economy grew in rapid and orderly fashion throughout the late 1820s. But growth was not rapid enough and was entirely too orderly for those who wanted to get rich quick or felt somehow thwarted. They included western developers, local bankers, indebted farmers, merchants and manufacturers hungry for capital, urban laborers and mechanics resentful of wealth, and New York financiers eager to crush their competitors in Philadelphia. In an era when people rent garments over Masonic lodges, the marbled BUS headquarters on Chestnut Street inevitably stoked suspicions that the oblivious Biddle could not dampen.[39]

The BUS ran twenty-nine branches around the country, did $70 million of business per year, issued 20 percent of the banknotes in circulation, held one-third of all American deposits, and was the sole depository for the annual surpluses accumulated by the federal government. It did not, however, have a death grip on local economies. In order to maintain leverage over state banks the BUS had to hold more of their dubious bonds than they did its own trustworthy notes. Hence the BUS was a guaranteed conduit funneling capital and credit to the West and South rather than squeezing them. Nor did the BUS have the power to expand or contract the money supply at its discretion. But the intricacies of the money market were lost on the president. Jackson had hated all banks since 1811, when he nearly lost everything on a botched land speculation. He hated debt. He hated paper money. He hated eastern "aristocrats." He hated foreigners, in whose portfolios lay a sizable share of BUS stock. As for politics, Jackson believed an assault on the BUS would stir the passions of his constituents, a belief reinforced by the shadowy fellows Biddle called Jackson's "kitchen cabinet." Chief among them was the "invisible" Amos Kendall, whom Harriet Martineau suspected of being the thinker, doer, and planner behind the curtain. Kendall, a Dartmouth man who went west, was the pasty-faced, prematurely gray, hired quill in charge of the president's newspaper claque. It was he who fed Jackson the words "The world is governed too much." Jackson's bank warriors also included Roger Taney, counsel and shareholder in a Baltimore bank slated for Treasury deposits if the BUS went down; David Henshaw, a customs official plotting to deposit government revenue in his own Boston bank; Churchill C. Cambreleng, the "Congressman from Wall Street"; and the publisher Francis Preston Blair, who made a fortune from government printing contracts in return for savaging the BUS in his *Globe*.[40]

Did Jackson have a clue what his assault on the BUS was about? If he thought he was saving westerners from exploitation, he misunderstood the cash flow. If he thought he was striking a blow against all banks, he got it backward, because destruction of the behemoth would just make it easier for less scrupulous state banks to proliferate. If he thought he was attacking speculation and greed in the name of honest industry, then he was blind to the profiteers in his own camp. Both houses of Congress reported, after lengthy inquiries, that the president's assaults on the constitutionality and effectiveness of the BUS were unfounded: "Amidst a combination of the greatest difficulties, the Bank has almost completely succeeded in the performance of this arduous, delicate, and painful duty." Hence, if Jackson's contempt had any rationale it must have been the political threat posed by an institution potentially able to "control the Government" and function as a "vast electioneering machine." After learning of Biddle's early request for a recharter, the president hissed, "The bank, Mr. Van Buren, is trying to kill me, *but I will kill it.*"[41]

Clay knew his Congress. He was sure he could push through a bill to recharter the BUS over the protests of the Jacksonian congressman James K. Polk and Senator Thomas Hart Benton. That would force Jackson to either swallow defeat or else risk a controversial veto in the midst of the presidential campaign.[42] Sure enough, the bank bill passed in June 1832 by votes of 107 to eighty-five and twenty-eight to twenty. Clay hugged himself. Webster strutted. Biddle rejoiced. But all they accomplished was to give frail Andrew Jackson a new reason to live. Styling this a heavenly call "to preserve the republic from its thraldome and corrupting influence," he put his advisers to work on the greatest (and most demagogic) state paper of his career. Kendall, the principal author of Jackson's veto message of July 10, ignored serial Supreme Court decisions, the will of the Congress, and Biddle's responsible record. Instead, he declared the monopoly "unauthorized by the Constitution, subversive of the rights of the States, and dangerous to the liberties of the people." The profits of the BUS, 30 percent of which flowed to foreign shareholders, "come directly or indirectly out of the earnings of the American people." The Supreme Court might rule in favor of the BUS, but the president had an equal duty to review the constitutionality of federal laws (as in the Cherokee case). It was unclear what "interest or influence, whether public or private, has given birth to his act" (implying that Congress was bribed), but this made it all the more incumbent on the president to stand against those who "too often bend the acts of government to their selfish purposes." Kendall and Jackson appealed to "the humble members of society—the farmers, mechanics, and laborers—who have neither the time nor the means of securing like favors to themselves." By vetoing the BUS, the president said, he had simply "done my duty to my country," relying for deliverance "on that kind Providence which I am sure watches with peculiar care over the destinies of our Republic, and on the intelligence and wisdom of our countrymen" (i.e., voters in the coming election).[43]

Webster damned Jackson's veto as an unprecedented, despotic usurpation of power, adding that Jackson presumed to annul federal law in the same manner as South Carolina. Clay, describing the veto as a capricious assault on representative government, nearly got into a fistfight with Benton. But not only did Clay lack the votes to override Jackson's veto; he misjudged the electorate. To be sure, the veto did "cause all the Election to be contested on the principle of Bank or no Bank" and invite journals such as the *North American Review* to warn of "wild and reckless speculation, ruined confidence, worthless currency, and prostrate and broken credit." But exhorting ordinary Americans to choose the "robber" Biddle over the saintly Jackson was like goading them to cry for Barabbas and crucify Jesus. The Democrats' party machines had all the ammunition they needed. Kendall and Blair reprinted the veto message in newspapers across the country, spinning the election as a showdown between democracy and aristocracy. Local Democratic committees staged barbecues, rallies, and parades of "Hurra Boys" waving hickory sticks on a scale exceeding that of the raucous campaign of 1828. They accused Biddle of trying to buy the election with huge infusions of money from the "Golden vaults of the Mammoth Bank." So tone-deaf was Biddle that he distributed 30,000 copies of the veto message himself, on the assumption voters would see through its bombast. Voters instead heeded the president's warnings about the political threat posed by the BUS's concentration of wealth.

Clay's 37 percent of the popular vote won him southern New England, Maryland, Delaware, and Kentucky. The Anti-Masonic candidate Wirt won nearly 8 percent and carried the quirky state of Vermont. Rebellious South Carolina voted "none of the above." The Jackson–Van Buren ticket swept all other states on the strength of a 54 percent majority. Philip Hone lamented that Jackson was more popular than George Washington. Wirt despaired that "he may be president for life if he chooses."[44]

Biddle and Jackson, knowing the BUS charter had four years to run, replenished their arsenals. But for the moment an even more pressing crisis eclipsed the war over the BUS. In 1832 Clay had brokered a mild reduction in tariffs in hopes of winning southern support for his candidacy. It failed to appease Calhoun. He stumped in South Carolina for candidates committed to nullification. After they gleaned 65 percent of the vote, the state legislature passed an ordinance banning the collection of customs duties, requiring loyalty oaths from state officials, and threatening secession if the federal government used force. Hayne resigned from the Senate to become governor and militia commander. Calhoun resigned as vice president to occupy Hayne's Senate seat. He was persuaded that northern and western business interests were engaged in a conspiracy to make the United States an industrial empire, reduce the South to colonial status, and then move against slavery. But Calhoun was no secessionist. His strategy was to pose as the only moderate who could restrain South Carolina and whose demands must therefore be met. That is why he urged Hayne to postpone execution of South Carolina's ordinance in order to buy time for a deal.

Jackson issued a ferocious ultimatum: "Disunion by armed force is treason. Are you really ready to incur its guilt?" He ordered General Scott to reinforce Fort Moultrie and dispatched navy cutters to Charleston. He sought to isolate South Carolina by reminding Georgia and Alabama of their reliance on him for Indian removal. Then Jackson asked Congress for authority to uphold federal law by coercion. Calhoun, daily expecting arrest by federal marshals, fulminated in the Senate against the "Force Bill" and "Bloody Act." Webster hurled back words dipped in hemlock. Both made a pretense of lambasting Clay for promoting some sort of compromise. After weeks of nervous backroom palaver the Congress arrived at a curious resolution. Clay patched together a mostly northern majority to pass the Force Bill, and a mostly southern majority to reform the tariff. It was not all Calhoun wanted, but it replaced over ten years the lofty "protective" tariff with a "revenue" tariff fixed at 20 percent, the same rate Calhoun had favored in 1816. On March 2, Jackson signed both bills into law. South Carolina's legislature answered in kind. It repealed its nullification act on tariffs while passing a new law spitefully nullifying the Force Act. So the dispute over substance dissolved, but the dispute over principle crystallized. Nullification had worked; "terrorism" was rewarded. This meant that South Carolina, or the entire South, could hold the Union hostage whenever it pleased in order to bargain for substance. Jackson belatedly understood and rued his acquiescence in this Compromise of 1833. "I can tell you, posterity will condemn me more because I was persuaded not to hang John C. Calhoun as a traitor than for any other act in my life!"[45]

The very day Congress acted to end the nullification crisis it reignited the war over the BUS by resolving the government's assets were safe in the BUS and ought to remain there. Jackson defied Congress again, fuming that Biddle "shan't have the public money. I'll remove the deposits!" All his own advisers except Kendall, Blair, and Taney agreed with Congress. It would be inopportune to withdraw Treasury funds before other arrangements, perhaps a national bank in the capital, had been established. So Jackson defied them as well. He sacked Secretary of the Treasury McLane in favor of William J. Duane, then sacked Duane when he demurred. At last he turned to the obsequious Taney, always eager to serve pocketbook and career. Over the last quarter of 1833, Taney ordered Biddle to transfer $6 million of the $10 million in federal funds it had on deposit to Baltimore's Union Bank, Philadelphia's Girard Bank, New York's Manhattan Bank, and twenty other "pet banks" of the Democrats. That forced the BUS to contract its own lending, but as usual Biddle proved his own worst enemy. He might have assisted local banks as best he could while blaming a contraction on Jackson. Instead, he hoarded the remaining BUS capital on the foolish theory that "ties of party allegiance can only be broken by the actual conviction of existing distress in the community."[46]

That was just what hard-money Jacksonians were taught to expect by the most widely read treatise on economics in American history. In February 1833 the editor William M. Gouge of Philadelphia published *A Short History of Paper Money and Banking in the United States*. It quickly ran through several editions, including a pulp version priced at twenty-five cents. Its Jeffersonian old-time religion preached that banknotes in general circulation were an unremitting evil. Greedy or unruly banks invariably issued too much credit and money. That drove prices up, creating a false sense of prosperity and tempting ordinary people to spend and speculate. That caused paper dollars to depreciate to the point where foreign creditors insisted on payment in specie, whereupon U.S. banks were obliged to squeeze their own debtors, spreading bankruptcy and panic throughout the economy. Gouge's remedy was to permit only coinage to circulate and limit paper transactions to commercial houses. Hard money would eliminate periodic panics and financial manipulations beyond the people's control, and would prevent moneyed aristocracies from taking advantage of the "humble members of society." That made sense to the "humble," who joked that "Gouge on Banking" might oblige Biddle to write "Banks on *Gouging*." After all, had he not brazenly lectured Congress, "Until the nature of man is changed, men will become speculators and bankrupts—under *any* system"? Citing Gouge's authority the hardest of hard-money senators, "Old Bullion" Benton, asked whether "PEOPLE, or PROPERTY, shall govern?" In June 1834 he carried a bill to expand the circulation of coinage by boosting the value of gold relative to silver from fifteen-to-one to sixteen-to-one. Benton boasted of success when the volume of gold flowing into the mint more than doubled. He scratched his head when inflation and speculation only sped up.[47]

What really happened to the U.S. economy over Jackson's second term? Whatever one thinks of the assault on the BUS, it seems logical to blame the crunch and inflation that followed on the actions of Jackson and Biddle. In the absence of laws requiring banks to hold sufficient reserves of specie and or specie-backed bonds (or both) to back up their paper instruments, evisceration of the BUS all but invited "wildcat banks" (a term invented in frontier Michigan) to issue all the loans and "rag money" the market would bear. And the market bore plenty because pioneers and speculators hungered to purchase millions of acres of land in the West. More plentiful, less reliable paper money helped create inflation and a land bubble certain to burst when the chickens came home to roost. Only that isn't what happened. The statistics reveal (1) no panic or recession in 1834 caused by Biddle's contraction; (2) no irresponsible lending by banks with inadequate reserves; (3) a land bubble that relieved inflation rather than contributed to it; and (4) an eventual panic that had less to do with American policies than with global trends.

There was no immediate panic, because the Treasury assets did not evaporate—they simply moved into other banks eager to invest—and because a flood of imported gold and silver made possible a responsible expansion of the money supply. Much of this specie was European gold attracted by higher interest rates, the apparent boom in America's "emerging market," and the prospect of purchasing silver at sixteen to one. But most consisted of Mexican silver that piled up in U.S. banks because an efficient new method was found to finance foreign trade. Up to that point merchants trading in China had been obliged to ship silver bullion halfway around the world in order to purchase opium from the British East India Company; they then traded the opium in China for tea, silk, and porcelain. In the 1830s the maturing of business ties between New York and London allowed merchants to finance their Asian purchases with bills of exchange drawn on English commercial houses. As a result, the silver Mexicans used to finance their own imports stayed in the United States, boosting the nation's hoard of precious metals by $20 million over the decade. That allowed banks to triple their issue of loans and currency and still increase the reserves in their vaults. Inflationary pressure grew with the money supply, but not because of the land boom. Rather, the millions spent on abundant real estate sold at low fixed prices drew off funds that otherwise would have chased scarce goods and fueled worse inflation.[48]

No American politician and certainly no "common man" knew anything of these trends. Clay just called Jackson a despot and rammed through the Senate an unprecedented censure of the president. The White House just declared Biddle's contraction proof positive why the malicious BUS had to be sunk. In Congress, Polk scored four direct hits on the listing BUS with resolutions that upheld the veto, removal of deposits, and transfers to state banks, and launched an investigation of the complicity of the BUS in the contraction. Biddle then scuttled his own ship by impudently refusing to release BUS records, thus ending forever any chance of a recharter. The United States would not have a central bank again until the Federal Reserve Act of 1913. What the "common man" got instead was the swift concentration of far greater, wholly unregulated financial power on Wall Street.[49]

The land bubble propelled the Jackson administration's final flights of economic pretense. Sales of public land in the Mississippi Valley soared from 4 million acres in 1834 to 20 million in 1836. That revenue, plus the income from tariffs, allowed Jackson to realize his dream of paying off the national debt for the only time in American history. But Jackson's scruples about states' rights did not permit him to countenance a federal surplus. Accordingly, Jackson sponsored the Distribution Act of June 1836 authorizing the Treasury to lend the surplus to state governments promoting internal improvements. All that did was to encourage boosters in Indiana, Illinois, and elsewhere to launch impossibly ambitious canal and railroad projects. It almost goes without saying that the loans were never repaid. Jackson also worried about homesteaders' being seduced into debt. Accordingly, he ordered the Treasury to issue a specie circular requiring that public lands be bought with hard money. All that did was to make cash-poor farmers more dependent on capital-rich speculators, magnify the fraud the president meant to expunge, and oblige his loyalists to fight little "bank wars" in every state of the Union. The Panic of 1837 soon rendered the Distribution Act and the specie circular ludicrous relics of a bygone boom. Jackson, rued the old Jeffersonian Albert Gallatin, "found the currency in a sound and left it in a deplorable state." The rich adjusted, the poor could not adjust, and the gap between them broadened on Andrew Jackson's angry watch.[50]

Imagine a man whose forebears were among the 100,000 Scots-Irish who sailed to Philadelphia in the 1700s, trod the Wilderness Trail into western Virginia, and began to acquire land. His father fought with Daniel Morgan, the "old wagoner," in the Continental Army; later went broke; lit out for Tennessee; and died. His gritty mother started over, with the help of five slaves and nine children, on a farm south of Knoxville. In 1809 the fifth child, now age sixteen, ran away to live with the Cherokee for three years. He preferred "the wild liberty of the red man." Chief Oo-loo-te-ka (He Who Puts Away the Drum) became a surrogate father to the boy he called Colonneh (the Raven). Experimenting with manhood, the youth drank to excess, went deeply into debt, then prospered in good American fashion by founding a school despite his own lack of education. When Jackson mustered militiamen to punish the Creeks in 1813, the six-foot-two-inch, muscular youth asked his mother's advice. Courage, she replied, was forever welcome in her home; cowardice was not. He proved his courage at Horseshoe Bend. When an arrow pierced his thigh he ordered a soldier to yank it out, then led a hobbling charge on the Creek ramparts. When musket balls shattered his shoulder and arm, he kept fighting until comrades hauled him away. Surgeons removed one ball, but left the other, since it seemed he must bleed to death anyway. Instead, he survived to become a lifelong friend of the general who witnessed the action. Biographies like that molded eminent Jacksonians such as Sam Houston.[51]

In 1818 Houston resigned his commission to study law in Nashville.

He joined the Masonic lodge, passed the bar, and became the most recognizable man in Tennessee because of his outlandish hats and embroidered shirts, sashes, and stockings. As a charter member of Jackson's Nashville junto Houston swept into Congress and then, in 1827, into the state house, where he distinguished himself among southern governors by promoting free public schools. The next year he helped manage Jackson's campaign and served as a pallbearer at Rachel's funeral. Then something terrible happened. The thirty-five-year-old Houston fell in love with nineteen-year-old Eliza Allen, the lovely blond daughter of a prominent planter. Her parents encouraged (perhaps forced) the match, made in January 1829. The next morning, Eliza was already muttering how she wished her husband were dead. She implied that there was another man; her husband questioned her chastity; they evidently never consummated the marriage; within three months she fled home. One can imagine what gossips conjectured about that wedding night. Houston might as well have been castrated in public. He resigned as governor, nearly drank himself to death, then joined his Cherokee father in exile in Arkansas.[52]

Houston went native, dressed in deerskin leggings and headdress, and between bouts of drunkenness and malaria served as ambassador of the Cherokee. Folks on the Mississippi threatened him for daring to defend Indians. At the White House Jackson welcomed his friend, but refused to halt Indian removal. On Capitol Hill an Ohioan accused him of fraud, whereupon Houston pinned the congressman to the ground and caned his nether regions. The House of Representatives made a monthlong circus of the case until Houston, sweeping aside the attorney Francis Scott Key, delivered a semi-sober oration so lofty that the thespian Junius Booth cried, "Take my laurels!" He was fined $500, but the trial healed his soul. "I was dying out, and had they taken me before a Justice of the Peace and fined me ten dollars for assault and battery, they would have killed me. But they gave me a national tribunal for a theatre, and set me up again." What would he do with his new life? Sam Houston had a notion to "repair to TEXAS."[53]

A trackless U.S.-Mexican border hundreds of miles long over which streamed thousands of legal and illegal immigrants whom the host country despaired of assimilating: that predicament was just as real in the 1830s as it is today, only the roles were *reversed*. The treaty John Quincy Adams concluded with Spain in 1819 fixed the limits of the Louisiana Purchase, and hence the U.S.-Mexican boundary, along the Sabine and Red rivers. Between those rivers and the Rio Grande lay an empire as large as France, watered by rivers spilling into the Gulf of Mexico, and almost vacant save for Native Americans. Under Spanish rule just 4,000 white or mestizo *tejanos* dwelled at San Antonio de Béxar, La Bahía (Goliad), and Nacogdoches. Mexico's violent war of independence reduced that number to just 2,500. Empire-building desperadoes noticed the vacuum. Aaron Burr's associate Philip Nolan (the "man without a country"), the pirate Jean Lafitte, the exiled Napoléonic general Charles Lallemand, the Virginia-born Dr. James Long, and the slave trader Jim Bowie (inventor of the "Arkansas toothpick" that disgusted Fanny Trollope) repeatedly tried to steal parts of "the Texas" with or without support from "the Texians." When Mexico became independent its interim government realized that grasping Anglos could not be kept out. How then could it secure the border, treaty or no treaty? [54]

The least bad answer, or so it seemed at the time, was to invite yanqui pioneers into Texas on condition that they embrace Mexican citizenship and the Catholic church. A modern-day Moses accepted those terms in return for permission to lead an exodus of 300 families across the Sabine. Who was Moses Austin? The only way to label him is *American*. Born in Connecticut, he traded his Puritan heritage for capitalism and Freemasonry; moved to Quaker Philadelphia, where he married; then moved to Cavalier Virginia, where he invested in mining and slaves; and finally went west to Missouri, where Austin developed the richest deposits of lead on the continent and founded the first bank west of the Mississippi. Alas, the War of 1812, deadly competition from the gunslinger John Smith T, a failed speculation in Arkansas, and the Panic of 1819 bankrupted Austin. So the fifty-eight-year-old striver started over again in 1820. Borrowing fifty dollars, a horse, and a slave, he fastened a shotgun to his saddle and rode for San Antonio. The governor, persuaded of his sincerity, recommended that Mexico City approve his petition. On the way home a rustler stole Austin's mounts, forcing him and the slave to trudge hundreds of miles on foot. He caught pneumonia and died five months later, but not before imploring family friends to persuade his reluctant son Stephen to take up the Texas land grant. [55]

Stephen T. Austin had been sent to Bacon Academy in Connecticut and Transylvania University in Kentucky with stern paternal instructions to dress, behave, and write like a gentleman; practice self-reliance; and always shun bigotry. Evidently he listened. Stephen matured into an equable, judicious man who made people feel comfortable whatever their background: just the personality required of a colonial proprietor on foreign soil. Once committed to the fantastic project of inventing Texas from scratch, he did better advance work than any promoter since William Penn. He traveled to Mexico City to secure clear title to his father's 15,000-square-mile land grant, a chore eased by his Masonic ties to generals Agustín de Iturbide and Anastasio Bustamante. The "fanaticism" of the Mexican Catholic church put him off, but he wisely kept silent about it. In the grant itself, Austin surveyed a capital at San Felipe de Austin and additional towns; offered lots ranging from 177-acre farms to 4,428-acre ranches at a price of 12½ cents per acre; promised settlers their titles would be "perfect and complete for ever"; agreed to take deerskins or beeswax from families short of cash; and ran advertisements throughout the American West and South touting the fertile bottom land along the Trinity, San Jacinto, Brazos, Colorado, Lavaca, and Guadalupe rivers. The enticements had the virtue of being true. Contrary to Yankees' impressions, the Gulf of Mexico coastal plain beyond the East Texas thickets received adequate rainfall and was excellent fruit, cotton, and cattle country. This was not western desert, but a natural extension of the American South. Finally, like Penn, Austin assured the local tribes that "it is not our wish to deprive the Indians of their hunting or fishing grounds." In practice, reciprocal raids marred Austin's first years until the number of whites grew sufficiently large to persuade the tribes to migrate.

According to a census taken in 1825, 1,800 free and 443 enslaved Americans were settled in Texas. By 1831, when Austin received his fifth and final grant as a Mexican *empresario*, Texians numbered more than 22,000 whites and 1,100 slaves. Who were they—the brave pioneer families of one fond American legend, or the gunslinging outlaws of another fond legend? It goes without saying that Texas had room for both, but the most accurate generalization describes the settlers as young, married southern men fleeing debt. Court clerks in Louisiana, Arkansas, Mississippi, and Missouri developed a shorthand for vanished residents sued for default: GTT (Gone to Texas). Let William Barret ("Buck") Travis of Alabama stand for them all. Having failed at four careers before reaching age twenty-two, he pleaded his case before a jury that literally hooted him down. To avoid debtors' prison, Travis kissed his wife and baby good-bye, then hightailed it for Texas. As for lawlessness, Texians simply displayed the same pride, honor, individualism, white supremacy, land hunger, and cussed distrust of authority celebrated by Jacksonians in the states of their origin. What is more, the *tejanos* shared their attitude toward distant authority as expressed in the adage *Obedezco pero no cumplo*: "I obey but do not comply." They had resisted Spain's mercantilism as naturally as the thirteen colonies had resisted Britain's, and joined their new Anglo neighbors in happily ignoring *pronunciamentos* from Mexico City. Nor did provincial authorities take seriously the requirement that Anglos practice Catholicism. Austin was torn. He needed to retain the confidence of the Mexican government, yet appease settlers whose own motto was "Obstinacy right or wrong." Texas, like Pennsylvania 140 years before, was both a booming success and a "labyrinth of trouble" for its sponsor.[56]

The politics of Mexico are too tangled, and the story of Texas is too familiar, to be told in depth here. But it is worth pointing out how closely events paralleled the run-up to 1776. Texas prospered under the "salutary neglect" of a faraway government whose occasional stabs at regulation could be safely ignored. Then the "mother country" began to get tough. In Mexico's case, Iturbide's bid for a centralized monarchy prodded the provinces into rebellion. Austin's friend Erasmo Seguín served as a delegate to their convention, returning with good news and bad. The Mexican constitution of 1824 blessed provincial autonomy and individual rights, but abolished Texas by merging it with populous Coahuila to the south. Some of the 5,000 *tejanos* were as angry about this as the 30,000 Anglos. In 1830 a second blow fell when the Mexican congress banned U.S. immigration and importation of slaves, discouraged U.S.-Texan trade, and posted more soldiers at the fort of Béxar. Those "intolerable acts" outraged Anglo-Texans. Austin pleaded for peace, but he could not prevent Jim Bowie from leading an ambush on Mexican infantry near Nacogdoches and marching the captives to San Felipe like a cat displaying mice to his master. Nor could Austin prevent his people from imitating the First Continental Congress. In October 1832, fifty-five delegates met in the capital to demand redress of grievances, organize committees of correspondence, and plan another convention in April. By then General Antonio López de Santa Anna had seized power in Mexico City and Sam Houston had reached Nacogdoches and been elected a delegate to the Texas convention.

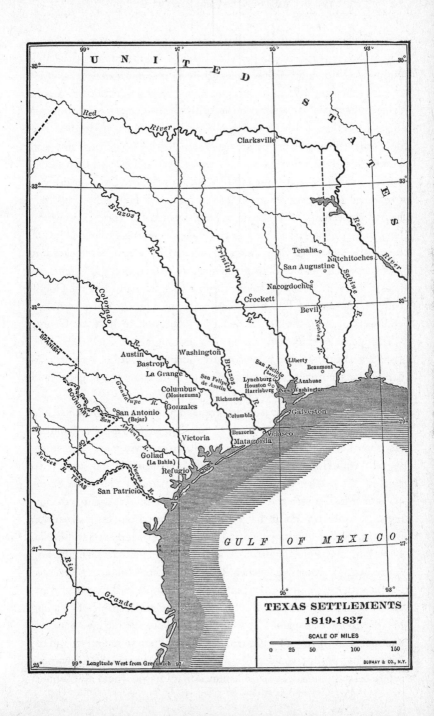

TEXAS SETTLEMENTS
1819-1837

SCALE OF MILES

0   25   50   100   150

BORMAY & CO., N.Y.

After drafting a constitution, the Texians dispatched Austin to Mexico City in hopes that he could persuade Santa Anna to restore their autonomy. Austin not only failed but spent the year 1834 in Mexican jails.

Like Benjamin Franklin after his public berating in London, Austin came home a rebel. Over fourteen years his dream of a free and prosperous Texas had taken on the "character of a religion." Now, satisfied he had done everything required by honor, Austin cried, "No more doubts—no submission." It was time to "go for *Independence*, and put our trust in our selves, our riffles, and—our God." Like Tom Paine, the ne'er-do-well newcomer Travis beat the drum for independence. He told Texians they had no choice but "liberty or death," and warned all "Tories, submission men, and Spanish invaders to *look out*." Like the redcoats who marched to Lexington Green, Mexican dragoons rode to Gonzalez in September 1835 to seize a small cannon. Like minutemen, the defenders cried, "Come and take it," then drove off the soldiers with a volley. Finally, like the Continental Congress, the Texian convention formed a provisional government at Washington-on-the-Brazos and asked Houston to be their George Washington. But Houston had no army, just ragtag militias, freebooters, some *tejano* patriots, and some friendly Indians. Nor did he have time to drill soldiers or impose a chain of command. This war was a long shot, and the early going confirmed it.[57]

Where was President Jackson throughout these exciting events? Did he not give a whit for Texas, or was he in cahoots with Houston to foment just such a rebellion? The truth is that Jackson craved Texas and corresponded with Houston regularly. But there is no evidence that he plotted a revolution, much less a war between the United States and Mexico. Suspicious of federal power and stingy to a fault, Jackson was a pure embodiment of American "cheap hawkery." His preferred modes of expansion were the frontier filibuster such as he waged against Indians and Florida, or the bargain-basement purchase such as Jefferson pulled off with France. Accordingly, Jackson sent an envoy to Mexico in hopes of buying Texas for not more than $5 million. The person he chose (on Van Buren's awful advice) was Colonel Anthony Butler, a Mississippian incapable of cloaking his contempt for Hispanic Catholics. Now, bribery and intrigue were second nature to Mexican politicians, but Butler's were so indiscreet and offensive that no one could do business with the United States without appearing a traitor. Santa Anna assumed that Jackson would now support the Texas rebellion, and Houston appeared to confirm his suspicions by pleading for help throughout the American South. This was when his native Tennessee became known as the "Volunteer State." To Santa Anna it was imperative that he crush the rebellion swiftly and savagely before more help arrived.[58]

For the time being all was chaos. Down on the Brazos there was a rumor that Santa Anna meant to provoke a slave rebellion. So even as Goliad's militiamen cried for liberty they swept through the cotton fields to torture or hang 100 suspected blacks. Greed inspired a second disaster when James Grant made a gambit toward Matamoros to seize disputed land grants on the Rio Grande. Grant and all but six of his men were gunned down in a Mexican ambush on the Nueces. Even the sole Texian triumph was ominous. When a deserter reported the garrison at San Antonio de Béxar had no stomach for war, Edward Burleson, a veteran of 1812; Ben Milam, a dashing old filibusterer; and "Colonel" Travis rallied 300 Texians to assault the pueblo. Houston thought the plan mad, but a disputed chain of command, sheer distance, and the bottle rendered his judgment irrelevant. Burleson's men invaded San Antonio in a brilliant house-to-house battle in December 1835, forcing the Mexicans to surrender and capturing 800 muskets and twenty cannons. Burleson did not seek reinforcements. Thinking the war all but won, he left Travis behind with 150 men.

In February 1836, 6,000 *soldados* crossed the Rio Grande in a freakish blizzard as white as their uniforms. General José Urrea marched one column up the road near the coast. Santa Anna drove another straight to San Antonio de Béxar. When Travis's lookouts spotted the Mexican cavalry, he ordered his men into the old mission now called the Alamo, cried "God and Texas," and swore no retreat, no surrender. Erasmo Seguín galloped off to get help. But James Fannin at Goliad and Houston at Gonzalez refused to order anyone on a suicide mission. Nor would help have arrived in time, because Santa Anna refused even to wait for his artillery train. On March 6, he raised a red banner designating no quarter and ordered his most seasoned infantrymen to make a frontal assault. He would later regret this costly engagement, but he believed he must make an immediate, unmistakable statement. Not only did the Alamo's 187 defenders perish; their corpses were burned and their cannons, including the iconic "come and get it" gun, were melted down. Two weeks later a worse disaster occurred 150 miles to the southeast. Fannin had reason not to reinforce the

Alamo, because Urrea's column was already threatening Goliad. But Fannin did not have good reason to fight. In the absence of orders from Houston, it would have been wise to retreat. Instead, he ordered his men to dig fortifications that became death traps when Urrea did deploy his artillery. Fannin was forced to surrender and Santa Anna again lowered his thumb. Reluctantly, Urrea gave orders that all 342 prisoners were to be shot or bayoneted, then burned.

The disasters were severe mercies. Hatred and fear unified the fractious Texians, and the several weeks bought by the doomed forward garrisons were put to good use by the Texas convention. On March 2, its delegates approved Houston's motion for a declaration of independence modeled on Jefferson's. Two weeks later they approved an American-style constitution. But just as in 1776 and 1787, the overwhelming need to ensure unanimity required that Texas bless chattel slavery. The convention also strengthened Houston's authority, permitting him to exert some semblance of control over the rowdy, surly, scattered volunteers and their self-appointed commanders. Houston called for a strategic retreat covered by Seguín's skirmishing rear guard of cavalry. This "runaway scrape" traded land for time, turned evacuees into recruits, allowed for some drill, and not incidentally put Houston in contact with the U.S. general Edmund Pendleton Gaines, commander of the Louisiana military district. Gaines just happened to deploy troops along the Sabine on the pretense of Indian trouble. He could, if needed, offer the Texians sanctuary and intervene should Mexicans violate the border.

Few understood Houston's retrograde motion; he was not sure himself how it would end. At the legendary "Which Way Tree" he at last decided to make a stand near Harrisburg, only to find that Mexican cavalrymen had incinerated the town. Houston gave his men a long look at the smoldering ruins, then chose his battlefield. It was a grassy meadow enclosed by the San Jacinto River and the Buffalo Bayou. When Santa Anna encamped, Houston ordered the only bridge out burned down. He meant to provide no retreat to the enemy or his own men. Still, he hung back, even as Mexican reinforcements arrived, until four o'clock on the afternoon of April 21. Many Mexicans, worn out from their march, were taking a siesta. Santa Anna thought it too late in the day to attack and never imagined the outnumbered Texians would take the offensive. By contrast, Houston's 783 men were well rested and twitching with fury. They had no uniforms, a

motley collection of weapons, one battle standard (a bare-bosomed goddess of liberty), one piper who knew only one naughty song, and scarcely
enough training to form a skirmish line. Indeed, after delivering one orderly volley they ignored Houston's command to reload. Using rifles as
clubs, drawing pistols or bowie knives, the frontiersmen charged the Mexican camp crying "Remember Goliad!" and "Remember the Alamo!" The
attack took the enemy utterly by surprise (Santa Anna himself was asleep
or "otherwise engaged"). At the cost of just nine of their own, the Texians
killed 650 Mexicans and captured 730, including Santa Anna, whose identity was betrayed by his own *soldados*. Sam Houston was a hero again (except to some rivals who claimed that they should get credit for the victory).
He was also a cripple again, courtesy of a musket ball that shattered the
tibia above his right ankle. For the rest of his life Houston suffered, but he
proudly displayed his "San Jacinto foot." The 4,000 Mexicans elsewhere in
Texas were leaderless, short of supplies, and bogged down by torrents of
rain. Those that did not desert straggled home.[59]

Houston and the interim president, David G. Burnet, spared Santa
Anna so that he could sign a treaty granting the province independence.
Infuriated Texians had wanted him hanged, but this did not dissuade
them from electing Houston their first president with 5,119 votes to just 587
for Austin. The old *empresario* felt rejected. During the war he led a delegation east to plead for help. In Washington City, Jackson told him the
United States must remain neutral, something the Texians should have
realized before their "rash and premature act." In Philadelphia, Biddle refused Texas a loan. In New York, Austin grew desperate enough to violate
his lifelong vow against bigotry. "A war of extermination is raging in
Texas," he cried, "a war of barbarism and of despotic principles, waged by
the mongrel Spanish-Indian and Negro races, against civilization and the
Anglo-American race!" All that got him was $10,000 from Wall Street
gamblers willing to wager a little on the future of Texas. Disgusted with
his former countrymen and himself, Austin made the long journey home,
only to contract pneumonia. Bedridden for months, he awoke from a
dream on December 27, 1836, and wheezed in excitement, "The independence of Texas is recognized!" Thirty minutes later he died.

In fact, it was not. Houston sent to the White House his own delegation conveying Santa Anna and a letter pleading for annexation by the
United States. Jackson haggled with the generalissimo over how many

millions Mexico might accept in exchange for Texas, then shipped him to Veracruz on a U.S. Navy cutter. On his own soil Santa Anna repudiated his treaty and insisted that Texas was still a rebellious province of Mexico. Jackson was also thwarted by Congress, where John Quincy Adams railed against any expansion of slavery in North America. Accordingly, Jackson's annual message in December called recognition of Texas impolitic. At last the Texian envoy William H. Wharton persuaded Old Hickory that without recognition by the United States, Texas would be unable to establish the credit and trade it would need to defend itself. Jackson relented; Wharton worked Capitol Hill; and on March 3, 1837, Congress resolved to recognize the "Lone Star Republic." It was Jackson's last evening as president, and for once he let go of his anger. Passing pipes and wine to his guests, he raised his glass in a final toast: "Gentlemen, the Republic of Texas!" [60]

The Twenty-Fifth State: Arkansas 1836 [61]

"I've traveled this wide world over, some ups and downs I've saw, but I never knew what misery was till I hit Arkansas." That talkin' bluegrass ballad sums up Arkansas's career as the state that just don't get no respect. The New Yorker Henry Rowe Schoolcraft hiked the Ozarks in 1818, astonished by the region's "unvaried sterility, consisting of a succession of limestone ridges, skirted with a feeble growth of oaks, with no depth of soil." Its settlers wore buckskins "abundantly dirty and greasy," and lived in a condition "not essentially different from that which exists among the savages." Washington Irving could not fathom the laziness of dirt-farming families content with dilapidated one-room shacks (albeit "a fiddle is a joy to their hearts"). In 1821 Mary Toncray Watkins called Little Rock "a wilderness of sorrows" where "the Sound of the Gospel of Jesus is not heard." The Englishman George Featherstonhaugh thought only that forgers, bankrupts, cutthroats, horse thieves, and gamblers were likely to settle in Arkansas, given the "tolerant state of public opinion which prevailed there in regard to such fundamental points as religion, morals, and property." The most indelible image (prior to "L'il Abner" and Dogpatch) was Edward P. Washburn's rustic painting *The Arkansas Traveler*. Civilization in the form of an elegant horseman arrives at the stoop of a squatter's cabin. The coonskin-capped farmer, seated on a barrel beneath a drying pelt and a misspelled sign hawking moonshine, looks up with suspicion, while his wife (smoking a corncob pipe) and five children peer out of the shack's doorless adit. Nor was Washburn conveying an eastern stereotype. He was illustrating a tale written by the native Arkansan Sandford Faulkner in 1840.[62]

Arkansas could not even get its name right. When Louis Jolliet paddled down the Mississippi in 1673 he encountered a Quapaw village that sounded like "Arkcansea." French cartographers rendered that as Arkansa and then pluralized it (as they did Illinois). But Congress incorporated the Arkansaw Territory in 1819. That angered William Woodruff, whose *Arkansas Gazette* was the first newspaper west of the Mississippi. A purist who scorned Noah Webster for blessing the usage "lengthy" (What's next, "strengthy"?), Woodruff successfully campaigned for the French spelling. Still, confusion continued for decades over pronunciation. In the 1840s the state's own senators disagreed, forcing the vice president to recognize one as "the Senator from ARKansas" and the other "the Senator from ArKANsas." The matter was not settled until 1881, when the legislature ruled for ARKansas even though its residents continued to be called ArKANsans.[63]

Yet another humiliation derived from Arkansas's origin as a leftover detached from Louisiana, then detached again from what became Missouri. The remaining block of 53,187 square miles was hemmed in by the Ozarks, by swamps and canebreaks along the Mississippi, and by Indian territory. Neither a favored destination nor a conduit to greener pastures, Arkansas began life "remote and restless." One might add, nearly empty. Smallpox had long since carried off 90 percent of the Quapaws, and just 1,000 whites traded and hunted around the French station Arkansas Post. As in Missouri, however, the end of the War of 1812 prompted an influx. The population surpassed 14,000 by 1820; 30,000 by 1830; and 98,000 by 1840. What attracted the newcomers? Imagine a line running from the state's northeast corner to its southwest corner. Above that line the Ouachita and Ozark massifs rendered life rugged. But below the line the Mississippi alluvial and Gulf coastal plains offered rich topsoil. More than 200 species of trees, especially pine and live oak, meant abundant timber. The Saint Francis, Black, White, Arkansas, Saline, Ouachita, and Red rivers, plus fifty inches of rain per year, supplied abundant water. Migratory birds, deer, quail, turkeys, rabbits, bears, catfish, gar, trout, and bass offered game in abundance. For all the sneering about poor land, fat mosquitoes, and boggy forests, Arkansas drew immigrants precisely because even poor, lazy, ignorant folks could get by, while men with capital could get rich on cotton and slaves. By 1840 Arkansas's per capita production of corn, wheat, oats, potatoes, chickens, horses, cattle, and swine exceeded that of any neighboring state. Most popular were the porkers, which outnumbered the people four to one and gave Arkansans their nickname: the Razorbacks.[64]

Politics in Arkansas Territory replicated the well-worn pattern of factionalism and land speculation woven with red threads of violence. The first governor was no factor, because he suffered "ague and fever" (locals called it "seasoning") and was absent most of the time. That left power in the hands of the twenty-two-year-old territorial secretary Robert Crittenden, a veteran of Jackson's Seminole war and brother of a senator from Kentucky. On his own authority Crittenden staged elections in 1819. Two years later he led one of the rival factions pressing to move the capital to Little Rock because each laid dubious claim to its real estate. But Crittenden lost his bid to become governor, then lost again when a former protégé, Henry Conway, was elected territorial delegate in 1827. Crittenden called Conway to pistols and killed him on the bank of the Mississippi. The ensuing feud prompted another duel, a free-for-all in which a man lost an eye to a bowie knife, and a mortal gunfight in an editor's office. Personal affronts were behind the shooting of a judge by another judge, and of the governor's nephew by a journalist. But power and land were almost always in the picture.[65]

Crittenden's mouthpiece, the *Arkansas Advocate*, first broached statehood in 1830. It warned Arkansas might become Indian territory if it did not escape federal oversight. His new rival Ambrose Sevier reminded readers of the *Arkansas Gazette* that heavy taxes would result if Arkansas chose to fend for itself. But Sevier changed his mind when Michigan sought statehood in 1833. Congress was sure to pair that free state with a slave state, and if Florida got its bid in first, Arkansas might have to wait many years. When a convenient census in 1835 turned up 52,240 Arkansans, well above the required minimum, the territory convened a convention dominated by planters from the cotton counties. Hillbilly delegates threatened to bolt unless they received equal representation in the state senate. So the constitution wrote into law the northeast-to-southwest "imaginary line." In Washington City Sevier lobbied hard with Democrats, assuring them that Arkansas would give its three electoral votes to Van Buren in the fall election. The new state joined the Union on June 15, 1836.

Arkansas was Jacksonian in almost every respect. Its folks made a pretense of being egalitarian even though a mere third of adult white males owned their own farms, the rest living with family, working as help, or squatting on public land. When the Choctaw and Cherokee began to move through the territory, the *Gazette* ranted about "turning loose a ferocious band of bloodthirsty marauding savages." The Quapaws were dispossessed as early as 1824. Except for some excellent flood-control levees promoted by planters in Chicot County, the state ignored internal improvements. Arkansans willingly hosted slavery, not least the poor whites whose self-esteem required a black underclass. Arkansas showed little interest in education. By 1850, when its sister state, Michigan, had 2,714 public schools and 280 libraries, Arkansas had just 353 schools and one library. Its industrial college (parent of the University of Arkansas) was not founded until 1871.

Finally, Arkansas leaped into the wonderful world of state-chartered banks. Its real estate bank ignored "common men" in favor of cotton kings, engaged in corrupt profiteering, and defaulted on bonds. When English creditors appealed to the honor of Governor Archibald Yell, he barked, "Arkansas is free to act, as she shall deem fit and correct." The second state bank collapsed after the accounts of its Fayetteville branch mysteriously disappeared. Most shameful was the spoliation of James Smithson's bequest to the United States for "the increase and diffusion of knowledge." Secretary of the Treasury Levi Woodbury put all $500,000 in the hands of a crooked broker who paid himself a fat fee for investing the gold in worthless Arkansas bonds. If not for John Quincy Adams, who eventually shamed Congress into replenishing the fund, the Smithsonian Institution might not exist.[66]

To be sure, the benighted frontier sucked in the evangelical clergy, who preached temperance, law and order, and literacy in more than 1,000 churches by 1860. As the saying went, "If you hear something lumberin' through the canebrake, it's either a b'ar or a Methodist preacher, and either one's bound to be hungry!" But it was hard to escape the stereotype when Thomas Bangs Thorpe published tall tales parodying his state in New York magazines. It was hard to escape the stereotype when even the wife of a future senator thought nothing of butchering nine hogs in a Little Rock smokehouse, then walking home in the snow to give birth. Yet that admixture of religious revival, insecure pride, and maternal backbone forged the character of such Arkansan achievers as Scott Joplin, Dizzy Dean, Eldridge Cleaver, Johnny Cash, Sam Walton, and Bill Clinton.

Philip Hone's diary entry for April 8, 1835, describes New York City's spirited mayoral race. He had hopes of turning out the Tammany Hall Democrats because "the Whig Party . . . are active, zealous, and confident of success." The election was marred by the usual riots in which "respectable persons were beaten and trampled in the mud," and the Democratic candidate squeaked by in the end. But the strong challenge mounted by Hone's "Whigs" and the majority they won in the city council were cause for national celebration. Within days Daniel Webster wrote of the "Whig" opposition, Davy Crockett addressed rowdy "Young Whigs," the *National Intelligencer* touted "Whig candidates," and spontaneous "Whig rallies" erupted in Portsmouth, Buffalo, Philadelphia, and Baltimore. Henry Clay made it official. In a stirring speech in the Senate he hailed the triumph of law and liberty in New York "over clubs and bludgeons. . . . Go on, noble city! Go on, patriotic whigs!" Just as Whigs defied tyranny in 1688 and 1776 so "the whigs of the present day are opposing executive encroachment," opposing what Philip Hone called an "imperial president."

Jackson's press agents knew at once this meant trouble. The National Republicans were burying their old label and reinventing themselves with an evocative name implying that Jackson was a monarch (King Andrew I) and the Democrats were Tories. Francis Blair gave the Democrats' newspaper chain the party line when his *Globe* chided an "Alias Party" that assumed a new name because its crimes had disgraced the old one. Everyone knew that this "organized incompatibility" of tariff men, supporters of the Bank of the United States (BUS), nullifiers, and former Federalists constituted the real Tories. But Blair failed to quash the excitement generated in a "republic of words" by a brawny word toting heavy historical baggage. By summer 1835 Whig organizations and newspapers sprang up in almost all states, creating a second great political party and the nation's second party system.[67] More will be said about what these Whigs stood for and whom they attracted. But at birth the second party system amounted to little more than a divide between Americans who loved or loathed Andrew Jackson and accused their rivals of pretense. To Democrats the Whigs were anything but Whiggish. To Whigs the Democrats were anything but democratic.

The Democratic convention in 1836 was certainly not. Billed as an open forum of delegates "fresh from the people," the conclave in Baltimore was a holiday for party hacks. Jackson's anointed successor was their unanimous choice, and Van Buren hoped Jackson's blessing would prove as effective with the voters at large. But those results in New York worried him, as did a growing rift between New York's conservative Democrats and radical Loco-Focos who wondered when Jackson's "common man" rhetoric was going to translate into help for the workingman. The insurgent mechanics and tradesmen demanded abolition of all banks; they also demanded the right to strike, free schools, and lenient bankruptcy laws. Van Buren also worried about southern Democrats whose loyalty might not extend to a northern dandy with a vulpine reputation. So Van Buren's strategy was to make the right noises to appease northern workers and shill all he could for the slave power. In the Senate he presided over adoption of the gag rule of 1836 against antislavery petitions. In New York, he ordered every Democratic official to persecute abolitionists. Postmasters refused to mail their tracts, politicians stirred up riots against them, the attorney general denounced their "pernicious" vision; and all of this was duly reported in Democratic newspapers in the South.

Van Buren's best hope derived from the immaturity of the opposition. Unable to agree on a single leader or even organize a convention, the Whigs tried faute de mieux for a repetition of 1824, when multiple candidates had denied Jackson a majority in the electoral college. So it was that Daniel Webster of Massachusetts, Hugh Lawson White of Tennessee, William Henry Harrison of Indiana, and Willie Mangum of North Carolina courted regional support while concentrating their fire on Van Buren's "corruption." But that stock accusation satisfied no mass appetite. Davy Crockett, the Whigs' own Tennessee leatherstocking, served up juicier meat. He wrote (or had ghost-written) a campaign biography depicting "little Van" as a fop. Van Buren, it said, turned up his nose at the common people he presumed to lead, rode about in a carriage with servants in uniform ("I think they call it livery"), dressed like a lord, and (here's the kicker) "is laced up in corsets, such as women in town wear. . . . It would be difficult to say, from his personal appearance, whether he was a man or woman, but for his large *red* and *gray* whiskers."[68]

Thanks to a bubble economy and Jackson's mantle Van Buren polled 51 percent of the popular vote and 58 percent of the electoral college. The forbidding Webster got less than 3 percent, which killed his national aspirations. But Harrison, alleged hero of the alleged battle at Tippecanoe (it was really more of a massacre), surprised everyone by winning nearly 37 percent and seven states stretching from Kentucky to Vermont. Like Jackson, Harrison was an aging frontier chieftain who appealed to multiple geographical and cultural constituencies. Like Jackson in 1828, Harrison ran a campaign safely devoid of substance. "Let him say nothing—promise nothing," advised an older but wiser Nicholas Biddle. "Let no Committee, no Convention, no town meeting ever extract from him a single word about what he thinks now and will do hereafter." Henry Clay, who always said and promised too much, spied in Harrison an alarming competitor. Eastern Whigs spied in him a welcome alternative to Clay. On Inauguration Day in March 1837 they all spied the fragility of a "monarchy" without Jackson as monarch. A crowd of 20,000 listened politely as President Van Buren droned through a platitudinous speech noteworthy only for describing forbearance toward slavery as "humane, patriotic, expedient, honorable, and just." Only after Van Buren shut up and the former president Jackson descended the Capitol steps did the crowd explode in deafening, affectionate cheers. It seemed, wrote Benton, "the *rising* was eclipsed by the *setting* sun." Just days later, when the Panic of 1837 wrecked the nation's economy, Whigs had their issue and four full years to master the electoral tricks honed in large part by the "little magician" Van Buren himself.[69]

The Twenty–Sixth State: Michigan, 1837

Michigan, whose name derived from the Ojibwa (Chippewa) word for "big waters," was a western, northern, and international frontier all at once. This explains why it remained insecure and nearly vacant years after Ohio, Indiana, and Illinois achieved statehood. Following American independence, the British continued to occupy their Great Lake fortresses at Detroit and on Mackinac Island until 1796. In 1805, when Michigan was carved out of the old Northwest Territory, the Ottawas, Hurons (Wyandots), Miami, and Potawatomis still roamed the entire peninsula save for a sliver around Detroit, which burned to the ground that year. In 1806 New

England hucksters floated a Bank of Detroit ostensibly to finance the fur trade, but really to peddle $1.5 million in unsecured scrip that made the name of Michigan mud. Then redcoats recaptured the forts in the War of 1812. So the territory was starting from scratch when President Madison appointed Lewis Cass governor in 1813. Over 18 years in office, Cass, a devoted Freemason, proved a master builder, shrewd entrepreneur, and "consummate politician, upon the machiavelian model (softened perhaps & *americanized*)." At length Cass made good the motto he coined for his empire: "*Si quaeris peninsulam amoenam CIRCUMSCIPE*" (If you seek a pleasant peninsula, just look around).[70]

Michigan's southern tier offered rolling, low-lying land traversed by gentle rivers and covered with the hickory, oak, and hemlock that to New Englanders betokened good soil. Northern Michigan's great stands of white pine were sure to attract loggers. Beaver, raccoon, mink, opossum, badger, weasel, otter, and skunk lured trappers. Trout and other game fish leaped in hundreds of lakes. The climate was rugged—the commandant at Mackinaw told Harriet Martineau, "We have nine months winter, and three months cold weather"—but Yankees were used to that. The trick was to get them to "look around" Michigan at all, given that attractive frontiers beckoned throughout what we now call the Middle West.[71]

Cass's first move was to serve as a veritable agent for John Jacob Astor, defending his monopoly of the fur trade as simply "the influence of capital, skill & enterprize." That won Michigan a rich and powerful friend in New York and Washington City. The first federal land office opened in 1818, federally funded surveys began, and a series of federal treaties liquidated Indian claims on the Lower Peninsula. The Treaty of Saginaw of 1819 was typical. Cass got himself appointed U.S. commissioner with a budget of $10,000. He loaded schooners with liquor and supplies, bribed a trapper whom the Indians trusted to serve as go-between, then called the Ottawas and Ojibwa to a council. Five great barrels of whiskey were prominently on display, but not to be opened until a bargain was struck. The thirsty Indians quickly agreed to sell 6 million acres around Saginaw Bay for $3,000 plus $1,000 per year in supplies and the services of a blacksmith. As the Indians retreated, land offices opened at Monroe, White Pigeon, Kalamazoo, Flint, and Ionia. Cass and the territorial delegate Gabriel Richard, the only Catholic priest to serve in Congress, lobbied successfully for federally funded roads linking Detroit with Fort Dearborn (Chicago),

Ypsilanti, Ann Arbor, and Saginaw, opening huge swaths of land preco-
ciously organized into thirty-eight counties. Cass sponsored a township
law modeled on the Puritans' famous regulation of 1647. He personally
explored the Great Lakes in the company of the geologist Henry Schoolcraft, whose reports on terrain advertised the territory's potential and
whose notes on the Indians inspired Longfellow's *Hiawatha*. Solid Yankee
that he was, Cass founded public schools, a "Catholepistemiad or University of Michigania," a library, and a Young Men's Society to promote virtue
and the arts. He pressed for democratic elections at an early stage. Above
all, Cass launched a publicity campaign stressing Michigan's docile Indians, rich soil, and salubrious climate—and its accessibility, thanks to
steamboat service on Lake Erie and the opening of the Erie Canal in 1825.
A ditty once ran, "Don't go to Michigan, that land of ills; the word means
ague, fever, and chills." Now the advice was: "Come all you Yankee farmers who'd like to change your lot . . . come follow me and settle in
Michigan-i-a." The land boom had just begun when Cass left to become
Jackson's secretary of war, but the groundwork was laid. Michigan's population, 31,640 in 1830, grew sevenfold over the ensuing decade, the fast-
est percentage growth in America.[72]

Most newcomers were of New England stock, with admixtures of
Virginians, Germans, and later the Dutch, who founded the town of Hol-
land. The first thing new arrivals did was acquire a wagon and hie themselves to one of the land offices. There they stood in long lines, perhaps
paying for the privilege of moving ahead. But woe to anyone who dared
bid more than $1.25 per acre minimum. He risked being run out of town on
a rail. Did that mean no speculators were sullying the process? On the
contrary: speculators moved among newcomers with elaborate brochures
depicting fictitious towns with churches, commons, roads, mills, and improved farms available at just twenty or twenty-five dollars per acre. One
land jobber likened Michigan to colonial Virginia, with the difference that
a fortune "ripens into maturity in a few years in Michg. whereas several
generations passed away in Va. before those large bodies of land were
productive to their holders." Cheated or not, many settlers suffered from a
poor diet the first year, frostbite the first winter, rattlesnake bites, and ill-
nesses such as the "ague" or "shakes." The ague was malaria, a scourge
before drainage of the swamps (or "bogs" as New Englanders called
them). But whether Yankees, Germans, or Dutch, Michiganders were com-

munal, helping each other raise houses and barns, plant a first crop, stage a husking bee, found a church, or look to each other's needs in town meetings. That ethic, plus government help and soil good for grain and dairy herds, ensured the future state's commonweal.[73]

Why wasn't Michigan already a state by 1835? That was what Michigan's "boy governor" Stevens T. Mason wondered. He inherited leadership in Detroit when Cass left and the territorial secretary John Mason resigned to claim land grants in Texas. President Jackson promptly named Mason's son Stevens to the post even though Stevens was only nineteen years old. When 160 citizens signed a protest, the youth issued a public statement assuring the people that he had learned the job from his father, always sought the advice of older and wiser men, and in any event would soon be subordinate to a new governor. In private, Mason researched the political affiliations of the protesters and forwarded the enemies' list to the White House. When the newly appointed governor then died of cholera, again putting Mason in charge, he pressed Congress to authorize statehood proceedings. Congress refused, pending resolution of a long-standing border dispute. It seems Ohio insisted its state border be adjusted to include the mouth of the Maumee River, where developers planned the city of Toledo and a canal linking Lake Erie to the Ohio River. Michiganders screamed, but the Ohioans enlisted support from their colleagues in Indiana and Illinois, states that had also received dubious northern adjustments of their borders. Young Mason was ready to fight. He called for a census, a constitutional convention, and election of state officials, all of which were completed over a few months in 1835. Then he took command of the territorial militia, occupied the disputed wedge, and arrested the Ohioans on the Maumee. The elections were a landslide. Michiganders approved the constitution by more than five to one, elected Mason their governor by a nine-to-one margin (over Nicholas Biddle's brother, no less), and cheered when the new legislature voted a war chest to match that authorized by the Ohio legislature. This "Toledo war" was what moved Ohio's buckeyes to refer to their feisty neighbors as wolverines.[74]

Lewis Cass leaned hard on Senator Benton, chair of the territorial committee, to arrange a truce. Southerners led by Calhoun stalled him until Arkansas was admitted. At last, in June 1836, Congress offered Michigan statehood on condition it cede Toledo and accept the so-called Upper Peninsula in consolation. Michiganders expected Mason to spit on this veiled surrender. But the boy governor was apprised that immediate statehood would earn his government 5 percent of future land sales plus a hefty portion of the federal surplus Jackson meant to distribute to states. The "Frost Bitten Convention" Mason summoned to Ann Arbor in December ratified the proposal. On January 26, 1837, Michigan entered the Union.[75]

The Wolverine State was perhaps the purest product of the great Yankee exodus. Its constitution, drafted by farmers rather than lawyers, was democratic and simple. It limited legislators' pay to just three dollars per day (a provision that would not be changed until 1948). It promoted public works and banned capital punishment. It was the first state to create a superintendent of public instruction. Its free public schools, university at Ann Arbor (1841), normal school at Ypsilanti (1852), and agricultural college at East Lansing (1857) became models for almost every subsequent state. The government also promoted science and engineering, a policy that struck pay dirt when Schoolcraft persuaded Douglass Houghton, a graduate of the Rensselaer technical school in Troy, New York, to become Michigan's state geologist. Fascinated by a two-ton "nugget" of pure copper found on the Upper Peninsula, Houghton did extensive fieldwork and filed annual reports on the methods of finding and mining the ore. He perished at age thirty-six when a sudden storm on Lake Superior swamped his boat in 1845. But by then his research had triggered the first American metals "rush." Michigan's Keweenaw mines yielded about 90 percent of the copper Samuel Morse used to knit the nation together with wires strung on poles cut in large part from Michigan's Saginaw lumber.[76]

The magical powers attributed to Martin Van Buren deserted him the moment he entered the White House. Federal receipts fell 50 percent in 1837, reviving the national debt briefly liquidated by Jackson. Money in circulation fell by two-thirds, forcing Congress to pass a debt relief act because the jails could not house all the bankrupts. There was little the federal government could do about the deepening depression even if Van Buren had jettisoned his laissez-faire principles. Instead, he wasted his term pushing for a bill to deposit federal assets in an independent treasury rather than rickety private banks. Van Buren also paid the political price for the Trail of Tears and the Seminole war. His refusal to press for the annexation of Texas, though prudent, angered expansionists throughout the South. His neutrality with regard to a Canadian revolt against Britain in 1837, though prudent, angered idealists throughout the North. His dispatch of General Scott to prevent loggers in Maine and New Brunswick from going to war over a boundary dispute, though prudent, angered spread-eagle nationalists everywhere. His incarceration of the mutinous slaves who seized the schooner *Amistad* in 1839 and washed up on Long Island gratuitously outraged genteel opinion. Lewis Tappan and John Quincy Adams launched court challenges on behalf of the refugees that embarrassed the president throughout the campaign of 1840. Van Buren even lost Andrew Jackson's confidence by refusing to dump Vice President Richard M. Johnson. Jackson himself had chosen Johnson, an old Kentucky colonel and a hero in the War of 1812, but Johnson's cohabitation with the mulatto mother of his two daughters was political poison. In the end the Democratic convention endorsed no running mate at all for Van Buren.[77]

Yet the Jacksonian machine remained so formidable that Van Buren polled 400,000 more votes in 1840 than any previous presidential candidate. How was it then that he got *trounced* 234 to sixty in the electoral college? First, the electorate burgeoned through population increase and state laws expanding the franchise. Second, the 80.2 percent turnout of eligible voters (up from 57.8 percent in 1836) was the highest in U.S. history. Third, leaders of the Whig Party, having failed in 1836 to replicate the results of 1824, managed in 1840 to mimic Jackson's 1828 campaign. In other words, they launched a demagogic crusade against demagogy.

Henry Clay thought he deserved the Whig nomination. But in order to sell himself as a viable "omnibus candidate" with national appeal he had to overcome all the controversial votes, "corrupt bargains," changes of positions, and enemies made during his long career in Congress. Clay tried to do that in February 1839 with a stunning speech in which he paired his old words of distaste for slavery with new words damning abolitionist "ultras." Time and Providence alone could resolve the issue; abolition was a "mad and fatal course" threatening "to deluge our country in blood." A friend suggested that his speech might just alienate "ultras" on both sides of the debate, whereupon Clay replied, "I had rather be right than President." That brilliant line took wing, but was laughably insincere. Clay intended to capture moderates on both sides and so build a nominating majority. He almost did. When the Whigs met in Harrisburg,

Pennsylvania, Clay garnered 103 votes to ninety-one for William Henry Harrison and fifty-seven for Winfield Scott.[78]

Eastern party bosses wanted no part of Clay, a two-time loser, preferring an "above the battle" military hero like Jackson himself. Scott was the first choice of most northern Whigs and the second choice of most southerners. But Pennsylvania's Thaddeus Stevens put Harrison over the top with a ruse. While visiting the Virginia caucus he "accidentally" dropped an indiscreet letter in which Scott praised abolitionists. Southern defections stalled Scott's bandwagon, and this persuaded Thurlow Weed to throw New York's big delegation behind Harrison. The convention then balanced the ticket by nominating John Tyler of Virginia for vice president. He was not really a Whig at all, but a states' rights man who hated Jackson and would play in the South. Finally, the convention tied ribbons around its pretty package, having put nothing of substance inside. It adjourned without even discussing a platform. Far from being irresponsible, incompetent, or hopelessly divided, the Whigs were fiercely united in their determination to win. American democracy was about winning, and the Whigs had learned through hard experience that the way to win is through comfortable imagery, not uncomfortable ideology. James Kent, New York's great jurist, advised, "I find from the Experience of 40 years in Politics that the more leveling, violent, democratic & unprincipled side of the electoral Contest for Power, is generally successful." Henry Clay, whose campaign tour for Harrison drew 100,000 howling men and swooning women, regretted "the necessity, real or imagined, of appealing to the feelings and passions of our Countrymen, rather than to their reasons and their judgments." A fellow Kentuckian was even more candid: "*I have no faith in this democracy, but it is the road to success . . . and no fellow shall out democrat me.*"[79]

William Henry Harrison was sixty-seven and retired for ten years. He was not a pioneer born in a log cabin, but a gentleman born on an estate in Virginia. His war record against the Indians and British was mediocre at best. As territorial governor he had tried to introduce slavery in Indiana. As a representative of Ohio in Congress he achieved virtually nothing over twelve years. None of that mattered. Harrison bridged the South's Cavalier culture, the West's Scots-Irish culture, and even New England's culture insofar as he was known (or billed) as a churchgoing family man and social reformer. His handlers forbade him to take a firm position on anything except when he could steal the Democrats' populist thunder. So it was that

William Henry Harrison, of all people, solved a political riddle that had stumped Jackson, Clay, and everyone else. He blessed Hamilton's capitalist system, but divorced it from Hamilton's contempt for democracy and wedded it to Jefferson's dream of equal opportunity. To Harrison, democracy and cheap credit were "the only means, under Heaven, by which a poor industrious man may become a rich man without bowing to colossal wealth." Finally, Harrison intimated how uncomfortable he was with Jackson's executive usurpations (to which he would devote his long, philosophical inaugural address). Clay and Webster loved this, because it implied that they themselves could chart the course for a pliable president.[80]

Harrison was the first fully packaged presidential candidate. If Jackson was Old Hickory, Harrison was Old Buckeye. If Jackson was the hero of New Orleans, Harrison was the hero of Tippecanoe. If Van Buren balanced Jackson's ticket, the Whigs offered "Tippecanoe and Tyler, too!" ("There was rhyme, but no reason in it," quipped Philip Hone.) Best of all was the slogan inspired by a bitter Clay man: "Give [Harrison] a barrel of hard cider and a pension of two thousand a year and, my word for it, he will set the remainder of his days in a log cabin, by the side of a 'sea-coal' fire and study moral philosophy." A Democratic editor in Baltimore cheerfully printed the slur, never guessing that "log cabin and hard cider" was exactly the image the Whigs needed to chisel in place of their own image as rich men who supported Biddle's bank. If Jacksonians gathered by the hundreds to wave hickory sticks, supporters of "Old Tip, the Farmer's President" paraded by the thousands in coonskin caps, torches in one hand and hard cider in the other (or sweet cider for women and those "on the wagon"). Nor was it just an affair of rubes in places like Vincennes, Indiana. A distillery in Philadelphia owned by Mr. Booz marketed whiskey in log-cabin bottles, thus popularizing the word "booze." In New York City a log cabin pavilion nearly forty yards long was erected on the corner of Broadway and Prince Street. Hone marveled at the "capital speeches" delivered there by gentlemen of the West, and the "hurrahs" elicited for Tip and Ty. The campaign seemed the civic equivalent of a religious revival, not least because of the ubiquitous clergy and women looking to the Whigs to restore order and decency to society. Harrison became the champion of hearth and home, morals, religion, and domesticity. Women joyously cried, "Whig husbands or none!" and carried brooms to the log cabin rallies in order to sweep out corrupt Democrats.[81]

Whig songs, cartoons, and journals (including Greeley's *Log Cabin*) contrasted their "man of the people" to the foppish Van Buren. What kind of "democrat" was it who paraded about in Belgian lace and sipped French wines from golden goblets? Democrats in their O.K. Clubs might chant for the man from Old Kinderhook, but Van Buren was not "OK" to the Whigs. He was "Martin Van Ruin" and "Van, Van, the Used-Up Man," slogans implying that he was to blame for the sour economy. If there was anything else resembling an issue in 1840, however, it has eluded historians. That is why the campaign is so very important. An opposition party won the White House for the first time since 1801, and did so by stealing its enemy's clothes. The Whigs bested the heirs of Jackson in the arts of imagery, spin, organization, and hoopla, proving you don't need a platform or even a "real human being" as candidate to seize the brass ring. You need only offer as many American voters as possible an incomplete syllogism whose conclusion they can draw for themselves. Nothing against Harrison, who was a decent enough nonidentity. He just happened to be the one with "availability," which, Benton sneered, was "the only ability sought by the Whigs." The *Democratic Review* saw what was coming when it mourned, "We have taught them how to conquer us!" But most Democrats pushed that truth far off,

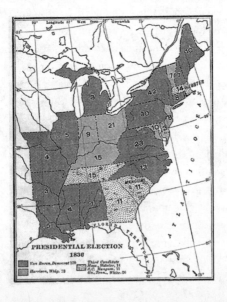

PRESIDENTIAL ELECTION
1840

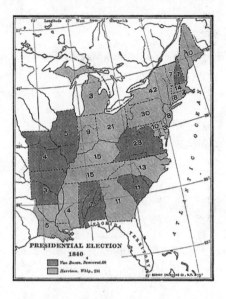

blaming their loss on "fraud, hallucination, and excessive democracy."
How ironic was that?[82]

Harrison's plurality was less than 1 percent, but so well distributed
that he carried nineteen states to Van Buren's seven and dominated every
region east of the Mississippi. The Whigs had proved that they, too, could
dance without tripping over the Mason-Dixon Line. This made them a
bona fide national party. How then could they explain the inscrutable
Providence that cheated them of their triumph when Harrison died of
pneumonia a month after his inauguration and the afterthought Tyler assumed the presidency? Whig preachers wailed that heaven must judge the
American people unworthy of redemption. Democratic preachers thought
it a judgment on the dishonest campaign. Whig senators led by Clay pressured Tyler to defer to Congress. Harrison's appointees led by Secretary of
State Webster insisted that Tyler defer to the cabinet. When instead Tyler
coldly insisted on his full constitutional prerogative, vetoed the Whigs'
legislative agenda, and cleaned out the cabinet except for Webster, politicians in both parties recognized this as an interregnum. All eyes looked
ahead to 1844. But before the term of the first "accidental president" ended,
the obscure Tyler bade Americans dream of manifest destinies as vast and
Romantic as Texas, the Pacific Ocean, and China.

* * *

Why dwell at length on mostly familiar political history? Because politics was the principal force holding together a disparate republican people. Americans were rendered American by virtue of their citizenship in states bound together by the verbiage of a Constitution fixing the rules under which the old power game might safely be played. Citizens could, and constantly did, argue about the meaning of words and bend the rules to their advantage, and sometimes they defined their identities by denying others (Indians, Negroes, Irish, women) a chance to define their identities through participation in the arena. But party politics in the nineteenth century was the national pastime, the rooting interest, the source of totems and comfortable myths by which citizens endowed their lives with meaning, coexisted with differently-minded people, and pretended politics made a difference. Sometimes it did.

Indeed politicking in the Jacksonian era was so persistent and fanatical that historians cannot resist disassembling the machine in hopes of discovering what made Americans tick. Thus, Arthur Schlesinger, Jr., interpreted Jacksonian democracy as a lower-class protest against monied elites, and the Jackson administration as a laudable antecedent to Franklin Roosevelt's New Deal (despite the fact their policies could not have differed more). A class conflict model would seem to explain why Jackson ran so well in the urban Northeast (and also why Schlesinger had so little to say about Jackson himself or his attitude toward Indians or Negroes). Recently, Charles Sellers depicted Jacksonians as crusaders vainly resisting capitalism. That would explain the "chants Democratic" of incipient labor leaders and publicists such as William Leggett in New York. But how then to account for the laissez-faire economics promoted by Jackson and Chief Justice Taney, or for the Democrats' excommunication of Leggett when he defended Loco-Focos and abolitionists? Another interpretation, as old as Frederick Jackson Turner and updated by Charles Beard, imagined that pioneer farmers were the core of Jackson's constituency. This would explain why Old Hickory routed Clay in states such as Illinois, Michigan, Missouri, and Arkansas, but not why the Democrats ran strong in cities. Marvin Meyers linked mechanics and farmers by defining Democrats as nostalgic Jeffersonians damning monopolies and corruption in the name of republican virtue. Robert Remini agreed that anticorruption was a thread sewing together what otherwise were local political machines. But what then of those "men on the make" who profited from the war over the BUS or demanded cheap credit to get ahead at their workbench or plow? Richard Hofstadter found that most Jacksonians were not anticapitalists, but rather frustrated capitalists hoping to expand opportunity as Jefferson's loyalists had sought to do a generation before. Lee Benson and Ronald Formisano analyzed the ethnic and religious composition of the Jackson coalition, and David Hackett Fischer portrayed the democracy as an alliance of the libertarian Scots-Irish and hierarchical Cavalier cultures against the "ordered liberty" upheld by heirs of the Puritans.[83]

All these interpretations have more or less merit, but their authors would probably agree there is a "blind men and the elephant" quality about Jacksonian scholarship. How can there not be, given that national parties could win elections, and thus survive, only by cobbling together disparate and sometimes contradictory blocs of voters? This means it is futile to apply such phrases as "*the* age of Jackson," "*the* Jacksonian persuasion," or "*th*e concept of Jacksonian democracy." If some single phenomenon did unite most of the people who waved hickory sticks, it was not ideology, group cohesion, or a legislative agenda, but rather moods of the sort that shaped the contemporaneous Romantic era in culture. Americans of many sorts were in one bad mood or another between 1819 and 1850. They were against many things and sensed that Jackson was, too. They displayed a psychology born of democracy and its discontents that was unimagined by the founders, but in retrospect seems an inevitable consequence of free markets in power, wealth, and pursuits of happiness.

Americans, Tocqueville remarked, are naturally suspicious of others' success. To attribute their rise to superior virtue demeans the self. Hence those left behind in a freewheeling, free-market democracy tend to assume the rich and mighty became rich and mighty through subterfuge, corruption, or favoritism. And if this were so, then the fact that such men had the power to constrain liberty and limit opportunity for the common man was simply unbearable. The real or imagined contrast between the myth of equality and the very unequal outcome of free competition in an era of blistering change was a torment to many Americans. Jackson himself was a rich planter, slave owner, land speculator, lawyer, politician, and military hero. But his orphaned childhood and tumultuous frontier life rendered him anxious, suspicious, and hostile toward institutions, and sympathetic toward all whites who were subject to others' authority. By instinct or accident Jackson discovered the secret of American politics, which is to rally the largest possible number of voters to oppose the smallest and vaguest of enemies. Leaders of the Democratic party caught the drift and rode it for thirty years, using rhetoric that damned corruption, corrupt bargains, conspiracies, "monster" banks, "satanic" mills, monopolies, aristocrats, usurpers, speculators, stockjobbers, abolitionists, and meddling self-righteous reformers who meant to "enslave," "shackle," "enchain," or "fetter" the naturally free and equal American workers and farmers. Democrats invited any voter with a grievance to assume he had been cheated, thwarted, exploited by powerful men who rigged the game in their favor. They apotheosized all that was natural, simple, and intuitive, implying that complex institutions of all sorts were artificial, oppressive, and undemocratic. Jacksonians wanted to rise in the hectic, industrializing market economy, yet at the same time flee from the impersonal human relationships it required. Their world was a volatile mix of aspiration and fear.[84]

Who then were the Whigs? A loose congeries of office-seekers devoid of ideas, as Henry Adams later dismissed them? A party of plutocrats, bluenoses, and their dupes, as the Democrats' propaganda suggested? A failed experiment that lasted just twenty years? The Whigs surely attracted the usual opportunists, many of the wealthiest southern planters, most northern merchants and manufacturers, almost all evangelicals, Unitarians, and anti-Masons preaching temperance, education, and restrictions on slavery. But the Whig Party was well endowed with ideas, beginning with Clay's American System; it survived almost as long as the Federalists had; and it found unity in a psychology at least as strong as that of the Democrats. The difference was that the anxiety felt by Whigs was internally directed, whereas the anxiety felt by Democrats was externally projected. Whigs looked for the source of society's ills and found it inside individuals whose duty it was to purge their vices and serve the public good. Democrats looked for the source of individuals' ills and found it in a society violating the public's rights. Whigs wanted a people as good as their government. Democrats wanted a government as good as the people.

In practice, Whigs such as Henry Clay and Daniel Webster worshipped the American Union. Trusting its institutions, they believed government should assist families, churches, and schools to mold citizens devoted to "sobriety, industry and frugality; chastity, moderation and temperance." Democrats such as Thomas Hart Benton and James K. Polk, though equally patriotic, were jealous of states' rights, distrusted self-styled moral authorities, and resisted "sage Doctors and sager Doctresses" who made it their business to tell others how to live. Whigs welcomed the world of stocks, bonds, and paper money, and considered internal improvements and social reform the true engines of opportunity whereby "the poor, in one generation, furnish the rich of the next." Democrats clung to hard money, and believed federal promotion of material growth inevitably favored some regions and industries at the expense of others. Whig journals such as *Hunt's Merchants' Magazine* preached a "gospel of commerce" that described business as a moral calling akin to the missionary's. Democrats damned this as blasphemy born of greed. Whigs were nervous about hasty territorial expansion, especially if it risked foreign war. Democrats hailed an idealistic doctrine of Manifest Destiny implying that expansion might go ahead without war. Whigs welcomed the industrial revolution. Democrats were at best of two minds. Above all, Whigs valued order, uniformity, and predictable behavior born of individual self-discipline that banished passion from human relations. Democrats valued individual liberty and local distinction, even if these meant excusing such passionate *dis*order as riots, duels, feuds, gang wars, strikes, corporal punishment, and armed resistance to federal authority. In sum, Whigs were anxious to realize the American dream through qualitative improvement; Democrats, through quantitative expansion. Whigs were anxious to pull Americans into a glorious future; Democrats, to give them the choice of preserving their present and past. Whigs offered a society "economically diverse but culturally uniform; the Democrats preferred the economic uniformity of a society of small farmers and artisans but were more tolerant of cultural and moral diversity."[85]

Not least, Democrats found their identity by dwelling on the Whigs' pretenses; Whigs found theirs by dwelling on the Democrats' pretenses. The real purpose behind the Whigs' profession of the "public good," said Polk, was "to build up an aristocracy of wealth, to control the masses of society, and monopolize the power of the country." If Whigs attracted votes, wrote a Democratic editor, it was because thousands of menial workers "think as their employers think, vote as they vote, and dare not call their souls their own." Whigs were nothing but "purse-proud swindlers." Ha! cried a Whig editor sick of Democrats' invoking the sacred rights of individuals and states. "There is no shallower cant than this cry of 'leave us alone.'" Who are the Democrats? asked Horace Greeley. They are a "race of hollow demagogues whose heaven is a majority."[86] In American politics only regional or single-issue quixotes can afford to be steadfast. Candidates and parties competing in national politics must constantly trim their sails. That in turn invites voters to choose sides not just on the basis of whose positions they share, but on the basis of whose hypocrisy offends their temperaments more.

Who were the conservatives and who were the liberals in this second party system? If one adopts twentieth-century definitions it might appear that the libertarian Democrats were the conservatives and the statist Whigs the liberals. But in the parlance of nineteenth-century Britain, where the labels originated, the reverse would be true. In regard to slavery, free-soil Whigs would appear the liberals and the Democrats supporters of a racist status quo. But in regard to workers' rights as understood later in the century, neither party was "progressive." In regard to ethnic and religious tolerance the Democrats would appear the liberals, since they embraced Catholics and immigrants. But in regard to education and social reform the reverse would be true. The only way to get a grip on the growing divide among Americans in the mid-nineteenth century is to purge our contemporary notion of the political spectrum and try instead to imagine the ambivalent anxieties of a freewheeling people with one foot in manure and the other in a telegraph office.

The important characteristics shared by the parties were their national appeal and their devotion, within limits, to compromise. Both appealed to significant blocs in all sections of the country. Both offered leaders who either cherished the Union or feared the consequences of division enough to broker, paper over, or agree not to talk about their sharp divergences. The ever more rancorous divide over slavery, however, would weaken the national appeal of the parties by placing unbearable pressure on northern Democrats and southern Whigs. Should those pillars crumble, the second party system must tumble, and with it the whole Solomon's temple revered as the Union.

Chapter 3

MIGRANTS, FARMERS, MECHANICS, AND CLOWNS

The Anxious, Exciting Birth of an Industrial People, 1830–1860

*A*nd Satan stood up against Israel, and provoked David to number Israel. And David said to Joab and to the rulers of the people, Go, number Israel from Beersheba even to Dan; and bring the number of them to me, that I may know it. . . . And God was displeased with this thing; therefore he smote Israel." Why King David's census was an abomination has been disputed for millennia, not least because two different accounts appear in the Bible. Did David not trust the Lord's promise to multiply Abraham's seed as the sands of the sea? Did his desire to know the number of men he could call to arms betray pride or a lack of faith in the Lord of Hosts?[1] Whatever the case, the precocious founders of the United States were not deterred. They wrote the census into their Constitution several years before the French revolutionaries did the same. The decennial census was imperative given the need to apportion state representation in Congress. But the census also inspired enlightened Americans, especially in the North, to imagine the nation's progress and character could be measured, and if measured, then stimulated and shaped through federal policy. That prospect horrified defenders of states' rights and laissez-faire markets, especially in the South. The rapid and decidedly uneven growth of the United States as documented every ten years was perhaps the greatest source of anxiety in the years of the second party system.

Today our brains are cluttered with account numbers and PINs, telephone numbers, stock indexes, interest and inflation rates, mortgage and insurance tables, cholesterol and blood sugar counts: numbers that rule ordinary people while enriching those who exploit them. The march of numbers began in the nineteenth century, thanks to the spread of the market economy, the craze for "statisticks" introduced to America in 1803, and primary education designed to equip children for lives of buying and selling. In the eighteenth century arithmetic was a lowly subject learned by rote and aimed at boys apprenticed as clerks, midshipmen, or surveyors. In the 1820s the first "new math" burst on the scene when Harvard's Samuel Goodrich and Warren Colburn published texts based on the theories of the Swiss pedagogue Johann Pestalozzi. They touted "mental arithmetic" by which youngsters were taught the underlying principles of mathematics so as to solve problems in their heads, without paper or slate. Goodrich and Colburn insisted that such skills were invaluable not only to merchants and bankers but to every mechanic and farmer seeking to maximize and invest the fruits of his labor. By mid-century, the publisher of Colburn's *First Lesson, Or Intellectual Arithmetic on the Plan of Pestalozzi*, was selling 100,000 copies per year to public schools from Maine to Wisconsin. Inasmuch as American men were busy in the marketplace, public schools hired young women to teach, thus altering gender perceptions. Until 1820 it was thought that mathematics was beyond girls' capacity and of little use to them anyway. Twenty years later numerate girls were in demand.[2]

A thirst for "authentic facts" accompanied the rise of arithmetic. When James Madison and the commercial promoter Tench Coxe urged Congress to collect data "marking the progress of society," few members paid any attention. But authors of gazetteers and almanacs compiled elaborate statistics (however inaccurate) on economic activity; and earnest reformers devoted to prison reform, education, and temperance bombarded the public with numbers revealing the scale and effects of crime, incarceration, recidivism, illiteracy, and alcohol. (One text asked, "How many tears have the wives of drunkards shed in the United States since 1790, supposing the number of drunken husbands to be 15,000 per year?" Answer: "Enough to float the U.S. Navy.") The spread of factories and rail-roads, especially in the Northeast, made a growing number of people dependent on Connecticut clocks, so much so that a British observer in the 1830s expressed disgust with a "reckoning, expecting, and calculating people" obsessed with punctuality. All this transformed the census. As late as 1830 it remained a simple head count. But over the following decade New England's scholars observed how British, Prussian, and French bu-reaucrats applied quantitative methods to statecraft and urged Americans to do likewise. In 1831 the native New Hampshireman and Yale graduate Joseph Worcester began compiling an annual *American Almanac and Repository of Useful Knowledge*. A Scottish transplant, Archibald Russell, provided a text, *The Principles of Statistical Inquiry*. In 1839 the American Statistical Association was founded in Boston. All demanded that Con-gress make the U.S. census of 1840 an accurate, multifaceted portrait of the nation and, incidentally, hire experts such as themselves.[3]

A more complicated census eventuated, but not as the experts in-tended. Secretary of State John Forsyth farmed out the levy to a fellow Virginian, William A. Weaver, who had been dismissed from the Navy for malfeasance. Weaver in turn hired his brother to do the grunt work. The result was a monstrous form with seventy-four columns in tiny print requesting data on all free white, free black, and enslaved males and fe-males. Federal marshals did their best, but the returns were a maze of numbers that drove Weaver's clerks crazy and pleased only the Democrats' patronage printers, Francis Preston Blair and John Rives. That caused more turmoil and delay in 1841, when the new Whig secretary of state, Webster, protested the exorbitant fees charged by the Weaver brothers and the printers. At length they were forced out and the census was released. It contained all manner of errors or impossibilities, not least the finding that Negroes in the North were ten times more likely to be "insane" or "idiotic" than those in the South. Advocates of slavery had a field day with this, but northern preachers delighted in statistics indicating that illegitimacy, crime, and poverty were far more prevalent in the South. Suffice it to say that the experience in 1840 was "a story of innocence lost." Americans em-braced statistics only to learn how easily they could mislead, or be used to mislead, citizens on pressing national issues. One result was a determined campaign to ensure that the census of 1850 would be both comprehensive and accurate. Another result was a determined protest by southerners against the collection of data useful mostly to northern businessmen, lawmakers, and agitators. How else could they react to tracts praising the "unique, mysterious little Arabic sentinels on the watchtower of political economy, 1, 2, 3, 4, 5, 6, 7, 8, 9, 0," that proved how slavery dragged down the southern economy? "If left alone, we have no doubt the digits themselves would soon terminate the existence of slavery."[4]

Raw population figures confirmed what many Americans felt about the swiftness of change in their town, county, or state. Today we wrinkle our brows over porous borders and exurban sprawl or, conversely, rusting factories and children moving away. Imagine a time when the nation grew by one-third every ten years: a time when it was believed the very character of the United States was being etched forever. No wonder politicians, preachers, journalists, and businessmen fretted over how to discipline and indoctrinate a teeming people intent on democracy. No wonder common folk in all walks of life fretted over how to prosper or at least not go bankrupt in explosive, unpredictable markets.

Americans took for granted their destined greatness; indeed, they flung numbers in the faces of their snide European critics. Between 1815 and 1850 the French increased from 29 million to just 36 million; and Britons (not counting the Irish) numbered just 21.5 million in 1850. Americans, by contrast, grew from 13 million in 1830 to 17 million in 1840 and 23 million in 1850. But the increase was disturbingly uneven. New York state grew 56.5 percent in the 1830s alone, plus another 27.6 percent in the ensuing decade, to surpass 3 million people. Pennsylvania expanded by 27.4 percent and then 20.8 percent, to reach 2.3 million. Yet those figures were modest compared with those for the Middle West. Ohio leaped 62.4 and 30.3 percent in those decades, to close in on 2 million people. Illinois more than tripled in the 1830s and grew another 78.8 percent in the 1840s. All told, the portion of Americans dwelling in eastern states dropped from 55 percent to 40 percent between 1810 and 1850; that in southern states dropped from 32 percent to 27 percent, while the share of western states grew from 13 percent to 33 percent. To be sure, southwestern states such as Mississippi attracted newcomers (thanks to cotton and Indian removal), but on the whole the South stagnated. Virginia expanded by just 17 percent over the twenty-year period, to reach 1.4 million; South Carolina, with only 15 percent growth, numbered 668,000 in 1850. Distribution of white citizens was even more skewed, inasmuch as one-third of Virginians and more than half of South Carolinians and Mississippians were enslaved blacks. All told, by 1850 free states contained 58 percent of the U.S. population and 67 percent of whites. Thus, northern states enjoyed a heady and permanent majority in Congress and the electoral college in addition to their predominance in foodstuffs, commerce, and industry. This meant that Democrats had to compete vigorously for votes in the free states in order to win the presidency and control judicial appointments. Their relative success in turn meant that the Whigs had to compete for votes in the slave states. But with the electorate growing and diversifying so quickly, just feeling comfortable on the playing field was a challenge to all parties.[5]

Not least, this was due to an influx of foreigners, which vexed northerners as much as overall trends vexed southerners. The end of the Napoleonic wars and the advent of Atlantic packets with fares as low as thirty dollars revived immigration. Over 10,000 newcomers docked in 1825; 23,000 in 1830; and 84,000 in 1840. Annual numbers on that scale troubled scions of earlier immigrants such as Philip Hone. But concern bordered on panic in the decade after 1845, when 2.4 million additional foreigners washed up on American shores. They accounted for one-seventh of all white residents. Over half of those people were Irish.

"The poor Irishman, the wheelbarrow is his country," wrote Ralph Waldo Emerson. The lot of the Irish was indeed the worst of any European ethnic group to date. Nearly 1 million Irish came to America between 1815 to 1845, but most of them were of the traditional sort: single males of some education and means from Protestant Ulster or the anglicized counties of "the Pale" around Dublin. The 1.5 million who emigrated between 1845 and 1855, by contrast, were almost all Catholic families from Ireland's destitute western counties. Some Irishmen ventured inland from Boston, New York, and Philadelphia to sell their muscle on railroads and canals. Others were drawn south to labor on levees and other dangerous projects because (a planter volunteered) "niggers are worth too much to be risked here; if the Paddies are knocked overboard . . . nobody loses anything." Thousands of Irish killed by accident and disease found rest in collective, unmarked graves. But mostly the Irish squatted in seaboard cities, too poor to move or buy land, contemptuous of the farm life that had impoverished them in the old country, and content to be "urban pioneers" in slums graced with a Catholic parish, if little else. Over time they became one of the most successful immigrant groups in American history. Ironically, they never intended that. The Irish considered themselves exiles, not emigrants. They saved and sent back to Ireland millions of dollars from their weekly pittances. They meant to go home someday and felt ashamed when they didn't. Who were these Irish, and why did they come?[6]

The contempt English colonists displayed toward peoples they perceived as indolent, such as Indians and Spaniards, was learned initially in Ireland, England's first overseas conquest. After the last major Irish revolt was crushed in 1690, Parliament imposed penal laws prohibiting Catholics from owning land, practicing a profession, engaging in commerce, bearing arms, riding a horse, living in towns, and patronizing church schools. The Act of Union of 1800 obliterated Irish nationality altogether. Accordingly, 90 percent of the population had no alternative to scratching out a meager sustenance as tenant farmers on two or three acres of indifferent soil. Catholic Irish became "strangers at home," exiles in their own country. No people west of Russia were so oppressed. Yet both people and land answered to the whip of their commercial-minded masters. From 1700 to 1840 English landlords drained bogs and fens, rationalized agriculture, and exported crops and wool, while Ireland's population, thanks to the highly caloric potato, increased eightfold. That was the era when Irish folkways later familiar to Americans took root.

They were partly a product of Gaelic semantics. English-speakers mostly gave voice to an active independence, as in "I was sad," "I went to America," or "I sued the bastards!" The native tongue of the Irish was passive, as in "Sadness fell upon me," "Going to America became necessary," or "You shall answer to his honour for this." Such linguistic resignation, reinforced by the Protestant ascendancy, bred in the Irish a fatalistic, long-suffering temperament that Englishmen mistook for torpor. They thought the rural Irish childlike, lazy, and stupid. Stripped of incentive, Irish commoners in fact affected the false humility characteristic of slaves. They were courteous, even obsequious, but their courtesy was a mask: they doffed their cap at a passing lord, while quietly farting. All the Irish had was one another, and they cemented those bonds in church, at fairs and festivals on saints' days, and at family gatherings where tenors sang soulful ballads of lost battles and lost loves. Insofar as Irish Catholics asserted themselves it was in the context of factional feuds. These arose mostly from disputes over land, giving birth to the gangs so familiar on the streets of American cities. The Whiteboys, Rightboys, and Hearts of Oak battled each other or, when possible, their Protestant tormentors. Finally, the Irish genius for politics later evident in American cities was born after the Act of Union. Catholic bishops gradually realized that the Irish could improve their lot only by playing the English political game. Their champion was Daniel O'Connell, a secular liberal turned pious Catholic by his wife Mary. In 1828 he mobilized County Clare's voters, previously pawns of their landlords, to elect him to Parliament. Over ten years O'Connell deployed enough carrots and sticks (threats of revolution) to push through an act emancipating middle-class Catholics. But he failed to win relief for poor cottagers or repeal of the Union. In 1840 he wrote with disgust, "It is vain to expect any relief from England. All parties there concur in hatred to Ireland and Catholicity; and it is also founded in human nature that they should, for they have injured us too much to forgive us." In short, political agitation backed by mob violence went only so far. Ireland could be fully redeemed only through a revolution, which all the sons of Hibernia had a duty to further by alms if not arms. That is why Irish-Americans felt themselves exiles. Like the Puritans after 1630 they meant to return someday and cleanse their isle of its oppressors.[7]

Emigrants carried those folkways, resentments, and lessons to America, especially when the scale and character of the emigration changed radically after 1838. By then Ireland contained over 8 million people. Impoverished Catholics from southern and western counties began to depart at a greater rate than Ulstermen. The bog-trotters, their wives and children in tow, paid a few shillings for transport from Cork or Limerick to Liverpool in hopes of scraping together the fare to New York. Profiteering captains of rickety old scows crammed families belowdecks almost as densely as slavers. Several dozen "coffin ships" foundered in storms or were stricken by cholera, typhoid, or yellow fever. Some docked with only half their cargo alive; families of emigrés took to staging "American wakes" prior to their departure. Then, in 1845, the North American potato fungus *Phytophthora infestans* struck Ireland full force. The blight killed every crop but one through 1850, in what became known as *An Gorta Mór* (the great hunger). Starvation, scurvy, disease, and despair took 1.5 million lives over those years and drove 1.5 million people to emigrate. The Irish were quick to blame their coldhearted English masters for cheering rather than trying to alleviate the calamity. Did not Charles E. Trevelyan's *The Irish Crisis* (1848) depict the famine as a divine act to relieve Ireland's overpopulation and prepare its moral regeneration through anglicization? Did not the *Quarterly Review* write of Ireland that "all her absurdities, errors, misery she owes to herself"? In truth, British insouciance was born of a dogmatic laissez-faire mind-set and imperial hauteur. Had such a famine struck England's Midlands, generous "poor laws" would surely have been enacted. But Ireland was considered a colony whose Catholic poor were literally "beyond the Pale."[8]

The plight of refugees from the famine was even worse than that of their predecessors, a fact to which Yankees testified. Nathaniel Hawthorne wrote, of the area near Liverpool's docks, "The people are as numerous as maggots in cheese; you behold them, disgusting, all moving about, as when you raise a plank or log that has lain on the ground." In *Redburn* (1849) Herman Melville described the Liverpool "land-sharks, land-rats, and other vermin" preying on the Irish, whose "women, children and the elderly lay huddled together, starving and dying." Some Irish committed suicide; some prostituted themselves to board a ship. Those who did stagger, emaciated and seasick, onto the East River docks fell into the hands of "Yankee tricksters" (often Irish themselves) ready to "help" newcomers in exchange for the few pennies they had left. Once ensconced in Five Points, South Boston, or some other ghetto, the Irish learned that most Americans were "in heart and soul anti-Catholic, but more especially anti-Irish." At least America's Protestant ascendancy was mitigated by a free labor market, democracy, and religious liberty.[9]

That made a world of difference because it encouraged two additional characteristics of the Irish diaspora that were *not* carried over the sea: a "devotional revolution" and a secular self-help movement. The first was wrought by New York's Bishop Hughes and Philadelphia's Bishop Francis Kenrick, who were appalled by the demoralized refugees. These refugees were Catholic only in name, as a consequence of British repression and their own superstitions. It was to make proper Catholics of the Irish-Americans that the hierarchy mandated a rigid catechism; iconography; high liturgy (chants, bells, and incense); devotion to saints (especially the Blessed Virgin); disciplines such as rosary beads, novenas, and stations of the cross; and strict sexual mores. To alleviate material suffering Hughes helped to found the Society of Saint Vincent de Paul (1846) and recruited

Sisters of Charity and Sisters of Mercy to open schools, hospitals, and hostels for children, widows, former prostitutes, and the homeless. Serving as nurses, teachers, and administrators, those nuns constituted the United States' first corps of professional women. To promote social mobility, Kenrick sponsored the founding of Villanova and Saint Joseph's universities. In all things, "The task of the church was not only to preserve the faith of the immigrants; in many instances it was to change nominal Catholics into practicing believers."

The second Irish-American trait was derived from the lesson Hughes learned in New York's school of hard knocks: that democracy and equality were a "pious lie." The Irish had to close ranks and look out for each other while proving themselves good Americans. They did so through lockstep adherence to the Democratic Party, Hibernian clubs, Saint Patrick's Day parades (an American innovation), loud paeans to civil and religious liberty, and fierce resistance to the abolitionist movement. For reasons of self-esteem (at least the Irish were white), secure employment, and hatred of the British Protestants who invented abolitionism, Irish-Americans supported the status quo on slavery. In that alone they broke ranks with the great O'Connell, who conflated the Irish and Negro struggles and inspired the name of William Lloyd Garrison's journal *The Liberator*. Otherwise the Irish, unlike other immigrant groups, never ceased to identify with the nation they left behind. When James Stephens and John O'Mahoney founded the Irish Republican—or Fenian—Brotherhood in 1859, Irish-American "exiles" opened a flow of money, guns, and volunteers to old Eire that continues to this day.[10]

The influx of Germans to the United States—over 1.5 million between 1830 and 1860—nearly equaled that of the Irish. In 1815 the Congress of Vienna replaced the Holy Roman Empire with a German Confederation of thirty-eight sovereign states. But the local princes followed the lead of Austria's Klemens von Metternich, who was determined to stamp out the ideologies born of the French Revolution's *"Liberté! Égalité! Fraternité!"* Enlightened Prussian reformers promoted a customs union (1834) for much of the German economy; an exemplary state educational system; and a *Rechtstaat*, or judicial system, based on equality under the law. But the Austrian and Prussian police suppressed agitation for representative government, especially after the French Revolution of 1830 provoked uprisings in southwestern Germany. The "hungry forties" followed, during which crop failures in overpopulated provinces and competition from cheap English factory goods ruined farmers and craftsmen alike. Finally, the French Revolution of 1848 started another wave of insurrections in central Europe. Americans cheered as the German people's tribunes gathered in the Frankfurt assembly to draft a constitution for a unified, liberal national state. (Some New Yorkers got their news from the dispatches of a foreign correspondent named Karl Marx.) But confusion and conflict among revolutionaries pressing national, liberal, or social agendas aborted the movement. By late 1849 the monarchs' professional armies restored authoritarian rule. Many Germans abandoned their *Vaterland* for economic or political refuge in America.[11]

Oscar Handlin's classic depiction of nineteenth-century immigrants as "uprooted" villagers crammed into tiny apartments in American cities and bedeviled by freewheeling capitalism may be valid for many Irish and later for eastern Europeans. It is not at all apt for these Germans.[12] Most were skilled husbandmen who settled, not in port cities, but in rich agricultural counties around Cincinnati, Milwaukee, Saint Louis, and San Antonio. Those who did take up urban residence were master craftsmen, merchants, and professionals familiar with the vagaries of the market economy. One small contingent, the "Latin farmers," so called because of their *Bildung* (classical education), did fail in an effort to plant utopian colonies on the Mississippi. Otherwise, Germans prospered, nurtured their cultural bonds, and assimilated into the American mainstream all at the same time. Three contemporary immigrant letters tell the story. One from Missouri reported settlements "remarkable for their completely German appearance and their purely German atmosphere." Another from Ohio described an immigrant's plan to buy 140 acres in Illinois for his children. "This costs $1.25 per acre. . . . If they do not move there, they can sell it in 8–10 years. Then it will be worth 20–30 dollars per acre." A third, from a former German bureaucrat, was posted from the heart of Texas. "Our life here might be quite tolerable," he wrote, "if only we had a bowling alley!"[13]

The sole disappointment most Germans felt stemmed from the inferiority, in their eyes, of American culture to their own. Hence the "Little Germanies" that sprang up in the Ohio and Mississippi valleys and on the western shore of Lake Michigan were more than ethnic defense mechanisms. German theaters introduced Goethe and Schiller to the American stage. Musical societies added the Romantic music of Beethoven, Schubert,

Schumann, and Mendelssohn to the works of Bach and Handel introduced by eighteenth-century immigrants. *Turnvereine*, begun in Cincinnati in 1848, tripled as athletic, intellectual, and social centers. *Bauvereine*
were America's first credit unions for farmers and workers. Needless to say,
German communities fostered *Gemütlichkeit* in beer halls and gardens, all
but inventing the American brewing industry. Maximilian Schaefer was
the first to age lager beer in 1846; after this, production soared with the
immigration. In 1840 Saint Louis counted three breweries making 3,000
barrels per year. Two decades later forty breweries were rolling out 210,000
barrels per year, with Cincinnati and Milwaukee close behind.[14] As for
German cuisine, it was (and is) looked down on because most Americans
forgot the origins of such beloved treats as pretzels, pickles, potato chips,
and coffee cake. The mild cylindrical *Wurst* known as the frankfurter was
standardized in 1857 by the butchers' guild in Frankfurt-am-Main. Ten
years later Charles Feltman thought to wrap the sausages in buns for the
convenience of day-trippers to Coney Island. Americans soon dubbed
them hot dogs. A similar story may be told of the chopped beefsteak-on-
a-bun named for the port of Hamburg.

The immigrant Karl Follen bestowed another enormous gift during
his tenure at Harvard as the United States' first professor of German (he
was later dismissed for his vocal opposition to slavery). In 1832 Follen put
up a *Tannenbaum* in his parlor. It caught the eye of his friends Ralph
Waldo Emerson and John Quincy Adams and inspired the first magazine
illustration of a Christmas tree. When Americans learned that Queen
Victoria blessed the custom (out of love for her German husband, Prince
Albert), the Christmas spirit even conquered dour New England. Most
important were the contributions Germans made to education and engineering. The Round Hill School founded near Northampton, Massachusetts, by the former Göttingen professor Johann Friedrich Herbart won
high praise from Horace Mann and inspired Michigan's schools. Friedrich
Knapp's German and English Institute, founded in Baltimore in 1850, received applications far beyond its 700-student capacity. In Wisconsin,
Carl Schurz's wife founded the first American kindergarten in 1856. By
then the Studebaker company was marketing sturdy and elegant wagons,
and the Prussian emigrant Johann Augustus Roebling had completed the
world's first suspension bridge at Niagara Falls.

Roebling, a favorite student of the philosopher Hegel, had emigrated to Pittsburgh with the notion of founding a technocratic utopia. He was diverted when he observed, on Pennsylvania's canals, how ropes made of hemp burst when pulled taut by winches. Roebling promptly invented a machine mighty enough to twist metallic wire into cables. That suggested the possibility of building suspension bridges flexible and strong enough to support railroads. Roebling spanned the Ohio River at Cincinnati, then expanded that design to fashion his masterpiece, the Brooklyn Bridge. Another name demanding mention is that of Heinrich Steinweg. His ship docked in New York in 1850, when he was fifty-three. For two years this failed revolutionary earned just six dollars per week as a craftsman while he learned English and American business methods. In 1853 he altered his name, founded Steinway and Sons, and began marketing the finest pianos in the world.[15]

Earlier German immigrants and their descendants referred to themselves as *die Grauen* (Grays) and newcomers as *die Grünen* (Greens). The Grays made immense contributions to colonial America, but took little interest in politics and clustered in their ethnic, mostly Pietist, churches. The Greens not only swamped them numerically but displayed a keen interest in public life.[16] Whether Protestants, Catholics, or radical free-thinkers, they were not wholly content with the communities they found in the New World. Lutherans thought their American brethren lax, uninformed, and disorganized. Accordingly, Carl Ferdinand Wilhelm Walther founded *Der Lutheraner*, a journal devoted to orthodox faith and the German tongue, then summoned like-minded churchmen to Saint Louis to found the German Evangelical Lutheran Synod (Missouri Synod) in 1847. Catholic arrivals protested against Irish domination of the church hierarchy. In time, they nudged Bishop Kenrick and other prelates into building churches and appointing German-language priests to serve the second large Catholic contingent in the land of religious liberty. Socialist Forty-Eighters blessed the democracy, free speech, and free press they found in the United States, but bristled when Americans damned them as wild-eyed atheists. All three groups reflexively voted Democratic, judging Whigs to be nativist, until slavery made them think twice. The journalist Franz Josef Grand, who staged a German-American convention and lobbied Pennsylvania to print documents *auf Deutsch*, exited the Democratic Party by 1850. Professor Francis Lieber, a veteran of Waterloo, left South Carolina College for Columbia University, where he wrote for the War

Department *A Code for the Government of Armies in the Field*. Gustav Ko-
erner quit Saint Louis in disgust over slavery to become a justice in Illinois, lieutenant governor, and consultant to the lawyer Abraham Lincoln.
Friedrich Hecker, who had led a workers' uprising in Baden in 1848, settled
in Belleville, Illinois, where he promoted *Turnvereine* and agitated for abolition. His comrade in arms Gustav Struve was invited to Long Island by
a wealthy brewer who put him to work on a nine-volume history of freedom. The most celebrated German in American politics was Senator Carl
Schurz of Missouri, but he was not nearly as lonely in that role as textbooks
usually suggest.[17]

The Forty-Eighters included some prodigious women. Mathilde
Anneke, rejecting the traditional spheres of *Kinder, Küche, und Kirche*
(children, kitchen, and church), leaped into the nascent suffragist movement. When German-language journals rejected her translations of Elizabeth Cady Stanton and Susan B. Anthony she started her own *Frauenzeitung*
(1852), presenting a theology mandating sexual and racial equality. The-
resa Albertine Louise von Jakob (pen name: Talvj) wrote historical essays
admired by William Cullen Bryant. Anna Ottendörfer helped her husband start the *New Yorker Staats-Zeitung*, managed the journal for thirty-
two years after his death in 1852, and founded a retirement home and
hospital for women. The New York Public Library's German-American
collection is named in her honor.[18]

German immigrants were soon to play a pivotal role in America's
sectional crisis. That is why the president of the University of Chicago
paid them this homage in 1906: "They saw what was right, and they
planted themselves firmly and distinctly on that side with no hesitation
or wavering. . . . They turned the balance of power in favor of unity and
liberty."[19] Perhaps the most poignant illustration of that was a gesture
made by Abraham Kohn, a dry goods merchant and city clerk in Chicago.
Following the election of 1860 he gave Lincoln an American flag on which
he had stitched, in Hebrew and English, a verse from Joshua: "As I was
with Moses, so I will be with thee; I will not fail thee, nor forsake thee. Be
strong and of good courage." Kohn was German, but also Jewish, one of
100,000 Ashkenazim in the mid-nineteenth-century exodus.[20]

As late as 1830 a mere 6,000 Jews lived in the United States. Almost
all were Sephardim whose ancestors migrated from the Mediterranean,
often via the West Indies. They were urbane, commercial people who kept

religion private, comfortably interacted (and sometimes intermarried) with gentiles, and were nearly invisible. When the first Jew arrived in Cincinnati, a woman evidently expected a robed shepherd with beard and staff. "Thee looks no different than anyone else!" she exclaimed. To be sure, entrepreneurs and utopians had tried to plant rural communes for distressed Jews from Europe. The most famous, the Tammany Hall sachem and publisher Mordecai Noah, promoted a 17,000-acre retreat named Ararat by donning ermine and anointing himself "judge of Israel." *Niles' Weekly Register* chuckled over this Jewish scheme to "get money, honestly if they can, *but get money*." The scheme really could not have been more American, and it inspired Leopold Zunz of Berlin to contact Noah, inasmuch as European Jews "have nothing to look for but endless slavery and oppression."

Zunz's gloom reflected social conditions after 1815, when the emancipation wrought by the French Revolution gave way to reaction. Most German Jews were once again subject to restrictions, discriminatory levies, and occasional pogroms. When "letters from America" began to circulate, young Jews were excited to hear "A German is gladly accepted as a workingman in America [and] the German Jew is preferred to any other." The Viennese novelist Leopold Kompert launched "On to America," a movement that the failures of 1848 accelerated. A newspaper in Prague reported in 1849 that "the desire for emigration, especially to North America, increases here day by day." Ships departing Bremen, it exulted, were even offering kosher food.[21]

America's Jewish population rose to 20,000 by 1848 and 150,000 by 1860, by which time the original six synagogues had increased to seventy-seven. Newcomers were not always welcomed even by their coreligionists. For just as in the late nineteenth century sophisticated Jews of central European descent would look down on rustic refugees from Russia, so in mid-century established Sephardic Jews looked down on the threadbare Germans. In fact, Ashkenazim leaped ahead. As Professor Lieber noted, the Jew from his homeland "gains in influence daily, being rich, intelligent and educated or at least seeking education."

The Jews did it through peddling. This was hard and risky work on the frontier in a strange land whose language one was just learning. Nevertheless, one-half to two-thirds of male Jewish arrivals acquired wagons and inventory to sell door-to-door in farm areas and small towns. They offered sewing supplies, tools, household amenities, and perhaps an array of cheap jewelry for women. Throughout the Midwest, the Southwest, and even New England, Jews replaced the familiar Yankee peddlers and inherited their reputation for sharp dealing. They were called "wagon barons" or "jewelry counts" until they earned enough to settle down as "store princes." The peripatetic peddlers achieved this goal so rapidly that Jewish Americans were concentrated in cities as early as 1850. Over 25 percent lived in New York, Philadelphia, and Baltimore alone, where they excelled in the bakery, bookbinder, jewelry, tailor, shoemaker, hatter, and printing trades. Closely observing supply and demand, Jewish merchants invented new products, most famously the riveted denim Levi Strauss jeans worn by miners and the ten-gallon hats that protected cowboys and farmers from sunstroke. They also invented numerous marketing tools such as installment purchases, money-back guarantees, and the mail-order catalog conceived by Julius Rosenwald. After relocating in America, Rosenwald earned enough to purchase a small watch enterprise called Sears, Roebuck, and Company.

From peddler to shop owner to wholesaler—to banker, partly because financial houses in New York and Philadelphia were among the few places Jews met discrimination. So they founded their own firms named for Kuhn, Loeb, Seligman, Goldman Sachs, Lehman Brothers, Guggenheim, and Rothschild. They forged national and international business ties through kinship networks, much like the Quakers before them. By 1860, 374 Jewish firms were large enough to earn business ratings, a number that tripled over the next decade.[22]

What Jews failed to achieve was uniformity in one thing they all had in common: Judaism. Free exercise of religion, the absence of a hierarchy, and the lack of American rabbis perhaps made that inevitable. But the importation of Reform Judaism by the Ashkenazim complicated matters immensely. Born of the Enlightenment and suckled by the French Revolution, the Reform movement inspired numerous western European and central European Jews to reconcile their faith with the demands of citizenship in national states otherwise dominated by gentiles. In 1819 Leopold Zunz, Abraham Geiger, and Heinrich Heine met in Berlin to form the Verein für Kultur und Wissenschaft der Juden. Their purpose was to arrest a trend among Jews toward nonobservance or conversion to Christianity. Reformers advanced a scientific, historical reading of scripture, and accused the rabbinate of stifling Judaism's humanitarianism with ossified legalism. Their pragmatism was evident in their abandonment of strict dietary laws and Sabbath observance that inhibited full participation in the secular world. To do so the Reformers turned Jewish eschatology upside-down. Whereas the Orthodox interpreted the diaspora as divine punishment calling Jews back to Mosaic law so that they might return to the promised land when the messiah arrived, Reformers viewed the diaspora as a divine blessing by which Jewry might become "a light to lighten the Gentiles." Thus did Rabbi Gustav Posnansky of Charleston teach, "*This* country is our Palestine, *this* city our Jerusalem. *This* house of God our temple."

Reform Judaism was tailor-made for Americans. By the 1840s and 1850s Leo Merzbacher, Max Lilienthal, Isaac Wise, David Einhorn, and other immigrant Reformers gave them plenty of choices, and choose they did. Congregants quarreled over Sephardic versus Ashkenazi customs, the proper order of prayers, the permissibility of music, separate pews for women and men, the strictness of legal observance, the authority of rabbis and the laity, and even architecture (some synagogues were indistinguishable from Protestant churches). Philadelphia's Mikve Israel and Charleston's Beth Elohim, for example, mothered broods of new synagogues due to serial schisms. Hence the Reform movement domesticated American Judaism while causing a crisis of authority. Who spoke for the law and the prophets? In 1838 Isaac Nordheimer took up a chair in Hebraic studies at New York University and found that the only scholars with whom he could converse were Protestant divines. Isaac Wise summoned to Cincinnati a national Jewish convention that promptly disintegrated. Not until 1875 did the first seminary and the first prayer book, the *Minhag America*, emerge. What Jewish men really needed was an ethnic but secular organization promoting fellowship, charity, and American identity. A dozen young men met that need in 1843 when they convened in New York's Little Germany to form B'nai Brith.[23] Organized like the Masonic lodges to which many Jews belonged, these "Sons of the Covenant" clubs swept the nation. Families in turn needed a shul to prepare boys, and even girls, for success in America while preserving their heritage. New York's B'nai Jeshurun, founded in 1842, provided a prototype copied in many American cities.

In politics, Jews kept their heads down, following the Talmudic injunction, "The law of the land is the law [for Jews]." Thus, Jews in the

South generally upheld slavery; and Jews in the North skirted the issue by voting Jacksonian until the 1850s, whereupon many Ashkenazim followed the drift of their Protestant *Landsmänner* away from the Democrats. In Chicago, Jews even took a leadership role, as proved by Abraham Kohn.[24]

Able young European arrivals were of tremendous value to a continental nation whose demand for busy hands, quick minds, and strong backs was insatiable. Yet immigration even on the scale of 1845–1855 was not the most decisive demographic shift of the time. That trend may be grasped through some simple statistics. While the U.S. population grew 144.4 percent from 1830 to 1860, that of the six New England states grew only 60.4 percent despite the historic fertility of their Puritan stock. Had New England kept pace with the nation, it would have numbered 4,776,000 by 1860 instead of 3,135,000. What became of those "missing" 1,641,000 Yankees? They pulled up stakes and left to subcolonize the upper Midwest and Pacific Rim in a veritable Yankee exodus. Thus, rather than shrinking, New England's national influence continued to grow, thanks to "missionaries" who replicated its culture across the American heartland.

"Fifty years ago," wrote a melancholy Vermonter, "this hill echoed to the axes of a hundred lumberjacks. One hundred years ago these pastures rippled with the bell-music of many hundreds of sheep." The balsam and birches were felled, the clearings put to the plow, the birds routed. Then all was reversed. "The jays are back, and so the crows. The old clearings are fast closing. The Yankees have gone. Why did they leave this New Canaan where they had tamed the savage forest and learned to live with the savage winter? . . . I am thinking of the kind of people who could and did form a *Vegetarian* Emigrating Society and head for Kansas; and who could and did conjure up Mormonism, Perfectionism, Christian Science, and in their lighter moments invent the friction match, the steam calliope, the normal school, the dollar watch, horsehair furniture, and barbed wire." It was a classic case of push-pull. Children of Yankee farmers saw no future at home, but prospects galore in the West. Still brimming with cultural confidence dating from Plymouth Rock and Bunker Hill, New Englanders embraced a new "errand into the wilderness" that helped make "Yankee" and "American" synonymous across the nation's whole northern tier.[25]

The fate of the U.S. experiment, insisted Lyman Beecher, depended not on winning the West—that was ensured—but on *who* won the West. Slavery forces or free-soil forces? Passionate Democrats or sober Whigs? Evangelicals, Catholics, or violent, whiskey-drenched heathens? In the early going the hillbillies led. After the War of 1812, the settlement of Ohio, Indiana, Illinois, and Missouri was largely the work of Scots-Irish from Kentucky, Tennessee, and Virginia. A generation later the Germans arrived. But in between the Erie Canal and Great Lakes steamboats opened the West to New Englanders whose numbers, ambition, and public-spiritedness swept all before them. Ohio's "Western Reserve" on Lake Erie was pioneered by General Moses Cleaveland of Connecticut. Philander Chase of New Hampshire became Ohio's first Episcopal bishop and the founder of Kenyon College. His nephew Salmon P. Chase emerged as the state's leading lawyer. The Vermont-born missionaries Philo Stewart and John Shipherd founded the coeducational college at Oberlin (and invented an excellent kitchen stove). Caleb Atwater, a graduate of Williams College, designed Ohio's public school system. The Harrison, Hayes, and Garfield migrants spawned four U.S. presidents. James Whitcomb of Vermont became governor of Indiana, architect of its school system, and the namesake of the Hoosier poet James Whitcomb Riley. Caleb Mills, a graduate of Dartmouth, modeled Wabash College on his alma mater.

Northern Illinois rapidly filled up with Yankees, delighting Governor Thomas Ford. Southerners, he complained, were personally generous but penny-pinching on public works. Yankees readily taxed themselves if there was "a schoolhouse to be built, a road to be built, a teacher or minister to be maintained." The pedagogue J. B. Turner of Massachusetts helped plant the University of Illinois. Benjamin Godfrey of Cape Cod and Anna Peck of Connecticut opened early female academies in Illinois. Fort Dearborn was named for a general from New Hampshire, and no one did more to tame the town that grew up around it than a man who was a product of Dartmouth and mayor of Chicago, "Long John" Wentworth (he was six feet six inches tall). Walter L. Newberry, a real estate mogul from Connecticut, bequeathed to Chicago one of America's premier libraries. An emigrant from Massachusetts, John Stephen Wright, the "great-grandfather" of boosters, promoted Chicago as a convention center. The Vermonter Gordon Saltonstall Hubbard began the city's meatpacking industry, paving the way for Philip Armour and Gustavus Swift, who were also transplanted Yankees. Stephen A. Douglas of Vermont became the state's most prominent politician. The Illinois Central Railroad, in which Harriet Beecher Stowe held a large block of stock, was directed, built, promoted, and advertised almost exclusively by New Englanders. Not least, a young man from Massachusetts worked and slept in his boss's store until he saved enough to set up his own shop in Chicago. His name was Marshall Field. The New Hampshireman Lewis Cass all but invented Michigan. Wisconsin, Iowa, and Minnesota (where the Pillsbury family settled) were also shaped by Yankee settlers, investors, and educators. Josiah Grinnell of Vermont, who founded Iowa's progressive Grinnell College, stood for all the young men Horace Greeley advised to go West because "I was the young man to whom Greeley first said it, and I went."

The moral impact of Yankees was mighty, but ambiguous. Reformers went everywhere, affecting the rage of Old Testament prophets. Moses Ordway, a graduate of Middlebury College in Vermont, fulminated against the 4,000 denizens of the timber town Green Bay in 1836. They "seemed to be agreed in only one thing and that was to blaspheme God." Every evening, it seemed, they lit bonfires, drank whiskey, and had dances "so wicked and wild that many of both sexes would lie drunken on the ground the next morning." Preachers, encouraged by the fact that more than one-third of Wisconsin's 305,000 residents in 1850 were Yankees, lobbied for prohibition. They ran afoul of Germans who wryly noted that Yankee nostrums such as Davis's Painkillers and Mrs. Pinkham's Compound were up to ten times more alcoholic than beer. Nor did the whitewashed Congregational churches that sprang up in the Midwest keep pace with the revivalist Methodists, Baptists, and Presbyterians. Nor were all New England crusaders on the side of the angels. A "profane, impious, and scandalous" Unitarian, Abner Kneeland, incurred Boston's wrath with a "how-to" tract on contraception, then fled to Iowa, where he founded the free-love community of Salubria. Radical utopias of all sorts speckled the Midwest in those decades, most inspired by heterodox Yankees. More typical was Brown University graduate Eli Thayer. When Kansas opened to settlement in the 1850s he founded the New England Emigrant Aid Company, ostensibly to fill the territory with free-soil supporters. Thayer won investors and migrants with a more mundane pitch: big money was to be made on the prairie; service to Mammon was service to God; glory plus 10 percent was the road to Millennium. Unfortunately for his speculations, the fanatic John Brown of Connecticut went to Kansas with blood in his eye and ruined the real estate market.[26]

Ever since 1776 four transplanted British cultures competed to shape American society. The Quakers of the Delaware valley spoke loudly in matters of conscience but planted only a few spotty subcolonies in Ohio, Indiana, and Iowa. The Cavalier planters of the Chesapeake begat the new "cotton South," but their hierarchical notions of liberty and slavery had no appeal to the nation at large. The wild Scots-Irish spread their libertarian ethic all over the West, but a continental republic could hardly be sustained by rugged individualism alone. That left the seed of the Puritans. Between 1820 and 1860 they steadily spread their notion of ordered liberty; tamed frontiers with law books, primers, and Bibles; then bade the Midwest stand with New England against Dixie's recalcitrant Cavaliers and Scots-Irish. Call it John Adams's and John Quincy Adams's revenge for their losses to Jefferson and Jackson in 1800 and 1828.[27]

In spring 1836, the season when real estate in Manhattan soared in the aftermath of the fire, the whole upper Midwest became a bazaar. Old Hickory may have rued the spirit of speculation, but his administration stoked it by rousting the Indians, opening new land offices, and placing them in the care of patronage hacks. These land agents raked off a percentage of sales that peaked at 20 million acres in 1836, conspired with land companies, and purchased the best soil on their own account. The companies, usually partnerships among capital-rich Bostonians and New Yorkers, bribed federal agents to set aside watered and wooded tracts prior to opening public sales, then resold the $1.25 acres to migrating Yankee farmers for up to thirty or forty dollars. James Fenimore Cooper bitterly lampooned the business in his *Autobiography of a Pocket-Handkerchief*, starring the vulgar huckster Henry Halfacre, Esq.

The most fabulous frontier auction occurred on the shore of Lake Michigan, where a town named for its pungent wild onions suddenly appeared on the map. Harriet Martineau "never saw a busier place than Chicago," where a scarlet-draped Negro on a white horse cried out to speculators the hour and place of the next land sale. As late as 1833 the Potawatomis far outnumbered the whites who drifted into the "chaos of mud, rubbish, and confusion" outside Fort Dearborn. According to their chiefs, "Their Great Father in Washington must have seen a bad bird which told him a lie," for they had no wish to sell their land. A palaver ensued during which rambunctious whites appeared "more pagan than the red men." But liberal bribes and whiskey jugs soon persuaded the chiefs to sign treaties relinquishing title to all northern Illinois. Within three years 5,000 boomers, many still housed in Conestoga wagons, fought to purchase not acres at $1.25 but mere lots at $10,000 and up. The *Chicago American* boasted that property had appreciated 100 percent per day, every day for five years. Manic investment peaked after January 1836, when the state legislature approved a canal to link Lake Michigan with the Illinois and Mississippi rivers. Touts called the corner of State and Madison "the germ of an immense city," predicting that "the place would one day have a population of 50,000." The middle run proved them conservative. Chicago surpassed 100,000 by 1860. But in the short run most people went bust in the Panic of 1837, leaving half the town's property in the hands of just 1 percent of its people. Desperate losers resorted to gambling, prostitution, and mob politics. In 1855, Mayor Levi D. Boone, grandnephew of pioneer Daniel, vowed to shut down Irish pubs and German beer gardens on Sundays (while leaving "American" bars untouched). So much violence ensued that he had to invoke martial law and repeal the ban. Two years later Mayor Wentworth's cops scoured the notorious Sands district (where the famous Water Tower now marks the start of Michigan Avenue's "magnificent mile"). The gambling dens and whorehouses simply relocated south of the river. Yet those were also the years when city fathers undertook to elevate every downtown street and building five to ten feet out of the mire. The project left behind subterranean haunts that popularized the term "underworld" but also established the town's reputation for engineering: "government in Chicago at times rivaled the feats of the Old Testament God." Vice and the vision to move heaven and earth combined to create the greatest city of the North American interior.[28]

Martineau thought it mad that lots in Chicago cost more than rich farms in the Mohawk valley. Such boomtowns had no future at all unless their hinterlands filled up with prosperous farms. They did fill up, thanks to an agricultural explosion that was far and away the most important U.S. development of the century's second quarter. During those years the number of American farms tripled to 1.5 million, land under tillage increased to

300 million acres, and all but the most remote rural families made the transition from mostly subsistence to mostly commercial production. So large and fertile were the inland farm belts that middlemen could ship wheat, corn, hogs, and beefs back east via rivers, canals, and lakes and undersell New England growers on their own markets. Already by 1840 Genesee flour and livestock shipped on the Erie Canal from western New York forced New Englanders to specialize in perishable vegetables, eggs, dairy goods, and hay for nearby urban markets. The flood tide of migrants to the Midwest completed the picture. From 1830 to 1855 the tonnage of vessels on western waters rose from 35,000 to 350,000; bushels of grain shipped east from Chicago rose from 2 million in 1850 to 30 million in 1860. These figures speak volumes about pioneers and the technology, transport, livestock, and markets that enabled them to drape America with amber waves of grain.[29]

What did a young eastern man and his wife have to do in order to make good out west? First they had to raise at least $500 to cover travel, forty or more acres of land, tools, implements, livestock, and seed. Given that wages averaged around twelve dollars per month in the 1840s, the only alternative to borrowing from kinfolk or banks was several years of toil and saving. Next, the migrants had to hire a "breaking team," usually two men and six oxen, who turned over the stubborn, deep-rooted clay sod of the prairie for two dollars to five dollars per acre. Rails for fencing cost a penny apiece, but miles of them added up. Materials for a log cabin could be had for fifty dollars, but the wife would soon want a farmhouse costing six or more times that. Digging a well, erecting a barn and sheds, hiring hands, and buying or renting a wagon drained what cash was left over. That explains why cheap credit was a boon and any drop in food prices a disaster. But a struggling capitalist on a farm had no hope of profit unless he could ship harvests to market. This explains why the farmers who made up the bulk of the electorate in new western states insisted that their governments sponsor ambitious canal, road, and rail projects. Politicians were eager to please, since they and their friends stood to gain from internal improvements as land speculators, commodities brokers, investors in transport, construction contractors, or lawyers serving all the above. But even the farmer blessed with a canal or railhead some twenty-five miles from home had a problem. Since a bushel of wheat weighs sixty pounds and a 100-acre farm might yield 2,500 bushels, a farmer might need to haul (or hire teamsters to haul) seventy-five tons of "truck" over rutted, muddy country roads. This explains why Illinois and Indiana, by 1850, stabled more oxen, horses, and mules per capita than any other American state.

Above all, farm families needed to work. Except in winter, developing a sizable homestead required unremitting toil from sunup to sundown. Even when crops and livestock were cared for there were always more trees and stumps to clear, outbuildings to raise or improve, tools to sharpen or fix, cows to milk, chickens to feed, hogs to butcher, butter to churn, clothes to mend, candles to make, and Bible stories to read to the children. Men at least could take pride as providers, participate in the community, and punish the jug after dark. A woman's life, recorded the *Ohio Farmer*, was "nothing but mend and patch, cook and bake, wash and iron, and make cheese, pick up chips [firewood] and draw water, bear children and nurse them." Male writers in turn complained of nagging, dissatisfied wives. Hence men and women alike, their eyes open for any laborsaving device, echoed the cry of the land speculator: "All hail the inventors! They do more benefit to their fellow men than a thousand theorizers on guano." The reference to guano was somewhat unfair. Tobacco farmers from Kentucky to North Carolina were restoring the acid content of their soil with tons of aquatic bird droppings harvested on Pacific islands. Their yellow burley, a mild leaf cured in flues, hooked a new generation of smokers at home and in Europe. But the speculator had a point: inventors of all sorts were busy improving farm life, enriching the nation, and thus confounding the prejudice of Abraham Lincoln and Karl Marx against the "idiocy" of rural life.[30]

Mechanical implements were a transatlantic phenomenon. But farms were on average much larger in America than in Europe, farmers were more numerous every year, labor was in much shorter supply, and sticky topsoil in the Midwest was a much tougher "row to hoe." The wooden moldboard plows with cast-iron shares manufactured by the tens of thousands in Pittsburgh and Worcester thus proved inadequate farther west. Even the all-iron "prairie-breaker" weighing 150 pounds cut furrows only three inches deep. So blacksmiths in Illinois began in the 1830s to tinker with steel shares and moldboards. John Lane was the first, but John Deere, who arrived from Vermont with seventy dollars, four children, and a pregnant wife, made the commercial breakthrough. His trapezoidal steel-plated, then all-steel "singing plows" manufactured in the humbly named town of Grand Detour, Illinois, whined through virgin prairies with ease, expanding cultivation to Iowa and beyond. Another Yankee mechanic, Joel Nourse, experimented with longer, sharply curved moldboards that turned the deepest furrows of all in fields already "busted" by Deere's device. Spring or fall planting of grain was another obnoxious chore. Diligent farmers who worked the seed in by hand could not plant more than ten acres per season; lazy farmers who just strewed seed around risked an indifferent harvest. Then in 1841 Moses and Samuel Pennock of Chester County, Pennsylvania, invented a drill to sow in furrows a controlled volume of seed from a tube. Twelve years later George Brown of Illinois built a double-row seed drill pulled by a horse. Summer weeding likewise ceased being a curse, thanks to horse-drawn mechanical hoes.

All this preceded the most famous innovation of all, the mechanical reaper. Obed Hussey of Ohio patented one in 1833, six months before Cyrus McCormick. But McCormick, a Virginian ironmonger, wisely spent six years in experimentation before throwing his machine onto the market and patenting his improvements. Pulled by two horses or mules, the McCormick reaper offered a comfortable seat, a drive wheel propelling a swivel cutter, and a platform from which the harvested crop could be easily raked off and bound by a child or farmhand. For thousands of years human beings had wielded scythes to harvest the fruits of the earth one flail at a time. Suddenly American farmers could reap ten acres per day at their ease. McCormick's Chicago factory opened in 1847, but the 1,000 reapers it turned out per year could not keep up with demand. Cut grain still had to be threshed and winnowed. So in 1837 Hiram and John Pitts of Maine designed a machine that separated wheat from chaff with a thresher pulled by a brace of horses on a treadmill. That helped Hiram Moore of Climax Prairie, Michigan (another delightful name), to imagine the "ultimate weapon," a combine. By the late 1840s, his gigantic prototype pulled by twelve horses could harvest twenty-five acres per day. That proved too much of a good thing. The combine's capacity and expense were beyond the needs and means of American farmers until the 1870s. All told, the U.S. Patent Office reported 659 agricultural inventions in the 1850s ranging from plows, seeders, corn huskers, and reapers to butter churns, yokes, and beehives. Production of farm implements in the year 1860 alone exceeded $34 million.[31]

How did isolated farmers, especially in the Midwest, learn of new technologies, seed strains, fertilizers, useful knowledge of all sorts, and the commodities markets on which their income (and mortgage payments) depended? They learned from books, such as Thomas Green Fessenden's best seller of 1834, *Complete Farmer and Rural Economist*; from touring lecturers such as Professor John P. Norton of Yale; and from journals and almanacs, whose broad circulation proves how many American farmers—especially the transplanted Yankees—were literate. In 1840 thirty such journals had a combined circulation of 100,000; twenty years later sixty farm publications reached 300,000. A surprising number of rural mailboxes were also filled by the semiweekly edition of the *New York Tribune*. Horace Greeley, who was interested in everything, knew that the destiny of the United States was born aloft on the shoulders of farmers. At the cost of just three dollars per year they could read the detailed, expert articles Greeley commissioned on soil and climate, land and crop prices, animal husbandry, veterinary techniques, and efficient use of one's time throughout the seasonal cycle. Regional newspapers such as the *Cleveland Plain Dealer* and the *Chicago Tribune* reprinted such articles or ran their own farmers' columns.[32]

Governments promoted agriculture, too, belying the image of the Jacksonian era as laissez-faire. Northern states led by New York sponsored agricultural societies, state boards of agriculture, and county fairs to disseminate knowledge among farmers. Through such media farmers and stockmen learned to appreciate the superiority of Durham, Devon, Jersey, Shorthorn, Guernsey, and Hereford cattle from England; Ayrshire cows from Scotland; the German Holstein beloved by dairy farmers; Merino sheep; and Berkshire, Hampshire, and Suffolk hogs. Thanks to improved breeding, the doubling of U.S. livestock between 1840 and 1860 yielded a tripling of beef, pork, wool, and dairy production. When threats arose to livestock or crops, state governments moved with alacrity. In the 1840s Massachusetts offered a bounty to anyone with a remedy for the potato blight and a dreaded pneumonia killing Shorthorns. New York appointed an entomologist to combat insect damage to crops. New Hampshire even put a price on the head of crows (then stopped when sly Yankees began raising birds for the bounty). Several states established boards of inspection to ensure the quality of their signature produce. The federal government imposed tariffs to protect domestic producers of commodities such as

Kentucky hemp, so vital to the Navy and merchant marine. Agriculture was such a concern to the U.S. Patent Office that farm implements were given a separate division in 1849. Only opposition from southerners concerned about states' rights prevented the creation of a cabinet-level department of agriculture prior to 1860. Recalcitrant southerners also blocked plans to found a national university in Washington, D.C., and to use James Smithson's bequest for the promotion of agriculture. The hectoring of the Yankee congressmen John Quincy Adams and Rufus Choate at last persuaded Congress to found the Smithsonian Institution in 1846, whereupon its director, Joseph Henry, went to work compiling valuable agricultural data on temperature, rainfall, frosts, dew points, and soil composition. In the 1850s national demand for vocational training spurred North Carolina, Georgia, New York, Pennsylvania, Michigan, Illinois, and Iowa to break ground for agricultural colleges that served as models for the great land-grant universities soon to come. In all these ways, the United States (especially in the North) embraced the same improvement ethic that had hastened the original spread of rural capitalism in England two to three centuries before.[33]

The phenomenon that Arnold Toynbee called the "industrial revolution" in 1880 was really a mix of four artifacts reinforcing one another: (1) machinery to work faster and more reliably than human beings; (2) factories to rationalize production through divisions of labor among hands and machines; (3) steam engines to transfer thermal energy stored in wood and coal into kinetic energy; (4) competitive, open markets to promote mass consumption. Historians used to conflate these phenomena under the term "industrial takeoff." Nowadays they are more apt to speak of a "market revolution." But everyone dates its coming-of-age between 1829, when Jacob Bigelow's *Elements of Technology* taught Americans a new word; and the 1850s, by which time few Americans could avoid noticing "the machine in the garden."[34]

The entrepreneurs, mechanics, and engineers who began to industrialize the United States responded to market incentives of which the chance to get rich was only the most obvious. As in every era, those with a genius for business had to be sensitive not only to economic conditions but to political and cultural values as well. Whether in transportation industries that required governmental involvement or in manufacturing industries that dominated the lives of whole towns and regions, business leaders had to double as community leaders in order to win support, or at least tolerance, for their enterprise. This meant that they had to contribute, or seem to contribute, to the wealth, opportunity, and security of the people and nation at large, even (or especially) when their methods appeared corrupt. To be sure, the Constitution established a uniform national economy. John Marshall's Supreme Court upheld the sanctity of contracts and struck down state monopolies. Chief Justice Roger Taney expanded protection for free competition in the name of community rights (the latter in the Charles River Bridge case of 1837).[35] But it is simplistic to think Americans made a fetish of capitalism at the expense of their commonsense notions of fair play. On the contrary, wage workers, consumers, and political spokesmen for specific regions and sectors were quick to protest perceived exploitation. By the same token, few businessmen were the heartless Gradgrinds of Charles Dickens's novels. For reasons both pragmatic and moral, most imagined themselves doing good by providing employment, affordable goods, and civic leadership according to the dictates of charity, philanthropy, and faith. In short, they wanted to feel good about doing well. Who doesn't?

Machinery appealed to entrepreneurs eager to enhance the productivity of labor and increase market share. Protective tariffs inspired by nationalism (and resentment of British imports) encouraged investment. Mechanical skills honed at American workbenches or carried to them by immigrants provided the knowledge base. But Yankee know-how would not have become a byword by 1860 without an additional fillip: strong patent law. That flew in the face of Justice Taney's Jacksonian worldview because patents were legal monopolies intended "to promote the progress of useful arts by the benefits granted to inventors; not by those accruing to the public." But the framers of the Constitution provided for patent protection on the common-law model, the primary purpose of which was to encourage English pirating of foreign inventions. Hence, the Patent Act passed by Congress in 1793 served Alexander Hamilton's program to stimulate American manufacturing. It recognized no foreign rights, while authorizing patents to any American who applied and paid the fee. The courts soon found this an invitation to fraud. Secretaries of state beginning with Thomas Jefferson were incompetent to judge issues of technical priority; hence myriad patents were granted that infringed on existing ones, and this in turn encouraged fast-buck artists to ignore patents altogether. The widespread theft of the cotton gin that vexed Eli Whitney was so much the norm that in Jefferson's words, "the whole was turned over to the judiciary." Justices Bushrod Washington and Joseph Story were not pleased to legislate from the bench, but perforce they did. In a long series of cases they established that a "discovery must not only be useful, but new; it must not have been known before in any part of the world," and even "the first inventor cannot acquire a good title to a patent, if he suffers the thing invented to go into public use . . . before he makes application" or fails to put the invention to public use in a reasonable time after obtaining a patent. Thanks to such judgments, U.S. patent law was bent to the purpose of rewarding not only novelty but social utility and not only inventions but their prompt delivery to the marketplace, while encouraging free use by all of imported techniques.

One thing remained: a genuine patent office staffed by people competent in both law and technology. To the shame of Congress it dithered for decades before replacing the law of 1793. Thanks to John Ruggles of Maine, it finally did in 1836. Citing the 100 or more infringement suits clogging the courts, Ruggles's committee called it evil that the Great Seal of the United States was being used to gloss over promiscuous piracy. His solution was to subject patent applications to rigorous tests of novelty, keep "promotion of the arts and the improvement of manufactures" foremost in mind, and "put an end to litigation before it begins." The resulting Patent Act of 1836, signed by Andrew Jackson on July 4, established a Patent Office staffed by experts empowered to reject applications, and stipulated that parties found guilty of infringement were liable for triple damages plus court costs. Some problems remained, not least the irksome provision allowing inventors to petition for extensions of patent protection on the grounds that they had not earned sufficient money over the initial seven years. (Thus did the Philadelphia sawyer William Woodworth and his assignees preserve a monopoly on a mechanical planer for twenty-eight years.) But the reform of 1836, as new magazines such as *Scientific American* were quick to point out, created an efficient, predictable legal environment for Yankee know-how to flourish. Ironically, Democrats may have thought that the new law would restrict issue of patents, hence the monopolies they despised. In fact, applications soared from an average of 544 per year in the 1830s to 6,480 in the 1840s and 28,000 in the 1850s. Those numbers, more than any others, reflect the American "takeoff."[36]

Technical revolutions cause enormous stress. They force old dogs to learn new tricks (or even ask their children for help). They consign to oblivion old manufacturers, old business models, and their trusting investors. They sow doubt about the moral impact of material change. Imagine the anxiety felt by the first generation to be accosted by inventions that altered human routines familiar for thousands of years. Elias Howe's sewing machine, first patented in 1846, not only took needle and thread out of female hands, save for minor repairs, but also showed that one could make a career of invention. Howe was content to let others mass-produce and market machines while he grew rich on royalties and infringement suits. The timeless ways women prepared meals were also transformed by canning, food processing that made Gail Borden a household name, and wood-fired ovens and ranges. Mason's screw-top jars, patented in 1858, soon lined the shelves of every American pantry. In the 1830s William Procter and James Gamble realized that a fortune could be made processing industrial waste: in the first instance, hogs' hooves, hides, blood, bones, and bristles discarded by Cincinnati's slaughterhouses. Their trademarked soaps, oils, cosmetics, dyes, lard, tallow, glues, fertilizers, and brushes soon became staples. By mid-century every respectable parlor displayed a family portrait, thanks to the chemical and optical experiments of the Frenchman L. J. M. Daguerre. In 1839 he mounted a silver-coated copper plate treated with iodine inside a box with a lens. Mercury vapor developed the plate; sodium thiosulfate "stopped" the development and thus "fixed" the image. Within months the dental supplier Alexander Wolcott and professors Samuel F. B. Morse and John William Draper of New York University were refining the daguerreotype, applying for patents, and launching publicity campaigns. Within a decade photography studios in every sizable town were selling millions of portraits at one dollar to two dollars each.[37]

What made Morse the "American Leonardo" was another miracle he did not conjure: the telegraph. Electrical experimenters such as Alessandro Volta, Humphrey Davy, Hans Oersted, and André Ampère had inspired inventors such as Carl Gauss and Charles Wheatstone to build working telegraphs by 1832. Morse's genius lay in turning science to practical use, which U.S. patent law was supposed to reward; and in getting government backing, which U.S. politics was *not* supposed to reward. First, he made the technology efficient over long distances by designing a single-circuit telegraph whose current was boosted by relays. (That derived from Joseph Henry's work on voltage and resistance.) Next, he replaced the cumbersome galvanometers used by the British with a receiver that sketched dots and dashes to symbolize letters. (His partner Alfred Vail crafted the instruments.) Finally, Morse persuaded the House Commerce Committee to cough up $30,000 for an experimental wire from Washington City to Baltimore. When he used it in 1843 to transmit instant results from that year's political conventions, the stunned Twenty-Fifth Congress declared, "Space will be, to all practical purposes of information, completely annihilated." The military, the post office, state governments, newspapers, and stockbrokers threw so much money into the business that in just six years telegraph lines connected cities from Boston to New Orleans. By then Morse had added his pièce de résistance: a sounder that chirped out the "dits" and "dahs" of Morse code. Telegraphy was the greatest communications breakthrough since the invention of movable type. In a free market anyone without access to it was dead. By 1860, six regional corporations controlled 23,000 miles of wire stretching as far west as the Missouri River. They in turn merged, in 1866, to form Western Union.[38]

Morse was a Massachusetts boy educated at Andover and Yale. Samuel Colt grew up in his father's textile mill in Connecticut, then ran off to sea. According to legend he was staring at a ship's capstan when it dawned on him that a cylinder might be crafted to revolve multiple bullets in the chamber of a pistol. He patented the revolver at age twenty-two in 1836, but flopped as a businessman until the government made his fortune as well. It was 1847, the nation was at war, and the cavalry badly wanted revolvers. Colt resumed production, promptly fulfilled his contracts, and plowed the "war bucks" back into the firm. His facility at Hartford, opened in 1855, was the largest private armory in the world. There Colt perfected the methods of mass production, interchangeable parts, and precision tools first developed in U.S. arsenals and known abroad as the "American system of manufactures." It was Colt's factory that taught Francis Pratt and Amos Whitney the machine tool trade. It was Colt who served as Mark Twain's model for *A Connecticut Yankee in King Arthur's Court.* It was Colt folks had in mind when they quipped that God had only made men, but the revolver made men *equal.*[39]

One industry already industrialized by 1830 was textiles, thanks largely to British machines and machinists put to work on the Merrimack River from Lowell to Manchester and in mills scattered through southern New England (Samuel Slater's "Rhode Island–style" factories). Why New England led the United States into the industrial age is an interesting question. Tempting answers include the region's Puritan work ethic, its close ties to Britain, its pool of capital built on trade, its solid financial network managed by the Suffolk Bank, the availability of labor (resulting from rural overpopulation), the highest educational level in the world, and a concentration of skilled tinkerers, woodworkers, and other artisans. Less appreciated is the critical role played by women in America's first industrial spurt. At a time when maidens, mothers, and widows elsewhere did housework and farmwork exclusively, tens of thousands in New England either worked outside the home as mill hands, servants, and teachers, or else produced for the market at home by sewing garments and footwear or plaiting straw hats from Cuban palms. The huge volume of yarn spun and woven in factories expanded this "putting-out" system until the mass-produced sewing machines of Brown and Sharpe, Willcox and Gibbs, and I. M. Singer and Company moved needlework into sweatshops (greatly reducing, alas, the volume of touching "samplers" on which American farm daughters learned their stitches). By 1860 one-third of New Englanders lived in towns, compared with 14 percent nationwide; one-third of Yankee women were employed outside the home, compared with one-fifth nationwide; and New England was making 75 percent of America's cotton goods and 60 percent of its boots and shoes.[40]

The early industrial revolution conjures images of sooty brick mills in which hundreds of barefoot women and children were chained for life to machines driven by clanging, hissing steam engines, their torment relieved only by rotgut gin or Methodist Sunday schools. In fact, the vast majority of American mills before 1860 were modest in size, semirural, and powered by stately waterwheels. Cotton mills employed on average ninety-seven people, glass plants sixty-five, shoe factories thirty-four, paper mills twenty-two, iron foundries twenty-four, and tanneries just two. Except in experiments such as Lowell (itself benign by the day's standards), workers went home to their own cottages. Still, there is no gainsaying the harshness of a sixty-hour to seventy-two-hour workweek (depending on sunlight) in the midst of noisy machines that had to be minded constantly to avoid losing a limb. The long hours were dictated by the fierce price competition among firms, but also by the technology. A three-story textile mill had an elaborate array of frames, carders, bobbins, mules, jennies, spinners, bleachers, warpers, winders, reeders, weavers, and dyers that—once set in motion by gears from a waterwheel—could not be stopped and restarted at will. This meant mill hands had to shrug off fatigue, boredom, swollen ankles, and lungs beset by clouds of lint while they performed their tedious tasks. To maintain efficiency and safety under such conditions, owners strictly forbade alcohol, tobacco, obscenity, fighting, uncleanliness, and slacking.

The "age of improvement" moved the orator Edward Everett to exult: "What changes have not been already wrought in the condition of society! What addition has not been made to the wealth of nations and the means of private comfort by the inventions, discoveries, and improvements of the last hundred years!" Militant artisans (and apologists for the South's brand of slavery) more likely shared the sentiments of the poet Thomas Mann: "For liberty our fathers fought, which with their blood they dearly bought, the factory System sets at nought. A slave at morn, a slave at eve, it doth my inmost feelings grieve." In most cases, however, mill workers were not "wage slaves" forced to take arduous jobs for meager wages in order to stave off starvation. Detailed studies of textile mills in New England and Delaware County, Pennsylvania, reveal an amazing turnover in the workforce (up to 95 percent over twelve years) and no "typical" factory worker. Young people willingly took employment in mills to escape the farm, augment their family's income, troll for husbands or wives, earn money for pretty clothes and a dowry, or save enough from their wages to buy land in the West. Thus did industry stimulate agriculture even as the growth of agriculture increased markets for factory goods.

The interests of owners and workers certainly clashed. A cotton or wool manufacturer invested heavily to start up a mill; struggled to maintain profit margins in an era when the retail price of cloth fell by more than one-third; and faced the prospect of instant ruin through fire, flood, or financial panic. Long hours and wages no higher than what hands earned on a farm were the mill owner's principal means to expand his margin. Tension between mills and towns was also intrinsic. Farmers, shopkeepers, and artisans might prosper from local industry, but they invariably resented the stranglehold mill owners wielded over real estate, livelihoods, politics, the environment, and sometimes even the conscience (through their patronage of churches). Hence, one might expect these early mill towns to have spawned trade unions, socialists, Luddite assaults on machines, and the "making of the American working class." This did not happen. Instead, in places like Rockdale, Pennsylvania; Dudley, Massachusetts; and Rochester, New York, most factory workers were docile, most agitators were better-paid mechanics, and most of the fires the agitators kindled soon flickered out. Class conflict on the scale that beset early industrial zones in Europe simply did not occur in the United States.[41]

Abundant, cheap, fruitful land, plus a shortage of labor, is the most obvious explanation. More than any other people on earth, Americans had the option of saying, "Take this job and shove it." But they were just as sensitive as Europeans to issues of moral economy. That they responded to these issues in different ways can be explained in great part by religion. Three main schools of thought competed to answer the question, "Who should control the machines?" Utopians such as Robert Owen imagined communal factory towns providing for all in a perfect harmony based on rational technocracy. But utopian communes invariably collapsed in dissension unless they were founded on rugged faith (like the Shakers). Militant trade unionists such as William Heighton, editor of the *Mechanic's Free Press*, wanted to reinvent industrial order on the basis of cooperation rather than competition, reason rather than legal and religious tradition, and equality rather than "artificial" distinctions of family, matrimony, education, rank, and wealth. But a revolutionary program that turned society into warring factions, plowed underfoot the consolations of faith and family, and by definition killed young people's hope that they might someday get rich appealed to few American workers. Evangelical preachers of the Second Great Awakening, by contrast, offered a far more appealing agenda. They imagined a society that blessed work and material progress while mitigating its abuses through faith-based reform, charity, and virtuous family life. They imagined an America liberated from strife and want once the hearts of workers and capitalists were purged of envy and greed. Did not the Lord say that from those to whom much is given much will be required? Of course, it is impossible to plumb the hearts of strangers, especially at a remove of 160 years. But testimonies such as that of the textile magnate John B. Crozer of Brandywine abound: "My mind was so thoroughly engrossed that for months and years no other subject than my business could engage or at least hold my attention. I feel, in review, that this was very sinful." He was humbled by a wave of strikes in 1842 and a disastrous flood (an act of God?) that washed away his mills the following year. Crozer and his fellow magnates, urged on by their sympathetic wives, dug deep to provide shelters, schools, and churches for the unemployed, widows, orphans, and waifs. Call it paternalism, but "giving back to the community" became a moral imperative for businessmen by the 1840s.

The Jacksonians' own pretense of equality among white males reinforced the notion of the "factory as republican community." In 1835 members of the Mechanics' Library and Benevolent Society of Poughkeepsie cheered Professor Alonzo Potter of Union College, whose lectures condemned "absurd and pernicious" distinctions between workers and bosses: "I give but utterance to the spirit of our institutions, and to the views of all good and wise men, when I say, that in this land we are, or at least ought to be, *all workingmen* and *all gentlemen*." In a free, inventive marketplace one's occupation ought not to be "mechanical drudgery," but rather a ladder climbing from the throstle frame (a spinning machine) to the engineer's office to the proprietor's estate. Potter became an Episcopal priest and was made bishop of Pennsylvania in 1845. Professor John McVicker of Columbia likewise taught economics as "moral philosophy." President Francis Wayland of Brown University was an exponent of "clerical laissez-faire" in his *Elements of Political Economy* (1837). In Philadelphia, Pastor Gregory T. Bedell asked his parishioners to make religion their "paramount business." Charles Grandison Finney taught, "A man's business ought to be part of his religion." The Unitarian Nathaniel L. Frothingham described God as the great taskmaster who calls all men to labor in the vineyard. The Reverend Henry Ward Beecher insisted that "the poor man with industry is happier than the rich man in idleness." If, as the historian Perry Miller wrote, business became dominant in America "at the expense of other value systems," then it behooved purveyors of political and religious values to hitch their carts to that horse.[42]

Democratic evangelism split an incipient labor movement already fragmented by divisions between skilled artisans and factory hands, native-born workers and immigrants, men and women, and temporary and permanent workers. Thousands of working-class men and women swarmed into revival tents to be "reclaimed, civilized, Christianized." The preachers were not put up to this (as some suggest) by capitalists seeking to "tame" an industrial workforce. But they might have had that effect. Evangelists defined work as a holy calling that should be done with humility and thanksgiving to the glory of God. They imparted to workers a zeal to improve their learning and skills, save and invest, spurn gambling and booze, and rise in the world. They damned the infidelity of radical agitators, while co-opting their social issues. Christian reformers attempted to shame business and government into restricting child labor, providing free public schools, abolishing debtor prisons, relieving pauperism, and banning alcohol. Between 1847 and 1853 New Hampshire, Pennsylvania, Maine, New Jersey, Ohio, and Rhode Island did pass laws to forbid employment of children under twelve and make ten hours "a legal day's labor" (albeit employers could still pressure workers to "agree" to work longer hours).

The evident might of industrial capitalism, the feeble resistance to it, and faith that democracy and Christian charity would suffice to minimize its abuses made for some heady dreams. In 1850 Henry C. Carey, son of the famous publisher Matthew Carey of Philadelphia, issued a manifesto entitled *The Harmony of Interests, Agricultural, Manufacturing, and Commercial.* Looking to a millennium when the Stars and Stripes would unite the globe under the doctrine "Do unto others as ye would have them do unto you," Carey held that the American system of industry (as opposed to the British system) would elevate and equalize the condition of all men and women through progress guided by "true Christianity" rather than "the detestable system known as the Malthusian." He held capitalist enterprise to be an act of divinely inspired sub-creation, wealth a reward for good stewardship, society a harmony of interests, opportunity a natural human right, and improved standards of living the natural partner of social, moral, and spiritual elevation. His book could not have appeared at a better time. During the 1850s, opportunity expanded for people in all stations in life, and the economy "seemed to be working in just about the way that the theory of Christian capitalism said it should."[43]

Why was there no American working class on the English model? Because no true American as yet believed himself or herself consigned for life to a "class"—unless it was the middle class.

* * *

The United States differed from Britain in another important aspect of the early industrial revolution: exploitation of coal. Since England was both largely deforested and rich in soft, bituminous coal, its forges pioneered the use of coke in the smelting of pig iron. Americans, by contrast, enjoyed a seemingly endless supply of wood, whereas the hard anthracite coal lining the hills of eastern Pennsylvania was difficult to ignite. Hence the use of coal was confined mostly to heating, cooking, and gas lighting; and "iron plantations" remained on the banks of streams (for water mills) in the middle of forests, because a single year's consumption of charcoal ate up 150 acres of trees. Then, in 1828, the hot-air, reverberatory blast furnace invented in Britain transformed all cost-benefit calculations. Put simply, the process involved an enclosed furnace fifty feet tall that reflected heat back down on the fuel while fanning the cauldron with preheated blasts from a steam-powered bellows. Mine owners in Pennsylvania were quick to report that such a process would easily ignite anthracite coal, produce purer iron than charcoal, and free iron makers from their slavery to woods, and to waters that froze in the winter anyway. By 1850 American investors and their engineers were operating sixty anthracite-fired iron furnaces in eastern Pennsylvania. Forty more using bituminous coal sprang up around Pittsburgh and Wheeling, Virginia. More than half used steam engines. The concomitant boom in mining was accompanied, in good American fashion, by profligate waste. About 45 percent of a coal vein served to support the underground shafts, 8 percent consisted of crumbly pebbles, 7 percent stuck to the shale, and 16 percent was lost during crushing and picking. All told, 76 percent of the mineral never made it to market and 30 percent was just left in heaps. As early as 1848 an observer mourned a "picturesque region, beautiful in all its seasons," now stripped, despoiled, perished. Nor could rainfall renew the terrain, because water reacted with slag to form sulfuric acid that killed flora and fauna miles downstream. But accountants had no line for pollution. Instead they recorded that U.S. iron production rose from 130,000 tons per year in 1830 to more than 700,000 tons by the 1850s, providing a bustling people with plowshares, girders for bridges and buildings, 300,000 wood-burning stoves per year, millions of horseshoes, and hammers and nails at half the price paid a generation before. No wonder *Hunt's Merchant Magazine* crowed in 1854, "Commerce is President of the Nation and Coal her Secretary of State."[44]

The relatively tardy exploitation of coal and the plethora of water mills also help to explain why the United States lagged behind Britain in stationary steam engines to power machinery. The simple principles underlying steam power were known in ancient times, but not until 1776 did James Watt design a reliable system of valves and condensers to alternate pressure and vacuum in a cylinder and thus drive a piston. Boulton and Watt's machines developing twenty or thirty horsepower quickly became the main power plants for British textiles, metallurgy, and mining. As late as 1830, by contrast, American manufacturers found their market confined mostly to the steamboat industry begun by Robert Fulton in 1812. It was during the presidency of Andrew Jackson—again—that the wheels of industry began to spin in earnest. The fall lines of gentle eastern rivers were already saturated with water mills, wild western rivers were unsuited to mills, and nobody wanted his power interrupted by frost or drought. By 1838 the United States counted 1,860 stationary steam engines, compared with 800 in use on steamboats. Clever mechanics fashioned gears, cams, and driveshafts to the machinery in paper mills, printing presses, gunpowder plants, mines, foundries, glassworks, lumberyards, and gristmills. Philadelphia and Pittsburgh led in the manufacture of engines, and Pennsylvania in overall number. But the state in second place was, surprisingly, Louisiana. So energy-intensive was the refining of sugar that bayou planters found it much cheaper to purchase a steam-powered mill for $6,000 than to support a small army of horses and slaves. By 1860 over 90 percent of the mills in Louisiana, 100 percent in Texas, and 72 percent in the Ohio Valley were steam-powered, as compared with 46 percent in the nation at large. Indeed, the largest steam engines in this era powered great flour mills in Cincinnati and Saint Louis. Such facts belie the notion that industry percolated westward only gradually or was choked in the slaveholding South. They also suggest that the water mills which powered New England's leadership became, in one generation, a symbol of backwardness.[45]

Nathaniel Hawthorne captured the most Romantic and wrenching employment of steam when he wrote of "the whistle of the locomotive" that "tells a story of busy men" and "brings the noisy world into the midst of our slumbrous peace."[46] If any novelty, technological or other, deserves the overwrought name "revolution," it is surely the railroad. When Jackson entered the White House in 1829, people, goods, and information—even in the most advanced countries—could not travel overland any faster than they did in the time of Julius Caesar. Then all concepts of space, time, and volume exploded. Where last year horse-drawn wagons might carry six people or one ton of goods twenty miles in a day, this year a railroad train could carry sixty people or ten tons of goods 200 miles in a day. Thanks to steam locomotion the potential command of human beings over the environment increased by three orders of magnitude (i.e., 1,000 times) with the promise of more and faster technology to come.

Once again, the technology was born overseas. In 1829, a quarter century after Welsh coal mines begin to experiment with crude locomotives, George Stephenson's *Rocket* amazed 15,000 spectators at Manchester's famous Rainhill Trials by tugging twenty tons over sixty miles at ten miles per hour. Just a few months later, Horatio Allen fired up an engine imported by the Delaware and Hudson Canal Company. "I had never run a locomotive or any other engine before; I have never run one since; but on the eighth of August 1829, I was the first locomotive engineer on this continent—and not only engineer, but fireman, brakeman, conductor, and passenger."[47] Merchants in Charleston, Baltimore, Philadelphia, and Boston immediately chartered rail corporations, obliging their competitors in Savannah, Norfolk, and New York to do likewise. Once again, U.S. politicians overcame Jacksonian scruples to give the new industry public money, liberal charters, land grants, and monopolies. Where private investment was inadequate, as in Virginia, Georgia, Michigan, and Indiana, state governments undertook construction directly at a cost of $43 million from 1830 to 1838 and another $90 million from 1844 to 1860. Needless to say, railroad boosters, lobbyists, and lawyers descended so thickly on state capitals that they became a virtual branch of government. Federal assistance included a reduction of tariffs on imported iron (repealed in 1843 in deference to domestic foundries) and grants in the 1850s of 18 million acres of western land that subsidized forty-five railroads including the mighty Illinois Central linking Chicago and New Orleans.

The argument should not be taken too far. Nearly 90 percent of the $1.25 billion spent on railroads by 1860 was private investment (much of it British and French); and the "stimulation" provided by governments was so haphazard, exuberant, local, and crooked that American railway construction was, in retrospect, a fiasco. Indiana, Michigan, and Illinois went belly-up when the Panic of 1837 exposed their premature, grandiose plans for internal improvements. Favored western railroads won charters, not because of existing demand, but in order to increase real estate values around towns that did not yet exist. Back east, politicians sensitive to charges of monopoly blithely approved multiple lines, heedless of what the traffic would bear. So great was the demand for civil engineers that poorly trained (or politically favored) men wasted huge sums by reason of poor design and management. State legislatures made a pretense of angry investigations into shoddy construction, sometimes firing up to 90 percent of their professional staffs (usually after elections put the other party in charge). Most scandalous was the willful refusal of railroads and state agencies to coordinate gauges of track. Hence passengers and freight on long hauls might have to be unloaded and reloaded several times. Only the four major trunk lines—Baltimore and Ohio, Pennsylvania, New York Central, and Erie—displayed an appreciation of interstate commerce. Their integrated routes cut travel between New York and Chicago from three weeks in 1830 to less than three days in 1857, while moving cargo three to five times more efficiently than water transport. Otherwise, most northern roads, and all southern ones, were conceived as artificial "rivers" connecting jealous ports with limited inland markets. In retrospect, what the United States needed was more, and more efficient, government, especially at the federal level. The numbers are always impressive—3,328 miles of track laid by 1840; 8,879 by 1850; and 30,636 miles by 1860—but the whole was less than the sum of its parts.

No two railroads emerged in identical fashion, but the Camden and Amboy line illustrates a passel of common characteristics. Lawmakers in New Jersey imagined rapid transit from New York to Philadelphia earning so much money that they promised their constituents a future entirely free of taxes. This led to a bitter feud between canal and railroad lobbyists: some people even packed pistols on the streets of Trenton. The legislature's equitable judgment was to issue charters to both in exchange for $100,000 in stock and a transit tax on passengers and freight. The railroaders raised $1 million in twenty-four hours. Canal stock went begging. So the canal company decided to use its right-of-way to build a second railroad. The legislature, fearing that its cash cow would be gored by competitive pricing, promptly merged the companies in exchange for another $100,000 in stock.

The newly christened Camden and Amboy then did something right. It dispatched Robert Stevens, son of steamboat magnate John Stevens, overseas to investigate the technology. He shipped a Stephenson locomotive to America, reassembled it, and named it the *John Bull*. It proved to be the prototype for all future U.S. designs. Stevens also oversaw tracklaying, completing the entire line by 1833. Counting rails, land, steamboats, and wharves on the Hudson and Delaware rivers, plus rolling stock, the initial investment was $3.2 million. In its first year alone the Camden and Amboy carried 105,000 passengers and 8,400 tons of freight; earned $500,000, against $287,000 in expenses; and emerged as one of America's largest corporations.

As with the telegraph, American engineers rapidly domesticated and improved railroad technology. England's first rails were made out of wood covered by iron straps that broke and buckled, especially in extreme heat and cold. Yankee engineers turned quickly to all-iron rails. English rails were not contoured; hence wheels easily slipped. Robert Stevens solved that with a T-shape design that gripped wheels to prevent derailments. English rails were laid over stone block that shifted or shattered in America's torrents and freezes. Roadbeds of crushed stone or gravel solved that. English locomotives had just two or four drive wheels that had difficulty on sharp curves. John B. Jervis fixed that in 1831 by extending in front of his *Brother Jonathan* a swivel or "bogie" truck with four small wheels. Thus was born the standard 4-4-0 design. Since early American engines burned wood, passengers were bedeviled by embers. Spark arresters in the smokestack relieved that problem. Since American farms were not enclosed by hedgerows and fences, animals wandered over the tracks. That problem inspired the cowcatcher. Other American safety features included bells, steam whistles, headlights, air brakes, and the reverse gear, especially vital on single-track lines. In 1836, when the engineer William Norris of Philadelphia demonstrated that a locomotive could easily mount grades approaching 400 feet per mile, not even America's mountains stood in the way of progress. The following year, when Jackson took the first presidential train ride, U.S. ironworks manufactured fifty locomotives; by 1855 that number had grown to 500.[48]

How important were railroads to the U.S. economy? Walt W. Rostow, a champion of "modernization theory," thought railroads such a tremendous stimulus to iron and coal production, machine tooling, manufactures, and national distribution of goods that he dated American industrialization from their inception. Robert Fogel, by contrast, ran counterfactual data sets on the assumption that railroads never existed, and arrived at the surprising conclusion that expanded canal and turnpike systems could have sustained equivalent growth. Albert Fishlow considered both these arguments extreme. To be sure, many railroads proved to be speculative ventures based on little genuine need, but it is absurd to deny the stimulus that swift, year-round steam transportation delivered to supply and demand in virtually every sector of industry and agriculture. Railroads hastened the settlement of the West, created vast new real estate markets, inspired new ways to marshal capital, transformed Wall Street financial markets, accelerated the progress of thermodynamics and civil engineering, and caused the public and private sectors to improvise new ways to flirt with (and sometimes seduce) each other. Far more than a mere technology, railroads were a "social invention."

Hence, the more interesting bottom line for historians is not what would have happened had no railroads been built, but why the supposed network of American railroads was really a crazy quilt. That did not happen in France, where the strong central government's bureau of bridges and roads built a nationally integrated rail system (albeit centralization of power also permitted stagecoach interests to retard railroad construction). Nor did it happen in Britain, where railroads were privately owned, but crown and Parliament regulated their routes, financing, and operation in the national interest. Chaos triumphed in the United States thanks to the combination of a weak central government, the dispersion of power among state governments, and a laissez-faire ethic empowering capitalists to do (or "get away with") whatever they wished. A persuasive argument has even been made that corruption in railroad planning, construction, and finance was a benign necessity. That is, visionary entrepreneurs had to play politics, bribe officials and judges, make outlandish promises, and engage in "creative" financing in order to get the job done at all. The result often resembled mercantilism more than free enterprise.[49]

In retrospect, one is tempted to say that the irrationality attending the railroad boom exposed the inadequacy of federalism in the Jacksonian era even more than did the rift over slavery. If the nation was going to grow in the manner everyone thought destined, then the role of the federal government would have to grow whether the proponents of states' rights liked it or not.

Because this technological revolution fell into many private hands, railroad directors, in turn, were obliged to make a managerial revolution that helped shape U.S. society for a century. Before 1840 almost all businesses, even merchant houses with far-flung interests and agents, were family affairs or partnerships run out of one place. (Even the Second U.S. Bank made do with two clerks and a boss: Biddle.) The report on U.S. manufacturing issued in 1832 by Secretary of the Treasury McLane revealed that the number of specialized firms had grown with the economy, but their scale and complexity had not, with the sole exception of textile mills. Railroads, however, simply could not be run out of one office by a few proprietors. They required coordination of dozens of activities performed hundreds of miles apart, more capital than the richest partners could muster, a wide range of technical and managerial talent, a large diversified labor force, and a pool of reliable contractors and suppliers in every phase from construction and procurement to operations and maintenance. The obvious model for such huge, dispersed organizations was the army. Not surprisingly, some 120 West Point graduates found lucrative careers as managers or civil engineers on the railroads. The "turnkey" contracting firm was pioneered by Horatio Seymour of New York and Joseph Sheffield and Henry Farnum of Connecticut, who provided all the workers and equipment needed to build railroads in exchange for fixed fees. The railroads themselves invented organizational pyramids just to operate safely. In 1841 a series of fatal crashes prompted the Massachusetts legislature to require the Worcester-to-Albany line to appoint managers with clear functional and geographical responsibilities. By 1846 the rapid growth of freight traffic on the Baltimore and Ohio (due mostly to Pennsylvania coal) caused its president (Louis McLane) and its chief engineer (Benjamin Latrobe, Jr.) to issue a *Proposed New System of Management* based on two chains of command under chiefs of operations and finance. In 1851 Daniel McCallum of the Erie Railroad drew up a detailed organizational chart defining the responsibilities, authority, accountability, standards of efficiency, and disciplinary procedures for all of its 4,000 employees. The following year the Erie again transformed the industry by establishing continuous telegraphic communications throughout its system. Henry Varnum Poor's influential *American Railroad Journal*, a British parliamentary commission, and a feature story in the *Atlantic Monthly* all urged universal adoption of McCallum's business model.[50]

J. Edgar Thomson of the Pennsylvania Railroad added another formidable weapon to the corporate armory: complex accounting. He obliged managers at every level of every division to submit monthly balances of receipts and expenses. Those data permitted accountants to compile annual reports on flows of passengers and cargo; gross, net, and marginal profits and losses; capital costs; and depreciation. These in turn enabled directors to make confident judgments about future expenses, investments, and dividends. Last, but not least, the railroads filled their managerial pyramids with salaried professionals who might own no company stock, but who expected promotion in return for their experience, performance, and loyalty. The divorce between ownership and management begun by railroad corporations soon characterized large corporations in most other sectors of American industry.[51]

Railroads obviously became a major source of employment, not only for civil engineers, managers, sales representatives, accountants, and other white-collar employees, but for skilled, semiskilled, and unskilled workers. Who was "workin' on the railroad" in the 1840s and 1850s, and how was it that these employees were no more successful than textile mechanics in advancing the cause of labor? To begin with, the "shortage of labor," so frequently cited as a reason Americans chased laborsaving technology, did not exist by 1850, at least not in the Northeast, thanks to the tide of immigrants. Railroads and their construction companies had no problem recruiting Irish and native-born workers. Their problem was keeping their workers. Just as in the textile industry, railroads suffered from wrenching turnover as workers trained at company expense quit after earning a nest egg, jumped to other railroads, got fired for indiscipline, or just "moved on down the line." Some railroads experimented with contracting out for services—for instance, hiring independent locomotive drivers, conductors, and maintenance foremen in the belief that they would be more cost-conscious and less prone to pilfer tools or embezzle fares. But by 1860 the regimented military model triumphed in the name of control. Railroads put their employees in uniforms; banned smoking, drinking, gambling, philandering, and fighting; imposed fines and punishments; and blacklisted repeat offenders with the help of Allan Pinkerton's pioneering detective firm. Resistance from workers was not as widespread as one might expect. After all, a railroader's pride and safety were imperiled by drunken, disorderly comrades. Most complaints and occasional strikes were born of resistance not to rules per se, but to inequitable enforcement, bullying superintendents, and a lack of fair grievance procedures.

Finally, railroad workers were discouraged from making common cause because they, no less than the managers, occupied a military-style hierarchy of ranks. Even menial labor in the new and exciting business of railroads raised a man well above the mill worker or farmhand, and a man willing to make a career on the railroad had genuine prospects. Promotion from baggage handler to conductor brought not only more wages but also the middle-class status that went with a top hat and kid gloves, gold pocket watch and burnished lantern, dignified interaction with passengers, and responsibility for the promptness and safety of the journey. Promotion from fireman to locomotive driver (or "engineer") made one a labor aristocrat of a different sort—and a traditional rival of that "dandy," the conductor. Engineers prided themselves on their mechanical skill and deft judgment born of harrowing plights. They polished and painted their engines in gaudy hues like the circus trains of a later era, and proclaimed their approach with distinctive toots of the whistle. Engineers became militant only when they were unfairly blamed for the mistakes made by switchmen, or when "their" engines were driven by others on their day off, or when management began insisting that engines be painted in no-nonsense black. Later in the century railroads became bloody battlegrounds between labor and management. But before 1860 work stoppages were rare and railroading was a privilege. "I've heard of the call of the wild," wrote a veteran, "the call of the law, the call of the Church. There is also the call of the railroads."[52]

What most people don't know is how well the train whistle harmonized with the whinny. The progress of the iron horse, far from rendering obsolete the flesh-and-blood variety, was accompanied by a rapid increase of the equine population in all regions and sectors of the economy. Freight and passenger cars, after all, were confined to their rails. So the more cargo they carried, the more wagons, coaches, and horses were needed to carry the goods and passengers to their final destinations. Likewise, the more people moved west by rail and then shipped their crops east by rail, the greater the horsepower required. Likewise, the larger cities grew on the strength of factories and trade, the more horse-drawn (or really, because of the harness, horse-*pushed*) streetcars, Conestoga wagons, Downing and Abbot "Concord coaches," hansom cabs, and trolleys they needed.

Oxen and mules, the sterile get of a horse and a donkey, did good duty in strenuous, plodding assignments such as pack trains, canal barges, and sod busting, but were too slow, stupid, expensive, ill-adapted, or uncouth for most tasks. ("What! the ladies will say, would you have us . . . be dragged about by oxen?") By mid-century, when nearly two-thirds of the nation's work animals were horses, Americans made calculated decisions based on trial and error, word of mouth, scientific husbandry, and studbooks as to which breed was "the right tool for the job." Manuals taught farmers and teamsters to understand "the nature of the machine—for such the horse is," and to give beasts the same care a devoted mechanic gave his engine. Only a fool would fail to provide his horses with good fodder, rest, and the latest harness design, or fail to assemble them into convivial teams (horses, too, have their friends and natural leaders), or fail to minimize draft (friction or drag) on roads and fields.

Thoroughbreds, trotters, pacers, cavalry mounts, and other prestigious breeds appealed to society's upper crust. Draft horses did democracy's heavy lifting. The large workhorses bred by Pennsylvania Germans in the eighteenth century were reinforced by the chesty breeds that nineteenth-century German immigrants imported for their breweries and farms. Most common, however, were the "Vermont drafters," a breed that appeared in the 1830s. They probably derived from a cross between the native Morgan horses and the Norman *cheval de gros trait* introduced a century before into Quebec. In the burgeoning Middle West they were known as Canucks, and much prized. All told, the demand for horses was so great in the United States that their numbers grew more than 70 percent in the 1850s, the height of the railroad boom, and they did not reach their peak of 27 million until the year 1910. Henceforth, whenever you contemplate the industrial revolution, think of horses as well as machines.[53]

Americans of the Jacksonian persuasion did not oppose progress. They just feared that Henry Clay and his ilk meant to exploit the nation's progress to impose "European" structures of power, class, and capital. Indeed, Democrats took as much pride as Whigs in the swift advances America made after 1830 despite the democratic, decentralized chaos disparaged by foreign observers. Americans were especially gleeful when the British themselves provided a stage on which to gloat. It happened in London, in

1851, when Queen Victoria and Prince Albert opened the Crystal Palace Exhibition, designed to impress the world with the wonders of British industry. American manufacturers, mechanics, and artists, with only modest help from the Smithsonian Institution and the Patent Office, stole the show. The most popular U.S. attraction was Hiram Powers's sculpture *The Greek Slave*, a fetching white nude that titillated Victorians of both sexes. Otherwise, the U.S. exhibit burst with unadorned but utilitarian manufactures that forced Europeans to credit American ingenuity. Connecticut clocks, Boston-made navigational instruments, Goodyear rubber goods, Steinway pianos, Colt revolvers, Robbins and Lawrence rifles, Borden foods, Singer sewing machines, and McCormick reapers won medals. The locksmith Alfred Hobbs embarrassed his hosts by opening an "unpickable" British device. On a percentage basis, Americans just held their own against British, French, and German techniques, but that was triumph enough, especially since most U.S. entries were production models, not prototypes. Yankees burst their buttons when the news arrived from their "mother country." Nathaniel Currier sold thousands of lithographs in which a stout, sour John Bull muttered, "Gammon! Don't believe it! Some Yankee trick! Wooden nutmegs!" P. T. Barnum capitalized on the Americans' popularity by charging admission to his Crystal Palace panorama, selling a souvenir book to go with it, and collecting fees from the manufacturers featured in it—an early, perhaps the first, example of "product placement."

The republican dream was coming true, remarkably fast, or so it seemed to Christian Schussele. Twelve years after the fair in London, he completed an imposing canvas that allegorized the march of progress in the United States. It depicted Benjamin Franklin presiding over nineteen inventors. Places of honor went to Morse, McCormick, Goodyear, Howe, and Richard Hoe (who invented the rotary press). They were flanked by Samuel Colt, William Morton (surgical ether), James Bogardus (iron architecture), Joseph Saxton (coin minting and coastal surveys), Peter Cooper (locomotives), James Mott (iron foundries), Joseph Henry (electricity), Eliphalet Nott (anthracite stove), John Ericsson (screw propellers and ironclads), Frederick Sickels (steam engines), Henry Burden (horseshoe machines), Erastus Bigelow (carpet loom), Isaiah Jennings (friction match), and Thomas Blanchard (copying lathe). Schussele called his work *Men of Science*, but his pantheon illustrates how thoroughly Americans equated science with technology, which they considered the handmaiden of democracy. "We have no populace, no rabble, but free and independent citizens. What has made them so?" asked the Jacksonian Benjamin Hallett. His answer was industry that placed the means of wealth within the reach of all and permitted all to "move forward evenly, as on the level of a railroad." Industry and democracy, thought Edward Everett, made America uniquely suited to produce the "civilized, cultivated, and moral and religious man." Salmon P. Chase praised the "new and almost infinite power" of machines that took from none and gave to all. New York's observer at the Crystal Palace bragged that Americans invented, not luxuries, but "articles of utility . . . for the advantages of the middling classes." The philosopher F. H. Hedge anticipated Marx's technological utopia where the "sacred frame of man shall no longer be bent and seamed with servile tasks." *Harper's Magazine* celebrated America's calling to feed and clothe the poor, and fashion a society "beyond the theorist's dream . . . the progressive order decreed by the God of Ages."[54]

Americans had cause to think of their nation as a laboratory testing whether material plenty could transform human nature. Between 1810 and 1860 U.S. manufactures grew tenfold to about $2 billion, 45 percent of which was "value added" beyond the costs of capital, labor, and raw materials. The United States' foreign trade grew from an annual average of $186 million around 1820 to $1.6 billion before 1860: a rate of increase greater than Britain's. The tonnage of shipping engaged in domestic commerce rose from 600 in 1825 to 2,200 by 1855. National wealth as valued by taxable property increased 400 percent over those decades; per capita income and purchasing power increased more than 50 percent. Economic historians estimate that the U.S. gross domestic product grew at an annual rate of 4.2 percent over the first half of the nineteenth century despite the severe contractions of 1808–1812, 1819–1822, 1837–1843, and 1857–1858. Assuming an average of zero growth in the lean years (in fact, it was probably negative), the economy must have averaged an expansion of 6.2 percent in all the fat years. To be sure, wealth was very unequally distributed, even in states without slavery. But so often did opportunity knock between 1815 and 1860—in new settlements and new industries—that farmers and tradesmen with little money or schooling "stood a better chance of business success than at any time before or since in American history." No wonder the American dream, the myth of a classless society in which all might get rich, exercised such "immense moral power."[55]

What use did people make of that power, and what morality did it serve? The answers voiced by a spectrum of social critics demand their own chapter. For the vast majority of white men and their female dependents, the American dream meant pursuing one's happiness through mass consumption and mass amusement. Hence, the final two industries to capture the national soul by 1860 were advertising and entertainment.

The tow-headed boy in a flannel cap crying "Extra! Extra! Read all about it!" is properly an American image. By 1828, the year when the Democrats buried voters in pro-Jackson broadsheets, the United States already contained more periodicals (850) and paper mills (150) than any other country on earth. By 1860, the periodicals surpassed 4,000 and the paper mills 550. Steam-powered rotary presses made possible bulk printing and sharply reduced prices. In 1833, Benjamin Day's *New York Sun* launched the "penny press," inspiring James Gordon Bennett's *New York Herald* (1835), Greeley's *New York Tribune* (1841), and at least one mass-circulation daily in every major city. In western towns a newspaper invariably opened for business before any church, school, or bank. What news did the journalists peddle? At the high end it was politics, literary criticism, and the latest in European art and philosophy. On the popular level it was politics, crime, catastrophe, scandal, more politics, and especially advertising. To Americans obsessed with business, consumer goods, or both, advertisements themselves were news; this is why they often dominated the front page. Americans needed to know right away what their neighbors were thinking and buying, what was "in" or "out," what spectacles were in town, and what their political party's "line" was on this issue or that.

By the 1850s magazine publishers explored every niche in the market. *Niles' Weekly Register*, *Hunt's Merchants' Magazine*, *De Bow's Review* (in the South), and specialized trade and industry journals covered business. The *Democratic Review*, *American Whig Review*, and *Globe* debated politics. The *North American Review*, *American Quarterly Review*, and *Southern Literary Messenger* tracked culture. In 1826 the Philadelphian George Graham invented the general-interest magazine, a genre perfected by *The Saturday Evening Post*, *Harper's*, and *Atlantic*, which literate, aspiring men and women could read and discuss together. But the magazine that did more than any other to dictate national norms of purchasing and behavior was *Godey's Lady's Book*. Founded by Louis Godey in 1830 and edited for forty-one years by the formidable Sara Josepha Hale, it became the template for all subsequent magazines packed with romantic stories, household ideas, beauty and fashion tips, and flashy advertisements. Hale defined what American womanhood meant, mediated "the conflict between senti-mental sincerity and fashionable hypocrisy," and convinced women that they must keep up with style lest they suffer the "*shame of being thought poor*." Finally, in the 1850s, the first magazine tycoon built the first pub-lishing empire. Henry Carter, a striving young English immigrant, ad-opted the name "Frank," made it a brand, and saturated the market with *Frank Leslie's Illustrated Newspaper*, *Frank Leslie's Lady's Magazine*, *Frank Leslie's New Family Magazine*, and *Frank Leslie's Budget of Fun*.[56]

"Advertising space in a newspaper," ventured a New York publisher, "is somewhat like a lady's favors, which are valued very much as she values them herself." Paid commercial notices had been the lifeblood of printers since colonial times, but the advent of mass marketing and mass circula-tion vastly complicated the question of how much advertising was worth beyond the paper it was printed on. That spelled opportunity for a would-be profession of agents who either peddled periodical space on commis-sion or else bought pages for resale to marketers. By the 1850s New York City, Boston, Philadelphia, and Chicago housed a demimonde of slick, fast-talking admen who sold middle-class domesticity by day and pur-chased prurient pleasures by night. Many were street hucksters, circus barkers, or confidence men who decided to "go straight." Since their cus-tomers could only guess at circulation numbers, the going rate for a col-umn, or the sales it would be likely to generate, advertising (confessed the agent George P. Rowell) was "one of the easiest sorts of business in which a man may cheat and defraud a client without danger of discovery." Knowing that a newspaper would settle for $300 to run an ad of a certain size, the agent assured the manufacturer of corsets or patent medicine that he *might* be able to secure the space for $400. He would then regale the manufacturer with visions of greater profits if he purchased a larger ad, assure him that "numbers never lie," and offer to write the loqua-cious, eye-catching copy himself. Thus Greeley denounced the "large class who delight to shine in newspapers as wits and poets, and announce their wares in second-hand jokes, or in doggerel fit to set the teeth of a dull saw on edge. . . . The fewest words that will convey the advertiser's ideas are the right ones."

Over time the advertising business went somewhat legitimate. Agents turned into consultants promising to protect advertisers from scams and assist them in targeting customers. The quality of advertisements improved. But the purposes of the enterprise never changed. The first purpose was to catch a reader's eye by word or image (sex and humor work best), then pitch the product by subtle indirection or a hard sell. The second was to convince consumers that the way to pursue happiness was to buy stuff which promised to make them look better, dress better, feel better, dine better, travel better, and impress or at least keep up with the Joneses. The third was to assure Americans that they didn't just want a product; they *needed* it.[57]

Advertisements aimed at females soared in the mid-nineteenth century. Though a "woman's sphere" was largely confined to the household, the outpouring of factory-made clothing, processed food, furniture, linen, drapery, tableware, ornamentation, and finery of all sorts made home economics and cultivation of tastes a big business. A few women began to rebel against the regnant "cult of domesticity." But most learned at their mother's knee, or by observing their neighbors, or by reading magazines, what respectable middle- or upper-class society expected of them. Women also knew how to get their husbands to pay for it all, their task eased by rising real incomes. One prosaic but telling example was the swift decision by American women that they must have fine china, preferably as a wedding gift. By 1820 every genteel hostess in eastern cities had an English tea set with which to perform her afternoon ritual. By 1840 no wife with social pretensions dared entertain without a complete, matched set of imported dinnerware with a floral or Chinese landscape motif. Suffice to say that the largest market in the world for Wedgwood and Staffordshire china, accounting for close to 50 percent of all sales, was the United States. How many wives had social pretensions which in turn required their husbands be good providers? Except for hillbillies, new pioneers, and immigrants just off the boat, the answer is nearly all. Even rural New England, that region of crowded, poor farms whose girls went off to the mills, surrendered to the ethic of domestic display. Every cottage, it seemed, required a clock, a bedstead, a patent stove, a parlor with upholstered furniture, oil lamps in every room, glass windows, additional wings, and a second story. As early as the 1830s Philip Hone rued the disappearance of the rustic New Hampshire he loved. The *New England Farmer* faulted its readers for building "houses too large for comfort, convenience, or beauty." Preachers disparaged how simple folk lusted for idolatrous displays of frippery even as the market for religious merchandise ballooned. Catholics had their crucifixes, icons, and devotionals, but now even Protestants advertised their faith with lithographs and amulets of Calvin, Wesley, or Luther, and display cases for the family Bible opened to some favorite verse. The American home was turning into something that people (usually women) designed for effect as well as function. It was a shame, thought Theodore Sedgwick, a lawyer in rural Massachusetts, how "our villages, which should glory in pure manners, ape the very silliest fashions of the cities," and once flinty Yankees now strove "to be alike" and to "disguise real differences in wealth and status."[58]

Disguise real differences in wealth and status? Therein lay the dilemma of pretense in the age of Jacksonian democracy. Most white Americans of Protestant lineage did not really want to be equal. Not only did they assume their superiority to Negroes, Indians, and Irish; they wanted to get ahead, and by definition this meant acquiring more wealth or status than others of their own sort. That clash between the myth of collective equality and the reality of individual ambition threatened to create a society in which nobody knew his place. Nothing cast that threat in sharper relief than the culture war that broke out in the 1840s and 1850s over control of popular entertainment. It galled urban sophisticates, the self-appointed guardians of taste and refinement, that the most famous American of their day was a clown.[59]

Mark Twain must have grinned to recall the "real bully circus" in *Huckleberry Finn*, especially the clown who "carried on so it most killed the people. The ringmaster couldn't ever say a word to him but he was back at him quick as a wink with the funniest things a body ever said." Chances are that the clown was Dan Rice, whose circus played Hannibal, Missouri, during Sam Clemens's boyhood. So far as one can believe anything Dan said about himself, he was born in 1823, the love child of Elizabeth Crum, a "shouting Methodist" from New Jersey; and Daniel McLaren, a crooked entrepreneur who lived hard by Five Points. If a marriage occurred (her father sued McLaren for seduction), it was quickly annulled. Elizabeth then wed a dairyman; this is why young Dandy learned about life while driving a milk wagon through Manhattan streets teeming with hogs, dogs, hucksters, and whores. At age fourteen he ran off to Albany, followed the Erie Canal to Buffalo, learned fisticuffs in Pittsburgh, jockeyed racehorses in Ohio, and everywhere proved himself "extravagantly fond of female society." At seventeen he wed the first of three wives. Then, in 1841, he met Sybil and found his métier. Sybil was a pig who appeared to tell time and answer questions by pawing the sawdust or shaking her head in response to Dan's secret cues. As this act, called "Porky and Presumption," moved about, drawing guffaws and handfuls of coins, Dan perfected his "method." The sure way to score a success was to sneak into a town incognito and pick up some local color to use in his act. Who burned down Jaacks's barn? Sybil would snout out the culprit! Who was the biggest fool in town? Sybil would hold the rubes in suspense until pointing at Dan himself! Throughout the performance he kept up a hilarious patter of earthy gags. In those days rowdy audiences were part of the cast: having paid for the privilege they felt free to hoot, holler, and harass performers. But Dan turned their gibes around, "quick as a wink," until people turned out just to hear their neighbors get zinged. In 1843, beneath Pittsburgh's smoky sky, Dan graduated to the circus.

The Latin word *circus* just meant a circle or round arena. The familiar definition denoting a traveling show with animal acts and clowns emerged in 1824, when James W. Bancker took his New York "Circus" on the road. The addition of bareback riding, female acrobats in skintight costumes, and the canvas tent introduced in 1825 by J. Purdy Brown (did he get the idea from religious revivals?) completed the spectacle summed up in the exultation "The circus is coming to town!" Dan McLaren was a natural, given his comic and equestrian talents, knack for publicity, and fighting trim. (Brawls were frequent among well-lubricated patrons, especially when they were cheated by pickpockets, gamblers, vendors, and ticket agents who oversold the house.) Dan was also a gorgeous fellow with twinkling, piercing eyes, and a Vandyke beard (no mustache) on his jutting chin. The only thing he had to change was his name. So he stole the name of the reigning "most famous" American showman, the blackface minstrel T. D. Rice. The newly minted Dan Rice added the "nigero" dialect to his own routines, as well as Shakespearean spoofs and topical humor. When Stephen Foster made himself America's first professional songsmith, Rice promptly put naughty lyrics to "O Susanna," "Camptown Races," and tunes of his own composition. Soon he was a manager, partner, and star attraction in Gilbert "Doc" Spalding's circus. By 1850 it was

"Dan Rice's Circus" because his was the four-star brand name in the brand-new industry of mass entertainment. He even took credit for making peanuts a signature snack. (But save some for the elephants!) Southern newspapers anointed him their region's favorite. Pittsburgh, Cincinnati, and New York claimed him as their favorite son. "Puff" pieces (some planted by Rice himself) screamed, "THE DAN RICE CIRCUS TRIUMPHANT!" In Washington City, boasted Rice, Congress adjourned so that Whigs and Democrats alike could relish his impersonations of the foul-mouthed Stephen Douglas, pompous Daniel Webster, and crabbed Charles Sumner.

Over the years Rice had financial reverses, quarrels with partners, and family troubles. But the performing animals, riders, and acrobats led by America's "only original humorist" were a smash everywhere. Everywhere, because Rice not only knew how to trim his politics to suit northerners, southerners, or westerners but could also lampoon whoever it was a particular audience wanted to laugh at, be it farmers, peddlers, philosophers, dandies, politicians, planters, "darkies," or competing performers. By the 1850s he was the first national show business celebrity by dint of "Wit so Whimsical! Jokes so Jocular! Puns so Pointed! Satire so Severe! Remarks so Ready! That all will say he is truly entitled to the rank of the Great American Jester!"

That was what troubled the bluenoses. For every newspaper heralding his arrival there was an editor, a pastor, or a local stuffed shirt damning Rice as a charlatan and a lowlife who played to the ignorant, unwashed, and immoral. How dare he mock American politics, religion, commerce, propriety, art, and culture! How dare he pander to audiences whose behavior offended and endangered any respectable man, let alone woman or child! Rice, sensitive as always to the nuances of public opinion and expert in the manipulation of crowds, fought back. He did so to reassure customers and head off those who insisted the circus *not* come to town. But the orphan who scaled the heights through audacity and native wit also fought to win acceptance for himself and his craft. Rice joined a Masonic lodge where he mingled with men of substance. He introduced serious readings into his act, styling himself a man of learning and decorum (inventing a brilliant career at Princeton in the bargain). He insisted he was a businessman. He made liberal contributions to charity. He boasted of his domesticity. He claimed that his audiences consisted of educated folk seeking "splendor, refinement, training, novelty, and true wit," all the while joking about the self-righteous critics he spied sneaking under his circus tent. When the *New York Tribune* called him a "grammatical assassin" under whose hand "the King's English nightly dies a thousand deaths," Rice bought front-page advertisements in Greeley's paper inviting the "RE-FINED AND INTELLECTUAL COMMUNITY" to buy tickets and judge for themselves. Finally, Rice literally wrapped himself in the American flag. As sectional tensions peaked in the 1850s he assumed a Stars and Stripes costume in which to preach Union throughout the North and the South. More Americans knew Dan Rice than knew their own congressman; there was talk of his running for president. As the man who fought to make clowns authentic interpreters of all that is inauthentic in America, he achieved success "more brilliant, except in one instance, than was ever attained by any member of his profession."[60]

That "one instance" of even more brilliant success was Phineas T. Barnum. He kept the name his parents gave him at his birth in Bethel, Connecticut, in 1810. Otherwise, he pioneered the celebrity's art of making and remaking one's image. His two fat volumes of memoirs are thus considered authoritative precisely because they are a whimsical grab bag of half-truths, misleading facts, and outright embellishments revealing deeper truths about Barnum and "Barnum's America."

Evidently he did not coin the phrase "There's a sucker born every minute"—the gambler Mike McDonald of Chicago did that—but it rang more true of Barnum because his method of bilking suckers was "legitimate" showmanship. P. T. Barnum was a peddler of pretense and found a huge market for it. He understood before anyone else that folks enjoyed being tricked. Perhaps you couldn't fool all the people all of the time, in Lincoln's words, but to fool all the people some of the time, and some people all of the time, was more than enough to make one famous and rich.[61]

The child of a farmer-shopkeeper, Phineas was portly, energetic, and mirthful. He could not remember a time when he wasn't fascinated by money and numbers (his later "Rules for Money-Making" became a model for all who get rich by professing to teach others how to get rich). While his schoolmates talked about fishing or made eyes at girls during recess, Phineas hawked sweets for pennies. But his great epiphany was the work of a puckish grandfather who filled the boy's head with visions of Ivy Island, a great estate he was destined to inherit. At age ten Phineas trudged for miles through vines and mosquitoes to discover that Ivy Island was a desolate bog. (He later cashed in on this practical joke by using the "valuable real estate" to secure the loan that got him started in business.) In 1825, when his father died penniless, P. T. worked as a store clerk in Brooklyn, observing how merchants sold short lots and customers paid in debased coinage. In Philadelphia as a teenager he ran lotteries, which earned him enough to go home and marry the aptly named Charity Hallett. They would have four beloved children. In 1833 Barnum started a newspaper, only to get sued and imprisoned for libel. His crime was telling the truth about a notorious usurer. Taking that lesson to heart, Barnum moved to New York, where a strange creature taught him his "true vocation." In the case of Dan Rice it was a sow; in Barnum's case it was Joice Heth, a wee, wizened Negress ancient of days. Her owner insisted that Joice Heth was 161 years old and had nursed the infant George Washington. Did Barnum credit the alleged bill of sale dated 1727? Or was it enough that dollar signs flashed in his eyes as the crone squeaked out "recollections" of Washington throwing a coin across the Potomac and telling the truth about the cherry tree? Barnum purchased the woman's services for $1,000, displayed her at Niblo's Garden, and pocketed $750 per week. He then took "The GREATEST Natural and National CURIOSITY in the World" on the road, benefiting from publicity both good and bad.

One can only imagine the tempestuous cultural crosscurrents stirred up by the sensational Joice Heth. First, there was the freak-show aspect. At forty-six pounds she was indeed "a mere skeleton covered with skin" like "a mummy of the days of the Pharaohs." She was blind, toothless, and paralyzed. Her fingernails were enormous curls. One skeptic surmised that she was a dummy made of Goodyear's india rubber with a ventriloquist supplying her voice. Second, there was the patriotic appeal of a living link to the father of his country. Heth reminded Americans from whatever party, religion, or section of their common origins and ideals; she was a fraudulent symbol of a genuine, much-needed truth. Third, there was the racial embarrassment caused, just four years after Nat Turner's rebellion, by a reminder of how many founders of the United States were suckled at the breasts and enriched by the labor of enslaved Africans. Fourth, there was the ambiguous tension between voyeurism and fraud. If Joice Heth was genuine, then paying money to gape at her was morbid. If she was fake, then paying money to gape was stupid. Was she really human? Was she really alive? Was she a slave illegally held in free states, or a free woman illegally exploited? In either case, was she being used more cruelly by northern hucksters than she ever had been in Virginia? Were Americans dupes? Did they need to be duped, in which case their measure of authenticity was simply whatever sells? When "Aunt Joice" died in February 1836 and the autopsy revealed that she was no more than eighty, New York's penny dailies went to war over all these questions. Barnum claimed that he wept over her corpse, but he went to the bank on her notoriety. She also helped take New Yorkers' minds off the fire that had consumed downtown just eight weeks before.[62]

"This is a trading world," wrote P. T. Barnum, "and men and women, and children, who cannot live on gravity alone, need something to satisfy their gayer, lighter moods and hours, and he who ministers to this want is in a business established by the Author of our nature. If he worthily fulfills his mission, and amuses without corrupting, he need never feel that he has lived in vain." After Joice Heth's death Barnum launched more promotions, including a circus tour in the South, that intensified his feel for the national market and the power of advertising. In 1841 Scudder's New York Museum went up for sale. Peale's Museum issued stock to raise capital for the purchase. Barnum scotched that by planting anonymous squibs asking who would be dumb enough to invest in "an exhibition of stuffed monkey and gander skins." Peale's dropped out, Barnum scooped up Scudder's cheap, and his American Museum opened for business. In addition to stuffed monkeys it offered such "transient novelties" as "educated dogs, industrious fleas, automatons, jugglers, ventriloquists, living statuary, tableaux, gipsies, albinos, fat boys, giants, dwarfs . . . dioramas, panoramas, models of Niagara, Dublin, Paris, and Jerusalem . . . the first English Punch and Judy in this country," etc. The most infamous exhibit was the Fejee Mermaid that men learned too late was a scaly grotesquerie, not the bare-breasted sailor's fantasy depicted on Barnum's 10,000 handbills. The showman knew that customers would come back no matter how many times they were gulled. "The titles of 'humbug' and 'prince of humbugs,'" he wrote, "were first applied to me by myself. I made these titles a part of my 'stock in trade,' and may here quote a passage. . . . : 'It's a great thing to be a humbug. . . . It means hitting the public in reality. Anybody who can do so, is sure to be called a humbug by somebody who can't.'"[63]

Barnum branched out in the 1840s, growing fabulously rich. He toured Europe with Tom Thumb, the midget who won Queen Victoria's heart. He blanketed the United States with advance publicity to ensure that the concerts by Jenny Lind, the "Swedish Nightingale," would cause a sensation. (Only in New Orleans did someone give Barnum and Lind a run for their money: the Dan Rice Circus was in town). But the *Struggles and Triumphs* of his memoir's title had only begun, because Barnum was not fooled by his own "Philosophy of Humbug": to become socially respectable he must massage his image. So he built Iranistan, a gaudy palace in Bridgeport, Connecticut, modeled on England's Royal Pavilion, then bought huge tracts of land in East Bridgeport to found what we would call an industrial park. Barnum meant to provide technology, jobs, and a healthy environment for thousands of workers. He was almost ruined when the directors of the Jerome Clock Company, claiming $500,000 in assets, swindled him into underwriting their loans. When the company folded in 1855, Barnum temporarily lost the American Museum. In 1857 a fire enveloped Iranistan. But Barnum recouped with profits gleaned from his original persona (the autobiography of 1855 and the pamphlet "How to Get Rich") and profits gleaned from a new persona as an apostle of thrift and clean living. He toured the nation inveighing against tobacco, strong drink, crooked businessmen, and any man who "takes no interest in politics," for such were unworthy to live in a self-governing nation. There was some irony in this, given Barnum's adage: "If I might provide the amusements of a nation, I would not care who made its laws." Once his capital was restored, of course, Barnum launched yet another career by lending his name to the circus billed as "The Greatest Show on Earth." That was ironic as well, because Barnum had snubbed Dan Rice. To socialize with a circus man was beneath him.[64]

T. D. Rice, Dan Rice, P. T. Barnum, Stephen Foster, and their imitators pioneered genuine American show business. Playwrights certainly did not. Most scripts performed in the United States in that era were European, knockoffs of European plots, European historical dramas, or comedies spoofing the British aristocracy. To be sure, theatrical troupes found it commercial to stage sentimental plays about Indians, farces about social climbers (e.g, Anna Ogden Mowatt's *Fashion* of 1845), and morality plays about wicked city life that cloaked prurient content in a respectable gloss (e.g., William Henry Smith's *The Drunkard, or the Fallen Saved* of 1844).

But none was the least bit memorable.[65] Real American entertainment was born in tents and arenas where clowns satirized pretense, jeering crowds threw rotten eggs, and Shakespeare was mangled for laughs. The Bard was far and away Americans' favorite scriptwriter because actors interpreted his comedies and tragedies in the same rollicking, bawdy, melodramatic, or slapstick styles that prevailed in Elizabethan times. Even the out-of-work English actor reduced to playing to "nobs and snobs" in America soon learned that he need only head west and "his pockets in a night or two were amply replenished." Yanks thought it hilarious when an actor in *Othello* stepped out of character to ask: "When was Desdemona like a ship? When she was *Moored*!" Famous Shakespearean lines recited in Negro or Irish lingo were also guaranteed showstoppers.

All this would seem to befit the Jacksonian age of the common man. But recall the dilemma of democratic pretense: white Americans getting ahead in the race for status and wealth did not want to be common. They wanted to think of themselves as refined, and they assuredly did not want the stages and loges of public theaters overrun by unsavory clowns and smelly yahoos. So from the 1830s to the 1850s, theater critics and their clientele, society folk such as Philip Hone and George Templeton Strong, drew a line between "highbrow" art and "lowbrow" entertainment. Highbrows insisted on segregated seating and paid top dollar for subscriptions to boxes. They surrendered some lowbrow venues to the mob while insisting that managers of serious theaters police their patrons' behavior. Even the bard of democracy, Walt Whitman, preferred the Park Theater "because the audiences there are always intelligent, and there is a dash of superiority thrown over the Performances," whereas at the Bowery or Chatham "bad-taste carries the day." Highbrows circulated "rube stories" meant to humiliate spectators into proper behavior. One told of a rube who grew impatient on hearing the Moor cry to Desdemona, "The handkerchief! The handkerchief!" So he shouted at the stage, "Damn it, my friend, why don't you blow your nose with your fingers and let the play go on!"[66]

The stakes in this contest involved not only manners but the nature and purpose of performance. All Americans agreed the theater was sacred. They diverged over why it was sacred and what cultural thirst its rites were supposed to slake. For lowbrows the minstrel show, theater, and circus were the only refuges in which people could tell the truth, puncture hypocrisy, or express their own anxiety, resentments, and (yes) bigotry in defiance of the oppressive conformity imposed by middle-class notions of deference and reserve. Herman Melville attested that "as, in real life, the proprieties will not allow people to act out themselves with that *unreserve* permitted to the stage, so in books they look not only for entertainment, but, at bottom, for more reality than real life itself can show." For low-brows the show had to be real because "all the world's a stage," and "the play's the thing" because audiences could participate and talk back to authority as they could nowhere else. For highbrows, by contrast, taste and decorum dictated that art be revered in silence, politely applauded, and discussed over tea in mixed company. To them, the tendency of mobs to drag culture down rather than let it uplift had scary implications for the republican experiment.[67] This is not to say that a comedy skit is superior to *Richard III*, or that "Suwanee River" is superior to Beethoven's C-Sharp Minor Quartet. It is simply to say that authenticity was ambiguous in a nation awash in pretense. Status-conscious Americans might snore through a play or an opera just to "make a show" of their refinement. No one ever made a show of attending the circus.

The culture war turned hot in the 1840s, when ubiquitous riots over race, religion, and class poured into theaters. Lowbrow mobs jostled for seats in contested theaters; violence erupted when managers tried to discipline audiences. America's upper crust grew especially alarmed in 1848, when it was learned that the Parisian revolution of the same year had been sparked by a theater riot. When the popular actor Frédéric Lemaître invited the "dangerous classes" to a benefit performance of his radical play *Le Chiffonier* (*The Ragpicker*), the crowd stormed out to assault the king's palace. In New York, Yankee nationalism was the impetus. In May 1849, the famous British actor William Charles Macready went head-to-head against America's favorite, Edwin Forrest, in rival performances of Shakespeare. Highbrows made a fuss over Macready. Lowbrows considered that treason. The actors themselves foolishly rallied their camps, thinking the rivalry good for publicity. At the Astor Place Opera House only a few "b'hoys" were admitted. Still, they hooted so loudly that the play could not be heard, then drove Macready off the stage with a volley of eggs and vegetables. "This cannot end here," vowed Philip Hone; "the respectable part of our citizens will never consent to be put down by a mob." So a delegation of city fathers persuaded Macready to resume his performances, under armed guard if needed. Lowbrows turned out, 10,000 strong, vowing to "burn the damned den of the aristocracy." They even drove off the mounted militia and refused to cease and desist when the Seventh Infantry marched up. The soldiers fired, killing at least twenty-two people and wounding 150. Clearly, the unity of a pluralistic republic required not unity but strict segregation of those who patronized mere entertainment from those who patronized art. That truth did not exactly square with the myth of Jacksonian democracy.[68]

Irish Catholics, Germans, and Jews pouring into America; Yankee migrants filling the Middle West; inventors, entrepreneurs, and number-crunching railroad executives revolutionizing transport and manufacturing; mass publications brimming with advertisements promoting a consumer culture; and national purveyors of hokum and humbug thwarting elite pretensions to leadership: the material achievements of the post-1830 generation were simply magnificent. But were the society and culture springing spontaneously to life realizing the founding fathers' ideals? A host of critics said no, or not yet, or at most maybe. They included reformers of every sort, artists, authors, philosophers, preachers, and prophets. In their various ways they damned the materialism, pretense, and vice that spoiled the dream of an honest, equal, hardworking, virtuous republican people. They asked what was worse: people who put on airs in a supposed democracy; people who refused to put on airs and mocked those who did; or people who did not know the "right" airs to put on? Every citizen wanted to believe the United States was the best the world had to offer. But that hope posed some troubling questions: "Is *this* the best the world can offer? Is *this* what liberty spawns? Is *this* what God intended America to become?"

Chapter 4

ROMANTIC REVELATORS, REFORMERS, AND WRITERS

Social Perfectionism and Cultural Pretense, 1830–1860

*T*here are more things in heaven and earth, Horatio, than are dreamt of in your philosophy," quoth Americans' favorite playwright. Americans felt that keenly, for the material progress so far described all happened during the Romantic era. As an adjective the word *Romantic* accurately describes North Atlantic culture in the first half of the nineteenth century. As a noun, Romanticism is doubly misleading. There were as many Romanticisms as there were Enlightenments; and the suffix *-ism* implies the sort of ideology or coherent body of thought that the Romantic temperament anathematized. All Romantic movements in literature, philosophy, religion, and art were reactions against the perceived hyperrationalism and classical formalism of elite eighteenth-century culture. Yes, human beings possessed logical minds able to grasp the laws governing nature and perhaps society. Yes, western science, technology, and industry marched forward in harness to explore, measure, and conquer the earth. But were men and women just thinking machines? Was human behavior as predictable and liable to manipulation as minerals, soil, plants, animals, magnetism, and electricity? Romantics cried, "Madness!" Human beings were also creatures of sentiment, emotion, and dreams that no logic could explain or satisfy. Reason could not tell why people loved, wept, or yearned for something beyond themselves. Reason could not say why a sunset, an Alpine vista, or a ruined castle on a misty moor thrilled yet saddened the spirit. Nor could people remake their society according to a rational blueprint, as the bloody, abortive French Revolution amply proved. As Edmund Burke taught, human communities are organic, not artificial. A wooden building might be razed and rebuilt to one's liking; a tree can only be pruned, fertilized, grafted. However much human beings strove for a better future, they could not sever their roots in the past. So Romantics looked backward as well as forward, sought truth in feelings, revered inscrutable nature, and wondered in awe what supernatural forces lurked beyond the frontiers of sensory impressions, measurements, and taxonomies.

Clever scholars traced the Romantic mood (for such it was: a *mood*) back to the Arthurian legend of broken hearts, ephemeral dreams, and unfulfilled quests; or to Plato, who imagined the human race chained and walking in file through a dark cave, never to know, except perhaps after death, the light of true love, justice, beauty, and peace; or to Homer, whose archetypal *Odyssey* depicted life as an endless, perilous journey; or even to the Garden of Eden, where humanity's parents were tempted by the ultimate Romantic lure: transcendent divinity. But the immediate sources of the Romanticisms that arose over the course of a century after 1740 were artists and writers cloyed by the unimaginative sureties of the Enlightenments. The English poets William Wordsworth, Samuel Coleridge, John Keats, Percy Bysshe Shelley, Lord Byron, and William Blake shattered the rules of verse to sing of wild nature's sirens and wild human longings. The Scottish novelist and historian Sir Walter Scott wrote paeans to the medieval chivalry and faith that rationalists maligned. The Swiss philosopher Jean-Jacques Rousseau wallowed in sentimentality, blamed private property for the heartbreaks of society, and apotheosized noble savages. The French essayist François René de Chateaubriand called the revolutionary generation of the 1790s back to humility and the "genius of Christianity." The German poets of *Sturm und Drang* ("storm and stress") such as Johann Wolfgang von Goethe and Friedrich Schiller asked why God had given man the power to dream impossible dreams and tried to define that ineffable holy grail called the sublime. The philosopher Immanuel Kant's

Critique of Pure Reason offered rational proof of the limits of rationality. German masters of the gothic novella such as Friedrich von Hardenberg (Novalis), Ludwig Tieck, Clemens Brentano, E. T. A. Hoffman, and the Grimm brothers wrote eerie tales of young men and children led to their doom by enchantresses lurking in forests and glens. They explored, in anticipation of Poe, the haunted moods known as *Sehnsucht*, *Lebensmüdigkeit*, and *Weltschmerz*, for which the translations "yearning," "weariness with life," and "world grief or sorrow" hardly suffice. Not least, titanic, rule-breaking composers such as Ludwig van Beethoven, Franz Schubert, Robert Schumann, and Felix Mendelssohn captured nature in sound while exalting and mourning the dashed hopes of mortals. No art conveyed the Romantic better than music.

All the Romantic moods rejected faith in reason and its Deistic "clockwork God." There was more in heaven and earth than was dreamt of in that philosophy. But if the Enlightened project of endless progress, social perfection, numerical rules for beauty, and mastery over the globe was itself exposed as *Romantic* in the sense of quixotic, then what could replace it? The Vatican's answer, institutionalized in the monarchical reaction symbolized by the Austrian minister Klemens von Metternich, was to restore the traditional authorities of throne and altar, and repress the liberalism, egalitarianism, and nationalism born of the French Revolution. Needless to say, such backward-looking Romanticism appealed to no Americans except those southern planters who found in Sir Walter Scott a champion of natural hierarchy led by chivalrous knights and ladies. Nor had ambitious Americans any use for the pessimistic Romanticism of Arthur Schopenhauer, who concluded that the best human beings could hope for was to find solace in Sisyphean labors they could never complete. Hopeful liberals and nationalists, by contrast, made the French Revolution a lost cause whose dreams might yet be fulfilled by common folk bound in cosmopolitan love, and in joy—Schiller's "divine spark" that made all men brothers. This spirit moved Byron and the philhellenes to volunteer in the Greek revolt against Turkey in the 1820s. It moved artist Eugène Delacroix to depict *Liberty Leading the People* in the Parisian uprising of 1830. It moved the revolutionaries of 1848 to imagine a "springtime of peoples" in which moral suasion alone toppled oppressive regimes. This sort of Romanticism melted American hearts because it suggested that Europeans yearned for the democratic nationhood Americans already had. But Romantic rebellion was a mood most Americans dared not encourage (beyond torchlit campaign rallies) lest it infect workers, the Irish, abolitionists, or southern fire-eaters.

Predictably, the Romantic mood most relevant to Americans was English, not continental; individual, not communal; futuristic, not nostalgic; and focused on sin, not institutions. The personal journey of Coleridge, who exercised a greater influence in the United States than American authors wished to admit, is illustrative. The young poet had lost his Christian faith and dabbled for years with the rational "necessitarian" doctrine (implicit in John Locke) that everything existing in time and space is subject to iron laws of cause and effect. If that were so, then sin did not exist, but neither did human free will. Coleridge rebelled. If the soul is not free then life is absurd. Besides, he *felt* free and yearned to *be* free in defiance of external constraints. That feeling and yearning must therefore originate from an impulse beyond space and time, from a supernatural realm empowering man to make free moral choices. Yet, being free, all people sinned by commission or omission: doing those things they ought not to do or leaving undone those things they ought to do. Thus did Coleridge's faith in freedom lead him to belief in a sin that was "original" in all human beings. "Wherever the Science of Ethics is acknowledged and taught," he wrote, "there the Article of Original Sin will be an Axiom of Faith in all classes." Nor was this a tenet imposed by Christian authorities, for if that were so then anyone who dismissed the gospel might dispense with sin, too. Rather, the sinful nature of man was an empirical "fact acknowledged in all ages, and recognized, but not originating, in the Christian Scriptures." Coleridge was a brooding, introspective, honest man, as all good poets are. He felt Romantic religious impulses, but did not know what to make of them. Then he read Kant. The philosopher's case for dual worlds—the *phenomenal* (material) and *noumenal* (spiritual)—convinced Coleridge that the seat of freedom, and thus of sin, lay in man's noumenal constitution. Ergo, a fallen supernatural being required a supernatural rescue. The application of reason to his Romantic feelings accordingly carried Coleridge back to an evangelical concept of grace and atonement. His bulldog belief in freedom, and hence "a renascence of the sense of sin," was a signature trait of the Romanticisms.[1]

The natural response of American evangelicals confronted with ubiquitous sin was ecstatic religious revival. Hence the era's cultural centerpiece was the Second Great Awakening. But Americans' heterodoxy, sectarianism, Romantic pride in their wilderness, Romantic disappointment with the fruits of 1776, Romantic longings ("We're free to pursue happiness yet somehow aren't happy"), and Romantic belief in their nation's eschatological destiny inspired a broad spectrum of social pathologies and prescriptions. They fell into three overlapping categories: religious movements purporting to reveal God's plan for America and so hasten the millennium; social reform movements aiming to purge specific sins that frustrated America's promise; and literary movements extolling the promise or exposing its pretense. The revelators, reformers, and writers exploited their liberty to argue about liberty and its stepchild, sin. Yet all were Romantic in at least one European sense plus a uniquely American sense. The language used, the promises made, and the calamities feared were invariably excessive, "burdened from the start by an overblown national rhetoric." [2] That excess has been attributed to the inferiority complex of a former colonial people desperate to get Europeans' attention. It might also be attributed to the excessive rhetoric in U.S. political and economic life: cultural authorities had to shout just to be heard over the din. A simpler, more troubling explanation is this. If in America the personal was political, the political was religious, and the religious pilgrimage of this new promised land bore the hopes of the whole human race, then how could the high priests of American culture talk about *anything* in hushed, measured tones?

American Protestantism in the Romantic era was characterized by loud disputations over fine points of theology, raucous revivals in country and town, and bewildering sectarianism. How was it then that Tocqueville, Trollope, and others thought the United Sates remarkable for its oppressive religious conformity? The answer is that sober theologians from Boston, New Haven, and Princeton; stump preachers imitating Charles Grandison Finney; and self-appointed prophets of splinter denominations all identified their nation with God's unfolding plan for humanity and considered sin the "enemy within" threatening to wreck, or at least delay, the culmination of history. Some zealots imagined Americans "preparing a place" for the Lord's Second Coming and the millennium (that mysterious thousand-year reign of the saints alluded to in the Revelation to John).

Others imagined Americans were "called unto liberty" in order to fashion utopia. Thus did the Shakers, or United Society of Believers, founded by "Mother Ann" Lee in the 1700s, plant utopian communes from New England to Kentucky. They specialized in plain, lovingly crafted furniture, and urged sexual abstinence. The Oneida Community—founded by one of Finney's converts, John Humphrey Noyes—was a perfectionist cult promoting social and gender equality. Its members specialized in animal traps, silver spoons, and sexual abandonment (everyone was "married" to everyone else). The Universalist minister Adin Ballou ran a heaven on earth called Hopedale until his "communist" followers sold off their 500 acres in Massachusetts. The socialist spirit of an intellectuals' retreat known as Brook Farm literally went up in flames when its new headquarters burned in 1846.

Most disappointed were the disciples of William Miller, a farmer in the "burned-over district" in upstate New York. Like Cotton Mather long before, he scoured reports of world news for signs that the end of days was near. At length he decided that Jesus was scheduled to return in 1843, whereupon thousands of Americans prepared for the rapture. When Jesus did not appear, Miller decided October 22, 1844, was the correct date. When that, too, proved false, a loyal remnant went off to found the Seventh-Day Adventist church.[3] It is tempting to dismiss such groups as the crackpot fringe of the Second Awakening. But they only took to literal, if extreme, conclusions what mainstream Protestants from Jonathan Edwards to John Adams believed. America was indeed said to be chosen by God to serve as the scene for the climactic act of the human drama. America was said to be uniquely blessed with a mission to build a virtual heaven on earth through unceasing material and spiritual progress. No foreigner could frustrate that mission, for if God be for us who can be against us? Only the people, through willful apostasy, could spoil the mission. It followed that Americans, free as no other people in history from the sinful impositions of monarchy, aristocracy, and priesthood, must hearken to the Lord's injunction, "Be ye perfect."[4]

By the standards of orthodox theology this conflation of biblical and civic religion amounted to heresy. Not since the fall of the Roman empire was it possible to imagine Christendom as coterminous with a political entity. The heavenly, eternal city of God, as Augustine of Hippo explained, was forever sundered from the mundane, ephemeral city of man.

All human kingdoms were more or less "of the devil," a fact obvious on the face of it and implicit in Satan's temptation of Christ in the desert. Rather, the church is the mystical body of Christ, the holy worldwide fellowship of all faithful people who are temporarily exiled from the kingdom that Jesus told Pilate was "not of this world." Human beings were alienated from their heavenly father because of their own rebellion born of a fallen nature. People were not sinners because they sinned; they sinned because they were sinners. And no one knew that better than saints who really tried not to sin in thought, word, deed, or omission. To be sure, this dogma outraged human notions of justice. Why would a loving God impose on his children commandments of perfection they were incapable of obeying, and then consign them to death and damnation when they failed? It would seem that God himself is the "author of sin," or at least most to blame for the human predicament. Orthodoxy replied that God so loved the world he indulged those very accusations by taking on human flesh and dying as a propitiation for the sins of mankind. Christian soteriology, a term for the economy of salvation, understood Jesus of Nazareth to be the spotless Lamb of God whose blood delivered the human race from bondage and death just as the blood of the Passover lamb delivered the Israelites from bondage and death. All God asked was that people accept his free gift of justification, aspire to sanctification even if it was not fully achievable in this life, and preach the good news to all peoples. Catholics and Protestants might argue forever about the roles of faith, works, scripture, and sacraments in the salvation of souls. But nowhere did New Testament prophecy identify salvation with a secular nation. Nowhere did it suggest the Parousia, or Second Coming of the messiah, depended on human activity. Nowhere did it say that human beings could change their nature or perfect society on their own. As for the "last days," the brief passage about a thousand-year reign of the saints (before or after Jesus's return?) was a weak reed on which to build a teleology. Yet insofar as Americans embraced these positions, they elevated their secular destiny, and the personal and national sins that might thwart it, to cosmic importance.

In Americans' defense, it must be said the temptation to imagine themselves a new chosen people with a messianic calling was (and still is) hard to resist. The separatist, antinomian character of the Puritans, Quakers, Presbyterians, Baptists, and Methodists that shaped the religious culture of the thirteen colonies inclined Americans toward heterodoxy and antiestablishmentarianism. They damned the Roman Catholic and Anglican creeds as superstition imposed by a venal priesthood and embraced instead the individualistic theology of George Whitefield and the "new lights" of the eighteenth-century Great Awakening. Americans were captains of their own souls, free to choose or reject God's grace. Americans believed their liberty, prosperity, and miraculous independence to be experiential proofs of the special providence scripting their history. Americans took it to be republican common sense that self-government depended on citizens willing to choose virtue over vice, public-spiritedness over selfish corruption, and the future over the past. Finally, Americans' precocious free marketplace of religion all but obliged theologians of whatever denomination to bless the material aspirations of a hustling, self-confident people and start "looking for God's will within the human lifetime rather than fixing exclusively on future rewards and punishments." Americans banished Tom Paine from their pantheon on account of his atheism, but they embraced his formula of 1776, which made America itself a sort of religion destined to remake the world or else perish in shame, depending on her citizens' virtues. The United States became an all-or-nothing proposition.[5]

In sum, whatever the "original intent" of Jefferson, Madison, and its other promoters, the First Amendment to the Constitution raised no wall of separation between church and state. How could it, when the state itself was a sort of church? All the "nonestablishment" clause did was ban Congress from setting up an official church like that of England; and the "free exercise" clause was motivated in part by Madison's belief that only liberty of conscience kindled in citizens the genuine faith and self-discipline without which self-government could not survive. In the ensuing free-for-all touched off by the Second Great Awakening, "peddlers in divinity" quickly learned that their consumers demanded spiritual products which met not only the need for private consolation but the desire for social progress. That is why all the successful religious movements of the nineteenth century competed to occupy, not vacate, the public square.[6]

Unitarianism certainly did. Indeed, it drained Christianity of so much spiritual content that little remained except a catalog of urgent social agendas. Unitarianism is often thought of as a mere spillover from Enlightenment rationalism that peaked around 1805 (when Harvard appointed Henry Ware, Sr., as Hollis professor of divinity) and amounted to nothing but the "fatherhood of God, brotherhood of Man, and neighborhood of Boston." Unitarians and Universalists were said to have rejected judgment and hell because they thought mankind too good for damnation and God too good to damn. But however much it was rooted in a "reasonable Christianity" that dispensed with miracles, original sin, and bloody sacrifice, Unitarianism came of age at the height of Romanticism and adjusted easily to its moods. Not until 1819 did William Ellery Channing, pastor of the Federal Street Church, fix the movement's creed. Not until 1825 did the 125 congregations of the American Unitarian Association fully secede from the Congregationalists. Channing denied the Bible was altogether revealed and infallible; denied the divinity of Jesus; denied an inscrutable, vengeful divine monarch as incompatible with common sense and republican morals; rejected original sin and hence the need for atonement; displayed no patience with human frailty; and dismissed Calvinism as an "attempt to exalt God by denying man freedom of will and moral power." By contrast, Unitarians affirmed the power of free Americans to purify the public square by hearkening to their own conscience and quickening their countrymen's. What quest could be more sublime than that?[7]

Calvinists at Yale and the Andover Seminary argued that Unitarianism comported neither with revelation nor with common sense. But insofar as they, too, exalted free will, republican religion, and the human capacity for reform, their "new school" theology amounted to a new "halfway covenant." Nathaniel W. Taylor, Yale's Timothy Dwight professor of didactic theology, implicitly jettisoned the Calvinist doctrine of prevenient grace (sinful man must wait for the Holy Spirit to quicken his soul), and hence predestination (God determined before all worlds whom the Holy Spirit would touch). Yale's declension in turn moved "old school" Calvinists to found the Hartford Seminary. But Taylor's creed, as expressed in his Sermon "Concio ad Clerum" of 1828, told Americans what most believed in their gut must be right. "The Bible is a plain book. It speaks, especially on the subject of sin, directly to human consciousness; and tells us beyond mistake, what sin is, and why we sin." No curse of Adam but rather a "moral depravity" resulting from man's "free choice of some object other than God" caused people to sin. Moreover, Taylor preached, what was the point of the gospel if it did not "reach the conscience with *its charge of guilt* and *obligations to duty*"—if, in short, it did not move men to pursue "the moral perfection of God"? As his mentor Dwight taught, the only government worthy of God is one based on voluntary obedience and devoted, as God is, to human happiness.

That was almost exactly what Finney's battalion of heterodox Presbyterians hammered home to sinners quaking on the anxious bench at their revivals. Don't use the excuse of a sinful nature or the need to wait for God. Admit the truth about yourselves and do something about it! Sin was selfishness, pure and simple. Everyone understood that. But Americans, a free people in a bountiful land, were called to do more than repent because, Finney insisted, "politics are a part of religion in such a country as this." Converts and their churches dared not "rest satisfied till they are as perfect as God" because "God cannot sustain this free and blessed country, which we love and pray for, unless the church will take right ground." At first the New England clergy condemned revivalists for their theatrics, unsystematic theology, and questionable mass conversions. Boston's Lyman Beecher warned Finney not to set foot in Massachusetts: "I'll meet you at the State line, and call out all the artillerymen." Yet Beecher's *The Spirit of the Pilgrims* (1831) soon echoed Finney to the effect that there was "no safety for republics but in self-government, under the influence of a holy heart, swayed by the government of God." Beecher invited Finney to preach at his own son's church, then joined the evangelicals himself when they offered him the presidency of Lane Theological Seminary in Cincinnati. There he wrote his *Plea for the West*, arguing that the heavenly destiny of liberty in America awaited the outcome of the spiritual war whose decisive theater was the frontier.[8]

The heirs of John Witherspoon—patriot and president of Princeton—split down the middle. Conservatives led by Charles Hodge upheld Calvinist orthodoxy, whereas "new lights" embraced the emerging American orthodoxy. Thus Albert Barnes wrote in his sermon "Way of Salvation" (1829) that according to common sense, God would not ask more of people "than *in any sense* they are able to perform." Quakers, whose belief in their divine "inner light" otherwise gave them unusual latitude in matters of faith, had been committed to social reform since the seventeenth century. Even the Episcopal church was injected with a dose of social activism when John Henry Newman's Oxford movement crossed the ocean in the 1830s and 1840s. In addition to calling parishes back to reverence for high-church liturgy, these Anglo-Catholics stressed charity, education, and ministry to the urban poor. Methodism—the largest U.S. denomination, thanks to Francis Asbury's tireless circuit riders—completed its domestication under Wilbur Fisk. As the first president of Wesleyan University in Connecticut, Fisk dismissed the Calvinist God as a "merciless tyrant" incompatible with the freedom of a republican people. He believed Christ's atoning death showered all people with prevenient grace; hence all were free to discipline their will, passions, and judgment ("the faculties") and obtain total sanctification. The Methodist Phoebe Palmer experienced that in 1837 after her baby burned to death in her crib: "my heart was emptied of self, and cleansed of all idols, from all filthiness of the flesh and spirit, and I realized that I dwelt in God . . . my ALL IN ALL." Her memoir, *The Way of Holiness*, called Americans to become "Bible Christians" for whom social reform was no more than faith in action. Hundreds of thousands of American women and men were touched by her books, lectures, and disciples.[9]

If one biography can stand for all the mainstream divines who wrestled with angels for the secret of America's destiny, it may well be Horace Bushnell's. Call him representative or just ordinary, but he sampled most of what American life had to offer while seeking a moderate Christianity compatible with civic religion. Born in Litchfield, Connecticut, in 1802, Horace worked as a child for his father, a farmer with a textile mill on the side; and for his mother, who ran a fabric shop. After graduating from Yale in 1827 he did stints as a teacher and junior editor for New York's *Journal of Commerce*. Still searching for his profession, he returned to Yale for a law degree, which he completed in 1831, the very year revivals swept through New Haven. Bushnell enrolled in the divinity school under Nathaniel Taylor, accepted a call in 1833 from a Congregational church in Hartford, and married. He remained in Hartford until 1859, living as fully as his circumstances and nagging tuberculosis allowed. He loved his mother, respected his father (whose sole flogging, he later remarked, was the best thing that ever happened to him), and had five children by his dear wife. He led a vigorous outdoor life, drank in moderation, and enjoyed a cigar while fishing. "Let's sin a little," he winked at his friends while lighting up. His Romantic temperament showed in his adulation of heroes, reverence for nature, and love of music, which he imagined as "a kind of divine power, enlisted in the interest of virtue and religion." The only thing Bushnell could not abide was pretense in politics, business, society, or the church.

Democrats and Whigs alike left him disillusioned. Out of "excessive honesty" (his own daughter's words), he turned down opportunities to make money. He flinched at the "pernicious sway" of fashion over his highbrow parishioners. Genteel society was a farrago of "hypocrisy, flattery, and fictitious feeling; an exquisite drawing room world of show and self-adulation." He advocated education for women, but thought the suffragists were combative and artificial. The good woman, he wrote, should adorn her intelligence with "a mantle of sweetness and charity" since "God made the woman to be a help for man, not to be a wrestler with him."

Not least, Bushnell disliked theological strife. He agreed with the experiential preachers that seminary theology was barren, but came to deplore the "artificial fireworks" they used to stoke their revivals. Having learned at Yale to reject Unitarianism, he shed a Trinitarianism based on sacrificial atonement for depraved human nature. He rejected both camps' efforts to prove their positions through argument, since the Romantic in him insisted religion was more akin to poetry than to science. Indeed, he concluded that all creeds were inadequate to describe the divine mystery and that they only bred disunity unhealthy for a republic. What heavenly star did Bushnell fix on to steer through the winds and currents of doctrine? The United States of America, whose creation he firmly believed was the reason God had tutored Anglo-Saxons in Protestantism and self-government. Accordingly, he supported a Christian alliance to combat the spread of Catholicism and the home missionary movement to redeem the frontier. In *Discourses on Christian Nurture* he argued that the family must be a nursery of piety, congregations a family of families, and the nation at large a family of congregations. Conversely, any movements threatening to rupture the unity of the United States were guilty of the paramount sin against God, who had watched over the nation while it was still in the womb (its "ante-natal" stage). So he urged northern abolitionists and southern nullifiers alike to repent and instead open their hearts to this sublime vision:

> *I see the nation rising from its present depression with a chastened, but good spirit. I see education beginning to awake, a spirit of sobriety ruling in business and in manners, religion animated in her heavenly work, a higher self-respect invigorating our institutions, and the bound of our country strengthened by a holier attachment. Our eagle ascends and*

spreads his wings abroad from the eastern to the western ocean. A hundred
million of intelligent and just people dwell in his shadow. Churches are
sprinkled throughout the whole field. The sabbath sends up its holy voice.
The seats of philosophers and poets are distinguished in every part, and
hallowed by the affections of the people. The fields smile with agriculture.
The streams and lakes and all the waters of the world bear the riches of
their commerce. The people are elevated in stature, both mental and bodily;
they are happy, orderly, brave, and just, and the world admires one true
example of greatness in a people.

Walt Whitman never sang a more beautiful hymn to the promise of American life.[10]

All Bushnell needed—the thing he searched for in sermons, lectures, and essays—was a bridge between his civic and biblical faiths. It came to him on a frigid February evening in 1848. His wife noticed, and asked what he had seen. "*The gospel*," he answered, "a revelation from the mind of God himself." His inspiration was Coleridge's *Aids to Reflection*. It taught Bushnell the inadequacy, even the deceptiveness, of human language when applied to spiritual and metaphysical concepts. Language, so applied, could never be taken "literally." Mere understanding sufficed to comprehend the natural world; it was human beings' endowment of intuitive reason that could apprehend the divine. Human beings were thus partly supernatural themselves, albeit sundered from God. Suddenly all was light. The next Sunday morning Bushnell preached, "What form is to body, character is to spirit. Sin takes away the image or form of God, and makes the soul a truly deformed creature." How did God remedy this? He sent his divine son (as Calvinists held him to be), not to perform "sacrifices and other Jewish machineries," but to present human beings with a "clear image of the divine beauty and goodness" (as Unitarians held) so their intuitive reason could repair their souls and rebind them (as in the Latin *religio*) with the divine "Triunity." Here was a vision that might reconcile Protestant theologies and rebind Americans into the holy nation God meant them to be.

In the short run it didn't. Bushnell's treatise *God in Christ* of 1849 was attacked from all sides. Dissidents in his own congregation wanted him tried for heresy. His health deteriorated. But the revelation, his daughter wrote, "is the key to Horace Bushnell" and all his subsequent work on behalf of the gospel and the American Union. In the long run Bushnell, more than anyone else, located the center of gravity in American Protestantism. He made national virtues of toleration (except for Catholics), relaxation of creeds, and social perfection through gradual reform that did not threaten the Union. And he promised, to a Union steeped in those virtues, the most Romantic destiny imaginable.[11]

If one biography can stand for the prophets who claimed "revelation from the mind of God himself" but stood *outside* the mainstream, it surely must be Joseph Smith's. He was simultaneously an eminent Jacksonian, a scion of the Yankee exodus, a creature and critic of the Second Great Awakening, a Romantic reformer, a charismatic utopian, a mystic nationalist, and a hustler in the manner of Barnum. What is more, his Church of Jesus Christ of Latter-Day Saints, unlike the other cults founded before 1860, not only outlived the environment of its birth, but continues to grow by leaps and bounds in the early twenty-first century. Prophet, genius, con man, crackpot, or all four in some proportion, Joseph Smith, Jr., founded what no less a figure than Leo Tolstoy called the "American religion."[12]

The tale is so dramatic, eerie, violent, erotic, and steeped in the pioneer spirit—so downright American—that it would surely have become a staple of pop culture but for the disconcerting insistence of Mormons that it all *really happened*. Even scholars trip over the tale because of the surfeit, not dearth, of evidence about Smith's life and times, and because spooky questions remain in spite of the evidence.[13]

His father and his mother, Lucy Mack, had moved from Massachusetts to Vermont only to lose their nest egg to land speculators and crooked merchants. In 1816, "the year without a summer" when everyone's crops failed, the family moved on to Palmyra, New York. Thus Joseph, Jr., grew up in the heart of the "burned-over district," where truth and illusion came at a discount. Revivalists touting this or that strain of Methodism, Presbyterianism, Baptism, and nonsectarian primitive Christianity wrestled for the souls of displaced Yankees whose own hoary folkways included divining, money digging, crystal gazing, magic amulets, counterfeiting, and necromancy. In 1820, the fourteen-year-old Joseph was vexed as to where truth, salvation, and riches lay hidden. So he did what adults seldom do (for fear of getting an answer). He took the apostle at his word: "If any of you lack wisdom, let him ask of God, that giveth to all men liberally, and upbraideth not, and it shall be given him" (James 1:5). On a lovely spring morning Joseph went into woods and prayed. A thick darkness descended; then came a pillar of blinding light encompassing God the Father and Jesus. When the quivering boy found his voice he asked which Christian sect he should join. Their answer was none of them, since all had the form of godliness but denied the power thereof, just as the Bible prophesied would be so before the last days.

Three years passed. Joseph grew up handsome and mischievous. He claimed to have found a "seeing stone," hired himself out as a treasure digger, and confessed that he "fell into many foolish errors, and displayed the weakness of youth." One night in September 1823, while Joseph prayed in bed for forgiveness, the angel Moroni appeared in glory. "He said there was a book deposited, written upon golden plates, giving an account of the former inhabitants of the continent. . . . He also said that the fulness of the everlasting Gospel was contained in it, delivered by the Savior to the ancient inhabitants." A breastplate containing two stones—the Urim and Thummim of the Hebraic priesthood—would give him the power to translate the book. The boy unearthed the box on the hill of Cumorah, but Moroni said he was to do four years of penitence before he could open it. So it was not until September 1827 that Joseph extracted the "pearl of great price" and began his career as "revelator."

Martin Harris, a neighbor, transcribed the first 116 pages Smith translated with the aid of the oracle stones. But Smith foolishly let Harris show the pages to his skeptical family. The text was stolen and translation ceased until the schoolmaster Oliver Cowdery volunteered to take dictation. "These days were never to be forgotten," he wrote. "To sit under the sound of a voice dictated by the inspiration of heaven awakened the utmost gratitude in my bosom." David Whitmer observed them: "Joseph Smith would put the seer stone into a hat, and put his face in the hat. . . . A piece of something resembling parchment would appear, and on it appeared the writing . . . and under it was the interpretation in English. . . . Thus the *Book of Mormon* was translated by the gift and power of God, and not by the power of any man." Smith and Cowdery went to work on April 7, 1829, and completed a manuscript of 275,000 words in just seventy-five days. It does not seem humanly possible, especially considering that Smith was barely literate and no expert on ancient Hebraic culture.[14]

"Other sheep I have, which are not of this fold" (John 10:16) is one of Jesus's most puzzling remarks. The "golden Bible," as the *Book of Mormon* came to be known, purported to locate those sheep in America. It told of Lehi, a patriarch of the tribe of Joseph who fled besieged Jerusalem around 600 BC with his sons, their wives, and the messianic prophecies of Isaiah inscribed on brass plates. The younger son Nephi received a vision in which the Lord told him to build a great ship and sail for a new world. But no sooner did they arrive than Lehi died and his older son Laman contested Nephi's authority. So the colony split into two warring tribes: the righteous Nephites and the wicked Lamanites whom God punished by turning their skin red. Hence the American Indians, "an idle people, full of mischief and subtlety." A thousand years of spiritual warfare ensued during which the Nephites built great civilizations, succumbed to sin and defeat, and then rebuilt again under holy prophets. All the while they awaited the Christ, whose servants were destined to thrive in "a chosen land of the Lord," "a land of liberty," and "the place of the New Jerusalem" (2 Nephi 10:10–14). At the moment of Christ's crucifixion, earthquakes transformed the American landscape. After the resurrection, Christ appeared to the Nephites, whose rekindled faith and ordered self-government created a golden age. Alas, by around AD 400 decadence set in again, compounded this time by the secret society of Gadianton, "which is most abominable and wicked above all, in the sight of God" (Ether 8:18). And so, in the climactic Lamanite war, Nephite civilization disappeared without a trace save for some gigantic burial mounds and the chronicles written in "reformed Egyptian" on golden plates. Mormon, the last prophet, instructed his son Moroni to bury the plates at Cumorah. There, in God's good time, the "stick of Joseph" (*Book of Mormon*) would rejoin the "stick of Judah" (the Bible) in fulfillment of Ezekiel's prophecy, and latter-day saints would restore the true church.[15]

In spring 1830 Harris mortgaged his farm to pay for the printing of 3,000 copies of the *Book of Mormon*. Locals hooted it down. The *Palmyra Freeman* called it "the greatest piece of superstition that has come to our knowledge." Others charged the unschooled youth with fraud or plagiarism. The book seemed a clumsy parody of the King James Bible. Every verb ended in *-eth*, and every other sentence began, "And it came to pass." (Mark Twain later said the book was "chloroform in print.") If the author did accurately depict some ancient customs, he could easily have picked them up from Ethan Smith's *View of the Hebrews*, published in 1823. That book also revived an old Yankee tradition to the effect that American Indians were descendants of the "lost tribes" of the Hebrews. Moreover, the *Book of Mormon* made howling mistakes such as naming Jerusalem as the birthplace of Jesus and populating America with horses and other animals unknown before the Spaniards brought them. The sibling rivalries among the Nephites would seem an obvious projection of the tensions in Smith's own family, and the *Book of Mormon*'s coded attacks on sectarianism, Catholicism, and Freemasonry (those "Gadianton robbers") were just what one would expect from an upstate New Yorker around 1830. "This prophet Smith," wrote the leader of the Disciples of Christ, Alexander Campbell, recorded "every error and almost every truth discussed in New York for the last ten years. He decided all the great controversies:—infant baptism, ordination, the trinity, regeneration, repentance, justification, the fall of man, the atonement, transubstantiation, fasting, penance, church government, religious experience, the call to the ministry, the general resurrection, eternal punishment, who may baptize, and even the question of free masonry, republican government and the rights of man."[16]

That was precisely the point. Other revivalists talked of restoring the primitive Christian church, but could not agree how to do it. Other preachers warned that salvation depended on questions to which they gave contradictory answers. Smith offered certainty, clarity, authority. He was no overrefined seminarian, but a simple farm boy whom God had chosen to reveal truth to a democratic people. Nor was it fishy that his revelations spoke to provincial concerns. Wasn't that just what the other revivalists told Americans to expect: a call from God to reform so they could usher in the millennium? So even though nobody purchased the *Book of Mormon*, copies distributed by disciples (mostly members of the Smith family) converted dozens, then hundreds. Perhaps the book's sharpest hook was its invitation to "ask God, the Eternal Father, in the name of Christ, if these things are not true; and . . . he will manifest the truth of it unto you by the power of the Holy Ghost." The Methodist pastor John P. Greene, the Campbellite preachers Parley P. Pratt and Sidney Rigdon, and ordinary rural folk like Brigham Young did so, and felt their hearts burn. William Richards declared that the book was either of God or of the devil, because no man could have written it. He decided it was of God.[17]

Smith had already started a church. Since the Lord taught Nephites to practice adult baptism by full immersion, he and Cowdery waded into the Susquehanna River in May 1829 to seek divine guidance. They received it from John the Baptist himself, who descended from heaven to ordain them in the Aaronic priesthood. Theosophies multiplied at a harrowing pace. The apostles Peter, James, and John descended to make them priests in the New Testament church. Jesus bestowed the priestly order of Melchizedek. Harris and Whitmer signed affidavits swearing they had heard God's own voice attest to the *Book of Mormon*. Eight other men claimed to have seen and handled the golden plates. By autumn 1830 the six original "Mormons," as the "Gentiles" nicknamed them, grew to more than 1,000. That provoked local persecution, so Smith sent missionaries to found a colony in Missouri and preach to the Lamanites (Indians), while he led the main flock to Kirtland, Ohio. There, on a terrible night in 1832, a mob pummeled Smith and Rigdon nearly to death. But for seven years the church thrived. Smith studied English, Hebrew, the Bible, and ancient history, while experiencing serial transfigurations. In a vision of the heavenly throne he learned that God is a human personage at whose right hand the Savior literally sits. He revealed that Christ's atonement covered all men and that a just God did not assign souls to heaven or hell for eternity. Rather, the biblical statements "All shall be judged according to what they have done in the flesh" and "In my Father's house are many mansions" meant heaven contained numerous levels to ensure that everyone got just what he deserved. In 1833 Smith inveighed against alcohol, tobacco, tea, coffee, and meat, while blessing education, since "the glory of God is intelligence." He designed a trellis of authority descending from himself through a council of twelve apostles to seventy "elect," then to elders, bishops, priests, deacons, and evangelists. Finally, Smith declared that saints must hold all things in common and contribute to building the town and temple. Rigdon, a veteran church builder; and Young, a natural leader, helped to mobilize saints for those tasks. It all came to an ecstatic climax in spring 1836, when the temple was dedicated. Smith and Cowdery prayed behind a veil while the congregation sat in suspense. Then the prophet emerged to proclaim that Jesus had descended to bless the temple, Moses had dispensed the keys to heaven, and Elijah had commanded baptism for the dead (to fulfill the prophecy of Malachi 4:5–6 and "turn the hearts of the fathers to the children and the children to the fathers."

Kirtland collapsed in the Panic of 1837. The Mormons' bank failed, sparking allegations of fraud, foreclosures, and warrants. Smith ordered his people to liquidate their property, follow him to Missouri, and build the real Zion. That was a curious judgment, given the fate of the Mormon colony founded by the missionaries near Independence. In 1832 mobs backed by sympathetic magistrates torched the Mormons' houses and farms, forcing a winter retreat during which dozens perished. But augmented by the Ohio exiles the colony started over in a town called Far West. By 1838 local gentiles were again vowing a "war of extermination" with silent approval from the governor. In October of that year militiamen shot thirty-two Mormon men and boys ("Nits will make lice") in cold blood. Smith and Rigdon offered themselves as hostages to prevent a slaughter and spent five months in jail, during which Young agreed to abandon Far West. When a timely bribe freed Smith he began his fourth effort to build Zion at a swampy Illinois river landing absurdly named Commerce.[18]

Smith renamed it Nauvoo ("beautiful location"), and had his Council of Twelve invest the tithes of the faithful in a "marvelous work and a wonder." In just five years the Mormons drained fens, planted orchards and fields, laid out streets, erected brick houses and schools, and raised a temple that at the time was the most elegant building in Illinois. Missionary work organized by Young brought in so many converts from Britain and Canada that Nauvoo's population swelled to 20,000. Distinguished visitors praised the colony. James Gordon Bennett of the *New York Herald* wrote that Smith "seems to have hit the nail exactly on the head, by uniting faith and practice—fancy and fact—religion and philosophy—heaven and earth, so as to form the germ of a new religious civilization, bound together in love and temperance—in industry and energy—that may revolutionize the whole earth one of these days."

Illinoisans were afraid of just that. They could no more abide the Mormons' success than Georgians could abide the Cherokees' success. Competing towns resented Nauvoo's prosperity. Whigs and Democrats realized the Mormon bloc had the power to swing state elections.[19] But it must also be said Smith courted his martyrdom. Contradicting the *Book of Mormon*, he embraced Freemasonry, made himself grand master of the Nauvoo lodge, and revamped temple liturgies to conform with Masonic rites and symbols such as the "all-seeing eye" and the "beehive."[20] He declared that the "literal gathering of Israel" was at hand, implying that doomsday was about to swallow up all the gentiles. He declared Nauvoo a province of God immune from Illinois law. He fancied running for president in 1844 on a platform of "theodemocracy." He unleashed a secret bodyguard known as Danites, who were immediately suspected of arson, murder, and the attempted assassination of the governor of Missouri. Most damningly, the prophet informed his elders of a new revelation instructing the church to practice polygamy. Hints of this dated from 1835, when Smith had an affair with a serving girl while translating an Egyptian papyrus said to be a lost book of Abraham. The Hebrew patriarchs had wives in profusion, so why not latter-day saints? This was a hard teaching, especially when Smith began quietly "sealing" to himself the wives of his closest followers. Dissension and paranoia gripped Nauvoo while vicious rumors about Mormons' lawlessness stoked envy, hatred, and fear in neighboring towns.[21]

It almost seems as if Smith were driven to reify the *Book of Mormon* apocalypse wherein Nephites succumb to sin and secret societies even as Lamanites move in for the kill.

In June 1844 six excommunicated Mormons started a newspaper to malign the Nauvoo establishment. Smith had their press destroyed and their papers burned. This provoked legal proceedings that reached the desk of Governor Thomas Ford. He proposed that Joseph and Hyrum Smith meet him in Carthage, guaranteeing safe passage. They went, Joseph said, expecting that "we shall be massacred, or I am not a prophet of God." Hence his last sermon, preached two months before for a martyred Mormon, may stand in retrospect as his own eulogy. In it he revealed "the great secret" that "*God was once as we are now*, and is an exalted Man, and sits enthroned in yonder heavens." He did not create the world, because matter and spirit have no beginning or end. Rather, he became divine in the process of bringing order out of chaos. Moreover, since Adam was fashioned in the image of God, men have it in their power to climb the same ladder to godhood. "This is good doctrine. It tastes good. I can taste the principles of eternal life, and so can you." He closed with a great farewell: "You don't know me; you never knew my heart. No man knows my history, I cannot tell it; I shall never undertake it. I don't blame anyone for not believing my history. If I had not experienced what I have, I could not have believed it myself. . . . I add no more. God bless you all. Amen."

The magistrate in Carthage clapped Smith and his brother in jail.

Governor Ford decided to let the legal process unfold, evidently trusting that the mob's fear of Mormon vengeance would stay its hand. But the militiamen guarding the jail let it be known that their guns were loaded with blanks. On the morning of June 27, 1844, the mob charged, shot Hyrum, then burst upstairs to find Joseph. The prophet fired first—his stalwarts had smuggled in a pistol—but received three wounds. His only hope was that some of the mob leaders were fellow Masons. So he dived through the window bellowing the Masonic cry for mercy, "O Lord, my God, is there no help for the widow's son?" A torrent of bullets silenced him after the word "God." [22]

What accounts for the numbers and zeal of the Mormon saints? Certainly Joseph Smith had charisma. But his death only propelled the church to make greater leaps of faith. Social and religious anxieties in the 1830s and 1840s provided a large pool of potential converts. But missionaries had even more success in decades to come. The Mormons offered structure, authority, a whole way of life. But separatist utopian sects with similar traits died out. So perhaps we should take Brigham Young at his word. When asked if his leadership had saved the church after Smith's death, he replied, "You know nothing at all about the revelations of the Church of Jesus Christ of Latter-Day Saints. It didn't depend on Brigham Young or any other man to lead the new Israel. A child or a fool could have done it." Why? Because the power of the church was its priesthood, said Young. Smith's church was all clergy and no laity, a true priesthood of all believers, a nation of saints in which every man spoke for God; served God through labor, money, and witness; and was sealed to God, wives, ancestors, descendants, and other adherents for all eternity. Once the church was established, said Young, "nothing could stop it." [23] By comparison the Mormons' zigzagging theology might seem of minor importance. But it, too, wielded power born of an adamant literalism. Other theologians interpreted biblical parables, metaphors, types, and shadows of things unseen in oblique language or else (like Bushnell) discounted language. Smith insisted that the scriptures meant exactly what they said in plain English. Literalism appealed to some folks' common sense. Literalism implied salvation and justice for all in the afterlife. Literalism yielded an anthropomorphic theology that made God a man, Jesus an American, America heaven on earth, and his followers immortal gods.

The prophet's death stunned Nauvoo. The mob at Carthage meant to seize Nauvoo; the Illinois legislature wanted its charter revoked. The fractured Council of Twelve did not know whether to dissolve the church or pray for a new prophet. Sidney Rigdon pressed his case, but he was eclipsed when a steamboat docked bearing Brigham Young. Young's stature, devotion, and leadership were recommendation enough. But Young settled the matter with a stirring speech that some elders exclaimed was the voice of Joseph Smith himself. Endowed with prophetic authority, Young expelled or suppressed all dissidents and prepared the faithful for their greatest exodus yet. It seemed beyond the powers of an army to organize a march of 10,000 people deep into the wilderness (remember the Trail of Tears). Somehow Brigham Young got it done in about fifteen months. On a frozen morning in February 1846, his advance guard of saints, bound for Utah, poled over the Mississippi. They believed, just like the nation that spurned them, that destiny lay in the West.

By the 1840s evangelical preachers sensed a loss of momentum. That wasn't supposed to happen. Finney, Beecher, and others had made revivalism a science whose lessons and laws they recorded in handbooks. They expected revivals to spawn more revivals until America became one vast church. Instead, other concerns arose to compete for people's attention. Nicholas Murray attributed this to impatience and pride. As early as 1836 he identified pitfalls that "every lover of Zion should be anxious to know." They included the distractions of politics, sectional strife, territorial expansion, a "haste to be rich," and a shift in attention from personal to national sanctification. Did not Jesus teach his disciples to pray, "Thy kingdom come, they will be done, *on earth* as it is in heaven"? Did not preachers urge Americans to "come out" of sinful lives, stagnant churches, and corrupt institutions? Indeed, but the more American Protestants cut to the chase in pursuit of social perfection, the more they noticed the mote in their neighbor's eye while ignoring the beam in their own. In short, *revival* gave way to *reform*. Preachers themselves were by no means immune. In 1834 the Reverend George Bush said that "the genius of Christianity" lay in its call to address "the temper, the pursuits, and the institutions of the world" and promote "the temporal and secular well-being of mankind."

The Reverend Calvin Colton said that the real purpose of revivals was to mobilize public opinion because "there is no power on earth so stern in its character, so steady, so energetic, so irresistible in its sway." The Presbyterian Robert Baird gave thanks that the triumphant Protestant faith empowered Americans "to make public opinion, not only strong, but *right*, on all points." And so, over the same years when Andrew Jackson defined reform as getting government out of citizens' lives, private crusaders poked their noses into every corner of life. Emerson put it succinctly: "The religious party is falling away from the Church nominal, and . . . appearing in temperance and non-resistence societies; in movements of abolitionists and socialists . . . all the soldiery of dissent." [24]

The first "religious party" coalesced in the anti-Masonic hysteria. Freemasonry, rumored to be a secret, elite, non-Christian fraternity that worshipped science and reason in occult, Catholic-like liturgies, would have been an irresistible target for reform-minded Yankees even if the scandal involving Morgan had never occurred. But none of those rumors contained much truth. Unlike the revolutionary, anticlerical cabals of Europe (e.g., the Bavarian Illuminati), American Masons and lodges were on proud public display. They were not exclusive, but open to all who professed and advanced in the craft. Discussion of religion and politics was forbidden inside the lodge; hence anyone able to swear oaths in good conscience could comfortably combine lodge membership with church or synagogue. Masons advanced through a hierarchy of degrees defined by merit, service, and virtue, not birth or wealth. In fact, no civil association had better patriotic credentials. From Franklin and Washington down to Jackson and Clay, many (perhaps most) of the nation's leadership cohort were Masons. They led the Continental army; designed the American flag, seal, and capital city; drafted the Constitution; permeated the executive branch under most presidents; and thoroughly dominated politics and society in such states as Tennessee and Kentucky. Masons, more than anyone else, romanticized the eighteenth-century dream of a fraternal republican people living in tolerance because they served a deity who transcended all sects and bade Americans be "Master Builders" of an unfinished pyramid towering heavenward to the "All-Seeing Eye." Today we think of the Masons, like the Odd Fellows, Elks, Moose, Kiwanis, Mummers, and Rotary, as just a philanthropic society. They used to be much more than that, as this ode to Grand Master DeWitt Clinton's Erie Canal conveys. Imagine a lilting melody sung a cappella by a soprano or tenor:

> *Let the day be forever remembered with pride*
> *That beheld the proud Hudson to Erie allied*
> *Oh the last sand of time from his glass shall descend*
> *Ere a union so fruitful of glory shall end.*
> *Yet it's not that wealth now enriches the scene*
> *Where the treasures of art and of nature convene*
> *Tis not that this union our coffers may fill*
> *Oh, no it is something more exquisite still.*
> *Tis that genius has triumphed and science prevailed*
> *Though prejudice flouted and envy assailed*
> *It is that the vassals of Europe may see*
> *The progress of mind in the Land of the Free.*
> *All hail to a project so vast and sublime*
> *A bond that can never be severed by time*
> *Now unites us still closer, all jealousy cease*
> *And our hearts, like our waters, are mingled in peace.*

High-minded visionaries? Or a heathen club where well-connected men conspired to manipulate politics and business to each other's advantage while indulging in who knew what beyond locked doors? Women, pastors, and political opportunists asserted the latter in language that ironically echoed the Jesuits' old assaults. Freemasonry was corrupt and foreign. It enforced its code of silence with death threats. It subverted politics and the courts. It escaped exposure by intimidating the press. Its charities served only its own members and their widows and orphans. Its creed promoted a watered-down ethics that said nothing of death, judgment, heaven, or hell. Masonry lured young men to perdition with wine, women, and song in the company of decadent older men. It was a mother's worst nightmare. Like the blogs of today, anti-Masonic newspapers sprang up to circulate horror stories about Masons' sins, their crimes, and their stranglehold over institutions. A pastor in Massachusetts, Nathanael Emmons, called the order the "darkest and deepest plot that ever was formed in this wicked world against the true God, the true religion, and the temporal and eternal interests of mankind." The faculties and students at Amherst, Williams, and other colleges founded by Calvinists angry over Harvard's Unitarian apostasy flocked to the movement. Masonry, said Myron Holley of Williams, promoted a sexual segregation that obliged husbands to embrace pretense and hypocrisy: "If the average family is a sham, so will be the republic." Denominations split; women voiced outrage; hundreds of clergymen quietly resigned from their lodges. When the Reverend Lyman Beecher, leader of Connecticut's Congregationalists, joined the crusade the balance was tipped. An Anti-Masonic Party, waging "a struggle of republican equality against an odious aristocracy," swept local and state elections in the Northeast and then went national. In 1831 its leading organ, the *Masonic Mirror*, vowed to "put down Masonry by the sword, if we cannot put it down without."

Not all women were hostile. Leading journalists such as "America's virago" Anne Newport Royall and Sarah Hale of *Godey's Lady's Book* defended Freemasons. Both had experienced Masons' gallantry and charity at difficult passages in their lives. They also observed plenty of hypocrisy on the other side, and took pleasure in publicizing the scandalous affairs of various clergymen in New England. If wives thought it reckless to allow their men out at night, how much more dangerous was it for men to leave their wives in the unsupervised company of syrupy "pious men" and "black-coats"? The Anti-Masonic candidate William Wirt, on the other hand, blessed his pious Presbyterian wife for rescuing him from private sin and a public life full of "faction, corruption, intrigue, and impenetrable and imperturbable stupidity and infatuation." He supported female emancipation in the belief that women alone could save the American republic. But Wirt's sudden death in 1834 deprived the party of a leader with name recognition. By 1836 most Anti-Masons migrated to the Whigs or to new splinter parties targeting slavery. Freemasonry paled beside that social malady, and everyone knew it.[25]

What did the Anti-Masons accomplish? Little besides disruption. Hundreds of lodges closed in the Northeast, and several states passed laws banning Masonic oaths. But the South was immune to the fever, lodges sprang up all over the West, and even New England's Masons reconvened after a decade or two. By 1850 some 2,000 lodges dotted the United States; and there were more than 5,000 by 1860. Over those years Greek-letter fraternities with their own secret mottos and handshakes, skulls and bones, blood oaths, and mystic symbols, conquered New England campuses and then spread nationwide.[26] It might be said the growing authority of women over domestic American culture gave husbands and sons even more incentive to flee the parlor for the company of males.

Reformers had rather more success against specific masculine vices, such as alcoholism, although how much that was due to their excessive rhetoric is impossible to say. They began by damning Americans as a "nation of drunkards" who consumed beverages containing almost four gallons of alcohol per person per year. That was the highest rate in U.S. history and higher than the rate for any Europeans save the Irish and Scots. Beer, wine, and rum accounted for some of the alcohol consumed, but most was quaffed in hard cider and whiskey. Since women and children, the enslaved, and perhaps half the men drank only in moderation, it follows that about two-thirds of the alcohol was consumed by one-eighth of the population. Call it a subnation of drunkards. Why several million men escaped their economic or social anxieties by crawling into bottle or jug is a matter of speculation, but the factors enabling them to do so are no mystery. Most Americans (often correctly) considered water from wells and streams contemptible and unhealthy. Fruit juice, unless fresh, was best spiked with rum. But the rum trade, so flush in colonial times, languished after 1783 when the British closed the West Indies to U.S. commerce. Tea was costly and unpatriotic. That left cider and whiskey that sold for just twenty-five cents per gallon, thanks to the Midwest's inexhaustible apple and cereal harvests. Social pressure made tippling hard to avoid. Morning, noon, and evening; at home and in shops; on ships and docks; in taverns, newspaper rooms, groceries, and clubs, Americans toasted each other in convivial republican solidarity. Any bluenose who declined the cup was assumed to be a snob thinking himself "holier than thou." Even children welcomed the shot that helped them sleep or fight off a cold. Did not Dr. Benjamin Rush, who first warned against overindulgence, attest to the medicinal qualities of temperate alcohol use?

Temperance: that was the word Boston's clergy and women were seeking in 1826, when they first rallied against demon rum. In truth they promoted total abstinence, symbolized by a T worn on one's vest—hence the term *teetotaler*. But abstinence sounded too harsh, too Catholic, so the movement went national in 1833 as the American Temperance Union. Its very success exposed the pretense. Recruits who did not partake at all criticized temperate drinkers, who in turn envied those somewhat less temperate. Preachers insisted that a man born again in the Spirit had no need of profane spirits, whence "we may set it down as a probable sign of a false conversion, if he allows himself to taste *a single drop*." Sarah Hale was on board this time, assuring her ladies that recipes printed in *Godey's* would never require a drop of liquor. Millions of temperance tracts paid for by teetotaling businessmen, such as the Tappans, warned young men of the slippery slope that ruined careers and marriages. Total abstinence was the high road to happiness, fortune, health, and longevity; the least indulgence put one on the low road to poverty, crime, disease, and death. In novels and plays, confidence men, gamblers, prostitutes, and thieves hung about taverns to prey on good boys and girls just arrived in the city. Greeley estimated that "not less than five millions of dollars are annually won from fools and shallow knaves" in New York alone and that "not less than a thousand young men are annually ruined," and the money was "wasted on harlots, strong drink, and extravagant living." Such behavior was loathsome enough when engaged in by swells. (Martin Van Buren's son bragged of "losing" his European mistress in a card game.) For everyday boys to waste their God-given talents through gambling and drink was subversive of everyone's American dream. Drunkenness was a form of slavery and certainly no private matter. One temperance group issued a new "declaration of independence" substituting the tyranny of "Prince Alcohol" for that of King George.

On the surface the movement was irresistible. By the mid-1830s some 7,000 local temperance clubs counted 1,250,000 people who had taken the pledge. Washington City climbed on the wagon, admitting that "the heavy perfume wafting up from the legislative chambers was a good third Old Monongahela rye, the other two ingredients being chewing tobacco and unbathed statesmen." William Goodell attributed the intemperate rhetoric of politicians and poets alike to wicked indulgence: "Here the disgusting secret is developed. Authors drink and write; readers drink and admire." Under the surface, however, the movement was exposed as "a thing more and more *in*temperate" because every man who could hold his liquor belied the propaganda depicting drunks in the gutter with delirium tremens. In 1836 the American Temperance Union split asunder when teetotalers pushed through a resolution mandating total abstinence.[27]

The revolt of the temperate against the temperance movement inspired Neal Dow of Maine to change tactics. Called the "father of prohibition" by

a later generation and a "moral Columbus" by his contemporary Horace
Mann, Dow was a self-made businessman, a grandson of a Puritan named
Hate-Evil Hall, and a bantam rooster on a mission. He guided tours through
Portland's slums, crowing, "Rum did that!" He single-handedly shut down
the city's Fourth of July festivities. He lobbied the state legislature until, in
1851, it passed "the law of Heaven Americanized": prohibition. As mayor he
led raids on grog shops and boasted of pouring $2,000 in booze into the
street. It seemed the law might achieve what persuasion could not. Yankee
do-gooders elsewhere followed Dow's lead until, by 1855, legislatures in thirteen northern states had banned the sale of alcoholic beverages. This first
experiment with prohibition ended badly and quickly. In Dow's own state
taverns violated the spirit of the law by selling the crackers, nuts, and sausages while giving away the whiskey and beer. Massachusetts tried to suppress intoxication among the lower classes by banning purchase of less than
fifteen gallons at once. Topers just pooled their dollars and divided the
hooch. Every state made an exception for "medicinal uses," with the result
that all one needed was a note from a doctor. Most damagingly, the attempt
to ban antisocial behavior caused a sharp increase in private binges that were
even more disruptive of families. Save in northern New England, all the
laws were off the books by 1860.[28]

What did the temperance crusaders achieve? If the scattered, often
ascribed statistics are at all accurate, Americans' consumption of alcohol
did drop severely in the 1830s and 1840s and, despite a rise in the 1850s,
never again approached its peak. Clearly a sharp swing occurred in public
opinion. But how great a role was played by temperance agitation, laws,
and taxes; religious exhortation; or new middle-class standards of propriety is impossible to judge. It became de rigueur for hosts to offer sweet cider and toddies along with the whiskey. Women made it known that they
thought manliness implied sobriety, domesticity, and success, not boorish
display. No doubt the association of hard drinking with the Irish and
Germans was an added disincentive for native-born Protestants. But market forces again helped determine cultural habits. In the decade after 1825
the price of coffee tumbled from twenty-five to fifteen cents per pound,
and then to ten cents after repeal of the Tariff of Abominations. Imports
of South American coffee beans quadrupled by 1840. In other words,
farmers, laborers, and middle-class folks got hooked on a new sort of
"high" conducive to morning and afternoon labor, and waited until eve-

ning to sample the spirits. Save for the 1920s, that has been the American pattern ever since.

Sylvester Graham graduated from temperance agitation to found another U.S. tradition: the health craze. Ordained a Presbyterian minister, this Connecticut Yankee rallied instead to Dr. William Alcott's gospel: "a vegetable diet lies at the basis of all reforms." Graham mastered the American art form of doing well by doing good. He went on the road selling the notion that a man is what he eats. Hence the best way to combat illness was not to imbibe patent medicines, but simply to eat vegetables, fruits, and above all Graham's own crackers, made from coarse wheat. The list of proscribed substances grew with Graham's fame on the lecture circuit to include alcohol, tobacco, tea, coffee, meat, spicy foods, and anything fried. He branched out to pronounce on hygiene (cold showers were good), sensible clothing, exercise, sex, and mental gymnastics. He became a franchise as Grahamite hotels, spas, health clubs, and magazines proliferated. By his death in 1851 (at age fifty-six), Graham had inspired a host of entrepreneurs hawking this or that "secret" to long life, health, or virility. Perhaps the most influential was Dr. James Caleb Jackson, whose Granula inspired the breakfast cereals later made famous by the Kellogg brothers and C. W. Post.[29]

Horace Mann was yet another moral athlete born on a rocky New England farm. "I believe in the rugged nursing of toil," he recalled, "but she nursed me too much." Only on Sundays was Horace left free to read, dream, and reflect on the meaning of life. Hence he remained a strict supporter of blue laws even after he denied the Sabbath any religious significance. His ladder up was education at Brown University and Tapping Reeve's law school in Connecticut. There he observed (as did P. T. Barnum) that "the most common thing imaginable is to rise from your knees at prayer to cheat a man out of two pence." While practicing law in Dedham, Massachusetts, Mann traded his Calvinist faith for a secular faith in America's progress through ordered liberty, economic stimulus, and moral reform. Elected to the general court in 1827, Mann sponsored bills to charter canals and railroads, restrict liquor sales, and build a humane asylum for the mentally ill. Following his first wife's death and financial woes stemming from his brother's failed textile mills, Mann resumed his career when the Whig landslide of 1834 swept him into the state senate. There he led the fight to pass Governor Edward Everett's proposal for a Massachusetts board of education. Every good deed gets punished. When Everett pleaded with Mann to serve as the board's secretary he agreed "in the spirit of a martyr." At least, that's what he wrote for posterity. What of his political and legal careers? asked his friends. Mann answered, again for posterity, with heroic words, "Let the next generation, then, be my client."

He was a Mann on a mission and no mistake. He inspected schools at home and in Europe, did assiduous research, and published annual reports rich in data and ideas about pedagogy, textbooks, classroom design, teaching aids, and curricula. They laid the foundation for a U.S. educational industry in the same way Douglass Houghton's annual reports created Michigan's copper industry. Like the preachers who taught that every person was capable of moral rebirth, Mann believed every child was capable of mental achievement if liberated from rote memorization enforced by corporal punishment. Like the reformers who identified some healthy habit as the "key to success," Mann considered education a cure-all if it was adequately funded and staffed. His twelfth and final report waxed utopian. "Education then, beyond all other devices of human origin, is a great equalizer of the conditions of men—the balance wheel of social machinery." Schools should teach not only the three R's, so necessary to commerce and culture, but "those articles in the creed of republicanism which are accepted by all, believed in by all, and which form the common basis of our political faith." In short, schools must be the churches imparting the American civic religion because "in a republic, ignorance is a crime." Mann held it to be the "will of God" that every child born into the world had the right "to such a degree of education as will enable him, and as far as possible will predispose him, to perform all domestic, social, civil, and moral duties." Indeed, the survival of republican government depended on schools to educate voters and tame their passions. What is more, education was an investment that paid for itself. The *American Journal of Education*, founded by Mann's counterpart in Connecticut, Henry Barnard, explained: "The very taxes of a town in twenty years will be lessened by the existence of a school which will continually have sent forth those who were so educated as to become not burdens but benefactors."[30]

Within five years of Mann's appointment public education was sweeping the nation. Within ten years every state in New England had its board of education headed by a paid commissioner. Within fifteen years—that is, by around 1850—normal schools to train teachers were springing up across the North; some 80,000 elementary schools opened their doors to

3.3 million students; and about 6,000 secondary schools opened theirs to 250,000 students. In the 1830s alone thirty-five new colleges sprouted; by 1860 the original "colonial nine" colleges had grown to almost 200. Presbyterians led with forty-nine schools, followed by twenty-five Baptist, twenty-three Methodist, twenty-one Congregational, fourteen Catholic, eleven Episcopal, and six Lutheran schools. The original nine and the new state universities of the Midwest began in these years to experiment as well with reformed curricula stressing science, political economy, and modern languages.[31]

Quality varied enormously. The legendary little red schoolhouse was often a drafty shack open just three months of the year; wanting in slates, chalk, maps, or books; presided over by hastily trained teachers whose main task was boxing the ears of mischievous boys. What is more, a dispersed rural population, poor transportation, Jacksonian hostility to public expenditures, and the fact that just a few years of book-learning sufficed for most occupations combined to make elaborate school systems impractical in much of the nation. As late as 1860 only about 15 percent of children were enrolled. But the momentum Mann and his disciples imparted made it only a matter of time before public schools were an ornament, or at least an obligation, in every American town and village.[32]

Expansion of public education created a mass market for textbooks. Noah Webster had paved the way with his *Blue-Backed Spellers* and *Grammatical Institute* collating patriotic tracts, speeches, and poems. Textbooks, he preached, must "form the morals as well as improve the knowledge of youth." But the primers that rode the tide of the public school movement and inculcated the civic religion as taught in the North were William Holmes McGuffey's *Eclectic Readers* for grades one through six. McGuffey was born in 1800 in western Pennsylvania. A Scots Presbyterian trained for the ministry, he was licensed to teach at age fourteen and felt more at home at the lectern than at the pulpit. He founded a school in rural Ohio and taught in a smokehouse in Kentucky until 1826, when the Reverend Robert Hamilton Bishop hired him for the new College of Miami (Ohio). Through the temperance and educational movements he befriended the Beechers in Cincinnati. It was Lyman Beecher's brilliant child Catharine who steered a publisher to McGuffey. Since he agreed with Beecher that schools and churches must civilize the frontier, McGuffey took pains to assemble materials ranging from nursery rhymes to essays, oratory, and snippets from classical literature. Then he experimented to learn which selections clicked best with youngsters. That formula plus aggressive marketing made the first *Eclectic Reader* a hit. So he went on to edit a whole series of graded readers, each more popular than its predecessor.

The books made McGuffey "America's schoolmaster" who "taught millions to read and not one to sin." The American Book Company later estimated that it sold 7 million *Eclectic Readers* from 1836 to 1850 and 40 million more by 1870. The nation's most familiar books, save for the King James Bible, the primers taught moral lessons as well as fine points of grammar and rhetoric. McGuffey's principles of selection did not include diversity or objectivity; nor would school boards and parents have wanted it so. The readers were meant to portray American society as reformers hoped and expected it would become. Hence the world that pupils encountered in their schoolbooks was overwhelmingly white, middle-class, upwardly mobile, Protestant, patriotic, and moral. The values they learned included education, temperance, honesty, thrift, diligence, patience, integrity, loyalty, generosity, and nonsectarian faith in God and country. Even the lessons in grammar and elocution carried moral baggage because McGuffey knew that words were the currency of self-government. Whether spoken or written, words should be like "beautiful coins, newly issued from the mint, deeply and accurately impressed, perfectly finished . . . and of due weight." McGuffey's snippets of speeches from Patrick Henry to Daniel Webster venerated the Union, abominated war except when it was a necessary evil, and offered models of "true manliness" and female domesticity. "Next to the fear of God implanted in the heart," McGuffey lectured, "nothing is a better safeguard to character, than the love of good books. They are the handmaids of virtue and religion." Conversely, "Bad books are the public fountains of vice."[33]

Geographies, histories, civics books, and other textbooks for secondary schools and colleges reinforced the values taught by McGuffey. They exalted nature, especially the American wilderness, even as they praised the material growth that tamed the environment. They implied that all Americans traced their heritage to the Pilgrim fathers and considered Protestant Christianity "the only basis" of a healthy, self-governing society. They extolled leaders who served the common good: "Franklin is the apotheosis of the great man; Washington of the hero." History texts invariably described God as the author of the United States. As one text of

1834 held, "The marks of divine favor shown to our nation, the striking interposition of divine PROVIDENCE in our behalf, cannot fail to enliven the patriotic sentiments of a pious mind." Nor did more advanced texts shrink from alerting students to the un-American values promoted by others. Popery headed the list. Catholicism was an idolatrous, oppressive, priest-ridden religion whose fruits were superstition, persecution, and poverty. That explained the decline of Spain and France, and the current torpor of Mexico. The only two Catholics granted honorary membership in the American pantheon were Columbus and Lafayette. Schoolbooks praised England for developing representative government, founding the thirteen colonies, and combating Catholic empires. But the schoolbooks chided the British for clinging to monarchy, aristocracy, and empire. The only stereotype likely to stir controversy was that of the American South. Histories and geographies written by Yankees, as the vast majority were, either deplored slavery or else indirectly condemned it by describing southern whites as languid, indigent, violent people, in contrast to industrious northerners.[34]

Such nineteenth-century political correctness reflected, no less than today's, the interpretations offered by the nation's most prestigious scholars. That first generation of U.S. historians included Jared Sparks, George Bancroft, Francis Parkman, Edward Everett, George Ticknor, William Prescott, and John Lothrop Motley. All were Yankees. All embodied the Romantic, nationalist historical philosophy of Georg F. W. Hegel. Most came from a clerical background. Most made a second career of politics in hopes of "making history" themselves. All affected a style of historical writing that can only be deemed excessive. Hegel imagined the stages of history unfolding in obedience to a dialectic between opposing ideas, usually exemplified in the lives of nations and empires, the outcome of which was continuous human progress. If God oversaw human affairs, as Hegel believed was the case, God was to be found in history. One might almost say God *is* history. American savants of advanced views and Romantic spirit felt their hearts leap at the notion. They imagined the past as mere prologue to the discovery of America and the birth of the United States, which undeniably incarnated Hegel's "spirit of the age." They celebrated this flattering epic with gaudy panoramas in words depicting John Smith's arrival at Jamestown, Wolfe's death at the siege of Quebec, and Washington's crossing of the Delaware. They might bow to the courage of Spanish and French explorers and sing paeans to noble Indians, but they made it clear those people's ages were over. Make no mistake: these were archival scholars adhering to the best German historiography of their day. But just as Finney urged preachers to pour their manhood into revivals as if they were going to war, so Motley wrote, "I now set myself *violently* to the study of history." The market responded to histories that read like adventure novels and that told readers what they wanted to believe about themselves and their era. Between 1800 and 1860 over one-fourth of the most popular books dealt with historical themes, and 111 historical societies catered to the sentimentality of a people otherwise hell-bent on their future. Therein lay a pretense. Americans embraced the past in order to fling it away.[35]

McGuffey's academic career misfired. His colleagues at Miami accused him of greed, vainglory, and coddling of students. But matters were brought to a head when an abolitionist frenzy hit the state's campuses like a tornado. Beecher opposed radical abolition as dangerous to the Union. So did McGuffey, in deference to Beecher or perhaps to the southern book market. His colleagues and the college president reviled him for that. He resigned in 1836; failed twice as president of other colleges; then repaired in 1843 to the University of Virginia, where he spent his last thirty years churning out readers and suspiciously scanning his royalty statements.

In retrospect, the most ghastly national sin of the era was slavery, yet the most feeble reform movement turned out to be abolitionism. That was something Americans could not have foreseen in 1833, when ships docked with electrifying news: Parliament had abolished slavery throughout the British Empire! A triumph of public opinion shaped by skilled agitators inspired by Romantic religious faith, it seemed to imply that the extinction of slavery in the United States was only a matter of time. Instead, it proved no precedent whatsoever.

The British movement began as a protest against the slave trade during the years when the thirteen American colonies were in revolt. Baptists, Quakers, William Wilberforce's Methodists, and exponents of natural law as expounded by William Paley of Cambridge blanched when they read Thomas Clarkson's *Abstract of the Evidence*, about the hellish traffic in African souls. They persuaded 400,000 Britons to boycott sugar and sign petitions. When Parliament at last defied the West India interests and criminalized the trade in 1808, abolitionists expected to touch the hearts of the sugar planters and abolish slavery as well. Fifteen years later, the movement's now elderly leaders admitted that would not happen. So they formed the gently named London Society for Mitigating and Gradually Abolishing the State of Slavery throughout the British Dominions, and went to work again on public opinion. Wilberforce damned slavery as "a national crime," "the foulest blot that ever stained our national character," and "a load of guilt, which has long hung like a millstone about our necks, ready to sink us to perdition." Members of the society preached and published, rallied church groups and women's groups, and trained more agitators to tug the heartstrings of the masses. Still, their moral, religious, and economic arguments got no serious hearing until the largest slave revolt in history erupted in Jamaica on Christmas Day 1831. More than 20,000 rebels seized large parts of the island, crying, "No watchman now! Brimstone come! Bring fire and burn massa house!" The British garrison restored order after a month in which 200 blacks and fourteen whites perished, but their commander reported that more insurrections were certain if the slaves were not freed. Westminster awoke to the fact that the costs of slavery outweighed its profits. When the Reform Bill of 1832 fortuitously cut the planters' representation in Parliament from thirty-five to fourteen, a torrent of abolitionist petitions washed away the resistance. In 1833 Parliament voted to free all slaves as of August 1, 1840 (later, this date was advanced to 1838), and to compensate their owners with £20 million raised by government bonds.[36]

The liberation of 800,000 enslaved Africans, almost all in the nearby West Indian islands, horrified whites in the American South and thrilled the nascent abolitionist movement led by Garrison and the Tappans. All parties were mistaken to take the British events as an omen. To be sure, humanitarian, activist, evangelical Protestantism was a populist force in both the United States and Britain, but the largest American denominations were sharply divided over slavery. In 1843 the Southern Baptist Convention seceded from the national church; in 1845 Southern Methodists did the same. Moreover, none of the conditions that favored the British Anti-Slavery Society prevailed in the United States. First, the United Kingdom was relatively compact and truly "united" rather than divided into federal states claiming their own jurisdiction. Second, the British were emancipating fewer than 1 million slaves on far-flung islands, not

3 million in their own midst. Third, Parliament was empowered to make law at its whim without fear of executive or judicial veto. Fourth, British public opinion could be aroused in a way that U.S. opinion could not, because northern Americans recoiled from the notion of coughing up money to abolish slavery with compensation just as much as southerners recoiled from the notion of giving up slavery without compensation. Fifth, the British Anti-Slavery Society was able to tap practical arguments in a way that the American movement could not. The burden of the revolt in Jamaica was borne by the nation and weakened the will of the West Indian planters. The cost of Turner's revolt fell on Virginia alone and steeled the will of its planters. Sixth, the British abolitionists were united and focused, whereas American opponents of slavery argued about almost everything.

William Lloyd Garrison's father was a drunken sailor from Massachusetts who abandoned his family to poverty when the boy was twelve. His mother was a pious, long-suffering Baptist who doted and depended on William after his older brother also surrendered to rum. You can guess the rest. He grew up alienated, self-righteous, and angry—like the sea lawyer who is always mouthing two grievances and a right. Garrison apprenticed himself to a local printer and started his own newspapers, but failed over and over again because readers choked on his polemical editorials. Then he learned how to play the victim. In 1830, when he was twenty-four years old and was working for the Quaker abolitionist Benjamin Lundy in Baltimore, he accused a merchant in print of murder for plying the slave trade. Garrison was convicted of criminal slander and jailed when he could not pay the fine. But his publicized martyrdom was noticed by the Tappans, who sprang him from prison and put him on their payroll. Garrison went back to Boston at once to publish *The Liberator*, taking care to ensure his continued persecution by declaring in its first issue: "I will not equivocate—I will not excuse—I will not retreat a single inch—AND I WILL BE HEARD." He certainly was, but not through *The Liberator*, whose subscribers (mostly free blacks) peaked at about 400. Garrison's real vehicle was the American Anti-Slavery Society founded in 1833. Its declared purpose was to persuade Americans of all sections that "slavery is a heinous crime in the sight of God"—indeed, one "unequalled by any other on the face of the earth"—that it must be abolished at once, without expatriation of blacks or compensation for slave owners, and that all the rights and privileges enjoyed by white citizens be extended to persons of color.[37]

The society's goals and methods echoed those of the British abolitionists whom Garrison sailed over to consult. He meant to organize local societies, wallpaper the nation with pamphlets and tracts, mount petition drives, and generally raise the roof in hopes of touching the hearts of slave owners and northern whites. Such "moral suasion" proved futile. Southern leaders accused Garrison of inciting revolts (in fact, he was a pacifist), blocked the mails, outlawed abolitionist speech, and imposed the gag rule in Congress. Northerners rioted against equal rights for Negroes. Garrison was pummeled through the streets of Boston in 1835. A mob in southern Illinois burned down the abolitionist Elijah Lovejoy's print shop in 1837, then shot and killed him as he ran from the flames. Abolitionism was repressed on numerous campuses, especially that of Yale, which had a huge stake in the cotton, shipping, and insurance industries. A growing number of northerners might grant that slavery was immoral and economically unsound, but they shuddered to think what might happen if the "immediatists" got their way.

Nonetheless, the number of radical abolitionists (as opposed to gradualist, back-to-Africa abolitionists) increased, especially among women and members of the clergy in New England and New York's "burned-over district." Theodore Dwight Weld was exemplary. He was the son of a pastor in Connecticut, and he was managing the family farm at the age of fourteen. Educated at Andover, he was such a brooding Romantic that he found himself looking out the window to see what time of year it was. By sheer force of will he trained his concentration and memory. (An artist said of his visage: "Its severity is like a streak of lightning.") Dr. Thomas Gallaudet, the physician to the deaf, introduced Weld to Lyman Beecher, who in turn got him into Hamilton College. A few months later Finney preached there. Weld was transfixed by Finney's insistence that sinners must reject self-interest and choose a "preference for disinterested behavior." Weld jumped into reform movements, worked with the Tappans, and then enrolled at Lane Seminary, where he collided with Beecher over what to do about slavery. In 1834 Weld led seventy-five conscientious objectors out of Lane in favor of abolitionist agitation at the grassroots. He helped found Oberlin College, trained scores of touring lecturers, and barnstormed through the nation gathering evidence for *Slavery as It Is*. Modeled on Clarkson's British exposé, it sold 100,000 copies. During his travels Weld met and married a woman from a decidedly different background.[38]

"I will lift up my voice like a trumpet and show this people their transgressions," said Angelina Grimké, who meant to re-create her nation, her world, and herself. Who was she? In religion, an "Episcopalian-Presbyterian-Quaker-nondenominational mystic. Regional affiliation? Charleston-Philadelphia-New York-Massachusetts. Gender roles? Un-married Public Speaker-Author-Debater-Married Mother-Women's Rights Activist-Educator. Class? Carolinian slave-holding aristocracy-Northern urbanite benevolence worker-utopian visionary." Angelina's father, John Grimké, a South Carolinian of Huguenot origins, was a planter and judge whose patriarchy included a wife, fourteen children, numerous slaves, and political protégés. The sixth child, Sarah, grew to despise the privileges and constraints of a Charlestonian lady. She wanted to go to Yale like her brother, but that was out of the question. She wanted to teach slaves to read, but that was illegal. Then in 1819 her father went north for his health, taking Sarah along, and promptly died. She fell in with some Philadelphia Quakers who believed in female equality, joined their meeting, and stayed. Her sister Angelina, her junior by twelve years, was even more rebellious. She broke with her family's Episcopal parish to follow an evangelical preacher. She witnessed the abuse of an elderly female slave and felt life-long guilt for not interceding. In 1829, she fled north to join Sarah (much to her mother's relief). Angelina made friends with Catharine and Harriet Beecher at their schools for women, then joined Philadelphia's Female Anti-Slavery Society. Why "female"? Because propriety did not allow "promiscuous" (mixed-gender) audiences, much less women speaking in public before men. But Angelina's *Appeal to the Christian Women of the Southern States* (1836) so impressed Weld that he invited the sisters to at-tend his boot camp for abolitionist lecturers. Angelina's eloquence and spunk were astounding. Audiences of 1,000 or more did not faze her. She welcomed hostile mobs in the belief that persecution purified and ad-vanced the cause. Her experience inside a slaveholding family gave her unmatched credibility. She also had the revivalists' flair for drama. Ange-lina's climactic address of 1838 in Pennsylvania Hall boldly began, "Men, brethren, and fathers—mothers, daughters, and sisters, what came ye out for to see? A reed shaken with the wind?" Thus did she personify John the Baptist, calling Israel to repentance and implying the kingdom of God was at hand if slavery was abolished. Northerners believed they were in-nocent of the sin of slavery, but Grimké cried, "They know not what they do." That turned insouciant northerners into the Roman soldiers on Calvary and the suffering slaves (or herself) into Christ.[39]

Historians once took for granted the religious grounding of the abolitionists' fervor, then dismissed it because the radicals all fell away from churches to pursue secular goals. Recently scholars have decided anew that many abolitionists were moved by a fiery faith—it just wasn't orthodox faith. By 1838, when Sarah Grimké published her *Letter on the Equality of the Sexes and the Condition of Women*, she and Angelina were drifting into a mystical communion with God that validated their growing identification of Negro emancipation with women's emancipation and their rejection of biblical verses that seemed to condone or tolerate slavery and patriarchy. Lucretia Coffin Mott trod a similar path. The Quaker child of a Nantucket sea captain who was absent for years, she imitated her self-reliant, shopkeeping mother. After moving to Philadelphia she married and had six children. But her spontaneous eloquence in Quaker meetings, mostly about the evils of slavery, launched her on a second, public career. She joined Garrison's abolitionist circuit, organized the first Anti-Slavery Convention of American Women, and became a staunch feminist. Convinced by her own intense feelings that divine truth was apprehended through intuition, Mott rejected original sin, declared "pure" human nature capable of perfection, considered monogamy an unnatural restraint, and insisted that the Bible had been mistranslated if it implied anything else. (In all these ways she took exactly the same positions as Joseph Smith.)

Male pastors, including Weld himself on occasion, warned the women against turning secular causes, however holy, into religions. But Angelina Grimké insisted that abolition was the gateway to the millennium: "God deigns to confer this holy privilege upon *man*; it is through his instrumentality that the great and glorious work of reforming the world is to be done." Sarah Grimké told friends to "be not troubled" about scholastic doctrines such as the Trinity and to seek religion through the heart, for it sufficed to know only that God is love. When friends suggested that she tone down her voice, Sarah likened their advice to "the powerful effort of the Jews, to close the lips of Jesus." Her friend Sallie Holley called the Trinity no mystery, but rather a flat contradiction. Abby Kelley wrote, "Abolitionism is Christianity applied to slavery."[40]

One must admire the zeal of these crusaders. But in the context of the time, they wandered beyond heterodoxy into eccentricity, fanaticism, and treason. Many abolitionists took up fad diets and health crazes; spurned conventional clothing and grooming; and dabbled in phrenology, séances, free love, and utopias. In "New England Reformers," Emerson wrote that "the fertile forms of antinomianism among the elder puritans seemed to have their match in the plenty of the new harvests of reform." Thomas Wentworth Higginson observed the "tendency of every reform to surround itself with a fringe of the unreasonable and half-cracked" (he approved). Henry B. Stanton peopled a "representative array" of abolitionists with a "crazy loon," a "lunatic" with wild hair to her waist, and a man dressed in a Revolutionary War outfit because he thought he was George Washington. John Murray Spear claimed to commune with the spirit of Benjamin Franklin and withdrew from the movement to be tutored by Thomas Jefferson.

The abolitionists' avant-garde notions, aberrant lifestyles, and precocious women repelled ordinary Americans, just as the behavior of hippies damaged their peace movement in the 1960s. But whence that allegation of treason? Its germ can be found in Garrison's *Thoughts on African Colonization* (1832), in which he accused anyone shrinking from immediate abolition of willful rebellion against God. "The slaves are men; they were born, then as free as their masters; they cannot be property; and he who denies them an opportunity to improve their faculties, comes into collision with Jehovah, and incurs a fearful responsibility." His colleague Henry Wright supplied the main premise: "God has a Government & Man has a government. These two are at perpetual *War*. . . . I regard all Human Governments as usurpations of God's power over Man.*" John Humphrey Noyes completed the syllogism. Educated in Yale's "new divinity," he embraced the perfectionist doctrine of "total abstinence from sin, and immediate emancipation from the chains of the devil." His notion of sin did not include promiscuity: the "complex marriage" he practiced at Oneida might have inspired Joseph Smith. (Why should other prophets have all the fun?) But Noyes reasoned that since slavery existed by dint of the government's power and law, anyone who pledged allegiance to government was an accomplice. "Every person who is, in the usual sense of the expression, a citizen of the United States . . . is at once a slave and a slaveholder . . . I must therefore separate them and renounce the last." Noyes boasted that his "hope of the millennium begins where Dr. Beecher's expires—VIZ., AT THE OVERTHROW OF THIS NATION."[41]

Garrison institutionalized that logic in 1838 when he founded the New England Non-Resistance Society. It went beyond traditional nonviolence to condemn citizenship itself as a form of participation in violence and oppression. That amounted to anarchism, but with a twist. The Non-Resisters judged men's sinful rule to be the real anarchy because it rebelled against God's law and order. Hence they were called to "come out" of the United States. Once in league with Noyes and Angelina Grimké, Garrison went utterly around the bend. True Christians, he preached, must oppose all civil and religious institutions. If the Bible admonished subjects to render unto Caesar, slaves to obey their masters, and wives to submit to the "godless enslavement" of marriage, then the Bible was bogus. His new motto for *The Liberator*—"No Union with Slaveholders!"—invited *northern* states to secede because the U.S. Constitution was "an agreement with Death and a covenant with Hell!" [42]

Moderate radicals, if that term makes sense, rued Garrison's anarchism and blasphemy. James K. Paulding accused Garrison of grabbing a "burning brand from the throne of God, to set fire to our institutions and consume our Union in ashes." [43] But Garrison's feminism was what precipitated the split. Mott, the Grimkés, Maria Chapman, Elizabeth Cady Stanton, and other women fed up with condescension insisted on the right to vote and hold office in the Anti-Slavery Society, and speak on an "open platform." A bloc led by Arthur Tappan opposed those demands in the belief that they would dilute the society's message and offend potential supporters. But the Garrisonians supported the women's agenda at the convention in 1840 and elected Abby Kelley to its business committee by a vote of 557 to 451. The dissenters walked out to found a rival abolitionist society. Disillusionment and secondary schisms split the movement asunder. The British model had clearly failed.

One faction, led by James G. Birney, gave up on moral suasion in favor of politics: "Vote as you pray and pray as you vote" was his slogan. Birney—a penitent slaveholder in Kentucky converted by Weld—ran for president on the Liberty Party ticket in hopes of obliging one or both major parties to embrace his positions. That has been third-party strategy ever since. Another faction, led by Weld himself, cried a pox on all houses and went back to the grassroots. The female faction focused increasingly on women's rights. Thus, Margaret Fuller took the classical education her father's home schooling drummed into her and rose in Yankee intellectual circles. Her manifesto

Woman in the Nineteenth Century (1845) demanded equal opportunity in education and the professions. In 1848 Lucretia Mott and Elizabeth Cady Stanton marshaled the first women's rights convention at Seneca Falls, New York (the "burned-over district" again). Their "declaration of independence" held that all men *and women* are created equal; hence females should enjoy equal rights in such matters as voting, property, marriage, and divorce. Interestingly, they followed up with a writ of nullification stating that any law which put women in an inferior position was "contrary to that great precept of nature, and therefore of no force or authority." Susan B. Anthony, another veteran abolitionist, became Mott's loudest, most fearless spokesperson. But as it happened Tappan was right: so long as issues of race, slavery, and sectionalism gripped the nation's attention, women could not get a hearing.[44]

The most important faction orphaned by the split in the Anti-Slavery Society was that of free Negroes. Wasn't all this supposed to be about them and their enslaved brothers and sisters? Didn't they donate their coins to keep *The Liberator* in print? Weren't dynamic former slaves such as Henry Bibb, William Wells Brown, and Frederick Douglass the most compelling presence on the lyceum tours? Free blacks never ceased to revere Garrison for his witness. But they noticed when he invited only three people of color to the founding of the Anti-Slavery Society, offered blacks no meaningful jobs in the organization, mingled women's demands with their own, and insisted that "celebrity" blacks speak and behave, like puppets, on his instructions.

A week after the schism in the American Anti-Slavery Society, free blacks gathered in Boston to address what the black pastor Stephen Gloucester called "a solemn crisis for people of color." By overwhelming numbers the caucus chose to stand with Garrison, but its members began to voice their disgust with the hypocrisy of white reformers who hated slavery in the abstract, "especially that which is 1,000 miles or 1,500 miles off," but were as prejudiced as anyone else when confronted with "a man who wears a colored skin." Blacks were invisible except as objects of pity and charity. In northern black communities "the Church is the Alpha and Omega of all things," yet black pastors and speakers were welcome in few white churches. The hundreds of African-Americans who had risked torture and death to flee slavery, or had labored to purchase their freedom for as much as $300, cringed to hear abolitionists such as Wendell Phillips say, "My friends, if we never free a slave, we have at least freed ourselves, in the effort to emancipate our brother man." Moreover, blacks in the North found their rights shrinking, not growing. A Pennsylvania state convention in 1838 inserted the word "white" before "males" in the suffrage clause of its constitution. New York's constitution barred most blacks from voting through a property qualification. Despite (or because of) intense lobbying by the colored community, white New Yorkers upheld the restriction in 1846 by a vote of three to one.

There were many exceptions, of course. Devoted Quakers such as the Motts played risky roles in the Underground Railroad sheltering fugitive slaves. Samuel A. Smith, a white carpenter in Richmond, Virginia, who packed the enslaved Henry Brown into a crate and shipped him to freedom, spent eight years in prison. White editors and publishers flooded the market with more than 100 heart-tugging narratives of escaped slaves. (Harriet Beecher devoured these accounts.) But free blacks took matters more and more into their own hands. The Philadelphia Vigilance Committee, founded in 1838 to coordinate assistance to fugitives, became an all-black operation within a year. At its peak, it helped 300 slaves per year escape to Canada. The New York Vigilance Committee of 1847 soon became over 50 percent black. Field operations of the General Vigilance Committee were managed by the African-American William Still, whose 780-page compilation is the richest primary source on the Underground Railroad. Most famously, Harriet Tubman fled slavery in Maryland and then returned fifteen times to assist runaways. Black communities from Boston to Rochester to Chicago staged regular ceremonies to honor those who did not just talk, but acted to liberate individuals. After 1838 they celebrated Independence Day not on July 4 but on August 1, when West Indian blacks got their freedom.[45]

Garrisonians believed that all inequality could be overcome if men just changed their hearts. The sincere attempts four men made to do exactly that proved it was an impossible dream.[46] The first, Gerritt Smith, inherited almost 1 million acres of upstate New York from his father, a partner of John Jacob Astor. At Hamilton College young Gerritt read Byron's poetry and became a hopeless Romantic. Then, in 1819, his mother died; his new wife died; and his father moved away, leaving the youth in charge of the entire estate. He overcame his despondency when long hours of staring at his wife's portrait unleashed a torrent of tears. There followed a religious conversion deepened by Smith's second wife, Ann "Nancy"

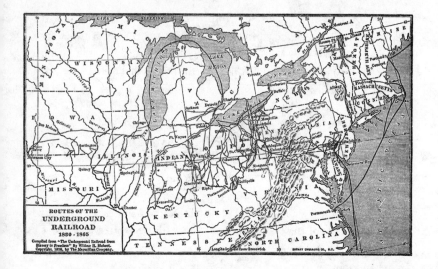

Fitzhugh. The daughter of a Maryland planter, she counted Robert E. Lee among her relatives. Nancy also veered from revival into reform, whereupon Smith gave lavishly to temperance societies and Sunday schools for the poor. He joined the abolitionist cause and spent large sums to buy freedom for the enslaved (he considered these sums "ransom," not payment for property). The loss of most of his fortune in the Panic of 1837 only made Smith feel more cleansed. But for charity's sake he labored to restore his wealth until, on August 1, 1846, he was able to realize a long-cherished dream. He announced his intention to give away plots of forty acres each near Lake Placid, New York, to 3,000 African-American families. He wanted them to escape the prejudice in the city, become self-reliant, and live side by side with himself and other sympathetic whites in a race-blind community.

James McCune Smith, the second dreamer, was the child of an unknown white father and an enslaved mother. Raised as a bonded blacksmith in New York City, he became legally free in 1827. After graduating from the city's African Free School and learning Latin from a black Episcopal priest, the boy attended medical school in Glasgow and joined a Scottish antislavery society. Back in New York at age twenty-eight, Dr. McCune Smith won instant respect and a socially prominent black wife. At the American Anti-Slavery Society convention of 1838 he shared the podium with Gerrit Smith, and they struck up an interracial friendship. When Gerrit launched his land scheme, McCune became the trustee in charge of selecting recipients. McCune also taught Gerrit his evangelical theory of politics. Government was like a sacrament, the outward and visible sign of an inward and spiritual grace. Absent internal conviction, government degenerated into politics serving the strong and oppressing the weak. Hence, for political action to have good effect, "the heart of the whites must be changed, thoroughly, entirely, permanently changed." White people must not feel sorry for blacks or purchase cheap grace through charity. Rather, they must feel as if they themselves were black, for only then could they endow blacks with dignity and regard them as equals.[47]

Frederick Douglass, the third seeker, found it hard to cross racial barriers because he was angry, proud, and a fighter. When he first heard Gerrit Smith's offer, he did not believe it sincere. Yet Douglass was also half-white, and so was another natural bridge. When, in 1855, Douglass published his second autobiography, *My Bondage and My Freedom*, McCune called him "a Representative American man—a type of his countrymen" and attributed his extraordinary talents to the "grafting of the Anglo-Saxon on good, original, negro stock." That is not how Douglass saw himself. No American of his era (except perhaps Walt Whitman) was more eager to sit for portraits and daguerreotypes. The camera loved Douglass's grim, determined eyes and thick halo of hair, and he knew it. Douglass was even more impressive to the ear. Having taught himself to read, write, and speak from old newspapers and a purloined primer of rhetoric, he talked "white" better than most white orators. (Was it partly from envy that Garrison told Douglass to stick to the plantation idiom?) After beating up his cruel owner in Maryland and escaping in 1838 with papers borrowed from a free black sailor, he settled in New Bedford, married, and joined the Anti-Slavery Society. His *Narrative of the Life of Frederick Douglass, Written by Himself* (1845) and hundreds of public lectures ("I appear before the immense assembly this evening as a thief and a robber. I stole this head, these limbs, this body from my master, and ran off with them") made him the best-known colored person in North America. Yet he still was not equal, not by a long shot; and that was ironic because his religious awakening began when he was a youth and he heard a Methodist circuit rider preach that all men, free or slave, were equal because all "were *sinners* in the sight of God." Equal—not because human nature is pure or perfectible, but precisely because it isn't. Yet he, Frederick Douglass, the handsome six-footer far better educated than most whites, was still not treated equally, even by abolitionists.

The dangerously famous fugitive slave toured Britain from 1845 to 1847. On his return he started his own journal in Rochester, thus ensuring Garrison's lasting enmity. He also accepted Gerrit Smith's stunning offer of land in the colony its black residents now called Timbucto. "In this, your new home," wrote Gerrit, "may you and yours, and your labors of love for your oppressed race, be all greatly blessed of God." Gerrit Smith's subsequent correspondence with McCune Smith and Douglass would grow into the most voluminous interracial dialogues of the century.[48]

The fourth seeker was a white who tried very hard to feel with a black man's heart. That was John Brown. A Connecticut boy whose family went to Ohio, he suffered poverty after 1837 and sampled all the sects before hearing Garrison's call to "come out" of wicked institutions. Far from being the old-fashioned Puritan of legend, Brown was an evangelical millenarian who believed his nation was in a "*very short* season of trial." When his business ventures all failed and his infant daughter was tragically scalded to death, Brown made up his mind to settle at Timbucto and live as a Negro. Douglass was stunned by Brown's evident sincerity, impatience, and belligerence. As early as 1847 Brown told him, "The true object to be sought is first of all to destroy the money value of slave property; and that can only be done by rendering such property insecure." His plan was to smuggle arms to Virginia and form a fugitive band of black guerrillas. As years passed during which southern politicians campaigned to expand the empire of slavery and stiffen enforcement of fugitive slave laws, the other three friends were seduced by Brown's vision. As Dr. McCune Smith put it, "The reforms which have swept over this land, have been, after all their noise and fury, mere acts of intellection. . . . They lacked heart and will." Only political action backed by the threat of violence would suffice to end slavery and, someday, give white people the hearts of black people.[49]

Timbucto failed. Only some thirty black families answered Gerrit Smith's call to "come out" and be free. Smith blamed that on blacks' own servility as much as on whites' prejudice. In fact, he had failed to provide the capital that families needed to move, establish a working farm, and pay property taxes. By the 1850s many of his black settlers faced foreclosure. By then John Brown was acting on his own plan to trade intellection for gun-play, killing even the dream that was Timbucto. Gerrit Smith forsook friendship with blacks, then went mad. McCune Smith's heart failed. Douglass went into exile. John Brown went to the gallows.

"I believe that man can be elevated; man can become more and more endowed with divinity; and as he does he becomes more God-like in his character and capable of governing himself. Let us go on elevating our people, perfecting our institutions, until democracy shall reach such a point of perfection that we can acclaim with truth that the voice of the people is the voice of God."[50] Believe it or not, that was the creed of Andrew John-son of Tennessee, a politician whose own whiskey-fueled career fell some-what short of perfection. But then, those were Romantic decades when all ideas and dreams were excessive. Some genteel captains of American cul-ture noticed, and pandered. The real prophets noticed, and recoiled.

New York City's commercial preeminence and its grip on the publishing industry persuaded its literati they ought to be captains of American cul-ture. The Knickerbockers James Fenimore Cooper, Washington Irving, and James Kirke Spaulding had already blazed a trail for a domestic Ro-mantic mythopoeia. The nation once thought lacking in history, ruins, and tragedy in fact abounded in them. Development was the American epic, wilderness its stage, rugged pioneers its heroes, and the disappearing frontier the stuff of tragedy for wild animals, Indians, and at length pio-neers themselves.[51] In the 1830s and 1840s a tight coterie of writers, poets, and artists embodied New Yorkers' creativity in the Bread and Cheese Club, Lunch Club, Sketch Club (aka the XXI), American Academy of Fine Arts, and National Academy of Design. Their output included es-says, novels, and such familiar Romantic poems as "The Old (Oaken) Bucket," "Home Sweet Home," "A Visit from Saint Nicholas" ("The Night Before Christmas"), and "Woodsman, Spare That Tree!" The *New-York Mirror* edited by George Pope Morris and *The Knickerbocker* edited by Lewis Gaylord Clark were the nation's leading reviewers of domestic and European literature and art. In 1839 John James Audubon, the brilliant portraitist of American wildlife, settled in New York. Wealthy patrons such as Luman Reed, Philip Hone, and Stephen van Rensselaer III subsi-dized a burst of creativity in letters, visual arts, and architecture. But the blazing binary star at the core of the galaxy was the friendship between William Cullen Bryant, editor of the *Evening Post*, and the young landscape painter Thomas Cole. Between them they all but invented the allegorical genre known as the Hudson River school. Its unique combinations of darkness and light, immediacy and vastness, tameness and wildness, the sublime and the picturesque have never been matched before or since.

Americans' "natural" empathy for the natural wonders of their virgin continent became nearly obsessive under the influence of English and German Romanticism. But as in all else, they insisted their sublime sensibilities express destinies, even truths, unique to the New World. Cole, the son of a failed English mill owner, grasped this intuitively. After his family emigrated in 1818, the lad daubed the odd portrait and studied in Philadelphia's fine arts academy before moving to New York in 1825. Bryant took one look at Cole's early sketches; began to promote Cole in print; and (when their respective travels permitted) took long hikes in the Catskills, where Cole eventually settled in 1836. The landscapes Cole captured, embellished, or conjured in the imagination of his disciples Frederic Church, Asher B. Durand, Jasper F. Cropsey, Albert Bierstadt, and Thomas Moran were more than celebrations of nature or of a pristine America. They were icons of the intuitive, self-referential religion that the Romantic revelators, reformers, and writers were all straining to appropriate for America.[52]

Romantic American artists, poets, and philosophers took it for granted that nature was God's real holy book. Without repudiating science as a valid means of capturing limited truths about the creation, they felt a potential for unlimited truth in the wild, the expansive, the unspoiled, and the excessive. Why waste time listening to theological debates when a field of wildflowers, a chasm in the Adirondacks, or a vista in New Hampshire's White Mountains put one's heart and soul in direct communion with the spirit in which we all live, breathe, and have our being?

In 1835, James Brooks wrote in *The Knickerbocker*, "God has promised us a renowned existence, if we will but deserve it. He speaks of this promise in the sublimity of Nature. It resounds all along the crags of the Alleghanies. It is uttered in the thunder of Niagara. . . . His finger has written it in the broad expanse of our Inland Seas, and traced it out by the mighty Father of Waters! The august TEMPLE in which we dwell was built for lofty purposes. Oh! that we may consecrate it to LIBERTY and CON-

CORD, and be found fit worshippers within its holy wall!" According to Asher Durand, landscape painting is "great in proportion as it declares the glory of God, by representation of his works." Cole wrote, "Art is in fact man's lowly imitation of the creative power of the Almighty. *We are still in Eden*; the wall that shuts us out of the garden is our ignorance and folly." As early as 1830 Bryant published a sonnet, "To Cole, the Painter, on His Departure for Europe," in which he urged his friend to "keep that earlier, wilder image bright," an American image of "savannas where the bison roves" and "skies, where the desert eagle wheels and screams." Fourteen years later, at Cole's funeral, Bryant thanked the departed spirit for "the opportunity of contemplating pictures which carried the eye over scenes of wild grandeur peculiar to our country, over our ariel mountain-tops with their mighty growth of forest never touched by the axe, along the banks of streams never deformed by culture." Durand committed his eulogy to canvas. His *Kindred Spirits* of 1849 placed the friends Cole and Bryant on a lofty crag in the Catskills. The figures are dwarfed by nature's expanse, but still somehow capture the eye as if the quality of soul that empowers human beings to drink all this in renders them sublime and universal as well. They are partaking of holy communion.

Willis Gaylord Clark wrote of Niagara Falls: "THERE IS A GOD!—and this vast cataract, awful, overpowering as it is, but a play-thing in his hand." Alfred Billings Street wrote, "Nature is Man's best teacher. She unfolds her treasures to his search, unseals his eye, illumines his mind, and purifies his heart. . . . And—holier theme—she teaches us of God." The New Yorkers' essays and poems invariably suggested that science, art, and religion all pointed to "Man's oneness with Nature," which in turn revealed (even *was*) the divine. Hence, "Nature, God, and Man" formed an "infinitely mutable Trinity." T. Worthington Whittredge begged for a vision of "something distinctive in the art of our country." The landscape as icon depicting America as the new Eden and the American as the new Adam met his criterion.[53]

Of course, it was partly a pretense. The tragic despoliation of American wilderness was a central trope in Cooper's novels. The *Literary Review*, commenting on J. F. Cropsey's art, likewise lamented, "The axe of civilization is busy with our old forests, and artisan ingenuity is fast sweeping away the relics of our national infancy. . . . What were once the wild and picturesque haunts of the Red Man, and where the wild deer roamed in freedom, are becoming the abodes of commerce and the seats of manufactures." If nature was God, but nature was dying at the hand of man, then what was to become of God? One way to dodge that conundrum was to substitute art for nature. The reviewer quoted above did that by praising Cropsey for preserving the image of a vanishing Eden. Transcendentalists, as will be seen, had their own answer. But what to make of Cole himself, whose popular cycles *The Course of Empire* and *The Voyage of Life* seemed to evince despair? The former depicted civilization moving from the savage state through the pastoral and consummation states to destruction and desolation; the latter depicted the human voyage from childhood and youth to old age and death. Was America doomed, in the same way as ancient Rome, to despoil Eden and bow to nemesis? Cole never was sure. But in his last years he began a new allegory, *The Cross and the World*, with the intention of illustrating how a Christian empire might ward off decadence and make a garden of civilization.[54]

William Cullen Bryant's celebration of art and nature bestowed on the nation a priceless gift. He launched a campaign to persuade the municipal government to purchase a chunk of Manhattan island before pavement covered it all. It took him twenty years to overcome the resistance of business interests, mean-spirited politicians, and taxpayers, but in 1856 New York City at last reserved land for a Central Park and hired Frederick Law Olmsted to design it. In decades to come almost every city and town in the United States followed New York's example. Thomas Cole would have approved.

New York's literary pretensions, by contrast, were swept away by a ferocious northeaster that blew up around Boston, the self-styled hub of the universe. We may call it the American renaissance, the New England renaissance, the flowering of American literature, or just New England's day. In any case, the overlapping careers of Emerson, Thoreau, Hawthorne, Melville, Longfellow, Lowell, and Poe (born in Boston) made an impression on American literature as deep as the one made by Yankees on demography, technology, religion, education, and reform. The climax came in the years around 1850, when *The Scarlet Letter*, *The House of the Seven Gables*, *Moby-Dick*, *Pierre*, *Walden*, *Evangeline*, and *Hiawatha* all appeared. It is tempting, therefore, to lump these authors together as representative of that era's Romantic spirit and "devotion to the possibilities of democracy." But as we have seen, there were many Romanticisms, as

well as many contrasting ideas of what American democracy was supposed to become. An especially dangerous misconception is the popular
image of the Romantic artist as a heroic individual alienated from society
and tortured by frustrated longing for the sublime. We can thank the English aesthetes for that: they crafted an image of alienation either to suit
themselves or to appeal to the sentimental Victorian marketplace. Even
the most exquisite Romantic poetry, music, and painting can sometimes
betray a cultivated preciosity and a tone of adolescent depression born of
the discovery that life and the world were not all they were cracked up to
be. Nor should we let Emerson and Thoreau dupe us into thinking that
Romantic writers spurned the base worlds of business and politics. All
the American authors of this era took a keen interest in politics and all
were attuned to the marketplace, however much they rebelled, or pretended to rebel, against it. The American market demanded excess, which
all these authors provided. What distinguished the popular writers and
works from the failures was their subliminal message about Americans
themselves. Optimism and flattery sold; pessimism and judgment did
not.[55]

Consider Edgar Allan Poe, who would seem an excellent example of
the brooding, alienated artist. Orphaned at the age of three, he grew up in
Richmond, the ward of a wealthy merchant who rued the boy's habit of
wasting his opportunities and talents. Poe dropped out of the University of
Virginia after a year in which he ran up $2,000 in gambling debts. In 1827
he fled to Boston, published his first volume of poetry, then enlisted in the
Army. Poe rose so swiftly through the ranks that he won admission to
West Point. But he threw that career away out of pique when his now widowed foster father remarried. The final blow landed when the merchant
died in 1834, leaving Poe nothing. The embittered author found solace in
the bottle, the inkwell, and his thirteen-year-old cousin "Sissy." The
Southern Literary Messenger and *Burton's Gentleman's Magazine* gave him
plum editing posts, but he quit both after quarrels. In 1839 his two-volume
Tales of the Grotesque and Arabesque made his reputation as a master of
gothic horror tales, psychological thrillers, detective stories ("ratiocina-
tion"), and spooky odes to unrequited love. That landed Poe a lucrative
position with *Graham's* magazine. But he quit again, lobbied unsuccess-
fully for a political appointment, and then moved to New York in 1844.
While Sissy, who had tuberculosis, coughed up blood, Poe drank all the

more, chased other women, and once attempted suicide. In 1849 he died, just forty years old, in a Baltimore gutter.

This seemingly archetypal artist who fled from success and destroyed himself was nonetheless as shrewd an observer of the marketplace as McCormick or Barnum. His letters and journals reveal that he understood how "the whole energetic, busy spirit of the age" put a premium on magazine literature. That is why he specialized in short stories, poems, and reviews. He called his works "literary commodities" and "wares" crafted according to the "saleableness of literature." He referred to knowledge as gold that increased when invested and compounded in the same manner as money. In reviewing a scholarly book that exempted the humanities from the laws of the market, Poe declared that "truth and honor form *no* exceptions to the rule of economy, that value depends upon demand and supply. . . . [It] is clear that were *all* men true and honest, then truth and honor, beyond their intrinsic, would hold no higher value, than would wine in a Paradise where all the rivers were Johannisberger, and all the duck-ponds Vin de Margaux." In his essay "The American Drama" Poe identified novelty as the coin of the literary realm and imitation as its great, conservative adversary. Among authors (as opposed to publishers) Poe was perhaps the most self-conscious student of business. He also commented extensively on politics, taking care to appear neutral lest he offend part of his market. One can sift Poe's tales forever in search of clues to his inner demons, but the satirical "Some Words with a Mummy" renders an unmistakable judgment on democracy. The narrator, a count, describes how "Thirteen Egyptian provinces determined all at once to be free, and to set a magnificent example to the rest of mankind. . . . The thing ended, however, in the consolidation of the thirteen states, with some fifteen or twenty others, into the most odious and insupportable despotism that was ever heard of upon the face of the earth." Poe asks the name of the usurping tyrant. "As well as the Count could recollect, it was *Mob.*"[56]

Poe could not stand his birthplace. He called Boston a "Frogpondium" the denizens of whose lily pad treated writers elsewhere to "contemptuous silence" while croaking praise for each other in the "tone transcendental." He conducted a war of words with Harvard's Longfellow, whom he considered a panderer, fraud, and plagiarist. Mocking Emerson and his fellow "mystics for mysticism's sake," Poe explained that all one need do to seem elevated (today we would say evolved) is "Put in something about the

Supernal Oneness. Don't say a syllable about the Infernal Twoness. Above all, study innuendo. Hint everything—assert nothing." Whereas Poe took human terror, greed, love, obsession, and reason itself to excessive extremes in order to isolate them in an analytical quest for truths, Emerson made excess and solipsism the measures of truth. Then he peddled self-worship as philosophy to a nearly defenseless audience.

Waldo, as the seventeen-year-old Emerson took to calling himself, was one of eight children raised by a stern minister given over to Unitarianism, that "feather-bed to catch a falling Christian." When the *pater* died in 1811 the household was reduced to a genteel poverty over which ruled the sternly devout widow and a lugubrious aunt eager to die. Cold is the only word for it, as the nineteen-year-old Waldo confessed when he wrote, "I have not the kind affections of a pigeon" and "There is not in the whole wide Universe of God . . . one being to whom I am attached with warm & entire devotion." Of course, he had to study divinity at Harvard and be ordained for the ministry, but doing both caused him to suffer inexplicable seizures as if he were allergic to the cloth. He was also penniless until, in 1828, he made the acquaintance of Ellen Tucker. She was a rich merchant's daughter already dying of tuberculosis. In between pledges of his undying love, Emerson nagged her to tears over what she called "the ugly subject": her will and estate. They married in 1829. She died in 1831. Her father contested the will, but Waldo prevailed to the tune of $23,000, a small fortune in those days. He promptly claimed that conscience forbade him to continue as pastor of Boston's Second Church, and he sailed off to Europe. By the time he came home Waldo had reinvented himself as a comfortable prophet disparaging the worship of money.[57]

Emerson's status as self-reliant, self-appointed prophet of a uniquely American philosophy was at least as dubious. An insatiable reader, he knew all the ancient and modern classics and returned from Europe especially taken with Carlyle, Coleridge, Blake, Shelley, Wordsworth, Kant, Goethe, and the mystical Swedenborg. Even Emerson's "The American Scholar," a speech Oliver Wendell Holmes would hail as America's intellectual declaration of independence, was a "comprehensive raid on Romantic articulations" well established abroad.[58] One auditor thought the speech echoed the "misty, dreamy, unintelligible style of Swedenborg, Coleridge, and Carlyle. . . . I much question whether he himself would have written such an apparently incoherent and unintelligible address, had he not been familiar with the writings of the authors above named." Emerson's essays were written to be spoken aloud, like an Athenian peroration, for their effect. Glib *aperçus*—a disciple called them "dots of thought"—tumbled over each other in defiance of the rules of logic. Years later, Lowell wrote to a friend, "Emerson's oration was more disjointed than usual, even with him. It began nowhere and ended everywhere, and yet . . . it was all such stuff as stars are made of. . . . I felt something in me that cried, 'Ha, ha, to the sound of the trumpets!'" Exactly so: whether on the lyceum circuit or in print, Emerson told Americans that they were unique individuals in touch with divinity, infinite in their possibilities, laws unto themselves—and told them in language so abstruse and uplifting that consumers of culture suspended their critical faculties. Emerson pioneered the career of public intellectual.[59]

New England's Transcendentalist movement made a fetish of Kant's notion that certain basic realities transcend human experience and reason, and thus can be apprehended only through intuition. We have encountered this notion already: the preachers and pantheists of the era said the same. But devolved Unitarians embellished their personal sense of the divine with Buddhist, Hindu, and Muslim Sufi mysteries carried to Boston on merchants' ships. In 1836 they formed a Transcendental Club; and they later published *The Dial*, edited by Margaret Fuller and then by the cult's high priest, Ralph Waldo Emerson. No one gainsaid Emerson's way with words. Every day he recorded in the journal he called his savings bank everything interesting that he read, heard, thought, or felt; then he made withdrawals whenever he was called on to indulge his "passionate love for the strains of eloquence." He enriched the American lexicon with such turns of phrase as: *whosoever would be a man, must be a nonconformist; for what is man born but to be a reformer; to be great is to be misunderstood; things are in the saddle and ride mankind; genius is sacrificed to talent every day; foolish consistency is the hobgoblin of little minds;* and *hitch your wagon to a star.*

Behind the aphorisms lurked a brilliant, voracious ego blind to any truths the human race might have learned by revelation or hard experience. Emerson's first essay, "Nature" (1836), began: "Our age is retrospective. It builds the sepulchres of the fathers. It writes biographies, histories, and criticism. The foregoing generations beheld God and nature face to face; we, through their eyes. Why should not we also enjoy an original relation with the universe? Why should not we have a poetry and philosophy of insight and not of tradition, and a *religion by revelation to us*, and not the history of theirs?" He taught only two realities: the ME and the NOT ME or OTHER, which is nature. The ME was as free as Adam to access the mind of the Creator and unlock "the palace of eternity." As it stands, "man is a god in ruins . . . the dwarf of himself." But man need only gaze on the world with new eyes to answer the timeless inquiry "What is truth?" and build a "kingdom of man over nature." The following year Emerson proclaimed that "the world is nothing, the man is all; in yourself is the law of all nature . . . in yourself slumbers the whole of Reason; it is for you to know all, it is for you to dare all. . . . [The] unsearched might of man, belongs by all motives, by all prophecy, by all preparation, to the American Scholar. We have listened too long to the courtly muses of Europe." Once Americans walk on their own feet, a "nation of men will for the first time exist, because each believes himself inspired by the Divine Soul which also inspires all men." In his "Divinity School Address" (1838) Emerson came right out and pronounced that man's religious sentiment made him divine, beatified, illimitable. Jesus gave us holy thoughts, but to suggest he had to die for humanity's sins degraded all parties. The very word "miracle" is a "monster." Away with preachers, pulpits, churches! Away with a historical Christianity that obscures "the moral nature of man, where the sublime is." Preachers claimed that the truth gave life, but mankind is called "to convert life into truth." What! Shall we found a new cult with new rites and forms? Not at all. "Faith makes us, and not we it, and faith makes its own forms." And so on and so on in essays such as "The Over-Soul," "Spiritual Laws," and "Circles." Emerson taught the postmodern creed that truth is accessible to all through the divine over-soul, the assurance of truth being simply one's feeling of certainty. "We know truth when we see it, from opinion, as we know when we are awake that we are awake." Hence the only sin man can commit is to limit himself through a failure to realize that he is "masterless." No wonder Friedrich Nietzsche, who conceived of a superman "beyond good and evil," carried a volume of Emerson wherever he went. It was Emerson who warned in "Circles": "Beware when the great God lets loose a thinker on this planet. Then all things are at risk."[60]

In fact, Emerson put nothing at risk that his disciples had not already discarded; and his broader influence was so feeble that his most representative disciple was the bathetic David Henry Thoreau (he transposed his given names). When Emerson settled in Concord, he invited Thoreau—a timid, tubercular, Harvard-trained teacher—to join his household and pursue a literary career. Thoreau made a minor splash in the magazine trade, thanks to assistance from Horace Greeley, but won lasting fame by camping out at Walden Pond in 1845–1847. Thoreau's self-reliance was less than heroic. He went into town almost every day, sponged off friends, and hosted regular picnics at his cabin. Mountain men such as Jim Bridger would have guffawed at the pretense. But Thoreau's little book *Walden, or Life in the Woods*, tugged New Englanders' heartstrings: "I went to the woods because I wished to live deliberately, to front only the essential facts of life, and see if I could not learn what it had to teach, and not, when I came to die, to discover that I had not lived." Thoreau in fact experienced little of life. He never married. He never traveled save for brief junkets to Quebec and Minnesota. His greatest adventure was spending one night in jail for refusing to pay his poll tax. Even his signature essay "Civil Disobedience" was ignored until after his death.[61]

Thoreau was standoffish and timid. Was he talking about himself in *Walden* when he (brilliantly) observed, "The mass of men lead lives of quiet desperation"? Caroline Sturgis Tappan likened him to a porcupine. Emerson thought him fit only to lead a huckleberrying party. But no one was chillier than the Sage of Concord himself. Emerson did remarry and had four children, but confessed that this marriage was (another) practical match devoid of romance. His few friendships—for instance those with Amos Bronson Alcott (Louisa May's father)—were purchased with loans and patronage. George Santayana considered his Transcendental method peculiarly sympathetic to the American mind. "It embodied, in a radical form, the spirit of Protestantism as distinguished from its inherited doctrines; it was autonomous, undismayed, calmly revolutionary; it felt that Will was deeper than Intellect; it focused everything on the here and now. . . . These things are truly American . . . and they are strikingly exemplified in the thought and person of Emerson."[62] No doubt all that is so. But it is also hard to resist concluding that Emerson's Transcendentalism was a manic effort to transcend himself because he was never comfortable in his own skin.

What a relief to move on to an author of real fiction who told real stories rooted in history; an American who admitted his "imagination was a tarnished mirror" and had real doubts about his countrymen's headlong flight into the future. Nathaniel Hathorne (he added the *w* when he was in his twenties) was born in 1804 in Salem, Massachusetts. Sure enough, one of his Puritan forebears adjudicated the Salem witch trials. He had the usual childhood tragedies, including the death of his father when he was four and a serious athletic injury when he was nine. But thanks to the money and love of a large extended family, the boy grew up happy with plenty of chances to indulge his love for books and the outdoors alike. He graduated from Bowdoin College in Maine in 1825, having made lifelong friends of Henry Wadsworth Longfellow and Franklin Pierce. He had also made a decision. He would follow the pioneering steps of James Fenimore Cooper and try to earn a living as an author. For twelve years Hawthorne was "the obscurest man of letters in America," but between dancing and card games he redeemed the time poring over New England's history and literature. *Twice-Told Tales* (1837) made his reputation, thanks in part to a rave notice in the *North American Review* by Longfellow. But the royalties did not suffice, so Hawthorne tapped his Democratic Party connections (Pierce was already a U.S. senator) to win a post in the Boston customhouse. That allowed him to marry his "dove," the invalid artist Sophia Peabody, and settle in Concord. Needless to say, he fell in with the Transcendentalists, losing $1,000 in their utopian venture at Brook Farm, but gaining the inspiration to write such denunciations of self-obsession as "Rappaccini's Daughter" and "Egotism; or, the Bosom-Serpent." Ousted by his landlord in 1845, Hawthorne again tapped his connections for a job in the Salem customhouse. There he reconnected with his Puritan heritage. So when a Whig victory left him unemployed in 1849, he moved to Lenox ready to write *The Scarlet Letter* (1850), *The House of the Seven Gables* (1851), and *The Blithedale Romance* (1852), an exposé of the "phantasmagorical antics" at Brook Farm. In 1852 he wrote the presidential campaign biography for Pierce; he was rewarded with four exciting years as U.S. consul in Liverpool.[63]

Emerson had contempt for the past. Hawthorne swam in it. Emerson imagined the human soul as a sun radiating power and light. Hawthorne sensed the "dark side of the force" as keenly as Cotton Mather and John Bunyan. Emerson buried a crystalline gospel in obscure, prolix prose. Hawthorne excavated hellish recesses in transparent prose. Just roll this (from *The Blithedale Romance*) over the tongue: "Bewitching to my fancy are all those nooks and crannies, where Nature, like a stray partridge, hides her head among the long-established haunts of men. . . . [T]here is far more of the picturesque, more truth to native and characteristic tendencies, and vastly greater suggestiveness, in the back view of a residence, whether in town or country, than in its front. The latter is always artificial; it is meant for the world's eye, and it is therefore a veil and a concealment. Realities keep in the rear, and put forward an advance guard of show and humbug." Pride and its symbols were Hawthorne's fixation: pride born of status, power, wealth, moralism, and not least intellect. He felt such pride, or its temptation, in himself; that is why one finds in Hawthorne a humility rare among his New England contemporaries. He shared the era's faith in democracy and the individual—indeed, he was far more of a patriot than Emerson and Thoreau—but could not bring himself to believe Americans were somehow released from the human condition. He rejected the Puritan theology even as they did, but could not deny the unwelcome truths that made the Puritans anxious.[64]

The Scarlet Letter is rightly Hawthorne's most famous work (though he preferred *The House of the Seven Gables*). Even people who have never read it or have only skimmed it in high school know that it paints a scathing portrait of the authoritarian, guilt-ridden Calvinist culture. But not only was the real message relevant to his own culture; it also anticipated the twentieth century's obsession with sex and sublimation. The earthy Hester Prynne must wear a red letter A, not only because the child Pearl (of great price?) conceived in her husband's absence proves adultery, but also because she refuses to name the father. Her lover, the eloquent pastor Arthur Dimmesdale, eats out his soul with guilt. Her husband, Roger Chillingworth, eats out his with a lust for vengeance. Nor was Hawthorne just purging himself of the burden of his ancestry by implying that Calvinism, not sin, was the cause of their troubles. *Twice-Told Tales* and *Mosses from an Old Manse* are replete with themes pricking his own society. "The Great Carbuncle" (1837) tells of a legendary gem of enormous brilliance and size hidden away in the White Mountains (the sort of thing Joseph Smith would divine and dig for). Obsessed adventurers have thrown their lives away in its vain pursuit, never even sure if it really exists. The Seeker discovers the carbuncle only to perish in the effort to reach it. The Cynic cannot see it through his tinted spectacles, and is cursed to spend the rest of his life wandering in search of light. The newlyweds Matthew and Hannah come to their senses, turn their backs on the gem, and walk away bathed in the light of their love, not for riches or nature, but for each other.

"The Celestial Railroad" (1843), a hilarious allegory published by the *Democratic Review*, made it obvious that Hawthorne was damning his own time from the standpoint of the seventeenth century rather than damning the Puritans from the standpoint of the nineteenth century. The narrator dreams an up-to-date version of *A Pilgrim's Progress* wherein the souls bound for the Celestial City are provided with every convenience. Railroads and steamboats can speed them to heaven. Luxury hotels await them in the towns. But Mr. Take-it-Easy, his guide, has no desire to reach a destination where there is "No business doing, no fun going on, nothing to drink, no smoking allowed, and a thrumming of church music from morning till night. I would not stay in such a place if they offered me house room and living free." The greatest pitfall is the town of Vanity Fair. Persecution of Christians has ceased; indeed, clergymen such as the Reverend Shallow-Deep and Dr. Wind-of-Doctrine preside over the pleasure dome in league with its capitalist stockholders. None of its pilgrims complete their journeys; they are said to sell off large estates in the Celestial City in order to lease small tenements in Vanity Fair.

No humor disturbed the nightmarish mood of Hawthorne's masterpiece "Young Goodman Brown" (1835). We are back in colonial Massachusetts to observe an upright young husband deceive his wife, Faith, in order to sneak off into the woods after dark. He is bound for some rite of passage administered by the pastors. He arrives late ("Faith kept me back a while") to find the whole town gathered for a black mass during which he—and Faith—are to be initiated into the coven. "Welcome my children" says the dark leader, "to the communion of your race. . . . There are all whom ye have reverenced since youth. Ye deeded them holier than yourselves, and shrank from your own sin. . . . This night it shall be granted you to know their secret deeds: how the hoary-bearded elders of the church have whispered wanton words to the young maids of the households; how many a woman, eager for widow's weeds, has given her husband a drink at bedtime and let him sleep his last sleep in her bosom; how beardless youths have made haste to inherit their father's wealth; and how fair little damsels—blush not, sweet ones—have dug little graves in the garden, and bidden me, the sole guest, to an infant's funeral." Perhaps it was all a nightmare, but Goodman Brown could never rest easy, or lift up his heart to a psalm, for the rest of his life.[65] But the Romantics were always excessive! One must not jump to the conclusion that Hawthorne thought human nature as utterly debased as Emerson thought it sublime. Hawthorne just took everything to an extreme in order to make a point. Human beings all suffer temptation; all sin; all lie about it. So they had better temper their self-congratulations and utopian ambitions with some candor, truth, and love.

While residing in Lenox Hawthorne met and befriended Herman Melville, another young Yankee struggling with fiction, Puritanism, and finances. Melville's paternal grandfather was a veteran of the Boston Tea Party, his maternal grandfather a Dutch patroon who served as a general in Washington's army, and his father a prosperous importer in New York. Herman, the second of eight children (all of whom lived to adulthood, beating the odds), seemed destined for a good, conventional education and career until 1827, when his father was swindled by a confidence man. The family fled to Albany, and then the children were orphaned by the death of their father in 1832. The boy Herman worked as a clerk, farmhand, and teacher until 1839, when he decided to go to sea on a Liverpool packet (the voyage later recounted in *Redburn* in 1849). Upon his return he briefly roamed in Illinois, then signed on to the whaler *Acushnet* bound in 1841 for the South Seas.[66]

Melville returned after three and a half years of rollicking adventures. He deserted the whaler in the Marquesas Islands to escape its brutal skipper, took nervous refuge with the allegedly cannibalistic Typees, signed with an Australian whaler, and jumped ship again in Tahiti. Another American whaler carried him to Honolulu, where he worked as a clerk and pinsetter in a bowling alley before enlisting in the U.S. Navy just to get home. The first literary fruit was *Typee: A Peep at Polynesian Life*, initially published in England because Harper Brothers refused to credit Melville's "anxious desire to speak the unvarnished truth." *Typee* was both a success and a scandal because its praise for the taboo-ridden culture of tattooed Polynesians broke so many Yankee taboos. The sailors drop anchor off the island of Nukuheva to be greeted by a "picturesque band of sylphs" dressed only in flowers and eager to satisfy "the unholy passions of the crew." Melville takes care to deplore this "grossest licentiousness and most shameful inebriation." But not only does he relate serial orgies in *Typee* and its sequel *Omoo*; he also savages missionaries intent on destroying the unaffected good nature of Polynesians (they have no *money*!) and saddling them with guilt. Contrasting the so-called savage with the civilized man, he asks himself: "insensible as he is to a thousand wants, and removed from harassing cares, may not the savage be the happier of the two?"[67]

Melville's reviews were decidedly mixed. He was a wonderful spinner of yarns, mixing action with descriptive color and tongue-in-cheek humor. But what a terrible influence on the morals of young Americans! Wanton sex was shocking enough; sex with colored savages was intolerable. Greeley called the books morally diseased. George Washington Peck called Melville a "sharp scamp" of a Yankee whose saturnalias could not be true. First, no manly mind would boast of sexual conquests; only a fraud would do so. Second, absent an "Epicurean elixir" it was physically impossible to perform the sexual athletics described. Third, everyone knew that Polynesian females were not half so attractive as the book suggested. In fact, the stories were only slightly embellished. In fact, Melville was abashed by "the sinful propensities of his nature" and "had seen enough to doubt the chances that reason, that celestial power, could conquer sexual temptation."[68] In 1847 Melville married and set up house in New York with his bride, mother, sisters, brother, and sister-in-law. Three more sea stories followed with no prurient content at all. *Mardi* was a tale of heroic love, *White Jacket* an indictment of navy life, and *Redburn* a coming-of-age reminiscence of Melville's voyage to England. They sank like stones.

Then Melville met Hawthorne, devoured his stories, and lived with him for a time in the placid Berkshires. In Hawthorne, Melville discovered a mentor who pried for "that which is beneath the seeming" and bade him engage in "ontological heroics." Melville determined to pour all his experience and talent into a mighty book that told truth, as he saw it, about life and the psychology of his countrymen. He already had in mind its "hook": the familiar South Sea legend about a great alabaster whale named Mocha Dick. He worked furiously, not least because he was desperate for money, completing the whale-sized manuscript by fall 1851. Melville, who was then just thirty-three, dedicated it to Hawthorne, assuring his friend that it was "broiled in hellfire." The book destined to become a classic of world literature received generally favorable reviews, though no one quite knew what to make of a tale more complex than a brig's rigging. Clearly, Captain Ahab's obsessive hunt for the elusive white whale that had bitten off his leg doomed not only himself but the *Pequod* and its piebald crew of sinners, saints, pagans, and salty philosophers. Beyond that the meaning was any-one's guess, and still is. But so pervasive were Melville's allusions to current American traits and trends that it is hard not to think Ahab is Emerson's "representative man" playing God; chasing a millenarian utopia with all the ruthless, conquering passion of Andrew Jackson; enlisting the manifold virtues and credulity of the crew in his mad quest; and taking everyone down with him.[69]

F. O. Matthiessen summed it up with his usual eloquence. "Melville did not achieve in *Moby-Dick* a *Paradise Lost* or a *Faust*. The search for the meaning of life that could be symbolized through the struggle between Ahab and the White Whale was neither so lucid nor so universal. But he did apprehend therein the tragedy of extreme individualism, the disasters of the selfish will, the agony of a spirit so walled within itself that it seemed cut off from any possibility of salvation." Americans did not want to contemplate that. They had far too much on their plate in 1851, too many worlds to conquer, too many dreams to fulfill. *Moby-Dick* sold about 2,000 copies, many of which no doubt went unread. Melville's literary career never recovered. Since *Typee* he had fallen into the sin of "revolting against the reader." Lewis Mumford understood what happened. "It is hard to refute Melville's black words," he wrote, "difficult to find an anti-dote for this spiritual nightshade. Melville's contemporaries did not try. They applied to the book the same medicine that worked so well in life: they agreed to forget it. Most of the sweetness and decorum of society rests on an agreement to forget it."[70]

McCune Smith, that African-American doctor who dreamed of a race-blind society, wrote a long, admiring review of *Moby-Dick*. He had no difficulty decoding the allegory. The *Pequod* was the American ship of state, hell-bent in vain pursuit of whiteness. Just so. For the burst of great literature c. 1850 proved *not* to be a renaissance or a flowering. It was the last glow of a sunset draping long shadows over the republic declared in 1776.[71]

Chapter 5

CONQUISTADORS
The Glory and Fraud of Manifest Destiny, 1841–1848

*T*he American people, having derived their origins from many other nations, and the Declaration of National Independence being entirely based on the great principle of human equality, these facts demonstrate at once our disconnected position as regards any other nation; that we have, in reality, but little connection with the past history of any of them, and still less with all antiquity, its glories, or its crimes. On the contrary, our national birth was the beginning of a new history . . . and so far as regards the entire development of the natural rights of man, in moral, political, and national life, we may confidently assume that our country is destined to be *the great nation* of futurity." Thus did John Louis O'Sullivan, editor of the *United States Magazine and Democratic Review*, brush aside all doubts about the character and career of the American nation. Preachers, reformers, and authors like Hawthorne (whom O'Sullivan published) might wring their hands over sin and warn Americans they must become morally worthy of their material destiny. But O'Sullivan, a staunch New York Democrat, assured readers that their national greatness was not only certain, but imminent: "We are entering on its untrodden space, with the truths of God in our mind, beneficent objects in our hearts, and with a clear conscience unsullied by the past. We are the nation of human progress, and who will, what can, set limits to our onward march? Providence is with us. . . . In its magnificent domain of space and time, the nation of many nations is *destined* to *manifest* to mankind the excellence of divine principles; to establish on earth the noblest temple ever dedicated to the worship of the Most High—the Sacred and the True. Its floor shall be a hemisphere—its roof the firmament of the star-studded heavens, and its congregation an Union of many Republics, comprising hundreds of happy millions, calling, owning, no man master, but governed by God's natural and moral law of equality, the law of brotherhood—of 'peace and good will amongst men.'" [1]

Six years later O'Sullivan distilled "destined to manifest" into the oxymoron "manifest destiny," the catchphrase historians ever since have applied to the Romantic mood of American expansion to the Pacific during the 1840s. O'Sullivan believed American expansion would be natural and peaceful, in contrast to the bloody conquests that marred ancient and European history. Pioneers would simply colonize the frontier, prosper, and build such a glistening temple of liberty that even foreigners would beg for admittance. Such, at least, was his pretense. In fact, O'Sullivan admitted the Texans won their independence in a war, supported (with some misgivings) his own party's belligerent policies, and later sponsored an armed invasion of Cuba. The very context of his ringing phrase of 1845 warned about European interference "for the avowed object of thwarting our policy and hampering our power, limiting our greatness and checking the fulfilment of our *manifest destiny* to overspread the continent allotted by Providence." But O'Sullivan boasted that all the "bayonets and cannon" of Europe could not prevent Americans from becoming a continental nation of hundreds of millions. [2]

This was hardly a new idea. Tom Paine had predicted the same destiny in 1776. Jefferson imagined "our rapid multiplication" would "cover the whole northern if not southern continent, with people speaking the same language, governed by similar forms, and by similar laws." John Quincy Adams, whose treaty of 1819 with Spain forswore Texas, nonetheless held that "North America appears to be destined by Divine Providence to be peopled by one nation. . . . For the common happiness of them all, for their peace and prosperity, I believe it is indispensable that they should be associated in one federal Union." Andrew Jackson craved Texas because it would unlock the doors to the Pacific. Nor could any foreign power stand in the way, as the young Whig from Illinois, Abraham Lincoln, crowed in 1836: "All the armies of Europe, Asia, and Africa combined, with all the treasure on earth (our own excepted) in their military chest; with a Buonaparte for a commander, could not by force take a drink from the Ohio or make a track on the Blue Ridge, in a trial of a thousand years." Manifest Destiny was no spasm. It was explicit in the original spirits of English colonization, including rural capitalism and the improvement ethic, the Protestant crusade against Catholic empires and monopolies, and the displacement of indigent peoples who made no good use of their patrimony. It was explicit, too, in Americans' contempt for any authority presuming to hobble their liberty. Just imagine Indian tribes, Mexican juntas, British lords, or for that matter U.S. federal marshals telling Americans: "No, you cannot settle here or do business there—go back where you came from!" Finally, expansionism was implicit in the strategic principles Americans had already canonized: liberty and independence at home (Declaration of Independence), a unilateral foreign policy (Washington's Farewell Address), and a New World off-limits to imperialism (Monroe Doctrine). To preserve opportunity for future generations, prevent a balance-of-power system in North America, and preempt European efforts to claim its vacant regions, the United States itself must expand coast to coast.[3]

Nor did Americans need O'Sullivan to justify their hunger for territory. Ever since the Louisiana Purchase, expansionists had argued that the United States enjoyed eminent domain by dint of natural law; geography; demography; the excellence of their institutions; their energetic development, republican virtue, federated government, and liberty; and the blessing of Providence and their millenarian mission to regenerate the human race by their example. How about blatant greed dipped in racial arrogance? Did not Americans construct a racial paradigm to justify seizure of land from Indians and Mexicans? Of course! But the paradigm was hardly a matter of controversy, given the overwhelming evidence for the superiority of Anglo-Saxon civilization in law, government, agriculture, commerce, science, technology, art, religion, and every other garden of culture. Anyone disputing that evidence in the 1840s would have been thought a lunatic. What politicians like Cass, historians like Parkman, and phrenologists like Samuel George Morton were anxious to know was why Indians and Mexicans languished like dry bones in Ezekiel's valley, and whether life could be breathed into them.[4]

If expansionist ambitions and arguments were as old as the republic, why did the mania of Manifest Destiny sweep the nation in the 1840s rather than before or after? One answer was the near disappearance of doubt concerning the possibility of a continental republic. Even the expansionist Thomas Hart Benton of Missouri had once thought the Rocky Mountains the abode of "the fabled God, Terminus," and some New Englanders thought even the Great Plains too remote to govern. The westward march of constitutional government, supported by steamboats, railroads, and telegraphs, dissolved such fears by the 1840s. A second answer was that American trappers, farmers, ranchers, planters, loggers, merchants, whalers, and missionaries were already infiltrating the Great North Woods, Great Plains, Rockies, and Pacific Rim. By 1840 the federal government had little choice but to extend its protection over them. A third answer, the one that caused Manifest Destiny to melt into "manifest design," was a fear that Britain and France were colluding in a sort of "containment policy" against the upstart Yankees. This made expansion a matter of urgency. A fourth answer was simply that Democrats beat the expansionist drum as an electoral ploy in 1844. Needless to say, Jacksonians preferred the second answer, since it fit their laissez-faire ideology and their myth of the rugged pioneer. Thus, Benton wrote, "It was not an act of government leading the people and protecting them, but like all the other great emigrations and settlements of that [Anglo-Saxon] race on our continent, it was the act of the people going forward without government aid or countenance, establishing their possession and compelling the government to follow with its shield and spread it over them." Benton was candid only up to a point. It was true the U.S. government did not recruit colonists or sponsor mercantilist chartered companies. But Benton himself lobbied tirelessly for federal expeditions to explore and publicize new frontiers, not least those led by his son-in-law Captain John Charles Frémont. Maritime-minded Congressmen likewise pressed for support of overseas commerce. This suggests a fifth, lesser-known, source of the expansionist mania in the 1840s: the contributions made, on threadbare budgets, by the highly professional U.S. Army and U.S. Navy.[5]

The American republic, in sharp contrast to the French republic founded in 1792, embodied a deep antipathy to the military profession. To be sure,

Americans apotheosized George Washington, waxed bellicose whenever Indians or foreigners appeared to check their ambitions, and rewarded battlefield heroes with office. But republican ideology, formed in resistance to King George's redcoats, declared standing armies a threat to popular liberties and a superfluous expense, given the United States' favorable geopolitics. Not least, military virtues such as hierarchy, discipline, obedience, and self-denial could not have clashed more with the democratic, individualistic, pioneering, go-getter ethics of Jeffersonians and Jacksonians. Hence anyone who voluntarily chose the profession of arms was an object of suspicion, and Americans made civilian control of the military a hallowed tenet of their civic religion. For most of the nineteenth century a majority of voters and their representatives held that except in wartime the wise policy was to starve the armed services and rely on militias commanded by amateur colonels. Flushed with pride after fighting the British Empire to a draw in the War of 1812, Americans remembered the victories won by militia forces at Lundy's Lane and New Orleans, while forgetting their frequent humiliations and all but ignoring the decisive contributions made by sixty-five graduates of West Point. If not for a few sober heads, the United States might have stumbled into future crises with no army to speak of at all.

The first sober head was that of John C. Calhoun, in his first incarnation as a nationalist. After the Treaty of Ghent in 1815 the U.S. Army was cut to just 12,000 officers and enlisted men. Following the Panic of 1819 a penurious Congress slashed it again, to 6,000, in the belief that just a few soldiers were needed to overawe mischievous Indians. But Secretary of War Calhoun's study of history taught him that the potential for another war with a European power was ever-present. Accordingly, in December 1820 he submitted to Congress one of the most important state papers of the era. War, he instructed, was "an art, to attain perfection in which, much time and experience, particularly for officers, are necessary." Hence, even as the Army performed peacetime roles, it must maintain an officer corps sufficiently skilled to go to war. His solution, pioneered by the Prussian general staff after 1806, was a cadre system in which a disproportionate number of officers and sergeants would be retained on active duty in skeleton regiments that could quickly be fleshed out in case of emergency. The professional army could then serve as a "faculty" ready at any time to train raw recruits. Calhoun also centralized the Army's departments in Washington City under chiefs of staff reporting directly to the secretary of war. In time ten such departments emerged: Adjutant General, Inspector General, Commissary, Medical, Pay, Quartermaster, Subsistence, Ordnance, Engineers, and Topographical Engineers. None paid dividends faster than the Ordnance Department's federal arsenals at Harpers Ferry and Springfield. Their mass manufacture of weapons with interchangeable parts, the so-called American system, endowed the U.S. armed forces with more potential firepower than those of any other nation on earth.[6]

Another sober thinker, Winfield Scott, was the United States' foremost professional soldier of the era. The son of a Virginia planter and son-in-law of one of Richmond's wealthiest men, Scott made no effort to cloak his aristocratic pretensions. He was vain, ambitious, and a stickler for rank, rules, and spit and polish—hence his nickname "Old Fuss and Feathers." Scott's political career predictably misfired in the age of the common man. But no one did more to mold the American officer. As a colonel of artillery in 1812 Scott was forced to surrender at Niagara Falls when the New York militia across the river refused to advance. Scott disdained militias for the rest of his life. Back in action in 1813, he studied a Napoleonic manual and played drill sergeant to a whole regiment. Its coolness in battle promptly won the legendary praise of a British commander: "Those are regulars, by God!" In the ensuing decades Scott was repeatedly passed over for promotion, in part because of his irascibility and his Whig politics. Still, presidents invariably summoned Scott to handle such thankless tasks as Cherokee removal, the Black Hawk War, the Seminole War, and violent disputes on the Canadian border. In 1841 Scott at last was honored with appointment as commanding general. But his greatest peacetime contribution was to realize Calhoun's vision of a professional officer corps. Scott pored over all the orders issued since Washington's time on tactics, ordnance, discipline, and supply. Then he studied European manuals and armies, and adapted their procedures to American sensibilities. In 1821, after two years of unremitting labor on the "*cursed* book," he completed the nation's first *General Regulations for the Army*.[7]

What about West Point? Wasn't the U.S. Military Academy founded in 1802 to train a professional, skilled officer corps? Not exactly. President Jefferson meant to promote engineering, but he was no friend of professional armies. West Point began life underfunded, understaffed, and confused. After 1810 it nearly disintegrated on the watch of its acting commandant Alden Partridge, an irascible mathematician from Vermont who did all he could to prevent the academy from becoming an elite institution. Academic and military standards became lax. Disillusioned cadets resigned; enrollment dropped. Had the situation persisted just a few more years, antimilitarism in Congress might have shut down the academy. But in 1817 Partridge was ousted in favor of Brevet Major Sylvanus Thayer, who was a veteran engineer, a graduate of Dartmouth and West Point, and a student of French military science. Thayer presided like an abbot (he never married) sent by a bishop to recall unruly monks to their discipline. His purge of Partridge's faculty, design of a rigorous curriculum, and insistence on military bearing all but reinvented West Point. It soon acquired a reputation as a nursery for officers and gentlemen devoted to duty, honor, and country. Its admission and graduation rates soared, so that by 1830 the academy provided over 60 percent of the Army's officers. Thayer also inspired the War Department to open postgraduate artillery and infantry schools in the 1820s. Such professionalism offended Jacksonian Democrats, who closed down the advanced schools and attacked West Point as the military equivalent of Biddle's Second Bank of the United States. Why should a democracy support an elitist school? Why should the taxpayer educate other men's sons so that they could spend their lives on the public payroll or else make fat salaries as engineers and railroad executives?[8] Partridge, the disgraced commandant, even fed Congressmen allegedly damning evidence in pamphlets such as *West Point Unmasked*.

In retrospect there was little danger the academy on the Hudson River would be abolished. Jackson thought it "the best school in the world"; Van Buren was a New Yorker. But Jackson subverted West Point admissions for patronage, ordered the reinstatement of expelled cadets, and discriminated against career officers in favor of Democrats commissioned from civilian life. All this prompted Thayer to resign in protest in 1833, drove more West Point graduates into the private sector, and caused the 58 percent of officers who served out a twenty-year Army career to form a beleaguered subculture. They lamented the low pay, slow promotion, and lack of appreciation. Their wives hated the frequent transfers and life in rustic forts. Their only psychic defense was to imagine themselves a sort of monastic order devoted to the service of others. One captain, asked to explain his "insane attachment" to the army, replied: "I verily believe our little army contains a better body of men than can be found in any other profession." That sentiment was especially strong on the frontier,

where the Army's struggle to maintain law and order was undercut by
greasy civilians who cheated Indians, sold them liquor and guns, and then
came crying for military rescue when violence erupted. Perhaps tension
between civilians and the military was inevitable in a frontier democracy.
But it grew considerably in the Jacksonian era.[9]

Finally, the Army suffered from the officers' contempt for enlisted
men. The simple reason was that very few American men of good character could be induced to enlist. Why subject oneself to five years of harsh
discipline, poor grub, and danger for lower wages than a mill worker or
farmhand earned? Officers took for granted that any homegrown recruit
must be a wastrel, drunk, or fugitive, and thought even less of the 40 percent of the ranks filled by "paddies" and "sauerkrauts." Drunkenness was
rife among soldiers. Desertion rates reached 27 percent. Regiments were
consistently understrength. In 1833 Secretary of War Cass and Congress
thought to fix the problem by reducing enlistments to three years and raising a private's pay from five to six dollars per month, while repealing the
Army's 1812 ban against flogging in hopes of instilling discipline. Whatever marginal effects those measures had, however, were erased by the
demoralization caused by the Second Seminole War. By 1839 more than
4,000 soldiers were slogging through Florida, while another 1,000 guarded
the troubled border with Canada. That left just 1,000 troops to patrol
1,500 miles of frontier beyond the Mississippi, not to mention the trails
blazed through the Great Plains. Yet the Army managed, as one foreign observer colorfully wrote, because those West Point graduates were scattered
over the land "like productive corn."

Thomas Hart Benton, acknowledged leader of the western Congressional bloc, pleaded with his fellow Jacksonians to overcome their antipathy
to the military in the interest of pioneers. He lobbied for a string of forts
from Minnesota to Oklahoma, exploration of the Great Plains, and patrols to shelter the Sante Fe traders. His most important achievement was
the "Act for the more perfect defense of the frontiers" in 1833. It created the
U.S. Army's first regiment of light cavalry, known as dragoons, under the
command of colonels Henry Dodge and Stephen Watts Kearny. Unlike
infantry regiments, the dragoons recruited only robust native-born
horsemen aged twenty to thirty-five "whose imaginations inflamed them
with the thoughts of scouring far prairies on fine horses amid buffalo and

strange Indians." Their first sortie, a reconnaissance and peace mission to the Kiowa and the Comanche in Oklahoma, ended in tragedy. Temperatures there in July 1834 reached 105 degrees. Horses and more than 100 men perished, including General Henry Leavenworth. But over the next ten years Dodge, Kearny, and their dragoons explored, mapped, pacified, and planted forts from the Red, Canadian, and Cimarron rivers in the south to the Des Moines and Missouri rivers in the north. Kearny's climactic expedition of 1845 visited the Sioux along the Oregon Trail, crossed the Continental Divide, turned south through Colorado, and returned to Fort Leavenworth (Kansas City) via the Santa Fe Trail. They covered 2,200 miles in ninety-nine days.

Nothing impressed the Plains Indians with the reach of the U.S. government more than these dragoons, who were fast, savvy, well-armed, diplomatic, and trustworthy. But frontier deployments were still far too small. In 1838 Benton and Senator Archibald Yell of Arkansas proposed expanding the Army to 14,000. Their Democratic colleagues shouted them down. Secretary of War Cass persuaded Congress to authorize $100,000 for a military road connecting frontier garrisons from Fort Snelling in Minnesota to Fort Towson in Oklahoma. Van Buren's secretary of war, Joel Poinsett, dragged his feet. He believed (rightly) that communication lines running along the Indian frontier could be easily disrupted. The best way to move troops quickly to trouble spots was to establish a base in the rear and build roads fanning out to the frontier. The obvious base was Jefferson Barracks near Saint Louis. General Edmund P. Gaines went even farther. His elaborate plan of 1838 called for eleven frontier forts connected to the Mississippi River by railroads. That got nowhere in the penurious, antimilitary Congress. So in the end soldiers on the frontier were left to improvise as best they could in an effort to shelter pioneers, broker peace among Indians, and not least draft the maps both duties required.[10]

Maps were the job of the Army's most elite branch, the Corps of Topographical Engineers, founded in 1838 by Colonel John James Abert. Over two decades its seventy-two cartographers (sixty-four of them West Point graduates) roamed the West as self-conscious agents of Manifest Destiny. They did more than anyone else to make the frontier real and accessible to the American people. They had precursors, such as Lewis and Clark, Zebulon Pike, and Major B. L. E. Bonneville. But their early maps were half guesswork and left great mysteries unsolved, such as whether the Great Basin was really a basin or drained into the Pacific via the legendary "Buenaventura River." The topographical engineers were professional scientists armed with theodolites, Zenith telescopes to measure latitude, and barometers to discern altitude. They were also trained by such artists as Robert Walter Weir and Rembrandt Peale to be keen sketchers of landscape. By the mid-1850s the Corps had accumulated enough knowledge to make possible a complete, accurate chart of the western United States by Lieutenant Gouverneur Kemble Warren, a dashing twenty-five-year-old who was second in his class at West Point. Far and away the most influential topographical engineer, however, was John Charles "the Pathfinder" Frémont.[11]

The love child of a Virginian woman and the French expatriate Jean Charles Fremon, young John grew up in Charleston unsure of his niche in society but determined to make it American and Romantic. Having added a t and an *accent aigu* (acute accent) to his surname, he found a political patron. Joel Poinsett arranged for Frémont to serve as math instructor to midshipmen on a two-year voyage to South America, then serve a stint as a railway surveyor. In 1838, when Frémont was twenty-five, Poinsett commissioned him as a topographical engineer so that he could accompany the French scientist Joseph Nicollet—Frémont spoke French fluently—on an expedition to the Great Plains. Nicollet's tutelage, Frémont wrote later, was his "Yale and Harvard." While billeted in Washington to write the report of the expedition, Frémont fell under Senator Benton's spell. He came away from their first interview intoxicated with the role he might play in the nation's destiny: "the opening up of unknown lands; the making unknown countries known; and the study without books—the learning at first hand from nature herself; the drinking first at her unknown springs—became a source of never-ending delight for me." The dashing Lieutenant Frémont with the aquiline nose and the cavalry whiskers was also intoxicated by Jessie, Benton's brilliant, beautiful, savvy daughter. When Frémont was sent to survey the Des Moines River in 1841 he suspected Mrs. Benton had arranged it to get him out of town. If so, it availed not, for on his return John and Jessie eloped. The senator, perhaps recalling his own tempestuous youth, blessed the union. Besides, he suspected this lad might be just the lever he needed to impel the federal government into realizing his dream of empire in the West.[12]

* * *

Benton's dream dated from 1821, the year Missouri became a state and Mexico won its independence from Spain. Everyone knew about Zebulon Pike's bleak description, in 1810, of the western plains as a rolling ocean of sand thrown up by merciless winds. It helped to compound the myth of a great American desert fit only for nomadic Indians. But William Becknell, a shrewd frontiersman and a veteran of the War of 1812, was more interested in Pike's offhand remarks about prices in Sante Fe. Flour, salt, cloth, and manufactured goods sold at a substantial premium there. So when news arrived that Mexico might open its borders to trade, Becknell advertised for partners to join a lucrative expedition. He did not mention the destination. His wagon train crossed the Missouri River on September 1, 1821, and in just six weeks traversed the Kansas plains, surmounted the 7,800-foot Raton Pass, and descended to Sante Fe. That remote pueblo on the upper Rio Grande del Norte served as capital city and market for Nuevo Mexico's 60,000 hidalgos, mestizos, and Native Americans. Becknell's twenty men were so smitten by the picturesque town and its inhabitants' "grateful evidence of civility and welcome" that all but two chose to remain. Becknell trekked back to Missouri, where (legend has it) he answered queries about his success by pouring Spanish dollars onto a boardwalk. His profits approached 2,000 percent. The Sante Fe Trail, and U.S. penetration of the Far West, might be dated to that day.

Every spring for twenty years, 100 to 300 traders assembled wagons, inventories, and sturdy Missouri oxen and mules for the annual journey along the Cimarron River to New Mexico. They demonstrated, for all pioneers to come, how to maneuver Conestoga wagons over prairies, rivers, and mountains; travel in convoys under the absolute rule of a wagon master; circle the wagons against menacing Comanche and Pawnee, and enlist public support for their private enterprise. As early as 1825 Benton lobbied for U.S. commissioners to survey the trail. In 1829, 1834, and 1843, he persuaded the U.S. Army to provide armed escorts. When eastern capitalists horned in on the trade, the volume of goods thrown onto the Sante Fe market drove prices down. But frontier entrepreneurs found new markets in Chihuahua or among Native Americans. This was the business that interested Charles Bent, son of a judge in Saint Louis. He teamed up with his brother and a clerk named Ceran St. Vrain to trade with Indians on the

Arkansas River. A rival firm built Fort Cass at the present site of Pueblo, Colorado. But the Bents trumped all competition in 1835 when they finished a massive adobe castle farther east on the river to draw traffic from the Santa Fe Trail. Hundreds of pioneers, mountain men, and merchants traversing the plains knew they were near safety and comfort when the gigantic American flag atop Bent's Fort appeared like a sail on the horizon at sea. The Bents and St. Vrain made hefty profits selling livestock, tools, household goods, weapons, tobacco, and coffee to traders, mountain men, and pioneers. But they made fortunes after 1842, when Pierre Choteau's American Fur Company peppered eastern magazines with advertisements for buffalo robes. Two thousand miles westward, the Bents paid the Cheyenne and the Arapaho twenty-five cents for skins that sold for five dollars in Saint Louis.[13]

In 1822, just a few months after Becknell advertised for associates, William Ashley put out a call for 100 enterprising young men to ascend the Missouri River to its source and work there for one to three years. A native of Virginia who followed Moses Austin to the Missouri lead mines, Ashley rose to become the state's first lieutenant governor. That helped him raise $30,000 in capital. He was also fortunate that his company's first pool of recruits included Jim Bridger and Jedidiah Smith, both destined to become larger-than-life mountain men. The first expedition was a disaster. The Big Muddy swallowed up the supply boats, Indians attacked repeatedly, and as luck would have it the market for beaver pelts nosedived just as the trappers returned. But Ashley persisted. He cut costs by eliminating the trading posts used by other fur companies, sold supplies to his trappers on credit, gave them leave to sell half their pelts on their own accounts, and steered them away from the dangerous Missouri River grasslands into the Rocky Mountains. The Hudson's Bay Company, which monopolized the Canadian fur trade in 1821, had a similar system. Finally, Ashley founded the practice of holding a rendezvous every year, usually beneath the Grand Tetons, where mountain men shared gossip and lore, and exchanged their pelts for supplies. Ashley returned from the first meeting in 1825 with 3,200 beaver skins weighing two and a half tons and valued at $48,000.

The news traveled fast, as did Ashley's report that wagon transport across the Great Plains and even into the mountains was no worse than on transport turnpikes back east. As fur companies proliferated, their trappers mastered the arts of survival in the wilderness, including coexistence with Native Americans, and blazed the safest trails into and over the Rockies. But this lasted for only fifteen years. The beavers became nearly extinct. Beaver hats went out of fashion. One by one, the colorful, courageous mountain men went broke until, in 1840, they staged their last rendezvous. There Robert Newell voiced his lament to Joe Meek: "Come. We are done with this life in the mountains—done with wading in beaver dams, and freezing or starving alternately—done with Indian trading and Indian fighting. The fur trade is dead in the Rocky Mountains. . . . We are young yet, and have life before us. We cannot waste it here; we cannot or will not return to the States. Let us go down to the Wallamet and take farms. . . . What do you say, Meek? Shall we turn American settlers?"[14]

The "Wallamet" to which Newell referred was the valley of the Willamette River, a tributary south of the great Columbia River in the Oregon Territory. And Newell was careful to say "American settlers" because Oregon was, at the time, still up for grabs. Initially the Spanish empire claimed the entire Pacific coast, but it relinquished its rights north of California (the forty-second parallel) to the United States in the Transcontinental Treaty of 1819. The Russian American Company based at Sitka, Alaska, briefly claimed the entire Pacific Northwest, but lacked the men, money, and navy to make the claim stick. A Russian-American treaty of 1824 fixed the southern boundary of Alaska at fifty-four degrees, forty minutes north latitude. That left Britain and the United States contending for the vast empire of woods and water encompassing present-day British Columbia, Washington, and Oregon. The British claim rested on the maritime explorations of George Vancouver and trading posts built by the mighty Hudson's Bay Company, whose ubiquitous blankets emblazoned with HBC ("Here Before Christ," joked the trappers) were much in demand among Indians. The American claim rested on the overland explorations of Lewis and Clark and the short-lived Astoria colony founded in 1812. The sensible solution—repeatedly urged by Senator Benton—was to partition the territory. But the British insisted on making the Columbia River the boundary, and the Americans demanded a boundary at least as far north as the forty-ninth parallel. So John Quincy Adams agreed in 1818 and again in 1826 to a joint Anglo-American "occupation" of Oregon. He assumed American pioneers would someday inherit the territory by sheer weight of numbers. The British feared the same outcome (Lord Castlereagh predicted that the fate of Oregon would be decided in the bedchambers of New England). That is why Sir George Simpson, governor of the Hudson's Bay Company, exe-cuted a grand strategy to enforce a British monopoly. He built a series of forts north of the Columbia; drove Yankee captains out of the coastal trade by offering Indians higher prices for furs; purchased exclusive rights to trade with Russian America; and ordered his resident superintendent, John McLoughlin, to discourage settlement of the Columbia basin.

McLoughlin was a frontier surgeon and trapper with the shaggy white hair and the temperament of a polar bear. He was infamous for whipping into shape (or just whipping) unruly employees, competitors, Indians, women, and even the priggish Anglican chaplain at Fort Vancou-ver. Simpson called McLoughlin "such a figure as I should not like to meet in a dark Night." Hence, when a sick and exhausted party led by the mountain man Nathaniel Wyeth straggled down the Columbia in 1832, it could be expected McLoughlin would banish them. Instead, he nursed the Americans back to health and invited them to plant farms along the Willamette! Did the big man have a sentimental streak? Was he a secret republican? (He later took American citizenship.) Did he disobey orders rather than provoke an international incident? Whatever the case, Wyeth went back east to spread word of the fertile, mild Oregon territory and the kind doctor of the Hudson's Bay Company. That publicity campaign coin-cided with another launched by the *Christian Advocate and Journal* in 1833. It published an appeal (most likely apocryphal) from a Nez Percé chief whose "poor blind people" craved the "White Man's Book of Heaven." Wyeth returned to Oregon the following year with two Methodist pastors in tow. By 1841 their Salem mission was surrounded by the farms and cab-ins of 500 American farmers, trappers, and traders.[15]

That was the moment Benton prevailed on Colonel Abert of the Topographical Engineers to send his son-in-law into the mountains. Ben-ton had failed several times to get Congress to promote settlement in Or-egon. Perhaps a sensational expedition might infect Americans, especially midwestern farmers ruined by the Panic of 1837, with Oregon fever. So Frémont kissed Jessie good-bye in May 1842, purchased supplies from Chouteau in Saint Louis, and boarded a steamboat on the Missouri River. On deck he met another man who would become a lifelong patron and friend—whom he described as "a man of medium height, broad-shouldered and deep chested, with a clear, steady, blue eye, and frank speech and ad-dress," not to mention an athletic horseman. His name was Christopher

Carson, but everyone called him Kit. With Carson serving as scout, Frémont veered off the Oregon Trail to Fort Saint Vrain. The party threaded its way through enormous buffalo herds and visited Indian villages whose warriors with burnished shields recalled the days of chivalry. When Jim Bridger warned of hostile Oglalla Sioux at South Pass, Frémont invited any man "disposed to cowardice" to go back. The whole party pressed on to the Wind River Mountains, where Frémont scaled the highest peak and stood with the wind in his hair as if he were posing for Thomas Cole. Then he unfurled a flag whose field displayed the American eagle surrounded by stars and clutching arrows and a peace pipe in its talons. Judging that he had sufficient "material," Frémont hastened home. His report, published in March 1843, not only advertised the Oregon Trail but fired imaginations the way the first satellite launches conjured up dreams of spaceflight. On reading Frémont one youth confessed, "I was no longer a boy," but inflamed with a passion for adventure and glory "away out yonder under the path of the setting sun."

In 1843 hundreds of families seeking a second chance set out in wagons on the 1,900-mile Oregon Trail. It stretched from Independence, Missouri; along the Platte and South Platte rivers; over South Pass; and into Fort Bridger. The wily Jim Bridger had located his trading post (in what is now southwestern Wyoming) on lush grasslands almost exactly halfway to Oregon. After seventy grueling days of travel the emigrants were in desperate need of supplies. According to Francis Parkman's eyewitness account, they "were plundered and cheated without mercy." But they were also refreshed for the last legs of their journey northwest to Wyeth's Fort Hall and along the Snake, Boise, and Columbia rivers. Frémont himself was among them, for Benton and Abert immediately hurled him back into the field to conduct a "great reconnaissance" of the entire Far West. As always, he recruited mountain men and Indians to serve as his scouts ("the Pathfinder" found few paths on his own). But the expedition would have been inconceivable without Frémont's vision, will, and belief in his destiny. He disobeyed orders and took enormous risks to explore the eastern ridge of the Rockies as far south as Pueblo, Colorado; the Powder River country and the Laramie Mountains; and the valley of Great Salt Lake, his first view of which Frémont compared to Balboa's first sight of the Pacific. (Brigham Young would pay close attention.) The party then followed the Oregon Trail to its terminus, where Frémont took

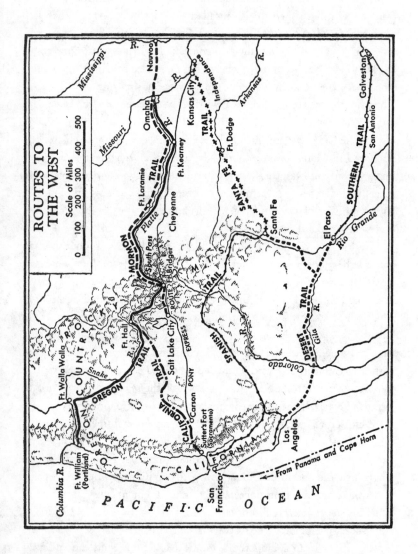

ROUTES TO
THE WEST

Scale of Miles

0   100 200 300 400 500

notes on the fertility of the Willamette Valley. That summer of 1843 settlers in Oregon formed a polity of sorts called the Champoeg Compact, elected officials, and petitioned Congress to annex them. In the Midwest spontaneous "Oregon conventions" shouted support. In Cincinnati Benton exulted, "Let the emigrants go on. . . . Thirty thousand rifles in Oregon will annihilate the Hudson's Bay Company, drive them off our continent."[16]

Frémont completed his official instructions when he rode alone to Fort Vancouver and paid a courtesy call on John McLoughlin. But instead of heading back east he led his apprehensive party south past Mount Hood and Klamath Lake into the barren desert east of the Sierra Nevada. There he proved once and for all that no "Buenaventura River" existed. His Indian guides pleaded with him to winter on the shores of Lake Tahoe. Instead, Frémont and Kit Carson drove the company through chest-high snow over the High Sierra to an altitude of 9,338 feet. Not one man was lost.[17] In 1844 the expedition traversed the San Joaquin Valley south to Tehachapi Pass, crossed the Mohave Desert to Las Vegas, marched through the exquisite buttes of present-day Utah, and then recrossed the Rockies to Bent's Fort, where they celebrated the Fourth of July. Frémont and his fellow scientists (chief among them the German topographer Charles Preuss) returned with bundles of priceless artifacts, samples, notes, maps, and sketches. Their 700-page report silenced anyone still doubting the fantastic potential of Oregon, Great Salt Lake valley, and the province of Alta California.[18]

California! That greatest of all western empires was virtual terra incognita when Mexican independence lifted the veil of secrecy the Spaniards had draped over their remotest, least populated, and most vulnerable colony. Concerned by reports of Russians on the Pacific coast, the Spanish crown dispatched José Gálvez with orders to occupy Alta (upper) California. He in turn recruited Franciscan missionaries led by Junípero Serra to found missions along El Camino Real (the "royal road"). Beginning in 1769 they established twenty-one bucolic adobe plantations from San Diego to Sonoma north of San Francisco Bay. In 1776 a great captain, Juan Bautista de Anza, led a column of soldiers and settlers over 1,600 miles of searing desert from Sonora to Monterey Bay. They established presidios at San Diego, Santa Barbara, Monterey, and Yerba Buena (San Francisco), and the pueblos of San Luis Obispo, San Jose, Carmel, and Los Angeles.

The colony did not thrive. Disease and demoralization killed 75 percent of the 72,000 native tribespeople along the coast, and many of the friars' 6,000 Indian converts fled from obligatory labor in fields and vineyards. Resupply from Mexico by sea was slow and perilous, owing to contrary winds and currents off California. Predatory Yumas, Apache, and Comanche overran the Sonora trail as early as 1781. Most enervating was the monopoly that Franciscans enjoyed over the best-watered land. As late as 1800 Alta California contained no more than 2,000 colonists of Hispanic descent.[19]

The Mexican government, chronically unstable and broke, was impotent to impose its will on the distant province. Any governor who took his authority seriously, especially if he did so to line his own pockets, was sure to be ousted by the Californios themselves. (That era inspired the legend of Zorro.) The Mexicans' only major achievement was to secularize the missions in 1833 and make generous land grants to favored rancheros. In just a few years they built, on the strength of their herds and Indian vaqueros, a robust trade in cowhides and beef tallow for New England's shoe and candle factories. Indeed, the American presence on the coast prior to 1846 was almost exclusively Yankee. William A. Gale, who arrived on the *Sachem* in 1832, represented the Boston firm of Bryant, Sturgis, and Company. He carried back news that the Spanish veil had lifted and Monterey was open to commerce. Richard Henry Dana of Cambridge, Massachusetts, sailed up the California coast in 1834. He reported, in *Two Years Before the Mast* (1840), not only that California was gorgeous and warm but that its "idle, thriftless" population happily purchased goods "at an advance of nearly three hundred percent upon the Boston prices." Several more Yankees married into the hidalgo elite. Their acknowledged leader was Thomas O. Larkin, also from Massachusetts, whose sole passion in life was money. He founded an all-purpose brokerage firm, clearinghouse, bank, and import-export business linking California with China, South America, and New England. In 1844 the State Department named him U.S. consul in Monterey.[20]

To reach California by sea meant a 12,000-mile, six-month journey around Cape Horn and then deep into the Pacific in order to catch westerly winds. To reach California by land meant following Frémont over hundreds of miles of moonscape and the daunting Sierra Nevada. The few who dared that in the 1840s found refuge at New Helvetia, better known as Johann Augustus Sutter's Fort. Rightly guessing a fortune could be made from the flow of emigrants, Sutter obtained 49,000 acres of land (plus Fort Ross when the Russians sold out in 1843), and erected an imposing stockade at the confluence of the Sacramento and American rivers. Sutter dreamed of building a "little Switzerland" on the Pacific, but the foreigners clustering around his fort were nearly all Americans. The same was true at Yerba Buena, where an English merchant observed with disgust, "Is there nothing but Yankees here?" In truth, no more than 800 Americans reached California by 1845. But many already anticipated playing the "Texas game" and staging an insurrection to shuck off Mexican rule. They faced, after all, just 7,000 or so demoralized Californios, of whom fewer than 2,000 were adult males.[21]

That an ambitious, commercial, maritime nation needed a sizable navy should have been obvious. The only serious threats to the United States were bombardment and invasion by sea, as the War of 1812 amply proved. Nor was nineteenth-century America blessed with "free security" by grace of the British navy. On the contrary, Americans rightly identified Britain as the only power on earth capable of choking their growth. Tugging the beard of the British lion was a quadrennial ritual for political candidates. Wars and war scares between the Anglo-Saxon rivals approached a dozen over the century. Americans who swore by John Adams's adage "Britain will never be our Friend, till We are its Master" should have been urging more sail and cannons. Instead, Congress shut down naval construction during the same decade when the Monroe Doctrine pretentiously warned the Spanish, French, Russian, and British empires that the Americas were off-limits. By the time Jackson entered the White House, the U.S. Navy had fewer than two dozen active warships and an annual budget of less than $4 million. The office of secretary of the navy had become a sop; its occupants (as Lieutenant Matthew Maury complained) were "usually selected from among politicians, who have never made Naval affairs any part of their study." Tocqueville knew that the American people, like their British ancestors, were "born to rule the seas, as the Romans were to conquer the world." But the American people were not ready to do what that required.

The U.S. Navy maintained stations on the Mediterranean, West

Indian, African, and South American littorals, but they consisted at most of one frigate and a few sloops or schooners. The navy's headquarters were notorious for being the "most rickety and stupid of all the Federal Departments." The bureau chiefs, mostly superannuated commodores "clinging like barnacles to their posts," squelched innovation. A rigid seniority system with no provision for retirement spoiled morale among officers stuck in junior ranks. The exhausting, claustrophobic, and brutal life on board ship spoiled morale among sailors whose sole escape was to quaff, brawl, and spew during shore leaves aptly called "liberty." Given desertion rates of 10 percent and an average annual turnover of 60 percent, captains and mates filled out their crews with anyone they could find, more often than not with foreigners. As with the Army, freeborn Americans had no economic incentive to enlist and nurtured a democratic hatred of hierarchy and discipline. Indeed, the U.S. Navy had a reputation for being even more oppressive than the British service. The Act of 1800 for the Government of the Navy (whose articles everyone nicknamed the "Rocks and Shoals") stripped sailors of the constitutional and common-law rights they had enjoyed as civilians. Of course, the same was true of Britain's articles of war and of the merchant marine (as Richard Henry Dana attested), because instant obedience to regulations and commands is a life-and-death matter at sea. What young officers and sailors despised was their enslavement, for months or years at a time, to capricious and tyrannical captains. On the "hell ship" christened the *United States*, Melville recorded 163 floggings during the voyage around Cape Horn. Yet most sailors preferred flogging—a manly punishment, quickly over—to losing their grog or liberty, being put on bread and water, being locked in irons, or being confined in the brig. In sum, just a couple of decades of peace sufficed to erase the navy's heroic image won by John Paul Jones, Stephen Decatur, Isaac Hull, Thomas MacDonough, and Oliver Hazard Perry.[22]

The navy did little more than show the flag even as American merchants, missionaries, and whalers skirted like schools of flying fish all over the Pacific. Beginning in 1819 Congregational missionaries led by Asa Thurston and Hiram and Sybil Bingham arrived in the Sandwich Islands (Hawaii) to conduct the first American project of nation building. Their party included a printer, doctor, and farmer, and their instructions were to teach the Polynesians not only the gospel, but literacy, agriculture, law, medicine, and every other blessing of New England civilization. They believed it providential that the great pagan warrior king Kamehameha I had died just before their arrival, and that queen regent Kaahamanu had abolished the *kapu* (taboo) system administered by the pagan kahuna priesthood. After 1825, when Kaahamanu and most of the *alii* (Hawaiian nobility) converted to Christianity, American missionaries became the powers behind the throne. They banned prostitution and grog, clothed the people, founded schools and private property, and all but took over the government. Their power alarmed the British and French, but no one hated the reign of the missionaries more than American whalers.[23]

After the War of 1812 the American whaling fleet increased dramatically. In 1816 sixty-one whalers were registered in U.S. ports, more than half in Nantucket. Thirty years later the fleet peaked at 722, with almost half the ships registered in New Bedford. That amounted to 75 percent of all the whalers *in the world*. To the uninitiated no occupation seemed more romantic. Rugged Down East captains, mates, and seamen kissed their long-suffering womenfolk good-bye and sailed around Cape Horn to roam the Pacific from the balmy South Seas to the icy approaches of Bering Strait. When the cry "Whale ho!" sounded from the capstan, men took to their boats to harpoon, bleed, and exhaust the leviathan until the master mate could deliver the fatal thrust of the lance at the heart and lungs. The corpse was then lashed in place to float on the surface while its tons of blubber were carved out in great cubic blocks. For days the blubber was cooked on board the main ship and the whale oil drawn off and stored in the hold. The stench was almost unbearable, but it smelled like money to the captain. "New England enterprise"—ran a whaling port's toast—"It grapples with the monsters of the Pacific to illuminate our dwellings, and with the problems of science to enlighten our minds." Before the first commercial development of petroleum in Pennsylvania in 1859 the only alternative to candlelight was the oil lamp fueled by the guts of the mighty sperm whale. In the peak harvest of 1837 American ships off-loaded 5,329,138 gallons of sperm oil. Ancillary markets included whalebone for women's corsets and ambergris for perfume.[24]

The reality of the whaling trade was anything but romantic. Whaling skippers were notoriously cruel, driven, and cheap. A deck hand might return from three years of terror, labor, and boredom to find that he actually owed money because his share of the cargo did not cover the advance and expenses the captain charged his account. Since the memory of a few drunken debauches on tropical beaches might be all a young man carried home, whaling crews were a mixture of the desperate, the resigned, and the ignorant, plus whatever adventurous Kanakas (Hawaiians) and Celestials (Asians) the captain could sign up. One can only imagine the fury of whaling crews who docked at Lahaina on Maui in the 1830s and 1840s, only to learn that missionaries from their native New England had banned booze and locked up the wahines (girls). So many Yankees deserted that the U.S. consul in Lima, Peru, lamented the loss of thousands of citizens who "either from shame or moral corruption" never went home. Whalers and sailors presented a decidedly different profile of American culture from missionaries, but their ubiquitous presence helped to Americanize the Pacific, and they contributed so much to the U.S. economy that even Congress had to address their concerns through tariffs, consular representation, and collection of data on winds, currents, and whale populations.[25]

The U.S. Navy began to recover in a "haphazard and unsystematic manner" under Andrew Jackson. He might despise merchants and military professionals, but not so much as he hated dishonor. When Malay pirates at a small pepper port on Sumatra murdered American sailors in 1831, Jackson dispatched the frigate *Potomac* halfway around the world to destroy their base. In 1835 the U.S. naval presence became permanent through creation of a Far East station. When the Argentines seized foreign vessels in protest over Britain's seizure of the Falkland Islands, Jackson sent a flotilla. He made repeated use of the West Indian station before and during the hostilities in Texas. In his annual message of 1835 Jackson called the navy "our best security against foreign aggressions." The naval budget, just $3.9 million in 1835, leaped to $6.6 million, 20 percent of federal expenses, by 1837. Those were the years when the steam engine, the screw propeller, exploding shells, and iron armor began to transform naval technology. Dynamic young officers including Matthew Maury, Sylvanus Gordon, Thomas ap Catesby Jones, and Matthew C. Perry (Oliver Hazard Perry's brother) lobbied for steam-assisted warships and vigorous support of American commerce. Perry, a veteran of the War of 1812 and an amateur scientist, commanded the Brooklyn Navy Yard, where he founded the Naval Lyceum and *Naval Magazine* in 1831. He would inspire the navy's own Corps of Engineers, revolutionize naval ordnance, and earn the title "father of the steam navy." He also began a publicity campaign for sea power sixty years in advance of Alfred Thayer Mahan. An important article in Perry's magazine, "Thoughts on the Navy," asserted that "all of our misfortunes as a nation, from the day we became one, have proceeded from the want of a sufficient navy." It was a scandal that the United States was only the eighth greatest naval power in the world. Perry judged that the fleet had to be tripled in size to be commensurate with U.S. trade and prestige.[26]

Jackson's initiatives died. Van Buren had no interest in naval reform, especially given the deficits born of the Panic of 1837. His secretary of the navy, the literary lion James K. Paulding, sided with the commodores who thought steamships ugly, noisy, inefficient (because of their need for fuel), and destructive of seamanship. Congressional pressure sufficed to get a few steamships under construction. (Paulding griped, "I am steamed to death!") Otherwise, naval expansion ceased. Van Buren's foot-dragging was most notorious in the matter of an exploring expedition. That was the brainchild of a former editor in Ohio who was a marine enthusiast, Jeremiah N. Reynolds. He swallowed the nutty theory of John Symmes that the earth was filled with hollow, concentric spheres with openings at the poles. Reynolds plumped for polar exploration (prudently dropping "Symmes Hole") and won over whalers, shippers, and young naval officers. When Congress authorized $150,000 for a Pacific expedition in 1836 the Navy Department put Thomas ap Catesby Jones in command. But Jones grew so weary of bickering over the cost, outfitting, and purpose of the expedition that he asked to be relieved. Van Buren then inexplicably put Secretary of War Poinsett in charge of the mission. Naval officers were livid, especially when Poinsett gave the command to a mere lieutenant, Charles Wilkes. But thanks to Wilkes's determination the 122-foot sloop USS *Vincennes* and five other vessels put to sea in August 1838 on the United States Exploring Expedition, nicknamed the "Ex. Ex."[27]

The inertia was at last overcome in 1841, when the commercial-minded Whigs took over the executive branch and Congress. Secretary of the Navy Abel P. Upshur was anything but a sea dog, but he wisely leaned on Maury and Perry. They made the eloquent case that every dollar spent on the Navy would repay the U.S. Treasury many times over through enhanced customs duties on foreign trade that in turn enriched America as a whole. Upshur's report of 1841 recommended a whole new generation of warships including steamers, new regulations modeled on Scott's Army manuals, professional bureaus to replace the stodgy Board of Commissioners, apprenticeships to serve as a nursery for seamen, an educational program for officers that gave birth in 1845 to the Naval Academy at Annapolis, and a greatly enlarged Marine Corps. He got only part of his program (landsmen led by Thomas Hart Benton persisted in their bias). But naval spending reached $8.4 million, one-third of all federal outlays, in 1842. The program included the iron-hulled *Michigan*; the ironclad *Stevens Battery*; and the *Princeton*, the world's first warship driven by a screw propellor. Within two years the Navy had thirty-eight vessels on active duty. Finally, Upshur named Maury chief of a new Depot of Charts and Instruments in 1842 and superintendent of the Naval Observatory that opened in 1844. Maury, shore-bound as a result of a stagecoach accident, poured his energies into oceanography serving American sea power.

The rise of the Whigs, however, was not the whole story. A solid majority for naval expansion emerged by 1841 because southerners of both parties displayed a surprising new interest in coastal defense. Thomas Butler King of Georgia drafted a report of the House Committee on Naval Affairs urging creation of a powerful Home Squadron. Senator Alexander Barrow of Louisiana called naval expansion no "local question" interesting just to the Northeast, but an issue vital to the South. Henry Wise of Virginia seconded Barrow. Senator William Preston of South Carolina described the Navy as "our glory and our strength." Upshur himself was a Virginian slaveholder. Why such naval-mindedness among southern agrarians? All spoke, of course, in terms of national defense and economics. But in fact, southern leaders were plainly worried about the consequences of the abolition of slavery in the British West Indies, finalized in 1838. In the event of another war, Great Britain could be expected to use its former slaves as shock troops. King warned that the British army meant to recruit 25,000 blacks. "These troops are disciplined and commanded by white officers, and, no doubt, designed to form a most important portion of the force to be employed in any future contest." Another southerner was convinced that Britain meant to "throw her black regiments from Jamaica into the Southern Country, and . . . proclaim freedom to the slaves." The result was a striking shift of the political winds. In the two decades before 1840 just one of six naval secretaries came from the South. In the twenty years after 1840, nine of twelve were southerners.[28]

In June 1842, just as Upshur's effort was gathering steam, the "Ex. Ex."

returned to port. The explorers commanded by Wilkes had circled the Pacific; discovered the Antarctic continent; charted the Tuamotu, Society, Fiji, and Sandwich islands; surveyed the western coast of North America from Puget Sound to San Francisco Bay; recrossed the Pacific; and sailed home via Manila, Singapore, and Cape Town. Over four years the "Ex. Ex." covered 87,000 miles; drafted 180 sea charts; and surveyed 280 islands, 800 miles of North American coastline, and 1,500 miles of Antarctic ice shelves. So vast were its collections of artifacts and specimens that Congress built the Smithsonian Institution's "castle on the mall" to accommodate them. But instead of a hero's welcome Wilkes returned to a spate of charges, countercharges, and courts-martial. His midshipmen testified to his favoritism, breaches of rules and etiquette, and love for the lash. The wondrous voyage to the ends of the earth was in fact excruciating; embarrassment rather than glory accrued to the Navy.[29] Yet Wilkes's account, published in 1844, not only enthralled readers such as Melville, Thoreau, and young Samuel Clemens but also made the case that America's destiny in the Pacific hinged on the fates of Oregon and California.[30]

The Tyler administration paid close attention. However barren of domestic achievements, it achieved peaceful triumphs in foreign policy greater than any since the days of Monroe and John Quincy Adams. They focused, directly or indirectly, on the Pacific: more proof that Manifest Destiny, pioneer myths to the contrary notwithstanding, was a matter of strategy at the highest levels.

Recall that the Whigs nominated John Tyler of Virginia for vice president in 1840 to provide ideological and geographical balance to their ticket. Their plan backfired when William Henry Harrison died just after his inauguration, making Tyler ("his accidency") president. The Whig congressional majority led by Henry Clay expected to steer the administration, only to rend their garments when Tyler vetoed two bills to resurrect some sort of national bank and two more that raised tariffs. This should not have come as a shock. Tyler's father, an anti-federalist judge and close friend of Thomas Jefferson, had inculcated in the son the eighteenth-century faith in classical republicanism. It was reinforced at the College of William and Mary where Bishop James Madison (cousin of the future president) taught Tyler that republics ancient and modern were at constant risk of decay from a want of public virtue. Most injurious was any concentration of power. By 1821, Congressman Tyler even thought his own party had succumbed by creating the Second Bank of the United States. "To my latest breath," he vowed, "I will, whether I am in public or private life, oppose the daring usurpations of this government." Not surprisingly, he feared Andrew Jackson because Jackson's solution to economic usurpation was political usurpation. The Whigs thought he was a convert. In fact, not only did Tyler cling to his archaic vision of a feeble central government; as president, he wielded Jackson's own favorite weapon, the veto. (That earned Tyler some cordial letters from the Hermitage, whereupon he and Jackson reconciled.) The Whigs in Congress formally expelled Tyler from their party. The Whigs in the cabinet all resigned, except for Secretary of State Daniel Webster, who was not about to cast away his long-awaited opportunity to make a mark in the executive branch and propel the presidential ambitions he still harbored.[31]

True to his Jeffersonian vision, Tyler believed that the expansion of America's agrarian "empire of liberty" could be achieved peacefully. This suited Webster, who inherited the task of preventing a third war with Britain when two crises roiled the U.S.-Canadian border. The first crisis dated from 1837, when revolts against perceived British tyranny erupted in Upper and Lower Canada (Quebec and Ontario). To escape the crown's soldiers, rebels took refuge on the United States' side of the border, while Patriot Hunter societies numbering as many as 200,000 New Yorkers and New Englanders helped the rebels in the belief that this was Canada's "1776." The U.S. and British governments rued the business, but could not ignore it when Canadian loyalists under an officer of the Royal Navy rowed across the Niagara River, killed an American citizen, and burned the steamer *Caroline*, known to have shipped supplies and volunteers to the rebels. The frontier was gripped by a war fever that General Winfield Scott was barely able to contain. Then a second crisis erupted along the Aroostook and Saint John's rivers, where the boundary between Maine and New Brunswick had never been fixed or surveyed. Millions of acres of choice timber were at issue. Loggers from both nations established armed forts; provincial and state governments called out militias and authorized war chests. Britain's bellicose foreign secretary Lord Palmerston dug in his heels. The British garrison in Canada reached 11,500 regular troops, more than the whole U.S. Army. Again Tyler summoned Scott in hopes that the

Mainers would respect federal authority. "Mr. President," said the general, "if you want war, I need only look on in silence. The Maine people will make it for you fast and hot enough. But if peace be your wish, I can give no assurance of success." [32]

Tyler's wish was "peace with honor" lest British hostility kill his hopes for peaceful annexation of Texas and Oregon. Webster rued the prospect of a war certain to scourge his native New England. But how could they appease the American public and Palmerston at the same time? The first answer popped up in May 1841. Francis O. J. Smith, a politician in Maine, beguiled Webster with what he called a "new mode" of diplomacy: an advertising campaign to educate Americans about the merits of British claims and peddle peace the same way agents sold soap and brocades. Smith added "it would not be unreasonable" if he received a commission for his work. Tyler paid him $3,500 from the president's "secret account" (established by Congress in 1810 to cover "contingent expenses in dealing with foreign nations"), whereupon Smith mobilized newspapers, churches, and pacifist organizations on behalf of a compromise. The second answer arrived in the autumn, when Sir Robert Peel's cabinet fell and Palmerston with it. The U.S. minister to Britain, Edward Everett, eagerly informed Webster that the new foreign secretary, Lord Aberdeen, wanted to liquidate the American quarrel. The British Empire had enough on its plate, having just experienced a war scare with France over the Turkish straits, a humiliating setback in Afghanistan, and the first Opium War in China. Aberdeen found the ideal envoy in Alexander Baring (Lord Ashburton), who had extensive business interests in the United States through Baring Brothers. He arrived in June 1842 with a £3,000 slush fund of his own to influence the press. In a matter of weeks Webster and Ashburton concluded a treaty awarding the United States 7,015 square miles of the disputed territory and Canada 5,012. To the west, in what became Minnesota, they agreed on a boundary at the forty-second parallel, thereby inadvertently bestowing on the United States the richest iron deposits in North America. Democrats such as Benton, Cass, and the Mainer Reuel Williams were loath to cede an acre of land or back any treaty crafted by Whigs. But after Webster made some judicious payoffs the treaty sailed through the Senate in August by a vote of thirty-nine to nine. The U.S.-Canadian border fell quiet. Had it not done so, the still more contentious matter of Oregon might well have become casus belli. [33]

In December 1842 an unusual delegation knocked at the door of the State Department. It seemed that a Hawaiian prince named Timothy Haalilio and his mentor, the Reverend William Richards, had come from the other side of the world to deliver an urgent request from Kauikeaouli, better known as King Kamehameha III of the Sandwich Islands. Webster was reluctant to see them until Richards hinted that Hawaii might become a British protectorate if they returned empty handed. Her Britannic Majesty's evil-tempered consul in Honolulu was urging Britain to act before Yankee missionaries and merchants completely dominated the islands. In 1836 and 1839 French warships had bullied the monarchy into legalizing Catholic missions and alcohol. The native Hawaiian population, diseased and demoralized, was shrinking. In hopes of earning respect for Hawaiian sovereignty the king and chiefs promulgated a constitution and elected a legislature in 1840. But most of all they craved recognition by Britain, France, and the United States.

Unbeknownst to them, the situation in Hawaii was falling·apart. The British consul's alarms had prompted the commander of Britain's Pacific squadron to intervene. In February 1843 Lord George Paulet trained the guns of HMS *Carysfort* on Honolulu until the king agreed to a protectorate. Suddenly British officers replaced American missionaries as de facto rulers. (Melville thought Paulet's maligned regime a decided improvement.) It was all a comedy of errors. Aberdeen had already decided, in October, to affirm Hawaiian independence against the apparent threats posed by the French and Americans. Then, on December 30, 1842, President Tyler used his annual message to denounce any attempts to colonize or dominate the Hawaiian islands, since the United States had a greater interest in them than anyone else. Finally, when Aberdeen repudiated Paulet's seizure, the prime minister joined the French in a declaration supporting Hawaii against U.S. encroachments. When the Hawaiian flag rose again on July 31, 1843, the king added a motto from Psalms: *ua mau ke aa o ka aina i ka pono* (the life of the land is preserved in righteousness). In fact it was preserved by mutual suspicions among the naval powers. But no one missed the significance of the so-called Tyler Doctrine. If the Monroe Doctrine applied to islands far out in the Pacific, it surely covered California and Oregon as well. Nor did anyone miss the fact that the king's new government included such "Hawaiian" names as Richards, Judd, Ricord, Lee, and Wyllie. In 1845 that government launched the great Mâhele reform that created a free market in land. American planters snatched up the rich volcanic soil. Caleb Cushing, perhaps Tyler's only friend in New England, was ecstatic: he thought Hawaii the key that unlocked the Pacific, and with it the China market.[34]

Hence, it was no accident that Tyler's annual message of 1842 paired the pronouncement on the Sandwich Islands with another request to Congress for funds to send a "messenger of peace" to China. European merchants for hundreds of years, and Americans for fifty years, had tried to pry open the Ching (Manchu) Empire but had been frustrated by China's strict commercial controls, onshore trading monopoly, protocols demanding that westerners kowtow like other barbarians, and sheer lack of interest in most western goods. For a few decades around 1800 the Chinese were eager to import sea otter pelts and sandlewood cut on Pacific islands, but those commodities played out. At last the British East India Company made a breakthrough. It smuggled in opium squeezed from poppies in India and made addicts of hundreds of thousands of Chinese. Merchants in Philadelphia and New York (including the Delano ancestors of Franklin Delano Roosevelt) followed in the wake of "John Company." What were the officials in Beijing to do? Legalize opium in order to stanch the outflow of capital, and tax the business themselves; or attempt to suppress the wicked drug traffic? In 1839 the moralist faction won out. Chinese officials confined all the foreigners in Canton, seized their ships, and flushed 20,000 chests of priceless opium into the harbor. At Lord Palmerston's urging, Parliament declared war in defense of "free trade." Britain's Royal Navy, using the first iron-hulled steamship in the Pacific, made short work of the emperor's junks. In the peace treaty of 1842 China ceded Hong Kong and opened five ports to British trade. American merchants hung around the edges to profit during the war, then lobbied the Tyler administration to obtain the same privileges for them. It wasn't necessary. Secretary of the Navy Upshur had already picked Captain Lawrence Kearny to lead the first U.S. Navy squadron to China. Kearny insisted that the United States sought only trade, not colonies or dominion. He displayed prudent patience while the Chinese took counsel, but urged the State Department to send a proper diplomatic delegation. Tyler asked Cushing to undertake the adventure. He was rewarded when the Chinese decided to admit the commerce of other powers if only to play them off against Britain. In 1844 the Treaty of Wangxia granted the United States most-favored-nation status, extended extraterritorial rights to American citizens, and outlawed the opium trade (a stipulation the merchants were pleased to ignore).[35]

Those were impressive achievements, but they paled beside Tyler's principal goal: annexation of Texas. The Lone Star Republic entered limbo after 1836 because Mexico's turnstile governments pledged to reconquer the rebellious province. Texas had a doughty little navy based in the old pirates' port of Galveston, but no regular army, an empty treasury, and a ballooning debt. Politics in Texas were a free-for-all among swaggering empire builders, none more outlandish than Mirabeau Buonaparte Lamar. Succeeding Sam Houston as president in 1838, Lamar expanded government debt tenfold, made genocidal war on the Cherokee to open central Texas to land speculators, and dreamed of extending Texas to the Pacific Ocean. Instead, his exhausted, starving column of 300 freebooters collapsed at the gates of Santa Fe and surrendered without a shot. Lamar did win treaties of recognition and trade from Britain, France, and the Netherlands, but only at the price of rescinding the Texans' request for annexation by the United States. Houston seemed "the only man for Texas," and proved to be so after his marriage to a twenty-year-old Alabaman named Margaret Lea. She coaxed him away from his bottle and into her Baptist church. She also gave him eight children born after his fiftieth birthday. In 1841, when Texans reelected Houston in a landslide, he took office determined to make peace with Mexicans and Native Americans, then renew his bid to merge with the United States. Even after a Mexican army briefly recaptured San Antonio and a Texan sortie to the Rio Grande ended in surrender and firing squads, Houston spurned cries for war. Instead, he negotiated an armistice in June 1843 and practiced such sly diplomacy that he earned the sobriquet "Talleyrand of the Brazos." He hoped to persuade Britain to persuade Mexico to recognize Texas, then dangle the prospect of an Anglo-Texas alliance to persuade the U.S. Congress to approve annexation. He was encouraged when Tyler, in May 1843, pressured Webster to resign and named the southerner Upshur secretary of state.[36]

A year of confusion ensued as rumors, false intelligence, plots, and secret emissaries crossed the Atlantic and the Gulf of Mexico. Everyone realized that the fate of Texas, and possibly the western half of North America, was going to be sealed soon, or sooner. But no one knew how.

Edward Everett, the cool New Englander serving as U.S. minister to Britain, assured Tyler that the British sought good relations and had no designs on Texas. Duff Green, the old Jacksonian hack acting as Tyler's unofficial agent in Europe, insisted that the British were conspiring with Texans and Mexicans to thwart U.S. expansion, abolish slavery, and split the Union. At the same time, Stephen P. Andrews, a lawyer and abolitionist in Texas (but a native of Massachusetts) attended the Second World Anti-Slavery Conference in London in hopes of encouraging British designs on Texas. Lewis Tappan and John Quincy Adams wished him well. The world's freedom, wrote Adams, "depended upon the direct, formal, open, and avowed interference of Great Britain to accomplish the abolition of slavery in Texas." Meanwhile, the U.S. chargé d'affaires in Austin, William S. Murphy, assured Houston that Congress would approve annexation, whereas the Texan chargé in Washington City, Isaac Van Zandt, feared the opposite. Dr. Ashbel Smith, the Texan in London, hinted that the British were raising a loan to compensate Texas slaveholders and cement an alliance. In August 1843 Lord Aberdeen himself promised the House of Lords to promote Mexico's recognition of Texas and the abolition of slavery there. Tyler believed one report, then another. Houston did not know what to believe. He knew only that the best way to court the Americans was to flirt with the British, while the worst outcome would be to plight his troth to the United States and then be jilted. That would strip Texas of all foreign support in the face of a still hostile Mexico.[37]

Tyler's annual message of December 1843 proclaimed, "It is time that this war had ceased." Tyler damned Mexico's obduracy and issued veiled threats against other parties tempted to meddle in American affairs. Andrew Jackson issued a public letter urging annexation lest the British make Texas a base from which to attack New Orleans and incite slave rebellions. Senator Robert Walker of Mississippi sustained the momentum with a widely circulated pamphlet arguing that Texas would enhance the security and prosperity of all parts of the country. His tortured but popular "safety valve" theory even suggested that cotton planters would gravitate, over time, to Texas until, at length, manumitted Negroes would migrate across the Rio Grande and depart the United States forever. Upshur sensed a favorable shift. He called Texas "the great measure of the Administration" and went to work on a treaty. He never finished it. On February 28, 1844, Upshur was blown to bits when a twelve-inch gun on the USS *Princeton* exploded during a demonstration on the Potomac. Who now would shepherd a treaty for Texas through the Senate? Tyler's surprising answer was John C. Calhoun. Perhaps he thought Calhoun would bring gravitas to his lightweight cabinet and rally southern support for Tyler's candidacy in the election that year. If so, he was terribly wrong. Calhoun sent the treaty to the Senate on April 22 along with a fat portfolio that included a letter he wrote to the British minister. It amounted to a long, fierce, and gratuitous defense of slavery. Tyler, Upshur, Jackson, and Walker had spent months spinning marriage with Texas as a vital national interest. Calhoun's kiss turned the Texas belle into an ugly sectional frog.[38]

No wonder the presidential front-runners in 1844 ran away from the issue as fast as they could. On April 27, Henry Clay and Martin Van Buren published simultaneous statements meant to remove Texas from the campaign. Clay opposed the treaty of annexation on the grounds that it would cause sectional conflict, provoke war with Mexico, and saddle the United States with the Texans' debts. Van Buren paid lip service to all points of view and then suggested annexation must await revision of existing international treaties. This "mortified" Jackson, who craved Texas and judged that only a border-state expansionist could beat the "old coon," Henry Clay. On May 13 Jackson told his fellow Tennesseean James Knox Polk, that Polk was "the most available man." Shrewd floor management would be needed. Van Buren remained the favorite of the powerful New York delegation, and Lewis Cass was the choice of expansionists led by Senator Walker of Mississippi. But neither prevailed, because the Democrats called to order in Baltimore's Odd Fellows Hall on May 27 clung to a two-thirds rule imposed by Jackson in 1832. After seven barren ballots Tennessee's chairman Gideon Pillow, George Bancroft, and Benjamin Butler proposed the convention choose instead between Polk and Senator Silas Wright of New York (knowing that Wright had already withdrawn in deference to Van Buren). The delegates swung behind Polk on the ninth ballot, hastily approved the party platform, then adjourned to pop corks at the bar. Only later did they learn that Walker, the night before, had inserted a novel plank in their platform. It stated that "the reoccupation of Oregon and the re-annexation of Texas, at the earliest practicable period are great American measures, which this convention recommends to the cordial support of the Democracy of the Union." Thus did the Democrats make 1844 the first presidential campaign fought primarily over foreign policy.[39]

Polk is usually considered the first dark horse candidate. His nomination was surely surprising. But no candidate touted by a patriarch as respected—and feared—as Jackson can be considered a dark horse. Likewise, everyone cites the Whigs' campaign gibe, "Who is James K. Polk?" But the Whigs just meant to suggest he was a pygmy beside Henry Clay. They knew very well that Polk had been Speaker of the House, Jackson's bulldog in the U.S. bank war, and governor of Tennessee. Born in North Carolina's rustic Mecklenburg County in 1795, Polk shared John Tyler's nostalgia for Jeffersonian idylls. In 1806 his family packed up their belongings in wagons and trudged through 500 miles of wilderness to middle Tennessee. Jimmy Polk was a sickly boy, the opposite of a Davy Crockett, but courageous. At age seventeen he underwent surgery without anesthesia for a gallstone. The following year the unlettered boy enrolled in college at Chapel Hill and excelled through sheer discipline. In 1823 he came out of nowhere to win a seat in the Tennessee legislature. That emboldened him to woo Sarah Childress, Murfreesboro's richest and most desirable young woman. Smart, loyal, and religious, she devoted her life to her husband's physical, spiritual, and political health. When Polk then defeated a bitter rival of Jackson in a race for Congress in 1825 his career was made and his profile complete. He was Young Hickory, the purest vessel of the Jacksonian spirit in the next generation. Whigs taunted him only because his narrow loss in the gubernatorial contest of 1843 seemed to confirm a Whig paper's prophecy: "Henceforth His Career Will Be Downwards." Nine months later Polk was a candidate for president of the United States.[40]

The political conventions coincided with the acrimonious debate over Texas during which some northern and southern senators alike threatened secession if annexation was approved or rejected. The question was called on June 8. Proponents could assemble just sixteen votes; thirty-five members voted nay. The split came on party as well as sectional lines. Benton was the only Democrat from a slave state to oppose the treaty (he feared war with Mexico). John Henderson, Walker's colleague from Mississippi, was the only southern Whig to vote in favor. Calhoun was humiliated, Clay ecstatic. The crushing defeat of "Mr. Tyler's abominable treaty" suggested Clay held the winning hand in the upcoming election. Sam Houston was furious. With Santa Anna swearing to drown Texas in its own blood he had no choice but to crawl, ten-gallon hat in hand, back to the

British, while he nervously awaited the outcome of the U.S. election of 1844.[41]

The campaign was devoid of certainties. Polk hoped his feisty support for "reoccupation" of Oregon up to latitude fifty-four-forty would reconcile northerners to the "reannexation" of Texas. (The prefixes were clever reminders that John Quincy Adams had bargained away prior U.S. claims to the territories.) Whigs countered by depicting Polk as an "ultra slaveholder." Democrats depicted Clay as a carouser who systematically shattered all Ten Commandments. As for his position on Texas, it was "a consistent straddle." Clay tilted against annexation until he sensed that this was hurting him in the South and West; then he tilted in favor of it, if a list of conditions were met. Clay was also bedeviled by the abolitionist Liberty Party, which showed unwelcome strength in the Northeast. Both candidates had to reckon with Tyler. He staged his own mini-convention (there were more spectators than delegates, reporters said) and stayed in the race until August, when Jackson convinced him that the honorable course was to throw his support to Polk. So many crosscurrents roiled the waters that it is impossible to say where the main channel flowed: one could say only that it was narrow. Polk tallied 1,337,243 votes to Clay's 1,299,062 and won in the electoral college 170 to 105. Clay held his own in the upper South (he won Polk's home state by just 250 votes), Ohio, and the Northeast, but Polk swept the southern and western tiers plus Virginia and Pennsylvania. It is tempting to argue the Liberty Party gave the "ultra slaveholder" his margin of victory because it drew off enough Whig support to award New York's thirty-six electoral votes to Polk. But they were decisive only because the issue of Texas put unbearable pressure on southern Whigs.[42]

In the 1830s wealthy planters and the merchants, professionals, and craftsmen who serviced the cotton culture rallied to the new party led by Clay. Proud to be called conservatives, southern Whigs stood for law and order, banks and internal improvements, moral reform, and social deference. They imagined themselves locked in a battle for civilization against poor white "crackers" who railed against privilege, mobbed, and voted Jacksonian. Their most effective weapon was ridicule. Augustus Baldwin Longstreet, a Georgian journalist educated at Yale and Tapping Reeve's law school, gave up lofty editorials attacking "corruption and filth" in favor of satirical sketches recording the hilarious, bombastic dialect of the frontiersman and the slithery pitch of the con man. His character Ransy Sniffle "was just five feet nothing; and his average weight in blackberry season, ninety-five," yet he "never seemed fairly alive except when he was witnessing, fomenting, or talking about a fight." Sniffle became the prototype for John Pendleton Kennedy's "distinguished loafer" Flan Sucker, William Tappan Thompson's cadaverous Sammy Stonestreet, Johnson Hooper's hustler Simon Suggs ("It is good to be shifty in a new country"), G. W. Harris's "durn'd fool" Sut Lovingood, and Thomas Bangs Thorpe's Jim Doggett, the antihero of tall tales such as "The Big Bear of Arkansas." These southwestern humorists invented the uniquely American genre that Davy Crockett epitomized and Sam Clemens immortalized. They were, to a man, Whigs. They knew that the braggadocio and pioneer virtue of Jackson's "common man" cloaked violence, ignorance, greed, and deceit.[43]

Southern Whigs had closed ranks with their northern colleagues to resist Jackson, Van Buren (an object of "aversion and even abhorrence"), and Tyler. But they withered under the barrage coordinated by Polk's campaign manager, Senator Walker. His incendiary pamphlet, "The South in Danger: Read Before You Vote" (1844), suggested the choice was between Texas and radical abolitionism.[44] Alexander Stephens even said, "The annexation project is a miserable humbug got up as a ruse to distract the Whig party in the South, or peradventure with even an ulterior motive— that is the dissolution of [the] present confederacy." But he joined other southern Whigs repudiating their party's platform on Texas. Clay tried to stop the bleeding with his "Alabama letters" stating that he had no personal objection to annexation but was unwilling to jeopardize the Union for the sake of Texas. Not only was that an obvious waffle; it impugned the patriotism of all parties. The Whig Thurlow Weed of New York called it an "ugly letter." Democrats chirped, "He wires in and wires out / and leaves the people still in doubt / whether the snake that made the track / was going South, or coming back." After the debacle William C. Preston of South Carolina lamented, "For the present the Whig party of the South is dispersed." Clay blamed it on extremist nativists, Catholics, Calhounites, and abolitionists. But the "old coon" himself failed to marshal a majority out of moderates.[45]

Anyone exposed to a textbook of American history knows what happened after the election of 1844. President Tyler chose to interpret Polk's paper-thin victory as a mandate for Manifest Destiny, and he petitioned

Congress to annex Texas through a joint resolution requiring only simple majorities. The lame-duck Congress obliged and the lame-duck president signed the law at the end of his term. There is rather more to the story. Tyler in fact sent the House of Representatives the Texas portfolio in June 1844, just three days after the Senate defeated the treaty. It was always his fallback plan. Nor was it obtuse for him to conclude that the voters had rendered a verdict in favor of annexation. The final tally was close, but that was because Clay shifted to a more pro-expansionist stance that helped him hang on in Ohio, Kentucky, and Tennessee. Even more tellingly, voters gave Democrats a majority in the Senate and an advantage of 143 to seventy-seven in the House. Vox populi spoke. Nor was Tyler just trying to ensure his place in history. President-elect Polk approved of getting the business done before he took office.

In his last annual message Tyler urged Congress to act swiftly lest Santa Anna lead a "war of desolation" sure to harm the poor Mexicans as much as the Texans. He also alluded to new evidence of British and French intrigues. Congressmen introduced a ream of resolutions in hopes of getting their names on the bill. The winner was the Whig Milton Brown of Tennessee, whose resolution passed, 120 to ninety-eight, on January 25, 1845. The federal government would admit Texas as one or more states, but reserve jurisdiction over its boundaries, refuse to assume its debt, and ban slavery north of the Missouri Compromise line. The Senate moved in the same direction until Benton (to block Calhoun) insisted annexation be accomplished by a new treaty to which Mexico would have to adhere. Luckily, Calhoun was sidelined with bronchitis over those critical weeks. That allowed Senator Walker and a silent partner named Polk to craft a compromise. Annexation might occur under the House plan or under Benton's plan, without Mexico's consent, depending on the wishes of the Texans and the new U.S. president. On the evening of February 27 the suspenseful roll was called. Three Whig senators voted with the Democrats; the Walker amendment passed by twenty-seven to twenty-five. The House accepted the formula the following day. Tyler signed it on March 1. The U.S. government had approved the first geographical extension of slavery since 1820 and had done so in a highly irregular fashion. It was a calamity, thought John Quincy Adams: "I regard it as the apoplexy of the Constitution."[46]

A popular misconception holds that Texas became U.S. territory at this time. In fact, whereas the resolution committed the United States to defend Texas, the transfer of sovereignty would not occur unless and until the Texans approved. A new state did join the Union on Tyler's last day in office, but it wasn't Texas. It was Florida.

The Twenty-Seventh State: Florida, 1845

From the day of the pirates to our day of offshore bank accounts, hedonistic resorts, and drug smuggling, Americans have found in the Caribbean an escape from their own laws and morals. The sandy spit that Juan Ponce de Léon baptized La Florida was no exception. In 1565 the Spaniards garrisoned Saint Augustine, the oldest European settlement on what became U.S. soil; and over a century Franciscans founded thirty-two missions to proselytize the Indians. But the province, which was 300 miles wide at the Panhandle and 400 miles long on the Atlantic coast, remained a derelict. The whole Spanish navy could not have policed its 8,426 miles of tidal coastline; nor could the army police its 54,000 square miles of jungle and swamp. Nor could either defend the Indians from European infectious diseases or from the renegade Creeks they called *cimarrónes* (whence "Seminoles"). By the nineteenth century the Native American Floridians were dead, the European population was measured in hundreds, and the whole peninsula from the Apalachicola River to Key West served as a refuge for Tampa Bay buccaneers, mutineers, deserters, fugitive slaves, Seminoles, and plunderers of shipwrecks (a frequent occurrence, especially during hurricane season). John Quincy Adams cited the anarchy as justification for the treaty of 1819 ceding Florida to the United States. But he was pretentious to think Americanization would ensure law and order. The mostly poor, mostly Scots-Irish "crackers" who spilled into the Panhandle had no patience for government. Hot blood, hot sunshine, laws so variable that even judges could not parse them, no jails, no constables, and plenty of places to hide encouraged "ingenious rascality." Florida was "a rogue's paradise."[47]

Advocates of western expansion such as Benton and Clay rued Adams's treaty because it relinquished the claims of the United States to Texas. But *Niles' Weekly Register* believed that Florida, once pacified, promised Americans "command of the gulph" [of Mexico] and valuable agriculture.[48] Most of the peninsula is less than 100 feet above sea level, and no point is more than sixty miles from salt water. Its fertile soil—composed of sand, seashells, limestone, peat, and muck; and drained by

1,700 streams and 30,000 lakes—supports a profusion of life. The 300 species of trees include pine, oak, cypress, palm, gumbo-limbo, and mangrove. The 400 avian species include the pelican, osprey, cormorant, egret, heron, ibis, and flamingo. Mammals include the manatee, bat, armadillo, otter, mink, puma, bobcat, bear, and panther. The 700 species of fish include pompano, flounder, snapper, tarpon, marlin, and shark. Alas, Florida also teems with malaria-bearing mosquitoes; bacteria that cause fevers; alligators; and coral, moccasin, and copperhead snakes as well as rattlesnakes. Rainfall averaging up to sixty-two inches per year mocks the nickname "Sunshine State," and central Florida is struck by violent electrical storms more often than any other spot on the globe.

Add to those hazards the Seminoles and yellow fever that emptied Saint Augustine in 1822, and one can appreciate Florida's reputation as a death trap. It was up to the military, which had no choice but to go where ordered, to begin taming the rogue's paradise. A U.S. Army post founded in 1824 planted civilization in Tampa. The U.S. Navy, after surrendering Key West to yellow fever, chose Pensacola in 1826. The Second Seminole War introduced thousands of men (and officers' wives) to Florida. Of course, that brutal conflict discouraged new settlements for six and a half years, but soldiers, sailors on "swift" riverboats, and civilian bounty hunters (who were offered $500 per scalp) explored the interior as far south as the Everglades, hastening development after peace was restored in 1842.

In 1822 President Monroe appointed as first territorial governor William Pope DuVal, a Virginian of Huguenot lineage who served for twelve years. At first DuVal maintained the Spanish division of East and West Florida along the Suwanee River, but that obliged the legislature to alternate meetings between Saint Augustine and Pensacola. Hundreds of miles of wild country separated them, necessitating a three-week sea voyage around Key West. So Duval appointed commissioners to select a central site for a capital. They chose Tallahassee, an old Indian campground on what for Florida was high ground. Its single log cabin was meant to be the temporary capital of a temporarily unified Florida in March 1824. Tallahassee remains the capital of a unified state to this day. Duval, a hard-core Jacksonian, vetoed early attempts to found banks lest they "create petty class distinctions and inflated sub-aristocracies which are ever at war with that plain and manly equality so essential to the preservation of virtue." But virtue was in short supply, not only among the murderers, gamblers, slavers, squatters, and drunks who poured over the border from Georgia, but among the erstwhile elite. One feud over banking provoked two duels, a murder, and a lynching that left all parties dead. In 1827 Ralph Waldo Emerson found Tallahassee "a grotesque place . . . settled by public officers, land speculators, and desperadoes." Eight years later Judge Robert R. Reid described the capital as "full of filth—of all genders" presided over by the rowdy Governor Eaton and his "drunk or crazy" wife Peggy. East Florida languished, owing to litigation over Spanish land grants and the death of its orange groves from the same freeze of December 1835 that beset New York's firemen. Saint Augustine actually lost population after the Seminole War; and years would pass before the pathetic village of Cowford near the mouth of the Saint John River became a commercial hub renamed Jacksonville. But the land office at Tallahassee sold 800,000 acres to eager farmers and planters of cotton, tobacco, rice, sugarcane, and livestock in Middle Florida's "red hill" country. Gravel, sand, fuller's earth, timber, pitch, and turpentine from the piney woods also made bustling ports of Pensacola and Apalachicola. By 1840, the territory's population surpassed 54,000, albeit nearly half were enslaved.[49]

Floridians first approved statehood in a referendum in 1837, and narrowly approved a constitution in 1839. But the Second Seminole War, the Panic of 1837, and factionalism took statehood off the table. Powerful planters wanted to wait until Florida could be split into two new slave states. East Floridians fearful of domination by populous middle Florida also wanted division. The former governor, DuVal, argued the infant community could not support the cost of state government. No one knew better than he how deftly the crackers evaded taxes. But Florida found an unlikely champion in a man denounced as "that little Jew politician." David Levy, born in the Virgin Islands and educated at Norfolk, moved to Saint Augustine when his father dreamed of founding a refuge for European Jewry. Young David was more interested in becoming American. He practiced law, bought a plantation, entered politics as a Jacksonian, and won the race for territorial delegate in 1841. In Congress he lobbied tirelessly for statehood. In Florida he countered Duval by explaining how the federal government would endow the new state with generous land grants and a share of future land sales. Levy carried the day when President Tyler named John Branch, Jackson's former secretary of the navy, territorial governor. Branch had moved his family to middle Florida and invested heavily in its future. In January 1845 Branch won the legislature over to statehood by promising that the state might be split in the future. That promise was bogus—northerners would never accept it—but it worked. Within a month Levy steered through Congress a bill to admit Florida together with a free state, Iowa. Tyler signed the law on his last day in office, March 3, 1845. The legislature at Tallahassee promptly elected David Levy (Yulee) the first Jewish United States senator.[50]

Duval was right about the crackers' antipathy toward government. The first state budget was $41,500, and annual spending averaged below $90,000 until 1860. The Jacksonian hatred of banks likewise prevailed. So stringent were the state's restrictions that no state banks were chartered until the legislature itself chartered one in 1855. Education? The same story. In 1851 the state founded "seminaries" to train teachers at Ocala (parent of the University of Florida) and Tallahassee (the future Florida State University), but as late as 1860 the state counted just ninety-seven schools with 8,494 pupils. The government showed vigor only in the enforcement of slave codes and the repression of free Negroes. As the state's population rose from 87,445 in 1850 to 140,424 by 1860, the percentage of slaves remained above 40 percent. Disciplining that underclass was everyone's business. Policing white people's behavior was pretty much left up to the women and the Baptist and Methodist clergy.

Florida's economy was less lethargic than its government, but only gradually assumed the shape that would be familiar a century later. Planters carpeted the frost-free central counties with citrus orchards. Cracker, Negro, mulatto, and Seminole cowboys drove Texas-size cattle herds through the sawgrass and palmetto of the lush Kissimmee Valley. The cigar king Vicente Ybor relocated his factories from revolution-torn Cuba to Key West, and then to Tampa in 1886. That was also the decade when Florida's phosphate began to be quarried, the Philadelphian Hamilton Disston purchased 4 million acres in the Everglades, and Yankee entrepreneurs competed to make Florida a winter playground. On the Gulf Coast Henry B. Plant built railroads to Pensacola and Tampa; on the Atlantic side Henry M. Flagler constructed Spanish-style grand hotels along his railroad to Miami and Key West. Vacationers, invalids, retirees, and workers in the construction and service industries boosted the state's population beyond 500,000 by the 1890s. Today David Levy's creation is home to Disney World, the space program, South Beach, and golf and retirement complexes.

But the original Florida will never die out so long as "darkies" gather in jook joints to dance the jubilee (jitterbug), bumper stickers proclaim "Redneck and Proud of It," policemen cruise with alcoholic "roaders" in hand, and transplanted Yankees are taught that "blacks is blacks, but there ain't nothin' sorrier than po' *white* trash." [51]

Americans held the Mexican republic in contempt. It began life in 1821 with a landmass as large as the United States, a population two-thirds as large, and economic prospects that titillated British investors and at least some American merchants. By the 1840s, Mexico was obviously a basket case. The population, 80 percent of which consisted of impoverished Indians and mestizos, did not grow. Neither did the economy. Mexico had abolished chattel slavery, but peonage on the great estates owned by Creoles amounted to much the same thing. The Spaniards left behind just one decent road linking the port of Veracruz with Mexico City, and even it was plundered by bandits. Otherwise the vast mountains and deserts were isolated. The only commodity worth the trouble to export was silver. The republic had been founded by patriotic revolutionaries, many of whom admired the liberal Spanish constitution of 1812. But liberalism in Mexico, as in Spain, was anathema to the Catholic church, hidalgos, and the army. Mexican governments were forever in debt and dependent on fickle, plundering generals to exert authority in outlying provinces. Their inability to protect life and property completed the wreck. Foreign investment was hardly encouraged when a mob crying "*¡Mueran los extranjeros!*" (death to foreigners) lynched five Frenchmen and an American in 1833.

Three political factions claimed they knew the answer to Mexico's woes, but their rivalry only made matters worse. Liberals (*puros*) led by the scholarly physician Valentín Gómez Farías favored a decentralized federal regime, disestablishment of the church, and reliance on civil militias, all of which smacked of Jacksonian democracy. But the *puros* were also fierce patriots bent on reconquering Texas. Gómez Farías even imagined that a war against yanquis was just what Mexicans needed to forge a nation. A second, conservative faction was equally hostile to the grasping Americans, but hated the idea of arming the masses. Conservatives wanted a strong, centralized state allied to the church and army. Finally, *moderados* shared both the *puros'* hope for reform and the conservatives' fear of social revolution. These moderates thought war might spark a lower-class uprising;

hence they alone favored peaceful relations with the United States. The
Mexican standoff was further complicated by personal feuds and shifting
alliances between politicians and generals. Thus, Santa Anna was deposed
during his captivity in the United States in 1836 by the conservative General Bustamente, who in turn was deposed when *puros* fomented a lower-
class revolt in 1840. That shocked conservatives and moderates into
restoring Santa Anna's dictatorial rule until *puros* and *moderados* accomplished another coup in December 1844, which established a parliamentary
regime led by General José Joaquín Herrera.[52]

For a few months it seemed the Mexicans might find their footing.
Herrera pledged reforms, promised stability, and accepted the fact that
Texas was lost. His government was prepared to make peace and recognize the Lone Star Republic so long as it remained independent. Accordingly, Mexico severed diplomatic relations with the United States in March
1845, and instead entertained a joint offer from Britain and France to mediate a Texan-Mexican treaty. The British and French chargés d'affaires
had been standing by in New Orleans. When their orders arrived they
sailed to Galveston and galloped inland to Washington-on-the-Brazos.
The new president of Texas, Anson Jones, informed them that public
opinion leaned strongly toward U.S. statehood, but he agreed to postpone
a convention in Texas for three months and suggested terms for a treaty
with Mexico. On March 30 the European envoys left in high spirits until,
just a few miles out of town, they met Andrew Jackson Donelson coming
the other way.

Polk had not wasted a day. At his very first cabinet meeting it was
decided Donelson must press the Texans into accepting annexation at
once. Polk reinforced him with two propagandists—Archibald Yell and
Charles Wickliffe—who promised the Texans federal protection, funds,
and patronage. Soon Commodore Robert F. Stockton swept into Galveston like a human squall. He claimed to have intelligence about a Mexican
army poised to invade and urged the Texans to strike first. Stockton was
doubtless exceeding instructions—neither Polk nor Anson Jones was ready
or anxious for war. But Stockton's alarms made Texans all the more eager
for a U.S. military presence. Even Houston, who had hoped to negotiate
with the United States for better terms, endorsed immediate annexation.
Then news arrived that Mexico had agreed to recognize Texas if it would
remain independent and would accept a boundary on the Nueces River

instead of the Rio Grande. Would the Texans be tempted? Donelson pleaded for U.S. troops to stiffen the Texans' resolve; Polk promised "prompt & energetic measures." He ordered General Zachary Taylor to lead a corps of observation into Texas and deploy it as close to the Rio Grande as "prudence will dictate." Polk was gambling this would deter, not provoke, Mexican belligerence. He won his bet when *puros* took to the streets on June 7 to overthrow the peace-minded Herrera, and failed. The war party was thus in temporary eclipse during the critical month when the special convention in Texas rejected the Mexican treaty and approved U.S. statehood. A state constitution was drafted, ratified in October by a margin of twenty to one, and ready to take effect as soon as the U.S. Congress convened in December.[53]

The ongoing border dispute ensured that the *puros* would get their war in the end, but at a time of Polk's choosing when the chance of British involvement was virtually nil.

The Twenty-Eighth State: Texas, 1845

Mirabeau Buonaparte Lamar, the transplanted Georgian who dreamed imperial dreams as president of the Lone Star Republic, commissioned the legend that has defined ever since what it means to be a proud Texan. In 1841 he paid the journalist Henry S. Foote to write a history glorifying Austin, Houston, the martyrs of the Alamo, and the heroes of San Jacinto, including himself. Henderson Yoakum's two-volume scholarly history of 1855 fixed the creation myth for all time. Yoakum, a Tennessee volunteer and fervent Jacksonian, believed that God had called courageous, freedom-loving Anglos to displace the Mexicans and Indians and make Texas the biggest land of opportunity yet.

He was right about its bigness. Even after the federal government pared down its borders, the state encompassed 266,807 square miles—an area greater than France—divided in due course into 254 counties, some bigger than Connecticut. Moreover, Texas was a "creation" because most of its soil arrived, over millions of years, as silty erosion washed down from the western mountains. Seismic upheavals completed the canvas by carving gigantic steps marching west and north and drained by rivers cutting east and south. The first step, the coastal plain, is a fertile crescent except for the Sabine River's Big Thicket. About 200 miles inland the Balcones Escarpment announces the onset of the hill country and central plains, where yucca, cactus, cypress, and ponderosa pines rise above a carpet of mesquite. Another 200 miles west the towering cliffs of the Cap Rock Escarpment herald the West Texas High Plains and Big Bend country. Stretching 750 miles on both meridians, Texas features climatic zones ranging from sultry to arid and torrid to frigid. Spring tornados, autumn hurricanes, 100 species of snakes, scorpions ("vinegaroons" to a Texan), and big cats defied human settlement. But land beyond measure and beyond authority was an irresistible lure to American desperadoes. As a jingle from the 1840s went, "The United States as we understand, took sick and did vomit the dregs of the land. Her murderers, bankrupts, and rogues you may see, all congregated in San Felipe." The "rogues" had a somewhat more flattering view of Texas: "great country for men and dogs, but hell on women and horses." [54]

Anglo-Texans imagined their history a triumph of the pioneer spirit over hostile Mexicans, fierce Comanche and Apache, and the tough blazing land itself. Triumph they did, but on "each step up the rocky plateaus, Texans left their blood, bones, and blasted dreams." On other western frontiers pioneers awaited the Army's protection before putting down roots. Only in Texas did civilians fight for their land and liberty before joining the United States. Only in Texas did "blood memory" forge a tribal consciousness not unlike that of the South African Boers. Only Texas embraced a local patriotism that struck other Americans as chauvinistic from the moment Anson Jones proclaimed, "The Republic of Texas is no more." This was because the boastful Texans in fact became ungrateful wards of the federal government! The terms of annexation dumped $5 million of Texas debt on the U.S. Treasury and granted the state ownership of all vacant land. But the state remained broke because its Jacksonian constitution mandated free distribution of land, outlawed banks, and rejected direct taxes. Debt and a dearth of revenue did not stop Texas from trying to wrest New Mexico ("Sante Fe County") from the control of the U.S. Army after the Mexican War even as officials in Austin demanded the feds pay the cost of the Texas Rangers' campaigns against the Comanche. In 1850 the Texans weaseled $10 million from Congress as compensation for their "lost" New Mexican territory; then they obtained $7,750,000 more in 1855. This encouraged such lavish, corrupt public works ($2 million meant for education found its way into railroad construction) that the state ran up another $1 million in red ink by 1860.[55]

Land laws in Texas seemed too good to be true. A family was entitled to 640 acres of land, married women and squatters had property rights, and veterans received large grants. In reality, settlement of the moving frontier was accompanied by the usual fraud, speculation, and lawsuits. Indians had no claims that the state was bound to recognize. Heirs to Spanish and Mexican grants were subject to intimidation or dispossession. More than 80 percent of the land in south Texas changed hands in the decades after 1846.[56] Huge disparities in wealth and power resulted. The census of 1850 counted 212,000 Texans. But the 58,000 enslaved Africans and 11,000 Mexicans were utterly submerged, and 16,000 European immigrants struggled just to survive. Most pathetic were the farmers and workers recruited by the *Adelsverein*, a colonizing society of Prussian nobility. In 1844 Prince Karl von Solms-Braunfels, a thirty-three-year-old cousin of Queen Victoria, was swindled into purchasing an invalid deed to 3 million acres of Indian country. The 7,380 Germans who sailed from Bremen to Galveston in 1845–1846 thus had nowhere to go. Within three years a third of them perished from disease and privation.[57]

The people who really shaped Texas were the Anglos, of whom 90 percent migrated from the American South. In the cotton and sugarcane counties of Texas, planters replicated the Cavalier culture born in colonial Virginia. They put on aristocratic airs, gambled, drank, dueled, raced horses, and honored hospitality and courtesy, especially toward women. They were also proud, violent, and terrified of slave revolts. One panic occurred in 1856, when the discovery of an arms cache in slave quarters led to the maiming or lynching of 200 blacks and, for good measure, expulsion of all Mexicans in the county. Meanwhile, in the interior of Texas, poor white pioneers replicated the rude, violent, hard-drinking Scots-Irish culture of Appalachia. They put up two-room dogtrot cabins, planted corn and wheat, and their men and women labored as jacks-of-all-trades on huge spreads that no family could cultivate. With the nearest neighbor several miles distant and no transportation except on the rivers, settlers in Texas were the loneliest people in America. Indeed, the state was so rural that even though population tripled in the decade before 1860, its metropolis of San Antonio numbered just 8,000. Houston was a struggling anchorage, Dallas a speck on the map, Galveston the only town with a railroad. Few Texans had access to schools, churches, or courts. What education and uplift they got were by grace of the family Bible; what justice they got were by grace of their Colt revolvers or the Texas Rangers.[58]

In legend the volunteer "ranging companies" that served as irregular cavalry in the wars against Mexico were chivalrous "white hats" cleansing the range of hostile Indians and outlaws. But not until 1874 did the legislature turn the Rangers into a sort of state police. Before that date they functioned as vigilantes ensuring white supremacy. Hence, the Rangers fought bravely against the Comanche, but also killed or expelled peaceful tribes. Their good, bad, and ugly qualities were all on display in the border war launched by the cunning bandito Juan Nepomuceno Cortina in 1859. He conspired with vaqueros and gringos alike to rustle *vacas* on one bank of the Rio Grande and sell them on the other. At times Cortina played Robin Hood for poor Mexicans. At other times he took payment from Anglos to "vote the Mexicans" the right way. Needless to say, he made enemies; this is why he invaded Brownsville with twenty caballeros, shot five men, and opened the jail. The citizens cried for the Texas Rangers, who only made matters worse by beating up innocent Mexicans and hanging one of Cortina's lieutenants. That turned a criminal feud into a race war. Cortina recruited a private army under the Mexican flag and spread the cry, "Death to the gringos!" For three months he terrorized Americans on the river until a disciplined detachment of Rangers under John S. "Rip" Ford defeated the rebels in pitched battles and chased them deep into Mexico.[59]

In legend Texas is western. Prior to 1860 it was utterly southern. The coastal planters and farmers, craftsmen, lawyers, and merchants dependent on cotton ensured a Democratic monopoly. A Whig faction arose after statehood, but the national Whigs' opposition to the Mexican War, rejection of the New Mexican claims, and resistance to the expansion of slavery made them anathema. Sam Houston tried gamely as senator and governor to fan the flame of Jacksonian nationalism, but in February 1861 Texans voted overwhelmingly to secede from the Union.[60] In the short run they doomed their state to the disasters that befell the rest of the Confederacy. Ironically, that bitter defeat also allowed Texas to escape the trammels of southern history and reinvent itself as a southwestern empire. In 1865, the year King Cotton died, King Cattle was born. Frederick Law Olmsted, who toured central Texas, tells why: "A great change occurred here in the prairie grass—we had reached the *mesquit grass*, of which we had heard much throughout Eastern Texas. The grass of the Eastern prairies is coarse and sedgy. . . . Where not burned, it lay, killed by the frost, in a thick, matted bed upon the ground. Our animals showed no disposition to eat it. This mesquit they eat eagerly as soon as we come upon it, as if it were an old acquaintance." [61] The shrub called running mesquite sinks roots seventy feet under the earth to drink from the aquifer. Its cream-colored flowers sprout bulging clusters of sweet, moist yellow beans. To grazing beasts it is simply ambrosia. When two shrewd entrepreneurs—John Chisum and Charles Goodnight—noted in 1865 that cattle costing them four dollars a head sold for thirty dollars in northern cities, they bred gigantic herds on the mesquite and drove as many as 300,000 cattle per year along the Sedalia, Chisholm, Loving, and Goodnight Trails to markets and railheads. The greatest cattle baron, Richard King, began by purchasing a land grant from a Mexican widow for less than two cents an acre and hiring a whole Mexican village for labor. The King Ranch soon surpassed 1 million acres.[62]

Texas filled up so rapidly that the state numbered 2.2 million by 1890. Its greatest city began to emerge in 1869, when the Buffalo Bayou Company began dredging the Houston Ship Channel. Dallas boomed as a railroad town in the 1870s. Yet, in the words of the historian Walter Prescott Webb, postbellum Texans were more "social debris" than a "well-ordered society." Racial antipathies hardened; violence increased. Seventy vigilante patrols roamed the state in search of rustlers, robbers, and gunslingers like John Wesley Hardin, who was credited with as many as forty murders. Judge Roy Bean, "the only law west of the Pecos," dispensed beer, whiskey, and "Hang 'em first, try 'em later" justice. Cattlemen and their hired guns executed Mexican shepherds because the sheep ate the grass down to its roots. They waged war on farmers who plowed up the mesquite and fenced the range with barbed wire. By the 1880s the heyday of the cattle drive was over.[63] But that was time enough for a new generation of Texans to apotheosize cowboy, horse, six-shooter, and steer, and link that romance to the state's original myth about grit and liberty on the frontier. The Civil War disappeared down the memory hole. Texas was not Georgia or even Louisiana writ large: Texas was protean and victorious.[64]

Finally, unlike the South, Texas got rich. Its first oil gusher spewed from the Spindletop drilling at Beaumont in 1901. By 1917 Texas sweet crude became the industry standard, the all-powerful Texas Railroad Commission dictated the supply and price of oil for much of the world, and the state was awash in revenue. But as much as Texas progressed in the twentieth century, its land and myths remained just as beguiling to T. Boone Pickens, J. Paul Getty, H. Ross Perot, and good old boys drinking Lone Star beer as they were to Mirabeau B. Lamar.

According to the historian George Bancroft, a Democrat from Massachusetts, President Polk told his cabinet, "There are four great measures which are to be the measures of my administration." They were tariff reduction, an independent treasury, settlement of the Oregon boundary, and the acquisition of California. Note that Polk did not say "measures I shall pursue" or "measures I hope to achieve"; rather, he said "measures which are to be." The domestic program was easy. Polk laid it out in his first annual message, and the Democratic Congress passed it by summer 1846. The foreign policy measures, by contrast, were risky and subject to so many variables that Polk could not imagine when, how, or on what terms they might be fulfilled. War against Britain over possession of Oregon, war against Mexico over the Rio Grande boundary and California, wars over both, or peaceful resolution of both all seemed distinct possibilities. Polk's campaign promise, iterated in his inaugural address, not to cede an inch of the U.S. claims further complicated the issues. So did events on the ground. The stream of American pioneers to Oregon and California and the tension along the Rio Grande were bound to trigger events, perhaps violent, whatever the wishes of the federal government. But nothing complicated diplomacy more than the ponderous pace of communication. It took weeks or months for news and diplomatic dispatches to arrive in Washington City from the Pacific, London, or Mexico, and weeks or months for Polk's orders to reach their recipients. So even though the president's goals remained fixed, his tactics were highly flexible, even experimental. Polk simply hurled diplomats, secret agents, soldiers, and sailors in every direction to increase the chances that some of his initiatives would pay off. That is what made his statecraft look like a grand design to some and a chaotic improvisation to others. In retrospect, Polk was lucky. But smart gamblers sometimes make their own luck.[65]

Historians usually discuss U.S.-British and U.S.-Mexican diplomacy separately. But these negotiations lurched back and forth simultaneously and sometimes influenced each other. Hence, only a chronological treatment can convey the tempo and confusion of events. In most cases Polk was reacting to out-of-date news or trying to anticipate future developments. But everything he did was deliberate. Polk rarely rested except on the Sabbath, and he left the capital only four times in four years. He kept a meticulous diary. He held cabinet meetings every Tuesday and Saturday morning. He strove for consensus because Bancroft (Navy), William Marcy (War), James Buchanan (State), Robert Walker (Treasury), John Mason (Justice), and Cave Johnson (Post Office), represented with distinction every wing of the Democratic Party. Finally, Polk regularly consulted key senators such as Benton and Calhoun. This all helps to explain why his brinkmanship never eroded his political base.[66]

Polk first went to the brink over Oregon.[67] The outgoing Tyler administration had offered partition of the territory along the forty-ninth parallel, whereas the British stuck to their old demand for a boundary on the Columbia River. But Congress, riding the crest of Polk's victory, passed a bill in February 1845 to establish a territorial government, build forts, and sell land all the way up to fifty-four degrees, forty minutes north latitude. Only Calhoun's assurance that a treaty on Oregon was imminent prevented the Senate from passing the bill. Britain's minister to the United States, Richard Pakenham, was only momentarily relieved. Polk's inaugural address in March asserted "clear and unquestionable" U.S. authority over all Oregon. British editorialists lambasted him for his arrogance. Even Calhoun called it "a profound blunder." In fact, Polk meant to stake out an extreme position just so the ultimate compromise would seem more favorable. "The only way to treat John Bull," he explained to a congressman, "was to look him straight in the eye." Andrew Jackson himself rose from his deathbed to endorse "a bold and undaunted course," because (he told Polk privately) "England with all her boast dare not go to war." Polk failed to reckon, however, on the bellicosity of western Democrats. His inaugural address elicited from Senator William Allen of Ohio the whoop: "Fifty-four-forty or fight!" Having heightened the tension with Britain, Polk sent William S. Parrott on a conciliatory mission to Mexico. Parrott reported that the Herrera regime was not amenable to resuming diplomatic relations, owing to a last-minute flurry of Anglo-French diplomacy over Texas and to hostile public opinion. But neither did Parrott credit reports of Mexican troop movements toward the Rio Grande.[68]

The final act of the Texan drama was about to begin in May when Pakenham received Aberdeen's instructions to submit Oregon to arbitration or else solicit an American proposal for a boundary on the forty-ninth parallel

THE OREGON BOUNDARY
DISPUTE

excepting only the southern tip of Vancouver Island. But Polk and Buchanan preferred to respond to a British proposal. Each government sought to bluff the other into making the first concession. In any case, nothing could be done until Polk appointed a new U.S. minister to the Court of St. James's, a process that ate up two months because the president's first four choices declined. Over those same weeks in late May 1845 disturbing news arrived from the American consul Larkin at Monterey. The Californios had risen to expel another despised governor and had apparently done so with covert British support. What would become of the province should the locals declare independence from Mexico? Polk immediately asked Bancroft to draft orders for Commodore John D. Sloat. He was to screen the Pacific coast, gather intelligence, and seize California's main ports in the event of a U.S.-Mexican war.

In July Louis McLane finally set sail to represent Polk in England. Known as a free trader and a man of compromise, he was a fine choice. Moreover, Polk and Buchanan had decided, in light of the rumors about California, to make an offer regarding Oregon after all. This offer suggested a partition on the forty-ninth parallel, but with no exception for Vancouver Island and no navigation rights for Britain on the Columbia River. Pakenham threw it back in their faces without even referring it to Aberdeen. Polk had no choice but to withdraw it, reaffirm the United States' demand for fifty-four-forty, and wait, especially since rumors were spreading again about a Mexican army moving north. In London both Aberdeen and McLane threw up their hands. Diplomatic malpractice by their countrymen across the ocean had led to an impasse over what should have been an easily brokered dispute.

The following month more out-of-date dispatches arrived from Parrott in Mexico. He advised the Polk administration to discount the bellicose rhetoric of the *puros* and generals. The Mexican army was ill-prepared and a military spending bill "sleeps in the Chamber of Deputies, where it will not soon be disturbed." Parrott thought a properly accredited U.S. diplomat could settle affairs with Mexico "over breakfast." Accordingly, Polk's cabinet decided to send John Slidell of Louisiana to Mexico City on a secret mission. While Slidell made leisurely preparations, however, more three-month-old dispatches arrived full of Larkin's alarms about British activity in California. Polk upped the ante. In October 1845 he drafted orders for Larkin and Commodore Sloat to shadow the British, foster pro-American sentiment, and arm the U.S. citizens on hand to resist any British incursion. One set of the orders sailed around Cape Horn with the USS *Congress* under the aggressive Commodore Stockton. Another set was borne across Mexico by Lieutenant Archibald Gillespie disguised (by Bancroft's Bostonian chums) as a Yankee merchant. His mission became even more dramatic when Benton reminded Polk that John Frémont was again exploring somewhere near California. Polk then secretly told Gillespie to locate Frémont and share Larkin's orders with him. What the president imagined Larkin, Sloat, Stockton, Gillespie, and Frémont might accomplish while utterly out of touch is anyone's guess. But it makes sense to assume that Polk meant to get Americans on the ground to contest any outcome in California *other* than U.S. annexation.

There is no reason to doubt that the president, as Buchanan put it, was "anxious to preserve peace, although prepared for war." Thus Slidell sailed from Pensacola in December with instructions to persuade Mexico to concede the Rio Grande and sell California and New Mexico for a price as high as $40 million. But just a few weeks after his arrival the *puros* began another revolt in hopes of unseating Herrera and mobilizing the nation for war. Knowing that urban mobs alone would not suffice, Gómez Farías allied with General Mariano Paredes y Arrillaga. On December 30, the outgunned Herrera resigned, whereupon Paredes betrayed the populists by making himself president, flirting with monarchism, and denouncing the "tyranny of the demagogues." Mexicans lost their last chance to unite in defense of their territory. Perhaps there never had been a chance. As a *moderado* later reflected, Mexicans could never display "a national spirit, for there is no nation." Paredes and the parliament did agree that Slidell must be snubbed, because to receive a credentialed diplomat would imply recognition of the United States' annexation of Texas. In any event, to cede even more territory would be political suicide. Not surprisingly, Slidell reported the Mexicans could never be dealt with until they received a drubbing.[69]

Polk started from scratch in his December message to Congress. He not only laid claim to the whole of Oregon but invoked the Monroe Doctrine to imply that British expansion anywhere (read: California) would be intolerable. As for Mexico, Polk ordered Zachary Taylor to begin shifting his 4,000 troops from Corpus Christi to the Rio Grande opposite Matamoros. All this made January 1846 the moment of maximum danger because Americans could not know if the British cabinet was considering war over California or Oregon. Indeed, the Tory government fell apart that month over the controversy surrounding the Corn Laws. Those tariffs protecting the home market from imported foodstuffs made the price of Britons' daily bread artificially high and had an especially cruel effect during the Irish famine. But the landowning squires nevertheless fought tooth and nail to retain them. So the queen invited the Liberal leader Lord John Russell to form a government. Had he done so, Aberdeen would have had to relinquish the Foreign Office to feisty Palmerston. But Russell failed to unite his own party, in what became known as the "Whig abortion." He was obliged (in Disraeli's words) to hand the poisoned chalice back to Peel. Aberdeen stayed in office, now more cautious than ever. He even planted editorials gently suggesting the wisdom of a compromise regarding Oregon, because the Hudson's Bay Company had suspended operations in the Columbia basin and the U.S. Congress was following Britain's lead by lowering tariffs. But Aberdeen, no fool, realized he must give Polk and the Congress an excuse for abandoning their announced policy. So he paired expressions of goodwill with a casual remark to the effect that the Royal Navy was sending thirty ships of the line to the North American station. On February 22 Polk's cabinet perused that little item in McLane's dispatch. On February 24 the cabinet agreed to entertain a partition on the forty-ninth parallel. At last, both partners were ready to stop their coquetting and dance.

Slidell's patience ran out that very week. He informed the Paredes regime that it had to choose peace or suffer the consequences. Foreign Minister Manuel de la Peña y Peña blamed the United States for the consequences. Slidell demanded his passports and embarked on March 15. Polk would not learn of that until early April, but he already thought it time to "take the remedy for the injuries and wrongs we had suffered into our own hands."[70] What injuries, Peña y Peña might well have inquired. But with the crisis over Oregon subsiding, Zachary Taylor deploying, and California up for grabs, Polk thought it time to start drafting a war message. Slidell's arrival reinforced that council of despair. The cabinet acquiesced, wishing only that the Mexicans would ease the task by firing the first shot. It learned, the next evening, they had already done so.

After Slidell's departure the Paredes regime was pressured to justify its existence by standing up to the hated yanquis. Reinforcements were at last ordered north. General Pedro Ampudia arrived to find American naval and ground forces training their guns on Matamoros. When Taylor refused his request to cease and desist, Ampudia ordered a reconnaissance in force to cross the Rio Grande some miles upstream. An American patrol blundered into this force on April 25, 1846, and all but a few of the sixty-three bluecoats were captured, wounded, or killed. Taylor's message— "Hostilities may now be considered as commenced"—traveled by horseback, coach, steamboat, and telegraph halfway across the continent to arrive on Saturday, May 8. Polk, Bancroft, and Buchanan spent all day Sunday assembling a dossier to prove they had bent over backward for peace. "It was a day of great anxiety to me," Polk wrote in his diary, "and I regretted the necessity which had existed to make it necessary for me to spend the Sabbath in the manner I have." His war message made what seemed an open-and-shut case. "After a long-continued series of menaces," he claimed, the Mexicans had at last "invaded our territory and shed the blood of our fellow-citizens on our own soil." In fact, Calhoun, Benton, and numerous Whigs had plenty of questions, which the president ignored or obscured. On what basis did Texas claim the Rio Grande boundary? What was Taylor doing way down on the river? How did the skirmish occur? Even Polk's loyalists questioned his cavalier use of executive power. Of what value was the constitutional power of Congress to declare war when it was presented with a fait accompli? But the Democratic leadership in the House closed down debate and rammed through a war measure by 174 to fourteen. The Senate was in pain—for over a day, Polk doubted that a declaration would pass—but party discipline, recollection of the Federalists' self-destructive opposition to war in 1812, and the absurdity of trying to argue Mexico's case before American voters silenced doubters. The Senate voted for war, forty to two.[71]

The *Union*, Polk's mouthpiece, screamed like an eagle. "We mean to conduct war with Mexico with all the vigor in our power. . . . *We shall invade her territory; we shall seize her strongholds; we shall even TAKE HER CAPITAL, if there be no other means of bringing her to a sense of justice*."[72] No wonder Whigs such as Congressman Abraham Lincoln suspected Polk had planned an aggressive war from the start. But no one could explain, if that was so, why the administration had made *absolutely no war preparations at all*.

A week later the steamer bearing Aberdeen's instructions on Oregon left Liverpool. What followed was pure anticlimax. Buchanan and Pakenham tied up a treaty, which Polk submitted on June 4 (having already buttonholed at least nine crucial senators). Northern Democrats led by Cass noisily protested the sacrifice of fifty-four-forty, but mostly to posture for their constituents. Whigs were content with the Columbia River and Puget Sound. Southern Democrats never cared much about Oregon anyway. Nobody wanted continued strife with the British when still larger prizes loomed in Mexico. The Senate ratified the treaty, thirty-eight to twelve, in executive session on June 11; and forty-one to fourteen a week later in public. The Oregon question had festered for thirty years. It was settled in thirty days.[73]

*   *   *

The familiar image of the Mexican-American War as the drubbing of a weak nation by a greedy, more powerful neighbor is valid insofar as the U.S. government nurtured huge territorial ambitions and Mexico had scant hopes of recovering Texas. It is also valid insofar as the United States eclipsed Mexico in every category of power. Paredes presided over an empty treasury, a corrupt bureaucracy, almost no navy, and an army demoralized, poorly equipped, and often unpaid. Lacking an arms industry, Mexican soldiers made do with obsolete British muskets and cannons. Yet the Mexicans also counted several advantages, so they were not loco to challenge the *norteamericanos*. The regular, battle-hardened Mexican army was several times larger than the American army. The Mexican soldiers could march quickly over enormous distances, and the Mexican cavalry was magnificent. Mexico could count on sympathy from Britain and France. Above all, the Mexicans would be fighting on the strategic defensive and on their own turf, where the same lack of infrastructure that impaired their economy would frustrate the invader. Finally, the Mexicans believed they held a trump card. If, just once, their army punched into Texas, the *tejanos*, Indians, and enslaved Negroes might rally to them as liberators. The United States, by contrast, were unready for war and split over the justice of "Mr. Polk's War." Any Mexican victories prolonging the contest, it was assumed, would surely erode Americans' will to fight.[74]

Zachary Taylor denied the Mexicans victories. Just two weeks after General Arista's forces ambushed the American patrol and surrounded Fort Texas opposite Matamoros, Taylor marched 2,220 regulars south from Corpus Christi. On May 8 and 9 they put to flight superior forces at Palo Alto and Resaca de la Palma, then linked up with the Navy to cross the Rio Grande and occupy Matamoros. Thanks to the withering fire of their "flying artillery," the Americans killed 1,500 Mexicans and lost just thirty-five of their own (including the war's first hero, Captain Samuel Ringgold). Lieutenant Ulysses "Sam" Grant wrote home: "I think you find that history will count the victory one of the greatest on record." It was not that, but it gave some strong hints about the nature of this war. The United States' firepower, combined arms tactics, and land-sea coordination by professional officers, proved glorious assets. But the U.S. volunteers, who poured into Matamoros only to fall sick, bully civilians, riot, or desert, proved shameful liabilities.[75]

Taylor's blunting of a Mexican invasion of Texas bought the Polk administration precious time to build up the armed forces and settle on a strategy. Polk's political biases ensured the first task would be botched. Begrudging even a modest increase in the regular army, he asked Congress for $10 million to raise 50,000 citizen soldiers, then appointed amateurish, exclusively Democratic, generals to command the recruits.[76] The volunteer regiments would vex the West Point graduates throughout the war. Polk was stuck with the irascible, aristocratic Whig Winfield Scott as senior general, but denied him a field command after they quarreled over who was most guilty of playing politics with the war. Instead, Polk left Taylor, "Old Rough and Ready," in charge of the Rio Grande. Taylor was also suspected of Whiggery, but at least he was a veteran frontiersman with a democratic demeanor and national admiration. Polk and Secretary of War Marcy did much better with strategy (thanks in part to Scott's input). They envisioned a three-pronged offensive to occupy all northern Mexico. Taylor was to await the volunteer regiments, then move as far as was prudent toward the city of Monterrey. Colonel Stephen W. Kearny's dragoons were to capture Santa Fe, then cross the desert to California. Meanwhile, commodores Sloat and Stockton were to secure San Francisco and Monterey, where the U.S. consul, Larkin, was based. Maybe even Frémont's troop would turn up.[77]

Polk meant to grab all he craved at the outset, expecting the Mexicans to accept the faits accomplis. But he dared not tip his hand. So the president asked the bishops of New York and Missouri for Catholic chaplains to reassure Mexicans that no harm would come to them or their churches. Then he asked Buchanan to draft a statement for Europeans' consumption. It was far too innocent: the secretary of state denied all territorial aims! Polk angrily edited the draft to say that Americans fought "solely for the purpose of conquering an honorable and permanent peace." That did not hoodwink Aberdeen, but he had written off California and was obsessed with repealing the Corn Laws. Nor was Congress fooled when, in August, Polk asked for $2 million to help resolve "all our difficulties with the Mexican Republic." In his secret diary Polk explained that $2 million plus cancellation of Mexico's $4 million debt should cover the cost of "Upper California, New Mexico, and perhaps some territory South of these provinces." Whigs and Van Buren Democrats howled against this "Two Million Bill" because they suspected an ulterior motive. Polk, in private life, was a full-throated proponent of slavery who put his money where his mouth was: he invested heavily in slave "property" in the belief that it had a profitable future in the Southwest. So the Democrat David Wilmot of Pennsylvania proposed an amendment prohibiting slavery in any lands taken from Mexico. The House passed the Wilmot Proviso, eighty-seven to sixty-four; the Senate adjourned without voting; Polk did not get his $2 million. He was perplexed by the "mischievous & foolish amendment," because he didn't see "what connection slavery had with making peace."[78]

At the start of the war volunteers lined up, cheered by patriotic crowds, in scores of cities and towns. Senator Calhoun might grumble that Americans, though vigorous and adventuresome, had as little wisdom or experience as an eighteen-year-old. But even most who believed the conflict unnecessary and self-serving felt obliged to support the troops in the field. Most churches, so vociferous on other social issues, kept quiet or were divided, north versus south. Catholics and Presbyterians openly supported the war. Only Quakers, Unitarians, and Congregationalists were outspoken in protest. But the aged John Quincy Adams and his mouthpiece, the evangelical Congressman Joshua Giddings of Ohio, damned the principle "Our country right or wrong" from the beginning. They accused Polk of lying to drag the country into "an aggressive, unholy, and unjust war" that any true patriot must oppose. After the Wilmot Proviso threw annexation and slavery into the debate, Polk's honeymoon ended. Moralists in New England grew into a faction known as "conscience Whigs" as opposed to the "cotton Whigs," and "barnburner" Democrats disassociated from Polk's southern base. Were Americans on the side of the angels or devils? James Russell Lowell's rustic "Mr. Biglow" had no doubt. "They may talk o' Freedom's airy / tell they pupple in the face / it's a grand gret cemetery / fer the barthrights of our race / they jest want this Californy / so's to lug new slave states in / to abuse ye, and to scorn ye / an' to plunder ye like sin."[79] The reviled administration clearly needed good news and finally got some on September 1. Americans had seized California.

John C. Frémont later faced a court-martial for his role, but the sheer romance of it all cleansed the drama of shame. It began in January 1846, well before war was declared, when Captain Frémont's latest "topographical expedition" brazenly toured California again. Whether he was freebooting or executing a "Benton family plot" is still a mystery, but General José María Castro's superior force convinced Frémont that a prudent retreat was in order. He rode slowly north into Oregon. The following month Polk's other spy, the U.S. Marine Lieutenant Gillespie, arrived in Monterey with the orders for Larkin to organize American settlers and conciliate Californios. Gillespie then tracked down Frémont at Klamath Lake to deliver another copy of Polk's orders, plus private letters from Benton warning of British intrigues. After battling snow and Indians in the mountains, Frémont arrived in California to find that U.S. citizens had already resorted to arms. Castro was cracking down on illegal immigration. A small band of settlers led by Ezekiel Merritt retaliated by placing the gentle governor, Mariano Guadalupe Vallejo, under house arrest at his ranch in Sonoma. Castro vowed to slaughter the rebels. After a skirmish was fought in the Marin headlands Frémont rode to the settlers' defense. He also bestowed a prophetic name on the magnificent strait entering San Francisco Bay. He called it the Golden Gate. Frémont urged the Americans to declare an independent Bear Flag Republic, and they did so at Sonoma on the Fourth of July. Three days later Sloat, having finally received confirmation of war, hoisted the Stars and Stripes in Monterey Bay. He was appalled to learn that Frémont had acted without orders, but happily turned over command when Commodore Stockton arrived. Stockton promptly folded the Bear Flag battalion into the U.S. Army and set up a fragile government pending the arrival of overland reinforcements.[80]

That Army of the West, as Brigadier General Kearny fancifully dubbed his 1,458 men, set out on its adventure in June. It consisted of regular dragoons from Fort Leavenworth, mounted volunteers from Missouri under Colonel Alexander Doniphan, and the promise of Mormon volunteers from Council Bluffs. In fact, the Mormons had been "volunteered by" Brigham Young, who hoped to obtain federal aid for the exodus of his people to Utah. Kearny's men made a grueling 856-mile march in just two months. They dragged their wagons and cannons over the dusty, hot plains to Bent's Fort and then over the Raton Pass, expecting a ferocious fight. They were relieved when a courier arrived from Santa Fe reporting that the Mexican garrison had disbanded. Since its governor, remembering the Texans' failure to capture the town just four years before, was full of fight, what had happened? One answer, maybe the only answer, is that another of Benton's spies had done excellent work. The senator dispatched James Magoffin, an old friend who was a trader in Chihuahua, to organize a "fifth column" in advance of the army. Magoffin assured merchants that

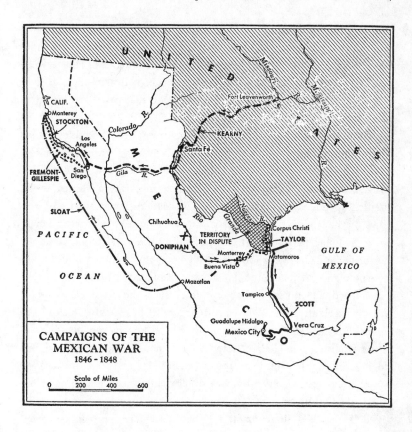

CAMPAIGNS OF THE
MEXICAN WAR
1846 - 1848

Scale of Miles
0   200   400   600

business on the Santa Fe Trail, shut down by Mexico, would boom under
the U.S. flag. The army officers he just bribed. So Santa Fe's governor
called a war council on August 16, only to find that nobody wanted war.
Kearny peacefully occupied the adobe town, entertained notables, and established a civil government respecting the residents' rights. He also announced the annexation of New Mexico to the United States.

The Army of the West resumed its march to the Pacific and learned
from Kit Carson, riding east, that Stockton had California in hand. So
Kearny sent most of his force back to Santa Fe. He regretted that decision
when his 100 men emerged from the desert. It seemed that easygoing An-
gelenos led by José María Flores and Andrés Pico had shrugged off the
martial law imposed by Gillespie. The rebels were formidable in the saddle.
In his first skirmish with them Kearney lost eighteen dead and thirteen,

including himself, wounded. Stockton and Frémont hastily dispatched re-inforcements that engaged Flores in a little last stand on the San Gabriel River. Their honor satisfied, the Californios came to terms in January 1847. The United States had pacified the vast province with the sailors off two frigates and four schooners, 100 dragoons, 200 marines, sixty "topogra-phers," and a few dozen armed settlers.[81]

More news that should have been good reached Washington City from Zachary Taylor. After three months of hassling with supplies, raw recruits, and Whigs who hoped to draft him for president, Taylor put his army on the road to Monterrey, the metropolis of northern Mexico. It was guarded by mountains, a river, and 10,000 soldiers in a semicircle of bas-tions. By the book, a successful assault would require at least 20,000 men. Taylor had 6,645, but thanks to his skilled engineers and artillerymen the bastions fell one by one, allowing his infantry to fight its way into the city from both ends. After a three-day slugfest costing 500 U.S. casualties Tay-lor accepted General Ampudia's offer to evacuate the city in exchange for an eight-week armistice. The armistice raised Polk's hackles. He ordered Taylor to rescind it and resume the advance as soon as supplies allowed. Taylor would, but in the meantime he had enough to do controlling his troops, winning the trust of a panicked enemy city, and preparing to blunt a new offensive threatened by none other than Santa Anna, home from exile and breathing fire.

Polk had himself to blame for that. Conquering provinces proved easy, but "conquering a peace" meant finding Mexican authorities willing to bargain. Since the Paredes regime refused, Polk recalled a feeler thrown out months before by an agent of Santa Anna. He hinted the general would gladly sell the land coveted by the Americans if they helped him regain power. So Polk ordered the Navy to permit Santa Anna to sail from Cuba through the U.S. blockade to Mexico, and dispatched a young offi-cer, Alexander Slidell Mackenzie, to talk terms with the general in Ha-vana. Mackenzie was a fiddle in his hands. He put the president's promises in writing, swallowed Santa Anna's assurances, and even took seriously the Mexican's recommendations for American military strategy. Santa Anna stayed put until August, when news from Mexico told him the moment was ripe. The *puros* and *moderados* had joined forces to topple Paredes. General José Mariano Salas and the liberal hawk Gómez Farías were in charge. They meant to wage total war against the *americanos*. Santa Anna sailed to Veracruz on a British steamer and proceeded to Mexico City, cheered by civilians and soldiers. Restored to command of the army in September, Santa Anna hotly repudiated the armistice at Monterrey: "Every day that passes without fighting in the north is a century of disgrace for Mexico!" He pledged to raise an army of 25,000 and drive the hated yanquis off Mexican soil. That same week the State Department received the Salas regime's rejection of peace overtures.[82]

That was enough to cancel out the United States' battlefield victories in the minds of many American voters. The two dozen seats gained by Whigs in fall 1846 would give them a narrow majority in the 1847–1848 session of Congress. Moral doubts explained some of the outcomes in the North, but Benton told the president most voters were simply impatient: "Ours is a go-ahead people." He urged Polk to make a "bold blow": seize Veracruz from the sea and then send an army on a "rapid, crushing movement" up the road to the capital. Benton even had in mind a commander—himself—who would deny Whig generals the glory. There was a snag. To outrank Scott, Benton would need an unprecedented commission as lieutenant general, which Congress was not likely to grant. Even Polk's cabinet thought it a bad idea. Yet the partisan president stubbornly pressed it, to the consternation of generals and congressmen, until Benton himself was embarrassed and recommended the appointment of Scott. The cabinet unanimously approved the Veracruz strategy on November 17. Two days later Polk placed Scott in command. All six feet four inches of "Old Fuss and Feathers" gave a shudder as he choked back his tears; then he raced out the door to plan the invasion that would make him a nineteenth-century Cortez. Not for the last time, protests and war weariness (after just six months) caused risky escalation of an American war.[83]

The Twenty-Ninth State: Iowa, 1846

"The whole country is good for nothing," thought Captain James Allen, because of the absence of timber. Captain Kearny, who founded Fort Des Moines in 1834, had more imagination. His column of dragoons marveled at prairie grasses up to ten feet high, vast potpourris of wild flowers, and strawberries so dense that their horses trampled out crimson tracks miles

in length. Surely these "handsome prairies" would someday sprout bountiful farms. Lieutenant Albert M. Lea's pamphlet recounting their expedition publicized the potential and fixed the name of this country. Marquette
and Jolliet, who explored its rivers in 1673, had bequeathed the impossible
sequence of vowels *Ouaouia*. English-speakers thought it sounded like
"Ioway." Thanks to Lea, the U.S. Congress chose Iowa.[84]

The first white man to settle in Iowa country was, as usual, a trickster.
Julien Dubuque came from Canada in the 1780s hoping to trade and mine
lead. The Meskawaki Indians living by Catfish Creek were less than cooperative until Dubuque threatened to incinerate their water supply. He told a
comrade to make an oil spill upstream, then hurled a fiery brand as the spill
floated by. Whoosh! The Indians decided to make the wizard welcome.
Under U.S. jurisdiction, however, the region was just a blank spot between
the Mississippi and Missouri rivers. Originally part of the Louisiana Territory, it was shunted to the Missouri, Michigan, and Wisconsin territories
before achieving its own territorial status in 1838. White settlement started,
literally overnight on June 1, 1833, when treaties made after Black Hawk's
War liquidated Indian claims to the western bank of the Mississippi. Hundreds of families poled over to claim bottomland. By 1838, whites numbered 23,000; by 1846, 96,000. The Sauk, Fox, Meskawaki, and Ioway (one
of the easternmost tribes to ride horses) moved sullenly west, ceding central Iowa in 1842, and all 56,290 square miles by 1851. With one hideous
exception, white settlement was achieved without bloodshed.[85]

Van Buren appointed as territorial governor Robert Lucas, a Methodist teetotaler from Ohio. Lucas pressed for immediate statehood because
he discovered most of the settlers were fervent Jacksonians and because
Florida needed a northern "sister state." But Iowans wanted no part of the
taxes required to support themselves until they sought to get out from
under a Whig territorial governor. In 1844, a heavily Democratic convention drafted a constitution providing for direct election of all state officials
and judges, paltry governmental salaries ("a fair price for services rendered but not a dollar for dignity," vowed a delegate), and no banks or
corporations without the voters' approval. Iowa also petitioned for boundaries that encompassed a chunk of southern Minnesota. Then Congress
got into the act. A northerner hoping to carve an additional free state out
of the region moved an amendment stripping Iowa of its westernmost
third. Voters rejected that out of hand. In 1845, with Polk and the Demo-

crats now in control, Iowa's congressional delegate Augustus Caesar Dodge revived the proposal for statehood, based on maximum boundaries. A young congressman from Illinois named Stephen A. Douglas interceded, thereby earning a reputation for compromise. He drew Iowa's boundaries along the Missouri and Big Sioux rivers in the west and latitude forty-three degrees, thirty minutes in the north. Iowans relented. Polk signed the state into existence on December 28, 1846.[86]

"No one who lives here knows how to tell the stranger what it's like, the land," wrote James Hearst, one of Iowa's many great authors. "It breathes dust, pollen, wears furrows and meadows, endures drought and flood, its muscles bulge and swell in horizons of corn, lakes of purple alfalfa, a land drunk on spring promises, half-crazed with growth—I can no more tell the secrets of its dark depths than I can count the bushels in a farmer's eye as he plants his corn." Cow Vandemark, a character of the novelist Herbert Quick, rhapsodized about his first gaze on Iowa's prairie. "I shall never forget the sight. The keen northwest wind swept before it a flock of white clouds; and under the clouds went their shadows, walking over the lovely hills like dark ships over an emerald sea." But the early years proved as hard there as anywhere else on the frontier, especially on women. A newcomer from New York state took one look at the prairie and sighed to her husband, "There are no trees, Ryal. I shall die of it." Mary Ann Ferrin Davidson moved from Vermont to Indiana, and then to Iowa in 1846. In her diary she wept over two-foot blizzards, an Indian assault, and a night spent alone with her feverish baby and husband while a violent thunderstorm extinguished the embers in the grate. Her husband recovered his health and together they planted twenty rich acres of corn, potatoes, turnips, and vegetables. But at length they took their five children and moved to Oregon. They were not alone. Some 10,000 Iowans joined the California Gold Rush in spring 1850. Several thousand more died in that summer's cholera epidemic.

Spring 1851 brought such torrents that the Des Moines River inundated farms miles from its channel. Severe drought and more cholera arrived in the summer. None of that deterred pioneers, who camped on the Illinois side waiting for river ferries. Steamboat ports at Davenport, Burlington, and Keokuk boomed as centers of supply, construction, and industries serving agriculture. In good years, Iowa did turn into an ocean of corn.[87]

That attracted the railroads. By 1856 the Rock Island, Northwestern, and Burlington lines reached the Mississippi and the first great bridge spanned the river (it gave rise to Abraham Lincoln's most famous lawsuit). Iowans went berserk, thinking "railroads were literally magic wands which had but to touch a community to create prosperity." One editor wrote: "The Railroad mania in Iowa presents some developments which to a quiet looker on must appear more ridiculous than any 'manifestation' of the spirit rappers. . . . So furious has the zeal of each locality become in favor of itself, and so venomous against every other point coming in competition, that even political ties are sundered and the 'cohesive power of public plunder' proves too weak for the repulsion of railroad mania." The laying of track all over the state was chaotic and corrupt, but railroad construction jobs and market access for farmers boosted the state's population to 675,000 by 1860.[88]

The people who became the Hawkeyes arrived in three waves. Most of the early birds migrated up the Mississippi valley to settle Iowa's southeastern counties. Some even brought slaves, thinking they were still in Missouri. They placed the original Jacksonian stamp on the territory. Next came immigrants, 200,000 of whom arrived between 1840 and 1870. Two-thirds were German or Irish, but the immigrants also included Scots, the founders of New Sweden in 1846, and Dutch (hence Van Meter, Bob Feller's homestead). The third wave dated from 1854, when Yankees, Yorkers, and Ohioans spilled into Iowa. The Amana colony arrived from Buffalo in 1855, plus pockets of Amish and Mormons. Quakers founded West Branch, an Underground Railroad depot frequented by John Brown and later the birthplace of Herbert Hoover. New Englanders alone quadrupled in only five years. Their ordered townships laid out on Iowa's grid of 160-acre quarter sections created the uniformity so striking today when seen from the windows of airplanes. Except where rivers impinged, the roads ran either north-south or east-west along property lines marked by rows of trees planted as windbreaks. Yankee influence also inspired a law of 1858 mandating tax-funded schools in every town and a high school in every county. Methodists founded Cornell College and Congregationalists founded Grinnell College. The coeducational University of Iowa opened its doors in 1855, and Iowa State Agricultural College opened in 1869. The state's literacy rate always ranked among the highest in the United States.

Education was no defense against prejudice. Iowa's territorial and state constitutions contained black codes restricting polling places and schools to whites. Lynching was not unknown. In 1851 the assembly passed a law excluding free blacks from entering the state. Five years later, when racial issues emerged again, the state was in turmoil. The Kansas-Nebraska Act had opened the possibility of slavery on the Great Plains. Iowa's Democratic Party shattered. A convention to amend the state constitution was held in January 1857. This time old Whig policies made headway in amendments to legalize banking, increase the state debt, and move the capital to Des Moines (Iowa City got the university as a sop). But the delegates passed the issue of Negro suffrage on to the voters, who rejected it by a margin of five to one. Just two years later Iowa's twig started to bend in more progressive directions. The state's founding father, Augustus Caesar Dodge, returned from the U.S. embassy in Madrid to run for governor. Despite the best efforts of the reunited Democrats, Dodge narrowly lost to a Republican. During the Civil War Iowa provided more soldiers per capita to the Union army than any other state. After the war a younger generation came gradually to embrace equal rights for the few colored people among them. By the 1880s they were rolling out carpets for George Washington Carver.[89]

Iowans, though ethnically diverse, shared certain traits. The Scots-Irish from the upper South, rural New Englanders, Germans, and Swedes all tended to be aloof, skeptical, and grudgingly charitable: they were perhaps closer than any other folk in America to a European peasantry. That is why the delightful lyrics of Broadway's Meredith Willson sing of a "chip-on-the-shoulder attitude" so "by-gone stubborn, we can stand touching noses for a week at a time and never see eye to eye." Yet they will give you a shirt and strong back to go with it "if your crop should happen to die." Willson—hometown, Mason City—knew his people.[90]

When Congress convened in December 1846 Polk finally got serious. He regretted the "misapprehensions" regarding the Mexican conflict, accused protesters of giving aid and comfort to the enemy, and pledged vigorous prosecution of the war. In February 1847 he pushed through Congress a bill to increase the regular army by ten regiments. However, these would not be available in time for Veracruz. So Scott peeled off the best units from Zachary Taylor's army. He and Taylor both assumed Santa Anna would move east in anticipation of Scott's invasion. Instead, Santa Anna marched 20,000 *soldados* due north, hoping to crush the 4,700 men Taylor had left and perhaps invade Texas. The whole U.S. war effort came to depend after all on a heavily volunteer force. Thanks to the engineer's eye of General John Ellis Wool, the Americans were deployed on a lattice of ridges south of a hacienda called Buena Vista. Those heights commanded the only pass leading out of a baked lava desert north toward Saltillo and Monterrey. Many a boy from Illinois or Arkansas must have wondered what fate brought him to that searing hell when a flash flood of Mexican uniforms turned the charcoal horizon blue and white. It was Washington's Birthday, 1847. Santa Anna sent Taylor a message: "You are surrounded . . . and cannot in any human probability avoid suffering a rout. . . . You may surrender at your discretion." Taylor removed his palmetto hat and replied, "I decline accepting your request." All the next day Santa Anna sent human-wave assaults that repeatedly threatened to turn the Americans' flank or break their lines. Jefferson Davis's Mississippi Rifles, sounding the banshee howl later known as the rebel yell, saved the flank. Batteries of "flying artillery" quickly maneuvered to defend the ridges with volleys of grapeshot whenever American units were near panic (hence Taylor's famous remark, "A little more grape, please, Captain Bragg"). When the inferno subsided Taylor counted some 700 casualties. He would surely be overrun the next day. Instead, the sun rose on a desert speckled with 591 corpses and hundreds more moaning wounded. Santa Anna had quietly withdrawn in the night.[91]

Two weeks later a lieutenant in the U.S. Navy greeted another brilliant dawn with the words, "We could not have selected a more propitious day." Commodore David Conner's reinforced West India squadron sailed from Tampico to the waters off Veracruz, a mosquito fleet of surfboats in its wake. Those shallow-draft gunboats designed by Winfield Scott and built by the Army's Philadelphia arsenal were the first amphibious craft in the New World. Their purpose was to land an army at the New World's Gibraltar. After noon the armada pummeled Mexican redoubts above the beach. It wasn't needed. General Juan Morales wanted no part of a battle within range of the naval guns. Even so, the landing of more than 8,600 soldiers from five until ten PM without the loss of a single man or surfboat was the greatest logistical feat in U.S. military annals to date. But that was only the curtain-raiser. Veracruz had two great onshore forts and the towering Castle San Juan de Ulloa on a reef opposite the harbor. Together they mounted 135 guns manned by 3,360 defenders. A long siege was out of the question because the fever season was nigh, but Scott spurned bloody frontal assaults: "We, of course, gentlemen, must take the city and castle before the return of the *vomito*." Storming the forts would cost "perhaps three thousand of our best men [and] although I know our countrymen will hardly acknowledge a victory unaccompanied by a long butcher's bill, I am strongly inclined to forgo their loud applause and . . . take the city with the least possible loss of life." So Scott's heavy artillery and Conner's warships unleashed a punishing barrage totaling 6,700 shells and shot over four days. The Mexican commander, afraid for the townspeople more than the forts, surrendered on March 27. Veracruz was seized at the cost of just seventy-three American casualties.[92]

Scott's expeditionary force quickly grew to 12,000 men while the quartermaster tried heroically to assemble nearly 1,500 tons of supplies and 9,000 wagons. The wagons required some 7,000 mules and twenty-five tons of fodder. Scott's own preparations included a stern proscription against mistreatment of civilians, orders to purchase rather than confiscate supplies obtained in Mexico, and his own conspicuous attendance at a Catholic Mass. On April 2 the army started its mountainous 250-mile journey into the Mexican heartland. Any number of narrow defiles might have to be forced. Anywhere Mexican guerrillas might be waiting in ambush or poised to cut Scott's supply line. The campaign could well have replayed the Afghan disaster that swallowed a whole Anglo-Indian army in 1841. That it did not explains why the Duke of Wellington later called Scott "the greatest living soldier." The first test was Cerro Gordo, a town squeezed between a river and mountains that Santa Anna deemed impassable. Scott dispatched his West Point engineers in hopes that something would turn up. Thanks to Lieutenant P. G. T. Beauregard and Captain Robert E. Lee, something did: a trail beyond the first range of mountains leading to undefended heights on the enemy's flank. The enemy's retreat moved Scott to crow, "Mexico has no longer an army"; then he repented of his exuberance as his own army dwindled daily from disease and expired enlistments. Soon he was writing to Secretary of War Marcy, "For God's sake, give me a reinforcement of 12,000 regulars."[93]

Adding to Scott's vexations was the arrival in camp of Nicholas P.

Trist with diplomatic orders he refused to reveal. It seemed that Polk's latest spies (the adventuresome journalists Moses Yale Beach and Jane McManus Storms) had found no Mexicans willing to sell out their country. So Polk ordered Trist, a Democrat who once clerked for Thomas Jefferson, to bring home a peace that would shove the Whig generals off the heroes' bench. Scott and Trist loathed each other until the latter received a démarche from Santa Anna promising again to make peace in return for a bribe. Since that meant dipping into Scott's army slush fund, Trist extended an olive branch to the general. Scott reciprocated with a jar of guava marmalade that cured Trist's diarrhea. Their resulting friendship would baffle Polk and Buchanan while permitting Santa Anna to dupe Americans for a third time. He pocketed Scott's $10,000 advance and used it to help recruit another army to defend Mexico City.[94]

After pausing at the lofty town of Puebla to rest his troops and await reinforcements, Scott made the risky decision to let go of his lifeline to Veracruz and leap forward to Mexico City. In August his 11,000 troops descended into the volcanic basin defended by 30,000 desperate Mexicans. The mountains and lakes surrounding the city forced a choice among three routes, all dangerous. Once again Lee, Beauregard, Captain James L. Mason, and Lieutenant George B. McClellan studied the ground and concluded that victory could be had if two crucial choke points were captured: a road junction near Contreras and a fortified former convent at Churubusco. Santa Anna lost 10,000 *soldados* in their fierce defense, but Scott prevailed at the cost of 1,000 U.S. casualties. Now all that remained were the castle of Chapultepec and the causeways into Mexico City. Scott and Trist offered an armistice, pleading, "Too much blood has already been shed in this unnatural war between the two great Republics of this Continent." When Polk later learned of the truce he angrily recalled Trist and sought an excuse to sack Scott. But events were far beyond his control. When the Mexicans still refused to talk peace, Scott's patience expired. On September 13, after another fierce fight, U.S. marines and soldiers raised their flag over Chapultepec. A Mexican officer was heard to lament, "God is a Yankee," doubtless crossing himself as he spoke. The following morning Scott was awakened by delegates from the capital. Santa Anna had fled. The city fathers wished to surrender.[95]

Trist's response to Polk's recall was a seventy-page letter explaining why he meant to ignore the president's wishes. The Mexican war parties were at last discredited; an interim regime under Peña y Peña was prepared to negotiate. Political chaos, the Mexicans' hope for British mediation, and the haggling itself ate up four months during which Scott and the Mexican elites alike grew increasingly nervous about guerrilla warfare. At last the Mexican commissioners met Trist on February 2, 1848, in the village of Guadalupe Hidalgo, to write a peace treaty. Its terms were within the limits originally laid down by Polk. Mexico ceded to the United States the Rio Grande boundary, New Mexico (including the future Arizona), and Alta (but not Baja) California in exchange for $15 million. In mid-January, on receipt of Trist's insubordinate letter, Polk thought the envoy had "acted worse than any man in the public employ whom I have ever known." In mid-February, on receipt of Trist's treaty, Polk sent it posthaste to the Senate. The vote was delayed for a week—Congress was mourning the death of John Quincy Adams—but on March 10, twenty-six Democrats and twelve Whigs voted to liquidate the war, whatever objections they had to the terms of the treaty. Only fourteen senators voted nay.[96]

For better or worse the Mexican War was the climactic achievement of the republic founded in 1776. It answered those preachers, reformers, and authors who anxiously asked what sort of nation Americans would make of themselves. Most obviously, the war displayed an impatient expansionism. Counting Texas, it added over 1 million square miles to the United States and realized the dream of continental empire. The war also exposed Manifest Destiny as a pretentious fraud. O'Sullivan prophesied peaceful, natural expansion through the efforts of pioneers. But the acquisition of Oregon and the Mexican Cession demanded shrewd, even cynical statecraft belying romantic notions of U.S. exceptionalism. The conduct of that war and diplomacy also disproved the Jacksonian conceit that the nation could do without professionalism. With few exceptions, the glorious battlefield victories were won by the regular Army and Navy. The unsung hero was Quartermaster General Thomas Sidney Jesup. He managed twenty-three federal arsenals that produced the tens of thousands of infantry weapons, artillery tubes, uniforms, boots, and tents the expanding army required, then transported those supplies to distant, primitive foreign locales. To mobilize the private sector Jesup wisely adopted open-market purchasing rather than tedious competitive bidding. The U.S. manufacturers, breeders, and merchants eagerly met the government's needs for ordnance, wagons, horses, mules, and victuals. Jacksonians' dislike for internal improvements notwithstanding, logistical support of armies and fleets south of the border depended heavily on railroads, steamboats, telegraphs, harbors, and levees (especially in New Orleans) constructed in the past fifteen years. Jacksonian pastoralism notwithstanding, the war proved Americans needed advanced technology. In every major action superior firepower canceled Mexico's advantage in manpower. Jacksonians' biases notwithstanding, the audacious U.S. offensive strategy relied on the excellent staff work of junior officers trained at West Point. "I give it as my fixed opinion," wrote Scott, "that but for our graduated cadets, the war between Mexico and the United States might, and probably would, have lasted some four or five years." Jacksonians' penury notwithstanding, the Treasury easily financed the war by issuing $76.5 million in debt.[97]

The Mexican War also revealed a people so steeped in politics, personal ambition, and bigotry born of ignorance that they hurt their own cause. Polk's partisan distrust alienated his senior commanders. The Democratic politicians that he turned into generals were (excepting the Mississippian John A. Quitman) incompetent. When Polk's belated expansion of the regular army opened 400 new billets, regular officers bitterly noted that only five were filled from their ranks. Grant, Meade, McClellan, and Robert Anderson were among those passed over. When Scott and Pillow feuded in Mexico, they planted their sides of the story with America's first war correspondents, George Wilkins Kendall of the *New Orleans Picayune* and James L. Freaner of the *New Orleans Delta*. As for Taylor, everyone knew the Whigs meant to run him for president. Morale was even worse among enlisted personnel. "An individual who enters the U.S. Army as a private soldier," wrote an Irish recruit, "must expect to be treated more like a vicious dog, than a civilized, intelligent, human being." The 75,000 volunteers whom Polk imagined "hardy pioneers of the West" mutinied, deserted, malingered, rebelled, and died in the deadliest war in American history. Just 1,548 soldiers were killed in action, but 10,970, 11 percent of the total muster, succumbed to dysentery, influenza, smallpox, measles, venereal disease, exposure, and even snakebites and tarantula bites. Bad water, exposure, poor diet, and crowded camps contributed, but army surgeons also suspected poor screening by recruiters anxious to fill their rosters.[98]

The Mexican War also exposed a self-righteous, self-confident nation. New Yorkers, noted Melville, were "in a state of delirium. . . . Nothing is talked about but the 'Halls of the Montezumas.'" Hawthorne spied a "chivalrous beauty" in the volunteers. The feminist Emma Willard praised U.S. soldiers for waging war like "knights of old." Polk invented an American trope by insisting that the war was aimed not at the Mexican people, but only at a corrupt, brutal regime. He called it a "war of reconciliation." Walt Whitman wrote, "Cold must be the pulse, and throbless to all good thoughts . . . which cannot respond to the valorous *emprise* of our soldiers." Maps and books about Mexico flew off the shelves. "Yankee Doodle," "The Star-Spangled Banner," and "Hail Columbia!" resounded on village greens. Inspired by visions of the Stars and Stripes rising above foreign soil, citizens rushed to purchase flags mass-produced, for the first time, by the Annin Flag Company and hawked by Whitman. They also hailed a new generation of heroes. Taylor became a "fabled personage" nurtured on "Kentucky's dark and bloody ground." Abe Lincoln recalled his "pride and sorrow" on hearing how "noble Ringgold fell." George Lippard left off writing prurient novels in order to romanticize war, "the great prompter of high deeds, the originator of noble impulses and generous actions . . . the great corrector of civil sloth, servile luxury, and national licentiousness." There is, he claimed, "a sublimity in battle." But whereas southerners and westerners felt no need to justify fighting an inferior race for land, northern intellectuals imagined the war an apotheosis of the republic's Anglo-Saxon moral mission. The *Democratic Review* claimed nations had "social duties" so that "when a nation keeps a 'disorderly house,' it is the duty of neighbors to interfere." The Boston Brahmin Caleb Cushing spoke in the same breath of chastising and uplifting Mexico. New England's *Uncle Sam* expressed the libidinous intoxication of a crusade amid lissome, dark-eyed *señoritas*:

> *The Spanish maid, with eye of fire*
> *At balmy evening turns her lyre*
> *And, looking to the Eastern sky,*
> *Awaits our Yankee chivalry*
> *Whose purer blood and valiant arms,*
> *Are fit to clasp her budding charms.*
> *An army of reformers, we—*
> *March on to glorious victory.*

By mid-1847 those reformers, stung by the indictments of Thoreau, Emerson, and the "conscience Whigs," carried the mission to its logical conclusion. If this was indeed a war of annexation, then the only way to redeem it was to annex all of Mexico and bestow on its unhappy people Yankee law, religion, and enterprise. The "All-Mexico" movement was no joke. Its rapidly growing appeal across party lines made Polk even more anxious for peace.[99]

Scott and the other U.S. commanders also feared that a prolonged occupation would escalate the guerrilla war and turn victory into disaster. (The Mexican revolt against French occupation fifteen years later proved they were right.) As for spreading civilization, officers knew their real task was to minimize the barbarism perpetrated by Americans, especially the volunteers and deserters or discharged men beyond their authority. After the fall of Monterrey some Texan cavalrymen burned, pillaged, and murdered 100 civilians. Taylor called Ohioan volunteers "a God damned set of thieves and cowards." Soldiers whose enlistments ran out formed gangs to rob banks, steal horses, and molest women. Gold-laden Catholic churches held special appeal. A volunteer wrote to his brother, a Protestant pastor, "I wish I had the power to strip their churches . . . to bring off this treasure hoard of gold, silver and jewels, and to put the greasy priests, monks, friars, and other officials at work on the public highways." Such greed helps to explain both the high rate of desertions among Protestant troops and the higher rate among Catholic troops, who found little to love in the American cause. The most famous deserters joined the Mexican army's San Patricio Battalion. When at last they were captured Scott ordered thirty-five of the traitors hanged at the same moment the Stars and Stripes rose over Chapultepec.[100] Professional officers knew it was vital to maintain good conduct in the midst of a dense, hostile population. They were only partly successful. It might even be said the Mexicans did more to civilize the invaders exposed for the first time to an alien, often appealing culture.[101]

To his lasting credit Polk spurned the temptation to annex densely populated portions of Mexico in the name of uplifting its people. To be sure, his statecraft sometimes bordered on farce and was saved in the end by Scott and Trist, the men he distrusted most. Yet the Treaty of Guadalupe Hidalgo rewarded Polk's determination to take just what the national interest required, no more. At the time nobody gave him credit except himself. According to the *American Review* "a civilized and Christian people regard an unnecessary war, in the middle of the nineteenth century, a spectacle of backsliding and crime over which angels may weep." According to the old Jeffersonian Albert Gallatin, "All these allegations of superiority of race and destiny neither require nor deserve any answer; they are pretences [sic] to disguise ambitions, cupidity, or silly vanity." According to Polk's last annual message, "The war with Mexico has thus fully developed the capacity of republican governments to prosecute successfully a just and necessary foreign war with all the vigor usually attributed to more arbitrary forms of government."[102] Only in America could all three views be valid, because all three pretended that morality, not *raison d'état*, was the only acceptable basis for policy.

The Thirtieth State: Wisconsin, 1848

In the lands encompassed by lakes Michigan and Superior and the rivers Mississippi and Saint Croix the receding glaciers left behind three regions, all wonderful. Hardwood evergreen forests blanket the northern tier, undulating prairies characterize the southeastern quadrant, and beyond the terminal moraine in the southwest is the hilly "driftless area." Their wonders include the fourth largest profusion of lakes on earth, the fluky Door Peninsula, yellow sand beaches, feral rivers such as the Wolf and Brule, and the gorges called dells on the Wisconsin River. Long, cold winters; forests of maple, birch, aspen, elm, ash, pine, hemlock, balsam, spruce, and cedar; deer, foxes, rabbits, skunks, chipmunks, squirrels, gophers, bears, beavers, eagles, geese, and seventy-four kinds of fish epitomize images of the Great North Woods. The frontier artist George Catlin wrote, "The Ouisconsin, which the French most appropriately denominate '*La belle rivière*,' may certainly vie with any other on the Continent or in the world, for its beautifully skirted banks and prairie bluffs." In 1838 the English writer Frederick Marryat called it "the finest portion of North America" because "nature had so arranged it that man should have all troubles cleared from before him, and have but little to do but to take possession and enjoy."[103]

Taking possession meant displacing the Menominees, Winnebagos, Chippewa, Fox, and Sauk, some 20,000 of whom flourished when the first French explorers arrived. European diseases and wars in which Indians consistently chose the losing side (the French versus the British, then the British versus the Americans) greatly reduced their numbers. So did the whiskey peddlers, "the avowed enemies of education and religion" in the words of a Catholic missionary. But Indians lost none of their land, even when Americans began to mine lead deposits around Prairie du Chien, until the drunken Winnebago chief Red Bird killed a white family in 1827. His death in prison made him a martyr. But the incident obliged the Indians to cede the lead mining region in 1829. The ensuing Black Hawk War forced more cessions, and Jackson's Indian Removal Act completed the rout. By 1837 almost the whole territory was open to miners, trappers, loggers, and settlers. The first to dominate were entrepreneurs who won federal leases in the lead mining country. Their profits on thousands of tons of ore shipped to New Orleans financed stately homes that are still on display in Galena, Illinois. Most of the original laborers were seasonal workers from downriver. But when the surface lead was picked over, proprietors recruited some 7,000 Cornish miners expert in shafting and drainage. Those who wintered in Wisconsin, often camping out in the tunnels they dug, were called "badgers." [104]

The politics of the Wisconsin Territory—created in 1836, when Michigan won statehood—were unseemly even by American frontier standards. Jackson appointed Henry Dodge governor. Dodge had been raised in Kentucky during its wild early years and had won laurels in the War of 1812 and the Black Hawk War. Lest any man challenge his honor, he wore pistols and bowie knives on his belt. No wonder Old Hickory liked him. Dodge's nemesis—they fought like a fox and a badger for years—was a Yankee hustler and raconteur to whom honor was a commodity: James Duane Doty. He was born on the New York–Vermont border in 1799, passed the bar in Detroit, paddled with Governor Cass around the Great Lakes, and won an appointment as Wisconsin's judge at age twenty-three. His sole pursuits in life were power through politics (it was said he had only one party, himself) and wealth through land speculation. Doty's first scam—embezzling federal funds to buy land—was undone by the collapse of prices in the Panic of 1837. But John Jacob Astor bailed Doty out and his next scam made him a fortune. The usual controversy had erupted over where to locate the territorial capital. Governor Dodge threw up his hands and said he would accept whatever the quarrelsome legislature decided. Doty quietly went to work on a town site between the lakes that were later named Monona and Mendota, and then bribed legislators with choice lots in the city of Madison, which was yet to be built. (Creative corruption indeed: it was an excellent choice.) Doty's bid for power also succeeded in 1838, when he won the election for territorial delegate in cahoots with a third candidate who split the Dodge party's vote. Doty used his ties to Cass and the Democrats to win federal funds for roads and canals in Wisconsin. But when the Whigs came to power in 1841, Doty just as easily charmed Clay and Webster with badinage, whiskey, and deals. He replaced Dodge as governor. Indeed, the only battle Doty lost was over the name of his adopted land. He wanted to spell it "Wiskonsan."[105]

"We are overrun here with land speculators, sharpers, &c &c," wrote an editor in Prairie du Chien. "Land speculators are circumambulating it and Milwauky is all the rage," wrote the *Green Bay Intelligencer.* Indeed, settlers arrived so fast that the population grew tenfold from 31,000 in 1840 to 310,000 in a decade. Still, voters in Wisconsin resisted statehood until 1846, when they realized that their share of federal land sales would exceed the paltry budget Congress provided. The admission of Texas made room for another free state. The Mexican War aroused interest in national politics. But the 103 Democrats (against just eighteen Whigs) who wrote a state constitution imposed such punitive clauses against banking that even some Jacksonians joined Yankee arrivals to vote it down. (No one minded that blacks were disenfranchised.) When a second constitution left bank law up to the people, a smaller turnout ratified it. That constitution of 1848 is still in force today. Congress and Polk approved it on May 29, making Wisconsin, the northwestern corner of the old Northwest, the last state east of the Mississippi to join the Union.[106]

In the 1850s Wisconsin more than doubled again, to 776,000, on the strength of bumper wheat crops and lumber shipped to mills throughout the Great Lakes. But issues of class, ethnicity, and corruption made those anxious years. Eastern investors bought up a third of the state in lots larger than 1,000 acres; the future general Cadwallader C. Washburn of Maine owned 130,000 acres. So resentful were voters of absentee landlords that the assembly nearly limited property rights to 320 acres or two city lots. Money, mocking, and the fear of "red socialism" combined to defeat it. "Resolved," opponents joked, "that all property in the United States, the State of Wisconsin, and the City of Milwaukee ought to be equally divided every Saturday night." Ethnicity was a more pressing matter. Immigrants, primarily Germans, surpassed one-third of the state population and half of

Milwaukee's 45,246 people in 1860. The city fathers lived on Yankee Hill; artisans and merchants in German Town, Tory Hill (Britons), Nauvoo (Mormons), or the Wooden Shoe District (Dutch); and laborers in the Bloody Third (Irish). Many of the Germans were Catholic; all ignored the state's blue laws. Even worse, their *Gemütlichkeit* and "strong, blooming" girls lured many a good "American" boy to the beer garden. An editor in Milwaukee noted a "significant transition of the foreigner and Romanist from a character quiet, retiring, and even abject, to one bold, threatening, turbulent." His solution was compulsory secular schooling: "then we are safe." But the Roman Catholic bishop John Martin Henni joined Protestant sects in establishing schools and colleges well ahead of the townships and the state. So Wisconsin tried, in vain, to prohibit doctrinal requirements for students or faculty members even at private schools.[107]

Attempts to suppress corruption proved just as vain. Crusaders as disparate as the German Republican Carl Schurz and the Irish Democrat Edward G. Ryan thought greed was making Wisconsin a "paradise of folly and knavery." When the gubernatorial election in 1855 ended in a virtual tie, the state canvassers, all Democrats, ruled the incumbent the victor by 157 votes while the chief justice, a Republican, inaugurated the challenger, Coles Bashford. In the ensuing lawsuits Ryan discovered sheaves of "supplemental ballots" had materialized long after the polls closed. Bucking his party, he exposed the fraud. But Bashford's vindication hardly cleaned up the state. Just two years later he fled to Arizona after pocketing a gigantic bribe from railroad promoters. Still, one can see in these early efforts to restrict the power of fat capitalists, mandate nonsectarian public education, and clean up state government the seeds of a Progressivism whose flower was the "Wisconsin idea" of the next generation. Robert La Follette, the great Progressive reformer, was born in Primrose the year of the disputed election.[108]

Moooo! Why is nothing said about cows? A headline from the *Milwaukee Sentinel* in 1861 gives the answer: "WHEAT IS KING" and Wisconsin its throne. Almost every farmer planted grain or hops for the breweries of Schlitz, Blatz, and Heilemann. As late as 1870 the census counted a mere twenty-five dairy farms out of 160,000 households engaged in agriculture. Of course, farm wives made some cheese on the side, but the state's 3 million pounds per year were piddling beside New York state's

100-million-pound mountain of cheddar. It was thanks to the vision of one man alone that Wisconsin became America's Dairyland. William Dempster Hoard, a minister's son, moved from the Mohawk Valley to Wisconsin in 1857. He worked as a farmhand, Methodist circuit rider, and pump sales-man before fighting with the Union army. In 1870, at age thirty-four Hoard founded a weekly to preach a new gospel of dairy farming lest the soil be depleted. In 1885 he followed his own advice—specialize, don't diversify—with a new journal called *Hoard's Dairyman*. Its circulation reached 70,000. Over five decades Hoard served a term as governor; chaired the university's board of regents; experimented to find the hardi-est, most nutritious strains of alfalfa for Wisconsin's soil; worked with the Department of Agriculture to popularize silos and scientific ensilage; and inspired farmers to replace Jersey cows with the more productive Holstein and Guernsey breeds. Milk could be shipped only locally before refrigera-tion, but high-powered marketing made Wisconsin synonymous with cheeses, including Colby, Swiss, Muenster, and the bland yellow product soon known as American cheese. By 1918, the year of Hoard's death, his adopted state produced almost two-thirds of all the cheese in the United States. He also bequeathed a proverb: "Speak to a cow as you would to a lady." [109]

The stillborn Wilmot Proviso haunted presidential politics in 1848. Would American citizens be permitted to revive slavery where Mexicans had abolished it? Could any candidate even answer that question without killing his chances of a nationwide victory? The Democrats' convention in Baltimore in May suggested the wisdom of silence or pretense. When Polk stuck to his vow not to seek a second term, three veteran Jacksonians threw their hats into the ring. Senator Lewis Cass of Michigan advocated "squat-ter sovereignty," meaning that western settlers should be free to decide for themselves about slavery. Secretary of State Buchanan proposed extending the Missouri Compromise line to the Pacific. Justice Levi Woodbury said nothing at all. Nor did the platform committee. But the nomination of Cass on the fourth ballot nevertheless rent the party. New York "barn burners" bolted to form a third party with some abolitionists and "Con-science Whigs." In high moral dudgeon, they met in a Universalist church in Buffalo to nominate Martin Van Buren and Charles Francis Adams. It seemed that Old Van ("the most fallen man I have ever known," snarled Polk) had repented of his lifetime of pandering to the South. The party's slogan—"Free Soil, Free Speech, Free Labor, Free Men"—was far too radical to carry even a northern state, but was sure to peel votes away from the moderate Cass.[110]

The Whigs' June convention in Philadelphia drew the proper conclusion. The Whigs unified behind Zachary Taylor, a war hero and Louisi-anan slave owner, balanced him with New York's Millard Fillmore, and offered *no platform whatsoever*. Nor did General Taylor, who apparently never voted or spoke out on issues, have any track record to attack. Cass, a lifelong partisan politician, was a mudslinger's delight. So the major parties colluded to stage yet another presidential race that ignored the nation's most pressing question. That allowed southern Whigs to stage a big comeback from 1844. Taylor, "Old Rough and Ready," polled 47.3 percent of the popular vote and 163 electoral votes from seven southern states and eight northern states. Cass polled 42.5 percent and 127 electoral votes. But Van Buren's nearly 300,000 votes (10.1 percent), heavily concentrated in the Northeast, might have tipped the balance against Cass in several states, especially New York.[111]

No president worked harder in office than James K. Polk. Even after the election, the outcome of which he deplored, Polk exhausted himself in order to leave "a clean table for my successor." He took special pains with his last annual message, crediting the "Sovereign Arbiter of All Human Events" for his administration's spectacular acquisitions, including respect from foreign powers who had "entertained imperfect and erroneous views of our physical strength." But everyone knew that. Polk saved the kicker for last. He confirmed "accounts of the abundance of gold" in California "of such an extraordinary character as would scarcely command belief were they not corroborated by the authentic reports of officers in the public service."[112] So his impatient pursuit of California had been justified after all, and the wealth of the province would more than pay for the war.

During his term Polk did not go home even once. Yet when he and Sarah boarded a train in Washington City on March 6, 1849, it was to make a triumphal tour across the South. His fatigue became illness, then incapacity. On returning at last to Nashville, Polk asked to be baptized, then died just three months after leaving the White House. He had fulfilled every promise made four years before. He proved a more effective

Jacksonian than Jackson himself. But he never fathomed the danger his feisty nationalism posed to his nation. Let Ralph Waldo Emerson rant about the nation swallowing arsenic and John C. Calhoun call Mexico forbidden fruit. They were extremists. Americans would never permit extremists to hazard the destiny none could deny was now manifest.

Chapter 6

FORTY-NINERS, FILIBUSTERS, FREE-SOILERS, AND FIRE-EATERS

Gospels of Slavery, Commerce, and Christ Sunder the Civic Religion, 1849–1860

Spiritual rhetoric bathed American politics, courtrooms, schools, and public festivities. The First Amendment proscribed a national religious establishment, but in so doing it invited citizens to embrace a civic faith that stood above sectarian creeds and guaranteed their free exercise. The last stanza of the popular hymn "America" began, "Our Fathers' God to Thee, Author of Liberty." The last stanza of "The Star-Spangled Banner" asked a "heaven rescued land" to "praise the Power that has made and preserved us a Nation." Both northern and southern Whigs and Democrats believed as fervently as the founding fathers that Americans were "called unto liberty" in a promised land destined to manifest the fruits of liberty—growth, peace, and prosperity—before the eyes of the world. Of course, Protestant evangelicals assumed that their God was the nation's protector, and they deemed Catholics, Mormons, and others "un-American." But high-minded patriots such as William Henry Seward, Jefferson Davis, Winfield Scott, and Robert E. Lee knew that sectarian bigotry was heretical to the civic religion, since it threatened the unity, liberty, and destiny of what G. K. Chesterton would call a nation with the soul of a church.

Civil religion broadly defined is a universal phenomenon. The ancient Greeks and Romans worshipped the divine patrons of their cities and empires. To chant "Great is Diana of the Ephesians" or burn incense to Caesar was both a political and a spiritual obeisance. The cults of Asian god-kings and god-emperors were civil as well as religious. Even Judaism had features of a civic cult in the eras of its monarchy and temples. Divine right conflated civil and religious loyalties in European monarchies; and the republics of the Italian Renaissance founded patronage cults around saints (e.g., Saint Mark in Venice). But a new sort of civil religion emerged from the Protestant notion of a republic as a holy covenant or social contract. Thus, James Harrington in England under Cromwell and Jean-Jacques Rousseau in Geneva asked what might sustain a government of the people. Their answer was a civil religion all the more powerful for being voluntary (not imposed by a priesthood), devoted to the commonwealth (not to a prince), and obedient to God or nature (not to corrupt human beings). Americans gave voice to such civic faith when they sang in their Revolution: "down with this earthly King; NO KING BUT GOD."

Who was this God of the American civic religion? In presidential inaugurals Washington called him the Almighty Being, Invisible Hand, and Parent of the Human Race; John Adams: the Patron of Order, Fountain of Justice, and Protector; Jefferson: the Infinite Power; Madison: the Being Who Regulates the Destiny of Nations; Monroe: Providence and the Almighty; John Quincy Adams: the Ark of our Salvation; Jackson, Van Buren, Harrison, and Polk: that Divine Being who had willed the birth and survival of the United States. As "high priests" of the national cult, presidents not only praised the deity but congratulated the people on the virtues that earned divine blessings. Those incomparable blessings included independence, civic and religious liberty, unequaled abundance, and a self-image allowing Americans to *feel good* about *doing well*. That was a precious birthright. But it required all citizens to honor, with religious devotion, the rule of law, republican virtue, the desire of others to feel good about themselves, and above all the national unity on which

everything else, including God's plan for history, relied. Hence, the unfor-
givable sin of the civic religion was schism. Journalists, preachers, and
politicians from all sections and sects habitually referred to the Union as a
sacred, holy temple whose rupture was as unthinkable as "overturning the
Christian religion."[1]

Alas, that temple of continental expanse was half slave, half free, and
crawling with money changers. The only way its worshippers could resist
schismatic temptations was by magnifying pretense until even Americans
grew weary of lying to themselves and each other. That is why, in the end,
their "better angels" became accomplices in the wreck of the Union.

Brigham Young led the first 2,000 Mormons across the frozen Mississippi
River in February 1846. His purpose was to lead the "camp of Israel" on an
exodus beyond U.S. territory so the saints would never again be murdered
and burned out of their homes. Eight hundred perished during that first
hungry winter in Iowa, but not before planting vast cornfields to supply
the thousands soon to follow the Mormon Trail blazed by the scouts. Over
the summer and fall Young prepared another winter quarters on the Missouri River opposite Council Bluffs. It was then that a captain of dragoons
arrived to recruit Mormons for the Mexican War. Young damned a nation
that expelled his people only to beg them to fight its battles. In fact, he
"volunteered" the Mormon Battalion in order to glean the soldiers' pay
and win favor with President Polk. In April 1847 Young resumed his march
westward, claiming not to know where the Lord would lead them. In fact,
he and his twelve apostles always aimed for the Great Salt Lake country
praised by Frémont. Thanks to timely guidance from a Jesuit and the
mountain man Jim Bridger, the Mormons found their way across Wyoming, then southwest into Zion. "This is the place," proclaimed Young on
July 21, "where I, in vision, saw the ark of the Lord resting." The place was
a valley, twenty by thirty miles, south of the saline inland sea and west of
the Wasatch Mountains' well-watered, well-wooded slopes. Within one
week of their arrival the pioneers had dammed streams to irrigate fields of
potatoes and corn and had platted an "instant city" with large lots and
broad avenues surrounding the site of Temple Square. In August, Young
and Heber Kimball rode back past huge herds of bison to shepherd their
human flocks. In Wyoming they were overjoyed to encounter a wagon

train bearing 1,500 Mormons and 4,000 animals. At the winter quarters on the Missouri they prepared 2,000 more saints for next year's trek. Young promised free land to all good stewards who would be "industrious and take care of it." He meant to build "the Kingdom of God or nothing."

Apparent miracles confirmed the Mormons' faith. In Iowa the advance guard had been fed in the wilderness, just like the Israelites in the Sinai, by a providential bevy of quails. At Salt Lake, in June 1848, the providential arrival of seagulls helped the Mormons rout a plague of black crickets devouring their crops.[2] They were initially spared trouble with Indians when Young pacified the local "Lamanites" without guns or whiskey. Not since William Penn, said Senator Salmon P. Chase of Ohio, had any governor done better by Indians than Brigham Young. Prudence, prayer, and iron discipline proved so successful that within three years the desiccated land amply fed 11,354 people. An army surveyor judged it "one of the most remarkable incidents of the present age." Of course, the kingdom of God was no democracy. Young reigned as a theocrat because every male was a member of the hierarchy of the Church of Jesus Christ of Latter-Day Saints (JCLDS), whereas no unbeliever had any privileges whatsoever. Young bolstered his spiritual authority through a brilliant legal adjustment. He boldly discarded the common-law principle of riparian rights (anyone owning land abutting a stream or lake had use of the water) in favor of the Spanish principle of appropriation (the community owned all water and distributed it as needed). That made possible an elaborate network of canals irrigating entire valleys. It also made Young a "pharaoh" controlling the lifeblood of the colony. Finally, the Church of JCLDS sent missionaries to plant dozens of subcolonies in Wyoming, Idaho, Colorado, New Mexico, Arizona, Nevada, and southern California that were in many cases the first white settlements there. The only goal Young failed to achieve was to escape U.S. territory! No sooner was Salt Lake City established than news of the Mexican cession arrived. Over the winter of 1848–1849 the Council of Apostles tried to preempt federal control by founding a state called Deseret with jurisdiction all the way to Los Angeles. That imperial assertion by a theocracy endorsing polygamy began fifty years of "cold war" between Utah and the United States of America.[3]

Another Mormon entrepreneur fired the starting pistol for the California gold rush. While still at Nauvoo the apostles had ordered a twenty-six-year-old elder named Samuel Brannan to sail with 200 converts around

Cape Horn. They arrived in July 1846 just a few weeks after the U.S. Navy raised the flag over Yerba Buena and changed its name to San Francisco. The Mormons tripled its population overnight; went into trade; and paid tithes, which Brannan used to build a personal empire. Brigham Young insisted he send "the Lord's money" to Utah. Brannan said he would only do so in exchange for a receipt signed by the Lord. Then the Mormon Battalion arrived with Kearny in 1847 and fanned out over the province when the members' enlistments expired. Several took jobs with James Marshall, the carpenter from New Jersey hired by Sutter to build a saw-mill on the American River. On January 24, 1848, they saw Marshall pull some glittering pebbles out of the millrace; then they accompanied him to Sutter's Fort, where Brannan ran a general store. Sutter tried to hush it up, fearing that his land grants would be overrun. Brannan knew just what to do. He quietly bought up every pick, shovel, pot, pan, and bag of sour-dough flour around. Then he rode into San Francisco on May 12 crying "Gold from the American River!" The Bay Area emptied of people. Yan-kees, Californios, and Indians, then Mexicans, Hawaiians, and Peruvians rushed to the western foothills of the Sierra Nevada: some 6,000 Forty-Eighters in all. After Polk confirmed the strike in December, they were joined by 40,000 Forty-Niners and 60,000 more gold prospectors by 1852. In 1852 alone placer-mining sourdoughs extracted $80 million from their pans, rockers, and riffle boxes, sharply increasing the money supply and propelling an investment boom unparalleled in American history.[4]

Feverish gold bugs chose one of three routes to California, all pro-tracted and perilous. The most direct route, especially for midwesterners, was overland. The 30,000 Forty-Niners who vied to outfit wagon trains made fortunes for merchants in the boomtowns of Independence, Saint Joseph, Kanesville (Council Bluffs), and Omaha. Transport companies sprang up at once, promising to provide all the animals and equipment needed for a safe, comfortable journey to Sacramento in exchange for $200. In fact, overloaded wagons, overworked mules, heat, cold, cholera, Indians, indiscipline, and ignorant trail bosses made many a journey hor-rific. The influx of customers was also a boon to proprietors of those Rocky Mountain trading posts. Fort Bridger enjoyed its flushest years until Mormons raised up their own posts on the trail to steer emigrants through Salt Lake City. Their arrival was more manna from heaven because many merchants had crammed their wagons with merchandise. When advised that ship-borne goods were already flooding the market in California, the overland merchants auctioned their stock on the dirt streets of Salt Lake City. The remote Mormon colony was thus showered with all the tools and luxuries of civilization at prices a small fraction of what New Yorkers paid.

The second route to the goldfields was by ship around Cape Horn, a voyage of 18,000 miles that before 1849 often took six to eight months. Its advantage lay in the fact that winter in the northern hemisphere was summer in the southern hemisphere. So tens of thousands of impatient argonauts paid up to $200 plus sixty dollars per ton of cargo on any tub willing to brave the Strait of Magellan or Drake Passage. Shippers might clear $70,000 on a single voyage: the price of a new vessel in Baltimore or Philadelphia. Their shipyards quickly met the demand for speed and more speed in the final heyday of wood and sail. Samuel H. Pook, William H. Webb, and Donald McKay were among the architects celebrated for their "sharp built" vessels with up to a dozen billowing sails, narrow hulls, and pointed prows that "clipped" through the wash. Equipped with the navy's invaluable charts of coasts, reefs, rocks, winds, and currents, Yankee skippers vied to double the Horn and reach the Golden Gate in record time. The romance was brief. By 1853 the clipper ship fleet was so overbuilt and competition from steamships was so stiff that cargo rates tumbled to $7.50 per ton. Those steamships serviced ports on both sides of Panama through which 20,000 people per year reached California in the 1850s. Depending on how long one had to wait for a mule train at Chagres on the Caribbean or a steamship berth at Panama City on the Pacific, an emigrant might reach San Francisco from the East Coast in just seven weeks. But tropical disease and violence in the spiny isthmus took so many lives that Panama was the riskiest route to the goldfields. William H. Aspinwall of New York had a technological solution. In 1855 his army of laborers completed a forty-eight-mile "transcontinental railroad" across the mountainous jungle.[5]

Why not a canal? Polk in fact instructed Nicholas Trist, even before the news about gold, to ask Mexico for transit rights across its narrow Tehuantepec isthmus. When the British threatened to occupy Belize and Nicaragua's Mosquito Coast, Polk's corollary to the Monroe Doctrine forbade European colonization in the New World even with the consent of the inhabitants. But more than bluff was called for once Forty-Niners be-gan streaming across Panama, turning Central America into a geopolitical stake of immense value. It was left to John M. Clayton, President Tay-lor's secretary of state, to compromise the Monroe Doctrine. Under the Clayton-Bulwer Treaty of 1850 the United States and Britain each agreed to dig no canals without the other's consent, and not to fortify a canal or colonize Central America. Democrats sarcastically suggested that Queen Victoria grant Clayton a knighthood.[6]

What might have become of California had the Mexicans (or British) learned of its gold a decade or two earlier is anyone's guess. In the event, so many Yankees poured into California that it qualified for statehood under the 1850 census. Therein lay a conundrum. California was likely to be a free state. The only prospective slave state to pair with it was New Mexico. But supporters of the Wilmot Proviso stood four-square against the re-introduction of slavery in the Mexican Cession. The upshot was that the richest prize won by the slaveholder Polk justified Calhoun's worst fear: California, for the South and the Union, was a poison pill.

Thomas Hart Benton likened abolitionists and nullifiers to the blades of a shears that did little damage alone but together might shred the Union. In the wake of the Mexican War their cutting edges were northerners' resis-tance to any expansion of slavery and southerners' resistance to any restric-tion on slave owners' property rights.[7] So long as those remained bargaining positions—opening positions from which the camps were prepared to retreat—the status of new territories and states might be brokered as in the past. But tolerance for the moral obfuscation this required became a scarce commodity by mid-century. Since demography, geography, and time all suggested that slavery had no future on the Great Plains and the Pacific coast, southerners found all the more reason to demand guarantees and northerners found all the less reason to grant them. Buchanan did not re-alize a corner had been turned when he suggested that Congress simply extend the Missouri Compromise line to the Pacific. Nor did Cass when he suggested that pioneers should decide for themselves about slavery. They were shocked when, in August 1848, a bill to create the free-soil Or-egon Territory provoked vicious speeches and the threat of a duel. Only the nationalist scruples of senators Benton, Houston, and Presley Spruance of Delaware carried the bill over the opposition of fellow slave-staters. The Mexican cession remained in limbo until the new Congress convened in December 1849.[8]

That Thirty-First Congress, forty-niners of a different sort, might have been history's most distinguished and contentious. The venerable Calhoun, Clay, Webster, Benton, and Cass were still present, if not in full voice, while some notable young men were in voice. Stephen Douglas, the five-foot-four-inch Vermonter trained in New York's rugged politics, had taken Illinois by storm. A Romantic expansionist and railroad promoter, the thirty-six-year-old Douglas, called the Little Giant, said he learned on the prairie that America was a young giant sanctioned by God to stretch its limbs in every direction. Since the principal threat to this civic faith was sectionalism, Douglas preached nationalism, especially after his marriage in 1847 to the daughter of a North Carolinian planter. He also aspired to Clay's mantle as compromiser and voice of the West. Jefferson Davis, the crippled forty-one-year-old hero of the Mexican War, expected to inherit Calhoun's mantle as voice of the South. Davis's family had moved from Kentucky to Mississippi, where his brother and he became planters notorious for their humane treatment of slaves. Davis meant to champion the peculiar institution against two young lions of far different persuasion. William Henry Seward, a forty-eight-year-old disciple of John Quincy Adams, had already served two terms as the first Whig governor of New York. Seward's unruly red hair belied his cold personality. He spent as little time as possible with his wife, Frances, and their five children in Auburn, New York. But her abolitionist piety tugged on his conscience. Seward remembered with shame the domestic slaves his own father had owned, and the misery he witnessed during a stint teaching school in Georgia. Much to the chagrin of his wily manager Thurlow Weed, Seward took risky political stands on behalf of emancipation of Negroes and civil rights for Catholics and Indians. Yet even he was outflanked by a new senator, forty-one-year-old Salmon P. Chase. A native New Hampshireman who moved with his widowed mother to Ohio, Chase was chiseled by three Christian sculptors: his uncle, Ohio's pioneer Episcopal bishop; Charles Grandison Finney, whose revival claimed Chase as a student at Dartmouth; and his legal mentor William Wirt, the Anti-Masonic candidate. During the years when Theodore Dwight Weld scorched Ohio with abolitionist sermons, Chase defended fugitive slaves and helped to found the Free-Soil Party. When that party won enough seats in the legislature to force Democrats into a coalition, Chase was elected to the Senate. He was said to be as ambitious as Julius Caesar. Southerners hated Chase even more than they feared Seward.[9]

The House of Representatives' class of 1849 was even younger. The average age was just forty-three, and more than half the members were freshmen. Nor did they defer to senior leadership, because there was none. The ten seats won by the Free-Soil Party denied a majority to either of the major parties divided in turn on sectional lines. That meant a vicious fight for the speakership. The Whigs' choice, forty-year-old Robert C. Winthrop, was a direct descendant of the Puritan founder of Massachusetts. He despised the South's insistence on sectional parity as injurious to the Union. The Democrats' candidate, thirty-four-year-old Howell Cobb, was a Georgian planter with 1,000 slaves. He despised the North's insistence on majority rule as injurious to the Union. The famous educator Horace Mann, who occupied John Quincy Adams's old seat in Congress, said slavery was Cobb's "politics and his patriotism, his political economy and his religion." Cobb himself wrote, "This Union is the rock upon which the God of nations has built his political church, and we have been summoned to minister at its holy altars." After fifty-nine futile ballots the House amended its rules and chose the speaker by a mere plurality. Cobb won, but the chamber was so divided that his power was slight.

Both parties looked to Zachary Taylor for succor. He was a Whig by adoption at least, a frontier soldier, and as militant a nationalist as Andrew Jackson. He selected a cabinet of moderates equally drawn from the North and South, but his chief adviser was none other than Seward. Surely Taylor would not let the "slavocracy" stand in the way of the nation's progress. Yet the president was also a Louisianan planter raised in Kentucky. Surely Taylor would not betray his fellow southerners. In fact, the president betrayed all factions a little bit so as not to betray "that Union which should be the paramount object of our hopes and affections" and the prosperity "to which the goodness of Divine Providence has conducted our common country." Accordingly, his first annual message simply suggested that Congress admit California under a constitution of its people's choosing, but refrain from taking a stand against the expansion of slavery on principle. Later he even promised to veto any Wilmot-type bill. The soldier-president just wanted to end military government in the Southwest, bestow "popular sovereignty" on its people, and otherwise face reality. He was altogether too sensible for a Congress so schooled in pretense that it assumed reality itself was a cloak. The stentorian planter and classical scholar Robert Toombs of Georgia spoke for southern Whigs when he vowed "in the presence of the living God" that he preferred disunion to the exclusion of slavery from lands bought with the blood of the whole nation. What made his threat palpable was the sad acquiescence of Alexander Hamilton Stephens. A frail intellectual beloved by Abe Lincoln among others, Stephens accompanied Toombs to the White House with a warning about secessionists. Taylor spat, "I will hang them as high as I hung spies in Mexico." But the president's own agenda was dead on arrival.[10]

"Ancient Henry" Clay yearned to spend his declining years as Jackson had done: resting on his plantation surrounded by a doting family and servants, and by politicos paying court to the sage. But Kentucky would not let him retire, so he hobbled back to his seat in the Senate, coughing from tuberculosis, while his colleagues gave him a standing ovation. Clay took the temperature of the new Congress and knew why a fickle Providence put him there: to serve once more as "one of the high priests officiating at the altar of the Union." So he drafted a litany all sections might be willing to pray and showed it to Daniel Webster on a raw night in January 1850. The march of the Union must not be impeded. But free western states would make the South an ever more beleaguered minority. It was imperative, therefore, to assure the South that its minority rights would be honored in perpetuity. Clay meant to introduce an "omnibus bill" satisfying each section's just demands and exorcizing every demon tormenting the Union. Webster promised to do all he could to bring his fellow northerners around.

Clay revealed his "great national scheme of compromise and harmony" to a packed Senate chamber on January 29. It included: (1) admission of California as a free state; (2) organization of the rest of the Mexican Cession into territories without restrictions, since slavery was unlikely to flourish there anyway; (3) denial to Texas of its claims to New Mexican lands; (4) compensation to Texas through federal assumption of its public debt; (5) legal slavery in the District of Columbia forever; (6) abolition of the slave trade in the District of Columbia; (7) denial of any federal authority to restrict the slave trade within states; (8) federal protection of southern property through a tough, enforceable Fugitive Slave Act. Every demigod in the Senate was heard from in the Olympian debate that ensued. Jefferson Davis, once the object of Clay's avuncular affection, accused him of siding with the "aggressive majority" that had "declared war against the institution of slavery." Calhoun, at death's door, crafted a fiery speech read by James Mason of Virginia. Ignoring Clay's grab bag of palliatives, Calhoun inveighed against abolitionism and insisted the responsibility for saving the Union lay with the North.

Webster gratified Calhoun and Clay alike with a mighty speech supporting the omnibus. He dared to say that in the matter of fugitive slaves "the South is right, and the North is wrong." He damned abolitionists whose agitation only forced planters to tighten the bonds of slavery, but also those who pretended "peaceful secession" was an option. Webster concluded with his signature line: "let us come out into the light of day; let us enjoy the fresh air of *liberty and union*." Businessmen in New York, fearing the consequences of disunion, sent Webster a gold watch. But opponents of slavery damned Webster as a Benedict Arnold or a Lucifer fallen from heaven. Those sentiments moved Seward to throw politics to the wind in the most stunning speech yet. Compromise was "radically wrong and essentially vicious" and the recapture of fugitive slaves "unjust, unconstitutional, and immoral." It was time Americans confessed "there is a higher law than the Constitution." Now it was southerners' turn to choke. In June nine slave states sent delegates to a convention in Nashville. They failed to form a united front, but a vocal minority called the "Fire-Eaters" proclaimed that slavery could never be safe inside the Union.[11]

The omnibus suffered the usual fate of that ilk. Shifting factions passed amendments to strike one and then another part of the bill until by August nothing remained but imposition of federal authority on Utah. Why Clay attempted it is a puzzle, because even if it passed it would probably be vetoed by the president. But the waste of six months proved a perverse blessing in that two actors and one illusion were killed off. Calhoun had died in March, muttering, "What will become of the South?" Zachary Taylor died of acute gastroenteritis in July.[12] The illusion of an omnibus died when Clay left to recuperate in Newport, Rhode Island. He realized that the best strategy was the same one he had used thirty years before to effect the Missouri Compromise: break the package into separate bills and cobble together discrete majorities for each. Clay was too old and tired for that. Stephen Douglas, chairman of the committee on territories, was primed for it. He had carefully given no one offense during the long debate; he sensed that Congress was anxious to adjourn; and he had assurances that President Millard Fillmore would look with favor even on bills unpopular in his native upstate New York. Douglas got it done in six weeks. The bills regarding the Texas boundary and money passed, thirty to twenty. California's statehood passed, thirty-four to eighteen (all the nays were southern). Territorial status for New Mexico passed twenty-seven to ten (all the nays were northern). A tough Fugitive Slave Law passed twenty-seven to twelve (northern Whigs voted nay; northern Democrats abstained). Abolition of slave markets in Washington City passed thirty-three to nineteen (all the nays were southern). Finally, the House of Representatives was on board thanks to Douglas's lobbying of northern Democrats and southern Whigs such as Toombs and Stephens.

It passed into history as the great Compromise of 1850 that appeased advocates of states' rights, extinguished the burning issues of the day, and gave the republic a new lease on life. In truth it *wasn't* a compromise, it *assaulted* states' rights, and it *fanned* the flames of disunion. Only four senators voted for all the bills. Only a few members, mostly from border states, compromised on anything. All the others voted their convictions and interests, and so hated half of the outcomes. The package threatened the rights of slave states by consigning them to permanent minority status, and it trampled on the rights of northern states by forcing them to assist in capturing runaway slaves on their own turf. Nor did the package contain any reference to the Missouri Compromise line or the Wilmot Proviso. Senator Chase grumbled, "The question of slavery in the territories has been avoided. It has not been settled." Even abolition of the slave trade in Washington City was a sham: it just drove the markets indoors. The only substantial results of what is more accurately termed the "armistice of 1850" were statehood for Forty-Niners and fat profits for speculators in Texas bonds. It all persuaded the "blades" of Benton's "shears" that abolition, secession, or both would be only a matter of time. Yet so tense was the session that Capitol Hill celebrated for twenty-four hours, during which "it was the duty of every patriot to get drunk." Angry waves receded from the Union's "holy citadel," said Senator Daniel Dickinson of New York. The settlement was "final and irrevocable," boasted President Fillmore.[13]

The Thirty-First State: California, 1850

"I looked on for a moment; a frenzy seized my soul . . . piles of gold rose up before me at every step; castles of marble dazzling the eye with their rich appliances; thousands of slaves bowing to my beck and call; myriads of fair virgins contending with each other for my love—were among the fancies of my fevered imagination . . . in short, I had a very violent attack of the gold fever." Unalloyed greed was the worst basis on which to found a frontier society. Had the vast, blessed province been settled gradually under orderly territorial government, Americans' pursuit of happiness would have been more successful and a great deal more peaceful. Instead, the "California dream" was sullied by plunder, hatred, fear, violence, and vice. About 250,000 people arrived in five years, almost all male, most armed, and half younger than thirty. As late as 1880 males still outnumbered females two to one. Courts, churches, schools, and other agencies of civilization lagged. Monopolists became a law unto themselves. They got very rich. But very few men in mining camps such as Poverty Hill, Devil's Retreat, and Hell's Delight ever found what the Mexicans called a bonanza; and those who did usually squandered it in bars, brothels, and gambling dens on San Francisco's "Barbary coast." Observing the mayhem, a correspondent from New York recorded, "the people of San Francisco are mad, stark mad." By 1853, when gold production began to decline, California was occupied by an army of thwarted fortune hunters whose only means of asserting their manhood was mobbing.[14]

Federal authority, weak to begin with, collapsed when soldiers and sailors deserted for the goldfields and the military governorship changed hands five times in one year.[15] When Congress then deadlocked over California's status, the governor invited the Forty-Niners plus a few Californios to petition for immediate statehood. Their convention at Monterey in September 1849 unanimously banned slavery (to keep gang labor out of the diggings) and drafted a constitution modeled on Iowa's. The state's boundaries occasioned lengthy debate. William M. Gwin, an erstwhile congressman from Tennessee with the ambition of Aaron Burr, wanted all the lands west of Salt Lake. He was seconded by Henry W. Halleck. Other delegates feared provoking resistance in Congress. They compromised on a "modest" 158,706 square miles bounded by the 120th meridian south to Lake Tahoe, and then a line southeast to the Colorado River. Finally, the convention adopted a busy state seal depicting the goddess Minerva and a grape-nibbling bear overlooking a miner, ships on the Sacramento River, and mountains capped by the motto "Eureka." [16] In November voters ratified the constitution by thirteen to one; elected Gwin and Frémont to the Senate; and convened a "legislature of a thousand drinks" (so named for its alcoholic haze), ten months before Congress approved statehood on September 9, 1850. A four-year floating crap game ensued among speculators; it ended when Sacramento was made the permanent capital.[17]

California lived up to a name given to an earthly paradise in an old Spanish romance. On the northern coast, where seals barked and whales spouted, ancient stands of redwood rose 300 feet into the sky. San Francisco Bay was the most capacious and lovely harbor in the world, and the bluffs of Big Sur were the most dramatic encounter of land and sea. From Santa Barbara to San Diego rocky spurs shared the surf with beaches of finely ground yellow sand. Marching along the state's eastern border were landscapes beyond the imagination of Thomas Cole: Yosemite Valley, Lake Tahoe, Kings Canyon, the Joshua trees of the Mojave Desert, and Death Valley. South of the Tehachapi Pass lay the arid beauty of the San Gabriel Mountains and Los Angeles valley. The Pacific coast's Mediterranean climate was equally enchanting to Yankees. No winter! There were just some wet months that quilted the hills with perky orange poppies. California also had a year-round growing season in soils hospitable to grains, vegetables, orchards, and livestock; and pale blue skies above green valleys amid brown hills crowned in the distance by white-capped purple mountains. California's clean desert air, blazing sunsets, and bracing salt breezes born somewhere beyond Hawaii made old people feel young and the young feel immortal. To be sure, the 450-mile valley drained by the Sacramento and San Joaquin rivers was torrid and tormented by Thule fog in the winter and the Santa Anna winds in the summer. But once irrigated, the Central Valley was a cornucopia.

Only the people were ugly. The corrupt Democrats who monopolized politics were torn between rival factions. Gwin, who floated schemes to divide California into a free state and a slave state, had a lock on federal appointments. David C. Broderick, an Irish saloon keeper trained by Tammany Hall, was the boss of a San Francisco machine with a lock on state patronage. In 1855 their feud so paralyzed the legislature that one of the state's seats in the Senate was vacant for two years. Broderick won out, only to be killed in a duel by the state's chief justice in 1859. The very word *justice* was a joke in Sacramento, in the self-policed mining camps, and in the streets of San Francisco, the "instant city" that made California the most urbanized state in the Union. It was bad enough that the hastily built, wooden town burned to the ground *five times* between 1849 and 1851 (arson was always assumed), but San Francisco was also a magnet for gangs, thieves, and swindlers from all over the world. Australian ex-convicts, the "Sydney Ducks," vied with the mostly Irish "Hounds" who hated all other foreigners. Fights over women and gold ended in murders, and murders ended in lynchings. Sam Brannan became a civic leader by rousing "respectable" merchants to form committees of vigilance. Too often the vigilantes were thugs and looters seeking to distract public outrage from themselves. In 1856 the committee of vigilance went so far as to defy the governor, arrest a state justice, and attack the militia sent to restore order. Vigilantes rightly accused officials of impotence and corruption, but their own contempt for due process set a dangerous precedent.[18]

Another legacy of the gold rush was racial conflict born of the whites' fear of losing their tenuous grip on the Pacific Rim. In 1850 the legislature passed a stiff tax on foreign miners that drove out two-thirds of the 15,000 Mexican Forty-Niners. The federal Land Act of 1851 inspired by Senator Gwin dispossessed 200 Californios holding Mexican land grants. Several hundred more rancheros sold out cheap rather than pay for interminable litigation. Oakland, Berkeley, and Alameda arose on the liquidated Rancho San Antonio. Angry Mexicans became bandits; local race wars broke out. When vigilantes rode down the dashing Juan Flores in 1857, he was hanged on a short rope to prolong his death throes. Even more threatening to whites were the 25,000 Chinese shipped from Canton by 1852. The docile, long-suffering "Celestials" took jobs from the Irish, drove wages down, and were despised as bearers of heathenism, filth, and disease. State laws stripped them of all civil rights until they "didn't have a Chinaman's chance." An even worse fate awaited the 150,000 Native American "acorn eaters" such as the Modocs. Half succumbed to disease, one-fourth starved, and one-twelfth were hunted and shot like wolves. By 1870 just 30,000 Indians remained on remote reservations in the deserts or lava flows around Mount Lassen.[19]

California's infant economy was a winner-take-all faro game favoring priority, capital, clout, hustle, and luck. Boston's Adams and Company thought it had won by dominating transport between the gold country and San Francisco. But the New York firm of Henry Wells, William G. Fargo, and Johnston Livingston swallowed up smaller express companies in hopes of competing. When Adams went bust in 1855 during a bank panic, the Wells Fargo Company became "the omnipresent, universal business agent of all the region from the Rocky Mountains to the Pacific Ocean . . . the Ready Companion of civilization." Its 147 branches cleared two dollars on every ounce of gold deposited, plus fees for assay, banking, credit, and shipment. The most fabulous real estate mogul was Heinrich Alfred Kreisler, a German butcher who reinvented himself as Henry Miller in 1848. He realized the way to get rich was to make himself cattle king of this booming new state. Over two decades the firm of Miller and Lux procured 1 million acres of range and bottomland along the San Joaquin River while supplying San Francisco with beef and wheat. Agoston Haraszthy, a refugee from the Hungarian revolution of 1848, reigned over the wine country. Appointed a state commissioner in 1861, he imported 200,000 cuttings of varietal grapes destined for Napa, Sonoma, and Santa Clara counties. The most famous monopoly was that of the Four Associates, who finagled the massive federal support needed to build the Transcontinental Railroad. In so doing, Leland Stanford and his partners gave Americans both their defining example of creative corruption and their worst Jacksonian nightmare. The Southern Pacific Railroad that controlled shipping and owned 11.5 percent of the entire state was a concentration of feudal power unmatched anywhere in the nation. The Four Associates exercised life-and-death power over towns such as Oakland, San Bernardino, and San Pedro (the port serving Los Angeles), and for good measure bought out their seaborne competition, the Pacific Mail Steamship Company.[20]

Californians rebelled during the "terrible 1870s," a decade of depression, unemployment, and falling wages and prices. Ambrose Bierce and Henry George were among the intellectuals pressing for social reform. But San Franciscans preferred the Irish "sandlot orator" Dennis Kearney, whose rowdy Workingmen's Party demanded the eight-hour day, regulation of corporations, and expulsion of coolies. The result was a new state constitution in 1879. But despite being the third-longest such document in the world (surpassed only by the constitutions of Louisiana and India), it had little effect, because the Southern Pacific "octopus" easily strangled the commission set up to control rates. Kearney's only lasting achievement came in 1882, when Congress passed the Chinese Exclusion Act. What did hurl the Golden State into a new era of booms and dreams was capitalist competition. After a lengthy court battle the Atchison, Topeka, and Santa Fe Railroad wrested rights-of-way into southern California in 1885. A decade earlier, Charles Nordhoff's best-seller *California for Health, Wealth, and Residence* had caused no mass exodus. But when fares from the Midwest tumbled to ten dollars or less and the railroads' advertisements lured snowbound families to sunshine and citrus fruit, sixty new towns sprang up to house 136,000 newcomers over the decade. "No happier paradise for the farmer can be found than Los Angeles County," boasted its board of trade.[21]

That boom revealed how artificial it was to place dense concentrations of Americans on the semiarid Pacific slope. There wasn't nearly enough water. Farmers and ranchers in the Central Valley had already sued for access to the San Joaquin River. When the state court upheld riparian rights in *Lux v. Haggin* (1886), the legislature passed the Wright Irrigation Act authorizing communities to create water districts and float bonds for canals. The intent was to help small farmers, but guess which "communities" profited most. Fred Eaton, an engineer who was the mayor of Los Angeles, conceived of a 233-mile aqueduct to divert the water of Owens valley and realize his dream of a city of 2 million people. Folks in Owens screamed, but Theodore Roosevelt judged that slaking the thirst of Los Angeles was more in the national interest. The aqueduct was completed in 1913, the same year San Francisco won its fight to dam the Tuolumne River in Yosemite Park. The naturalist John Muir founded the Sierra Club to fight that unholy plan to flood the exquisite Hetch Hetchy Valley, but was beaten when Woodrow Wilson made San Francisco's city attorney his secretary of the interior. Muir is said to have died of a broken heart.[22]

Great cultural and educational institutions gradually arose. The Jesuit universities in Santa Clara and San Francisco dated from the 1850s. Methodists founded the universities of the Pacific (1851) and Southern California (1879). But Leland Stanford, Jr., University, named for the governor's deceased son, did not open until 1891; and the University of California began its rise to excellence only under Benjamin Ide Wheeler in 1899. By then California's image as "a commodity that can be labeled, priced, and marketed"—and seductively robed in myth—was indelible. Bret Harte, Mark Twain, Jack London, and Hubert Howe Bancroft made fond legends of the "days of forty-nine" and a cosmopolitan El Dorado "linked imaginatively with the most compelling of American myths, the pursuit of happiness." But nothing helped white Californians feel good about doing well more than Helen Hunt Jackson's best-selling novel *Ramona* (1884). Jackson, the daughter of a stodgy Calvinist professor at Amherst College, despised Hispanic culture until, after some personal tragedies, her soul found rest among the surviving rancheros, Franciscans, and Indians around Santa Barbara. Transfixed by their piety and persistence, she crafted a Romantic parable replete with rosaries, rodeos, tortillas, mantillas, and mission bells, about the tribulations of old Californios in the crucible of Yankee civilization. The tears American readers shed over *Ramona* freed them to sentimentalize, even appropriate, the very past they were busy destroying.[23]

Obloquy and obscurity were all Millard Fillmore received for a life in the public service. Hard factory work and study by candlelight carried the youth from rural poverty to a law practice in Buffalo and four terms in Congress as a Whig expert on tariffs and banks. Like his fellow New Yorker Seward, Fillmore detested slavery, but believed it an inheritance that must be endured "till we can get rid of it without destroying the last hope of free government in the world." That rejection of Seward's "higher law" in deference to the civic religion made Fillmore an acceptable northern running mate for Zachary Taylor in 1848. But Fillmore was unacceptable four years later because the "cotton Whigs," angered by his patronage appointments, joined "conscience Whigs" angered by the compromise bills to deny Fillmore renomination by a mere fourteen votes. The convention of 1852 dragged on for fifty-three ballots before choosing another war hero, Winfield Scott. Democrats were just as conflicted. Cass, Douglas, and Buchanan traded the lead over forty-nine ballots until New Hampshire's delegates did for Franklin Pierce what Tennesseeans had done for Polk in 1844. They styled their man "Young Hickory of the Granite State," a dark horse able to unite the party. Southern delegates, convinced that Pierce was a "northern man with southern principles," tumbled to him in an avalanche.

Of course, the Democrats did not say much about slavery in the general campaign—nor did the Whigs, whose platform also paid lip service to the Compromise of 1850. So the American people again swept nasty truths under the rug and sat back to enjoy a mudslinging contest. Scott was deemed an aristocrat plotting a "reign of epaulets." Pierce was deemed a coward who had fainted twice during battles in Mexico. Voter turnout would not sink so low for the rest of the century. In the end enough southern Whigs deserted Scott and enough northern Free-Soilers returned to the Democrats' fold to give Pierce 51 percent and 254 electoral votes to Scott's 44 percent and just forty-two electoral votes. The party of Jackson swept the deep South, won every northern state but Massachusetts and Vermont, and controlled both houses of Congress. But the monolith was an illusion stemming from the fact that Whigs simply fractured on sectional lines in advance of the Democrats. Pierce himself turned out to be a guilt-ridden drunk. He and his wife, a tubercular invalid who had lost two children to illness, watched in horror as their third child was killed in a railway accident en route to Washington City. Fillmore's fate was equally bitter. His wife caught pneumonia during the wintry inaugural ceremony and died within days. The following year Fillmore lost his daughter. Finally, he lost his political legacy when it became evident that the only other citizen who really believed that the Compromise of 1850 was a "final settlement" was President Pierce.[24]

That vexed the Reverend Dr. Charles Colcock Jones, a Georgian and an icon of sorts for southern apologists. He reigned as a benevolent patriarch over his strong, placid wife; his accomplished, obedient children; three plantations; and 100 slaves in the rice country—and a church congregation renowned for its faith, erudition, and patriotism. As a youth Jones won laurels at Phillips Academy, Andover; and Princeton Theological Seminary. As a man he resigned as pastor of Savannah's prestigious First Presbyterian Church in order to devote himself to the "moral enlightenment of the slaves." Despite a lifelong pulmonary affliction Jones rode from sunup to sundown every Sabbath to teach and preach to Negroes. His voluminous letters refer constantly and by name to his "bondspeople," especially when yellow fever or cholera threatened. He married them, buried them, and tried to prevent the breakup of families. A Yankee observer, Frederick Law Olmsted, praised Jones's influence on local planters, who seemed "remarkably intelligent, liberal, and thoughtful for the moral welfare of the childlike wards Providence has placed under their care and tutorship." Whether Jones was all that saintly may be disputed, but there is no denying his faith and patriotism. He prayed and believed Americans would never betray their Christian calling so much as to break up the Union.[25]

That is what makes the letters he exchanged with his son, C. C. Jones, Jr., so poignant. In May 1854 one of the pastor's missives from Maybank plantation described in loving detail his wife's daily routine as she cheerfully worked, rested, and looked to the needs of family, visitors, servants, pets, and farm animals. Gentle sea breezes, spring flowers, warm sun, a "general concert of birds," and a hickory shade tree filled out the idyll. "You will recognize all this as very natural—what you have seen many times," wrote the father. "Sure our hearts should be full of gratitude to God for all His unnumbered and undeserved favors to us as a family." The son answered from Harvard, where he was studying law. He thanked his father for the joyful reminders of Georgia and his mother. Then his tone changed. "To me the picture, always so attractive, is rendered even more dear in consideration of the endless confusion, turmoil, and bloody scenes which are now hourly transpiring in our midst. Mob law, *perjury*, free-soilism, and *abolitionism* are *running riot*." A fugitive from slavery had been captured in Boston, but rather than honor the law of the land abolitionists bore false witness, bribed, and mobbed to prevent "a perfect Virginia gentleman" from reclaiming his property. "Do not be surprised," concluded C. C. Jones, Jr., "if when I return home you find me a *confirmed disunionist*."[26]

What prompted the violence was the arrest of Anthony Burns on suspicion of robbery. He was innocent of that, but it was discovered that his former master, a plantation owner, had obtained a warrant for his recapture. Abolitionists stormed the jail in a vain attempt at rescue. A policeman was knifed to death. A federal marshal called out soldiers and marines, assuring the White House he had matters under control. Pierce wired back, "Your conduct is approved." He and Attorney General Caleb Cushing meant to make an example of Burns in the most resistant northern city and thereby reassure southerners. Richard Henry Dana, Jr., argued Burns's case, but the commissioner ordered Burns to be marched under heavy guard to a U.S. revenue cutter for transport back to slavery. One of the 50,000 Bostonians who turned out to protest recalled, "We went to bed one night old-fashioned, conservative, Compromise Union Whigs & waked up *stark mad abolitionists*."[27]

A law providing for the return of fugitive slaves had been in effect since 1793. But it was rarely enforced, and it was rendered nugatory by the Supreme Court's ruling of 1842 in *Prigg v. Pennsylvania*. The Court held that northern officials were not required to arrest fugitives in violation of their states' due-process provisions. The Fugitive Slave Act of 1850, by contrast, empowered federal marshals to require all citizens to assist in the capture of runaways and placed adjudication of runaway cases in the hands of federal commissioners who were rewarded with a doubled fee for every Negro carted back to the South. "This filthy enactment," Emerson chided, "was made in the nineteenth century, by people who could read and write." He urged Yankees to resist, and a few dramatic rescues and escapes received widespread publicity. But the act was met with more apathy than rebellion. Over six years fewer than 100 fugitives were returned to bondage, as against perhaps 5,000 who escaped through the North to Canada. Along the Ohio River and the Mason-Dixon Line, reported Clay, most whites complied with the law. Incidents such as a shoot-out at Christiana, Pennsylvania, in 1851 were rare. After all, these were years when most northern states stiffened their legal restrictions on free Negroes. Northern investors, merchants, and mill owners made comfortable profits on the cotton trade. Northern ports dominated the coastal slave trade and even tolerated traffic in African slaves carried on in contempt of U.S. and international law. New York was the headquarters of this traffic because its legitimate commerce with West Africa and the West Indies plus its high turnover of shipping allowed dummy companies to operate easily undercover. Especially notorious were the Portuguese Company based at 158 Pearl Street and its legal counsel, Beebe, Dean, and Donohue, at 76 Wall Street. Greeley's *Tribune* credited reports that a slave ship departed from New York every two weeks in 1854. The only determined effort to suppress the piracy was made by an upstanding Virginian, Henry A. Wise, who served as minister in Rio de Janeiro during the 1840s. He was shocked to learn that one-fifth of the 30,000 slaves imported annually to Brazil sailed in ships flying the Stars and Stripes. Administrations of both parties in the United States deplored it, but the Navy's small South American squadron was impotent to suppress it. If Yankees shrugged at the illegal capture of thousands of Negroes abroad, why should they care about the legal capture of a few hundred at home?

That, at least, was the thinking of Clay, Douglas, and the other sponsors of the Fugitive Slave Act. What they failed to anticipate was the ammunition its enforcement or lack of enforcement handed the propagandists. Abolitionists and "Fire-Eaters" both had a stake in exaggerating the number of runaway slaves, and both deemed every escape—or recapture—to be an act of criminal theft. Governor John A. Quitman of Mississippi accused northern conspirators of having "stolen" 100,000 bondsmen over the years, and Josiah Henson, a Negro conductor on the Underground Railroad, boasted of 50,000 fugitives in the North (both were impossible figures). Thus did extremists collude to depict the centerpiece of the Compromise of 1850 as a practical failure. Harriet Beecher Stowe turned it into a moral calamity.[28]

Uncle Tom's Cabin, or Life among the Lowly began to appear in serial form in July 1851 and was printed in book form in 1852. It is still worth rereading. For despite its occasional mawkishness and its opaque Negro dialect, Stowe's melodrama retains the power to move readers to tears. It is also surprising because our popular caricatures of it are false. The author had more sympathy for planters than for businessmen in the North. Her southerners of both races comprise affectionate extended families, caring for each other in ways only close relatives did in the North. She implied slavery was repugnant in part because of its thralldom to the market forces revered in the North. Thus, Mr. Shelby, the kind Kentucky planter, is forced to sell two loyal slaves lest his whole plantation fall to heartless creditors. Eliza, a black exemplar of republican womanhood, flees across the icy Ohio River to prevent the breakup of her family. Uncle Tom, a black exemplar of Christian sacrifice, permits himself to be sold down the river. In Louisiana Tom is purchased by Augustine St. Claire, whose saintly, sickly child Eva prefers plantation life to her cousins' house in Vermont because "it makes *so many more round you to love.*" As Little Eva dies in Uncle Tom's arms, St. Claire vows that he will free his slaves, but he is killed before he can do so. The estate sale then hurls Tom into the hands of the sadistic Simon Legree and his two Negro slave drivers. They mean to break the spirit of the hymn-singing Tom by forcing him to take up the lash against fellow slaves. Tom refuses: like the apostle Stephen at his stoning, he has a vision of heaven that reconciles him to martyrdom. So Legree beats him to death, but Tom survives just long enough to bless Shelby's son, who had hoped to purchase his freedom, and to redeem the souls of the two Negro drivers.

Uncle Tom's Cabin is easily dismissed as a saccharine fable full of stock characters. In that sense it failed to make the victims of slavery into real people. But it was no ham-handed attack on the Fugitive Slave Act.

Rather, Stowe crafted a religious allegory exposing slavery as an institutional sin trapping *all parties*—white and black, southern and northern—by forcing them to muffle Christian truth in pretense. The Shelbys and St. Claires agonize over their moral duty to the slaves in their care. Tom and Eliza agonize over their duty to masters and kin. Eliza's proud husband George questions the existence of a God who would countenance slavery. Legree, swilling whiskey by his lonely hearth, is tormented by the thought that Tom's God of Love might really exist. The middlemen dealing in slaves duck conscience by denying the possibility of a "pious nigger." None can redeem a system in which the best intentions are thwarted by events beyond anyone's control. Even the most humane masters eventually die or go broke, whereupon their chattels are torn from their families, turned into cargo for heartless brokers, and bid on by the likes of Legree. But Stowe made it plain her real target lay in the North. "Do you say that the people of the free states have nothing to do with it, and can do nothing?" she asked in her conclusion. "Would to God this were true! But it is not true. The people of the free states have defended, encouraged, and participated; and are more guilty for it, before God, than the South, in that they have *not* the apology of education or custom." That is why Stowe followed up with "An Appeal to Women of the Free States of America, on the Present Crisis in Our Country." [29]

Harriet, one of Lyman Beecher's brilliant children and the wife of another theologian, was encouraged by her sister-in-law to "make this whole nation feel what an accursed thing slavery is." A vision in church gave her the plot. Her grief over the loss of a child to cholera gave her the hook. The way to get northern women to feel Negroes' humanity was to tell of children yanked from their mother's bosom in the unholy name of property rights. But Stowe's "great American novel" (in the mind of the man who coined that phrase) was no sudden burst. At age nine she learned from John Brace at the Hartford Female Seminary to examine all sides of an issue and have something to say before taking up her pen. At thirteen she amazed her father with a cogent essay on the "natural and moral sublime." At seventeen she wrote a satire inspired by Coleridge, quipping that the "modern" approach to language was "1st. To conceal ideas. 2d. To conceal the *want* of them." Great men used unintelligible prose "to convince common-sense people of their own insufficiency and ignorance." She went on to publish short stories and essays in national publications, developing a style that pleased her Romantic audience but in fact planted the seeds of Realism. She spurned happy endings, posed discomfiting truths, explored human dilemmas, and mastered description. (Few passages are more evocative than her long description of Uncle Tom's cabin itself.) She put all that together in what became the biggest per capita best seller and longest-running play in American history. The novel sold 300,000 copies in 1852 alone, or one copy per every four votes cast for Franklin Pierce in the North.[30]

How about its effects? Was Stowe "the little woman who wrote the book that started this great war," as Abraham Lincoln is supposed to have said later? Not exactly, and not just because there is no way to measure the impact of cultural fads. Doubtless many copies of *Uncle Tom's Cabin* were bought for display in the parlor or else were wept over once and forgotten. The Fugitive Slave Act, as noted above, caused no widespread outrage. As for slavery itself, nothing could be done about that. Or at least most northerners figured that nothing unsettling needed to be done until a series of new provocations pushed Free-Soilers into a corner. Perhaps, several years later, many a woman and man in the North decided that Stowe got it right. But the chronology proves that the event causing northerners to flock to antislavery political parties was not *Uncle Tom's Cabin*. It was a railroad scheme hatched by their fellow Yankee Stephen Douglas.

As early as 1844 the businessman Asa Whitney of New York floated a plan for a transcontinental railroad to run from Milwaukee to Oregon. When the Pacific coast fell to the United States two years later it became part of the national catechism that such a railroad must be rapidly built by private investors subsidized by federal land grants. But there was nothing sacred about Whitney's route. Douglas suggested the rails run from Chicago to San Francisco. Senator David Atchison of Missouri stumped for Saint Louis. Pierre Soulé and Judah P. Benjamin, who proposed a route from New Orleans, won the ear of President Pierce. He authorized James Gadsden, a railroad executive from South Carolina, to purchase additional Mexican territory south of the Gila River so tracks might be laid from El Paso through Yuma to California. Santa Anna drove a hard bargain, but Gadsden brought home a treaty in 1853, only to meet with resistance from the Senate. It was the first time Americans balked at grabbing new real estate. At length the Senate approved the purchase of 30,000 square miles of desert for $10 million, but no railway bill was likely to pass until two or more sectional interests cut a deal.

The deal that emerged in January 1854 was a bill to create a Nebraska Territory west of the Missouri River on terms that facilitated both a central rail route and the expansion of slavery if pioneers chose that option under the "popular sovereignty" provision of the Compromise of 1850. Douglas proudly claimed sole authorship. "There is a power in this nation," he preached, "greater than either the North or the South. . . . That power is the country known as the Great West. . . . There, sir, is the hope of this nation—the resting place of the power that is not only to control, but to save, the Union." Atchison, the leader of the F Street Mess (an elegant boardinghouse for southern senators), claimed that he inspired the bill as a way to get support from northerners for repeal of the Missouri Compromise. Both were disingenuous. Douglas meant to dupe southerners into funding his railroad in return for the hope (which proved vain) of new slave states on the prairie. Atchison meant to dupe Free-Soilers (including his rival Benton in Missouri) into selling their souls for the hope (which also proved vain) of a railroad act. Douglas put his conscience at ease with some tortured logic. The bill did not really repeal the Missouri Compromise line banning slavery north of latitude thirty-six degrees, thirty minutes, because admission of California had already terminated that agreement by creating a free state partly south of this latitude. In truth, no congressmen in 1850 imagined that they were erasing the sacred line, now thirty years old. Douglas claimed they had done so not only willingly but (somehow) without a moment's debate on the matter. By admitting the possibility of slavery north of the line, Douglas just shifted the "saddle to the other horse."[31]

None of this fooled Chase, Seward, and others who damned Douglas as an ambitious traitor. The wily Seward even encouraged southern Whigs to propose explicit repeal of the Missouri Compromise. That would smoke Douglas out and kill Democrats in the North. The Whig Archibald Dixon of Kentucky went farther still: he proposed an amendment affirming the liberty of citizens "to take and hold their slaves within any of the Territories of the United States or of the States to be formed therefrom." Then the Pierce administration got into the act, suggesting a formula asserting constitutional property rights without any mention of the Missouri Compromise. Sensing a fiasco, Douglas shocked all parties by introducing an entirely new bill calling for two territories—Nebraska and Kansas—subject to all federal laws except the eighth section of the act that admitted

Missouri as a state, since it had been "superceded [sic] by the principles of the legislation of 1850, commonly called the compromise measures, and is declared inoperative." Not for nothing did Harriet Beecher Stowe mock unintelligible prose meant to conceal a want of ideas! Shorn of obfuscation, Douglas's new bill offered Free-Soilers a future free state in Nebraska and a gamble on Kansas in exchange for repeal of the Missouri Compromise. It touched off a fifteen-week brawl during which the Little Giant huffed, puffed, bribed, and buttonholed with energy even his enemies had to admire. Thanks to Democratic discipline, support from the southern Whigs, Pierce's patronage, and railroad lobbies, Douglas prevailed. The Kansas-Nebraska Act passed the Senate thirty-seven to fourteen in March and the House 113 to 100 in May. Seward's handler Thurlow Weed called it a crime. The wall erected by Jefferson "to guard the domain of liberty is flung down by the hands of an American Congress, and slavery crawls like a slimy reptile over the ruins." Douglas boasted: "I had the authority and power of a dictator throughout the whole controversy in both houses. The speeches were nothing." One could just as well argue that he lost control the instant he deposed his first bill. He never did get his railroad. But thanks to Douglas a new game called popular sovereignty would be played out on the frontier. Too bad that nobody thought to draft rules.[32]

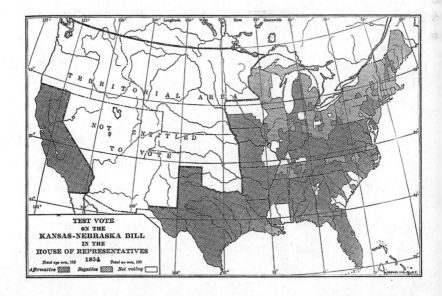

TEST VOTE
ON THE
KANSAS-NEBRASKA BILL
IN THE
HOUSE OF REPRESENTATIVES
Total aye vote, 113 1854 *Total no vote, 100*
Affirmative Negative Not voting

Common sense suggested that people petitioning to become a new state ought to be free to decide by majority vote whether or not to allow chattel slavery. This idea had worked in California and seemed likely to work in New Mexico and Utah (Brigham Young was as yet noncommittal). Every new state since Missouri had submitted a constitution of its own drafting and won swift approval from Congress. But in no case had the status of a potential new state been cast into doubt while its land was still nearly without settlers. What was to be the regime under which a territory became populated? To ban slavery before a petition for statehood would deprive slaveholders of property rights and make a farce of popular sovereignty. But to permit slavery during territorial years would require the federal government either to sanction its spread into lands even the South forswore in 1820, or else expel slave-owning settlers at the moment of statehood. Kansas was especially vulnerable because pro-slavery Missourians risked facing free states at three out of four points on the compass. Nor were Free-Soilers above reproach: many opposed slavery because they opposed blacks. "I kem to Kansas to live in a free state and I don't want niggers a-trampin' over my grave," exclaimed a clergyman. In sum, popular sovereignty was an invitation to chaos, but Douglas was evidently blinded by the pretense that American elections were free and fair. It never occurred to him that partisans might fight for new turf under rules of their own making, or no rules at all.

Even before passage of the Kansas-Nebraska Act a New England Emigrant Aid Society was organized to fill Kansas with Yankees and "Beecher's Bibles" (Sharps rifles and two field guns). Colonel Jefferson Buford of Alabama bankrolled recruits from the South, and Atchison encouraged Missourians to claim Kansas for slavery. A low-level guerrilla war soon crackled over the plains, and the territorial governor, Andrew Reeder, busy with his own land speculation, did nothing to dampen it. Reeder did call for elections, but the campaign allowed Missouri's "border ruffians" to add a new wrinkle. They crossed over to vote in such numbers that the turnout was 200 percent. Their "bogus legislature" promptly mandated slavery and criminalized the expression of antislavery opinions. Reeder retreated to Washington City, crying, "Kansas has been invaded." When Free-Soilers set up a rival government in the village of Lawrence the pro-slavery regime in Leavenworth called it treason (which it technically was). Violence escalated for a year until Senator Charles

Sumner of Massachusetts could stand it no longer. On May 19–20, 1856, he fulminated against the violence and fraud done in the name of a wicked cause, singling out a South Carolinian, Andrew Butler, for taking "the harlot, Slavery" as his mistress. Two days later Butler's kinsman Congressman Preston Brooks, bludgeoned Sumner nearly to death with a cane made of gutta-percha. Meanwhile, in Kansas, the pro-slavery government also lost patience. It raised a small army under command of a U.S. marshal, razed the rival capital of Lawrence, and put Free-Soilers to flight. That atrocity, in turn, completed John Brown's metamorphosis. Believing himself "God's chosen instrument," he recruited four of his sons, a son-in-law, and two settlers for a "mission" at Pottawatomie Creek. Under cover of night they broke into the cabins of pro-slavery settlers and dismembered five men with swords. Kansas continued to bleed until Pierce at last dispatched a hands-on governor and enough federal troops late in 1856. All told, combatants on both sides murdered some 200 of their fellow Americans. This was no "war between the states." It was civil war, pure and simple.[33]

That horrible fact should have humbled all parties. But it didn't. Southern journalists and congressmen cheered the sack of Lawrence as a "Glorious Triumph of the Law and Order Party Over Fanaticism in Kansas." They blamed abolitionists for inciting the Pottawatomie "massacre" and extolled Preston Brooks for defending the South's honor. Northern journalists and spokesmen escalated the Lawrence affair to Napoleonic proportions, vilified Sumner's assailant, and excused or even denied the murders inspired by Brown. That took some doing, since Kansans knew very well who led the killing spree, and newspapers in Chicago ran fairly accurate reports of it by June 5. But the antislavery press led by the *New York Times* and the *New York Tribune* chose to trust no sources other than Free-Soil newspapers in Kansas, their own correspondents (one of whom shot a pro-slavery judge), and the accounts read into the record by like-thinking congressmen. So most northern readers were told to blame the trouble in Kansas on boozing, cursing border ruffians and attribute any violence Free-Soilers committed to self-defense. As for the murders allegedly ordered by Brown, they might not have happened at all, or else (given the mutilations) might have been the work of the Comanche. John Brown himself kept silence to shield his accomplices from prosecution and his fund-raising efforts from bad publicity.

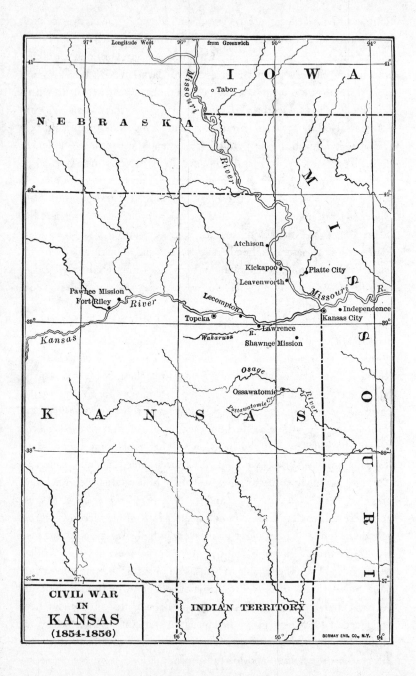

Longitude West 96° from Greenwich 95° 94°

IOWA

o Tabor

NEBRASKA

Missouri River

MISSOURI

Atchison

Kickapoo   Platte City
Leavenworth

Pawnee Mission   River
Fort Riley   Lecompton   Missouri   R.
                    Topeka   Independence
Kansas   Kansas City
                    Wakarusa   R. Lawrence
                              Shawnee Mission

                    Osage
          Ossawatomie
                    Pottawatomie Cr.   River

K A N S A S

CIVIL WAR
IN
KANSAS
(1854-1856)

INDIAN TERRITORY

BORMAY ENG. CO., N.Y.

He headed east, convinced more than ever that only blood sacrifice could atone for his nation's sins.[34]

"Twenty years ago the South had no thought—no opinions of her own," wrote the fifty-three-year-old Virginian George Fitzhugh in 1857. "Then she stood behind all christendom, admitted her social structure, her habits, her economy, and her industrial pursuits to be wrong, deplored them as a necessity, and begged pardon for their existence." But soon he thought all would be forced to admit how "wise and prudent" the South had been in retaining slavery. Fitzhugh was a provocateur: he went so far as to deduce that northern white factory employees would fare better as slaves. But Fitzhugh was right to note that the self-conscious defense of slavery was a recent phenomenon. It sprang to life in 1819–1820, in the debate over Missouri's statehood, when northern congressmen first called slavery demonic. It matured in the 1830s under the pressure of the Hayne-Webster debate, the nullification crisis, and the Anti-Slavery Society. It peaked in the 1850s in reply to *Uncle Tom's Cabin* and "Bleeding Kansas." Southern ideology is a good place to start in hopes of explaining how the American civic religion, cherished by citizens everywhere, cracked like the Catholic church when Martin Luther refused to recant.[35]

Asked if she knew her maker, Harriet Beecher Stowe's trickster child Topsy replied, "I spect I grow'd. Don't think nobody never made me." Nobody made the Old South. It just grew from the Cavaliers' tobacco, rice, and indigo cultures in the Chesapeake and Carolinas through Scots-Irish Appalachia and the cotton states of the Gulf to the Mississippi River and out yonder in Texas. That sprawling South was not the least homogeneous. As two generations of scholars have shown, it was diverse and complex. Slavery was barely in evidence throughout much of Delaware, the western counties of Maryland and Virginia, eastern Tennessee, Kentucky beyond the bluegrass, and southern Missouri. The tobacco country of tidewater Virginia and North Carolina survived, but had long since conceded leadership to younger elites in the Deep South. There King Cotton reigned because British textile mills could not get enough of the stuff. Prices doubled to fifteen cents per pound in 1857, and production soared to 3.8 million bales by 1860, by which time cotton accounted for 57 percent of all American exports. Needless to say, that gave the lie to any notion that the South was provincial. Except for New York and a few other ports, the "cotton South" was the most *globalized* sector of the U.S. economy.

Nor should the prevalent cash crops obscure the abundance of a region that grew almost one-third of the nation's wheat; half of its corn, horses, pigs, and oxen; and 90 percent of its mules. Likewise, the 2,292 planters who owned more than 100 slaves should not obscure the breadth of the slave interest. The number of people owning twenty or more slaves reached 46,274 in the census of 1860; the 383,637 who owned at least one slave represented about one-fourth of all white families in the South. Would plantations and farms have been more profitable if worked by free labor? Perhaps, but the 10 percent average return on investment in slaves drove the price of strong field hands from about $1,000 in the mid-1830s to nearly $2,000 by the mid-1850s. For all their prattle about paternalism and contempt for materialism, southern slave owners were capitalists. Indeed, they were just as fixated on the bottom line as a Wall Street plunger, because agriculturalists lived on the margin, borrowing against next year's crops to pay this year's expenses or mortgage. Nor were plantations opulent islands dotting an ocean of poverty. In alluvial parts of the South the income per capita (blacks included) was higher than that of the United States as a whole. Finally, enterprising southerners made belated efforts to diversify through manufactures and trade. Richmond; the Cumberland River towns in Tennessee; and Prattville, Alabama, had large ironworks staffed in part by contracted slave labor. Hemp factories in Kentucky and sugar mills in Louisiana were as high-tech as any northern factories. Coal, lead, iron, and copper mines speckled the South; railroads created boomtowns. Atlanta grew from a few shacks in 1845 to a city of 10,000 people by 1860, with foundries, mills, warehouses, carriage shops, tanneries, banks, and four rail junctions. Development in the South surely compared badly with that in the North. But looked at another way, the South's 20 percent share of U.S. industry belied the image of a vast rural sink inhabited only by planters, slaves, and slovenly "crackers." Its commodity markets, trade, manufactures, and service industries supported a middle class of professionals often educated in the North.

Southern distinctiveness nevertheless struck people of both sections as true. In the North dynamic promoters of industry had their hands on the throttle of the economy, and even the majority still on the farm needed railroads and canals to move harvests. Hence the weight of northern political interests was on the side of protective tariffs, internal improvements, and banks. In the South dynamic promoters of cotton exports had their hands on the tiller of the economy, and even the minority urging diversification relied on those exports for capital. Hence the weight of southern political interests was on the side of free trade and opposition to subsidies for the North and the West. The coastal South's semitropical geography, climate, flora, and fauna were unlike anything in the North; and the southern uplands from the Blue Ridge to the Ozarks had no northern counterpart except perhaps upper New England. But whereas impoverished Yankees hastened to improve their lives in places like Iowa, southern hillbillies either stayed put or else replicated their lives in places like Arkansas. That suggests a cultural divide. Northern society was a stew of Puritans, Yorkers, and Quakers seasoned with Scots, Irish, and Germans. The people born in or adapting to those "cradle cultures" defined liberty in terms of law, equal opportunity, and self-reliance. The recipe for southern society, which attracted few immigrants or transplanted northerners, remained to a remarkable extent what it had been in colonial times: Cavalier planters in the low country; Scots-Irish "Bordermen" in the highlands; and Africans, mostly enslaved. Scions of Cavaliers and the cotton lords aspiring to their dignity clung to an aristocratic notion of liberty whereby the obvious inequality of human beings meant that anyone who was not a master was a dependent—a slave of sorts—to somebody else. The Scots-Irish clung to a libertarianism hostile to all external authority. Both southern white cultures bristled at the suggestion that the federal government, much less self-righteous Yankee reformers, had any right to tell them how to manage their households.

The Constitutional Convention had in its wisdom accommodated all the American cultures and notions of liberty. The social contract that emerged was a powerful glue that kept the Union together for over half a century. Free exercise of religion was another glue because the mostly Protestant denominations united Americans of all regions and cultures in the worship of a God they believed was the author of their glorious Union. The party system perfected by Democrats and Whigs was a third, tacky glue because it obliged aspirants to national power to build coalitions across sectional lines. How vexatious it was, therefore, when a handful of northern agitators began to shout that the American civic religion was false! The founders, they insisted, had sinned by accommodating slavery; hence God would smite America unless the evil was abolished at once. The first glue turned into a solvent by such agitation was religion. Consider the Methodists' Baltimore convention of 1844 that "rent the body of Christ." Wesleyans in the North thought it high time the national church condemned slavery. Bishop James O. Andrew of Georgia protested in a voice that "cracked like thunder." The northern clergy might preach as they wished in their pulpits, he argued, but a resolution declaring slavery a sin would turn southern pastors into pariahs. That would "leave us no option (God is my witness that I speak with all sincerity or purpose toward you) but to be disconnected with your body." The resolution passed. Southern Methodists promptly seceded in order to save their ministries to free and enslaved souls alike.[36]

Southern Unionists needed all the glue they could find during the decades when their section's representation in Congress fell from 46 percent in 1800 to just 35 percent by 1860. As Alex Stephens wrote, southerners looked "to the security which rests upon principle, rather than upon numbers," because the "citadel of our defense is principle sustained by reason, truth, honor, and justice." He could take heart from the fact that the South won the constitutional argument. Slaves were property protected by federal law. The South also won the racial argument. Most northern states also discriminated against blacks on the grounds they were inferior beings. But once Free-Soilers used political and moral arguments against slavery, another glue began to dissolve. Southern apologists were forced to plead for toleration, diversity, and the minority rights without which democracy becomes liberty's bane. For three decades, from the Missouri Compromise until 1850, they advanced Calhoun's doctrines of states' rights and "concurrent majorities" limiting the federal government to measures affirmed by most people in free and slave states alike. But as the southern Whig Joseph G. Baldwin observed, states' rights "will not always, or even often, prevail, when they come in contact with the impulsive and eager utilitarianism and impatient wishes of the people." By the 1850s the hustling, expanding North and West were clearly impatient with sectional strife that hampered the nation's growth (especially the deadlock over western railroads). So the South devised a new strategy based on new guarantees and new slave states. Secession was always a last resort. Southern kingpins really spent half a century searching for ways to remain in the Union with their honor intact.[37]

Later historians had trouble comprehending this. They assumed pro-slavery propagandists must have been wicked reactionaries, scribblers pandering to the planters, or else crazy. One scholar confessed to being "astonished" by their ideology. Another likened the southern intellectual to "an alien child in a liberal family, tortured and confused, driven to a fantasy life." The truth was that the "theologians" of slavery chiseled their orthodox creed to ward off what they deemed modernist heresies. Their learned republic of letters included such authors as James H. Hammond, William J. Grayson, Edmund Ruffin, Nathaniel Beverley Tucker, Joseph Glover Baldwin, Johnson Jones Hooper, and William Gilmore Simms (the "father of southern literature"). Their essays, satires, and letters published in *De Bow's Review*, the *Southern Literary Review*, and elsewhere debated the merits of various societies: industrial and agricultural; free and slave; ancient, medieval, and modern. They engaged in what E. N. Elliot called a "sober and cautious reflection" on the purpose and ethics of human society, exposing their peculiar institution to "the lights of History, Political Economy, Moral Philosophy, Political Science, Theology, Social Life, Ethnology, and International Law." Lamenting the "misguided spirit of sickly sentimentality" that caused their forebears to be ashamed of slavery, they judged it the "cornerstone" of a social edifice "more manifestly consistent with the will of God" than that of the North. "Slavery is no evil," concluded Senator Hammond. "On the contrary, I believe it to be the greatest of all the great blessings which a kind Providence bestowed on our glorious region."[38]

How on earth did they reason themselves to beliefs so out of touch with the progress of western civilization? They began (and usually ended) with the Bible because nowhere in the Old or New Testament was slavery proscribed. Scripture simply admonished masters to care for their servants as the Lord cared for them, and servants to obey their masters without feigning, as they obeyed the Lord. The stories of Hebrew patriarchs and the parables of Jesus contrasted slaves with mere hirelings whose link to the employer was purely financial. There was "no love lost" between them. The bond between masters and servants was an invitation to affection, loyalty, and mutual service. Grayson, a Charlestonian lawyer and staunch Unionist, made the distinction in an epic poem, "The Hireling and the Slave" (1854), rebutting *Uncle Tom's Cabin*.[39] Northerners who oppressed free Negroes and white hired "help" preached a rarefied morality requiring nothing of them. The Irish mother and child starving in the gutter of a northern slum were literally nobody's business. "Much of the benevolence of our time is malignant," said the southern *Biblical Recorder*; "much of its philanthropy selfish; much of its charity cruel." By contrast, southerners surrounded by Negroes lived out their morality and had a personal stake in the health and contentment of their "people." Grayson even compiled data purporting to show that slaves on the whole were better fed, better housed, and better nursed than workers in the North.

Were miscreant slaves punished? Of course, but no more severely than criminals in the hands of the northern police. Were some planters violent, impetuous men? Of course, but southerners were well acquainted with sin. Only northern perfectionists claimed sin was eradicable by human effort. Southern preachers and planters had no use for such theological novelties. Under the impact of the revivals they had long since abandoned the enlightened deism of their Jeffersonian grandfathers in favor of an evangelical ethos that "won official preeminence as the arbiter of 'civil religion' in the South." This ethos inculcated a simple faith in divine love and judgment, providential guidance of human affairs, the power of prayer, the stain of sin, and the hope of salvation. To condemn slavery because some masters were cruel made no more sense than to abolish marriage because some husbands or wives were violent or shrewish.[40]

Recent history favored the abolitionists. By 1850 slavery existed nowhere in the Atlantic world except Brazil, Dutch Guiana, Cuba, and the American South. But southern apologists cited the whole counsel of history. Since slavery was once universal, it must be an expression of natural law. The glory of ancient Egypt, Greece, and Rome rested on the shoulders of slaves. Abolition was a modern heresy inspired by the French Revolution. Citing Edmund Burke, southerners argued for the health of organic societies rooted in the past and based on hierarchy and obligation. By contrast, northern society was at best an anthill and at worst a war of all against all. Southerners justified singling out Africans as unfit for liberty by citing the Bible's curse on the children of Ham and the mostly northern anthropologists who placed the Negro race at the bottom of their ranks of intelligence. It followed that slaves under the care and tutelage of whites were better off than their savage cousins in Africa. Hammond and Fitzhugh argued from economics and sociology. Just as a temple rests on a sturdy foundation, wrote Hammond, so every civilization rests on a "mud-sill" class that performs its menial labor. In practical terms, slavery was sturdier than a fractious, dissatisfied working class whose members envied those towering over them. In moral terms, slavery ensured that the "mud-sill" was treated humanely; this explained why the enslaved population in the South had quadrupled since 1800 almost entirely on natural increase. Northern employers considered their endless supply of immigrant labor expendable. Southerners paid handsomely for workers whom they supported for life. Indeed, northerners implicitly granted the point when they urged Kansans to choose freedom because land was dearer and labor cheaper in the North. Fitzhugh retorted: "Now, is it possible that they are simpletons as not to see that they are asserting that the white laborers of the North, as slaves to *capital*, get less wages than our slaves . . . ? There is a vast deal more of knavery and hypocrisy than idiocy about these men."[41]

Finally, the southern apologists were not reactionary in the sense of rejecting progress. They simply believed that English and Yankee society had jettisoned the faith, chivalry, and deference of the medieval era along with its superstition and stagnation. The result was a social jungle posing as democratic and Christian but steeped in bigotry and privation. "All the steam power in the world," chided Simms, could not create happiness; nor could railroads carry "one poor soul to heaven." The South progressed without embracing the demons that tormented the North. Calhoun called it the "conservative portion" that served as America's ballast and "may, for generations to come, *uphold* this glorious Union of ours." The scientist Henry Ravenel said that his region was "the breakwater which is to stay that furious tide of social and political heresies now setting toward us from the shores of the old world." Southerners habitually credited piety, paternalism, and commonsense education for their immunity to the Millerites, Mormons, Shakers, rappers, utopians, and free-love cults to which the North was uniquely susceptible.

The purpose of such advocacy was to boost confidence and stiffen spines in Dixie. But a few northerners were stung by the critiques. James Kirke Paulding, the old former colleague of Washington Irving, was one. He came to believe that the seventeenth-century English civil war raged on between the heirs of the Roundheads and Cavaliers in America. Yankees, like Cromwell's Puritans, were aggressive, self-righteous, avaricious, and given to fanatical witch hunts, most recently abolition. "When the love of self becomes the ruling passion," he wrote, "and the golden calf the only divinity . . . then will this majestic fabric of Freedom . . . crumble to pieces, and from its ruin will arise a hideous monster with Liberty in its mouth and Despotism in its heart." In 1850 Paulding wrote an open letter from New York to the people of Charleston urging South Carolina to secede so that some part of the American Eden might escape enslavement by the almighty dollar. But therein lay the "slaveholders' dilemma." To stay in a Union dominated by Yankees meant slow strangulation enabled by "compromises" reflecting southern cowardice and northern fraud. ("I have no hope, and no faith in compromises, of any kind," wrote Simms in 1850, "and am not willing to be gulled by them any longer.") But to bolt from the Union meant renouncing the civic faith of Washington, Jefferson, and Jackson. It seemed the only way southerners could be true to themselves and their country was to increase their leverage through the expansion of slavery. This explains why the South, in the 1850s, began to pursue its own foreign policy.[42]

Once upon a time, *filibuster* referred not to a parliamentary ploy by a minority aiming to block a vote, but to an insurrectionist ploy by a minority aiming to take over a country. The Spanish *filibustero* derived from the Dutch *vrijbuiter* (freebooter, or pirate) in the seventeenth century. But no one mastered the art like Americans, who sequestered west and east Florida, Texas, Oregon, California, a portion of Maine, and numerous Indian lands through spontaneous occupations or coups d'état or both. As one historian wryly noted, "Infiltration and internal subversion were virtually the only original American contributions to the technique of worldwide imperialism." Southern expansionists embraced that technique the moment the Wilmot Proviso revealed opposition to new slave states in the West. Believing that slavery must expand, "Fire-Eaters" like Quitman and the South Carolinian Robert Barnwell Rhett unfurled the banner of Manifest Destiny in the Caribbean. Northerners like Douglas, Buchanan, Dickinson, and O'Sullivan encouraged them. Not coincidentally, all of the above were Freemasons.[43]

Between 1848 and 1860 American filibusters hatched some two dozen plots against portions of Mexico. But the main targets of filibusters were Central America and Cuba. By mid-century the Spanish grip on Cuba was tenuous, conjuring southerners' fears of a takeover by the abolitionist British or, worse, a Haitian-style black republic: "If the slave institution perishes in Cuba, it perishes here." So O'Sullivan hatched a plot through Masonic lodges in New York, New Orleans, and Havana where his brother-in-law Narciso Lopez was grand master. Lopez recruited 500 mercenaries outfitted by Yankee investors and offered command to Robert E. Lee and Jefferson Davis. When they declined, Lopez himself led an abortive invasion in May 1850. He fled for refuge to the Masonic lodge in Key West and beat criminal charges thanks to Masons in the U.S. judiciary. In 1851 Lopez invaded Cuba again but was captured and executed beside fifty U.S. citizens. Fillmore did not protest. The Whig president said he would be glad to have Cuba were it populated by a kindred race, but he had no use for an empire of slavery.

Franklin Pierce did see the utility of acquiring new lands for slavery. Proving himself a true "doughface"—the Whigs' slur for a northern Democrat in league with the slave states—Pierce entertained Quitman's scheme to seize Cuba while authorizing the U.S. minister in Madrid, the Louisianan Pierre Soulé, to offer $130 million for Cuba. Queen Isabella II was not amused. Soulé, Buchanan (minister to Britain), and the Virginian John Y. Mason (minister to France) met in October 1854 to trumpet the Ostend Manifesto. It urged the purchase of Cuba at any price, but it held that if Spain refused, the United States was justified "by every law human and divine" in seizing the island. A few years before, either ploy might have worked. Enough northern congressmen had crossed sectional lines to annex Texas and pass the Fugitive Slave and Nebraska acts. But the Ostend Manifesto shared the front page with news of the bloodshed in Kansas. Horace Greeley called the diplomats brigands. Congressman Gerrit Smith cried shame: "Never has there been so self-deceived a nation as our own. That we are a nation of liberty is among our wildest conceits."[44]

Filibusters in Central America proved just as vain despite the stubborn exertions of a "grey-eyed man of destiny" heralded as the South's savior. Born in Nashville in 1824, William Walker grew up slight and studious, but he had a hypnotic gaze, a romantic soul, and inexhaustible energy. By the age of twenty-three he had graduated from the University of Pennsylvania medical school, interned in Paris, studied law, and passed the Tennessee bar. Yet he chucked it all in the belief that destiny meant him for greatness. In 1848 he courted a beautiful mute in New Orleans, agonizing when she slowly perished of yellow fever. Walker then joined the gold rush to California, where the poet Joaquin Miller sensed "A dash of sadness in his air / Born maybe, of his over-care, / And, maybe, born of a despair / In early love."

Walker spent a few frustrating years as a journalist fighting corruption in San Francisco and then embraced it himself on a grand scale. He led filibusters in Baja California and Sonora, where he pompously declared himself "president." Mexican soldiers expelled him, U.S. soldiers arrested him, and jurors in California acquitted him—stealing more of Mexico was no crime to them. In 1855 Walker repaired to Nicaragua, which was torn by civil war and was a likely locus for both slavery and an isthmian canal. Financed by expansionists in New Orleans and by northern tycoons such as the shipping magnate Cornelius Vanderbilt, Walker recruited filibusters to invade Nicaragua. He managed a violent overthrow of the government, then reintroduced slavery. Stephen Douglas insisted Walker's government was "as legitimate as any which ever existed in Central America." Wall Street staged a rally on his behalf. The Pierce administration recognized Walker's regime. Just weeks later an angry coalition of Central Americans amassed to expel the usurper. Walker vindictively burned his capital city, Granada, and then fled to the American South, where he peddled an "empire of slavery" and his memoirs. He met his real destiny in 1860, when the British captured him in Honduras. Local authorities put Walker to death. Of all the overwrought obituaries, the one in *Harper's Weekly* best captured the man: "He never displayed any constructive ability; his energies were wholly destructive. He was brave, persevering, and energetic, but had little or no foresight. . . . His works, from first to last, have been injurious rather than beneficial to the world." Northerners busy codifying their own American gospel decided those judgments neatly personified the South as a whole.[45]

Most interpretations of the 1850s suggest that the rise of two hostile nationalisms in the United States was either inevitable (hence "the irrepressible conflict") or accidental (hence "the blundering generation"). An older generation of progressives argued that northern capitalists in league with western farmers were sure to collide with the feudal South. That idea fell flat before evidence of how intertwined the regions' economies were. Revisionists impressed by the Union's durability argued that the clash arose because leadership passed from the likes of Clay, Webster, Calhoun, and Benton to bunglers like Douglas and nonentities like Pierce and Buchanan. But contingencies alone could not have broken the Union if not for the protracted conflict that exhausted compromise after 1850. A younger generation of progressives argued that conflict became irrepressible once slavery posed intractable moral issues. That hypothesis ran afoul of northerners' prejudice: did anyone other than radical abolitionists mean to risk the Union, not to mention his life, over the plight of Negroes? Still other historians stressed a cultural divide that made separate nations of the North and South. As we have seen, there was a divide, but so were there deep affinities shared by Americans, especially in the border states.[46]

A synthetic interpretation stressing the conflict over the expansion of slavery would seem most persuasive. That issue would never have arisen but for the Mexican War and the Wilmot Proviso (absent the war, the whole Louisiana Purchase north of thirty-six-thirty north latitude would probably have come in as free states with nary a murmur). That issue alone exacerbated economic, cultural, moral, and constitutional conflicts by forcing the contestants to confront the contrary sources of their patriotism. Southern partisans loved the Union because they imagined its *institutions* embodied a civic religion affirming the principles of economy, culture, and law peculiar to each of its united *states*. Northern partisans loved the Union because they imagined its *spirit* expressed a civic religion affirming principles valid for all its *united* states. Northerners believed God and the founders intended the Republic to become a temple of liberty built by the free labor of free men on free soil. Slavery was a false note in its choir, but disharmony was tolerable if muted and destined someday to fall silent. What northern partisans could not tolerate was the prospect of slavery's expansion into lands of opportunity reserved for their own children and grandchildren. In other words, the North's affirmation of self turned into condemnation of the other the moment southerners reasoned that their own affirmation must grow or die. Economic tension between the sections was real. Statesmen in the 1850s were mediocre. Passions ran high. But the various glues of Union dissolved because each section's rational measures of self-defense looked to the other like mad provocations. All that remained was for men in both sections to turn the dispute into a virility contest played out before the eyes of God and their women. Honor demanded that southern gentlemen, sooner or later, hurl a gauntlet at the feet of their rivals, confident the prissy men of the North would back down. Honor demanded that northern men, sooner or later, say no to the South's latest demand, confident it was another bluff. Almost no factions intended to force a choice between the Union and truth as they understood it.[47]

Yet Americans ended up doing just that, because northern Whigs, Free-Soilers, abolitionists, and Democrats alienated by Kansas-Nebraska felt obliged to rebut southern assaults on their way of life just as southerners rebutted assaults made on them. Had the free states bowed to "modernist heresies" breeding greed, vice, corruption, and oppression of workers? On the contrary! Northern farmers, mechanics, merchants, manufacturers, and factory hands lived off their own labor, not that of others. Call it the Protestant ethic, the American dream, or just the free market; the northern economy transformed Adam's curse to live by the sweat of his brow into triumphant service to self, family, community, nation, and the whole human race. Horace Greeley called work "a blessed boon of God, to alleviate the horrors and purify the tendencies of our fallen state!" The free labor market offered every man the chance to study, work, save, invest, buy land or open a shop, employ others hoping to scale the ladder, and rise as high as his talents and moral fiber could take him. "You are destitute because you have idled away all your time," Horace Bushnell, the North's most representative theologian, lectured his stepbrother. "Your *thousand pretences* for not getting along better are all nonsense. . . . *Go to work* is the only cure for your case." The North's prosperity, technology, and diversification proved how consistent its society was with the will of God and the principles of political economy. In the South masters forced people to work under the lash. In the North employers and workers voluntarily served their joint interests. In the South government was held in suspicion. In the North government helped make people's dreams come true. In the South masters muffled free speech and censored the mails—proof enough that their regime was artificial. In the North people of all opinions debated and voted in freedom.

Why was the South "morally unjust, politically unwise, and socially pernicious," as Seward put it in 1856? Reams of statistics in northern magazines, newspapers, and travelogues proved it was all due to slavery. Southern apologists congratulated themselves for owning their workers, but that was precisely the point. Even if Negroes were not expensive, wasteful, and lazy (as northerners were quick to believe), a system based on coercion was inefficient and dehumanizing to all parties. "Enslave a man and you destroy his ambition, his enterprise, his capacity," wrote Greeley. Slavery enervated whites as well by making honest work degrading instead of ennobling. Slavery also bred violence, ignorance, and disease. Northern visitors reported in shock that "pistols, dirks, bowie knives, or other instruments of death, are generally carried throughout the slave states" and "deadly affrays . . . matters of daily occurrence." Northerners were quick to believe that men who whipped slaves for a living were just as likely to beat their wives, especially when drunk. Southerners might be "more civilized than the Seminole," sneered Ralph Waldo Emerson, but only "a little more." The apparent lack of education in the South was easily explained because there one looked in vain for the public schools that were taken for granted in the north. The illiteracy rate among rural whites was appalling. So, it seemed, was their health. Yellow fever, cholera, and malaria were more severe in the South. Pellagra; hookworm; tetanus; respiratory and intestinal diseases; eye, skin, and teeth disorders; and "female afflictions" made the South a literally sick society. Northern experts attributed that situation to torpor, ignorance, and poor hygiene. Most scandalous was the southerners' sex life. Northerners noted the many mulattoes and concluded (perhaps projecting their own fantasies) that planters kept harems. Northerners blushed at their own bawdy districts; but in Nashville, New Orleans, and Charleston, men's social life seemed to revolve around elegant bordellos. The illegitimacy rate was far worse down South. Did not George Bourne, a Presbyterian minister from Virginia, admit to "an erotic society" indulging in "all the vicious gratifications which lawless power and unrestrained lust, can amalgamate"? George Weston dared draw the conclusion: "to destroy slavery is not to destroy the South, but to change its social organization for the better."[48]

How was it that a pernicious and uncompetitive system had not only survived but become so mighty as to threaten the whole nation's future? Senator Salmon P. Chase offered an answer. Abundant evidence proved the founders had expected slavery to disappear and surely did not intend the federal government to promote it. Chase thought natural law, common law, and the Fifth Amendment (specifying that no person could be "deprived of life, liberty, or property, without due process of law") should have made slavery illegal from the beginning. But the tyranny had been sustained by statutory laws imposed on behalf of a tiny minority. In 1850 Chase summarized his findings: "1. That the original policy of the Government was that of slavery restriction. 2. That under the Constitution Congress cannot establish or maintain slavery in the territories. 3. That the original policy of the Government has been subverted and the Constitution violated for the extension of slavery, and the establishment of the political supremacy of the *Slave Power*." Northern Democrats denied this, but the truth seemed obvious to Chase, Seward, Sumner, Greeley, and even southerners such as Hammond and Stephens. Which states were overrepresented by 30 percent in Congress and the electoral college, thanks to the constitutional clause counting a slave as three-fifths of a person? The South. Who controlled the White House for all but four years (John Quincy Adams's term) after 1800? Southerners and their "doughface" northern allies. Who controlled the Supreme Court, important cabinet posts, the House speakership (for twenty-eight out of thirty-five years), and won all the showdowns eventuating in "compromises"? The South and its northern Democratic toadies. How dare southern advocates claim their section was an exploited colony when the slavocracy ruled the North like a conquered province? How dare they accuse the North of embracing British-style industrialism when it was clear that the South planted British-style aristocracy on American soil?[49]

Short of *northern* secession, which only a few Garrisonian zealots desired, the only way to break the grip of the slavocracy was to form a new majority opposed to the expansion of slavery. But the quick demise of the Free-Soil Party proved that a negative platform was not enough. What Chase, Seward, Sumner, Thaddeus Stevens, Benjamin F. Wade, Joshua Giddings, and other Free-Soilers needed was a mass party that exploited fear of the slave power the way Jacksonian Democrats exploited fear of the money power, but was otherwise devoted to Whig principles and to a civic religion of hope rather than fear. It took a few years for that party to coalesce, but the principles were available in Clay's American System. So were the principles that inspired the North to pursue its own foreign policy in the 1850s. Seward derived that policy from what he called the twin gospels of Christ and commerce.

In 1848 Seward's eulogy for John Quincy Adams brought tears to the eyes of Albany's hard-boiled state legislators. His subsequent biography of "Old Man Eloquent" persuaded Charles Francis Adams that Seward was his father's political heir. Adams had faith that Americans, if moved by morality and aided by government, would cover the continent with agriculture and industry; conquer the commerce of the Pacific; and then carry the torch of democracy, science, and Christianity to benighted Asia. Adams also believed that slavery must be abolished lest the "Author of history" judge Americans unworthy of their high calling. Seward agreed. But he was a shrewd enough politician to know that a curmudgeonly, one-issue scold had no chance of becoming president. So even as he invoked a "higher law" with regard to slavery, Seward summoned Americans to be missionaries for the two great agencies of civilization.

The first, the gospel of Christ, needed no introduction. The American Board of Commissioners for Foreign Missions had been growing for decades, thanks to recruits and contributors moved by Romantic hymns and appeals concerning the heathen. The "gospel of commerce," by contrast, was conceived in 1839 by *Hunt's Merchants' Magazine*. British economists led by Richard Cobden convinced Freeman H. Hunt that free, peaceful trade among nations not only maximized everyone's wealth but inculcated morals as well. "Commerce," wrote Hunt, "is now the lever of Archimedes; and the fulcrum . . . wanted to move the world is found in the intelligence, enterprise, and wealth of the merchants and bankers, who now determine the questions of peace and war, and decide the destinies of nations." By the 1850s all this seemed prescient. The United States' foreign trade grew 144 percent over a decade. American clippers and steamships circled the globe. China, the largest market for merchants and missionaries, lay open to western influence. France, the Low Countries, and western Germany were quickly industrializing in imitation of Britain. But Seward assured the Senate that there was "no seat of empire so magnificent" as America, because it was now in position to intercept the commerce of the Atlantic and Pacific alike. The United States "must command the empire of the seas, which alone is real empire." He imagined an empire built by free men, free soil, and free labor; an empire governed by merchants and missionaries preaching trust, tolerance, and progress; an empire in which "one nation, race, or individual may not oppress or injure another, because the safety and welfare of each is essential to the common safety." Seward exhorted Americans to "take up the cross of republicanism and bear it before the nations."[50]

Such notions were not limited to the Whigs. A Young America movement (imitating the Young Italy and Young Germany republican movements of 1848) also touted Americans' mission to reform the world by example. But Young America appealed mostly to Democrats hoping that the rhetoric of Manifest Destiny might win northern support for southern filibusters while distracting attention from slavery at home.[51] Seward hoped an American mission to promote freedom abroad would reinforce freedom at home. In any case, the maritime interests he represented were not about to let foreign opportunities slip just because sectional issues had not been settled. Four contingencies convinced them the chief opportunity lay in Japan, the mysterious empire whose Tokugawa shogunate had kept it isolated from the world since 1638. First, the desk-bound naval scientist Lieutenant Matthew Maury explained the geopolitical imperative. In 1848 he showed Congress how a flat Mercator projection suggested that the shortest route between San Francisco and Shanghai was a straight line. Then he replaced the map with a globe and string, proving that a great-circle route skirting the Aleutian Islands was far shorter. Sailing ships could not take advantage of that because of contrary winds, weather, and currents, but steamships could so long as they had a coaling station en route. Since Japan was known to be rich in coal, Maury pressed for a mission to persuade the shogun to open his ports. Second, the steamship magnate Aaron H. Palmer of New York lobbied for a mission to win for U.S. merchants the same advantages in Japan that the British had won in China. Palmer made converts of Seward and Senator Hannibal Hamlin of Maine. Third, congressmen from New England hotly protested the torture and death that the xenophobic Japanese imposed on whalers shipwrecked on their forbidden shores. Fourth, missionary societies longed for the chance to civilize a race deemed barbaric.[52]

Democrats, especially southerners, shrugged at all this. But Daniel Webster, the Whigs' Secretary of State, called Japan's coal "a gift of Providence, deposited, by the Creator of all things, in the depths of the Japanese islands for the benefit of the human family." In 1851 Fillmore approved two expeditions: one to bear his personal letter to the Japanese emperor, and another to explore the North Pacific as Wilkes had explored the South Pacific. The naval secretary's first choice for commander proved unfit, but his second was ideal: Matthew Calbraith Perry, "Old Matt" from Rhode Island. In addition to being an apostle of steam and commercial empire, Perry was a consummate diplomat. By the time his twelve warships, including three smoke-belching steamers, reached the Lew Chew (Ryukyu) island of Okinawa in May 1853, the Democrat Pierce was president. But the Whig initiative ran its course. In July Perry's "black ships" anchored off Edo (later renamed Tokyo), spreading panic among the city's million residents. Perry neither threatened nor reacted to threats. He just waited until the shogun's officials agreed to bear Fillmore's letter to the mikado; then he sailed away, promising to return in the spring. During that interlude Lieutenant John Rodgers began a lesser-known but equally important expedition that surveyed the Japanese archipelago. This helped to convince the *bakufu* (bureaucracy) that Japan must either let the Americans get a foot in the door or risk having their door kicked in like China's. Perry returned in March 1854 to negotiate the Treaty of Kanagawa granting the United States a consulate and access to two coaling ports. An ironic exchange of gifts ensued. The Japanese bestowed on their guests exquisite lacquers, porcelain, silk, samurai swords, four little dogs, and an exhibition of sumo wrestling. The Americans bestowed on their hosts a miniature locomotive, a telegraph, a daguerreotype, firearms, and a black minstrel show.

When Perry docked at Hong Kong, American merchants in China elevated him to the ranks of "Columbus, DaGama, La Perouse, Magellan—these inscribed their names in history by striving with the obstacles of nature. You have conquered the obstinate will of man . . . and brought an estranged but cultivated people into the family of nations." When the news reached New York the chamber of commerce showered Perry with awards and money. In fact, he obtained no trading rights. That task fell to Townsend Harris, a dour Puritan bachelor appointed consul to Japan in 1856. He suffered two years of lonely indignities until the *bakufu* made up its collective mind to grant a commercial treaty and send seventy-seven samurai across the ocean to study the barbarians known as Yankees. San Francisco lionized those first Japanese emissaries in the belief that Americans were now poised to export their civilization to Asia in exchange for Asian riches: the gospel of Christ and commerce fulfilled. But when the samurai reached Washington City in May 1860, America's mission abroad was the last thing on anyone's mind.[53]

What sort of edifice had Americans built on the cornerstone laid by the founders? Ultras in both sections agreed it was a "house of Usher" based on two clashing architectures. They just disagreed over which wing was comely and which ugly. Southern spokesmen apotheosized their way of life, warning, "You northerners may boast of your pieties, factories, farms, and free labor in the belief that God and the Constitution favor your cause. But so long as we're in this Union we'll ensure that you don't feel good about doing well." Northern spokesmen apotheosized their way of life, warning, "You southerners may boast of your chivalry, cotton, and smiling slaves in the belief that God and the Constitution favor your cause. But so long as we're in this Union we'll ensure that *you* don't feel good about doing well." The schism was even harder to bridge in the 1850s because almost everyone on both sides was lying.

Did planters sacrifice financially by caring for slaves? No; the institution proved increasingly profitable. Were Negroes unfit for freedom because they were not fully human? No; voluminous essays in *DeBow's Review* and the *Southern Agriculturalist* taught planters how to maximize efficiency through the same incentives and discipline to which white employees responded. Were Negroes nonetheless inferior? Their growing employment as artisans and mechanics, the economic success of some free blacks, and the contempt planters and slaves shared for "white trash" (a term possibly coined by slaves) suggested otherwise. Were paternalistic masters the rule? Perhaps, but many masters were absentee landlords and all delegated the driving of slaves to overseers, whose numbers doubled in the 1850s alone. Did planters not claim a sexual droit du seigneur over slaves? Enough did so that one-fifth to one-third of colored Americans in 1860 had some white ancestry. Did blacks have no notion of family? In fact, wise planters fostered family life among slaves as a source of stability and children, a bountiful "interest" on their investment. Did planters therefore take pains to keep black families together? Often they could, but sometimes would not because they in turn were "slaves" to the market. Even the esteemed Reverend C. C. Jones broke up some families when affairs required.[54] That is why the heart of the system was not the mint julep or magnolia, but the slave market (and number-one tourist attraction) in New Orleans. Every year thousands of blacks passed through pens stinking of smoke, sweat, excrement, and lye disinfectant. Sellers and brokers auctioned them off like horses, often falsifying their age, ancestry, and health in defiance of Louisiana's elaborate "redhibition laws." Sharp buyers scrutinized the naked Negroes, mulattoes, quadroons, and griffes to examine musculature and check for scars, deformities, and disease before pronouncing them "likely." Most slaves sold in New Orleans came from out of state; half of interstate sales broke up a nuclear family.

Were slaves at least content when governed by a kind "massa" and "missus"? Common sense says they probably were, if one assumes they knew or imagined no other life. But weighty evidence suggests freedom meant a great deal to slaves in spite of all efforts to keep them docile and ignorant. Just as the Underground Railroad moved fugitives to the North, a word-of-mouth "underground telegraph" transmitted news gleaned from the North, from southern journals, or just from conversations overheard in the plantation's "big house." Olmsted's tour convinced him that knowledge of abolitionist agitation "has been carried among the slaves to the most remote districts." That slaves could imagine a better life was equally proved by their bargaining ploys. They knew how to resist overwork and ill-treatment by playing dumb, feigning illness, lollygagging (hence their reputation for laziness), or sabotage. Their "Br'er Rabbit" trickster stories celebrated the guileful triumph of prey over predator. But nothing proved slaves' discontent more than their religion. White pastors presiding over sanctioned services told them God willed that servants obey their masters. But "the ante-bellum Negro was not converted to God. He converted God to himself." This happened in slave quarters and fields, where blacks sang of their own identity as a chosen people, longed for an exodus from "Egyptian" bondage, and prayed to a Lord as intimate as their African deities had been. "Gwine to write to Massa Jesus / To send some valiant soldier / To turn back Pharoah's army / Hallelu!" Slaves claimed for themselves the promises in the Old and New Testaments about divine deliverance of the downtrodden and poor in spirit. They found dignity, power, and spiritual freedom in scripture. They made Psalm 68:31 the most quoted passage in African-American worship: "Princes shall come out of Egypt; Ethiopia shall soon stretch out her hands unto God."[55]

Southerners posed as proud Cavaliers, masters of their domain and all who dwelled therein. That may have been their most painful prevarication. If the planters were secure and benevolent, why did they need strict slave codes, censorship, and armed posses to maintain their system? (South Carolina even forbade slaves to have pet dogs lest they get a feel for being the master.) Why did they stoop at election time to stirring up fears and prejudice among whites who did not own slaves? Why did their "plantation novels" drop copious hints that southern patriarchs weren't even masters of their own household? George Tucker's *The Valley of Shenandoah* (1829) founded the genre. William Alexander Caruthers wrote several such novels, including *The Cavalier of Virginia*, in the 1830s. Many novelists of the same sort, led by the Harvard-educated Virginian Daniel

Hundley, responded to *Uncle Tom's Cabin* in the 1850s. All extolled the traditional southern gentleman's devotion to family, hospitality, honor, agrarian life, and service to country. But almost all lamented that few such gentlemen remained! Southern novels told of decline, debt, and expiring lineages giving way to "new-rich swells" who resembled northern hustlers more than southern gentry. Their most telling revelation was that the real rulers of the plantations were *women*. "In this her proper sphere," Hundley confessed, "woman wields a power, compared to which the lever of Archimedes was nothing more than a flexible blade of grass. She it is who rules the destinies of the world, not man . . . a power behind the throne greater than the throne itself." Of course, planters' wives made a pretense of deference, but they governed family and slaves as deftly as belles made puppets of beaux. Everyone knows the responsibility that fell on the shoulders of the fictional Scarlett O'Hara. The South Carolinian diarist Mary Boykin Chesnut insisted it was that way even before the war. While the master amused himself with sport and politics, management of his sprawling estate and dependents frequently fell on the matron; hence Chesnut's complaint: "there is no slave like a wife." This is not to suggest that the South was a matriarchy, but it does suggest that patriarchy took unusual, burdensome forms in a plantation economy where the household and workplace were coterminous and distinctions of race complicated distinctions of gender. It is tempting, therefore, to suspect that insecurity mingled with pride drove planters and even white yeomen to fight for the Cavalier myth with all the masculinity they could muster.[56]

What of the North? Was it really a haven of free labor and land where an honest government and a democratic citizenry partnered in progress? When Senator Robert Toombs of Georgia pronounced the U.S. government the most corrupt under the heavens, the New Hampshireman John P. Hale agreed. When Earl Gray warned Britons that democracy in America bred demagogy, dishonesty, graft, bribery, and incompetence, the *New York Evening Post* agreed. When a Virginian inveighed that "the majority of northern members of congress are as corrupt & destitute of private integrity, as the majority of southern members are the reverse," the *Philadelphia North American* agreed. When an editor in South Carolina accused the North of hypocrisy, his interlocutors in New York agreed. "The fact is," they confessed, "our public men are all rogues, honest men are driven from the polls—the ballot boxes are in the hands of ruffians—the very men elected . . . are so many swindlers, stock-jobbers, liars, even forgers and robbers." Honest opinion was "awed and confounded, silenced and incapable." The North, it seemed, was a fraud, and its institutions were the playthings of a "plundering generation."[57]

Corruption—however it is defined—exists wherever laws exist. Edward Gibbon wryly called corruption the infallible symptom of liberty, and Samuel Huntington argued that corruption was especially prevalent during dynamic social transitions. So it should surprise no one that hustling Americans endowed with layers of law, unprecedented liberty, and the most dynamic civilization in history practiced corruption as a matter of course. What is more, Americans were usually pleased to overlook corruption in the service of public projects that expanded opportunity for the many. However embarrassing to exponents of republican virtue, not to mention evangelical preachers, the evolution of U.S. business and politics had, by the 1830s, built corruption into the system.[58] Jackson's Democrats purported to fight corrupt concentrations of wealth, but they did so through a political machine based on the spoils system, sweetheart contracts, a bribed partisan press, intimidation, fear-mongering, ethnic blocs, and vote fraud. The Whigs, in order to win, followed suit. Likewise, economic interests purported to fight corrupt concentrations of power, but did so by co-opting officials through graft, kickbacks, and campaign contributions. What ploy became the first reflex of anyone hoping to win office, shape law, and dip into the public purse? Why, accuse one's opponents of corruption, of course.

Sleaze took center stage in the politics of the 1850s for at least four reasons. First, a booming economy that nearly doubled the volume of money made Yankees more eager than ever to get rich quick. Since railroad and canal building, land speculation, bank charters, and patents all required some government imprimatur, corruption was tempting and often necessary to get big things done. A second factor was the explosion in journalism. Before 1850 only a few newspapers had permanent correspondents in the national capital or state capitals. After that date swarms of mercenary reporters hungry for scandal made corruption seem out of control as never before. Third, native-born northern workers grew increasingly angry about fortunes made at public expense because their own wages were being depressed by the spike in immigration in 1845–1855. Many laborers considered protest against foreigners and protest against corruption as two sides of a coin. Fourth, the escalating debate over slavery ignited the old "paranoid style" of American politics by conjuring fears of a corrupt slave conspiracy.[59] Still, where there is smoke there is fire: corruption was pervasive and centered in the North.

In the 1850s the Democrats' Tammany Hall machine ceased running front men and unabashedly took over New York City. The grand sachem Fernando Wood won the mayoralty—on a reform platform, of course—whereupon his "forty thieves" at city hall plundered the municipal coffer; bought judges, aldermen, and policemen; and made fortunes for dummy corporations, favored contractors, real-estate insiders, and themselves on every public works project. Greeley campaigned against growth in the city's debt because "They will steal enough without it." Young Milwaukee's tax base was small, so its urban machine financed graft by expanding the city's debt by a factor of twenty from 1851 to 1858. Mayor "Long John" Wentworth's machine ran Chicago, where policemen closed down only those dens of prostitution and gambling that competed with their own. Mayors in Philadelphia, Cincinnati, and Saint Louis (where liquor licenses were the medium of exchange) ran similar shops. Nativists blamed it all on ignorant immigrants. But ward heelers were the ones who lured Paddy to the polls with booze and walking-around money, handed him phony naturalization papers, then carted him to the next ward to vote again. If the result was still in doubt, control of the ballot box fixed it. San Franciscans joked that elections began only after the polls closed. Did city machines corrupt pristine state governments, as nativists also claimed? In fact, Thurlow Weed and his Democrat counterparts in Albany functioned as literal power brokers, using their clout with the national party to buy and sell offices, influence, and press coverage. Nor did state legislators need instruction from the big city about what to do when railroad and canal promoters spread around millions in land grants or bonds that a favorable vote would release. Lawmakers in Wisconsin and Michigan were notorious for pocketing payoffs after adjournment, when they "were no longer a legislature"—hence "acceptance could in no way be considered or regarded as a bribe." Maine was the most bluenosed of states, yet its Democrats swept into office in 1858 by "enfranchising" thousands of French Canadians where the census showed that only a few hundred lived. "My God, we have voted like hell," bragged a ringleader and candidate. "I have not worked so long for nothing."

Shenanigans at the state level were small potatoes compared with deals done on Capitol Hill. Such was the federal government's power to give and to take away that "borers" (lobbyists) more numerous even than journalists haunted the halls of Congress. They spread around money, women, and gifts on behalf of clients seeking: subsidies for steamship lines; gigantic land grants to railroads like the Illinois Central; patent extensions; contracts to Indian agents (a license to steal); or juicy appointments as customs officials, land-office agents, public printers, and postmasters. It all added up to tens of millions of dollars in political capital. Douglas and Pierce dipped into that pool to purchase votes needed to pass the Kansas-Nebraska Act. Enough said. Before 1776 leaders in the thirteen colonies railed against such "royal patronage" and profiteering. By the 1850s it seemed that Americans had democratized corruption and put democracy itself up for sale. The system had its defenders. Payoffs and patronage greased the wheels of government. Internal improvements served the public good. Urban machines assimilated and pacified immigrants. Young America disparaged "old fogies" whose notions of civic virtue impeded progress. But as the manic decade went on, a growing number of northerners recoiled against corruption, and southerners feared that union with the "arrogant, aggressive, mercenary and unprincipled" North must end in their own corruption.[60]

* * *

By blue Ontario's shore, a Phantom gigantic superb, with stern visage accosted me. Chant me a poem, it said, of the range of the high soul of Poets . . . chant me, before you go, the Song of the throes of Democracy. . . . A nation announcing itself . . . we are executive in ourselves . . . we stand self-pois'd in the middle, branching thence over the world, from Missouri, Nebraska, or Kansas, laughing attacks to scorn . . . nothing is sinful to us outside of ourselves . . . all is eligible to all, all is for individuals, all is for you . . . if you would be freer than all that has been before, come listen to me . . . these States are the amplest poem, here is not merely a nation but a teeming Nation of nations . . . Others take finish, but the Republic is ever constructive . . . O America because you build for mankind I build for you . . . for the great Idea, the idea of perfect and free individuals, for that the bard walks in advance . . . I listened to the Phantom by Ontario's shore, I heard the voice arising demanding bards, by them all native and grand, by them alone can these States be fused into the compact organism of a Nation.

He was destined to be the poet laureate of American civic religion, but somehow Walt Whitman did not belong. He surely did not belong among people like Hawthorne and Melville who looked to America's past and prophesied doom. Whitman looked to America's future, sensing glory he never won for himself.

Whitman's birthplace was Long Island; his roots were English Quaker and Dutch; his father was a failed real estate developer. Walt's education was minimal, but he worked as a law clerk, printer, and school-teacher until age twenty-two, when he leaped into Gotham's inky news-paper swamp. No good. He was bounced (or bounced himself) from a dozen journals in New York between 1841 and 1848, quit a job in New Orleans in 1849 because he was disgusted with slavery, then founded a Free-Soil paper in Brooklyn on a shoestring. All the while he was "sim-mering, simmering" until Emerson brought him "to a boil." In 1855 Whitman set the type for a thin volume called *Leaves of Grass* dominated by the epic later entitled "Song to Myself." Nobody noticed. So the next year Whitman issued a second edition including "On Blue Ontario's Shore," quoted above. It was secretly bankrolled (by a phrenology firm), graced with an unauthorized blurb from Emerson, and promoted by anonymous raves that Whitman planted himself. Still, his revolutionary free verse, operatic and biblical cadences, and sublime message remained a mere curiosity. Whitman did land the editor's post at the *Brooklyn Times*, but he was fired in 1859 for advocating sexual freedom for women. That year a third edition of *Leaves of Grass* flopped when the publisher went belly-up (the same fate suffered by Melville's *The Confidence-Man* the previous year). Whitman suffered a nervous breakdown simultane-ous with that of the Union. Was it caused, as some critics suspect, by repressed homosexuality? Who cares? Whitman transcended himself as Emerson never did. He imagined himself as America ("I contain multi-tudes"), hugged all its regions and peoples, claimed that bards were the nation's salvation, and never stopped learning. That is why he rewrote and reissued *Leaves of Grass* like pungent, newly harvested hay for the rest of his life. Perhaps what happened in 1860 is that Whitman realized his genius had led him astray on one awful point. The chants Ontario's Phantom demanded were *war chants.*[61]

Did Whitman sing of the "throes of Democracy" because he realized American politics were in chaos by 1856? Southern Whigs defected en masse after their debacle in 1852. Northern Democrats were clobbered in 1854 when two-thirds of their congressmen supporting the Kansas-Nebraska Act lost their seats. Northern Whigs then might have emerged stronger, as an omnibus Free-Soil Party; but they did not, because of the red-herring odor of immigrants. All the votes cast for Winfield Scott did not match the number of aliens, mostly Catholic recruits for the Democrats, who arrived between 1852 and 1856. Various Whigs soldiered on locally, but could no longer compete at the national level. Accordingly, Protestant Free-Soilers, nativists, and temperance advocates experimented with alternatives. The front-runners in 1854–1855 were the Know-Nothings, a Masonic-style organization that claimed 1.5 million members favoring strictures on naturalization. Their political arm, the American Party, won stunning victories in the Northeast and showed strength in the upper South. But their secrecy and bigotry also repelled potential supporters. Free-Soil politicians in various states hoped that "fusion" with anti-Nebraska Democrats or nativists or both might lay the basis for a new majority.

One such party took the bland but venerable name Republican. Four nondescript gatherings claimed to have rocked its cradle: a conclave in Ripon, Wisconsin, in March 1854; congressional caucuses in May and June; and a state convention held "under the oaks" of Jackson, Michigan, in July. But Republicans lacked likely leaders. Strong Whig organizations, such as Seward's in New York, were loath to gamble on a new brand name. Chase, the ex-Democrat in Ohio, was loath to break bread with ex-Whigs. No smart politician believed any party defined primarily by opposition to slavery stood a chance of winning, even in the North. Only in Wisconsin and Michigan did Republican organizations mount strong campaigns in 1854. So the bewildering upshot of the midterm elections was a polyglot Congress that included 121 supporters of the Know-Nothings. Ninety-two of them were also Free-Soilers. In sum, the crack-up of the old system in no way determined what new one might coalesce. As for the Democrats, the northern defections they suffered while Kansas bled sharply increased southerners' clout. But they carried on as a national party, thanks to the immigrant vote, northern doughfaces ambitious for office, and Douglas, who knew that the Democratic Party was the only thing left holding his beloved Union together.[62]

In 1855 some of the clouds blew away. The Know-Nothings, having broken down party loyalties in such key cities as Philadelphia and Cincinnati, broke down themselves. Three times the American Party mustered to plan a run for the presidency; three times its northern delegates walked out rather than endorse repeal of the Missouri Compromise. Prominent Free-Soilers and fusionists sniffed an opportunity. One by one they decided to hitch their wagons to the Republican Party. Seward, safely reelected to the Senate, hoped to win the Republican presidential nomination. In Albany he delivered a rousing speech that Richard Henry Dana, Jr., hailed as "the key-note of the New Party." Rather than bashing immigrants, who were only pawns in the Democrats' game, Seward focused his ire on a "privileged class" of slave owners out to trample the freedom of 25 million Americans. "The Republican Party is sounding throughout all our borders a deep-toned alarum for the safety of the Constitution, of union, and of liberty. Do you hear it?" Chase also embraced the Republican label in hopes of becoming president. He brokered a complicated fusion with the Know-Nothings, then double-crossed them by extending an olive branch to Germans and anti-Nebraska Democrats. The Republican slate swept Ohio's elections of 1855. In December the American Party was co-opted permanently when its northern congressmen agreed, after a long fight, to help elect the Republican Nathaniel P. Banks of Massachusetts Speaker of the House. Southerners of all persuasions were mortified; northerners of most persuasions were ecstatic. Still, no interstate Republican organization existed as yet; hence there was no understood platform. A few missteps or maladroit speeches and the Republicans would have reverted to a fringe party. To become truly "available" Republicans had to focus northerners' frustrations on the Democrats fronting for slavery without being branded abolitionist or disunionist. They had to attract Know-Nothings without conceding the whole immigrant vote. And they had to promote a Whiggish program for economic development without offending old-fashioned Jacksonians.[63]

They filled that tall order in critical Illinois, where fusion movements had stalled, owing to the state's deep party loyalties. Indeed, the main contest in 1855 was decided when Whigs backed an anti-Nebraska Democrat in order to block the Senate candidate whom Douglas endorsed. Only after the Little Giant's machine retained its hegemony over state government did moderate Whigs, Free-Soil Democrats, and Know-Nothings realize the urgent need for a new party. The 270 delegates who convened in Bloomington were all the more eager to lay aside differences because they met on May 29, 1856, just days after the army of ruffians sacked Lawrence, Kansas, and Brooks bludgeoned Sumner on the floor of the Senate. The wise Illinoisans drafted a radically moderate program. It bowed to nativism, but condemned discrimination and said nothing about temperance. It demanded restoration of the Missouri Compromise, but shunned abolition and made no explicit objection to new slave states. It slated Know-Nothings, Democrats, and a German for prominent posts. It did not even use the Republican label. But its delegates carried to that party's first convention a formula for winning the Middle West. The realignment in U.S. politics was complete.[64]

In February the Know-Nothings' southern rump wooed its northern brethren one last time by nominating Fillmore for president. It was not even known whether Fillmore (who was overseas on a diplomatic mission) supported the nativist cause. But he certainly had name recognition. In June the Democratic convention in Cincinnati (the first held west of the Alleghenies) battled for seventeen ballots before rejecting the discredited Pierce and the controversial Douglas in favor of Buchanan, Pennsylvania's veteran doughface. Buchanan, a dapper old bachelor (he had been jilted in his youth), was another "northern man with southern principles," and was unstained by Kansas-Nebraska by virtue of having served Pierce as U.S. minister in London. By contrast, the Republicans meeting in Philadelphia later that month swiftly constructed an "Illinois" platform. It called on Congress to ban from the territories "those twin relics of barbarism—Polygamy and Slavery" (thus smearing the South with the Mormon brush), admit Kansas as a free state, and reject southern plots to annex Cuba. The platform was silent about slavery where it already existed. Likewise, Republicans endorsed a transcontinental railroad, but in deference to former Jacksonians said nothing about tariffs and banks. They skirted nativism by affirming the liberty of conscience of all *citizens* but leaving open the issue of *aliens*. Choosing candidates was a thornier task, but with managers as subtle as Weed, Greeley, Francis Preston Blair, Joshua Giddings, and Thaddeus Stevens working the floor, the party was not going to do anything stupid. Seward bowed out because Weed did not expect to win this

first contest and wanted to save his man for 1860.[65] Chase was too abolitionist. Justice John McLean was too old and nativist. So the managers
took a page from the Whigs and selected a soldier with "no past political
sins to answer for." To the embarrassment of Senator Benton, a Democrat,
it was his son-in-law John "the Pathfinder" Frémont. The *New York Tribune*
credited this choice to a "spontaneous instinct" among delegates. In fact,
the instinct was that of the managers and hardly spontaneous. The hero not
only distracted attention from the party's abolitionist base but gave it the
perfect slogan: "Free Soil, Free Speech, Free Men, and Frémont."[66]

Mudslinging, hokum, and candor in 1856 left Americans unsure
whether to be smug or alarmed. Democrats and Know-Nothings questioned Frémont's military record, hinted that he was a closet Catholic, and
baited their opponents by using the term "Black Republicans." Democrats
and Republicans said laughingly that the Know-Nothings' nightmare concerning a papal conquest of America was as likely as a Norman invasion.
Know-Nothings and Republicans dismissed Buchanan as a tired Old
Hunker Democrat with no platform beyond spoils. "There is no such person running as James Buchanan," announced Thaddeus Stevens. "He is
dead of lockjaw." All very funny. But what to make of Hannibal Hamlin's
slap about the Democrats' having no issue left besides slavery, "the standard on which it measures every thing and every man"? What to make of
Robert Toombs's prophecy that the "election of Frémont would be the end
of the Union, and *ought* to be"? Wall Street feared this so much that it
dumped money on Democrats. (Greeley griped, "We Fremonters of this
town have not one dollar where the Fillmoreans and Buchaneers have ten
each.") The fear was not idle. In the heartland, evangelicals turned the
campaign of 1856 into just what party managers hoped to avoid: a crusade.
They rallied for Frémont in churchyards where pastors such as the Congregationalist Charles Boynton blessed them for choosing God's side in
the eternal battle between "truth and falsehood, liberty and tyranny, light
and darkness, holiness and sin." The Methodist Benjamin Adams said
that stumping for Frémont was equivalent to preaching the gospel.
Hymns were composed, such as "Think that God's eye is on you / Let not
your faith grow dim / For each vote cast for Fremont / Is a vote cast for
Him."[67]

In light of the outcome, the Georgia pastor and planter C. C. Jones
chose to be smug. "The election has ended well," he wrote to his son. "May

it be the first act of divine goodness in our deliverance from the political and religious heresies which have been disorganizing and destroying the country." Despite dreary fall weather across the North, voter turnout was the largest in history. Buchanan won 45 percent of the vote and 174 electoral votes; Frémont won 33 percent and 114 electoral votes; Fillmore won 22 percent and Maryland's eight electoral votes. The good old electoral college had once again translated an iffy plurality into a comfortable victory. But cogent observers raised an alarm. Republicans swept the northern tier of states but were not even on the ballot in most slave states. Democrats swept all but one slave state plus Illinois, Indiana, Pennsylvania, and New Jersey, but nowhere in the North did they surpass 50.4 percent. It took little imagination for Republican bosses to see all they needed to do was capture most of Fillmore's voters in the lower North and perhaps a few Germans to win the presidency "with no pretense to bisectionalism" at all. As a Unitarian in Philadelphia observed, the Republicans "have not got a President, but they have what is better, *a North*." John Greenleaf Whittier asked, "If months have well-nigh won the field, what may not four years do?" [68]

American voters "hire" their presidents to solve some pressing problem or else make it go away. Buchanan's temperament decidedly favored the dodge over the fix. The sixty-five-year-old lawyer from Lancaster County had wealth, distinction, experience, and a certain charm. But he was plodding and meticulous to a fault, and this helps to explain why his long diplomatic career produced bulging files but no signal achievements. Polk had likened him to an old maid. So Buchanan was in character when he declared in his inaugural address that the feud over slavery in the territories wasn't the president's business at all, or that of Congress. "This is, happily, a matter of but little practical importance. Besides, it is a *judicial* question, which legitimately belongs to the Supreme Court of the United States, before whom it is now pending, and will, it is understood, be speedily and finally settled." He was referring to *Scott v. Sandford*, the climactic litigation of a case that dated from 1842, when Dred Scott first sued for freedom. He claimed his master, an Army doctor (since deceased), had illegally kept him in bondage while stationed in Illinois and the Wisconsin Territory. A court in Missouri found in Scott's favor, but a federal court reversed the decision, whereupon Scott's attorneys appealed to the Supreme Court in February 1856. Chief Justice Taney and his colleagues had numerous options. They could uphold the lower court, rule a noncitizen had no standing, decide Scott's status on the merits without making a precedent, or pronounce on the constitutionality of all federal and territorial laws regarding slave property. President-elect Buchanan interfered by informing the justices that he hoped they would settle the case promptly and render a broad decision. On March 6, 1857, forty-eight hours after Buchanan's upbeat inauguration, they did.[69]

By a vote of seven to two the Supreme Court denied Dred Scott's standing, because citizenship under the Constitution was never meant for an "inferior class of beings" with "no rights which the white man was bound to respect." But Taney's book-length opinion went on to deny the authority of the federal government and, a fortiori, a territorial government to restrict citizens' property rights. At a blow the Court struck down the Northwest Ordinance, the Missouri Compromise, the Compromise of 1850, and the Kansas-Nebraska Act. It held that no majority in Congress or Kansas could prohibit slavery on a frontier. What next? Would the Taney Court strike down state laws against slavery as well, as litigants in New York had already petitioned? A brilliant minority opinion by Benjamin R. Curtis of Massachusetts revealed serious flaws in Taney's reasoning (e.g., U.S. citizenship predated the Constitution and had been granted to free persons of color in at least five states). But the political implications of the Dred Scott decision caused Greeley's *Tribune* to use words like "atrocious" and "wicked," and the *Chicago Tribune* to express "detestation" over "inhuman dicta." Seward thought that 350,000 slaveholders exerted inordinate power. Now just *"five* slave holders and *two* dough faces" legislated from the bench in favor of the potentially limitless expansion of slavery.

In one sense, Taney's judgment was a final yawp of Jacksonian anger. Just as Taney's mentor and patron Andrew Jackson had lost his wife weeks before his inauguration, so the seventy-nine-year-old Taney had been rendered despondent by the deaths (in 1855) of his wife from a stroke and his daughter from yellow fever. Just as Jackson flexed his executive muscles to crush southern threats to national unity, so Taney flexed his judicial muscles to crush what he imagined were northern threats to national unity. In 1856 Taney, a Marylander, confidentially seethed over northern "cupidity," "evil passions," and "aggression." Still, he could not have imposed his will unless the other southern justices agreed to render a "broad" decision. There is evidence that they did because of the stated intention of the two dissenters to file minority judgments upholding the Missouri Compromise. That obliged southerners to fall in behind Taney or else concede to Congress the power to legislate for the territories.[70]

Buchanan cheered the Dred Scott decision, mistakenly supposing that it would clear the path toward a fair resolution in Kansas under his newly appointed governor, Robert J. Walker. Kansans had their own notions of fairness. The ruffians convened in Lecompton to draft a pro-slavery state constitution. It was quickly approved in a rigged referendum, although Walker judged that a sizable majority of settlers preferred Free-Soil. Buchanan stood by him until southerners in Congress and his own cabinet pressured the president beyond endurance. On December 2, 1857, Buchanan circulated the text of his annual message. It endorsed Lecompton. Stephen Douglas, that human wrecking ball, now swung to the other extreme. He burst into the White House damning the mockery of popular sovereignty. Buchanan warned Douglas he might suffer the fate of those Democrats who bucked Andrew Jackson. "Mr. President," Douglas snapped, "I wish you to remember that General Jackson is dead." The Democrats' two most prominent northerners thus became bitter enemies. That was the last thing the party needed in the face of the Republican challenge.

Meanwhile, one referendum in Kansas ratified the Lecompton constitution by ten to one because Free-Soilers boycotted the voting; and another in January 1858 defeated it 100 to one because pro-slavery men stayed home. Buchanan stuck to his guns. On February 2 he recommended that Congress approve the constitution because Kansas was already "as much a slave state as Georgia or South Carolina." Douglas battled as he had in 1854, but this time in opposition. Buchanan battled with all the weapons in a president's arsenal. The fact that a petition for statehood from Free-Soil Minnesota was also on the docket made southerners all the more determined not to surrender Kansas. At last on April 30 a majority accepted a suggestion by Congressman William H. English of Indiana that a third vote on Lecompton be held under federal oversight. Buchanan signed the bill, gambling his administration on the outcome. Three months of preparation ensued. Federal troops and officials were deployed. On August 2 Kansans stopped bleeding and voted 11,300 to 1,788 *against* slavery.[71]

Buchanan's presidency was wrecked. Southerners felt cheated by Kansas; northerners by Dred Scott. Meanwhile, the financial markets had crashed and even tough businessmen were falling on their knees in prayer. It seemed the tribulation had begun.

The Thirty–Second State: Minnesota, 1858

"You ask how we manage to protect ourselves from the snow and cold," wrote a Swiss immigrant to her in-laws. "As for the cold, we have two pot-bellied stoves, and we get them red-hot with pleasure or anger; that keeps us warm, and [Theodore] isn't stingy with wood even though it's painful for him to cut it. As for the snow, it comes in through the roof, melts, drips onto the upstairs floor and then trickles down to the lower floor. It's awful, but what's to be done about it—the house isn't worth repairing." Sophie Bost wasn't one to complain. Her family surmounted each day's travail with prayers and the Bible. Still, she confessed in March 1863 that Minnesota scared her, especially "the Indians who are said to be planning to make a new incursion in the spring . . . and the scarlet fever that's raging all around here." At other times she wrote of droughts, depressions, and insect plagues. But the Bosts persevered. They even saved enough to retire to a bungalow in southern California.[72]

French *voyageurs* were the first whites in North America's fountainhead, whence waters flow to the Gulf via the Mississippi, the Atlantic via the Great Lakes, and Hudson's Bay via the Red River of the North. Their posts and portages around Lake of the Woods served as headquarters for a fur trade stretching thousands of miles. The fountainhead's northern forests of birch, pine, and aspen were home to fierce tribes of Sioux until the equally fierce Chippewa drove them south to the prairie. Both regions glistened with 12,000 glacier-carved lakes filled with bass, walleye, pike, coho, and chinook salmon. Loons cried eerily overhead; red-winged blackbirds flocked by the thousands. In 1805 Zebulon Pike led the first American expedition into the region. In 1819 the army founded Fort Snelling at the confluence of the Mississippi and a river the Sioux called "Minnay-sotor" after its milky water. In 1820 the Cass expedition charted watercourses far to the north. In 1821 Major Stephen H. Long ascended the Minnesota and Red rivers to Canada. But the most Romantic prize eluded all parties until 1832, when Henry Rowe Schoolcraft found the source of the "Father of Waters" at a limpid lake he named Itasca.[73]

The human source of the future state arrived two years later. Henry Hastings Sibley, the son of a Yankee migrant who served as a justice in frontier Detroit, craved the "active and stirring life." So he ran away to join the fur traders at Sault Sainte Marie and rose to become the American Fur Company's factor west of Fort Snelling. A "lithe, athletic, intellectual, upright, and rebellious" youth with a high forehead and drooping mustache, Sibley spent fifteen years trading among the Sioux and serving as sole civil authority within 300 miles of Mendota. But the idyll ended in 1848 when Wisconsin became a state and Minnesotans demanded territorial status. They numbered fewer than 4,000, but had hopes for boom times ahead and elected Sibley to realize them. He proved deft, perhaps because of the diplomacy and florid speech he learned from the Sioux. Not only did Congress honor Sibley's premature credentials; it approved a new territory, the name Minnesota, and a capital at Saint Paul. Alas, the man President Taylor appointed governor proved to be Sibley's nemesis. "Bluff Alex" Ramsey, a "heavy, phlegmatic, unreflective, morally compliant" fellow with muttonchop whiskers, was a Pennsylvania Whig who had hoped to be named customs collector in Philadelphia. Frustrated in that, he determined to squeeze a fortune out of his proconsular post. "The frontier, to Ramsey, was an invitation to exploitation. . . . To Sibley it was a lost Arcadia." [74]

When Ramsey arrived in 1849 Saint Paul consisted of some twenty half-built structures. Yet he promised the legislature that Minnesota would be "peopled with a rapidity exceeding anything in the history of western colonization." All that this required was Indian removal and advertising. Ramsey accomplished the first in 1851 when the Lakota Sioux, indebted to traders and threatened with a cutoff of rations, relinquished the whole southern half of the territory, except for a ribbon on the Minnesota River, for seven cents per acre. Most of the $1,665,000 purchase price was placed in a trust; most of the rest was paid to the creditors (Sibley included). The Sioux got almost nothing. In 1854–1855 the Chippewa bowed to similar pressure. The ostensible parties to such treaties, the missionary Henry B. Whipple said scornfully, were the chiefs and the U.S. government, but "the real parties are the Indian agents, traders, and politicians." Ramsey and his successors met the second requirement by boosting Minnesota as a "green tree of empire." In 1853 the territory had its own exhibit at New York's world fair. In 1854 it was host to a blue-ribbon delegation, including Fillmore, Bancroft, and Weed. In 1855 a commissioner of emigration was named. In 1856 a New England Society of the Northwest toasted the conquest of Minnesota by Yankee enterprise, thrift, education, and faith. Catholics founded churches over which the famous Bishop John Ireland would preside. Whipple founded an Episcopal diocese; Eric Norelius spread Lutheran churches, and Saint Paul's Jewish congregation dated from 1856. Given that Minnesota had to compete with California for settlers, its success was phenomenal. Between 1850 and 1857 over 5 million acres were sold to a white population that grew 2,500 percent, to 150,000. Over half were of Yankee origin; one-third came from Germany, Scandinavia, and Great Britain. In the seven ice-free months of 1858 1,000 steamboats docked at Saint Paul, boosting its population to 10,000. Its rival Minneapolis emerged when the first bridge spanning the Mississippi River opened in 1855.[75]

Though ripe for statehood, Minnesota was sharply divided. German and Irish pioneers fell in behind their Democratic governor, but Scandinavians and Yankees rallied to the Republicans. The boundary issue raised more uncertainty. In 1856 speculators "persuaded" the legislature to incorporate a company with vast powers to develop real estate, mills, and a new capital farther west. Their plan was to create a state on an east-west axis stretching deep into Dakota, thereby enriching themselves and leaving room for another free state to the north. A committee chairman foiled the plot by absconding with the boundary bill until the legislature adjourned. Meanwhile, the territorial delegate Henry M. Rice did serious business in Washington City. He lobbied for a single state bounded on the west by the Red River, adding as a sop to the hustlers in Saint Paul a bill promising federal grants for railroad construction. Rice's enabling act passed in February 1857. That shifted attention back to the prairie, where fraudulent elections resulted in a polarized constitutional convention of fifty-nine Republicans and fifty-five Democrats. Sibley, the chairman, could not even get them to sit in the same room. So separate caucuses drafted two constitutions, then ironed out the discrepancies. Then they cleverly ensured ratification by tacking the referendum onto their own party ballots so that anyone voting for candidates ipso facto approved the constitution. In the race for governor, Sibley defeated Ramsey by just 240 votes, but the constitution passed sixty to one. The scene shifted back to Capitol Hill, where Minnesota's petition collided with that of Kansas until the compromise offered by William English broke the logjam. In the vain hope that Kansas would become a slave state, southerners acquiesced in Minnesota's admission on May 11, 1858.[76]

The state paid for the circumstances surrounding its birth. The Republican landslides of 1859–1860 ousted Sibley in favor of Ramsey, who exploited his power as governor and senator over fifteen years to wheel and deal in railroads and other schemes. In 1861 the national dispute over slavery dragged Minnesota into civil war.[77] In August 1862 Ramsey's Indian predations came home to roost when Red Cloud led the great Sioux revolt in which nearly 500 whites in the Minnesota Valley were massacred. Whom did Ramsey summon to crush the uprising and "exterminate" the Sioux? Sibley, of course, who defiantly disobeyed. His militia did push Red Cloud into Dakota, but otherwise he spared nonhostile Sioux, pacified the Chippewa, and even defended Indian captives against lynch mobs, giving President Lincoln the chance to commute the sentences of all but thirty-four of the renegades. (Lincoln had taken to heart Father Whipple's judgment that "as sure as there is a God, much of the guilt lies at the nation's door.") Sibley's reward was a reputation as an Injun-lover that ended his political career.[78] Nor did Minnesota's woes end with the coming of peace. Year after year grasshopper swarms devoured the wheat until Governor John S. Pillsbury called for fasting and prayer in 1877. That year farmers harvested a 30-million-bushel bumper crop.[79]

Minnesota grew up in the decades that followed. Bishop Ireland began the "most successful Catholic colonization program" in U.S. history to resettle impoverished Irish people on farms. Swedes, Norwegians, Danes, and Germans reinforced the communities founded before 1860. Hard-driving Minneapolis boomed on the strength of its factories, powered by the Falls of Saint Anthony, and absorption of neighboring towns. So fierce was the Twin Cities' rivalry that each falsified census returns in hopes of besting the other (Minneapolis pulled ahead in the disputed count of 1890 with 182,967 against Saint Paul's 142,581). Northern Minnesota grew even faster on the strength of timber and iron. In 1871 a German immigrant, Frederick Weyerhauser, founded the Mississippi River Logging Company. By 1882 Minnesota's lumber mills were cutting 400 million board feet per year. In 1875 George Stone sold the Pennsylvanian millionaire Charlemagne Tower on the potential of Minnesota ore, and arranged for tax breaks from the state under the guise of "an act to encourage mining." Tower then built the Minnesota Iron Company and Duluth and Iron Range Railroad, which by 1887 was extracting 400,000 tons per year. Even that was chump change after 1890, when Leonidas Merritt discovered that the Mesabi Range was laced with red powder assayed to be 64 percent hematite. Andrew Carnegie, John D. Rockefeller, James H. Hill, Jay Cooke, and other eastern captains of industry poured so much capital into the Mesabi that Minnesota shipped 43 million tons of iron ore in the 1890s and over 200 million the following decade: one-fourth of the world's production. The port of Duluth jumped to 30,000 people. Finally, the University of Minnesota grasped for greatness in 1884, when beloved president Cyrus Northrup arrived from Yale. Over twenty-seven years he founded colleges of law, education, engineering, mines, agriculture, nursing, pharmacology, and medicine, and a graduate school affiliated with the Mayo Clinic.[80]

The Republicans usually had a lock on state politics. But farmers, loggers, and mill hands—many of them socialist-minded immigrants—were bound to challenge absentee owners of extractive industries. It began in 1867, when Oliver Kelley, a farmer at the Elk River, founded Patrons of Husbandry to press for railroad regulation. That gave birth to the Granger movement. In subsequent decades Minnesotans flocked to the Greenback Party, Antimonopolist and farmers' alliances, the Populist Party, Prohibition and Nonpartisan leagues, and finally the Farmer-Labor Party after the Great War. Minnesotans also led movements for labor, female, and civil rights.[81] Its children may not all be above average, as Garrison Keillor quips, but it somehow makes sense that the progressive Gopher State spawned Sinclair Lewis, F. Scott Fitzgerald, Hubert Humphrey, and Robert Zimmerman, better known as Bob Dylan.

Buchanan rode to the White House on a boom the like of which even Americans had never experienced. Since 1846 their economy had received successive injections of adrenaline including Britain's embrace of free trade, the Mexican War, the California gold rush, railroad mania, and voracious European markets for cotton and then for commodities of all sorts during the Crimean War (1854–1856). Wall Street bulls binged on real estate, shipping, and especially railroad stocks until, by June 1857, the situation became scary. "What can be the end of all this but another general collapse like that of 1837, only upon a much grander scale?" asked James Gordon Bennett in the *New York Herald*. "Government spoliations, public defaulters, paper bubbles of all descriptions, a general scramble for western lands . . . hundreds of thousands in the silly rivalries of fashionable *parvenues*, in silks, laces, diamonds, and every variety of costly frippery are only a few among the many crying evils of the day. The worst of all these evils is the moral pestilence of luxurious exemption from honest labor, which is infecting all classes of society." Clever bears such as Winston Churchill's grandfather Leonard Jerome sold short and cashed in when a dip in exports caused by a summer sell-off. Then, on September 12, a hurricane capsized a steamship bound for New York with 426 souls and a half million ounces of California gold. The result was a liquidity crisis that crashed markets and ruined banks. The Panic of 1857 was on.[82]

Jeremiah Calvin Lanphier recommended sackcloth and ashes. He had tried for two decades to get rich in New York before giving up at age forty-eight. He decided that Wall Street must burn as it did in December 1835, but this time by the fire of the Holy Spirit. On a Wednesday in September 1857 he called businessmen to noontime prayer in an upper room of the North Dutch Church at the corner of William and Fulton streets. Just six stragglers peeked in. The following Wednesday twenty desperate traders showed up, and a week after that, forty. During those "melancholy days," as Whitman called them, prayer meetings spread all over New York; and then to Chicago, where Dwight Moody rededicated his life; and to Philadelphia, where the retailer John Wanamaker and the Quaker Hannah Whitall Smith evangelized high society; and then to "every nook and cranny of the great Republic." Southern businessmen prayed, too, but the revival was strongest in the North. It seemed the Lord was calling men in the free states to repent of materialism and stand up against slavery. Bestselling accounts such as Samuel Irenaeus Prime's *The Power of Prayer* and William Conant's *Narratives of Remarkable Conversions* suggested that the triumph of contrition over pretense and pride might be a sign of the coming millennium. By February 1858 the revival was front-page news, something even Finney never achieved. Banner stories tallied attendance at meetings and featured celebrity converts. The clergy and women got into the act, but the leadership of lay businessmen marked the inception of a "muscular Christianity" that defined the rest of the century. The YMCA, founded in England in 1851, took flight in America during the revival of 1858.[83]

If the Holy Spirit blows and lists so mysteriously that even theologians cannot explain spontaneous revivals, what can historians make of them? Certainly none would dare claim the spiritual burst after the Panic of

1857 helped to spark civil war. But it is at least plausible that the upper-class urban revival helped reinforce the North's growing political consciousness urged by the Republicans, the growing revulsion against corruption and vice, and the growing tolerance for moderate antislavery opinion. Among southern opinion leaders the panic itself reinforced contempt for the North. The *New Orleans Picayune* considered the stock market crash a judgment on the North's moral turpitude, while boasting that sturdy southern banks weathered the storm. *DeBow's Review* declared, "The wealth of the South is permanent and real, that of the North fugitive and fictitious." The *Charleston Mercury* asked, "Why does the South allow itself to be tattered and torn by the dissentions and death struggles of New York's money changers?" It seemed to the abolitionist Elihu Barrett that "the North and South were wholly occupied in gloating upon each other's faults and failings."[84] In sum, there is evidence that leaders in both sections were getting fed up with falsehood. It remained to be seen how much truth the Union could bear, and for how long.

Moral rehabilitation found immediate expression in two sensational episodes, one absurd (though bloody) and the other sublime. The absurd one was Buchanan's diversionary crusade against Mormonism. Utah's legislature (elders handpicked by Brigham Young) twice requested an enabling act for a state constitution. Twice Congress refused, because of Utah's polygamy and alleged encroachment on Indian lands. Pierce had tried clumsily to impose federal authority by appointing three territorial judges. Two were Mormon apostates; the third was a philanderer and profiteer ("Money is my God," he bragged to reporters). Utahans accused these judges of trying to steal their land and vowed to recognize no authority other than Young. One judge ended up mysteriously dead. Eight federal surveyors were "killed by Indians." When rumors reached Salt Lake City of federal troops on the march, Young prophesied death to gentile invaders. Death came to 137 innocent pioneers whose California-bound wagon train was assaulted in September 1857. Young claimed that Paiutes were responsible for this "Mountain Meadows massacre." Others accused Mormons of dressing like Indians or else putting the Paiutes up to it. The truth is unknown. But back east moral outrage fanned by the revival and a prurient exposé called *Horrors of Mormonism* gave Buchanan the chance to look tough and proceed against a scapegoat nobody loved. He appointed a governor to oust Young and ordered Colonel Albert Sidney Johnston to march on Salt Lake City. The Mormons' riposte in spring 1858 was to torch their own homes and fields, attack the army's supply trains, and prepare for a siege. Johnston's men were hot to slaughter the "modern sodomites" and might have done so had Thomas L. Kane, a sympathetic Pennsylvanian, not reached Utah first. Kane brokered a modus vivendi in which Young pretended to honor civil authority so long as he retained religious authority and no soldiers came within thirty-six miles of Salt Lake City. Buchanan washed his hands of the "Mormon war" by granting a blanket pardon. All he achieved was to puff up the legend of Brigham Young and make the federal government appear impotent in the face of odious sin. Republicans noticed.[85]

The sublime expression of the new moral climate occurred in the state that had driven the Mormons into the desert: Illinois. Stephen Douglas, having mended his fences after the storm over the Kansas-Nebraska Act, hoped that reelection to the Senate in 1858 would catapult him to the White House. The Dred Scott decision was a body blow to that hope because it seemed to strike down his precious principle of popular sovereignty. But Douglas parried with a new doctrine even less respectful of the rule of law. The Supreme Court might insist on a citizen's right to enjoy his property (i.e., slaves) in the territories, but that right would be a "dead letter" if territorial legislatures refused to enforce it through slave codes and police regulations. In short, the way to save popular sovereignty was nullification. That plus the Little Giant's defiance of Buchanan over Lecompton restored his popularity among Free-Soilers in Illinois. Some Republicans even heeded Greeley's advice that they not contest Douglas in 1858, in hopes that he would gravitate into their camp. At their convention in Springfield in June they fought off that temptation in the belief the Democrats' split and Douglas's tortured logic made him vulnerable. The decisive factor, however, might have been local: office-hungry Republicans could hardly expect to prevail in lesser races if they failed to contest the top of the ticket.

Their nominee's acceptance speech did not disappoint. Abraham Lincoln's words cut to the marrow, beginning with an allusion to Jesus from the gospel of Mark: "A house divided against itself cannot stand. I believe this government cannot endure, permanently half *slave* and half *free*. I do not expect the Union to be *dissolved*—I do not expect the house to *fall*—but I *do* expect it will cease to be divided. It will become *all* one thing, or *all* the other." How had the house been divided during five years when Democrats promised to end the strife over slavery? Timber did not turn into a house on its own, premised Lincoln. The frames were cut and joined by craftsmen named Stephen, Franklin, Roger, and James (Douglas, Pierce, Taney, Buchanan). Their carpentry fitted so well that it was impossible to suppose they worked without a plan. If the conspiracy was not checked, Lincoln warned, "We shall *lie down* pleasantly dreaming that the people of *Missouri* are on the verge of making their State *free*; and we shall *awake* to the *reality*, instead, that the *Supreme* Court has made *Illinois* a *slave* State."[86]

Who was Abraham Lincoln—and why? The ever-expanding library devoted to America's martyred high priest cannot be distilled into thumbnail biography.[87] All one can do is posit the familiar facts of a life that began in a log cabin in Kentucky in 1809. Abraham's unlettered parents, Thomas and Nancy Hanks Lincoln, were content to be dirt farmers fighting daily to stay warm and fed. Their faith as primitive "Hard-Shell" Baptists sustained them. It also made them hate slavery. In 1816 the usual disputes over land titles caused the family to start over in Indiana. When Nancy died from poisoned milk, Thomas married the widow Sarah Bush Johnston, who proved to be Abe's best friend and advocate. By day the boy worked with ax and plow under a father who imagined no other life. By night, and during rare months of schooling, the boy learned the three R's under his stepmother's wing. By 1830, when the family migrated again, to the Sangamon valley in Illinois, Abe itched to invent himself. But poling corn and pork to New Orleans on rafts stiffened his resolve never to work for the profit of another man (including his father, who sequestered Abe's wages), and jobs as a clerk and postmaster soured him on unremunerative tedium. So Lincoln tried politics, choosing as his beau ideal Henry Clay of Kentucky. The choice was automatic. Whigs stood for the education, internal improvements, commerce, and industry that young men like himself needed to rise in the world. Democrats extolled the stagnant, ignorant life of the yeoman that Lincoln was fleeing.

The lanky youth's hopeful message peppered with barnyard humor struck a chord with farmers and townspeople. From 1834 to 1841 Lincoln served in the legislature, mastering the art of logrolling on behalf of banks, steamboats, and railroads. He also kept reading everything within reach, including philosophy, science, mechanics—and law, a profession for which his experience and diligence uniquely suited him. After teaming with William Herndon in Springfield, Lincoln became one of the state's top trial lawyers. He handled some 5,000 cases, represented the biggest client around (the Illinois Central Railroad), and made a comfortable living. Yet Lincoln was somber, chronically so. He took little pleasure in possessions and spurned alcohol and tobacco. Around women he behaved like a rabbit. When Mary Todd at last caught the skittish Lincoln in 1842 ("a policy Match all around," judged his Whig mentor), Abe paid dearly for the social graces and political goad Mary imparted. Herndon likened her to a toothache. Not least, Lincoln seems to have had no spiritual source of comfort. While he was a boy in Indiana he rejected his father's religion. In Illinois he courted political damage by writing a "handbook on infidelity." Even after marriage he attended no church. Instead, Lincoln clung to a Calvinist "doctrine of necessity" whereby irresistible forces of nature predestined men's thoughts and deeds so that free will and moral responsibility could not exist. No wonder acquaintances considered Lincoln a morose, brooding man, and biographers imagine pathologies to explain him. But it only made sense that a lonely bachelor, then an unhappy husband, whose human relationships revolved around litigation and legislation might conclude that no man was free except to act in his short-term self-interest. More interesting is the evidence that despite his stubborn defense of this doctrine, Lincoln acted as if contingency and judgment, not necessity, shaped the course of human events. Nobody worked harder to fashion the outcome of a lawsuit or a political race. Nobody upheld more strenuously the sovereignty of "cold, calculating, unimpassioned reason." Nobody dueled hypochondria more effectively than Lincoln did through incessant, sardonic humor.[88]

Perhaps the riddle is better put by asking, not what the man believed at specific points in his life, but rather what faith he embodied throughout his life. That question at least has an answer. Abraham Lincoln embodied the American civic religion as Federalists, Whigs, Republicans, and most northerners knew it or came to know it through Lincoln himself. His address of 1838 to the Young Men's Lyceum in Springfield, entitled "On the Perpetuation of Our Political Institutions," began by anathematizing the passion, demagogy, and mob rule of the Jacksonian era. The republic could not long survive such decadence, but must succumb to a tyrant trampling liberty in the name of slavery or abolition. There was only one remedy. "Let reverence for the laws, be breathed by every American mother . . . —let it be preached from the pulpit, proclaimed in legislative halls, and enforced in courts of justice. And, in short, let it become the *political religion of the nation*; and the old and the young, the rich and the poor, the grave and the gay, of all sexes and tongues, and colors and conditions, sacrifice unceasingly upon its altars." But that early speech just hinted at the civic faith Lincoln embraced after years spent asking himself what made laws deserving of reverence. He drank in the rhetorical culture—Clay's buoyant nationalism, Webster's Romantic hyperbole, Calhoun's scorching syllogisms, Douglas's paeans to democracy, the Southern Whigs' satire, and the biblical quotes and allusions that struck Lincoln as expressive of truth and made politics sacred to the people. Then Lincoln drafted all those devices into the service of his American dream. It rested on law derived from the Constitution, but law was a mere instrumentality if it did not serve values derived from the Declaration of Independence. His congressional term in 1847–1849 planted more seeds that germinated during the five years his political career lay fallow. By the time the Kansas-Nebraska Act forced Lincoln back into the arena, his voice and faith were mature.[89]

In 1854 Lincoln moved around Illinois like flying artillery to bombard popular sovereignty. If the founders had intended pioneers to decide for themselves about slavery, why had they declared the whole Northwest Territory free soil in 1787? If popular sovereignty was not meant to encourage the spread of slavery, why had Douglas torn down the fence containing it? Even the hogs would know better, Lincoln joked. If popular sovereignty meant men should govern themselves, how could it guarantee some men a "fair chance" to enslave others? How, in short, could moral democracy be twisted into a tool of immoral tyranny? In Peoria Lincoln soared above rhetorical questions to confess a creed. He said, "My ancient faith teaches me that 'all men are created equal'"; he invoked "our ancient faith, that just powers of governments are derived from the consent of the governed"; he called voters back to "the national faith"; he called liberty the "first precept of our ancient faith." He alluded to Moses, the prophets, and Paul the apostle on behalf of natural law. He argued that 433,643 Negroes with a "market value" of $200 million would not be living in freedom "but for SOMETHING which has operated on their white owners, inducing them, at vast pecuniary sacrifices, to liberate them." That something was conscience. Lincoln's climax was millenarian. If Americans reaffirmed the faith of their fathers that all men are endowed with rights to life, liberty, and the pursuit of happiness, then "we shall not only have saved the Union, but we shall have so saved it, as to make, and to keep it, forever *worthy* of saving. We shall have so saved it, that succeeding millions of free happy people, the world over, shall rise up, and call us blessed, to the latest generations."[90]

Whether one calls this biblical republicanism, the American creed, Romantic nationalism, or the democratic faith, Lincoln's civic religion rested on mutually supporting propositions about natural rights, practical politics, and utilitarian interests. He never stopped being a Whig insofar as he believed in vigorous government to promote progress and expand opportunity for all. The sectional divide frustrated progress. Worse still, the move to spread slavery to the territories threatened to choke the opportunity of future generations of farmers and mill hands clambering for the next rung up on the ladder of self-improvement. But republics could not survive on material interests alone; that was the error of Douglas and the slavocracy. Citizens must defend each other's rights and dignity or else risk their own. If Americans did so, then Lincoln believed as firmly as any churchgoer that the author of natural law would bless their republic "to the latest generations."

Did Lincoln believe Negroes were equal in natural endowment? By no means. He shared the assumption of his era that Africans were innately inferior. Did Lincoln mean to assault slavery where it already existed? He denied this loudly and often, wishing only that slavery tread a "course of ultimate extinction." Did Lincoln believe in pure democracy? On the contrary; he respected states' rights to determine their own suffrage. Did Lincoln sanctify individualism? No, again. His obscure but telling "Lecture on Discoveries and Inventions" argued from the Bible and history that no progress was ever achieved except through human cooperation inspired by reason. Nor was he a determinist, for Lincoln wrote that speech to rebut a lecture by George Bancroft, "On the Necessity, the Reality, and the Promise of the Progress of the Human Race." The factors that drove progress in recent times, Lincoln held, were the printing press, the discovery of America, and patent law. None was inevitable; all were cooperative. What lurked at the core of Lincoln's faith was humanity itself. No man had a right to rule other men, own other men, or eat the bread other men earned by their sweat. Either America stood for that or it stood for nothing. Negroes were human beings and everyone knew it. Ergo, Americans North and South must cooperate to adjust to those truths lest they cast away reason, progress, and their future. That made Lincoln more radical than the hottest abolitionist or secessionist. Lincoln, the border state Whig, insisted on liberty and union, now and forever.[91]

* * *

Walt Whitman, an American, one of the roughs, a kosmos, disorderly, fleshy, sensual, eating, drinking, breeding . . . I speak the pass-word primeval, I give the sign of democracy, By God! I will accept nothing which all cannot have their counterpart of on the same terms . . . Have you outstript the rest? Are you the President? It is a trifle—they will more than arrive there every one, and still pass on.

Lincoln's partner "Billy" Herndon was a bibliophile. That is why a copy of the curious *Leaves of Grass* appeared in their law office in 1857. Lincoln picked it up and found it hard to put down. Mary Todd Lincoln thought Whitman disgusting, and her reaction no doubt tightened the poet's grip on her husband. No one knows whether Lincoln also read Whitman's plea of 1857 for a "Redeemer President of These States" to emerge from "the real West, the log hut, the clearing, the woods, the prairie, the hillside." But he must have been struck by Whitman's command, "Consider, you who peruse me, whether I may not in unknown ways be looking at you!" Within months the politician turned lawyer turned politician again.[92]

The race of 1858 in Illinois for the Senate deserves its fame, but not because of the crusty legends of later invention. Lincoln was no David slaying Goliath, but the most feared debater in the state (and almost a foot taller than Douglas). Lincoln did not sacrifice himself in order to boost his party's hopes for 1860. He desperately wanted to win in hopes of boosting his own ambitions. Lincoln did not skewer Douglas by asking in Freeport whether there was any lawful way to exclude slavery in territories against the wishes of a single citizen. The Dred Scott decision had already posed that question and Douglas had already answered, if feebly. Even the seven face-to-face encounters would be more accurately called the Douglas-Lincoln debates because Douglas, the Democrat, was more prominent, spoke first in four of the seven debates, and won. What really distinguished the campaign was its candor. Lincoln made sixty-three speeches and Douglas double that number in a searching dialogue that was not about popular sovereignty so much as pragmatism, principle, and the place of slavery in the nation's future.

Douglas appealed to the "destiny which Providence has marked out for us." He famously boasted that "this is a young and growing nation. It swarms as often as a hive of bees. . . . I tell you, increase, and multiply, and expand, is the law of this nation's existence." Did any white person want to risk that limitless destiny on behalf of inferior persons? Evidently, Lincoln did. Douglas insisted the "House Divided" speech threatened "a war of sections" and a national uniformity which could not be achieved in a large republic except by "erecting a despotism." Lincoln appealed to the morals that justified faith in America's destiny. "The real issue in this controversy—the one pressing upon every mind—is the sentiment on the part of one class that looks upon the institution of slavery *as a wrong* and of another class that *does not* look upon it as wrong." All the Republican Party asked was that no provision make the wrong larger. "That is the issue that will continue in this country when these poor tongues of Judge Douglas and myself shall be silent."

Their rebuttals degenerated as the debates wore on. When Douglas branded his opponent a "Black Republican," Lincoln retreated on the issue of Negroes' equality. When Lincoln implied that his opponent was conspiring to nationalize slavery, Douglas baited him by asking, "Do you desire to turn this beautiful State into a free negro colony?" and told the crowds that all "who believe that the nigger is your equal" were free to vote Republican. Neither candidate had a solution to the larger problems of race or precise policies to recommend for the territories or for free blacks. Douglas was no friend of slavery; Lincoln no friend of disunion. They really were not far apart. But neither man allowed nuance, stump tactics, or pretense to obscure the choice offered to northern voters. Douglas called for business as usual lest white America forfeit its destiny in a fight over slavery. Lincoln called for tough love lest white America forfeit its liberty through expansion of slavery.[93]

The debates ended on October 15. Three weeks later a statewide turnout even larger than that of 1856 cast 125,430 votes for Republican candidates, 121,609 for Douglas's Democrats, and about 5,000 for Buchanan's loyalists. But the distribution among districts was such that Democrats retained a majority in the legislature empowered to elect the state's senators.

Lincoln said he was too hurt to laugh and too big to cry. He in fact accomplished three things. First, his competition for votes in a free state forced Douglas to make statements so poisonous in the South that they killed any chance of his heading a unified Democrat ticket in 1860. Second, Lincoln's well-publicized speeches clearly distinguished the Republican stance from that of Free-Soil Democrats, while suggesting that northerners' moral and material interests coincided. Third, the campaign elevated Lincoln himself to the status of sectional leader, a status he reinforced by campaigning for Republicans in other states throughout 1859. He would rather have been in the Senate, but his ambitious engine had a full head of steam.

The Thirty-Third State: Oregon, 1859

Residents of the Beaver State tickle their children with the tale of two signs where the California Trail veered off from the Oregon Trail in what is now southern Idaho. One arrow was marked with a lump of gold quartz; the other with the words "To Oregon." Literate emigrants followed the latter. There was something to it. Single males from all over the world made Californian society turbulent and cosmopolitan. Young Midwestern families made Oregon orderly and provincial—but only by comparison. Oregon got its share of hustlers, ruffians, and frontier violence, and even a little gold rush. Oregon's precocious self-government under the Champoeg Compact was more pretense than fact. Its very name was an error most likely derived from Wisconsin. French explorers in the Great Lakes region reported a river flowing west, which they rendered as Ouisconsink. In 1715 a map engraver mistakenly rendered this as Ouaricon—sint. In 1760, the famous ranger Robert Rogers was told by Indians that the "western river" ran all the way to the ocean. He begged the crown's leave to search for the northwest passage that he styled Ourigan. At last, in 1778, Rogers's assistant Jonathan Carver published a (mostly imaginary) log of their exploits that fixed the name Oregon on the unknown province where a great river flows into the Pacific.[94]

The Oregon Territory shrank by half when Britain and the United States partitioned the Pacific Northwest in 1846, then by half again when Congress spun off the Washington Territory in 1853. The 97,703 square miles that remained still encompassed such startling diversity, beauty, and richness that pioneers referred to the end of the Oregon Trail as Eden. The rugged Pacific strip from Astoria to Coos Bay was virtual rain forest. Inland from the coastal range lay the fertile Willamette. East of that valley loomed the volcanic Cascades anchored by 11,235-foot Mount Hood in the north and Crater Lake in the south. East of the mountains lay forbidding "moon country" and the Harney High Desert, a lava plateau pocked with craters and cinder cones. The central plateau just south of the Columbia River, however, proved to be good wheatland, and the Blue Mountains of the northeast good pastureland. Most striking were Oregon's 30 million acres of ponderosa pine, western juniper, and Douglas fir, where deer and antelope played amid bears, foxes, beavers, ducks, and wildflowers. The human ecology that this terrain would shape was obvious. Over 60 percent of the people would settle on the Willamette's north-south axis, and most of the rest on the Columbia's west-east axis, leaving the southeastern quadrant of Oregon as barren as northern Nevada.[95]

According to H. H. Bancroft's romantic account, Methodist missionaries and brave pioneers plied the Oregon Trail bent on replicating New England's peaceable kingdoms. When threatened by Indians, the Hudson's Bay Company monopoly, and newly arrived squatters, the pioneers gathered at Champoeg in 1843 in a "solemn conclave of democracy," then petitioned for annexation by the United States. In truth, the motley fifty or so in attendance mostly looked out for themselves. The mountain man Joe Meek hoped to raise enough rifles to defend his sawmill at Oregon City against John McLoughlin's claim for the Hudson's Bay Company. The Methodist Jason Lee was there to procure land grants for missions that did not exist. Others hoped to declare Oregon independent. Thus, the petition for annexation sent by Robert Shortess was meant to preempt the Champoeg initiative. The convention did fashion a government, but in order to win broad allegiance it promised all white males the vote, 640 acres to all settlers, legal and military defense, respect for Indians "except in just and lawful wars," and no taxes at all. That utopia was bound to crumble the moment it was required to do anything. Missionaries achieved even less. Those Indians who did not perish of white men's diseases were pushed out of the Willamette valley and thus segregated from the people they were supposed to imitate. The Methodists blamed their lack of converts on "unprincipled white men and Romanists," but at length decided that the Indians did "not seem to understand" the point of the gospel. Pioneers learned that corn and tobacco grew poorly in Oregon, whereas hardier cereals, vegetables, potatoes, apples, and pears did well. They also learned from the Chinook canoe masters how to harvest the Columbia's bounteous salmon. Otherwise, they did little to prepare for Oregon's growth under the American flag.[96]

That became painfully evident when Cayuse braves, angry over having been pushed beyond the Cascades, got the idea the Methodists meant to kill them and steal all the land. In November 1847 they murdered a dozen whites, including the missionaries Marcus and Narcissa Whitman, and took fifty-three hostages. Protestants accused priests of inciting the Indians and considered banning Catholics from Oregon. But they were reluctant to join a militia for fear that squatters would seize their dubious claims in their absence. A committee of females even threatened to ostracize any young man who refused to defend them from savages. Needless to say, the Champoeg regime could do nothing but borrow and beg for volunteers, prompting Oregonians to demand federal protection.[97] In August 1848 Polk finally dispatched officials and soldiers to establish territorial government. But Congress fouled the pitch by stating, "All laws heretofore passed in said territory making grants of land [are] declared null and void." Chaos resulted when the 11,500 "old Oregonians" were swamped by 30,000 new arrivals between 1849 and 1855. Feuds over land became violent. Lawsuits lasting decades began. The most poignant case was that of Nimrod O'Kelly, who trekked the Oregon Trail at age sixty-five hoping to make a new start. He killed a trespasser in self-defense, spent the rest of his life dodging the gallows, and never did get a deed to his farm.[98]

Another myth about Oregon is that its original settlers were stolid scions of the Yankee exodus. In fact, about 2,200 Missourians and 1,800 from other slave states were among the 9,682 white adults in the territory by 1850. That, plus the breakup of the party system back east, made for interesting politics. Joseph Lane, the state's unlikely founding father, was Carolina-born and Kentucky-bred. In 1846 he left his wife, ten children, and career as a Jacksonian to fight in the Mexican War. Polk rewarded him with an appointment as governor of Oregon. Less than a year later his Whig replacement arrived, but Joe Lane's cocky charm and blunt removal of Indians won him election as Oregon's delegate to Congress from 1851 until statehood. In Salem (the capital since 1853), Lane crafted a patronage machine called the "Clique." On Capitol Hill he won federal funds by courting the southern congressional bloc. But the dispute between Douglas and Buchanan splintered Lane's Clique, and Know-Nothings, Republicans, and a prohibitionist Maine Law Party added to the confusion. The settlers' abiding hatred of taxes also moved them to reject statehood twice. In 1856 everything changed. The events in Kansas suggested that slavery might be foisted on Oregon against its will if it did not achieve statehood soon. The territorial justice George H. Williams declared, "The true policy of Oregon is to keep as clear as possible of negroes, and all the exciting questions of negro servilure." When a constitutional convention met in August 1857, Democrats briefly rejoined on behalf of their Jacksonian principles. The constitution opposed provision for a public university, restricted corporations and banks, and mandated voice voting instead of a secret ballot. Racial issues were left to the voters. Lane's party fought for slavery, but blundered by threatening not to respect a contrary outcome. Oregonians ratified the constitution by two to one, rejected slavery by three to one, and opposed admission of free blacks by seven to one.

Lane's disappointment turned to frustration when he found Congress too distracted to pay attention to Oregon. When the statehood bill finally reached the House floor in February 1858, crazy crosscurrents tossed it like a boat daring the Columbia River's perilous bar. Southerners opposed a new free state unless Kansas came in as a slave state. Some northerners opposed Oregon's ban on free blacks. Republicans shied from creating a state with a Democratic majority. Democrats wondered if Oregon would elect Buchanan's men or Douglas's men. All opposed federal assumption of the $4.5 million in debt Oregon amassed by waging Indian wars without taxes. Thanks to some Republican defectors Lane at last piloted the statehood bill into port by the slim margin of 114 to 103. Buchanan signed it on Valentine's Day. On Saint Patrick's Day the news reached Salem, Oregon City, and Corvallis, where brass bands played "Hail Columbia" and celebrators consumed stocks of whiskey and gunpowder.[99]

Oregon's future hinged entirely on transportation. The falls of the Willamette blocked navigation above Oregon City. The Dalles, where the Columbia River pours violently through a gorge, forced emigrants to ply the treacherous Barlow Road over the Cascades and shippers to unload and reload their cargoes. Oceangoing vessels had to brave the Columbia bar. Indian wars sputtered until 1883, and as late as 1896 the state staged a "Good Roads Convention" because none existed as yet.[100] But the vortex of a great city and transport network came into existence in 1845 at an Indian

site called simply "the Clearing." It overlooked the confluence of the Willamette and Columbia rivers 100 miles from the sea. The promoter Asa Lovejoy wanted to name it after his native Boston; his partner Francis Pettygrove
after his hometown, Portland, Maine. They flipped a coin and Pettygrove
won. The town's first institutions—a Catholic church and a Masonic
lodge—fought losing battles against each other and human nature. By the
1870s Portland was notorious for vice; vigilantism; labor unrest; outrageous
port fees; and mobbing of Chinese, Negroes, and Indians. It was also the
riskiest Pacific port, given the twenty-foot draft at the Columbia's sandy
mouth and the ten-foot draft at Portland. The city fathers, having mastered
creative corruption, turned everything around in a decade. State-financed,
steam-powered dredging of the Willamette began in 1881; federally funded
dredging of the Columbia began in 1884. The Northern Pacific transcontinental railway reached Portland in 1883 and a railroad from San Francisco
in 1887. After Portland absorbed two neighboring towns in 1891 it housed
70,000 people working at bustling wharves, warehouses, sawmills, flour
mills, factories, and banks. Oregon's metropolis was always about business: not until 1911 did Reed College open its doors.[101] In the Willamette
valley, by contrast, the state founded schools at Corvallis (1868) and Eugene (1876) that were destined to become great universities. Rivers and
rails ensured the state's steady if unspectacular growth. Agriculture and
orchards, dairy farming and livestock, logging and mining, and shipping the
produce of all the above boosted Oregon's population to 413,000 by 1900
and 953,000 by 1930. The state remained mostly Protestant, but Catholics
and Mormons became strong minorities. Likewise, the state emerged from
the Civil War a Republican bastion, but its politics remained as idiosyncratic
as during territorial days. California likes to take credit for west coast populism, but tax and rate revolts, the Grange, the American Federation of Labor, voter initiatives, and recalls were pioneered on the Pacific slope by
Oregon. Ever since the days when their forebears spurned that gold quartz
on the trail, Oregonians have cherished the right to say no.

"From all I'd heard in the past three days," the fictional Harry Flashman recalled, "I'd formed a picture of John Brown as a towering figure
with flowing white locks, glaring like a fakir and brandishing an Excelsior
banner in one fist and a smoking Colt in t'other; what I saw was an elderly
man, spare and bony in an old black suit, like a rather seedy farmer come

to town for market . . . nothing out of the ordinary—until you met the gaze of those eyes, clear bright grey and steady as a rock"—the eyes of a man who couldn't be talked out of anything. Our own image is just as mistaken, thanks to John Steuart Curry, whose famous mural of Bleeding Kansas depicts "a gigantic figure of cyclonic rage" summoning Americans to a bloodbath. John Brown was many weird things, but insane he was not, however convenient his insanity would have been to people both North and South. What makes his fantastic exit from life so ironic is that none of those involved, except Brown himself, got from it what they wanted.[102]

Lincoln described the debate over slavery as part of the eternal struggle between right and wrong. If that were so, wondered John Brown, why did northerners betray their God and conscience by refusing to take part in it? We have already encountered Brown as a serial bankrupt, an ascetic moralist who whipped his sons and had them whip him, a Calvinist saint with a black man's heart at Gerrit Smith's utopia, and a grim reaper dismembering ruffians in Kansas. He was also a hustler for whom ends always justified means. Embezzling and swindling were second nature to Brown; he dodged twenty-one lawsuits over the decades. Yet softhearted (or softheaded) Transcendentalists refilled his pockets during fund-raising junkets in 1857 and 1858. The most generous came to be known as the "secret six." Later they claimed ignorance of Brown's intentions. That is hardly credible, given Brown's revelation to Frederick Douglass of a plan to incite slave revolts, his draft of a constitution for the Negro republic he meant to set up in Virginia, his recruitment of filibusterers, and his purchase of guns. Indeed, Brown would have launched his raid in 1858 had his accomplice Hugh Forbes not spilled the beans. Forbes had expected Brown's rich Yankee friends to drown him in dollars. When he judged his pay inadequate to the risk, he revealed the plot to senators Seward and Henry Wilson. The "secret six" flinched; Brown fled west. He did, however, work in a trial run that December by killing a slave master in Missouri, rustling his livestock, and spiriting eleven Negroes to freedom.

If his abolitionist supporters thought Brown would give up his quest, then they were the loony ones. If his supporters thought the fanatic was still bent on inciting revolt, then they were cowards of the sort who pay hoodlums to throw rocks through an enemy's window. Brown did not care what they thought: he pointedly strewed their damning correspondence all over the farmhouse he rented in Maryland in 1859. Five miles away lay the federal armory at Harpers Ferry, a spooky town on a bluff overlooking the Shenandoah and Potomac rivers. Late on October 16, Brown's twenty-two raiders (including five Negroes) crept to the bluff, severed the telegraph wires, and crossed the Baltimore and Ohio railroad bridge with a wagon loaded with pikestaffs (presumably to arm freed slaves). The armory fell into their hands without bloodshed. Then Brown did nothing except to invite swift attack by waylaying a train, only to let it chug on and spread the alarm. By the next afternoon local militiamen laid siege to the raiders. By evening Lieutenant Colonel Robert E. Lee arrived to command federal troops. On the morning of October 18 Lieutenant J. E. B. Stuart led marines in the final attack. All told, ten of the raiders were killed, including two of Brown's sons. Brown himself suffered painful sword wounds. The raiders killed four people, including a black baggage handler on the night train. The affair was so inexplicably botched that it seemed explicable only by madness. ("How do you know I'm mad?" asked Alice in Wonderland. "You must be," said the Cheshire Cat, "or you wouldn't have come here.") But why then did Lincoln regard the raid as merely "peculiar"? Why did Governor Wise of Virginia conclude from his interviews that Brown was "a bundle of the best nerves I ever saw" and "a man of clear head"? Why did the *Boston Post* write that if Brown were a lunatic, "then one-fourth of the people of Massachusetts are madmen"?[103]

The notion of twenty-two men overthrowing white rule in Virginia was not mad in the minds of northerners conditioned to hope (or fear) that slaves were sticks of gunpowder in need of a match. Nor was it mad in the minds of southerners who had accused abolitionists for decades of wanting to get up a war of the races. But if that was Brown's purpose—as his weapons, words, and target proved—why didn't he do anything after seizing the armory? Brown testified that his only object was to free slaves, not commit murder or treason or incite insurrection. Perhaps he was toying with language before a hostile court, because in his eyes slavery amounted to war against humanity; hence anything done in defense of the enslaved was a just act of war. But regardless of what Brown said before or after Harpers Ferry, his inaction suggests that he either expected the heavenly host to descend or else set himself up to be martyred. Either outcome would serve his objective: to free slaves. Suppose that Brown had managed to hole up in the hills with an army of fugitives and terrorized Virginia as he had Kansas. Fear, bigotry, and devotion to law and order would probably have caused most northern whites to rally behind threatened southern whites. The drift of the North toward moderate antislavery opinion and the Republican Party would have gone into reverse. The abolitionist cause would have been set back a generation.[104]

Instead, the wounded Brown promised his wife he would "make the utmost possible out of a defeat," and proceeded to do so with eloquence. "I see a book kissed," he announced at his trial, "which I suppose to be the Bible, or at least the New Testament, which teaches me that all things whatsoever I would that men do to me, I should do even so to them. . . . Now, if it is deemed necessary that I should forfeit my life for the furtherance of the ends of justice, and mingle my blood further with the blood of my children and with the blood of millions in this slave country whose rights are disregarded by wicked, cruel, and unjust enactments, I say, let it be done." When Brown was hanged on December 2, church bells from Boston to Chicago tolled dirges. Preachers and armchair abolitionists such as Thoreau went into mourning. Few besides the alleged pacifist Garrison blessed Brown's violent methods, but his pitiful failure and noble death unleashed torrents of sympathy more disturbing to southerners than the raid itself. Secessionist "Fire-Eaters" moved into the center. Now, when people like Robert Toombs cried, "Never permit this Federal government to pass into the traitorous hands of the black Republican party," people like the Reverend C. C. Jones listened. "The whole abolition crusade which has been preached for thirty years *ends in the sword*," wrote Jones. "There is no place left for forbearance—no ground for compromises."[105]

The congressional session that began three days after Brown's death exposed the wreckage. In a long battle for the House speakership southern Democrats decided that they hated their own pro-Douglas wing even more than they hated the Republicans. Civility vanished. Members went armed. Government ground to a halt. But amid the wreckage gleamed a nugget of hope. John J. Crittenden of Kentucky, the man who occupied Clay's old seat in the Senate, raised the border states' banner of Union and cried to all who had ears, "To me!" Fifty members of Congress seconded his motion denouncing both major parties. "Cotton Whigs" in New England rallied at Boston's Faneuil Hall in support of a Constitutional Union Party. In May 1860 delegates from twenty-three states convened to nominate the only candidates they believed could save the United States. But their most charismatic figures—Crittenden, Winfield Scott, and Sam Houston—were too aged. They settled on bland John Bell of Tennessee in hopes that he could appeal to southern moderates while his running mate Edward Everett stumped the North.[106]

Democrats were already busy destroying the party created by Andrew Jackson. They met in the "Fire-Eaters'" capital, Charleston (a stupid choice), and split down the middle when northern delegates rejected a platform endorsing slavery in the territories. In fact, matters had gone far beyond the dispute over popular sovereignty. During the winter, southern caucuses decided the Union would never be safe for them until the federal government guaranteed new slave states, banned abolitionist societies, enforced a strict fugitive code in the North, and resumed the African traffic in slaves. For a northern Democrat to acquiesce in any of that was political suicide. When the same deadlock paralyzed the party's conclave in Baltimore in June, the Free-Soil Democrats ran Douglas for president while the slave-state Democrats nominated John C. Breckinridge of Kentucky (Buchanan's vice president). What folly! Didn't the schism invite their nightmare, a "Black Republican" victory? Douglas suspected that some Fire-Eaters wanted exactly this. His first words as candidate were a warning: "Secession from the Democratic party means secession from the federal Union." But to the extent that southern managers were not just trapped by emotion and pride, they either expected Douglas to come begging for a last-minute fusion, or else hoped to throw the election into the House where the solid South might prevail. To risk the Union in such fashion might seem wildly irresponsible, but dueling and gambling came naturally to Cavaliers. Besides, they considered themselves the injured parties.[107]

Republicans powwowed in the Wigwam, a capacious hall built to host the first of Chicago's many conventions. It proved wholly inadequate for the 20,000 loyalists who camped outside to chant, cheer, and drink in the belief that the Democrats' confusion spelled victory for them.[108] All the party needed to do was to offer northern voters a positive platform and an unthreatening candidate. Greeley advised: "An Anti-Slavery man *per se* cannot be elected; but a Tariff, River-and-Harbor, Pacific Railroad, Free Homestead man *may* succeed *although* he is Anti-Slavery." Party managers agreed. They defied the abolitionists by drafting an even more moderate platform than that of 1856. The plank opposing slavery in the territories no longer contained the phrases "relic of barbarism" and "all men are created equal." It even upheld states' rights and damned John Brown for the "gravest of crimes." The rest of the platform consisted of progressive remedies for the recession begun in 1857, including protective tariffs, frontier homesteads, and a transcontinental railroad. Finally, to court Germans, the Republicans opposed laws impairing the citizenship rights of immigrants.

The convention mulled over candidates with the same objects in mind; as a result, the standard account goes like this. Seward, the front-runner, was too hot a potato. His impolitic speeches about the "higher law" and "irrepressible conflict" caused doubts about his ability to carry such battleground states as New Jersey, Pennsylvania, and Indiana. Chase was even more tainted with abolition. This left the door open for Lincoln, the host delegates' favorite son, assuming he could persuade eastern delegates of his stature and his manager David Davis could match the "oceans of money" dispensed by Thurlow Weed on Seward's behalf. Lincoln accomplished the first chore with his address at Cooper Institute in New York, where 1,500 Republicans were disabused of their notion that Lincoln was a hayseed or simian. Davis accomplished the second chore by ignoring his boss's order to make no backroom deals. "Lincoln ain't here," he reportedly barked, "and don't know what we have to meet." When Seward fell short on the first ballot with 173½ votes to Lincoln's 100, then gained only eleven on the second ballot compared with seventy-nine for Lincoln, the momentum was obvious. Chase dropped out and Ohio put Lincoln over the top. Weed was incredulous; Seward distraught. The convention then appeased New Englanders and ex-Democrats by nominating Hannibal Hamlin of Maine for vice president.

That story, though accurate, is incomplete. Recall the shame northerners felt when apologists for the South spat on their graft, plunder, and rigged elections. Recall the prayer meetings of northern businessmen in 1858. The great unspoken concern among Republicans was corruption, and no one was more vulnerable in that regard than Seward, veteran of the pork barrel; and Weed, "king of the lobby." Weed made things worse when he could not resist pushing one last deal through the New York state legislature, which was 80 percent "bought and paid for." At issue was a charter for a network of trolley car lines in Manhattan. The New York Central Railroad dispensed $500,000 for services rendered. Weed pocketed $60,000 in stock for the planned Third Avenue line. The stench was so bad that even a friendly newspaper asked, "Is it wise, is it safe, to give this Albany Regency an opportunity to transfer their corruption to the wider field of Washington?" It called "Grand Larceny and Plunder," not slavery, the great issue of 1860. To be sure, Seward's extremism on slavery (and his courting of Catholics) hurt him, but so did the issue of his integrity. One delegate recognized the Republicans' debt to Seward but shouted, "*We won't pay it in hard cash to Thurlow Weed.*" A party styling itself new and clean needed a candidate above reproach. That is why Republicans printed 100,000 copies of a Congressional exposé of the *Democrats'* corruption while calling their own man "Honest Abe."[109]

Like Jackson's "hurra boys" waving hickory sticks thirty-two years before, Republican "wide-awakes" brandishing torches organized evening marches down Main Street in towns throughout the North. To judge from their slogans about good government and prosperity the election had little to do with slavery, much less the fate of the Union. James Russell Lowell dismissed secessionist threats as the "old Mumbo Jumbo." Seward asked, "Who's afraid?" Carl Schurz recalled that the South had cried wolf before but had then taken a swig from its jug and crawled back. This time it might take two swigs. Greeley's *Tribune* reckoned that there was no more chance of southerners' bolting the Union than of lunatics' plotting an escape from their asylum. Lincoln thought southerners had "too much of good sense, and good temper, to attempt the ruin of the government." Whether they were naive, or whistling past the graveyard so as not to spook voters, insouciance worked. By October Douglas conceded defeat by suspending his campaign in the North: "Mr. Lincoln is the next President. We must try to save the Union. I will go South." Douglas, only forty-seven years old, but weakened by drink and fatigue, and sure of a hostile reception, now atoned for the prior damage he had done to the Union. He barnstormed from Memphis to Mobile, in effect pleading Lincoln's case before southerners. His health broken, Douglas would die the following year.[110]

The returns in the election of 1860 precisely profiled the American people. Lincoln's vote was 98 percent northern; Breckinridge's 85 percent southern. The biggest majorities were Lincoln's 76 percent in Vermont and Breckinridge's 75 percent in Texas. In most free states Lincoln battled Douglas; in most slave states Breckinridge battled Bell. The outliers were

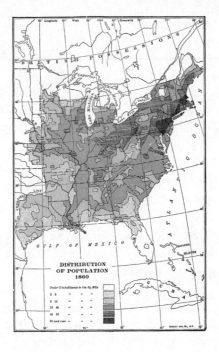

just as revealing. In Pennsylvania Buchanan's Democratic machine turned out 37.5 percent of the vote for the southerner Breckinridge and held the despised Douglas to just 3.5 percent. Breckinridge also finished a strong second in California and Oregon. Most tellingly, perhaps, urban voters in both the North and the South leaned toward the cross-sectional candidates, Douglas and Bell, reflecting their interest in a national market. But America was still mostly rural, so 1860 can be read as the triumph of northern country folk over their less numerous southern counterparts. All told, the Republican won 39.8 percent of the popular vote, the Democrat 29.5, the southerner 18.1, and the Constitutional Unionist 12.6. Might the latter three factions have fused somehow to frustrate the Republicans? The answer is no. A combined opposition on the west coast could have denied Lincoln seven of his 180 electoral votes, but in all the other free states his share of the votes—more than 50 percent—assured him a comfortable victory. Breckinridge's seventy-two electoral votes from the Deep South, Maryland, and Delaware; plus Bell's thirty-nine from Kentucky,

Tennessee, and Virginia; plus Douglas's twelve from Missouri and New Jersey added up to just 123.[III]

The reaction in the South was genuine shock. If the Republicans had dismissed talk of secession as bluff, so had the southern Democrats discounted the chance that the bluff would be called. It was time for everyone to sober up, but since the Republicans were too busy toasting themselves, only some southerners did. Lincoln's old friend Alex Stephens, the latter's Georgian colleague Herschel Johnson, Senator Judah P. Benjamin of Louisiana, the South Carolinian Benjamin F. Perry, and others tried to persuade their neighbors that Lincoln was moderate. In any event his powers would be checked and balanced, since the Republicans did not win a congressional majority and the Supreme Court was in friendly hands. It would be folly for the South to fling away its constitutional armor and march alone into a world hostile to slavery. At the very least the South should not cry until hurt. Even the ideologue James Hammond urged restraint lest precipitate acts divide the South. He likened the Fire-Eaters to "the Japanese who when insulted rip open their own bowels." To judge by the Unionist majorities in New Orleans, Vicksburg, Atlanta, and Richmond, southern businesspeople agreed. Unionists repeatedly begged Lincoln to help by offering words of assurance. He obliged them only once, and then indirectly, by sitting on the podium while Senator Lyman Trumbull of Illinois affirmed states' rights and denied Negroes' equality. The northerners' reaction was angry; the southerners' reaction nil. "This is just as I expected," said Lincoln. He fell mum, content to refer inquirers to his speeches and platform.

So be it. Lincoln's speeches gave southern editors all the ammunition they needed. This was the man who proclaimed that a house divided could not stand, but must become "all one thing." This was the man whose reading of the Declaration of Independence condemned slavery as a sin against the civic religion. This was the man who told the Cooper Institute that nothing would satisfy the South but "this only: cease to call slavery *wrong*, and join them in calling it *right*." This was the man whose abolitionist supporters damned southern patriots and wept for John Brown. Now the entire North had turned insult to injury by electing this man president. In the heat of the campaign of 1860 (and the summer was exceedingly hot that year) folks in the Deep South burned Lincoln in effigy and called Republicans the party of "free love, free lands, and free negroes." They spread rumors of insurrections among slaves who thought that "Black Republican" meant Lincoln himself was a Negro. Mobs, vigilantes, and "minutemen" cowed Unionists and called for the lynching of suspect strangers. They vowed to prevent Lincoln's inauguration or else secede from the Union. They preached resistance as a duty to God, fasted and prayed, and cheered when their clergy cited biblical and civil texts to justify slavery and secession. They likened the sectional strife to a poisoned marriage and Republicans to a castrating wife. In November they discovered their own rhetoric delivered them into the hands of the Fire-Eaters.[112]

The very day Lincoln's victory was certain, the South Carolina legislature scheduled elections for a state convention. "Cooperationist" candidates who hoped to bargain for guarantees from the North were buried. On December 20, 1860, the convention voted 169 to zero to restore South Carolina to its status of 1775 as an independent republic. In order to show proper respect for the opinion of mankind, the delegates issued an elaborate defense of a state's right to secede, a list of grievances against the North, and a specific accusation against Republicans for "having invested a great political error with the sanctions of a more erroneous religious belief" in order to wage war against slavery. Upholding the right to secede was easy. Southern jurists invoked the history of the Articles of Confederation and ratification of the Constitution to interpret the Union as a contract that states might freely cancel, especially if their rights were violated. The Declaration of Independence provided ample justification for dissolving political bonds, and the Tenth Amendment implied that the federal government had no right to coerce. Indicting the North was a harder assignment, since opposition to western slavery and noncompliance with fugitive slave codes were simply assertions of *northern* states' rights. But advocates of the South made the case, in advance of the evidence, that "Black Republicans" meant to throw the whole weight of federal authority on to the abolitionist side of the balance. Explaining what good would come from secession was their hardest challenge. Southern intellectuals had long surmised that soil depletion combined with a growing slave population meant the plantation culture must expand or die out, leaving millions of indigent, dangerous blacks behind. But just how the Cotton Kingdom might annex Cuba or parts of Mexico in defiance of the powerful North (not to say Britain) was skirted. Nor did the alleged fear of too many blacks jibe with southerners' calls for resumption of the African slave trade. Nor would turning the free states into a hostile foreign country do anything but make the problem of fugitives worse.[113]

Such objections could not compete with the propaganda, grassroots organization, panic, intimidation, and fear that combined to give secessionists additional lopsided victories in Mississippi, Florida, and Alabama. South Carolina celebrated Christmas in the certainty that this time it would not stand alone. Why did a large majority of whites in the Deep South rush to secede? There is no way to identify motivation when emotions, interests, pride, bigotry, and conviction could all be made to point in the same direction. But two measurements are suggestive. First, a breakdown of the returns of 1860 by county reveals that poor whites in regions where slaves were sparse (e.g., upstate Alabama) were more likely than whites in plantation belts to vote for Unionist candidates. Yet when it came to voting on secession, poor whites were less likely to vote for cooperationists. In other words, many crackers did not identify with rich planters and were lukewarm on sectionalism. But after Lincoln's victory, crackers felt more threatened than rich whites by the prospect of someday having to live and compete with free blacks. As an Irish editor in Alabama put it, "Submit to be governed by a sectional party whose grand aim is not to raise the negro in the scale of being, but to sink the Southern white men to an equality with a Negro . . . ! Far better ten thousand deaths than submission to Black Republicanism."

The second measurement is a content analysis of secessionist editorials. Complaints about the containment of slavery and the violation of fugitive codes appeared only sporadically. The ubiquitous complaints were legal and moral. For decades the South had watched its influence fade. The Republicans' triumph would hasten that trend until Dixie sank to the status of satrapy in a corrupt, heretical northern-run empire. The time to act was now or never. Likewise, all secessionists fumed over the hateful slanders heaped on them by people they had once thought their countrymen. They were fed up with being called sinful, licentious, barbaric, brutal, and, worst of all, un-Christian. An editor of the *New Orleans Bee* managed to make the point calmly: "Lincoln's triumph is simply the practical manifestation of the popular dogma in the free States that slavery is a crime in the sight of God, to be reprobated by all honest citizens, and to be warred against by the combined moral influence and political power of the Government. The South, in the eyes of the North, is degraded and unworthy, because of the institution of servitude." [114]

The Fire-Eaters called it a revolution akin to 1776, but they really thought of themselves as a righteous remnant clinging to the institutions and civic religion bequeathed by the founding fathers. "We propose to do as the Israelites did of old under Divine direction," vowed the Alabamian William Lowndes Yancey, "—to withdraw our people from under the power that oppresses them and in doing so, like them to take with us the Ark of the Covenant of our liberties." A delegate from Mississippi assured his wife, "Our cause is holy, and we fear not the arbitrament forced upon us of the Sword." The delegate C. C. Jones, Jr., of Georgia was not quite so cocky. "I do not apprehend any very serious disturbance in the event of Lincoln's election and a withdrawal of one or more Southern states," he wrote. "On what ground can the free states found a military crusade upon the South . . . there would be no *casus belli*. . . . But may God avert such a separation, for the consequences may in future be disastrous to both sections. Union if possible—but with it we must have *life, liberty, and equality*."[115]

Chapter 7

HORSEMEN OF THE APOCALYPSE

The Sanguinary Salvation of American Myths, 1861–1865

*I*n a span of just twenty years crises of coalescence gripped the whole northern hemisphere. In Europe they brought forth new nation-states or else forced existing ones to broaden their popular base. In Asia the wrenching impact of western commerce and empire confronted Confucian, Hindu, and Muslim civilizations with three dire options: subjugation, vain resistance, or rejection of ancestral ways if only to master the weapons and tools that made white barbarians so formidable.

The crises began in 1851 when a Chinese youth befuddled by missionary tracts proclaimed a "kingdom of Heavenly Peace" that turned into the T'ai P'ing rebellion, a thirteen-year civil war which killed 30 million Chinese. In 1854 Britain and France launched the Crimean War to defend the Ottoman empire against Russia and persuade the Turks to embrace reforms. The Sepoy Mutiny of 1857 challenged the East India Company's informal raj, prompting the British to impose direct imperial rule on India. In 1859 French and Piedmontese armies defeated Hapsburg Austria, allowing most of Italy's states to unite under King Victor Emmanuel. In 1861 Czar Alexander II reacted to Russia's defeat in the Crimea with progressive reforms. From 1864 to 1871 the Prussian prime minister Otto von Bismarck waged wars against Denmark, Austria, and France to bestow unity and universal male suffrage on the German states. Side effects of those wars included the federalized Austro-Hungarian empire and the democratic Third French Republic. In 1867 Parliament enfranchised the British working class with the Second Reform Bill and founded the Dominion of Canada. In 1868 the daimyo and samurai who overthrew the Japanese shogunate in the name of Emperor Meiji realized Japan needed a strong central government and broad replication of western technology to ward off western imperialists.

Nationalism, liberalism, industrialism, imperialism: those burly isms hammered like Olympian blacksmiths to forge larger territorial units, deeper political bases, broader markets, and complex bureaucracies in a climate of intense competition. How perverse it was, therefore, that in February 1861, just days after the unification of Italy and the abolition of serfdom in Russia, Jefferson Davis declared American unity fractured so that slavery might live. That retrograde motion condemned Americans to a crisis of coalescence bloodier than any save China's. Yet the apocalypse foreseen by Melville redeemed a pretense "four score and seven years" old. The Civil War made the United States a singular, indelible nation.[1] The war extinguished the danger that two or more jealous republics on American soil would expose themselves to European manipulation and cast away the independence, unilateralism, Monroe Doctrine, and Manifest Destiny sacred to the American civic religion. The war purged the sectional blackmail and strife that had paralyzed federal policy. Indeed, the war turned the federal government into a mighty republican instrument serving Yankee agriculture and industry and preparing for a great leap forward toward world power. The war even exposed reservoirs of virtue beneath the surface of a nation devoted to self-absorbed hustling, and of truth ("Glory, glory, hallelujah, His truth is marching on") that trampled pretense like grapes of wrath. At least, that is how northern Americans chose to interpret the ordeal.

But where does the Civil War fit in the context of world history? If abolishing slavery was its deepest meaning, then the United States was just belatedly catching up with the rest of Christendom. If preserving the Union was the war's deepest meaning, then it merely restored the *status quo ante bellum.* Nor was a "new birth of freedom" needed to foster industrialization, as the authoritarian Japanese, German, and Russian empires proved. Indeed, from a global perspective the real rebels in 1861 were the Yankees. After winning their independence, republics throughout the western hemisphere functioned comfortably as commodity producers for a free-trade empire led by Britain. Only the northern United States yearned for high tariff walls behind which to build a free-labor empire challenging British supremacy. Ironically, the Republican Party was at liberty to pursue that ambition only after southern secession.[2]

What made the Civil War special was the character of the nationalism it bequeathed. Lincoln did not love the Union for the sake of its power. He loved it because its liberty under law permitted people to enjoy the fruits of their labor and apply their genius in the service of human progress. Lincoln's constituents, not to say his opponents, rarely or barely grasped that. For decades the South had invoked the Constitution to defend its peculiar institution, while Free-Soilers invoked the Declaration of Independence to condemn expansion of slavery. In 1861 the South invoked the Declaration to justify secession, while the North invoked the Constitution to condemn it. Lincoln took a lonely stand on both documents, in the manner of theologians citing "the whole counsel of God" to refute a heresy based on a snippet of scripture. The Declaration established that all men are created equal as regards their endowment with natural rights. But this apple of gold, as Lincoln styled it, was secured by the silver frame of the Constitution. In the heat of passion Americans pitted their sacred texts against one another, each side accusing the other of claiming to fight for freedom while in fact fighting—in the name of God, no less—to impose slavery or abolition on fellow Americans. Only Lincoln (who Frederick Douglass said "knew the American people better than they knew themselves") refused to divide the word of truth even in the interest of peace. But the total war Lincoln was forced to wage ensured he would win few converts. The resurrected United States purged some old myths only to fuse nationalism even more inextricably with a cult of material progress disguised as a holy calling. That coalescence of Union and creed, power and faith, rendered Americans ever since uniquely immune to cynicism and uniquely prone to sanctimony.[3]

Most telling of all, the Civil War did not answer the question that triggered it: what's to become of the Negro?

Almost no one outside Massachusetts and South Carolina wanted a war. When it came anyway, James Buchanan was the obvious scapegoat. To this day historians rank him as one of the worst presidents: the man who botched Kansas, divided the Democrats, blinked at colossal corruption, then paced the floor sobbing while the Union dissolved. The last charge is a bum rap. Cabinet members testified to Buchanan's determination. He himself insisted, "All our troubles have not cost me an hour's sleep or a meal's victuals, though I trust I have a just sense of my high responsibility." His archives prove he was vigorous to the end. There just wasn't much Buchanan could do as a lame duck with no political base, meager executive powers, a hostile Congress, a mediocre cabinet of suspect loyalties, a tiny army scattered about the frontier, and a navy mostly in dry dock or stationed abroad. Accordingly, the president hoped his credibility as a prosouthern doughface might forestall secession. In his annual message of December 3, 1860, Buchanan attributed the crisis to the "intemperate interference of the Northern people with the question of slavery," while urging forbearance absent an "overt and dangerous act" (i.e., don't cry until you're hurt). He echoed Andrew Jackson by likening secession to revolution, but rattled no sword in fear that this would provoke the secession of the upper South. Instead, Buchanan pleaded for constitutional amendments to reassure slave states and deliver the sacred Union.[4]

For two months into 1861 Senate and House committees labored over bills to guarantee slavery, stiffen the Fugitive Slave Act, restore the Missouri Compromise line, or admit New Mexico as a slave state. A parallel peace conference met in Willard's Hotel, where the former president, Tyler, and delegates from twenty-one states chewed over packages similar to the "Crittenden Compromise" emerging on Capitol Hill. With capital markets tumbling for fear of disunion (and repudiation of southern debts), businessmen in New York, Philadelphia, and Boston lobbied Congress to compromise and forwarded peace petitions signed by 100,000 citizens. That inspired Buchanan to suggest a technique recently used to good effect by Napoléon III and the Italian duchies—a national plebiscite—in the correct belief that a vast majority of voters preferred compromise to disunion. Why could no deal be cut? Why could no majorities be found save for a stillborn amendment to shelter slavery where it already existed?

Fire-Eaters deserve much of the blame. Buchanan's pacific response to South Carolina's secession was to assert federal authority but not to exercise it lest alienation become irreversible. But that left the question of the

U.S. Army installations at Charleston. The rebels meant to oust them, but on the day after Christmas Major Robert Anderson evacuated his little command from Fort Moultrie on the mainland and took refuge inside the squat brick walls of Fort Sumter in the ship channel. Buchanan, bucking protests from southern friends, decided that Anderson must be provisioned and hired a civilian ship, the *Star of the West*, for the task. South Carolina's gunners would have none of it. On January 9 their "ring of fire" forced the *Star of the West* back to sea. The northern press erupted, especially when news arrived the same week that Mississippi, Florida, and Alabama had seceded. By February 1 Georgia, Louisiana, and (to Sam Houston's heartbreak) Texas followed suit. Their militias took possession of federal arsenals, post offices, and customhouses, plus the U.S. mint in New Orleans. Finally, all the seceded states sent delegates to Montgomery, a ragtag city of 9,000 people (half Negro), to found the Confederate States of America. The remarkably moderate assembly displayed a "mania for unanimity" led by Alex Stephens (Lincoln's old friend) and Howell Cobb (lately resigned from Buchanan's cabinet), while marginalizing Fire-Eaters such as Rhett, Toombs, and Yancey. As a result, the convention was able to found a new nation, elect Jefferson Davis and Stephens as provisional president and vice president, draft a constitution, reconstitute itself the first Confederate congress, design the Stars and Bars banner, and authorize an army of 100,000 volunteers—all in less than a month. In some respects the Confederate constitution bettered the one drafted in Philadelphia. It mandated a single six-year term for the president, empowered him with a line-item veto, required a two-thirds majority for spending bills not requested by the executive (to nail the pork barrel shut), and encompassed a bill of rights. But it prohibited internal improvements and (needless to say) sheltered slavery. In an unguarded moment Stephens even told a cheering Savannah crowd that the cornerstone of the Confederacy is "*the great truth that the negro is not equal to the white man*" and that slavery "is his natural and normal condition."[5]

From the northern perspective secession killed any incentive to appease. Indeed, the founding of the Confederacy transformed a debate on the future of slavery—over which northerners divided—to a debate on the future of the Union, over which northerners firmly agreed. That stiffened the spine of the other bloc responsible for thwarting compromise bills: the

Republican Party. Lincoln opposed tinkering with the Constitution, had little enthusiasm for congressional bargains, and in no event would retreat from his plank opposing slavery in the territories. Lincoln refused (rightly) to be made to "apologize" for prevailing in a legal election. He believed (rightly) that the fever must reach a crisis before breaking: "The tug has to come, and better now, than any time hereafter." He feared (rightly) that his young party would shatter if forced to recant its core principles. He sensed (rightly) that his fellow midwesterners would wage war on their own rather than let the Mississippi River fall into "foreign" hands. But above all he expected (wrongly) that once he was safely inaugurated southern Unionists would quash the secessionist coup-makers. To Buchanan's vexation, therefore, most Republicans voted against the Crittenden measures even as southerners understandably pleaded that they could trust no compromise Republicans did *not* bless.[6]

It is often said the archaic provision whereby a president-elect waited four months to take office (and a Congress-elect waited over a year) compounded the agony of "secession winter." In fact, it bought precious time. It is said that the peace talks were a "great nonevent." In fact, they had a critical impact on the eight slave states that had not bolted the Union. Frustrated Fire-Eaters could not understand the hesitancy of the upper South, since these states were most vulnerable to abolitionism, the Underground Railroad, and raids like John Brown's. But border folk knew northerners best and so feared them least. Their slave populations were sparser; their planters were less dominant. Most of all, the border states knew they would suffer the brunt of a war. No wonder delegates from the upper South sponsored the peace talks and voters in the upper South recoiled from secession the moment parleys began.

On January 17, 1861, Kentucky's legislature refused to summon a state convention at all. On February 4, Virginians slogged through winter slush to elect 152 convention delegates of whom a mere thirty-two favored immediate secession. The same month candidates favoring immediate secession were buried by a vote of four to one in Missouri, three to one in Tennessee, and two to one in North Carolina (where the people sang "to live and die for Union" instead of "Dixie"). Arkansas narrowly defeated secession. Governor Thomas H. Hicks of Maryland refused to call the legislature to session. Delaware's assembly condemned secession on principle.[7]

The fever seemed indeed to have broken. Credit Buchanan for that. He knew well that most voters in the border states, though loyal, recognized no federal right to coerce disloyal states. Hence the way to contain secession and perhaps lure the Deep South back was to shun provocations.

Moreover, the prominent wing of the Republican Party agreed with Buchanan. To be sure, a "war party" composed of radical politicians such as Samuel Chase, Thaddeus Stevens, and Charles Sumner; and cheerleaders such as Ralph Waldo Emerson, Horace Mann, Horace Bushnell, and Frederick Law Olmsted damned the president's passivity. They believed the greatest danger to Union lay in mollycoddling traitors. But moderates led by secretary of state–designate William H. Seward believed that the greatest danger lay in hurling threats and insults which could not be backed up. Seward stayed in close touch with leaders of the border states, got back channel reports from Washington via Attorney General Edwin M. Stanton, and dropped soothing hints about Lincoln's intentions. Seward urged the president-elect to adopt a "forbearing and patient" policy and draft a "wise and winning" inaugural speech. Lincoln brooded, then decided Seward was right. He must assert federal authority in "Secessia" but refrain from exercising it, while exuding goodwill. Thus, he would arrange for ships to collect import duties offshore rather than in southern ports. He would impose no postmasters, marshals, or judges. He would make no effort to recapture seized federal property or reinforce Charleston's Fort Sumter and Pensacola's Fort Pickens. He would try to lure southerners into his cabinet and give more reassurances about the survival of slavery where it existed. Only after the war would aggrieved rebels and proud Yankees alike embrace the myth of a Republican "firm policy." The real posture Lincoln assumed at the start was nearly identical to that of his maligned predecessor Buchanan.[8]

On a rainy February 11, 1861, the day before he turned fifty-two, Lincoln bade farewell to neighbors in Springfield whom he would never lay eyes on again. His twelve-day "victory lap" over the tracks of twenty-odd railroads was not ennobling. In town after town Lincoln's offhand remarks peppered with jokes seemed to trivialize the crisis and legitimize caricatures of an ectomorphic buffoon. Perhaps the thaw in "secession winter" did augur a "Unionist spring." Still, John Quincy Adams's grandson blushed to observe the man on whom all depended "perambulating the country, kissing little girls, and growing whiskers."[9]

The Thirty-Fourth State: Kansas, 1861

Legend would make Charles Robinson the ideal founder of Kansas. Born on a farm in Massachusetts, he studied diligently despite fragile health and established a medical practice in the Berkshires. After his wife and two infants died, Robinson joined the forty-niners and led the fight to save squatters' lots in Sacramento. He then went home and worked as a journalist until 1854, when he answered the call to save Kansas for freedom. Indeed, it was he who chose Lawrence as headquarters for the New England Emigrant Aid Society. Robinson's moral suasion failed to restrain violent Jayhawks and Border Ruffians, but he won such respect that voters would elect him the state's first governor. Ever the hustling Yankee, he speculated in real estate while pushing every progressive movement from women's rights and farmers' relief to education for Negroes and Indians.[10]

Only the legend is just that. In 1865 just 3,000 Kansans came from New England, compared with 12,000 from the Middle Atlantic, 35,000 from the Middle West, 43,000 from the Upper South (half from Missouri), and 15,000 from overseas. The imbalance increased after the Civil War when railroads blanketed Ohio, Indiana, Illinois, and Iowa with campaigns to attract settlers to central and western Kansas. By 1914, when the first native-born Kansan was elected, seventeen of the state's nineteen governors had roots west of the Alleghenies.[11] That means the archetypical founder of Kansas was really the "lean and hungry" Hoosier James Henry Lane. He decamped to Kansas in 1855, expecting that his record as a hero of the Mexican War and a Jacksonian congressman would catapult him to leadership of the Kansas Democrats. He sensed at once how the dispute over slavery was wrecking that party. Instead, he helped launch a Free-Soil party and ran men and guns into Kansas along "Lane's trail" through Iowa. His career hiccuped after he killed a rival land speculator and Stephen Douglas disowned him. So Lane played chameleon again. In May 1859 he staged a convention where Horace Greeley and a son of John Brown blessed the birth of the Kansas Republican Party. Its instant candidates won a margin of two to one in a convention to replace the Lecompton constitution. The Republicans banned slavery (but denied the vote to free Negroes), chose Topeka for the capital after a "committee on skulduggery" solicited the usual bribes, and refused to annex the portion of

Nebraska south of the Platte River because its settlers were said to be Democrats. In October Kansans at last conducted a fairly fair poll on this Wyandotte constitution. The *Lawrence Republican* crowed that whereas "office holders, Buchanan Democrats, old pro-slavery leaders" opposed Wyandotte, 65 percent of the voters approved. The following month Lincoln paid a visit to celebrate the triumph while Lane dispensed patronage. He meant to become a U.S. senator.[12]

As it turned out, everyone had to wait. Southern senators still damned a Free-Soil Kansas; northern Democrats held out for boundaries including the Platte country and Colorado's Pike's Peak region, where gold had been struck in 1858. Not until January 29, 1861—after eight southern seats became vacant—did the Senate pass and Buchanan sign a statehood bill. Then everyone waited another year while Kansans quarreled over the terms and their kingpins waged their own civil war. "On all public questions," groused a paper in Leavenworth, "there is a Robinson version and a Lane version, and neither is the truth." Over the months when the Union dissolved, the Missouri border bled anew, and drought struck thousands of homesteaders, Kansans again had no government. A private relief committee helped "hungry children and their grief worn parents," but foundered on charges that Samuel C. Pomeroy was skimming its funds. (Mark Twain would lampoon portly Senator Pomeroy in *The Gilded Age*.) And by the time the state was up and running in January 1862, Senator Lane was busy impeaching Governor Robinson.

The story still isn't done. After Fort Sumter, Lane patronized Lincoln to win appointment as a brigadier general and returned to Kansas a "grim chieftain." His Jayhawk invasions terrorized whites in Missouri while encouraging thousands of blacks to flee to his camp. A chaplain suggested putting the fugitives to work helping pioneers brace for the winter. "I'll do it!" said Lane. By spring 1862 some 6,400 Negroes were free and employed. Lane was also among the first to recruit black regiments, one of which fought with distinction at Chickamauga. White Kansans were not pleased that the blacks in their midst grew to 12,527 by 1865. Robinson even blamed Lane for provoking the worst atrocity of the war: the Confederate guerrilla William Quantrill's massacre at Lawrence in August 1863. But Lane remained defiant until charges of influence peddling broke his spirit. In July 1866 he leaped from a carriage and fired a pistol into his mouth. It took death eleven days to claim his restless soul.[13]

Kansas continued to be restless, its colorless image in *The Wizard of Oz* notwithstanding. The Kansa or Kaw Indians quietly died of smallpox and federal neglect, but the sophisticated Cheyenne and Kiowa horsemen resisted the whites' encroachments, slaughter of buffalo, and broken treaties until at last they retreated to Indian Country in 1878.[14] By then railroad companies, no fewer than fifty-one of which had been chartered by the territorial government, made raucous cattle-drive towns of Ellsworth and Abilene on the Kansas Pacific trunk line, and Dodge City, Newton, and Wichita on Cyrus K. Holliday's Atchison, Topeka, and Sante Fe. For a few years reality matched the Hollywood image of dusty streets lined with clapboard stores, stables, a jail, a bank, and false-fronted saloons for thirsty cowboys. There really was a gritty U.S. marshal who tamed Abilene, and a "saloon war" when Luke Short bought into Dodge City's Long Branch. The gambling, music, and "sporting women" Short sold provoked respectable citizens to crack down. When Short imported such firepower as Wyatt Earp, Bat Masterson, Shotgun Collins, Texas Jack Vermillion, and Doc Holliday, however, the town council prudently decided not to enforce its vice code until Short prudently moved on to New Mexico. Only in 1887, when a terrible winter killed thousands of cattle and Texas railroads ended the cattle drives, were Kansans wholly free to embrace Victorian progress and propriety.[15]

Progress in a 350-by-200-mile rectangle of largely treeless tableland rising to 4,000 feet above sea level and subject to climatic extremes required not only transport, but rugged people and a signature crop.[16] The Preemption and Homestead Acts offering federal land for virtually nothing, plus lusty advertising by railroads and speculators, increased the population to over 1 million by 1885. Wichita, in those years, was the fastest-growing city in the United States. Most of the newcomers were midwestern, German, or Scandinavian, but 50,000 black "Exodusters" found refuge in Kansas from persecution by southern whites after Reconstruction. Pioneers in central and western Kansas had to live in sod huts, pray weekly for rain, and burn buffalo chips to keep warm, but proudly claimed between forty and 360 acres each. The signature crop, according to an old wives' tale, arrived in 1873 in the baggage of Mennonites whose Ukrainian "Turkey Red" seed made Kansas the richest breadbasket on earth. What really happened was a twenty-year process during which Kansans quit competing with Iowa corn, learned to plant winter wheat,

and chose hardy strains over "soft" ones. The credit belongs to Edward
Mason Shelton, who patiently tested fifty Russian strains at Kansas Agricultural College; and Barnard Warkentin, a Mennonite miller whose flour
took European wholesalers by storm in 1888. Within thirty years Kansas
was harvesting one-seventh of the whole world's wheat.[17]

Propriety on the frontier—especially the promotion of schools,
churches, and government services—was women's work. And nowhere
did women demand public power more than in Kansas. In 1861 they won
the right to vote in elections for school boards, the right to independent
property, and equal custody of children. With support from a Republican
governor, John St. John, female crusaders in 1880 made Kansas the first
state to adopt a constitutional prohibition against alcohol. When enforcement later grew lax, the Kansan Carry Nation won national celebrity
swinging her ax in dram shops. Kansas approved women's suffrage in city
elections and bond referenda in 1887, and elected America's first female
mayors. Women were instrumental in the pioneering by Kansas of agencies to foster health, education, and welfare. Not least, feminists such as
the fiery Mary Elizabeth Lease and dogged Annie Diggs goaded farmers to
stand up against railroads and banks. In 1893 their Populist Party literally
brawled with Republicans over control of the legislature.[18]

Booms born of wars and bubbles, and busts born of plagues, droughts,
and depressions are the fate of any region reliant on commodities markets. In good years Jayhawks defined gold as a Kansas wheatfield. In bad
years they mourned, "In God we trusted, in Kansas we busted." In 1896
the editor William Allen White answered his famous query "What's the
Matter with Kansas?" by mocking its people's penchant for blaming out-
siders and passing a law against something. Even less gracious was the
wag for whom a Kansan's piety was "to thank God that you are not as
other men are, beer-drinkers, shiftless, habitual lynchers, or even as these
Missourians." But the historian Carl Becker praised the idealism found at
America's geographical center. "To venture into the wilderness," he wrote
of Kansas in 1910, "one must see it, not as it is, but as it will be. The frontier, being the possession of those only who see its future, is the promised
land which cannot be entered save by those who have faith." [19]

Lincoln's leisurely progress ended in Harrisburg on Washington's
Birthday. Seward's son Frederick arrived with intelligence from General

Winfield Scott and Pinkerton's detective bureau to the effect that an assassination plot was afoot. So Lincoln and a bodyguard slipped through hostile Baltimore and arrived in the capital in the wee hours. Seward virtually kidnapped Lincoln, much to the consternation of other Republican leaders. He installed Lincoln in Willard's Hotel, chaperoned him to briefings by Buchanan and Scott, took him to church, hosted him in his townhouse, toured him around Capitol Hill, vetted his inaugural address, and otherwise monopolized his time for three days. Seward failed to dissuade Lincoln from naming his rivals to the cabinet (if only to keep a close watch on them and repay electoral debts). Chase of Ohio got the Treasury, the party hack Simon Cameron of Pennsylvania got the War Department, Caleb Smith of Indiana got the Interior, and able Gideon Welles of Connecticut got the Navy Department. Border states were represented by Attorney General Edward Bates of Missouri and Postmaster General Montgomery Blair of Maryland. But Seward's own last-minute threat to decline the State Department did persuade Lincoln to make soothing noises in his inaugural address, delivered under the Capitol's scaffolded dome on March 4.

Lincoln, like Buchanan, called secession "the essence of anarchy," but denied federal power to interfere with private property (i.e., slaves). He pledged to hold federal property (i.e., the forts), yet promised not to use force to recover seized assets. He pleaded for patience, since there was "no single good reason for precipitous action." He appealed to Americans' intelligence, patriotism, and Christianity. He insisted the "momentous issue" rested with the dissatisfied parties because "the Government will not assail *you*." Then Lincoln soared: "We must not be enemies. Though passion may have strained it must not break our bonds of affection. The mystic chords of memory, stretching from every battlefield and patriot grave to every living heart and hearthstone all over this broad land, will yet swell the chorus of the Union, when again touched, as surely they will be, by the better angels of our nature." Knowing what was to come, we weep. Not knowing what was to come, some Republicans heard timidity; most northern Democrats heard vacuous rhetoric; and Confederates heard judgment and threat. Only border state Unionists heard mystic chords. That is because Lincoln's eloquence offered little besides Whiggish nostalgia for the status quo of the previous autumn.[20]

Lincoln's peace offensive collapsed within twenty-four hours when

Major Anderson, the Kentucky Unionist (and slaveholder) in command of Fort Sumter, reported that his eighty-five soldiers had to be revictualed in just four to six weeks or else evacuate or surrender. That much everyone knew, although the time frame was briefer than Lincoln expected. More shocking was Anderson's judgment that no resupply ships could get through unless an amphibious assault first knocked out the Confederate shore batteries (the "ring of fire"). That would require a force of 20,000 trained soldiers, twice the size of the whole U.S. Army. General Scott endorsed the gloomy prognosis. Withdrawal seemed inevitable, as Anderson no doubt meant to convey. Three weeks passed during which Lincoln dithered, cabinet members intrigued, and newspapers led by the *New York Times* screamed, "Wanted: A Policy." Charles Francis Adams despaired: "the man is not equal to the hour." So what happened to energize Lincoln on March 28? Did he rise to the challenge or crawl into his cocoon of "necessity" and let events take their course? Whatever the case, a flurry of news forced his hand. First, Scott shocked the cabinet by advising the desertion of Sumter and Pickens. He judged that the forts were of little military value but of great political value if their abandonment forestalled what the old soldier suspected would be a horrible war.[21] Second, just before the Senate's adjournment, Republicans led by Lincoln's colleague from Illinois, Lyman Trumbull, resolved that it was the president's duty to use every means to protect federal property. Third, the former naval captain Gustavus Vasa Fox reported that a stealthy naval operation after dark might easily resupply Anderson. Fourth, the Virginia convention spurned Lincoln's request that it disband in return for a retreat from Fort Sumter. The next day the whipsawed cabinet decided at last to hold Pickens and try to hold Sumter. Lincoln ordered the preparation of a flotilla. Seward, sniffing gunpowder, took up his pen.[22]

Few documents are less understood than Seward's "Thoughts for the President's Consideration," submitted on April Fools' Day 1861. The most notorious passages damned the administration's paralysis, called for declarations of war against all European powers seeking to profit from America's troubles, and invited the president to delegate policy to Seward himself. Historians used to describe the memorandum as reckless, fantastic, and wild. They imagined the fifty-nine-year-old statesman had gone off his rocker, or had reacted in pique when the cabinet foiled his bid to be Lincoln's "prime minister," or supposed a foreign war could magically undo secession. Such historians not only forgot Seward's lifelong devotion to the peaceful gospels of Christ and commerce but ignored the fact that he was secretary of state. During weeks when the crisis at Fort Sumter preoccupied everyone else, Seward was reading dispatches and conversing with foreign ministers. What he learned was appalling. Spanish soldiers had taken a pretext to invade the Caribbean republic of Santo Domingo. A French fleet had sailed, perhaps to assault Haiti or show its flag in New Orleans. The British minister Lord Lyons had warned of an intervention to protect cotton shipments. There was talk of Anglo-French-Russian recognition of the Confederacy. Meanwhile, the U.S. diplomats inherited from Buchanan sat on their hands or hummed the song of the South in foreign capitals. For Lincoln to plunge into civil war while Europe's sharks were circling seemed suicidal. All Seward could do was to ward off hostilities while bluffing the Europeans as brazenly as he could.

He had already done so. In February Seward gave Lyons cause to warn London that the Lincoln administration might "endeavour to divert the public excitement to a foreign quarrel." In March Seward disrupted a dinner at the British embassy with spitting "defiance of Foreign Nations." On April 1 he confronted Lincoln himself, albeit his "Thoughts" were crafted with multiple targets in mind. He hoped tough talk would satisfy impatient Republicans. He hoped the elevation of national interests above party would rally northern Democrats and border states. He hoped defense of the Monroe Doctrine would queer the Europeans' dalliance with the Confederacy. Seward's "desperate diplomacy" would largely succeed abroad. But he never reached his domestic targets, because Lincoln buried the memo, told the secretary who was in charge, and turned his thoughts back to Fort Sumter.[23]

If Lincoln expected war, why did he let Congress recess? Why did he assemble an even weaker flotilla than the one Buchanan had mustered, and then permit Seward to reassign its flagship, the steamer *Powhatan*, to Fort Pickens? Why did he assure South Carolina's governor on April 6 that no effort would be made to ship arms and men to Fort Sumter? Why did he appeal yet again to the Virginia convention with talk of trading "a state for a fort"? For that matter, why did he turn the War Department over to the venal, incompetent Cameron? If, on the other hand, Lincoln was bent on peace, why did he consult four northern governors, urging

Pennsylvania especially to rally militias? The best way of reading the riddle is that Lincoln reasoned himself at last into the policy most likely to see Republicans and the Union through their windows of vulnerability. Either the brave Major Anderson would be saved, thus cheering northern opinion, sinking Confederate prestige, and giving Europeans pause; or else the resupply effort would founder amid the rebels' shells, thus inflaming northern opinion, soiling the Confederate cause, and scaring Europeans into neutrality. Perhaps Lincoln absorbed Scott's and Seward's advice after all. The military value of saving Fort Sumter was trivial compared with the political value of losing it.[24]

Thus did Lincoln deliver the fate of the American people to Jefferson Davis, who did not want war either. Davis's own inaugural address described peace and free trade as the "true policy" of an agricultural people. In public he insisted the Confederacy just wanted to be left alone. In private he fretted a war might cost "thousands of lives and millions of treasure." As late as April 6 he insisted that the Confederacy and the United States had "many reasons to feel towards each other more than the friendship common among other nations." But he could not deny that Union garrisons mocked Confederate sovereignty. He could not help asking what earthly use Sumter and Pickens were to Lincoln, unless Lincoln intended aggression. He could not play for time because, as an editor in Alabama warned him, secessionist zeal was "oozing out under this do-nothing policy."[25] Finally, Davis was as persuaded as Seward that the Confederacy could not survive without reinforcement from some states in the upper South, and war was the way to make them choose sides. Indeed, to quiver while federal warships sailed into the Confederate ports would amount to autocastration. So Davis instructed General Pierre G. T. Beauregard to demand Fort Sumter's surrender. Anderson politely prevaricated. On April 12, before dawn, Confederate batteries erupted. Federal gunners replied until their shells were gone and their powder was aflame. On the afternoon of the following day Anderson struck the Stars and Stripes.[26]

> But God his former mind retains, / Confirms his old decree;
> The generations are inured to pains, / And strong Necessity
> Surges, and heaps Time's strand with wrecks.
> The People spread like a weedy grass, / The thing they will to pass,
> And prosper to the apoplex. —Hermann Melville, 1861[27]

Lincoln believed in necessity. He had let events take their course. Yet the news from Fort Sumter so angered him, not to say opinion throughout the North, that on April 15 he cited a law dating back to the Whiskey Rebellion of the 1790s to call on the states for 75,000 three-month volunteers. In retrospect that was the first flex in a long series of exercises that bulked up the constitutional powers of the commander in chief. At the moment it was an impatient blunder. Had the president "held his fire" for another twelve weeks, a planned convention of the border states might have hit on a formula to cement their loyalty and isolate the Deep South. If instead war proved inevitable, then the emergency session of Congress convened on the Fourth of July could have launched the full-scale mobilization needed to do the thing right. Instead, haste and cheapness prevailed. It might be said that Lincoln was desperate to get boots on the ground to defend the federal city itself. But Washington, D.C., was under no threat until Lincoln's pledge to repress rebellion by force plunged his Virginian neighbors into the apoplexy that Melville described. For months, secessionists in Virginia had mobbed and railed to no avail. But twenty-four hours after Lincoln's call for volunteers the women in the galleries of the state convention shamed their men into action. "Mothers, wives and daughters," one woman wrote to the *Richmond Daily Courier,* "buckle on the armor for the loved ones; bid them, with Roman firmness advance, and never return until victory perches on their banners." Students at Henry College, the University of Virginia, and the College of William and Mary also howled for war (thus signing their own death warrants). A diarist in Norfolk recorded, "The effect was magical!" and "Immediate Secession!, is the universal cry!" When Virginia's delegates voted, eighty-eight to fifty-five, to secede on April 17, the Confederacy was endowed at a stroke with Richmond's Tredegar ironworks, Harpers Ferry's machinery, Norfolk's navy yard, a boundary on the Potomac—and Colonel Robert E. Lee, the man Winfield Scott had hoped would command the Union forces. Moreover, the Confederate Congress reinforced Virginia's conversion— and defied Lincoln—by moving the capital to Richmond, just 100 miles south of Washington. In Nashville, Senator A. O. P. Nicholson wrote of a similar "revolution in public sentiment" in his state. Folks called Lincoln a "tyrant" and voiced the Scots-Irish cry, "Liberty or death." Tennessee seceded on May 7; Arkansas the next day; and North Carolina, surrounded by Confederate territory, on May 20.[28]

Had the northernmost tier of slave states joined this second secession-ist wave, the Union cause would have been fatally handicapped. But a delicate mix of coercion, forbearance, luck, and local initiatives spared Lincoln the worst. In Baltimore mobs stoned the "nigger thieves" of the Sixth Massachusetts regiment, killing four soldiers while they tried to change trains on April 19, then destroyed railroad bridges and telegraph lines. A "plug-ugly" Irish regiment from New York City heard of the vio-lence and boasted, "We can fix that Baltimore crowd. . . . We boys is so-ciable with pavin' stones!" But General Benjamin Butler rendered that unnecessary by ingeniously shipping troops via Chesapeake Bay. Still, an anxious week passed during which Lincoln gave daily thanks to the navy yard's commander, John A. Dahlgren, without whom the capital would have been utterly naked. Meanwhile, the president ordered General Scott to scour Maryland and arrest anyone suspected of treasonous motives. That disturbed Attorney General Bates, not to mention Chief Justice Taney, who ruled that the executive branch had no right to suspend habeas corpus. Lincoln ignored Taney, inasmuch as "the very existence of the Na-tion is assailed" and it would be absurd to jettison all the laws of the land out of respect for just one. Still, martial law could not have saved Mary-land for the Union if the state's loyal governor had not sequestered the state assembly until elections returned a Unionist majority.[29]

Kentuckians also went Unionist in their elections of spring 1861, but instead of supporting the war effort declared the state neutral. "To lose Kentucky," remarked Lincoln of his native state, "is nearly the same as to lose the whole game." He funneled secret shipments of arms to Unionists, but otherwise respected the state's sovereignty and seconded a resolution by Crittenden denying the war was waged "for any spirit of conquest or subjugation, nor purpose of overthrowing or interfering with the rights of established institutions of these States" (i.e., slavery). Fortunately for Lin-coln, the Confederate general Leonidas Polk violated Kentucky's "neutral-ity" in September 1861 in the belief that he was preempting a Union seizure of Columbus, Kentucky (which the federal army hardly needed, given its secure base at Cairo, Illinois). Polk's rashness invited federal forces to en-ter the state in the name of defending it, and to round up suspected seces-sionists.[30]

Missourians eschewed neutrality by going to war with each other. Governor Claiborne F. Jackson encouraged secessionist militias to seize the federal arsenal at Saint Louis, whereupon a rigged convention might take Missouri out of the Union. But Congressman Francis P. Blair, Jr., and Captain Nathaniel Lyon of the U.S. Army mustered a Unionist force consisting mostly of Germans, smuggled 20,000 muskets to Illinois, and took the rebel militia camp by surprise. When Saint Louis erupted in riots, Lyon's troops killed twenty civilians. The irate governor then prevailed on the legislature to appropriate funds for a Missourian army. Blair in turn persuaded Lincoln to make Lyon a brigadier general with authority to oust the government in Jefferson City. The federal and state armies skirmished for months, during which Lyon was killed and the impetuous John C. Frémont almost threw the game away. He placed the state under martial law, condemned the property of traitors, and freed their slaves. Lincoln was aghast. He insisted Frémont rescind the decrees and then removed him when the general refused. The congress at Richmond likewise tried to leverage public opinion by "admitting" Missouri to the Confederacy even though it never managed to leave the Union. In March 1862 Brigadier General Samuel R. Curtis at last ran the state army to ground at Pea Ridge in Arkansas, whereupon Missourians simply reverted to the border ruffian tactics they had learned in "bleeding Kansas."[31]

Union control of Maryland greatly reduced the vulnerability of Washington, D.C. Union control of Kentucky and Missouri, hence the Ohio-Mississippi river system, would greatly increase the vulnerability of the western Confederacy. That is why General Scott's so-called "anaconda" strategy for gradual asphyxiation of the rebellious states made perfect sense. His Virginian heritage taught him the Yankees were foolish to think southerners would not fight for the slave-owning oligarchy. His campaign in the Mexican War taught him that amateur soldiers like those flocking to Washington in parrot-like plumage would be useless without professional training. He also noted how many regular Army officers, including some of the best West Point graduates, were defecting to their seceded states (313 out of 1,105 resigned their commissions). Scott's reading of military history and theory also caused him to buck the conventional wisdom that this war would be brief. It was true that every international war since 1815 had ended in two years or less. But this would be a civil war waged in a theater as large as Europe. Finally, Scott assumed the Confederates would adopt George Washington's strategy, which was to fight on the strategic defensive, conserve resources, troll for foreign assistance, and hope to break the will of a superior foe over time. Given all this, Scott reasoned that the way the North could prevail with minimal bloodshed was to isolate the battlefield so the enemy got no foreign help, invade on several fronts to disperse the enemy and defeat him in detail, then occupy his sources of men and supplies. Scott's war memorandum of May 3, 1861, urged Lincoln to exploit the Union's naval, industrial, and manpower advantages to blockade the Confederate coasts, secure the entire Mississippi River, and hurl well-trained and well-supplied armies into the heart of Dixie.[32]

Scott's "anaconda" plan was destined to shape the entire Union war effort. But when initial rumors of a slow buildup and a long war leaked out, Republican congressmen and editors led by Horace Greeley demanded a sudden offensive to capture Richmond and quash the rebellion. The fact that the ninety-day volunteers would soon go home increased the pressure on Lincoln. Indeed, if there was even a slim chance of quick victory it was probably worth the gamble. But Brigadier General Irvin McDowell, a worthy enough staff officer in the Mexican War, had never commanded large formations. One-third of his 35,000 volunteers never even got into the fray that erupted on Bull Run near Manassas Junction, Virginia, on July 21. That allowed Beauregard's 21,000 recruits, reinforced just in time by Joseph E. Johnston's 12,000 men from the Shenandoah, to blunt the Union flanking attack (and bestow en passant the epithet "Stonewall" on Thomas J. Jackson and his brigade).[33] McDowell called for retreat, but an orderly withdrawal was beyond the capacity of his green troops. The "Forward to Richmond!" parade—and the lady and gentleman picnickers cheering it on—careened backward to Washington in the rain. That in turn panicked the capital, where people ran amok demanding that Lincoln resign. But if Bull Run shattered northerners' morale, it strangely did little for southerners' morale. Jackson claimed that he could have danced into the White House the following day if given just 5,000 troops. Richmond chattered over whom to blame for the fact he did not: Beauregard, Johnston, or Jefferson Davis. In retrospect, disarray in the Confederates' novice regiments plus the difficulty of storming the Potomac's bridges rendered a capture of Washington highly unlikely. But Manassas bequeathed a legacy of overconfidence, impatience, and recrimination that affected the Confederate cause for two years. Charleston's diarist-in-retrospect, Mary Boykin Chesnut, summed up the impact of the battle on rebels and Yankees alike. It "lulls us into a fool's paradise of conceit" but "will wake every inch of their manhood."[34]

Her priapic image was apt. The mere fact of the battle delivered a psychological shock challenging the virility of men in the North and South. The first Bull Run was no brawl on the Senate floor, mobbing of abolitionists, riot against Negroes and Irish, or bushwhacking on the frontier. The first Bull Run was a matter of tens of thousands of white Protestant American males, formed up in regular units under constituted authorities, leveling muskets and killing each other. Northerners had been wrong to think the secessionists would fold up their tents the moment the Union showed that it meant business. Southerners had erred in believing the Yankees would flee to their shops rather than face Confederate chivalry on the field of honor. The toll (fewer than 1,000 killed, about 2,600 wounded) was small by later standards, but it swept away Americans' last inhibitions against internecine slaughter. Hatred, fear, and anger ("How dare they fight us at all!") made the other side's cause diabolical, which made one's own cause holy, which meant that any man shirking his duty to God and to country was no man at all.

The campaigns of the Civil War, though momentous and thrilling, may seem too familiar to require retelling. But it is important to grasp how the war unfolded day by day, not sector by sector. Thus, it is artificial to chop up a narrative into chapters such as "The War in the West," "The War in the East," or "The Naval War." All sorts of news clattered over telegraph lines from all sorts of theaters in a chronology even more bewildering to Lincoln and Davis than it is to us today. It is also important to grasp that the ebb and flow of combat inspired judgments about what victory seemed to require. The process was jerky and gradual, but such was the escalation of violence that victory turned out to require political, economic, social, and cultural transformations beyond anyone's imagining at the start.[35]

The ratchet effect began during the nervous weeks after Virginia's secession. On May 3 Lincoln called on the states to raise an additional forty regiments of 1,000 troops each for a three-year hitch. When more than 200 regiments formed, he was more perplexed than pleased, because the federal arsenals could not begin to equip, clothe, and feed so many soldiers. Yet when Congress convened in July, it summoned 500,000 more volunteers; then it summoned another 500,000 after Bull Run. Most of these citizen-soldiers rallied to colors planted by their states, counties, or towns and served under "political officers" appointed by governors or elected by the rank and file. Incompetence, indiscipline, and confusion were often the rule.

President Davis and his congress likewise responded to the outbreak of genuine war by calling for 400,000 three-year volunteers to be enlisted by the Confederacy instead of the states. Some governors protested, most notably Joseph E. Brown, who claimed that every young Georgian was already enlisted in the state militia and insisted that a national levy was unconstitutional. Yet states' rights were only the first in a long list of vaunted southern principles sacrificed on the altar of Mars. The second to go was free trade, when the inexperienced secretary of the treasury, Christopher Memminger, asked the congress for duties on imports and exports so as to make a pretense of financing the war. The third to go was property rights, when the Confederate government ordered states to collect up to 1 percent of the value of all land, slaves, and other wealth. Rather than tax their own citizens, state governments fobbed off inflationary bonds on Richmond. More perverse was the encouragement, by Davis's cabinet, of a movement by "people's committees" to ban the export of cotton. This brilliant folly derived from the misconception that Britain and France were so in thrall to King Cotton they would break the Union blockade and recognize and perhaps ally themselves with the Confederacy rather than see their textile plants languish. All that the self-imposed embargo achieved (as with Thomas Jefferson's embargo of 1808) was to starve the economy of foreign exchange, wreck domestic producers, and spawn wholesale corruption. Finally, the Confederate treasury borrowed directly from state governments and banks (mostly in New Orleans), pledging to repay with customs receipts that never materialized. This drained precious liquidity from a private sector facing market disruption and soon armed invasion. It also invited the South to default on the millions its businesses owed northern creditors, thereby ending the influence the Cotton Kingdom had wielded on Wall Street. Since the Confederacy had invoked the code duello against an enemy with advantages of four to one in manpower, four to one in wealth, ten to one in manufacturing, and two to one in railroad mileage, it could scarcely afford a niggardly system of public finance dependent on citizens' old-fashioned republican virtue.[36]

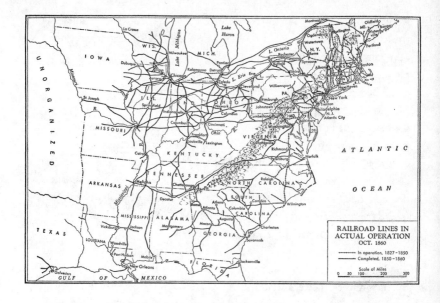

RAILROAD LINES IN
ACTUAL OPERATION
OCT. 1860

In operation, 1827–1850
Completed, 1850–1860

Scale of Miles
0   30   100   200   300

Instant armies dwarfing those fielded in 1776, 1812, and 1846 kicked off fierce competition for arms, and thanks to Josiah Gorgas the South got a leg up on the North. Gorgas resigned his command of Philadelphia's Franklin Arsenal (in deference to his wife, a native of Alabama) and won appointment as chief of ordnance for the Confederacy. He dispatched agents to Europe and the Caribbean with orders to corner the markets in weapons and saltpeter. He rehabilitated the cannons, small arms, and machine tools captured from Harpers Ferry and Norfolk, and expanded the Tredegar Works. Above all, he scrounged throughout the Confederacy for ores and chemicals to supply eight new arsenals located safely in the Deep South. Gorgas was ably seconded by the Confederacy's secretary of the navy, Stephen R. Mallory, who built almost from scratch a fleet of thirty-three ships in 1861, contracted for commerce raiders manufactured in England (most famously Captain Raphael Semmes's *Alabama*), and experimented with innovative weapons in hopes of breaking the Union blockade. These weapons included submarines, exploding mines, and ironclad warships, of which the *Merrimack* (rechristened the CSS *Virginia*) was only the first. All Mallory's effort and ingenuity, however, could not break the vicious circle traced by Confederate finances, priorities, and the cotton embargo. Without access to markets abroad the Confederacy would suffocate, yet over the course of the war it spent twenty dollars on the army for every one dollar on the navy. One scholar tersely concluded: "We are thus back to the starting point: no cotton exports, no money; no money, no ships; and no ships, no cotton exports."[37]

The Union quartermaster, General Montgomery Meigs, was resourceful and honest, but hamstrung by Cameron, the crooked secretary of war. Cameron's idea of mobilization was to authorize officials to buy anything from anyone at any price, thereby sparing himself the bother of reviewing competitive bids. To American hustlers that was a patriotic license to steal. The War Department spent millions on obsolete or defective muskets, decrepit horses, spoiled meat, moldy biscuits, and threadbare uniforms (some from the firm of Brooks Brothers). The *New York Herald* rued the fortunes made overnight and said that 1861 was "The age of shoddy . . . shoddy brokers in Wall Street, or shoddy manufacturers or shoddy goods, or shoddy contractors for shoddy articles for shoddy government. Six days a week they are shoddy businessmen. On the seventh day they are shoddy Christians." Cameron's own cut came from contracts with his railroad interests to ship all the shoddy goods to the troops. When it came to underwriting the war Lincoln was no help at all. "I don't know anything about 'money,'" he quipped. "I never had enough of my own to fret me, and I have no opinion about it any way." And the secretary of the treasury, Chase—a former Democrat suspicious of banks, loans, and federal spending—was not the right man to engage in creative financing. Indeed, his first instinct was to make an absurdly precise underestimate—$328 million—of the cost of saving the Union. Congress answered his call in August 1861 by passing an Omnibus Revenue Act drafted by John Sherman, the Ohioan named to replace Chase in the Senate. His bill called for the first U.S. income tax and authorized floating $250 million in bonds. The numbers embarrassed everyone soon enough, but the act set a "borrow, but tax" pattern that permitted the Treasury to tap the nation's wealth without setting off ruinous inflation.[38]

Lincoln was no mechanic either, but he was smitten with weapons after a visit to the Washington navy yard, where Captain Dahlgren showed off an eleven-inch naval gun he had worked for years to perfect. Flirtation turned into a hot affair as Lincoln attended numerous "champagne experiments" of new weapons at the Anacostia and Potomac arsenals commanded by George Douglas Ramsay. Could new breech-loading rifles or crank-operated "coffee mill" machine guns end the war quickly? Lincoln hoped for a "technological fix" so devoutly that he encouraged his secretaries to pass on letters from inventors even though ninety-nine of 100 were probably bogus. To be sure, the Civil War introduced or advanced various technologies, but none except ironclad warships proved decisive in battle. In the 1850s secretaries of war Jefferson Davis and John B. Floyd (also a southerner) were enthusiastic about a British breech-loading rifle designed by Christian Sharps, but their budget discouraged them from attempting to retool whole arsenals and refit the whole army. Had they done so, the war might well have ended at Bull Run. As for the crude machine gun invented by Richard Gatling, the War Department did not even give it a trial until 1864. A significant reason why the North's mechanical prowess yielded no wonder weapons was James Wolfe Ripley, aged sixty-seven in 1861, who considered it his job as chief of ordnance to nix every new idea. This wouldn't work; that cost too much; the other would take too long; the generals wouldn't like it; the troops would need retraining; in a word, no. Ripley survived in office because many generals shared his bias against novelty and Lincoln felt his own lack of expertise. Lincoln contented himself with purchasing some breechloaders and "coffee mill guns" out of his own pocket. Balloon observation to help artillerymen adjust fire, repeating rifles to multiply the Union cavalry's firepower, "triage" judgments of soldiers' wounds to maximize the lives saved by surgeons, and numerous other innovations certainly presaged the future of warfare. But the Union (and Confederate) infantry fought mostly with 1861 muzzle-loaded Springfield and Enfield rifles, and the artillery fought mostly with muzzle-loaded, smooth-bore twelve-pounder cannons. Those weapons were lethal enough in a long war of attrition that dictated close-quarters combat and favored mass over mobility.[39]

Lincoln was fortunate that the Navy at least was in excellent hands, because the blockade he declared made it responsible for sealing off 3,500 miles of Confederate coast with nearly 200 outlets to the sea. Speed was essential because the European powers had agreed that blockades were illegal unless they were effective (otherwise they amounted to hit-and-miss, state-sanctioned piracy). Moreover, a "blockade" ipso facto implied a state of international war rather than internal rebellion. In May 1861 the British government seized on that diplomatic faux pas to declare neutrality, and hence recognition of the Confederacy's rights as a belligerent. Seward and

Charles Francis Adams, the U.S. minister in London, hotly protested, but their plaints would be futile unless and until an effective blockade was a fait accompli. Secretary of the Navy Gideon Welles ("Father Neptune" to Lincoln) and Assistant Secretary Gustavus Fox were blessed by the fact that the U.S. merchant marine and shipyards were based almost entirely in northern ports, and gratified when only 373 of the Navy's 1,554 antebellum officers went south. Welles and Fox promptly commissioned more than 200 civilian ships in 1861 alone and laid keels for twenty-three steam-powered oceangoing gunboats. At the suggestion of the Smithsonian regent Alexander Dallas Bache, a veteran of the Coastal Survey, they also assembled an expert blockade board to study the southern coastline and recommend ship stations. Meanwhile, the steamboater James Buchanan Eads advised Lincoln that the weapons needed to crack Confederate power in the West were armored steam-powered gunboats of the sort the British had toyed with in the Crimean War. The Navy Department rushed Eads to Saint Louis, where he grabbed all the available iron and put 4,000 workers on round-the-clock shifts. In a matter of months the Union would launch a fleet of nearly invulnerable "floating batteries" or "Pook turtles," named for Eads's chief designer, Samuel Pook. Speed was just as essential on the Atlantic lest the Confederate ironclad under construction at Norfolk make every wooden ship in the U.S. Navy obsolete. Welles entrusted the task to John Ericsson, a brilliant Swedish-American recluse in New York. Ericsson's *Monitor*, a long, thin raft almost wholly submerged except for a revolving turret with two eleven-inch Dahlgren guns, would prove itself barely seaworthy barely in time. To worship this miracle, joked a naval inspector, would not be idolatry, for the vessel "is the image of nothing in the heavens above, or the earth beneath, or the waters under the earth."

That sea power was far and away the Union's most vital asset should have been evident to anyone resigned to a lengthy war. As early as August 1861 an amphibious assault on Cape Hatteras Inlet taught the navy a valuable lesson and sent the Confederacy an ominous signal. The old pirates' den on North Carolina's Outer Banks seemed impregnable because naval lore decreed, "A ship's a fool to fight a fort." But flag officers Silas Stringham and Samuel Francis du Pont suspected that the Navy's new rifled cannons had the range and accuracy to batter coastal forts in advance of the landing of troops for a ground attack. Troops were not even needed.

Stringham's seven ships and 141 guns were enough to cause the enemy at Cape Hatteras Inlet to raise a white flag. In November, Captain du Pont guided another flotilla into Port Royal Sound between Charleston and Savannah. This time even the naval barrage was not needed, except to alert slaves on the sea island plantations: "Son, dat ain't no t'under, dat Yankee come to gib you Freedom." Confederate ports fortified under Robert E. Lee's watchful eye would be harder to crack, but it was just a matter of time before the U.S. Navy visited all of them.[40]

Provocative deployment of sea power nonetheless ran a risk that became all too apparent on November 8, 1861. Captain Charles Wilkes, the irascible explorer now in command of the sloop *San Jacinto*, fired across the bow of the British mail ship *Trent* in the Bahamas and boarded it for the purpose of arresting the Confederate diplomats James Mason and John Slidell. This was a coup to cheer northern public opinion. It was also a violation of neutral rights every bit as outrageous as those perpetrated by the British navy on American ships before the War of 1812. Prime Minister Palmerston backed up his protests with threats, including the mustering of a Canadian militia and a fleet bearing 11,000 redcoats to North America. The U.S. minister in London, Charles Francis Adams, was beside himself for fear that Seward really meant to provoke a foreign war. The State Department had already soured relations by ordering the arrest of British consuls who were in touch with the Confederacy, a practice Lord Lyons deemed "monstrous." Lincoln and the rest of his cabinet simply felt trapped. To stick by their guns could provoke British belligerence, while to grovel before "rebel spies" would outrage northerners and undermine the Union's prestige. But it was Seward who carried a face-saving formula in the cabinet meeting on Christmas day. He had never desired a war that would hurl Britain onto the Confederate side: what nonsense! But he knew that a bluff stance was the only way to dissuade the British from meddling in America's troubles. So he persuaded Lincoln to release Mason and Slidell on the grounds that Wilkes had exceeded his authority, yet concede no point of principle. Seward also took the occasion to remind the British that as the governor of New York he had scrupulously prevented Americans from intervening in Canada's civil strife in 1838, and he now expected the British to do likewise. After all, how would Whitehall react if foreigners assisted a rebellion in Ireland? The prospect of European intervention remained real, but peaceful resolution of the *Trent* affair invoked the muse of White House secretary (and future secretary of state) John Hay: "And so, a generous people, at the last / Will hail the power they did not comprehend, / Thy fame will broaden through the centuries; / As a storm, and billowy tumult overpast, / The moon rules calmly o'er the conquered seas." Hay called this an ode "To William H. Seward."[41]

The man Lincoln appointed after Bull Run to whip 175,000 recruits into the proud army of the Potomac and suppress the rebellion was short, dashing, and brash. People called him "little Mac" and "the young Napoleon." He seemed forever in motion attending to organization, logistics, and drill. Around the District of Columbia he designed an elaborate ring of fortifications and staged endless reviews, saluting his soldiers from horseback and making them feel like winners. Journalists, generals, congressmen, and even the president doted on him. All this was enough to cause Scott to resign in disgust (pleading ill health). When Lincoln turned over command of the Union forces to a man less than half Scott's age, the martinet boasted, "I can do it all," His pride and petulance were on display in the indiscreet letters he sent to his wife. At first the letters boasted of his prestige and the great work God had given him. Later they whined about plots and called the members of Lincoln's cabinet so many geese, puppies, old women, idiots, and traitors. Prickly personalities and some genuine issues got his goat, but the main cause of discord was his own lethargy. As late as December, when the Thirty-Seventh Congress finally convened, the general still had no timetable for an offensive. Lincoln called it a bad case of "the slows." Benjamin Wade, chairman of the Joint Committee on the Conduct of the War, suspected it was a sign of disloyalty.[42] Finally, after the Army of the Potomac took the field, "little Mac" squandered a good chance to take Richmond by storm and blamed everyone else when he didn't.

That much everyone knows about George Brinton McClellan, a "vain and unstable" man "possessed by demons and delusions" or even a "messianic complex." Yet no less an authority than Ulysses Grant confessed, "McClellan is to me one of the mysteries of the war." The mystery is worth plumbing because its clues reveal much about northerners' opinion in the months before the rebellion became a "great civil war." In his annual message on December 3, 1861 (McClellan's thirty-fifth birthday), Lincoln angered Radical Republicans when he insisted that the general in chief was "in considerable degree, the selection of the Country as well as of the Executive." Lincoln soon regretted the selection, but his statement was true.[43]

McClellan, the middle of three sons of a doctor, grew up in a town house in Philadelphia across the street from the First United States Bank. His family was fervently Whig. George learned to restrain passion through self-discipline, value education and science, and espouse a respectable low-church Episcopalianism. After graduating second in the class of 1846 at West Point, he ably served Scott in the Mexican War. But the growing influence of passionate Free-Soilers caused the whole McClellan clan to defect to the Douglas Democrats in 1852. In their view, the Whig Party of Clay and Webster had ceased to exist. In 1857 George resigned his commission to manage two railroads: the Illinois Central (where he scheduled trains for the Lincoln-Douglas debates), and the Ohio and Mississippi. He married in 1860 and leased a home in Cincinnati, but he was not surprised by secession and war. In fact, he accepted the call to command Ohio's regiments so quickly that Pennsylvania and New York never got to bid for his services. A small but rugged mountain campaign in the Kanawha valley of western Virginia made him the Union's first victorious general.

McClellan was certainly vain and overly cautious. He managed the army like a railroad whose motto is "Safety first." But deluded, demonic? On the contrary, he thoroughly agreed with Douglas that hotheads were to blame for the crisis; agreed with Lincoln that a Unionist silent majority would emerge in the South; and agreed with Scott that a destructive war must be avoided if possible. McClellan's humane and patriotic purposes were to limit the carnage lest passions increase, persuade rather than compel the Confederates, and restore the Union to the health it had enjoyed in the 1840s. That profoundly conservative vision required making no issue of slavery. Thus, although the general hoped some "day of adjustment" might bring succor to slaves, he also hoped to be able to "dodge the nigger— we want nothing to do with him. I am fighting to preserve the integrity of the Union & power of the Govt." That, too, matched Lincoln's thought as expressed in his annual message of 1861: "In considering the policy to be adopted for suppressing the insurrection, I have been anxious and careful that the inevitable conflict for this purpose shall not degenerate into a violent and remorseless revolutionary struggle."[44]

Yet Lincoln squirmed under political pressures from which McClellan was partly insulated. The press wanted to know when the administration would begin fighting this war. On Wall Street, noted George Templeton Strong, "Nobody believes in [Lincoln] any more." Radical Republicans damned Lincoln's reticence in regard to slavery. The *Trent* affair risked a war on two fronts. McClellan got sick. The Treasury was at its wit's end. So was the president. On January 10, 1862, he pleaded with Meigs, "General, what shall I do? The people are impatient; Chase has no money; . . . the General of the Army has typhoid fever. The bottom is out of the tub." A month later typhoid fever carried off Lincoln's son, whose death in turn deranged the first lady.[45]

In fact, winter 1862 brought help beyond hope to the president. First, when Secretary of War Cameron finally resigned in January, Lincoln named Edwin McMasters Stanton to the critical post. That was a curious choice. Stanton, a lifelong Democrat, had served with Buchanan. All one can surmise is that he wheedled his way into the most critical cabinet post by scheming with Seward, befriending McClellan, impressing Chase with his abolitionist sentiments, and betraying Cameron by suggesting that the last-named endorse the heretical notion of arming the slaves. Once in office, the bespectacled, long-bearded Stanton did for the Union war effort what Lazare Carnot, *l'organisateur de la victoire*, had done for the armies of the French Revolution. Stanton epitomized probity, energy, and bipartisanship. He established the United States' first massive bureaucracy to administer its first national mobilization, and did so by working with, not against, private business. Thanks to Stanton, the North exploited the material superiority that allowed it to rally from setbacks in ways the South never could.[46]

Two weeks after the War Department changed hands, another unknown quantity proved his mettle. Ulysses Grant, an indifferent West Pointer known mostly for horsemanship, was another veteran of the Mexican War who grew bored in peacetime; he had resigned his commission in 1854. Unlike McClellan he failed in business, sometimes drank to excess, and hit bottom clerking at a store in Galena, Illinois. When the war broke out Grant volunteered to command a regiment (much to the consternation of his pro-slavery father-in-law). He suddenly found a purpose in life. His tough but affectionate discipline turned a mob of surly farm boys into the Twenty-First Illinois Volunteer Infantry, one of the Union's

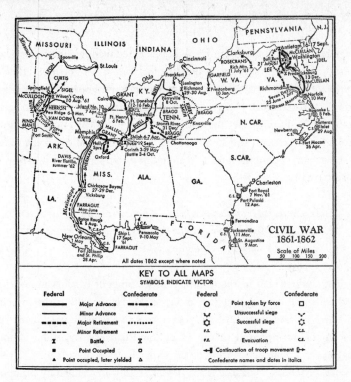

crack units. Thanks to the clout of Congressman Elihu B. Washburne, Grant was soon a brigadier in command of the troops swarming to Cairo. The maps he pored over told him the Tennessee and Cumberland rivers, which met the Ohio just fifty miles to his east, were the keys to invading the South. The Confederates knew that as well and were hastily building Fort Henry and Fort Donelson where those rivers crossed the Tennessee border. But Grant distracted their attention with demonstrations against their Mississippi River "Gibraltar" at Columbus, Kentucky. Then he begged for permission to strike Henry and Donelson. Such indirection would become Grant's signature.

Henry W. Halleck, the Army's "Old Brains" in command of the western district, balked at Grant's motion until it was seconded by another brilliant naval appointee. Flag Officer Andrew Hull Foote was the fifty-five-year-old scion of an old New Haven family of sailors. Having distinguished himself in blue-water combat, shipyard management, and moral reform (no drinking, swearing, or work on the Sabbath), he never expected to end up in charge of ironclad gunboats 1,500 miles from the sea. Despite "a work of almost insuperable difficulty," however, Foote launched the flotilla ahead of schedule and all but invented techniques for combined operations by the Army and Navy. Just four days after getting Halleck's approval on February 2, 1862, Grant's soldiers and Foote's "floating batteries" captured Fort Henry. Just ten days after that Grant earned a nickname by forcing the "unconditional surrender" of Fort Donelson's 14,000-man garrison and supply cache. That in turn allowed the federals to occupy Nashville and threaten the critical railroad junction at Corinth, Mississippi.[47]

> *Stern weather is all unwonted here. / The people of the country own*
> *We brought it. Yea, the earnest North / Has elementally issued forth*
> *To storm this Donelson.*—Melville, 1862[48]

An anxious wait in the telegraph office brought Lincoln more stunning news the day after he had received terrible news that the ironclad CSS *Virginia* had ventured from port to sink two wooden vessels and run a steam frigate aground in the Hampton Roads estuary. The Union's blockade and planned invasion of the James River peninsula were at risk. Then, on March 9, 1862, the USS *Monitor*, frantically hauled by a tugboat down from New York, proved the equal of the *Virginia*. The ironclads' all-day slugfest ended not so much in draw as a double technical knockout.[49] But that was all the Union required. The *Virginia* limped back to the Elizabeth River, where it sank two months later. That enabled the U.S. Navy to proceed with the most ambitious sea lift in history in support of one of the most brilliant operational plans of the war: McClellan's Peninsula campaign. Back in December he had first imagined a joint land-sea invasion of Virginia from the coast. A surprise attack from there would bypass the prodigious earthworks he imagined Joe Johnston had built at Manassas and would cut in half the distance to Richmond. But when Lincoln impatiently ordered a general advance in late February, McClellan confined his army to a tentative march back to Manassas. To his utter embarrassment he found that Johnston had skedaddled south to the Rappahannock, leaving behind dummy camps and "Quaker cannons" to gull the federals. Lincoln and Stanton promptly stripped McClellan of the post of general in chief so he could devote all his undoubted talent to the Army of the Potomac. He obliged them by overseeing the tremendous staff work required to ship the whole army to Fort Monroe at the mouth of the James

River opposite Norfolk. It was to be Veracruz, 1847, on a much larger scale; and the Navy proved just as efficient in 1862. It shipped 120,000 men, 14,000 animals, forty-four artillery batteries, siege mortars, and supplies and baggage for all of the above. The losses en route totaled six mules. A British observer called it "the leap of a giant." Unbeknownst to McClellan just 15,000 defenders barred his path, so he might have marched into Richmond before the azaleas dropped their blooms. But whatever pluck he possessed was trumped by his aversion to big battles that would harden hearts on both sides. With good reason, too, because a very big battle erupted in the West just a day after McClellan invested Yorktown.[50]

The 40,000 soldiers of Grant's Army of the Tennessee halted their southward march at Pittsburg Landing just twenty-five miles northeast of Corinth. There, the Mobile and Ohio Railroad crossed the Memphis and Charleston, the Confederacy's only east-west trunk line. Halleck wanted Grant to await Don Carlos Buell's 35,000 troops from Kentucky before challenging Albert Sidney Johnston, the general the Yankees feared most. The chesty, jut-jawed Johnston struck first, promising his 50,000 recruits clad in butternut "a decisive victory over agrarian mercenaries, sent to subjugate and despoil you of your liberties, property, and honor." At dawn on April 6 a regiment from Ohio, camped around a little church named Shiloh, awoke to the bloodcurdling rebel yell. The men fled in terror to the main Union lines, which the rebels assaulted all day in relentless, if disjointed, waves. Bodies piled up in the Peach Orchard, where Johnston was shot; the Sunken Road; the Hornet's Nest; and the Bloody Pond, where wounded enemies reunited in thirst and exhaustion. But the cool, fighting retreat of William Tecumseh Sherman's division protected the Union flank, and naval gunboats broke the attackers' momentum at Pittsburg Landing.

It had been "the devil's own day," muttered Sherman. "Lick 'em tomorrow, though," promised Grant. And he did, thanks to the arrival of Buell's "lost division." On April 7 fresh Union troops pushed the Confederates back until Beauregard sadly sounded retreat. The "complete victory" he had already reported to Davis turned into disaster. The Confederacy would never regain the initiative in the West, as war correspondents including Whitelaw Reid soon realized. But their initial reports sickened the home front with lurid descriptions of battlefield gore and army hospitals "more unendurable than encountering the enemy's fire." The dead, wounded, and missing numbered 25,000, nearly seven times the tally at Bull Run, where the war is said to have started in earnest. No, Americans' consciousness of the scale of this war was born in the same place their Romanticism expired: at Shiloh. Two weeks later the Confederate congress imposed three years of military service on all healthy men between ages eighteen and thirty-five: the first conscription in American history.[51]

> The church so lone, the log-built one, / That echoed to many a parting groan
> And natural prayer / Of dying foemen mingled there—
> Foemen at morn, but friends at eve—/ Fame or country least their care
> (What like a bullet can undeceive!) / But now they lie low,
> While over them the swallows skim, / And all is hushed at Shiloh.
> —Melville, 1862[52]

Nothing exposed the futility of the Confederate cause as did the appearance of the U.S. Navy, that same week in April 1862, at the sandy mouth of the Mississippi. Had Albert Sidney Johnston anticipated an assault on New Orleans, or just questioned the myth of its invincibility dating from 1815, he might have peeled off divisions for its defense. But to do so would have ensured the loss of Corinth's railroads to Halleck's combined armies. As it was, both prizes were lost when the Navy's marriage of human will and technology arrived in the person of David Glasgow Farragut. Though he was a Tennesseean married to a Virginian, the sixty-year-old flag officer could not imagine trampling the Stars and Stripes he had sailed under for fifty-one years. Gustavus Fox recommended to Welles that the wiry patriot be given the Gulf Coast command with orders to capture the lower Mississippi. On April 18 Farragut's nine sloops, fourteen gunboats, and nineteen mortar schooners opened a furious six-day barrage at Fort Jackson and Fort Saint Philip seventy-five miles south of New Orleans. No white flags appeared, so Farragut did not rest on the seventh day lest two rebel ironclads being built upriver enter the fray. Instead, he bade two brave lieutenants to ram the chain links blocking the channel so the fleet could run the gantlet. Now it was the rebel gunners' turn to launch "the greatest fireworks display in American history." All but four Union ships survived to overawe the mobs that gathered on the levees of New Orleans in hatred and disbelief. Their hatred grew after the political general Benjamin "Beast" Butler placed the city under stern martial law and gratuitously threatened its women. When the terrible news reached Richmond, Jefferson Davis "buried his face in his hands," then sought out an Episcopal priest to baptize him. Varina Davis said "a peace which passed understanding seemed to settle in his heart." Still Farragut's fleet kept chugging upstream to take the surrender of Baton Rouge and Natchez. Farther north Foote's gunboats and Halleck's soldiers had already captured the bastions at New Madrid, Missouri, and Island No. 10. When Memphis surrendered on June 6, the Union owned the whole river except for Vicksburg's bristling bluffs.[53]

To understand the cadence of the Civil War's drumbeat, consider that all those dramatic thrusts in the West occurred during the same weeks when the Peninsula campaign wilted. McClellan wasted a month besieging Yorktown because he fell for John Magruder's pretensions of strength. That gave Joe Johnston time to reinforce Richmond. After May 4 the federal host still advanced at a snail's pace because Pinkerton's detectives reported that Confederate forces were triple their actual number. McClellan the railroader knew and trusted Pinkerton, but simple long division should have told McClellan the general that the whole Confederacy could not concentrate and supply 200,000 men. His own numbers shrank when Lincoln recalled a whole army corps to secure Washington, D.C. He then lost all hope of reinforcements when Stonewall Jackson's 16,000 "foot cavalry" led 40,000 federal troops on a wild-goose chase in the Shenandoah Valley. The final handicap (or excuse) setting McClellan up for defeat was "the confounded Chickahominy" River swollen by downpours. Joe Johnston waited until a sizable mouthful of Union troops bobbed across at Fair Oaks, then ordered them chewed up and swallowed. Like Shiloh, the battle ended in a bloody draw that favored the Union—the bridges were held. As at Shiloh, the rebel commander was one of the casualties. But as Johnston gallantly testified, "The shot that struck me down is the very best that has been fired for the Southern cause yet." He referred to the man Davis chose to replace him: Robert E. Lee.

Trite as it is to repeat, Lee is an icon, a marble man, epitome of the chivalrous Cavalier. Devotees and debunkers still debate how much he deserves his postwar and posthumous veneration, but there is no question that Lee "posed" for his role in life as much as George Washington did. Born to one of Virginia's First Families, Lee dispensed hospitality to his visitors at Arlington in a sprawling classical manor fronted by Brobdingnagian pillars. Heir to a heroic tradition, Lee graduated with honors from the U.S. Military Academy without incurring a single demerit. Sworn to public service over private advancement, he remained in the Army while most of West Point's "long gray line" took lucrative jobs in the private sector. An enlightened patriot, he lamented slavery and opposed secession, but found he could not make war on his own neighbors and heritage. If a single attribute defined the tall, white-bearded patrician with the gentle tongue and leonine heart, it might be Lee's belief that he must not only live up to, but be seen to live up to, the most rigorous code of honor. Yet what mattered most to the story was Lee's way of war. Its operational results were stunning, but its tactical costs and strategic goals were beyond his nation's means.

McClellan clearly meant to reduce Richmond post by post with artillery rather than launch frontal attacks, so Lee seized the proffered initiative. He summoned Jackson from the valley and sent the plumed cavalry chief J. E. B. Stuart on a hell-for-leather reconnaissance around the whole Union army. He riskily divided his inferior forces, trusting lieutenants A. P. Hill and James Longstreet to do nothing stupid. Then he delivered a series of hammer blows at the enemy's flanks that quickly persuaded McClellan to withdraw to the navy's sheltering guns on the James River. Down South "Bobby" Lee was an instant hero; up North, a bête noire. But Lee's swashbuckling victories came at a price. The Army of Northern Virginia lost over one-fourth of its strength during the Seven Days Battles. The Confederates could not stand such attrition for long. Lincoln and Stanton knew that, and so they called Halleck east to serve as general in chief and ordered Major General John Pope to mount another offensive with the troops of the Washington garrison. Pope's progress ceased when J. E. B. Stuart raided his headquarters, bagging a $350,000 purse, and Stonewall Jackson bloodied his flank at Cedar Mountain, then circled around to sack the Union depot at Manassas Junction. Pope was baffled; "Lee's lieutenants" were wasps. But he still expected to overwhelm the Army of Northern Virginia once he was reinforced by McClellan's Peninsula veterans. To forestall that, Lee ordered Jackson to dig in around the old Bull Run battlefield and invite an attack. No sooner did Pope snap at the bait than Longstreet and Lee appeared out of nowhere. This time the federals fought hard despite wilting artillery barrages and rebel-yell charges, but the Second Bull Run ended on August 30 with the Union skedaddling again. Lee had inflicted 15,000 casualties on the Union and his own troops had suffered just 9,000. Still, he rued this war of attrition and urged Davis to let him take the war north. An invasion might rally Maryland and Kentucky to their side, capture precious provisions, and perhaps yield the climactic victories needed to win foreign help. Davis saw the logic: in September 1862 he bade Lee's army and Braxton Bragg's out in Mississippi to navigate by the polestar.[54]

Honest Abe was only half candid when he forswore "revolutionary struggle." The whole raison d'être of the Republican Party was to reinvent government through a long list of acts that some historians consider a second American revolution. Thanks to the absence of southern Democrats the Republicans enjoyed majorities of 105 to forty-three in the House and thirty-one to ten in the Senate. The 428 acts that the Thirty-Seventh Congress proceeded to pass made it the most prolific in history. Even those acts meant strictly as war measures had lasting consequences. Not least of those was the Legal Tender Act of February 1862. It authorized the Treasury to circulate an initial $150 million in paper notes nicknamed greenbacks. They amounted to promissory notes, but since the act declared them legal tender, they functioned as currency and gave every citizen clutching a few dollar bills a stake in the Union cause. Even more revolutionary was the Revenue Act, providing for an income tax and excise taxes on virtually every productive activity. In 1775, Americans had taken up arms to resist a few shillings in taxes imposed from abroad. Now they were asked to swallow theoretically limitless taxes for the purpose of taking up arms against one another. The bill, whose 20,000 words made it the longest to date, itemized imposts on everyone from locomotive manufacturers to circus barkers. It sparked a month of lively debate until Thaddeus Stevens struck the perfect false note. The taxes would so augment Union forces that the South would quit within ninety days, whereupon it would be right and proper "in accordance with the practice of nations, the dictates of wisdom and of justice, to make the property of the rebels *pay the expenses of the war* which they have so wantonly caused." Well, that made it an easy vote after all. The House approved, 125 to fourteen, and Lincoln signed the bill into law on July 1. Yet another war measure, the Militia Act of July 1862, imposed a veiled form of conscription by authorizing the president to raise 300,000 more "volunteers" in the event states could not fill their proportional quotas. Almost overnight the burdens of U.S. citizenship grew as weighty as the blessings.

The rest of the Republican revolution amounted to a headier version of the American System that the Whigs never managed to pass. The Morrill Tariff of 1862 jacked up duties on imports to record levels in the name of protecting domestic manufactures. Few "infant" industries needed the security blanket anymore, given America's advanced technology, Britain's flagging competition, and the Union's own war spending. But high tariffs remained on the books for the rest of the century. Morrill's even more famous bill, the Land Grant Act of July 1862, gave states 30,000 acres of federal land for every member they sent to Congress, on the condition that the proceeds be used to promote agriculture and industry. No fewer than sixty-nine land-grant colleges would emerge, including state universities throughout the Midwest and in California. The Homestead Act of May 1862 reduced the price of virgin soil in the West to zero for anyone who improved the land for five years. Congress created the Department of Agriculture the same month. In July it voted to stitch the continent into a single market through the Pacific Railroad Act, which bestowed huge land grants and loans on the Union Pacific and Central Pacific railroads. Most controversial was the National Banking Act, which established a system of federally chartered banks. It was designed to help float the government's war debt and stimulate growth. But fierce opposition from Wall Street, state banking interests, and old Jacksonians delayed passage until February 1863, when Senator John Sherman offered bribes under the guise of a compromise. One amendment killed proportional distribution of the Treasury's largesse in favor of the large eastern banks, and another ensured 100 percent redemption of federal notes only at national banks based in New York, Philadelphia, and Boston. The financial community, theretofore surly, found patriotism plus 5 percent more to its liking. All those initiatives were grist for the mills of hustlers and frauds. But the extraordinary Republican package made the partnership between the public and private sectors more than a motto. Indeed, it invited Yankees to gaze beyond the present conflict and imagine the glorious future in store for a reunited nation.[55]

Abolition was the only revolutionary struggle Lincoln was chary of waging. When Greeley chastised the administration in his column of August 1862, "The Prayer of Twenty Millions," Lincoln seemed to make his position crystal clear: "My paramount object in this struggle *is* to save the Union, and it is *not* either to save or to destroy slavery. If I could save the Union without freeing *any* slave I would do it, and if I could save it by freeing *all* the slaves, I would do it; and if I could save it by freeing some and leaving others alone I would also do that." What more ammunition do revisionists need to argue that the "Great Emancipator" was at best a pragmatic racist? But the evidence is suspect because Lincoln always spoke (in public) as a statesman for whom politics was the art of the possible. His letter to Greeley was simply written after he made up his mind to push for emancipation but before he was prepared to admit it.[56]

It is futile to debate whether a pragmatic Lincoln made a pretense of morals or a moral Lincoln made a pretense of pragmatism. He was a brilliant, subtle, troubled man feeling his way through a national identity crisis. Probably not a day passed without Lincoln's agonizing over the "right" thing to do, yet he was the last person to claim to be heaven's mouthpiece. As a friend of Mary Todd Lincoln attested, the president pored over his Bible "in the relaxed, almost lazy attitude of a man enjoying a good book." Doubtless Lincoln found the "good book" very good because of the truths it told about human nature. Lincoln's other favorite diversions—satire, poetry, and the theater—likewise suggest someone obsessed with probing man's powers of self-delusion. For decades his doctrine of necessity had served as a handy defense mechanism against the awful reality of reality, but mental tricks were a luxury Lincoln could no longer afford. Perhaps Providence and free will did coexist. When Senator Orville Browning of Illinois concluded from the First Bull Run that heaven would not bless the Union until it confronted slavery, Lincoln mused, "Suppose God is against us in our view on the subject of slavery in this country, and our method of dealing with it." By the time of Shiloh Lincoln appreciated how slavery united and buttressed the Confederate war effort: "we had about played our last card, and must change our tactics or lose the game." Gideon Welles remembered Lincoln as being "convinced that the war must be prosecuted with more vigor, and that some decisive measures were necessary on the slavery question, not only to reconcile public sentiment and to consolidate and make uniform military action, but to bring the slave element to our aid instead of having it turned against us." In other words, a war waged merely to save the Union not only could not be won but probably should not be won.

Think about that. Americans, free to function as their own theologians, are expert at persuading themselves that the thing they want to do just happens to be the right thing to do. Lincoln persuaded himself that the right thing to do also happened to be thing he wanted to do, which was win the war and save the Union. In September 1862 he meditated on the mystery in private: "God's purpose is something different from the purpose of either party. . . . I am almost ready to say this is probably true—that God wills this contest and wills that it shall not end yet."[57]

Why Lincoln's thinking evolved is less of a mystery. Tens, hundreds, and then thousands of fugitive slaves shadowed the Yankee invaders, begging for asylum, food, shelter, work, or a musket and uniform. Enslaved Negroes just freed themselves. It was they who made the war something more than a schism in the whites' civic religion. Union officers needed to know the legal status of these people and how to dispose of them. As early as May 1861, Ben Butler at Fort Monroe found the formula needed: he did not "free" the slaves in the imprudent fashion of Frémont; he declared them "contraband" property subject to the war powers of the commander in chief. Congress approved, through Confiscation Acts that by 1862 freed all persons in thrall to masters supporting the rebellion and ordered the Army to grant asylum to runaways. Of course, vocal spokesmen for the urban Irish and "butternut" Democrats of the Ohio Valley cried, "We won't fight to free the nigger." But Lincoln also felt the hot breath of Radical Republicans for whom the war was unholy unless it became a crusade against bondage. Congressman George Washington Julian of Indiana delivered their ringing keynote in January 1862 when he made the commonsense point that 4 million slaves could never be neutral in this conflict. "As laborers, if not as soldiers, they will be the allies of the rebels, or of the Union." The Second Confiscation Act authorized the War Department to hire as war workers "so many persons of African descent, as can advantageously be used." The Militia Act authorized Negro soldiers.

Lincoln and Stanton shrank from opening that Pandora's box. Their last attempt at a moderate course was a request that loyal slave states embrace the gradual, compensated emancipation Congress mandated for the District of Columbia. Only when delegates from the border states said, "No, thank you" by a two-to-one margin did the president decide he must invoke his war powers. Emancipation, he told Seward and Welles on July 13, was "a military necessity. . . . We must free the slaves or be ourselves subdued." Nine days later he apprised the cabinet of the preliminary proclamation he painstakingly drafted at his desk in the telegraph office. Postmaster Montgomery Blair feared it might cost the party the fall elections. Secretary of State Seward feared it might backfire if foreigners took it to be an admission of weakness. But the cabinet bowed to the president's will on condition that he postpone the decree until a battlefield victory.[58]

Foreign opinion mattered more than ever when the Union's winning streak of spring 1862 turned into a slump during the summer. In retrospect it seems highly unlikely that Britain and France would have formally recognized, much less fought for, the Confederacy. The benevolent hegemony known as the Pax Britannica was more bluff than battalions in the backwash of the Crimean War and the Sepoy Mutiny. Britannia ruled the waves thanks mostly to an absence of challengers; and ironclads might render the existing Royal Navy obsolete. Imperial crises demanded constant attention. The peace in Europe was fragile. Parliament was as tight-fisted as ever. Canada was more of a hostage than a threat to the Yankees. Most (not all) of Britain's middle-class and working-class people loathed the notion of risking war on behalf of American slavery. The government, with one eye always on Ireland, feared supporting a secessionist movement. As for the adventurous Napoléon III, he promised to act in harness with London and in any case decided that the way to exploit the Civil War was to collude with Mexican royalists to establish a French satellite in the New World. But Charles Francis Adams could be certain of none of that. He reported from London that British strategists exulted at the thought of their upstart rival splitting in two; that the gentry toasted the wreck of democracy in America; that businessmen fumed at the Union's blockade and tariff; that workers went hungry.

What is more, King Cotton was by no means all puff and no fiber in 1862. Nearly one-fifth of England's families lived off textiles one way or another, and the depletion of inventories from the 1860 bumper crop hurled Lancashire and Liverpool into depression. The *Economist* estimated that the Civil War made paupers of 35,000 heads of households, and increased tenfold the number of people on the dole. The stock exchange tumbled. Even more worrisome to the City of London, Britons had invested £50 million in the United States. If the price of a Union victory was American bankruptcy or payment of debts in despised greenbacks, then Britain would be far better off forcing a brokered peace. Over the summer, members of Parliament introduced a rash of resolutions to that effect. The Scottish journalist Charles Mackay filled the *Times* of London with plaudits to the gallant Confederates, which John Thadeus Delane, its peer-smitten editor, was eager to print. "The North and South must now choose between separation and ruin," he editorialized. Meanwhile, British shipyards built commerce raiders for the Confederate agent James D. Bulloch (a maternal uncle of Theodore Roosevelt). All this caused the Confederate envoy Slidell to be "more hopeful than I have been at any moment since my arrival." By late summer 1862 Palmerston ordered the British foreign office to arrange with the French for a joint diplomatic intervention if the tide of war ran in the Confederates' favor.

Three counterweights kept British opinion somewhat in balance even during those months of maximum southern influence. First, there was reason to hope that the textile industry's hard times soon would improve. Before the war started, British firms began to hedge their dependence on U.S. cotton by planting crops in Egypt and India where "free" workers, who were paid only a few pennies per day, proved far cheaper than slaves. After 1861 they sharply increased their acreage, beginning a geoeconomic upheaval that outlasted the war.[59] Second, it turned out the North had a "king" of its own: King Corn (as the British called wheat). Urbanization, the Irish famine, and free trade had made Britain vitally dependent on imports of food, one-fourth of which came from the United States. Smart Britons realized they needed close ties to the North more than the South. The third counterweight seems silly by comparison with cotton and grain, but it should not be discounted. Seward, hoping to match the Confederates' propaganda, dispatched to Europe the most skilled and experienced political operatives from his native New York: Archbishop John Hughes and Thurlow Weed. Adams thought private emissaries "of no value"; but in fact, they were free to speak off the record and were known to be Seward's confidants. Hughes played the Catholic card at Napoléon's court. Weed became a familiar guest in the country houses of lords, in the pulpits of evangelical churches, and in the back rooms of newspaper offices. Everywhere he peppered his benign case for the Union with veiled warnings about the danger of testing the "temper of Congress and our People." Still, one thing was lacking: an unmistakable commitment to the same abolitionist cause that had conquered the British public three decades before. Lincoln knew that. He was just waiting, like Palmerston, for a sign to appear on a battlefield, perhaps in some ripening cornfield near the Potomac.[60]

After the Second Bull Run, the Army of Northern Virginia disappeared into the Shenandoah Valley. Its 50,000 soldiers were tired and ravenous; some were even shoeless. Their commander was in worse shape. When his beloved horse Traveller shied, Lee injured both wrists breaking the fall. So Lee rode to Maryland in an undignified ambulance, trusting in Stuart's cavalry screen and the enemy's inertia (McClellan was back in charge) to escape detection. Lee's plan was for Jackson's corps to bag the federal garrison at Harpers Ferry and then reunite with him around Hagerstown. McClellan was slow to pursue, for good reason. If he marched into a trap or let Lee slip behind him and capture Washington, the war could be lost in an afternoon. But neither had he been idle. In just a few weeks "little Mac" restored the army's morale and led it on a risky retrograde march in hopes of forcing Lee to withdraw—assuming that Lee could be found.

Then chance, fate, luck, or Providence intervened—or so say the breathless accounts. At a former rebel campsite, Union soldiers found three cigars wrapped in a paper marked "Special Orders No. 191." These orders outlined Lee's strategy for the campaign. McClellan authenticated the handwriting as that of a Confederate staff officer he knew from West Point. He wired to Lincoln, promising victory. But sixteen hours spent poring over the orders, the map, and his other intelligence data suggested no plan. How large was the rebel army? The document did not say. When would the rebels rendezvous? The document indicated that Harpers Ferry should already have fallen, but McClellan knew it had not. Had the captured orders been superseded? There was no way to know. At last he ordered assaults in the gaps of South Mountain, Maryland, in hopes of locating Lee. But D. H. Hill's division put up such resistance that the Confederates gained a day, during which Jackson's men bagged the supplies and 12,000 soldiers at Harpers Ferry and marched to join Lee on Antietam Creek near a village called Sharpsburg. At last, on the misty morning of September 17, the armies squared off. McClellan's conception

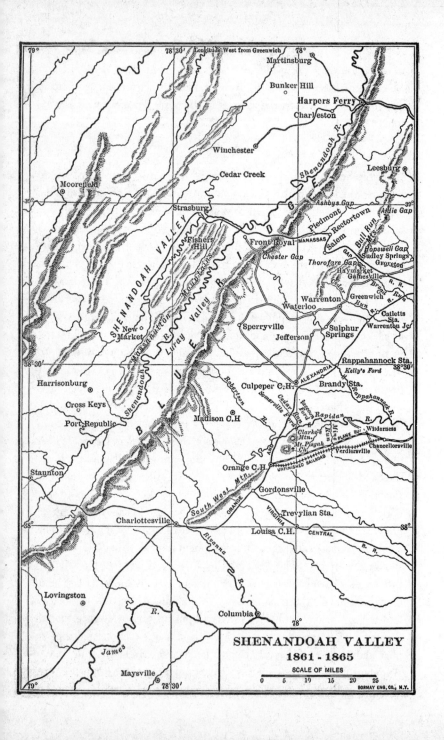

SHENANDOAH VALLEY
1861 - 1865

SCALE OF MILES

0   5   10   15   20   25

BORMAY ENG. CO., N.Y.

was sound. One Union corps was to cross the creek and threaten Lee's right while three corps overwhelmed his left flank. McClellan's lieutenants let him down. Instead of wading his men though the shallows, Ambrose Burnside launched wasting assaults on the bridge that still bears his name. That gave Lee precious hours to peel off the men needed to buttress his left and center. In those sectors Joseph Hooker and the other division commanders made poorly timed, misdirected attacks that piled up corpses at Dunker Church, the Cornfield, and Bloody Lane. A New Yorker remembered "a savage continual thunder that cannot compare to any sound I ever heard."

Sheer numbers still verged on breaking the rebel lines until late afternoon, when A. P. Hill arrived with the last contingent from Harpers Ferry to blunt the last Union assault. The bloodiest day in American history anticlimaxed despite 12,400 Union and 10,700 Confederate casualties. A slave woman asked a soldier in gray if he had a hard fight: "Yes, Aunty, the Yankees gave us the devil, and they'll give us hell next." But they didn't. The next day, the traumatized armies just sat, tortured by the moans of the wounded and the stench of the dead. Lee drew off, and McClellan, who wanted no more of such carnage, did not pursue. Lincoln wanted no more of McClellan. Seven weeks later Little Mac was packed off to recruit troops in lukewarm New Jersey.[61]

Lincoln chose to interpret Antietam's hell as heaven's will. Believing that "God has decided this question in favor of the slaves," he issued the preliminary Emancipation Proclamation on September 22. As of January 1 all enslaved persons in states rebelling against the United States "shall be then, thenceforward and forever free." For the moment the bombshell was a dud, since it would liberate no one in the loyal slave states nor anyone in the Confederacy not already occupied by Union troops. The real explosion erupted over Lincoln's apparent arrogation of Cromwellian war powers: "I think the constitution invests its commander-in-chief, with the law of war, in time of war. . . . Is there—had there ever been—any question that by the law of war, property, both of enemies and friends, may be taken when needed?" Everyone's predictions about the initial reactions proved out. Britain's Tory press damned the proclamation as an incitement to racial war, and the Liberal press mocked its empty words. Among British voters, however, Adams reported a groundswell of support. Palmerston persuaded the cabinet to remain "lookers-on" and shelve plans to mediate the American war.

Back home, the antiwar Democrats, called Copperheads, spat venom at the White House. Their Pennsylvania convention resolved that America was a government of white men; their campaign slogan in Ohio endorsed "the Constitution as it is, the Union as it was, and the Niggers where they are."

Lincoln expected this. It mattered more that he cemented his base. Thus did Charles Sumner exclaim, "The Administration belongs to us now, and we belong to the Administration." The *New York Herald* described Lincoln as "a man of fixed principles" in the mold of Old Hickory. The *New York Times* and *Post* used words like wise, necessary, and glorious; the *New York Tribune* spoke of "the beginning of the end of the rebellion; the beginning of new life for the nation"; the *Chicago Tribune* hailed "the grandest proclamation ever issued by man." Perhaps the shrewdest observation was made by a foreign correspondent, Karl Marx. "President Lincoln never ventures a step forward before the tide of circumstances and the call of general public opinion forbids further delay. But once 'Old Abe' has convinced himself that such a turning point has been reached, he then surprises friend and foe alike by a sudden operation executed as noiselessly as possible." Marx's insight was quickly borne out when Lincoln noiselessly reserved the right to suspend habeas corpus wherever evidence of treasonous activity warranted. To the Copperheads it was the act of a dictator; to Lincoln it was just "constitutional common sense" applied to an unprecedented constitutional crisis.[62]

Still, Republicans prayed for some miracle to reverse their slide before the elections. They got one in Kentucky, where for once the Confederate generals proved the worse bunglers. It took Braxton Bragg nearly a month to shift his army to Chattanooga because the siege of Corinth forced him to use railroads running south to Mobile and then northeast through Atlanta. His 34,000 men did not begin marching north until late July to join Kirby Smith's 21,000 men crossing the mountains from Knoxville. But how could their bold invasion succeed against Buell's 55,000 troops and an equal number of new recruits drilling at Cincinnati and Louisville? By raising Kentucky! Indeed, the rebel armies never merged, because Bragg insisted on making a pompous detour to install a "liberated" state government in Frankfort. Nor did the rebel armies swell, because although Kentucky's farm boys waved their hats as gray columns passed, they hid in their barns when rebel recruiters turned up. The irascible Bragg cussed them out. The more thoughtful Smith judged, "Their hearts are evidently with us, but their blue-grass and fat cattle are against us." When Buell and Bragg stumbled into each other at Perryville on October 8, the slaughter yielded naught but another draw and an unmolested retreat by the rebels. Lincoln sacked the dithering Buell in hopes that William Rosecrans, who was a Democrat and a Catholic, would show more spunk. But at least the border states were secure.[63]

The Democrats spun the elections of November 1862 as a rebuke of emancipation. In fact, the Republicans emerged with all but two governorships and all but three legislatures, and gained five Senate seats. Moreover, most of the thirty-four House seats they lost went to War Democrats, not Copperheads. Just as important—perhaps more so—Lincoln got to replace three Jacksonian justices on the Supreme Court in 1862 with loyal Republicans, and he expected the elderly Chief Justice Taney had not long to live. All this was enough to embolden the scrupulous Lincoln. In December he used his annual message to transcend legalities and claim at last the high priestly mantle of the civic religion. "The dogmas of the quiet past, are inadequate to the stormy present. . . . We must disenthrall our selves, and then we shall save our country. We, even we here, hold the power and bear the responsibility. In giving freedom to the slave we assure freedom to the free. . . . We shall nobly save or meanly lose the last, best hope of earth. Other means may succeed; this could not fail. The way is plain, peaceful, generous, just—a way which if followed the world will forever applaud and God must forever bless."[64]

The year ended badly for the Union cause. Burnside (whose name became synonymous with muttonchop whiskers) took charge of the Army of the Potomac too eager to prove he was the anti-McClellan, which is to say, a mover and brawler. He marched 122,000 men to the Rappahannock near Fredericksburg, where Lee expected to winter in peace, then decided the way to trick Lee was to try no tricks at all. Burnside instructed the Union's excellent engineers to throw six bridges across the river. He then ordered regiment after regiment to charge across a broad killing field below Longstreet's entrenchments on Marye's Heights, or else assault an impregnable stone wall and a sunken road defended by Jackson's corps. This was the first Civil War battle to offer a chilling anticipation of World War I. Burnside squandered 13,000 young men for the privilege of maiming or killing just 5,000 rebels. Meanwhile, Rosecrans lost a third of his army assaulting the Confederates' winter quarters at Murfreesboro,

Tennessee, managing only to nudge Bragg twenty miles south. On the Mississippi, even Grant and Sherman lost momentum when rebel cavalry raids in their rear frustrated their efforts to lay siege to Vicksburg. The war news deepened a pall which political intrigue cast over Washington that gloomy December. Apparently, the secretary of the treasury was leaking stories on Capitol Hill that questioned the secretary of state's devotion to the Union cause. In truth, Chase trumped up the rumors to promote his presidential ambitions. But to Radical Republicans, Chase was a reliable zealot and Seward a fox, so a Senate delegation banged on the White House door. Lincoln consulted his mother wit. The next day he summoned the senators to a meeting of the cabinet minus Seward, then coolly declared that the administration always sought a consensus. Wasn't that so? Chase blanched, but had to agree, leaving the senators scratching their heads. They had expected Seward to resign, but now it seemed that Chase must step down. In the event, neither one did, because Lincoln told both the country needed their services. In the wake of the tiff, however, the cabinet ceased to function as a collaborative body. In that way, too, the war made the presidency a mightier and lonelier office.

Despite the month's disappointments, Lincoln not only signed the Emancipation Proclamation on New Year's Day but also instructed the War Department to mobilize Negro regiments, the first of 188,000 African-Americans to don Union blue. General Halleck, with masterful understatement, observed, "The character of the war has very much changed within the last year." It had indeed, not because Lincoln declared total war, but because he now understood that a "people's war" between social systems was already a fact and could be won on no other basis.[65]

A people's war was the last thing Jefferson Davis could afford. His white people were much fewer in number, his economy was much less able to stand profiteering and waste, and his anachronistic cause was less likely to arouse zeal. The "Southrons" pretended to be staging a revolutionary sequel to 1776, while in fact they were attempting to make a swift counter-revolutionary fait accompli. By the time the secessionists realized what kind of war had to be fought, it was already too late. Not until 1863 did the congress of the Confederacy attempt to take measures that the better-endowed United States had already taken. In March it approved the monthly issue of $50 million in treasury notes, which recipients could use to purchase thirty-year bonds paying 6 percent. But inflation and loss of faith were already so far advanced that a mere $21 million would be sold. A draconian tax bill passed in April fared even worse. It obliged all producers to give the government 8 percent of their produce and to pay 10 percent on all profits from sales of produce, plus a graduated income tax topping out at 15 percent. Planters, not to say small farmers and tradespeople, hated all that, not least because the tax-in-kind (TIK) agents plundered the countryside like publicans in the Bible, and urban extortionists and smugglers either ducked taxes or bribed officials. All told, tax revenue covered just 5 percent of the budget, and bonds covered just 30 percent. Thus the Confederacy paid roughly two-thirds of its bills by printing over $1 billion in treasury notes. By late 1864 the inevitable hyperinflation drove the purchasing power of a Confederate dollar below a penny. Rich and poor alike went broke from the disruption of trade, ravages of war, absence of males, and laws that proscribed the planting of cash crops in favor of foodstuffs. (Planters cheated, needless to say, and as many as could slipped tobacco and cotton through Union lines.)

Transport was the ultimate bottleneck. Even when food was available soldiers and cities went hungry because the South's 113 jealous and inefficient railroads ran in the wrong direction (i.e., to the ports), could not replace worn-out rolling stock, and resisted the war department's feeble efforts at centralized management. Great quantities of corn and meat piled up at depots, where they fed only rats, insects, and mold cultures. Even horses and mules grew scarce when the Union invasions occupied or isolated the breeders of Kentucky, Tennessee, and Texas. Anyway, southern quartermasters were hard put to it to supply the 5,300-odd bushels of fodder the army's livestock ate every day in 1862. The following year some 20,000 Confederate horses died of hunger, fatigue, and battle. That was as telling a turning point as any. Shortages of leather, lead, zinc, coal, and other commodities also crippled the war economy. Even salt disappeared for all who could not afford extortionist prices. Citizens of the Confederacy hated their own speculators as much as they hated the Yankees, but the regime in Richmond was powerless to police price controls. Nobody suffered more than town dwellers. In spring 1863, crowds rioted and looted from Mobile, where they cried "bread or blood"; to Richmond, where Jeff Davis personally faced down angry women demanding "bread or peace."[66]

What about the Union blockade? Was it not the garrote strangling the Confederate war effort? Yes and no. By the end of 1862 the U.S. Navy screened all southern ports, but it would not capture another major port until August 1864 (at Mobile). Credit the Confederates' secretary of the navy, Stephen Mallory, who realized after the debacle at New Orleans that coastal defense was his highest priority. This allowed intrepid blockade-runners to slip out of Bermuda and Nassau, where the Union Jack gave them sanctuary, and into southern anchorages under cover of night or fog. One study estimates the Union stopped one in ten blockade-runners in 1861, one in eight in 1862, one in four in 1863, and one in three in 1864. Another study estimates that there were 1,000 successful voyages compared with just 300 captures. Suffice it to say that 330,000 European small arms found their way to the South, and the Confederate army never ran out of bullets and shells. That said, blockade-runners could have been more of an asset if not for their eye on the main chance. Stern government regulations to the contrary, many "Rhett Butler" types stuffed their holds with low-bulk, high-profit luxuries ranging from corsets and lace to ice and furniture lest southern gentlewomen go without. But the truly devastating effect of the blockade was on exports. In 1863 Davis repented of the cotton embargo, negotiated a $6 million loan in France, and asked the industrialist Colin J. McRae of Alabama to impose a modicum of control over the clandestine business. It was too little, too late. All told, the Confederacy lost some $700 million in revenue from lost exports. The overall reckoning? At the stroke of secession the South had a booming economy and no army. By 1864 it had an army and no economy, unless one counts smuggling, tax evasion, speculation, hoarding, trading with the enemy, scrounging, and subsistence farming.[67]

Almost as demoralizing to southerners' morale was conscription. Even at its peak the Confederate army's effective strength (not including the sick, the wounded, and deserters) never exceeded 40 percent of the Union's. So the compulsory service law was amended in September 1862 (just after Antietam) to collar every male of age eighteen to forty-five, and again in February 1864 to apply to males ages seventeen through fifty. Southern manhood could abide privation and the risk of death. It could not abide the law's exemptions for mechanics, professionals, skilled workers, teachers, state officials, overseers of twenty slaves, and anyone able to afford a $500 commutation fee. The exemptions fostered such widespread abuse that yeomen called the conflict a rich man's war and a poor man's fight. The congress of the Confederacy listened, but only tinkered with the exemptions. The stricken economy was already short of railroad workers, factory hands, and managers; white women could hardly be left alone with their slaves. Instead, the conscription bureau dispatched ruthless impressment gangs to scour the boondocks for white males to turn into cannon fodder. At that the Confederacy excelled. Nearly three-quarters of military-age males wore the uniform at some point.[68]

Everyone quotes William E. Gladstone's famous address in which he expressed "no doubt that Jefferson Davis and other leaders of the South have made an army; they are making, it appears, a navy; and they have made what is more than either, they have made a nation." Forget it. Those were the remarks of a British Liberal politician pandering to textile workers with words he hoped were not true. But were they? Some historians are dubious that a powerful nationalism can spring into being overnight, or else judge Confederate nationalism a failure because of its defeat in war. Such doubts, however, ignore the work of decades of sectional strife that prepared secessionists to think of themselves as a separate nation, as well as the separate identity nurtured by whites in the former Confederacy for a century *after* 1865. They also fail to consider that the Confederacy's fight to the bitter end in spite of all the problems noted above makes a strong prima facie case for its nationalism. Southern leaders certainly thought of themselves as fighting for a republican vision decidedly different from that of the North. Indeed, they imagined secession as nothing less than a rebellion against politics and political parties altogether. To some degree that was born of necessity. From the instant Lincoln called for volunteers, the Confederates fostered a fortress mentality. But Davis's cabinet, the Confederacy's congressional leaders, and most of the Confederate state governments already embraced the notion of a single-willed nation rejecting the partisanship, patronage, and corruption of the 1850s. Preachers called Confederates the "righteous remnant." Teachers rewrote history so that its march of liberty culminated in 1861, not 1776. Davis and Stephens ran unopposed, spoke for the nation rather than a party, and trusted in Providence because "Liberty is always won where there exists the unconquerable will to be free." For a while, democracy without discord seemed a done deal because the Confederate congress was a rubber stamp. It never did get around to setting up a supreme court.

Jefferson Davis, ever the dignified man of duty, knew what had to be done. Often disparaged as an ideologue, a poor judge of men, and a disappointing commander in chief (given his military background), Davis actually displayed remarkable energy, imagination, flexibility, and doggedness under the most trying conditions. And he did it all despite constant pain (he had facial neuralgia), despite blaming himself for every setback, despite hurtful betrayals, and despite the grief he and Varina shared when their five-year-old boy crushed his skull in a fall from the balcony. Davis made his share of mistakes, especially in his handling of generals; but then, so did Lincoln. If a mountain of evidence suggests that Davis was at fault for the Confederacy's defeat, another mountain credits him with "holding the fort" for four terrible years. After all, it was he, not Lincoln, who first anticipated a total war and tried to mobilize for it. It was he, not Lincoln, who first plumbed the war's meaning in mystical, biblical language. Indeed, the principal character flaw of Jefferson Davis might well have been his unshakable faith that he was doing (in the words of his Episcopal prayer book) "all such good works as Thou hast prepared for us to walk in." Davis declared ten days of fasting and prayer, compared with just three days in the North. Davis likened all doubters to Israelites pining for the fleshpots of Egypt. Lincoln, by contrast, was never sure where God stood. In the end, Lincoln won, only to die. Davis lost, only to live with that bitter fact. His portraits are simply beautiful.[69]

Davis enjoyed a long honeymoon during which hatred for the Yankee invaders hardened Confederate nationalism. Only dyed-in-the-wool advocates of states' rights or frustrated office seekers questioned the administration in Richmond. By 1863, however, measures accepted as a necessary evil just appeared evil. Davis, like Lincoln, was a tyrant sequestering private property, taxing and regulating, suspending civil liberties, and—worst of all—failing to win the war. His cabinet was always churning and with few exceptions remained undistinguished. The best of the lot was Judah P. Benjamin, a former senator from Louisiana, but his talents were partly wasted, his good ideas were mostly scotched, and his reputation was sullied by slurs, some anti-Semitic.[70] All this is not to say that the "rebellion against politics" miscarried. A formal two-party system emerged nowhere except in North Carolina. But twenty-six congressmen formed an antiadministration faction, and their number increased to forty-one (out of 106) in the elections of fall 1863. Since numerous districts were under Union control and local causes of discontent varied wildly, that sole wartime Confederate canvass was "a crazy quilt of idiosyncratic, almost apolitical contests conducted before a largely apathetic though sometimes angry electorate." Davis took off on a strenuous cheerleading tour and was gratified when the incessantly critical *Richmond Examiner* bade an "enthusiastic welcome home to our second Washington." But words could not dispel the tension between the desire for liberty and demand for unity. Voters insisted Richmond do *something* about *everything*, but resented in like measure its actions and its failures to act. This suggests that the war exposed a deep contradiction in Confederate nationalism: not between states' rights and a centralized war machine, but between the elite's conservatism and its pretense of rebellion. The contradiction was evident in the refusal of border states to espouse the cause, the choice made by as much as a tenth of southern men to enlist in the Union Army, and the way the Confederacy's war effort slowly but steadily undermined the very institutions and values that secession was supposed to preserve, not least slavery itself.[71]

So why did Johnny Reb suffer and risk life and limb? Historians who are also war veterans know that any answer is guesswork. Young men cannot sort out even their own motivations, and when pressed to say why they enlisted are likely to make something up, if only to ward off interrogation. "All's fair in love and war" says it best, because only in courtship do men pretend and play-act more than in war. That observation extends to the letters written by soldiers during the Civil War. How can one tell whether a youngster's expressions of patriotism and faith were sincere or just crafted to reassure Ma and Sis? The cynic might be inclined to give credence to letters expressing privation, despair, and revulsion, while discounting lofty, heroic sentiments. But soldiers are also skilled at fishing for sympathy, so even their expressions of woe cannot always be trusted. Consider also that soldiers rarely record thoughts in real time. Their letters and journals are composed during respites when memory becomes a toy. Consider, too, that war is like the elephant groped by the blind men. The experiences of an infantryman in Virginia differed markedly from those of a cavalryman in Tennessee, an artilleryman in a Savannah shore battery, and a supply sergeant comfortably posted in Raleigh. It is easy enough to sift letters, make a few generalizations, and choose poignant examples to dramatize them. But those examples may reflect only what historians or their audiences want to hear. The only sure wisdom resides in platitudes. War usually is 95 percent boredom and 5 percent terror. War usually does bring out the best and worst in people. The fear felt by soldiers in their first combat often turns into flight in the absence of leadership. The fear felt by veteran leaders often turns into lust to destroy the source of their fear, something onlookers deem to be courage. War is a psychic arena in which men make rolling revaluations of their own bodies and reputations, duty to comrades and units, devotion to families and homes, and commitment to country and cause. The Civil War was especially cruel for teenagers, who should have been sneaking off to a fishing hole or ball field after their chores, but instead pointed muskets at other American boys in places named Cheat Mountain, Savage Station, and Cold Harbor.

In the first flush of excitement, many a Johnny Reb doubtless felt called to the colors, or else draped in patriotism his surrender to community pressure and the desire to earn the respect of fellow boys, win the admiration of girls, make parents proud, and perhaps flee parents and farm in search of adventure. As ugly reality confronted him, more visceral motives emerged: *fear* (of that awful unknown "abolition"), *hatred* (for those who would shove it down southerners' throats), and *love* (of one's land, one's liberty, and commanders like Lee). Unit cohesion was unusually strong in the Civil War because so many regiments were raised in a single locale. A soldier's cowardice would follow him home. Not least, the Scots-Irish "cracker" or pioneer culture dominant in the southern backcountry affected Johnny Reb's fierce, personal style of war. In the broadest sense the rebel yell was a distant echo of the Celtic berserkers who charged Roman legions and English squares. In the narrowest sense it voiced the simple reply a hillbilly gave to his captors when asked why he fought: "Ah fit 'cause y'all are down heah." It was enough to sustain Confederate soldiers during three years of marching; fighting; foraging for anything other than stale cornbread; mending frayed uniforms; picking at lice and fleas; vomiting at the sight of hogs gorging on amputated limbs; and burning or dying from dysentery, malaria, typhoid, smallpox, measles, mumps, gangrene, and exposure. Stoic routine was broken by song, drink, gambling, other things one didn't write home about, and coffee whenever the Yanks would send some over the line in exchange for tobacco. By and by, even self-defense lost its power to motivate. The Yankees were seemingly "down heah" to stay and nobody wanted to be the last man to die in a lost cause.

Many rebels' devotion to family and home waxed as their commitment to country and outfit waned. Most of the 100,000 deserters went over the hill in 1864–1865. Like the slaves who just walked away after the white men rode off to war, deserters just walked away when the white women beckoned them home.[72]

Indeed, "It may well have been because of its women that the South lost the Civil War." Perhaps that's too melodramatic, but much evidence suggests that women were a reliable barometer of morale. In 1861 most bonneted belles cheered their departing knights as if the war were a medieval tournament. Jefferson Davis made the pretense official by locating the heart and soul of Confederate patriotism in its females: "To the women no appeal is necessary. They are like the Spartan mothers of old." For a couple of years many were. Southern women helped recruit volunteers, nursed and clerked, sewed and quilted, joined civic organizations, oversaw plantations or plowed fields themselves, juggled household budgets with the volatile currency, and thought it a virtue to do without. In spring 1862 Henry Ward Beecher remarked icily on "the strange part that has been played in this conflict by Southern women. . . . Women are the best and the worst things that God ever made! And they have been true to their nature in this conflict. Southern men have been tame and cool in comparison to the fury of Southern women."

By that autumn death, privation, worry, and work began to wear on the proudest of hearts. A woman in South Carolina wept to recall how "poor country women with babes in their arms" gathered in silence at the Spartanburg depot while the latest roll of the dead was recited. Every shriek meant another new widow. Southern towns were "thinned out of men," obliging the women to do all the work of men despite daily anxiety about restless slaves in plantation counties and "blue devils" in the spreading war zones. In Shelby County, Alabama, just 200 of 1,800 adult males remained at home. In Tennessee, a former editor of the *Southern Ladies' Book* complained, "My work is never done" as she struggled to keep her children fed and safe. "I fear the blacks more than I do the Yankees," confessed a woman in Vicksburg. "I am afraid of the lawless Yankee soldiers, but that is nothing to my fear of the negroes," wrote a Virginian. In the Mississippi Valley war made refugees of thousands of women, often with old folks and children in tow. Some turned beggar, prostitute, collaborator. Even quiescent regions suffered when desperate Confederate foragers cleaned out their larders. Governor Zebulon B.

Vance of North Carolina heard women's plaints. If God Almighty had another Egyptian plague in reserve, he fumed, "I am sure it must have been a regiment or so of half-armed, half-disciplined Confederate Cavalry." Yet women who cursed their men's absence cursed the presence of skulkers and profiteers even more.

By 1863–1864 southern women even began to let go of the Rock of Ages. Their stay-at-home pastors assured them that initial victories were signs of God's blessing on the Confederate cause, and intermediate defeats a summons to penitence as the price of victory. They had more trouble explaining climactic setbacks except by doubting whose side the angels were on. Chapels emptied as women prayed and ministered to each other in parlors and farmhouses. Those were the sort of women who wrote obituaries for sentimentality: "War has hardened us"; "My heart became flint"; "I cannot weep, I dare not pray." Cascading fear, loneliness, helplessness, rage, and resentment left them with the feeling that their men had deserted them for a foolish war, stripped them of the "separate sphere," then failed to live up to their part of the paternalist bargain, which was to defend the womenfolk from famine, fire, and foe. In 1864 a gritty Georgian charged that the men who abandoned Atlanta "have brought an everlasting stain on their name."[73] The longer the war raged, the more southern women concluded their mournful letters with pleas that husband, son, brother, or uncle come home. Imagine the angst such letters induced in soldiers whose nerves were already stretched to the limit. Imagine, too, how the cry of a damsel in distress could make the capital crime of desertion seem the more honorable calling.[74] Still, one must take care not to confuse cause and effect. The gray legions might have quit at the last because their home front collapsed, but the home front collapsed because the legions' endurance and courage could not repel invaders whose will to nationhood equaled their own and whose resources and management skills far exceeded their own.

The dirty little secret in Massachusetts was that the war of northern aggression damned by Confederates was just that. For years public intellectuals such as Emerson, Mann, Sumner, and Theodore Parker had been redefining American nationality in ways that simply excluded the slaveholding South. The portraits of Dixie by Olmsted, Bryant, and Hinton

Helper gave them the data to argue that Yankees were the true heirs of the founders and the southerners were bastard children. They imagined the Republican Party a switch to whip the South into conformity. They exulted when secession brought civil war. Emerson sensed "a sentiment higher than logic, wide as light, strong as gravity" infusing colleges, churches, banks, and farms in New England. "We are wafted into a revolution . . . an opportunity to bring about the moral revolution in American life, that final fulfillment of the covenant that Americans had made in undertaking the political revolution of the eighteenth century." At the outset most Republicans had no such intention. (When a delegation of preachers from Chicago told Lincoln he must abolish slavery, the president asked why the Lord's only channel to him was via "the roundabout route of that wicked city.") But the exigencies of war soon delivered the Republican Party into the hands of armchair hawks for whom "the Union would exist on northern terms, or not at all."

It is customary to speak of the Civil War as cementing American nationhood. Thus Lincoln referred to the Union twenty times in his first inaugural address, but to the nation five times in his Gettysburg address. Thus "United States" turned into a singular noun. Thus soldiers enlisted under the banners of states, but were mustered out of the Grand Army of the Republic. Thus the war and the Republican revolution vastly expanded the impact of federal power on individuals. But to speak of the Civil War as having cemented American nationhood is fallacious. In truth, Yankee men and women engineered the wartime transformations for a bewildering, intertwined mix of motives: to save the Union, to realize its manifest destiny, to preserve the American dream for their children, to prove to the world that self-government worked, to do God's will, to build heaven on earth, and if possible to win glory and wealth in the process. Most northerners also imagined that they fought in defense of their own liberty, Constitution, and holy flag. The rebel, after all, had seceded and then fired on Fort Sumter. The rebel was a traitor—"The Man without a Country" described in the Boston Unitarian Edward Everett Hale's stirring war propaganda. But a minority of influential Yankees considered this an offensive war, a righteous crusade not only to abolish slavery, but to abolish the South until, in the words of Wendell Phillips, "the ideas of Massachusetts kiss the gulf of Mexico."[75]

The Civil War started in Massachusetts, when Governor John A.

Andrew put the state on a war footing in January 1861—before the Confederacy was created. That explains why its two regiments responded like minutemen the day Lincoln called for volunteers. But Andrew did something far more important to the Union war effort. He enlisted Boston's leading bankers, manufacturers, and merchants to provide money, supplies, and their own expertise. The rest of New England, and then the whole North, followed Massachusetts in war as they so often did in law, education, and religion. New York, as always, was split because Manhattan's immigrants and financiers hated Negroes, market disruption, or both—in the fall of 1862 they would elect as governor the notorious copperhead Democrat Horatio Seymour. But for the initial two years of the war, New York's Republican governor, Edwin D. Morgan, threw the richest, most populous state into the lists of the Union. Morgan's bounty system helped to recruit 500,000 volunteers, the most of any state, and his elite managerial team was the first to crack down on "shoddy" procurement. Union League clubs sprang up in New York City, Philadelphia, and other hubs of business to raise private money and regiments for the cause.

Midwestern governors met resistance from antiwar "butternuts" in their southern counties, none more so than Oliver Perry Morton of Indiana. When the legislature became Democratic in 1863, Morton set up a private treasury supported by loans and ruled the state by executive decree. A grateful Lincoln sustained him with Union troops who all but occupied Indiana. Beyond the Mississippi River, frontier states contributed ardor. Iowa's 80,000 volunteers ranked near the top as a percentage of their state population. But there is something more to be said. Although New England, New York, and Pennsylvania were the Union's demographic and industrial core, they led a victorious war effort only because of the manpower, foodstuffs, railroads, horses, geography, and loyalty of the upper Midwest. Indeed, the nation's center of gravity shifted so sharply in the Civil War era that twelve of the eighteen presidential elections from 1860 to 1928 would be won by midwesterners.

The War Department began recruiting civilian experts even while Cameron was at its helm, but the relentless Stanton made the practice universal. He centralized pyramidal bureaucracies to oversee procurement, supply, and medicine, and appointed a War Board that functioned like the Prussian general staff. His civilian Ordnance Commission kept watch on the $1 billion spent by officers of the quartermaster, Montgomery Meigs, in

New York, Philadelphia, Cincinnati, Indianapolis, Detroit, Chicago, Saint Louis, and San Francisco. The North's industrialists and workers were ready to meet their needs because the panics of 1857 and early 1861 left many factories and workers idle. Over the course of the war Colt and other private manufacturers produced 650,000 fifty-eight-caliber rifled muskets to augment the 800,000 turned out by the Springfield Arsenal at a cost of just twelve dollars per weapon. Mass-produced artillery tubes made Robert P. Parrott's West Point Iron and Cannon Factory the biggest contractor. Scores of public and private shops molded 750 million rounds of small-arms ammunition. The du Pont chemical plants, heavily guarded against Confederate saboteurs, supplied gunpowder. Textile mills tripled their production of spun wool because of the cotton shortage. New England's shoe factories poured forth boots and belts. Forests in the Great Lakes region were felled at a frantic pace to fill the tenders of Union railroads.

Food was the least of problems. Northerners ate better during the war than ever before, and their soldiers rarely went hungry (albeit hardtack, salt pork, and desiccated vegetable chew, an experiment to ward off scurvy, were hardly palatable). Midwestern soil proved so fertile, farm implements so efficient, and foreign markets so insatiable that northern harvests increased despite the shortage of men. Travelers on the prairie invariably remarked on the women in long dresses and bonnets urging a horse from the seat of a McCormick seeder or reaper. Indeed, agriculturalists responded to wartime price increases so energetically that after feeding their own army and people they exported 200 million bushels of corn and wheat, 500 million pounds of pork, 400 million pounds of lard, and 380 million pounds of butter and cheese, while compensating for lost southern produce by sharply increasing production of fruits, cotton, tobacco, flax, and sugarcane. Illinois alone increased its cotton crop from 6,700 pounds in 1862 to nearly 1.6 million pounds in 1865. Only livestock herds shrank slightly, because the war consumed so many animals.

Bustling citizens moved by patriotism and profit joined with federal managers moved by duty and pride to make the war machine that drove old Dixie down like a steam-powered pile driver. Nothing illustrated this better than railroads. Recall Daniel McCallum of the Erie railroad, who helped make a revolution in corporate management. Stanton named him the transport czar in February 1862 and then prevailed on Lincoln to seize control of all American railroads in May. McCallum hired the Pennsylvania Railroad's president J. Edgar Thomson and its chief engineer Herman Haupt, who in turn hired an experienced staff. Much jawboning ensued with railroad executives, but in time all bowed to presidential authority in exchange for guaranteed rates. The profits of the Baltimore and Ohio railroad more than doubled, to $5.7 million, once Harpers Ferry was finally secure. So efficient were the Union's rails that Haupt shipped more supplies to the Army of the Potomac in four days in July 1863 than Confederate rails had provided to the Army of Northern Virginia in four months during the previous year. When Stanton needed to reinforce the Army of the Cumberland in September 1863 Haupt's trains transported 23,000 soldiers with horses and kit from the Potomac to Chattanooga in a little over a week. When McCallum's rolling stock was stretched thin early in 1864, the Baldwin Works and others turned out fifty-three new locomotives in three months. The "pugnaciously efficient" Haupt (who refused a salary) also managed the U.S. Military Rail Road (USMRR), which was responsible for reconstructing, maintaining, and operating captured southern lines. With Nathan Bedford Forrest's men tearing up track and blowing bridges at will, that was a Sisyphean task. But by 1864 the USMRR controlled 2,100 miles of track: the world's largest railroad. At a total cost to the government of $30 million it was the world's best bargain, too.[76]

The Union war effort as a whole seemed an improvised potlatch. But the Treasury's issue of unbacked greenbacks to the ultimate sum of $431 million was a gamble that paid off because "funny money" covered just 16.5 percent of the Union budget compared with 61.7 percent of the Confederate budget. The Bureau of Internal Revenue collected another 16.5 percent, so that the North's flush capital markets paid two-thirds of the cost of the war. That was what Secretary of the Treasury Chase intended when he tapped Jay Cooke of Philadelphia to market war bonds. Cooke's epithet, "the financier of the Civil War," suggests that he functioned as a personal lender and elite broker in the manner of Robert Morris during the Revolution or Stephen Girard in the War of 1812. But he did nothing of that sort. Cooke's national network of agents, his Barnum-like advertising techniques, and his enticing offer—6 percent bonds that could be purchased with greenbacks—tapped every American's wish to feel good about doing well. His firm sold "PATRIOTISM AND PROFIT," urged everyone with just fifty dollars to "MAKE THE U.S. GOVERNMENT YOUR SAVINGS BANK," arranged for newspapers to applaud subscribers by name, and cleared over $3.5 million in fees. Congress added a powerful fillip by requiring federally chartered banks to put a sizable chunk of their assets in war bonds. All told, northerners bought over $2 billion worth, with the result that the Union suffered just 75 to 80 percent inflation. To be sure, rising prices plus mechanization and the employment of women depressed real wages 20 percent. By 1863 a wave of strikes and new union movements hinted at what postwar America was to become. But the social costs of war in the North were as nothing compared with those in the South.[77]

That in turn made women a source of strength, not weakness, for northerners' morale. The rightly famous ones included nurses like Clara Barton, "the American Florence Nightingale"; Dorothea Dix, the "dragon" who battled the male surgeons of the U.S. Sanitary Commission on behalf of female volunteers; and the widowed Mary Ann Ball Bickerdyke, whose angelic care of soldiers brought forth an oath from Ulysses Grant: "My God, man, Mother Bickerdyke outranks everybody, even Lincoln." The famous women also included abolitionists such as Anna "Joan of Arc" Dickinson, an eloquent teenager who toured the North preaching love for the Negro; the feminists of Seneca Falls who founded the Women's National Loyal League; and Mary A. Livermore and Jane C. Hoge, whose Northwestern Soldiers' Fair converted Chicago into "a vast theater of wonders" and inspired "sanitary fairs" in almost every northern city and town. All these women were moved by a Unitarian, Quaker, or evangelical spirit and were convinced that "in this war, patriotism and holy charity are twin *sisters*." Indeed, all the antebellum reform movements revived during the war as men and especially women promoted temperance, sanitation, literacy, and the distribution of Bibles among the troops. They even raised $350,000 for stricken textile families in England: a small sum with a big payoff for public diplomacy. Still more important were the anonymous women who kept the home fires burning, wrote encouraging letters, did absent men's work, or followed the armies to cook, wash, clean, mend, and flirt for the war effort. No doubt there were tens of thousands like Emily Dickinson of Amherst, Massachusetts, a recluse whose poetry offers barely a hint that there was a war on. But even this is significant. Insofar as the war did not torment northern households it gave their women no cause to despair but instead a chance to contribute to nothing less than the kingdom of Heaven.[78]

So we arrive at the heart of it all: the mystical font of the Union war effort. Historians used to imagine Johnny Reb as more religious than Billy Yank. Others stressed how revivals swept through the ranks of both armies in the middle years of the war. Still others suggested the difference was theology rather than zeal, with rebels clinging to a more personal old-time religion and Yankees embracing a more corporate, Lincolnesque vision of God at work in America. But that was too simple. First, the Union army was a jumble of German and Irish Catholics, Protestants of all denominations and social classes, and blacks as well as whites. Second, the broad spectra of religious waves during the Civil War traced a veritable interference pattern across northern states. No generalization seemed plausible unless we surmise that faith and confusion coincided because apostles of *two religions* struggled to define the climactic event in the nation's providential career. The strugglers were mostly unwitting, since they pretended America itself was God's church and the Union cause was God's will. But a few clear-sighted observers such as the orthodox Lutheran Phillip Schaff, the "Old School" Presbyterian Charles Hodge, and the Catholic convert Orestes Brownson saw what was happening. They scoffed at the notion that sacrifices in war would make America either a Christian republic or a Republican church. They suspected that the North's mighty *civic religion* would plow sectarian faith underfoot, trample charitable impulses in a war of unchecked ferocity, and validate all manner of sin and corruption. Lincoln suspected that, too, and so he pleaded—though in vain—for civility and forgiveness even as he exhorted the North to wage total war. Mark Noll called this "the great theological puzzle of the Civil War": Lincoln clung to "a complex view of God's rule over the world and a morally nuanced picture of America's destiny," whereas the North's most eminent theologians peddled "a thin, simple view of God's providence and a morally juvenile view of the nation and its fate."[79]

Occupying the radical middle as always was Horace Bushnell, a devolved Congregationalist who equated the Christian dispensation with Manifest Destiny. Bushnell, like Emerson, welcomed the war in the belief that spilled blood would wash away all "feeble pretences of philosophy" and purge democracy of its sins: "we shall no more be a compact, or a confederation, or a composition made up by the temporary surrender of powers, but a nation—God's own nation." That was tame compared with the ideas of enthusiasts who so closely identified America with God's dispensation that they leaped to the conclusion this war must be Armageddon, the end of history. Gilbert Haven of Massachusetts, who was an army chaplain and Methodist bishop, saw the Civil War as ushering in a "world-Republic" dispensing liberty, equality, and perpetual peace to mankind. Hollis Read of the American Tract Society believed that the Civil War was calling humanity to its "final and high destiny" to serve as doorkeepers for the Second Coming of Christ.[80]

No rebel would have swallowed that. How many Yankee recruits swallowed it cannot be known unless one takes at face value every pious missive to loved ones back home. No doubt many young men brought up on McGuffey's readers, Whig values, evangelical sermons, and the era's flood tide of hymns could sing Julia Ward Howe's "Battle Hymn of the Republic" (1862) with utter sincerity:

> *I have read a fiery gospel writ in burnished rows of steel:*
> *As ye deal with my contemners, so with you my grace shall deal;*
> *Let the Hero, born of woman, crush the serpent with his heel,*
> *Since God is marching on.*

And God bless them for their courage and selflessness. But what happened when northerners discovered the war was no more a camp meeting with guns than it was the chivalrous charge southerners expected? What happened when Billy Yank could not forget how his whole unit turned tail in its first battle, how he made eye contact with that Johnny Reb before blowing his head off, how he languished in a fetid hospital terrified of infection from the stinking man dying in the next cot? What happened back home when civilians were exposed to Mathew Brady's stark photographs of the war's carnage? Doubtless many soldiers continued to pray for God's protection and assurance of some higher purpose in their ordeal. The 140,000 veterans who reenlisted for another three years in 1864 must have risked their lives for something more than Stanton's generous bonuses. "Never in a war before," wrote one private, "did the rank and file feel a more resolute earnestness for a just cause, and more invincible determination to succeed."

But an army doctor from Illinois was not alone when he said, "There is no God in war. It is merciless, cruel, vindictive, unchristian. . . . It is all the devils could wish for."

Such sentiments, combined with the usual wartime breakdown of conventional mores, posed a quandary for patriotic preachers. Most solved it, not by reassessing the morality or conduct of the war in light of Christian precepts, but rather by escalating the rhetoric of blood sacrifice, a holy cause, a demonic enemy, and a millenarian promise. All Methodists and almost all Presbyterians and Baptists eagerly blessed the sword with the cross. Archbishop Hughes flew the Stars and Stripes on his cathedral. Observance of the Sabbath collapsed ("There is no Sunday now," attested the reformer Dorothea Dix). Indeed, many clergymen blessed fighting and working for the Union as a manner of worship. Even some Quakers "opposed to cutting men down like grass" declared this "a holy war." Henry Ward Beecher set a record for tear-jerking imagery when he wrote that "a pitiful God" watched like a midwife over soldiers "in the throes and groans of the Mother" awaiting birth of a child whose names shall be "peace, concord and universal intelligence." Soldiers were not only Christlike, but Christ himself: a metaphor that for Walt Whitman never grew old (or blasphemous). One of the few dissenters was James Cruickshanks of Worcester. He challenged the mood in the North, not on constitutional, pragmatic, racist, or even humane grounds in the manner of copperheads, but on theological grounds. "In a word, the army is the people's God," he preached on a Union fast day. "They idolize it—they worship it" to the point that Americans' very religiosity was in danger of making them "the most idolatrous people on the globe." In terms of the civic religion, this made Cruickshanks a heretic. The Union had just made "In God We Trust" its national motto and emblazoned the words on its money.[81]

How to explain a righteous war nonetheless waged with spiraling violence, not only in battle but in rapacity toward civilians, plunder, atrocities, and guerrilla activity? Lincoln said that the two sides prayed to the same God, but neither embodied God's will. His conclusion was correct, but his premise was not. Civic gods in masquerade bade American souls on both sides to join cults of human sacrifice. The southern cult mandated slavery; that was why a Georgian woman could write, "I have lost my religion since the Yankees set my negroes free." The northern cult mandated the South's utter destruction; that was why Bushnell prophesied, "These United States, having dissolved the intractable matter of so many infallible theories and bones of contention in the dreadful menstruum of their blood, are to settle into a fixed unity and finally into a nearly homogeneous life."[82]

The Lincoln administration launched the first bold experiment in homogeneity when it amputated a limb from the state of Virginia.

The Thirty-Fifth State: West Virginia, 1863

No pithy description of Harpers Ferry surpasses the one Thomas Jefferson sketched: "You stand on a very high point of land. On your right comes up the Shenandoah, having ranged along the foot of the mountain an hundred miles to seek a vent. On your left approaches the Patowmac, in quest of a passage also. In the moment of their junction they rush together against the mountain, rend it asunder, and pass off to the sea." Yet "very high" Harpers Ferry was the lowest point in the counties destined to become West Virginia. Its western plateau and Kanawha valley drain into the Ohio River, on whose bluffs cling Huntington, Parkersburg, and Wheeling. In the north central hills the Monongahela River rushes past Morgantown to its rendezvous with the Allegheny River at Pittsburgh. In the east Appalachian runoff feeds the west branch of the Potomac. Everywhere spiny ranges chop the region into isolated valleys where loblolly pine bristles, bears lumber, foxes lurk, raccoons forage, opossums freeze, skunks spray, and squirrels scatter at a rifle's report. Forests covering 75 percent of the state and limy or sandy soil discourages commercial farming. Speculators such as George Washington dreamed that fortunes could be made in Virginia's northwestern counties. Instead, these areas became living museums where homesteaders preserved seventeenth-century dialects, music, and handicrafts. Long after folks elsewhere processed crops by machine, farmers in West Virginian dells still relied on scythes, smokehouses, and grist mills.[83]

Geography and arrested development made stepchildren of the ultramontane pioneers. As early as 1776 they talked of forming a separate state called Westsylvania. By 1841, when their numbers had grown, but their influence in Richmond had not, western counties threatened to found a state called Appalachia. A state convention that year expanded their representation, but ignored their pleas for internal improvements while weighting the tax system in favor of slave owners. Especially aggrieved was the town of Wheeling in the Panhandle, whose prosperity was linked to the industrial North.[84] Even the completion of the Baltimore and Ohio

Railroad brought no prosperity, because the Panic of 1857 hit the marginal mountain economy hardest of all. Such grievances later made it seem almost inevitable that the western counties would bolt from Virginia when Virginia bolted from the Union.

Unfortunately, history is messy. Yes, those counties were heavily Scots-Irish or German, Methodist, egalitarian, and innocent of slavery as compared with the tidewater's mostly English, Episcopalian planter elite. But the up-country was not so homogeneous or alienated as legend suggests. Consider the West Point graduate who had been born in Clarksburg in 1829. It would be heartwarming to think that "Stonewall" Jackson rose to fame from orphaned, barefoot boyhood. In fact, the Jacksons were part of a caste of frontier aristocrats who dominated local offices and imitated the Cavaliers. Nor did the Jacksons' fervent Methodism, imparted by Bishop Asbury himself, incline them toward abolition. According to Jackson's widow, Stonewall "died, as he was born, a Virginian." [85] Moreover, western Virginians eventually did achieve fair representation, obliging Richmond to court their favor. In the 1850s Whigs, Know-Nothings, and Democrats divided the vote in the hills. In 1860 Breckinridge outpolled Douglas in several counties. Most tellingly, although thirty-two western delegates voted nay at Virginia's secessionist convention, eleven voted aye and four abstained. Unionism was strongest among Whigs and in counties where slavery was rare, but the ratios were closer to 51 than 100 percent. Western Virginia was a border region twice over, condemned to its own civil war in the midst of the greater one. [86]

Mountain delegates who opposed secession crept home by circuitous routes, weighing their loyalties. Morgantown's Henry Dering was among the angriest. "Talk about Northern oppression, talk about our rights being stolen from us by the North—it's all stuff," he wrote, and warned of the day "when all Western Virginia will rise up." His neighbor Waitman T. Willey and Clarksburg's John S. Carlile agreed. In May 1861 they summoned a convention of Unionist Whigs to Wheeling. But no one knew how to proceed until Francis H. Pierpont, the "father of West Virginia," showed the way. Pierpont, who was then forty-seven, was a lawyer for the Baltimore and Ohio as well as an industrialist. He loved Virginia, western Virginia, the Union, and the law all at once. Separation might become necessary, but the only legal way to do it was for the parent state's government to agree to dismemberment. Accordingly, he persuaded the delegates at

Wheeling not to secede, but to declare Virginia's offices vacant and elect a loyalist state government. It proved a brilliant coup when Lincoln recognized "Governor" Pierpont and Capitol Hill welcomed "senators" Willey and Carlile.[87] Meanwhile, McClellan's troops from Ohio overcame mud and precipitous terrain to secure the Kanawha and Monongahela valleys, while local volunteers pushed east along the Baltimore and Ohio. To be sure, Stonewall Jackson reduced Harpers Ferry to rubble. But that foretaste of total war steeled the nerves of wavering Unionists. In November another convention at Wheeling met to design a new state with Virginia's blessing in the person of Pierpont.[88]

Colonel Rutherford B. Hayes of Ohio observed the cleavages among western Virginians: "The Secessionists in this region are the wealthy and educated, who do nothing openly, and the vagabonds, criminals, and ignorant barbarians of the country; while the Union men are the middle classes—the law-and-order, well-behaved folks." The latter dominated the conventions, since Virginian loyalists boycotted the statehood movement. Another distinctive feature of the convention was piety: nearly all sixty-one delegates were churchgoers, and fourteen were pastors. Still, the convention was divided between "liberals" favoring a New England–style constitution modeled on Ohio's and "conservatives" hoping to replicate old Virginia. The latter triumphed on issues of land tenure, but the former empowered townships at the expense of counties, replaced voice voting with a secret ballot, and provided for frequent elections. Almost all agreed that the new state must secure the Baltimore and Ohio railroad, and hence annex the counties that became its eastern panhandle.[89] Finally, the delegates rejected the names Kanawha, Allegheny, and Augusta in favor of West Virginia. That was descriptive and imperial, and it implied legitimacy. Congress had doubts. When Carlile introduced a bill for statehood in June 1862 some members objected to despoiling a state of 38 percent of its territory, or else feared that the bill would steel Virginians' will to fight. Radical Republicans made abolition the price of statehood and wrote it into the bill that finally passed, ninety-six to fifty-five, on New Year's eve. Lincoln hesitated, too, until he convinced himself that people who "have been true to the Union under very severe trials" must not be betrayed. The Whigs in Wheeling hiccuped, but agreed on Lincoln's birthday, 1863, to make a "gradual" end to slavery. That June West Virginia became a state: the only one besides Maine to be carved from an existing state.[90]

No one knows the true mind of western Virginians during those turbulent years. The 98 percent vote for the constitution meant little, because it was boycotted by opponents and was held amid Union troops.
More telling are the 15,000 West Virginians who fought for the South (as
against 25,000 for the North) or gave clandestine aid to Confederate bushwhackers.[91] Could the "war-born state" reconcile its populace after the
shooting stopped? It could not. The Republican-controlled legislature
stripped former rebels of their rights to vote, hold office, practice law,
teach, or plead in court. To add insult to injury, the Fifteenth Amendment
promised suffrage to Negroes. Soon mobbing by "rebel ruffians . . . yelling
and blaspheming like demons" swept over the state. Horace Greeley
warned the Republicans that they were losing their grip on West Virginia.
He was right. In 1871 the Democrats swept to power, promising whites
civil rights and segregated schools. Gerrymandering kept them in power
for twenty-six years.[92]

The obverse of West Virginia's optimistic state seal depicted a farmer
and a miner and bore the slogan *Montani Semper Liberi* ("Mountaineers
Always Free"). The reverse displayed fruit, cereal, a viaduct of the Baltimore and Ohio, and *Libertas E Fidelitate* ("Liberty from Loyalty"). As the
population rose to 442,000 by 1870, the founders of the state imagined it
becoming an industrial entrepôt. Instead, the postwar decades fixed the
state's image as a backwater. Census data revealed that West Virginians
were not only comparatively poor and illiterate but also (as they remain)
96 percent white and native-born. The national press ignored West Virginia except to report breathlessly, in the 1880s, on the feud between the
Hatfields and the McCoys over old Civil War wounds, the theft of a hog,
and a romance. Nobody grasped that the feud was the death rattle of a
whole way of life.[93] In 1880 President Hayes ordered the first crackdown
on moonshine distillers. Revenue agents invaded the Appalachians to
smash stills and establish an elaborate network of homeland surveillance.
In 1887 the state government imposed a firearms code. Far more important, two railroads—the Chesapeake and Ohio, and the Norfolk and
Western—penetrated the interior valleys to extract coal and timber. As
thousands of hillbillies mortgaged their lives to the pickax, steam drill,
sawmill, and coal car, West Virginia emerged as an industrial colony characterized by outside exploiters, outside agitators, and internal demands
for welfare and patronage. In 1899 a progressive, Senator Nathan B. Scott

of Wheeling, moaned, "I had not the remotest idea that every man, woman, and child in West Virginia wanted a government position, but I believe they do." Why not, if that was the only alternative to a lung-choking life in the mines? In 1897 Jerome Hall Raymond of the University of Chicago learned the same lesson. He was hired to make a university of the little state college in Morgantown, but the legislature deemed him "unsuited to West Virginia conditions" when he called for promotion by merit, strict academic standards, and curricular reforms. Nor did the state show any interest in the centerpiece of the progressive era, civil service reform.[94]

The state's symbolic coming-of-age was the election in 1912 of its first governor born after statehood: Henry D. Hatfield, a thirty-seven-year-old medical doctor and child of the feuding clan. The state was again a battleground because of violent strikes. For fifteen years the United Mine Workers had tried to resist "yellow dog" contracts that banned unions, imposed twelve-hour shifts, and obliged miners to sell their souls to the company store. Hatfield toured the mines and treated the injured, one of whom was the radical Mary Harris Jones. The men in the pits loved her foul-mouthed advocacy. They called her Mother Jones. She called West Virginia medieval. Hatfield settled the strike by persuading operators to stop treating their miners like serfs. But in the hard times following World War I striking miners waged such battles with the police and militia that the U.S. Army had to be summoned. Union membership dropped 50 percent while coal and rail executives toasted each other at the 6,000-acre Greenbrier resort built by the Chesapeake and Ohio in White Sulphur Springs. Not until the New Deal did miners win collective bargaining rights, regulation of working conditions, and unemployment and disability insurance.[95] After World War II the railroads' decline and the shift by electrical power plants to oil and nuclear energy depressed the third-ranking coal state all over again. Having defected from the Old South, West Virginia never found a home in the North.

Tactics involve how to win battles at minimal cost through judicious employment of the weapons, terrain, and forces at hand. Operations (campaigns) involve how to win important objectives, perhaps through a series of battles. Strategy involves how to break the enemy's will to resist (i.e., win the war) through multiple campaigns targeting his center of gravity.

All are exercises in economy—allocation of scarce resources—because no army or nation possesses unlimited resources. The challenge to commanders, especially if they are the underdogs, is to translate tactical expertise into operational breakthroughs, and operational brilliance into strategic success. Confederate commanders, try as they might, never surmounted that challenge, owing to tactical misfortune, operational nostalgia, and strategic contradiction. The misfortune was that the South undertook its war for independence in a decade when weaponry favored defense. The only way to destroy a field army was to close with it in mass infantry charges so lethal and disruptive that even the victors could not press the advantage without sizable reserves—something the rebels rarely had. Hence the southerners' élan and initiative produced tactical triumphs that did not translate into victorious campaigns. The damaging nostalgia was for Napoléon. Lee mounted deft, swift, and deceptive campaigns in hopes of a victory so shocking that the North would throw in the towel, in the manner of Europe's monarchs. But the American theater of war was too large and the enemy too democratic for glorified raids across the Mason-Dixon Line to force a strategic decision. Finally, the contradiction arose from the Confederacy's contrary imperatives: to rest on the defensive until the North said, "To hell with it," yet to attempt stunning offensives to gain the foreign assistance needed to prevail in a war of attrition. Put tactics, operations, and strategy together, and the ebb and flow of the Civil War—seasons of inertia alternating with seasons of spasm—make sense after all, especially the pivotal seasons of 1863.[96]

Winter was all inertia. The Union's blockade kept tightening its grip, but the Union suffered two embarrassing setbacks when the doughty Confederate navy drove the bluecoats out of Galveston and Charleston's batteries turned back a whole squadron of ironclads. Grant spent the winter cursing cotton speculators (he made scapegoats of Jews) and searching in vain for ways to get an army to Vicksburg through the bayous and thickets of the Yazoo River delta.[97] In Tennessee Rosecrans was inert save for blind swipes at Forrest's marauding cavalry. In Virginia in January, Burnside led the Army of the Potomac on a "mud march" that eventuated in nothing but his loss of command. It was now up to "Fighting Joe" Hooker (the nickname originated as a typographical error) to restore the morale of the often-beaten Yanks by upgrading their rations and, allegedly, their access to prostitutes. He was also the first Potomac commander

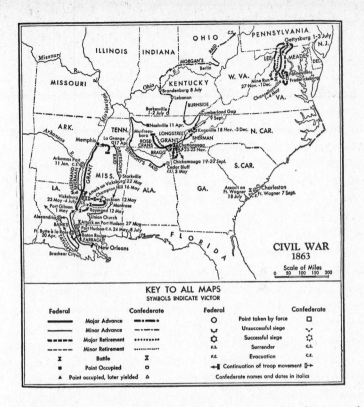

CIVIL WAR
1863

Scale of Miles
0   50   100   150   200

KEY TO ALL MAPS
SYMBOLS INDICATE VICTOR

Federal		Confederate		Federal		Confederate
———	Major Advance	—·—·—		O	Point taken by force	□
———	Minor Advance	—··—··		�	Unsuccessful siege	
-----	Major Retirement	·········		⍖	Successful siege	☆
-----	Minor Retirement	··········		F.S.	Surrender	C.S.
X	Battle	⅀		F.E.	Evacuation	C.E.
▪	Point Occupied	□		◄■	Continuation of troop movement	□►
▲	Point occupied, later yielded	△			Confederate names and dates in italics	

to stress cavalry training and convince Union troopers that they were the equal of J. E. B. Stuart's riders. Finally, the men loved it when the handsome, clean-shaven Hooker vowed, "May God have mercy on General Lee, for I will have none." [98] He turned south in April with confidence. A feint by the Union on the James River forced Lee to send Longstreet's corps to the Peninsula, leaving Lee just 50,000 soldiers to face 75,000 invaders. Hooker's bold plan was to fake another frontal attack on Fredericksburg while swiftly marching half his army across the Rappahannock into Lee's rear. All went well until Stuart located and routed a German regiment on the federals' right flank. Hooker choked. Instead of pressing ahead, he retreated to the village of Chancellorsville. Lee resorted at once to his old trick of splitting his army. Stonewall Jackson led 26,000 men on a risky flank march the whole length of the Union lines and then hurled them into battle with barely a moment to catch their breath. The surprise they achieved that evening of May 2 was so thorough that nothing stopped the rebels until they reached the artillery park at Hooker's own headquarters. The Army of the Potomac slunk north again, less 17,000 men. Lee lost 12,000 troops plus his "right arm": Jackson was mortally wounded by friendly fire while reconnoitering yet another attack.[99]

Lincoln despaired, expecting a surge of support for copperheads in the North. But the cobwebbed telegraph office, where the president marveled like Robert the Bruce at the patience of spiders, soon clattered with the news that Grant had run out of patience. Having eliminated every possible route of advance on the eastern bank of the Mississippi, he decided that the impossible must be the solution. He would ferry his men to the western bank of the mile-wide river, march them south to Louisiana, and then cross the river again. At least, he would if the dying Foote's successor Admiral David Dixon Porter could run the Union river flotilla past Vicksburg's batteries. Sherman thought the whole business mad. But Porter's gunboats, transports, and barges did survive the furious barrage on the night of April 16 and rendezvoused with Grant at the aptly named Hard Times Landing thirty miles south of Vicksburg. Still, the 30,000 veterans ferried to the eastern shore found themselves outnumbered deep in enemy territory and tethered to the flimsiest of supply lines. Grant had thrown his hat over the wall like the proverbial Irishman. He retrieved it by marching fifty miles *deeper* onto enemy soil to capture the state capital of Jackson. The Confederate general John C. Pemberton was doubly confused by Grant's ruse and his own conflicting orders. Davis told him to save Vicksburg at all costs; Joe Johnston told him to save his army at all costs. So Jackson fell on May 14 and William Tecumseh Sherman's troops burned everything they could not eat or cart off. Then Grant turned his columns around and chased Pemberton's sorties back into Vicksburg. After a failed frontal assault on May 22, he laid siege to the tormented city. Reinforcements swelled Grant's host to 75,000. "Wan, hollow-eyed, ragged, footsore, bloody" Johnny Rebs swelled Vicksburg's population by 30,000: impotent soldiers amid terrified, starving women and children pounded daily by Porter's mortars. It spoiled a beautiful spring. Bluebirds chirped in pungent magnolias, recorded one diarist, and "all save the spirit of man seems divine."[100]

Vicksburg's torment gnawed at Jefferson Davis: his plantation at Davis Bend lay just south of the city. So he summoned a war council in mid-May and put the question to Lee: should they send reinforcements to Johnston in hopes of breaking Grant's grip, or roll the dice on another invasion of the North? Not surprisingly, Lee made strong arguments for the latter. An offensive could fulfill the oath Davis swore back in 1861: "to feed upon the enemy and teach him the blessings of peace by making him feel in its most tangible form the evils of war." An offensive would indirectly relieve Vicksburg by drawing off federal reserves. It was certainly less risky than trying to move whole divisions from one end of the Confederacy to the other on overtaxed railroads. Above all, an offensive might at last pry London and Paris out of neutrality. When Davis consulted his cabinet, only the postmaster general, a Texan, pleaded for Vicksburg. On June 3 Lee bade his invincible Army of Northern Virginia to trudge once again through Harpers Ferry en route, he hoped, to Harrisburg.[101]

So the familiar drama played out. Lee's 65,000 troops crept into virgin Pennsylvania shadowed by the Army of the Potomac. The job of Stuart's cavalry was to locate and count the federal forces; he considered that this was best done by another showy gallop clear around them. But terrain pushed Stuart farther east than he expected, and then could not resist the temptation to sack bursting barns and smokehouses. His absence made Lee very nervous, especially when Lee learned on June 28 that the enemy was also across the Potomac and was commanded, not by Hooker, but by George Gordon Meade, a West Pointer from Philadelphia who did not make mistakes. Two days later a rebel detachment looking for plunder collided with Union cavalry at a road junction called Gettysburg. Not knowing where danger might lie, Lee ordered his columns to concentrate. The first to arrive captured the town on July 1, but stiff resistance saved the high ground south of the town for the arriving Union infantry. Lee waited until his whole army drew up by the next afternoon, and then assaulted the enemy flanks in hopes of a double envelopment. On the extreme southern end Longstreet's men won a hot contest in the rocks of Devil's Den. That exposed Little Round Top, the capture of which would render Meade's position untenable. But charge after charge by brave Alabamans broke against the lonely Twentieth Maine regiment commanded by a bookish colonel, Joshua L. Chamberlain. When his surviving troops ran out of ammunition—and help had still not arrived—Chamberlain cried "Fix bayonets!" and led a countercharge that carried the field.[102] The Union's right (northern) flank on Culp's and Cemetery hills held out against less determined attacks. So at last, on July 3, Lee ordered George Pickett's Virginians to pierce the guts of Meade's army on Seminary Ridge. A two-hour barrage by 150 cannons was supposed to hit the earthworks and batteries of the federal lines, but Confederate gunners firing in defilade were unaware that defective fuzes caused their shells to burst late. So over half of the 14,000 attackers were killed, wounded, or captured in the frontal assault that ensued. All told, Lee squandered 28,000 soldiers at Gettysburg, one-third of his army, compared with 23,000 Union casualties. Meade, like McClellan, let the chastened rebels slink away. But Lee's last, desperate wager was lost.

Is there anything new and important to say about Gettysburg? Perhaps so. Perhaps Lee's imagination and patience did not falter after the abortive attacks on the second day. Perhaps his plan for the third day was the most brilliant of all: he just kept the fact of its failure secret out of concern for his army's morale. That plausible plan concerns something everyone knows—J. E. B. Stuart's belated return on July 2—and something many people may not know: the action Stuart's men fought on July 3 well behind Culp's Hill. It is always assumed that Stuart just meant to disrupt the federal troops' supplies and reinforcements or harry their expected retreat. But strong circumstantial evidence suggests that Lee sent Stuart around the northeastern tip of the Union lines with orders to circle back west and charge Seminary Ridge from the rear in support of Pickett's charge from the front. Why didn't Stuart deliver that mortal blow? Because 2,700 cavalrymen from Michigan, in fighting trim thanks to Hooker's attention, defeated Stuart's 6,000 gray ghosts about four miles short of their goal. The Union general who rallied his men with the cry "Come on, you Wolverines!" was George Armstrong Custer.[103]

> Sloped on the hill the mounds were green, / Our centre held that place of
>   graves,
> And some still hold it in their swoon, / And over these a glory waves.
> The warrior-monument, crashed in fight, / Shall soar transfigured in
>   loftier light,
> A meaning ampler bear;
> Soldier and priest with hymn and prayer / Have laid the stone, and
>   every bone
> Shall rest in honor there.—Melville, 1863[104]

The following day, which happened to be the Fourth of July, Pemberton surrendered Vicksburg. Four days after that Port Hudson, the last rebel outpost on the Mississippi, sued for terms. Lincoln proclaimed, "Grant is my man and I am his for the rest of the war," and exulted that the Mississippi, Father of Waters, flowed unimpeded again. The Confederate cavalry commander stranded in the Red River country sadly agreed: "A bird, if dressed in Confederate gray, would find it difficult to fly across the river."[105] Just ten days after that, on July 18, the Union gained a moral victory no less important. Colonel Robert Gould Shaw led the Negro Fifty-Fourth Massachusetts Volunteer Regiment in a do-or-die assault on Battery Wagner, a strong rebel fort in Charleston harbor. It failed. But the courage and professionalism displayed by the soldiers quashed any doubt that colored units might tip the balance in the Union's favor. That was a timely development, because deadly draft riots had terrorized New York and other cities during the busy month of July. Last but not least, in Britain Charles Francis Adams raked in the chips after Lee's lost wager. He didn't realize it until September, when the British agreed not to make delivery of two fearsome "Laird rams" with seven-foot spikes on consignment to the Confederate agent Bulloch. But Palmerston in fact had decided to give the Confederates a cold shoulder upon hearing of Vicksburg and Gettysburg.[106]

Nothing exposed the South's handicaps like the struggle for the important rail junction at Chattanooga, the fulcrum of a third seesaw that wavered and tipped in 1863. It began in late June, when Rosecrans maneuvered his Army of the Cumberland around Bragg's flank at Tullahoma, Tennessee, forcing the outnumbered rebels to retire to Chattanooga on—that date again—July 4. Rosecrans then pried them out of the city by crossing the Tennessee River to the west and threatening to bottle them up. All the Union troops needed to do was fortify Chattanooga while awaiting reinforcements, the pacification of Unionist East Tennessee, and repair of rail lines by the USMRR. Instead, Rosecrans pushed on, expecting a third bloodless triumph across the border in Georgia. What he didn't know was that Davis and Lee had done what they chose not to do before Gettysburg: peel off two Virginian divisions to ride the rails west. On September 19–20 Bragg exploited his rare manpower advantage to rout the weary bluecoats at Chickamauga and lay siege to Chattanooga in turn. Lincoln, Stanton, and the railroad executives came to the rescue with two of Meade's army corps and a new high command starring Grant, Sherman, Philip Sheridan, James McPherson, and George H. Thomas, "the rock of Chickamauga." Grant did well to cobble together a supply line served by wagons, steamboats, and pontoon bridges. Otherwise Thomas's Cumberland infantry took the war into their own hands. On November 25 they cleared out the rebel gun pits at the base of Missionary Ridge and without waiting for orders drove their tormentors up and over the precipitous slopes back into Georgia. The Confederacy lost another critical hub in its transport network. The United States won another portal into the heart of Dixie.[107]

Six days earlier, Lincoln had delivered "a few appropriate remarks" at the dedication of Gettysburg's cemetery. He could not know how they would be received, so he added a disclaimer: "The world will little note, nor long remember what we say here." He was buoyed by the year's victories, but wounded by the copperheads' repetition of Confederate propaganda that called him a tyrant, fool, "royal ape," boor, hayseed, drunkard, henpecked husband, nigger-lover, coward, and bastard mulatto. "No President had engendered so much hate." Lincoln also had just 272 words with which to follow Edward Everett's florid two-hour address; and on their face Lincoln's words merely summarized rhetorical nostrums long used by Webster, Clay, and popular actors such as Edwin Forrest. Even "Four score and seven years ago" and "of the people, by the people, for the people" were borrowed. But Lincoln struck four chords as booming as the start of Beethoven's Fifth. First, he invoked the Declaration of Independence (1863 minus eighty-seven equals 1776) to pronounce that "all men are created equal." Second, he defined the "great civil war" as a test whether "*any* nation so conceived" can endure. Third, he declared that the soldiers' blood had made this ground "hallowed" (holy, set apart). Fourth, he exhorted "us, the living," to redeem their devotion by fighting on for "a new birth of freedom." In short, Lincoln implied that liberty was equality, America was humanity, the soldiers were martyrs, and the war must be won lest liberty perish from the earth. Like Ezra and Nehemiah of old, Lincoln called the faithful to rebuild Solomon's temple, this time on a stronger foundation. Ralph Waldo Emerson, a master of coded rhetoric, had Lincoln down pat: "Rarely was man so fitted to the event. . . . What pregnant definitions; what unerring common sense; what foresight; and on great occasions, what lofty, and more than national, what humane tone. . . . The very dogs believe in him."[108]

In whom did Lincoln believe? He knew that the northern public must glimpse light at the end of the tunnel in 1864, or he and—perchance—the

Union war effort were goners. But the stolid Halleck and Meade had tunnel vision only for caution. Grant, his best fighting general, was laden with baggage including battles with Halleck and the bottle. Grant was also suspected of being a Democrat, possibly with presidential ambitions. Imagine, therefore, Lincoln's relief when a federal marshal arrived from Illinois with a letter from Grant renouncing all interest in politics so long as Lincoln had a chance for reelection. The president asked Congress to grant Grant a third star. Halleck would become a deskbound chief of staff, and Meade would remain commander of the Army of the Potomac. But Grant would serve as the new Winfield Scott: commander in chief of the U.S. Army. He arrived at the Willard Hotel on March 8, 1864, alone except for his son Fred. At first nobody recognized him. After dinner he strolled to the White House and went unnoticed again until Lincoln recognized his grizzled face and beamed, "Why, here is General Grant!" When Seward made Grant stand on a davenport in the East Room so that everyone could get a look, he "blushed like a girl" and perspired freely while shaking hands. This introduction to Washington society, Grant recalled, was "my warmest campaign of the war." What mattered most was Grant's avid endorsement of Lincoln's strategy for 1864, which was to exert maximum and sustained pressure on every front to pin down and destroy the remaining Confederate armies. Grant and Sherman, the latter commanding the army at Chattanooga, were also in full accord. They resolved to pull like horses in tandem and plow the rebellion under.[109]

"Public sentiment is everything," lectured Lincoln like a professor in 1858. "*With* it, nothing can fail; *against* it, nothing can succeed." Six years later the war president facing reelection desperately needed to mold public sentiment. That explains his rare temper tantrum of May 18, 1864, when New Yorkers awoke to news of a "presidential proclamation." It called for a day of fasting and prayer to the "throne of grace," and for 400,000 draftees to make good the Union's terrible losses in Virginia. Most of the city's editors ascertained quickly that the alleged press release was a fraud, but the *World* and the *Journal of Commerce* ran it. The White House went into full damage control. The captain of an English mail packet was urged to delay his departure until the affair was cleared up. Seward sent a blanket denial over the nation's wires. Stanton ordered General John A. Dix, commander of the Department of the East, to get to the bottom of it. He did within twenty-four hours: the forgery was not a treasonable conspiracy, but a scam by a typical American hustler hoping to cause a run on the gold market. That did not stop Lincoln from ordering Dix to arrest the editors for providing aid and comfort to the enemy, and seize their presses. Why did Lincoln overreact? First, he had, by coincidence, sketched out a proclamation very like the fraudulent one while he was alone in the White House the evening before. Was this evidence of a treacherous leak? Second, he and Stanton had long monopolized war news to put the administration's "spin" on it. They nationalized the American Telegraph and Western Union companies in 1862. They granted the New York Associated Press exclusive rights to official news in exchange for the right to approve (i.e., censor) the copy supplied for front pages all over the country. Even independent war correspondents were policed because they served at the pleasure—and under the rules—of the Union Army's commanders. Had a rival news service recently started by Henry Villard hatched the plot? Just in case, Lincoln ordered his arrest, too, until Stanton got Villard to cooperate in exchange for an occasional scoop. Stanton assured the president that he had again "got the telegraph under his control."[110]

Suspension of civil rights by federal, state, and military authorities remains one of the most controversial issues of the Civil War. But before pronouncing judgment on Lincoln one must consider the facts and the context of a horrendous war that by 1864 the North could lose only by losing its nerve. This meant that Peace Democrats, defeatists, sympathizers with the South, haters of emancipation, and outright spies and conspirators threatened the northern cause as much as Robert E. Lee did. Given the prevalence of mobbing and intimidation in peacetime America, one might just as well marvel at how much law and order survived in wartime America. Nonetheless, abuses were legion. The Union Leagues that sprang up in the Midwest were no gentlemen's clubs like those in the East, but paramilitary auxiliaries that rallied Republicans and bullied critics of the war with the warm encouragement of governors such as Yates of Illinois. In Indiana, Kentucky, and Missouri soldiers broke up Democratic rallies, arrested alleged agitators, and guarded the polls. To be sure, in the state elections of fall 1863 Lincoln effectively pleaded with northern voters to subordinate their hatred of blacks to their love for the Union. But there is no question that strong-arm tactics helped Republicans recoup the previous year's losses. The Democrats' most devastating defeat occurred in Ohio, where Congressman Clement L. Vallandigham had denounced the war with such passion that General Burnside court-martialed and exiled him. Vallandigham repaired to Canada, relishing his role as a victim of an unconstitutional war. When the Democrats ran him in absentia for governor, however, nearly 110,000 new or formerly disgruntled voters cast their ballots for the Republican candidate. As happens so often in American history, the peace movement created a backlash that damaged its cause. Union soldiers, especially, swore they hated copperheads more than rebels.[111]

Subversive activities also discredited the Peace Democrats' stance as defenders of civil rights. Self-appointed leaders of "dark lantern" societies such as the Knights of the Golden Circle, Order of American Knights, and Sons of Liberty claimed to have tens of thousands of members ready to confront the Union Leagues. They were led for the most part by confidence men and crackpots who sought rebel gold with big talk about coups d'état. Federal investigators happily exaggerated the threat, and Republican officials happily used it to justify the persecution of copperheads. The most notorious witch hunter was Colonel John P. Sanderson. As a Union provost in Saint Louis, he hired a platoon of detectives, magnified every boast and lie they heard in taverns, and presented rumors and surmises as fact. He imagined the Order of American Knights had 500,000 secret adherents, and that they whispered the password Nu-oh-lac ("Calhoun" in reverse), aided the rebel bushwhacker William C. Quantrill, took their orders from the Confederate general Sterling Price, and conspired "to overthrow the government." What made Sanderson's fantastic report of June 1864 credible was that Jefferson Davis hoped it was true. That February the Confederate congress appropriated $5 million to fund insurrection and sabotage. The spymaster Clement C. Clay, Jr., and the spy Thomas Hines purchased arms for the Sons of Liberty in New York (bribing Mayor Fernando Wood, who was a kingpin of the Tammany Hall machine). They entertained plots to rescue and arm 10,000 Confederate prisoners of war from Chicago's Camp Douglas, form a secessionist Northwestern Union, burn down Broadway, storm the Democratic convention, raid New England from Canada, and abduct President Lincoln. The Republicans made the most of it. In August 1864 Colonel Henry B. Carrington, the Union commander in Indiana, tipped off by a double agent, captured arms caches and 300 alleged Sons of Liberty. In Indianapolis and Cincinnati high-profile treason trials dragged on past Election Day. The Confederates even gave Lincoln a handy "October surprise": a sensational bank robbery at Saint Albans, Vermont.[112]

What to make of the Lincoln administration's record on civil rights? Did Lincoln tear up the Constitution when he raised armies and waged war before Congress was even in session, declared a blockade contrary to international law, spent unappropriated funds, imposed martial law, suspended habeas corpus, censored the mails, controlled the news, sequestered private property, ignored court orders, and condoned military interference with local politics? Or was Lincoln correct in arguing that he must possess implied powers to "preserve, protect, and defend" the Constitution lest it become a dead letter? After all, how could the document's purpose—to form a more perfect Union—be served if the Union ceased to exist? The real test is whether Lincoln recognized limits on his war powers so as not to destroy the liberty he purported to save. A fair reading can only conclude that Lincoln passed this test with higher marks than the founding fathers (recall how the patriots treated the Tories after 1776). Lincoln carefully deferred to Congress, sought ex post facto approval when necessary, pooh-poohed the "dark lantern" panics, authorized few arrests without warrant north of the Ohio River, and even shrugged when Vallandigham reportedly sneaked home in disguise. Most of all, Lincoln was keenly aware that the victory he hoped was imminent by 1864 would bring all his war powers to a sudden, deflating end. That is why he started work early on legal rehabilitation of seceded states and on a constitutional amendment abolishing slavery. Indeed, to suggest that Lincoln meant to subvert the rule of law is preposterous. The man wasn't sure about God, but he had worshipped the law ever since his address to the Springfield lyceum in 1838.[113]

If a conflict of interest marred Lincoln's tenure, it stemmed from his role as both party leader and war president. The political reversal in 1863 could easily lead to another reversal in 1864 and topple him in favor of a Democrat willing to negotiate peace and eager to roll back emancipation. So Lincoln had to pretend not to notice when state officials and Union Leagues spread alarmist propaganda, staged show trials, and accused their neighbors of being copperheads. He had to rally the Republicans and at the same time appeal to the War Democrats. He even had to thwart challenges from inside his own party when Chase started telling tales out of school again. Hoping to supplant Lincoln at the top of the ticket, Chase quietly spread the word that the president was a terrible war leader and lukewarm on racial equality. Then he blessed Senator Sam Pomeroy of Kansas, who formed a Chase for President Committee. Lincoln pretended "to shut my eyes to everything of that sort" so as not to set off another crisis in the cabinet, but his eyes and ears were wide open. He called in his chits from Republican managers and appointees all over the North, not least in Chase's state, Ohio. Then he cashed them in February 1864, when "The Next Presidential Election," a mailing franked by Senator John Sherman, urged Republicans to dump Lincoln for "a statesman of rare ability." Irate state committees pricked that bubble at once, obliging Sherman to pretend that Chase's people had misled him, and obliging Chase to pretend he knew nothing about it. But the charge about Lincoln's being lukewarm on race inspired abolitionists such as Wendell Phillips, Elizabeth Cady Stanton, and Frederick Douglass to nominate Frémont for president on May 31. The following week Republicans gathered in Baltimore to endorse Lincoln, an amendment abolishing slavery, and the South's unconditional surrender. That appeased William Lloyd Garrison and sank the Frémonters. But the Republicans' choice of venue also symbolized their desire to reach out to the border states and the War Democrats. They even dropped the name "Republican" in favor of "Unionist" and replaced Vice President Hamlin as running mate with the Democrat Andrew Johnson of Tennessee. It all amounted to a smudging of ideological lines in hopes of dividing the northern electorate into two camps: those who were foursquare for the Union and victory, and those who weren't. That was good politics. But recasting his administration as an omnibus coalition made Lincoln's personal leadership even more indispensable.

The other ingredient of victory in November, the other maker of that public sentiment Lincoln called almighty, was out of his hands: "Upon the progress of our arms all else chiefly depends." As late as his renomination in June, little progress was evident.[114]

"The Americans are making war as no people ever made it before," wrote a correspondent for the *Times* of London in 1864. "Their campaigns combine the costliness of modern expeditions with the carnage of barbaric invasions. Grant squanders life like Attila, and money like Louis XIV." That was the point. In the first week of May the Union launched the all-points offensives Lincoln had wanted for years. The 120,000-strong Army of the Potomac, with Meade in command but Grant calling the shots, took aim at the Army of Northern Virginia. Ben Butler moved inland on the James

River. A German-American, Franz Sigel, attacked in the Shenandoah Valley. Sherman lashed out from Chattanooga in search of Joe Johnston (whom Davis reluctantly but inevitably chose to replace Bragg). The Union's goal was not so much to capture Richmond and Atlanta as to force Confederate armies to perish in their defense. The perishing took much longer than the generals and public expected. The carnage was almost intolerable. It began in the tangled Wilderness south of the Rapidan River. Grant meant to slip the army "over the river and through the woods" to the very outskirts of Richmond. Lee not only wasn't surprised but realized that a fight in the forest would favor defenders outnumbered two to one. When the armies engaged, there was no way to control isolated units that could not even see beyond fifty yards, especially when artillery set trees ablaze and turned underbrush into a hell. More than 25,000 men fell, not a few were burned alive, and two-thirds of them wore blue.[115]

> *The wagon mired and cannon dragged / Have trenched their scar; the plain*
> *Tramped like the cindery beach of the damned—/ A site for the city of Cain.*
> *And stumps of forests for dreary leagues / Like a massacre show. The armies*
> *have lain*
> *By fires where gums and balms did burn, / And the seeds of Summer's reign.*
> *Where are the birds and boys? / Who shall go chestnutting when October*
> *returns?* —Melville, 1864[116]

Veteran soldiers had shrugged at Grant's vow to stay on the offensive no matter what. Every general said that. Sure enough, when Grant surrendered the Wilderness on May 7 and marched his columns east to the familiar Chancellorsville crossroads, another retreat seemed certain. Then something amazing occurred. The soldiers were told to turn right, not left back to Washington. Grant was as good as his word. The Army of the Potomac broke into song. That was the first of a series of sidesteps by which Grant exploited his numbers to turn Lee's right flank and force him to fight or retire. At Spotsylvania Courthouse twelve savage days of fighting, sometimes hand to hand, left corpses piled in heaps at the "Bloody Angle" and elsewhere. "Many a man has gone crazy," recorded Oliver Wendell Holmes, Jr., the future Justice. "More desperate fighting has not been witnessed on this continent," recalled Grant. But still no decision. Lee retired south in good order. His cavalry blunted a gallop by

Union troops to the suburbs of Richmond (albeit J. E. B. Stuart was killed). The former vice president, John C. Breckinridge, humiliated Franz Sigel's superior force at New Market. The rebel fortifications at Bermuda Hundred were more than Butler could handle. Those victories allowed Lee to reinforce his main body, now entrenched at Cold Harbor, and withstand ten more days of frontal assaults in June. So Grant executed the boldest sideslip yet, southeast across the James and then back west to Petersburg, twenty miles south of Richmond. It seemed he was tenaciously pursuing Lee. In fact, Lee had to follow Grant, and now more than ever. Petersburg, perched on a tributary called Appomattox, commanded Richmond's rail links to the south. If it fell, Jeff Davis would have to relinquish his capital. Lee's loyal remnants dug in. Three more days of attacks pushed Union casualties for the campaign above 60,000. On June 18 Grant settled into a siege. It would last over nine months.[117]

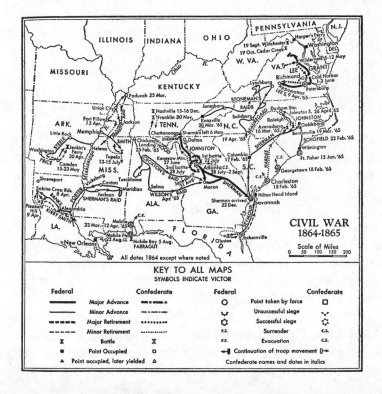

Sherman faced like frustrations minus the butcher's bill. Johnston's western rebels skulked through the north Georgia hills like Cherokees, giving battle only when Sherman lost patience and squandered 3,000 men in charges up Kennesaw Mountain. So all he could do was methodically nudge the enemy south while taking care his flanking units walked into no traps. Not until mid-July did the federals cross the Chattahoochie River to confront Atlanta's elaborate earthworks dug by slave labor. Johnston dared them to get bogged down in trench warfare at the end of a long supply line menaced by Forrest, while his own army was still at large. Johnston knew he needed only avoid a crushing defeat to make Sherman's campaign seem a flop in the run-up to the North's presidential election. Unfortunately, other political forces trumped that wisdom. With Richmond at risk and thousands of white and black refugees streaming into Atlanta, President Davis felt as pressured as Lincoln to score a battlefield win. He feared inaction would turn Atlanta into another Vicksburg. On July 17 Davis sacked Johnston in favor of John Bell Hood. A skilled and impetuous division commander who had literally given an arm and a leg for the cause, Hood was eager to give Davis (and Sherman!) just what they wanted: pitched battles out in the open.[118]

Meanwhile, Lincoln suffered what monks call the dark night of the soul. A flying Confederate column urged on by Jubal Early had exploded out of the Shenandoah Valley and made straight for Washington. It was a beau geste, no more, because Early's 15,000 men could hardly storm the ramparts McClellan had built. But on July 11 Lincoln was obliged to play the abashed spectator at a battle inside the District of Columbia. Congress and the public asked how that could happen. A week later, in light of Grant's losses, Lincoln issued the demoralizing proclamation calling for 500,000 more recruits to be conscripted if necessary. Again the public reeled. Then, on July 30, Union officers both drunken and dumb squandered a chance to escape the sweltering trenches at Petersburg. A regiment of miners from Pennsylvania had seeded a tunnel under enemy lines with 8,000 pounds of gunpowder. The explosion was unearthly; nearby rebels who were not blown to kingdom come staggered away like zombies. But confusion over which units would exploit the breakthrough (Grant and Meade had rescinded their order placing a black division in the van) and confusion over their route of attack (they wandered into the crater itself) allowed the Confederates to plug the gap. If polls had been taken in those days Lincoln's "approval rating" would have hit bottom. Thurlow Weed considered the president a dead duck; the Republican national chairman suspected the same. Greeley floated the notion of peace talks, which Lincoln took seriously for a time. By contrast, the Democrats' morale was sky-high when they met in August on Lincoln's turf in Chicago to nominate Lincoln's nemesis McClellan for president and Vallandigham's disciple George Pendleton for vice president. On August 23 Lincoln wrote a memo pledging the cabinet to make every effort to save the Union by Inauguration Day, since "it seems exceedingly probable that this Administration will not be reelected."

Summer had brought some flickers of hope. In June the USS *Kearsarge* finally sank the CSS *Alabama*, which under Captian Raphael Semmes had destroyed or captured sixty-four Yankee merchantmen. In late July Hood made his reckless attacks around Atlanta, lost 13,000 irreplaceable men, and still did not turn Sherman's flanks. On August 5 Farragut cried "Damn the torpedoes—full speed ahead," sank an enemy ironclad, and closed the port of Mobile. But no breakthrough, no public relations coup, was forthcoming until Sherman decided to *disobey* orders. He rejected the goal of bagging Hood's army—that could be done some other time—and instead made pincer movements south of the city that obliged Hood to vamoose or be trapped. On September 2 Sherman telegraphed, "Atlanta is ours, and fairly won." Then he ordered all civilians out of their homes in preparation for his razing of the critical railroad town.[119]

Two weeks later Philip Sheridan's Union cavalry pushed Early out of the Shenandoah Valley. Sheridan's orders were to ensure that it never again would serve as a rebel breadbasket and invasion route north. So Sheridan conjured the horsemen of the apocalypse—war, famine, pestilence, death—to visit one of America's sweetest valleys until, he vowed, even the crows went hungry.

Everyone highlights the striking swing in public opinion. Unionist newspapers mocked the Democrats' postulate of a lost war. The Democrats were divided. In states holding local elections in October, Republicans ran strong. Everyone says the fall of Atlanta decided the presidential election and thus killed the Confederacy's last hope of survival. Everyone is probably right, but in that case Lincoln was wrong. As late as October 13 Major Thomas T. Eckert, chief of the War Department's telegraphy staff, thought that Lincoln was "unusually weary and depressed." The president reached for a blank telegram and totted up the likely results in the electoral college. In McClellan's column he placed New York, Pennsylvania, New Jersey, Delaware, Maryland, Missouri, Kentucky, and even Illinois, for a total of 114 votes. That left himself a spare 117, thanks only to West Virginia's five votes. Eckert asked for the pen, then added to Lincoln's tally three votes from the other new state that the president had just finagled into the Union.[120]

The Thirty-Sixth State: Nevada, 1864

Twenty days by stagecoach from Saint Joseph, Missouri, carried Sam Clemens into sight of the dribble of shacks boastfully named Carson City. "Visibly our new home was a desert, walled in by barren, snow-clad mountains. . . . Every twenty steps we passed the skeleton of some dead beast of burden, with its dust-coated skin stretched tightly over its empty ribs. Frequently a solemn raven sat upon the skull or the hips and contemplated the passing coach with meditative serenity." That first afternoon Clemens witnessed a casual shooting and the daily visit of the Washoe zephyr, "a soaring dust drift about the size of the United States." A sleepless night ensued while his party stood tiptoe on chairs and bunks for fear of tarantulas on the floor. The man who became Mark Twain in Nevada was known to exaggerate. But Horace Greeley reported, "Here, on the Humboldt, famine sits enthroned." The first director of the state historical society confessed, "Nevada is scarred." And one of its leading scholars wrote, "The land between California and Salt Lake City was a virtual Land of Nod—east of Eden and west of Zion's religious settlements." [121] Anyone who has crossed the state on U.S. 50, the "loneliest highway in America," knows the terrible tranquillity induced by barren mountains and valleys so sere that the air vibrates. Imagine gazing in every direction and spying nothing man-made except the pavement and the odd lonely sign: NEXT GAS 91 MILES. All of Nevada lies in the Great Basin, except for a wedge in the northeast and the southern triangle on the Colorado River. Otherwise, the Humboldt, Carson, Truckee, and Walker "rivers" fed by snow melt disappear into sinks that turn into alkaline mudflats. The desert climate is so extreme that the growing season around Elko is just 103 days, compared with 239 days around Las Vegas. But so scarce is water that only about 1 million acres are cultivated even today. Sagebrush, greasewood, yucca, cacti, piñon pine in the high country, geckos, horned toads, jackrabbits, and rattlers define most life in Nevada. Its human population was the smallest of any state until 1960. The federal government still owns 85 percent of its land.

Forty-Niners exited Washoe (as Nevada was known) as fast as they could. Mormons planted some colonies, but in 1857 Brigham Young called them home to defend Salt Lake from Buchanan. A name they left behind in Carson Valley—Gold Canyon—attracted some sourdoughs who found little gold. Ethan and Hosea Grosh, however, suspected that the bluish mud, which other miners despised, might indicate silver. A secret assay confirmed a yield of $3,500 per ton. But Hosea died of gangrene after spearing his foot with a pick and Ethan perished in a blizzard. That gave a "loud-mouthed trickster," Henry "Pancake" Comstock, a chance to steal what he still assumed was a gold strike. Styling himself a partner of the Grosh brothers, he duped miners into paying for licenses to dig in his Comstock Lode at Virginia City, a shantytown that Virginny Finney baptized with booze. The rush began in 1859 when assayers from Grass Valley, California, confirmed the biggest silver strike since Solomon's mines.

Who were the losers? The Washoe, Shoshone, and Paiute Indians who hoped that the white man had no use for their ranges. The situation exploded in 1860 when Paiute warriors killed prospectors who had abducted two Indian girls. The ensuing "Pyramid Lake war" began with an ambush that killed seventy-six white vigilantes. It ended when militiamen stiffened by the U.S. cavalry shot 160 Indians and consigned the rest to remote reservations.[122] Who were the winners? Shysters from San Francisco, who bought out the claims of "pikers" for fifty dollars per foot, sold shares in mines real and imagined, and then dunned stockholders for assessments to "develop" their claims. Only three of the 135 Comstock mines that were traded on the San Francisco exchange paid dividends. In Nevada itself swindlers salted mines and planted newspaper stories of strikes. Mark Twain's rule of thumb was, "Get the facts first, and then you can distort them as much as you like." A reporter from San Francisco attested, "I have seen more rascality, great and small, in my brief forty days in this wilderness of sagebrush, sharpers and prostitutes than in thirteen years' experience in our not squeamishly moral state of California." But silver there was. Deep sideways tunnels contrived by the German engineer Philipp Deidesheimer yielded $300 million over two decades. Virginia City sprouted a wild-west vanity fair of mansions, hotels, and theaters featuring Sarah

Bernhardt, Lotta Crabtree, and Junius Booth. Its jails were filled to capacity—a sure sign of prosperity, in Mark Twain's opinion.[123]

In March 1861 Congress acknowledged the silver rush by carving from Utah a territory it named for the Sierra Nevada ("snow-clad mountains"). When Lincoln was told that the 20,000 fortune hunters included secessionists, he quickly ordered the Republican James W. Nye of New York to take charge as governor. However, since Nye came by sea, William Stewart—another man bent on becoming the founding father of a new state—intercepted him in San Francisco. Stewart, a lawyer educated at Yale, was a red-bearded, hot-tempered giant who faced down notorious killers. He persuaded Nye to locate the capital in Carson City, where Stewart owned property, and then sustained the decision (so the legend goes) by drinking Virginia City's lobbyists under the table. A scruffy legislature convened in an upper loft strewn with sawdust to hear Governor Nye start by lauding the Union and finish by asking for prohibitions on guns, gambling, and drinking on the Sabbath.[124] There was no chance of enforcing those. But territorial laws didn't matter, because Lincoln meant to rush Nevada to statehood even though it contained only one-fourth of the population required. In 1863 the president was frustrated because Stewart's rival, Judge John North, inserted a provision in the draft constitution to tax the inflated market value of mine property. The wily Stewart mobilized mine companies' money and pikers' votes to reject that. So in 1864 Stewart and Nye staged another convention, which agreed to tax only the actual proceeds of mines. When voters approved by a margin of nine to one, the 16,543-word constitution was telegraphed to Washington City at a cost of $3,416.77 so that Congress could approve statehood before Election Day. Lincoln bagged three more electoral votes. Stewart and Nye became U.S. senators. They arrived on Capitol Hill just in time to vote for the abolition of slavery and were rewarded in 1866–1867 when Congress transferred to Nevada all of Utah west of the 115th meridian and the southern wedge on the Colorado River.[125]

Best of all, the railroad was coming. After struggling for years in the High Sierra the Central Pacific crews laid 500 miles of track across the Great Basin to link up with the Union Pacific in 1869. The construction chief, Charles Crocker, selected the site for the spur to the silver lode and named the town after a Union war hero, Jesse L. Reno. An instant hub, Reno was a Babel of English, Irish, Welsh, German, French, Italian, Spanish, and

Chinese. Just a few years later those voices cursed the railroad's land monopoly and its discriminatory freight rates. A few years more, and the silver played out. The population of Virginia City fell from 11,000 to less than 3,000. Reno survived, thanks to the railroad, cattle ranches, and Basque sheepherders, but the state's population shrank from 62,266 to just 42,335 by 1900. There were even mutters on Capitol Hill about revoking Nevada's statehood.[126] The nearly empty interior counties recovered somewhat when a savvy ranch wife discovered more silver at Tonopah and prospectors found gold near Ely. But the cradle consisting of Reno, Carson City, and Lake Tahoe languished. Those were the years when Francis G. Newlands, a Progressive who married Comstock Lode money, imagined that irrigation, education, regulation, and the suppression of vice could make Nevada a "model commonwealth."

Newlands could not have been more wrong. Nevada became a "desert of buried hopes" because (or until) more colorful boosters realized that the way to restore Reno's fortunes was to peddle sin, not reform.[127] It started in 1897, when San Francisco canceled a ballyhooed prizefight featuring Gentleman Jim Corbett. Nevada legalized boxing overnight and booked Corbett for Carson City. Tens of thousands of fans rode the rails to Nevada in freight cars stuffed with food, liquor, and beds. Reno jumped into the fight game, specializing in "angles" such as interracial matches. In 1906 it began a new racket when the wife of the president of U.S. Steel came to town to take advantage of Nevada's lenient divorce laws. Entrepreneurs soon built dude ranches and swank hotels to cater to such lovelorn celebrities as the film star Mary Pickford. They also coined a nickname that blazed in electric lights on an arch when Governor Fred Balzar pulled a switch in 1929: "Reno: The Biggest Little City in the World." Two years later Nevadans battled the Great Depression with their own sort of New Deal: legalized gambling. By the end of the decade Harold Smith and William Harrah were marketing their casinos nationwide in imitation of the first self-service grocery chain, Piggly Wiggly.[128]

Relief came to southern Nevada, thanks to the Republicans' passage of the $50 million Boulder Canyon Project in 1928. California managed to obtain rights to almost all the water and power Lake Mead and Hoover Dam would provide, but the groundbreaking in September 1930 (in 120-degree heat) brought millions of dollars and thousands of workers to Boulder City, the new industrial town of Henderson, and a gasoline stop called Las Vegas. Momentum built during World War II, when the army made Nellis Air Base a major pilot-training facility and the mobster Bugsy Siegel made Las Vegas the site of his fabulous Flamingo Hotel. The cold war bestowed another gift: the Nevada test site, where more than forty atomic bombs were detonated during the 1950s. The outsize federal presence ensured that Nevada would get all the water and power it needed. That, plus air-conditioning and a recession-proof gambling industry, pushed the state's population past 160,000 by 1960. But the glamor and glitz obscured, like cosmetics, the blemishes of a callow, dependent society. Even after Howard Hughes bought up the Las Vegas "Strip" in 1967, residents of the city gave hints of profound discontent. One old-timer was sorry to see the Mafia go, for now Vegas was run "by corporate cats whose word isn't good for twenty-four hours. At least when a hood gave you his word, it stuck." Billions in revenue from "gaming" (the industry's euphemism for gambling) permit Nevada to boast of a poverty rate lower than that in all but three other states. Yet Nevada ranks near the top in alcoholism, suicide, crime, and divorce. That is why the state born of silver and the Civil War requires two final words. "Why did I move to Tonopah?" said a western Canadian, sipping a highball. "Because the only free countries left in North America are New Hampshire and Nevada." Why fret for Nevadans? asked a native. Because we never grew up, and "it is only possible to transcend our adolescence if we are willing to remove our masks *before* gazing into our mirrors." [129]

Lincoln's happiest day in the telegraph office was undoubtedly November 8, 1864. He won 55 percent of the popular vote and every state except New Jersey, Delaware, and Kentucky. The tallies in New York and Pennsylvania were razor-thin, but even without those giants the president would have won nearly two-thirds of the votes in the electoral college. Patriotism, patronage, party organization, propaganda damning the "treasonous" opposition and warning of plots, plus furloughs permitting some Union soldiers to vote (78 percent favored Lincoln) help to explain the success of the Republicans' efforts. The patriotism of George McClellan, it is usually said, helps to explain the failure of Democrats' efforts. Like Stephen Douglas before him, McClellan chose the Union over self when put to the test. He repudiated his party's platform at the convention and again just a fortnight before the election. "I intend to destroy any and all pretense for any possible association of my name to the Peace Party," he said, lest he be unable to look his former comrades in arms in the eye. Did that cause the Peace Democrats to stay home? There is no way of knowing, but the fact that turnout in states casting ballots in both 1860 and 1864 rose 4 percent—despite the death or absence of so many men—suggests otherwise. McClellan might have opposed peace at any price, but he shared the Peace Democrats' revulsion against racial equality—the principal reason they feared Lincoln's war. Suffice it to say that McClellan received 400,000 more votes than Douglas had. But Lincoln increased his total by 350,000, and Republicans rode his coattails to enormous majorities in the House (145 to forty) and the Senate (forty-two to ten).

Lincoln's occasional premonitions about dying in office ("This war is eating my life out") are not needed to explain his haste in pressing for an amendment to abolish involuntary servitude. The new Congress would not convene until December 1865, by which time his war powers would surely have expired. Readmission of the seceded states might invalidate the Emancipation Proclamation and invite federal courts to return Negroes to bondage. Thankfully, a Jacksonian demon had fled when Chief Justice Taney died in October. Lincoln graciously replaced him with Chase and was rewarded when Chase licensed an African-American lawyer to plead cases before the Supreme Court: a sort of Dred Scott decision in reverse. Still, Lincoln wanted to shove an amendment, "a King's cure for all evils," through Congress before the war ended so that he could make its ratification a condition for southern states' reconstruction. No one since Jackson was more up to the political hustle the business required. Lincoln cracked the Republican whip, lobbied the sixty-six Democrats in the House, divvied out patronage with both hands, and made horse trades. The next Congress would surely pass the act anyway, so why not get paid for your vote? Finally, antislavery societies (with women again in the lead) submitted petitions signed by hundreds of thousands of people. Only twenty-five Democrats helped (eight by abstaining), but that was enough to give the Thirteenth Amendment a two-thirds majority amid cheers and tears on January 31, 1865.[130]

By then the other agenda item so dear to Massachusetts—the utter effacement of southern society—was also being fulfilled. For two months after Atlanta fell, Sherman's army had hunted in vain for the rebel army while Hood looked in vain for a chance to dislodge the invaders. Tired of this, the commanders reached independent decisions to throw their text-books away. Hood linked up with Forrest and invaded Tennessee in a for-lorn stab at the Union supply line. After two needless and bloody battles (Franklin and Nashville), the proud Confederate Army of Tennessee no longer existed. Sherman linked up with no one and stabbed at no one. He reached the considered conclusion that while all war is hell, this all-American war was, fittingly, a new sort of hell. The enemy's center of gravity lay not in armies or cities, but in the will of the people; hence a victorious strategy must target what used to be called noncombatants. On November 16 Sherman and 62,000 soldiers left burning Atlanta behind and began a march to the sea that made Sheridan's apocalypse in the Shenandoah Valley seem paltry. Like an army of ants Sherman's men me-thodically devoured every morsel stored up from the rich harvest in Geor-gia and torched everything aboveground over a swath sixty miles wide. Worse things occurred on the fringes where "bummers"—bandits and deserters, both rebel and Yank—killed, raped, and pillaged for the hell of it. Sherman evinced contempt, but made no move to stop them. A month later the army sniffed salt water, teamed up with the U.S. Navy, and made Savannah a Christmas present for Lincoln.

> *I heard the bells on Christmas Day / Their old familiar carols play,*
> *And wild and sweet / The words repeat*
> *Of peace on earth, good-will to men!*
> *And in despair I bowed my head; / 'There is no peace on earth,' I said;*
> *'For hate is strong, / and mocks the song*
> *Of peace on earth, good-will to men!"*—Longfellow, 1864[131]

Sherman's men were especially hot to punish South Carolina, the cradle of secession, and they got their chance in February 1865. Columbia burned to the ground (by accident, it was claimed), and Charleston surely would have burned if the bummers had gotten their way. Instead, Ashley Cooper's sweet city was occupied from the nearby sea islands by black and white regiments who loved the South even if they hated its cause. By March 4, Inauguration Day, Sherman was in North Carolina facing mere dregs of an army again led by Joe Johnston. It seemed just a matter of weeks before Sherman's hammer met Grant's anvil at Petersburg. It was time for southerners to practice repentance and northerners mercy. Or so at least Lincoln believed.[132]

The brief text he read as sunlight pierced the clouds over the Capitol is as holy as any in the American canon. Lincoln reminded the nation that four years before, both parties had deprecated war, yet both had accepted it and prayed to the same God for succor. In fact, some people on both sides wanted war, many (especially in the border states) never accepted it, and rival civic deities were the recipients of the prayers. But Lincoln pretended otherwise in order to summon Americans to the true Almighty who "has His own purposes" and gave "to both North and South this terrible war as the woe due to those by whom the offense [slavery] came." Then he held out the prospect of justice and mercy in a paraphrase of the prophet Isaiah:

Fondly do we hope, fervently do we pray, that this mighty scourge of war may speedily pass away. Yet, if God wills that it continue until all the wealth piled up by the bondsman's two hundred and fifty years of unrequited toil shall be sunk, and until every drop of blood drawn with the lash shall be paid by another drawn with the sword, as was said three thousand years ago, so still it must be said, the judgments of the Lord are true and righteous altogether. With malice toward none, with charity for all, with firmness in the right as God gives us to see the right, let us strive on to finish the work we are in, to bind up the nation's wounds, to care for him who shall have borne the battle and for his widow and his orphan, to do all which may achieve and cherish a just and lasting peace among ourselves and with all nations.

How was it that Lincoln, a lifelong skeptic, came to meditate on moral theology more profoundly than any cleric? His antebellum speeches, the first inaugural, the Gettysburg address, and the second inaugural surely trace a spiritual journey. But his remarks on national days of prayer in 1861 and 1863 had already described the war as a divine punishment for "our own faults and crimes as a nation and as individuals." Is it sacrilegious to point out that in March 1865 Lincoln spoke not for the ages, but to shape public sentiment in the weeks and months just ahead? His hint that still more wealth might be sunk and more blood drawn was a *threat* to persuade Confederates to quit now lest they invite more "judgments" of the kind Sherman pronounced. His call for charity without malice was a *challenge* to vengeful northern ideologues to practice the pieties they loved to mouth and thus realize the millennium (justice and peace with *all* nations) that they claimed the war was about.

Frederick Douglass called it a righteous effort. That was an interesting term to choose: effort. Radical Republicans, flushed with their electoral landslide, were in no mood to meditate on their own sections' sins or forgive secessionists. Even Mary Todd Lincoln gagged on her husband's sweet words. The only Americans who seemed to respond were the generals.[133]

Johnny Rebs were slipping out of the trenches at Petersburg at the rate of 200 per day even as Grant's forces became overwhelming. Lee's only chance was to beat a stealthy retreat into Carolina and try to link up with Johnston. But massive assaults by the Union on April 2 stove in the Confederate flank and made stealth impossible. That night Richmond was burned by its own army's hands, as if denying the enemy its resources would do any good at this stage. The next day Lincoln himself paid a visit to ponder the wreckage, greet newly freed Negroes, and sit at the desk of Jefferson Davis. Lee's fugitives limped, as fast as their unshod feet and spent horses permitted, toward the Orange and Alexandria Railroad at Lynchburg. They fell twenty miles short because Custer's cavalry and Union infantry blocked their road at Appomattox Court House. Now Lee had to choose: order his men to run for the hills, join up with such feared marauders as John "Gray Ghost" Mosby, Quantrill, and Forrest, and wage a protracted insurgency; or accept honorable defeat and trust Lincoln's pledge to "bind up the nation's wounds"? Lee made the decision knowing that the leaders of the Confederacy might be imprisoned and possibly executed for treason. His answer was that "as Christian men" they had no right to consider anything but the effect their decision would have on the country. He begged for terms. Now Grant—"Unconditional Surrender" Grant—had to choose: arrest the officers, guard the soldiers, and await instructions; or figure out what Lincoln meant when he counseled "let 'em up easy"? Grant ordered generous rations for the famished rebels, then gave them leave to go home and take their horses with them so they might plant a crop for their families. "This will have the best possible effect upon the men," answered Lee, who promised in turn to lend his prestige "to pacifying the country and bringing the people back to the Union." Sherman later gave Johnston even more generous terms (which Stanton had to retract). Most tellingly, the fiery slave-trading millionaire Nathan Bedford Forrest told his loyal troopers that he meant to follow Lee's lead and urged them to do likewise. "You have been good soldiers, you can be good citizens. Obey the laws, preserve your honor, and the government to which you have surrendered can afford to be and will be magnanimous."[134]

But it couldn't. The Whigs' three-decade crusade against Jacksonian passion had delivered Americans of both parties and sections to unrestrained passion upon Lincoln's election in 1860. The Civil War, far from destroying antebellum America, was the completest expression of its political culture, racial fixation, paranoia, industrialism, mysticism, self-delusion, and anger. Even Lincoln could probably not have exorcized the demons unleashed by the war. (Second terms are usually debacles.) But he was not allowed even to try, because the angry demon repelled by the magnanimous generals lodged in the soul of John Wilkes Booth. Deeming his thespian talents more suited to intrigue than infantry, this twenty-nine-year-old mustachioed zealot had dabbled in Confederate espionage. When the Confederate government dissolved, Booth formed a conspiracy of fools, dullards, and drunkards to decapitate the U.S. government. On the evening of April 14 one thug failed to kill Seward when his revolver misfired. Another, who was to target Andrew Johnson, chickened out. But Booth got to his man at Ford's Theater and discharged a derringer into his skull. Lincoln gave up the ghost by morning. Steely Stanton wept. Grant blamed himself for having declined the president's invitation to the theater. Mary Todd Lincoln went mad. Parishioners nodded and sobbed during Easter sermons mourning the Union's savior and noting that he had been slain on Good Friday. Whether it was ironic or tragic, Booth's literal *coup de théâtre* was the perfect, excessive climax to a Romantic American drama in which performance was politics. Booth even bore a strange resemblance to Poe. Of course, all he achieved was to douse what embers of mercy Lincoln had kindled in the North and turn Lincoln himself into a martyred messiah.

They have killed him, the Forgiver– / The Avenger takes his place,
The Avenger wisely stern, / Who in righteousness shall do
What the heavens call him to, / And the parricides remand;
For they killed him in his kindness, / In their madness and their blindness,
And his blood is on their hand.—Melville, 1865[135]

Nor was Lincoln the last casualty. Three weeks after Appomattox, thousands of Union soldiers whittled away at Camp Fisk near Vicksburg. Most had been prisoners of war, some for as long as two years. Their discharge papers were in order. They yearned for some passage home. At last the great steamboat *Sultana* docked. Its owners made handsome profits moving soldiers and supplies on the Mississippi River at five dollars per enlisted man and ten dollars per officer. Needless to say, the crew crammed men onto the decks until they scarcely had room for a blanket or a pot to pee in. The *Sultana* was registered for 300 passengers, but on April 27, 1865, it carried 1,866 veterans plus eighty-five crewmen, seventy-five civilians, and 160 animals. About ninety miles south of Memphis, at three o'clock in the morning, its boiler blew. More than 1,500 men were blown to bits, scalded to death, or drowned in the swift black currents on that blackest of nights. The veterans' midwestern families received the mocking news around May 3, the day the funeral train returned Abraham Lincoln's body to Springfield.[136]

The worst pretense of all was to imagine the Civil War over. But that is what most Yankees, at least, preferred to believe. They embraced the myth that their nation's sins had been purged by the blood of their soldiers and president. They believed more strongly than ever the myth of America as a "new Israel" destined to regenerate the whole human race. They certainly believed Stanton's remark when the Grand Army of the Republic (minus the black troops) passed in review: "You see in these armies the foundation of our Republic—our future railway managers, congressmen, bank presidents, senators, manufacturers, judges, governors, and diplomats; yes, and not less than half a dozen presidents."[137] So believing, Yankees leaped back to hustling as fast as they could.

Chapter 8

RADICALS, KLANSMEN, BARONS, AND BOSSES

Reconstructed, the Great White Republic Flees to the Future,

1865–1877

*L*et no one persuade you that the American Civil War was anything but a catastrophe. Combat and disease claimed 620,000 lives, a percentage of the population of 1860 equivalent to 5.97 million in the year 2008. Half a million more came home wounded. The war carried off almost one in every twenty-five American males (compared with one in 160 during World War II); and since most were in the prime of their youth, the nation paid a staggering price in lost labor and children never conceived. More than 220,000 dependents, mostly widows, qualified for pensions, while hundreds of thousands more were doomed to the workplace and/or spinsterhood. To be sure, immigration revived with the end of the killing and return of prosperity. The federal government, not to mention railroads and land speculators, posted agents in Europe to foster immigration. The U.S. population nearly reached 40 million in 1870 and passed 50 million in 1880. Still, the growth rate of 36 percent in the 1850s fell to just 27 and 26 percent in the following decades. Statistics on the cost of the war are tougher to interpret. An official accounting in 1879 put the United States' expenditures at $6.2 billion and those of the Confederacy at $2.1 billion. After adding in $3.3 billion for pensions and adjusting for inflation, the result in 1990s dollars is over $100 billion. Another econometric measure—including expenditures, destruction, lost labor, and lost consumption—estimates that the economy took a hit equivalent to an average year's income for every man, woman, and child in the North and two and a half years' income for everyone in the South. Moreover, although the greenback experiment is usually deemed a success, economic historians confirm the suspicion of contemporary critics that the inflated prices for war matériel caused by paper money increased the Treasury's indebtedness by $300 to $600 million. The Morrill Tariff likewise obliged the War Department to pay higher prices for goods that might have been purchased abroad. In any event, the U.S. national debt increased by a factor of *forty-two* and peaked in 1865 at $2.8 billion.[1]

Did the Civil War at least stimulate industrialization? Historians of both Marxist and liberal bents once took this for granted, and it must be said that progressive optimism is a wonderful asset for a people to have. In retrospect, the Union's national mobilization and distribution of resources doubtless taught American business powerful lessons in how to achieve economies of scale, a phenomenon to be described in due course. But professionals in the dismal science of economics are not surprised when their numbers reveal civil war to be a very ill wind that blows good to *some* firms, industries, and regions, while it slams like a hurricane into everyone else. Americans pioneered no major civilian technologies between 1861 and 1865 and ceased doing pure science. They invented no new models of management and paid a huge cost in lost opportunities. To be sure, hot-air balloons for artillery spotting, the Gatling gun, submarines, and ironclad warships debuted in the Civil War, but only the ironclads had a significant impact on combat. Railroads and telegraphs, by contrast, made a huge impact, but they were mature technologies before the war. So the Union's impressive war effort really absorbed the energies of an industrial machine already in place. Production of pig iron had grown 17 percent between 1855 and 1860 and would grow 100 percent from 1865 to 1870. During the war it grew 1 percent. Railroads had spread 8,700 new miles in the five years before the war and would spread 16,200 miles in the five years after. During the war just 4,000 miles of track were laid. Data on river and harbor improvements, overall manufacturing, commodity production, and exports tell similar stories. Needless to say, the war wrecked or sharply depreciated the South's capital goods; and the end of the war flung the North into a brief but wrenching recession as government spending plummeted, industries reconverted, and demobilized men competed for jobs.

Distinguished historians such as Hunter Dupree, Robert Bruce, Alfred Chandler, Stanley Engerman, and Robert Sharkey even conclude that the Republicans' "new deal" (as a North Carolinian Unionist called it) retarded or distorted investment. High tariffs, high government spending, subsidies, land grants to state governments, and the lobbying and patronage encouraged by all of the above were no triumph for laissez-faire, but a mercantilist program that favored "rent seeking" over competitive advantage. That is, capitalists invested where a government benefice ensured a tidy return rather than where their money could be used most efficiently. The federal banks bequeathed by the war were a boon, but they also invited abuse by strengthening Wall Street's grip on the valves controlling the flow of Americans' money.[2]

The Civil War made losers of laboring men except insofar as they escaped the perils of combat. (One might even consider their decline in real wages due to inflation the "bounty" they paid to escape conscription.) Skilled workers made starts at organizing in some sixty trades during the war, but the 200,000 unionized by 1865 were merely a tithe of the North's industrial force. Federal troops broke strikes in arms factories, coal mines, textile mills, and railroads. Employers hired Negro scabs in New York. The Illinois legislature made work stoppage a crime for the duration. That was to be expected in wartime. But labor unions did no better after the war, because the resumption of rapid economic growth and opening of the Great Plains dampened militancy, and the renewed influx of immigrants weakened the bargaining position of organized labor. Defining what Republican "free labor" ideology meant in terms of a social contract was not only unfinished business, but a business barely begun when hard times returned in the Panic of 1873.

The Civil War likewise made losers of poorer women who lost their male providers or hopes for matrimony, and were forced into manual labor or domestic service. Nor should it be forgotten (although it usually is) how the war must never have ended for tens of thousands of wives, mothers, and sisters condemned to care for physically and psychically crippled veterans. Numerous middle-class women no doubt longed to reclaim the secure domesticity of antebellum life, but those who shared the excitement of the war effort must have been frustrated in this regard. Upscale feminists such as Elizabeth Cady Stanton and Susan B. Anthony expected that a war to liberate slaves would bestow rights on women as well. Had not Stanton considered marriage a form of slavery and denial of divorce rights a "sin against nature"? Instead, Radical Republicans told them that this was "the Negro's hour" and raising the "woman question" (as it was known) would only endanger the movement for black males' citizenship. In any event, female activists quarreled. Some believed suffrage the first priority, others reform of divorce and property laws, and still others the reform of *men* through preaching and prohibition. Harriet Beecher Stowe predicted that easy divorce laws would prove a calamity for women. So two rival suffragist associations emerged in 1869, followed by the Women's Christian Temperance Union in 1874 and a divorce reform league. Progressive women did better in education. The new land-grant state colleges embraced coeducation, and normal schools prepared young women to teach. The elite colleges Vassar (1865), Smith (1871), and Wellesley (1875) joined the ranks of what would be called the "Seven Sisters." On the Great Plains hardy pioneer women scored some victories for female emancipation after the Civil War, but suffragist movements remained stymied until World War I.[3]

One might expect a war between whites to take some pressure off Native Americans, but the latter were in fact the biggest losers of all. In Oklahoma the Cherokee chief John Ross (still going strong at age seventy) urged the Five Civilized Tribes to form a "United Nations of the Indian Territory" and remain neutral. But the whites would not let them, and for a variety of ironic reasons most sided with the southern heirs of Andrew Jackson. Troops from Texas and Arkansas scoured the Indian lands of U.S. Army outposts. The Confederate Indian commissioner bore generous alliance offers and reminded the tribes how Seward had said that all land "south of Kansas must be vacated by Indians." Slave-owning chiefs had an obvious stake in the southern cause. Anyway, the Union suspended its annuity payments, rendering past treaties dead letters. Several tribes split down the middle, or else changed sides after the federal victories at Pea Ridge, Arkansas, in 1862; and at Honey Springs (near present-day Muskogee, Oklahoma) in 1863. But rebel guerrillas including Quantrill's took vengeance on pro-Union tribes; a quarter of all Creeks and Seminoles perished during the war. Most of the farms, orchards, and towns planted by the Five Civilized Tribes during their twenty years' exile were ruined. The treaties foisted on the survivors by a vengeful United States in 1866 removed them to still poorer land and obliged them to swallow a "war guilt" clause accepting blame for the hostilities. Worse still, the severity of the treaties varied according to the tribes' residual strength; hence those who had tried to stay loyal suffered most, and the Confederates' Choctaw and Chickasaw allies suffered least.[4]

According to a Cheyenne legend, the Great Spirit warned that the apparent blessing of horses obtained from the Spaniards would change life forever. Instead of peaceful hunting and gathering in the company of their dogs, the Cheyenne would forage and hunt far afield, fighting incessantly with other tribes. So it was. Long before Manifest Destiny, the vast grasslands stretching from Texas to Canada were the site of an anthropological free-for-all among 250,000 diverse tribespeople whose fortunes depended on how they adjusted to the equestrian revolution. So much for the notion that Indian braves in their feathered headdresses were pretty much all alike. As the horse culture spread from south to north after the seventeenth century, Native Americans made strategic decisions whether to stick with settled agriculture like the Pueblos, turn pastoral nomads and raiders like the Apache, or attempt to strike a balance. To the extent that horses and horsemanship became measures of wealth and prestige, the social, political, sexual, and religious foundations of tribal life were shaken. More conservative tribes tethered to crops and villages became prey. Competition for lands blessed with water, forage, buffalo, and trade routes defined a balance-of-power system every bit as sophisticated as Europe's. Thus, the Apache destroyed the Jumanos' ascendency in Texas but in turn were driven into the desert by the more mobile Comanche. Kiowa on the northern plains imitated their success, forcing the Cheyenne and Arapaho cultures to adapt or die. Blackfeet allied with Gros Ventres to grab the horses and buffalo grounds of the Cree and the Crows. But no tribe reinvented itself as a nomadic, militant horse culture more successfully than the obscure Lakota, originating in present-day Iowa. They bought or stole enough horses to go on the warpath in the 1780s and 1790s just as smallpox was decimating the sedentary tribes around them. Over fifty years the Lakota crossed the Missouri, claimed broad buffalo ranges, allied with the Cheyenne and Arapaho, and conquered the holy Black Hills in what the

French called the Sioux country and the Americans named Dakota. Their fierce, Romantic resistance to whites would make the Lakota and their Oglala relatives the representative Plains Indians, but their empire rose on the strength of anomalous and ephemeral advantages.[5]

The Sioux empire, like all the other Native American patrimonies, collapsed during and after the Civil War. Strife began in 1862, when starving Sioux terrorized Minnesota and starving Cheyenne stole food and horses from whites who overran their reservation during the gold rush at Pike's Peak. Those first white Coloradans came by their violent proclivities honestly. Many were veterans of "Bleeding Kansas"; and many of the volunteer cavalrymen ordered to deal with the Indians were veterans of the recondite campaign of 1861 in New Mexico. After secession Texas revived its old claim to the whole Rio Grande valley and dispatched a mounted column that swiftly captured Albuquerque and Santa Fe. A regiment of Coloradan volunteers under Colonel John Slough routed the overextended Texans at Gloriéta Pass in the deathly Sangre de Cristo Mountains. But the hero that emerged was a mad Methodist preacher, John M. Chivington, whose 400 men picked their way over the mountains and fell upon the invaders' rear. On his return to Colorado, Chivington was promoted to colonel and ordered to stop the Cheyenne's depredations. It wasn't necessary. Chief Black Kettle journeyed east to consult with the "white father," Lincoln; then he returned to promise peace and more concessions in a powwow near Denver. He led his people to winter quarters at Sand Creek on their reservation at the Arkansas River. He even hoisted a white flag and the American flag over his tent. He did not understand (blame the interpreter) that his white interlocutors demanded restitution of stolen property and that Lincoln's territorial governor granted all whites the power "to take captive, and hold to their own private use and benefit, all the property of said hostile Indians that they may capture." This simply echoed the Confiscation Acts under which the Union waged war in the East. But Chivington, his zealous blood up, went farther, saying that it was "right and honorable to use any means under God's heaven to kill Indians . . . and damn any man that was in sympathy with Indians."

On November 29, 1864, Chivington's regiment crept up to Sand Creek under cover of night and attacked the tepees without parlay or quarter. Soldiers went berserk, cutting down children, raping women, severing men's genitals, and taking 200 scalps that earned them drinks on the house in Denver. Captain Silas Soulé, who forbade his men to join the atrocities, later testified against Chivington. For that he was shot dead in the Denver dust. Congress ultimately censured those responsible for the Sand Creek massacre, but the Civil War's validation of violence proved the stronger writ. In the wars that followed against the Cheyenne, Arapaho, and Sioux nations, generals such as Sherman, Sheridan, and Custer would conclude that assaulting Indian noncombatants in their winter camps was the most effective method of what is now called counterinsurgency. Sheridan put it rhetorically: "Did we cease to throw shells into Vicksburg or Atlanta because women and children were there?"[6]

The next most effective method was bloodier still: kill off the buffalo. It wasn't policy, just another spillover of violence onto the plains, when railroad workers, merchants, sportsmen, and then professional buffalo runners ("hunters" was a tenderfoot's term) deposited surplus Civil War cartridges from their Springfield and Remington rifles into the skulls of some 15 million buffalo. Sheridan once reported that it took him three days to pass through a single herd, and a conductor on the Kansas Pacific reported nonstop buffalo along a 120-mile stretch of the railroad. So there were beasts aplenty to satisfy the hunger of frontier whites for meat, the hunger of eastern whites for trendy robes, and the hunger of all the Plains Indians for food, clothing, and tools. But berserks—again there is no other word—escalated the killing until 5 million beasts fell in 1873 alone. Their putrid carcasses lined the trails; their bones were stacked in great pyramids at the railheads. Their swift disappearance drove ever more desperate Indians into war against whites and each other. When whites began crossing the still sacrosanct Powder River buffalo grounds on the way to yet another gold strike in Montana, the Oglala Sioux Red Cloud (and a young chief, Crazy Horse) rallied tribes to resist. Meanwhile, Congress and the War Department began to realize that feeding Indians was cheaper than killing them. The Fort Laramie Treaty of spring 1868 restored fragile peace to the northern plains by granting the Sioux and their allies expansive, exclusive hunting grounds "so long as the buffalo range thereon in sufficient numbers to justify the chase." The presidential candidate Ulysses Grant approved. He meant to enlist Ely Parker—a Seneca Indian, master of Freemasons, and veteran of Vicksburg—on behalf of a program called "concentration." The idea was to settle Indians on agricultural reservations secure from prospectors, railroaders, and buffalo hunters. The Sioux had cause to be skeptical, but agreed to give peace a chance and retreated to their sacred Black Hills.[7]

We rend our garments over the cruelty visited on Native Americans. But think: white Americans had been more than willing to mow down hundreds of thousands of other whites in pursuit of their national destiny. Is it realistic to think they would balk at dispatching far smaller numbers of "savages"? Indeed, Indian wars did not even rank at the top of the list of the Civil War episodes that made losers of many Americans. That honor belonged to Reconstruction, the template for all future occasions when the United States won a war only to lose the peace.

Abraham Lincoln did not know what to do with emancipated slaves. In his last speech, he suggested the wisdom of granting the vote to a few Negroes, including war veterans, but otherwise deemed them incompetent to function as full citizens. So did the chief of his Freedmen's commission, the abolitionist Samuel Gridley Howe, whose wife wrote "The Battle Hymn of the Republic." So did the anthropologist Louis Agassiz of Harvard. He went so far as to warn that assimilation of blacks would promote miscegenation, pollute the white race, and drag America down. Military experiments during the war suggested otherwise, but offered no clear-cut solution. On the Sea Islands of Georgia and South Carolina, Union officials hired blacks to work the rice and cotton plantations, but the confiscated land mostly found its way into the portfolios of Yankee speculators. Outside New Orleans, generals Benjamin Butler and Nathaniel Banks encountered resistance from freed blacks to whom work contracts looked like a milder form of slavery. Ulysses Grant's "negro paradise" centered at Davis Bend successfully endowed 3,000 former slaves with collective land grants and self-government. But that could not serve as a model, because Republicans' devotion to property rights precluded land redistribution except under martial law. The same disappointment awaited 40,000 blacks in Georgia who were promised forty acres of "Sherman land" and a mule. All these "rehearsals for Reconstruction" posed the question of what freedom meant: equal protection, suffrage, property, education, or just nothing left to lose? Frederick Douglass wisely said, "Verily the work does not end with the abolition of slavery, but only begins." To most whites, the work either ended at Appomattox or would end with the swift repair of the Union.

Lincoln craved swiftness, lest the end of the rebellion terminate his war powers and free Radical Republicans, conservative Supreme Court justices, and former secessionists to wage civil war by other means. As early as December 1863, Lincoln issued a Proclamation of Amnesty and Reconstruction for occupied states in the South. Except for high-level Confederate officials and former U.S. officials who had violated their oaths, all citizens could obtain a presidential pardon in exchange for pledging loyalty to the Union. Once a mere tenth of enrolled voters had taken the pledge, they could form governments and resume participation in Washington, D.C. To be sure, they had to amend their state constitutions to provide freedom and education for the enslaved, but they were otherwise free to make any regulations "which may yet be consistent, as a temporary arrangement, with their present condition as a laboring, landless, and homeless class." Louisiana, Arkansas, and Virginia eventually claimed reconstructed status under that gentle war measure. Radical Republicans thought Lincoln's measure far too gentle. They refused to seat Louisiana's new congressmen and in July 1864 passed a harsher plan inspired by Benjamin Wade and drafted by the Unionist Henry Winter Davis of Maryland. It required loyalty oaths from 50 percent of a rebellious state's whites and banned former Confederate officials and even soldiers from voting or holding office. To demand civil rights for blacks while denying them to the South's most prominent whites was to ask far more than human nature could handle. Lincoln gave this Wade-Davis Bill the treatment reserved for well-meaning mistakes: the pocket veto. Its sponsors replied with a manifesto accusing Lincoln of a "studied outrage on the legislative authority of the people." Clearly, the struggle between the White House and Congress over Reconstruction began well before Lincoln's death.[8]

What is more, Andrew Johnson carried forward the main lines of Lincoln's policy, just as Lincoln had followed the main lines of Buchanan's policy in 1861. (Presidential transitions are never nearly so cut-and-dried as historians and popular memory want them to be.) Johnson agreed with Lincoln that the Union was legally indissoluble—secession being a pretense. Hence loyal southerners needed only to get back into proper relation with the other states. Far from Congress's having power over them, they should be permitted to wield power in Congress. Southerners who had fought bitterly for the right to secede saw the virtue in reversing their stance if it meant no trials for treason, no confiscations, and no pressure to treat blacks as equals. Republican zealots, not trusting this application of the parable of the "prodigal son," groped for some other theory to parse an unprecedented situation. Thaddeus Stevens argued that the eleven naughty states had forfeited constitutional protection and thus were at the mercy of Congress no less than lands conquered from Mexico. Charles Sumner suggested that the rebels never ceased to be American citizens, but their states had committed suicide and reverted to the status of territories under congressional oversight. Theories begat theories until 1869, when the Supreme Court found in *Texas v. White* (a case involving redemption of Confederate bonds) that ratification of the U.S. Constitution had forged "an indestructible Union of indestructible states." But what really mattered was less getting one's head around theory than getting one's hands on power to mold a "new South."

That is what made Andrew Johnson pivotal: he represented so *old* a South that Republicans misunderstood. They thought they were adding a border state Unionist with mild abolitionist views to their ticket for 1864. In fact, they plunged their spades into a Tennessee grave, dug up the dry bones of their most hated nemesis, and breathed life into the still angry spirit of Old Hickory. Like Jackson, Andrew Johnson (who was born in Raleigh) grew up poor, uneducated, and fatherless. After apprenticing as a tailor he followed Jackson over the mountains to seek his fortune in Tennessee. A homely man with a clean-shaven moon face and bulbous nose, Johnson doted on his wife, a humble shoemaker's daughter who raised his five children. Along the way he mastered the rough-and-tumble patois of frontier stump speakers and entered politics a foursquare disciple of Jackson. In 1843, he joined Congress at age thirty-four; and he would serve for ten years before being gerrymandered out of his district. It seems Johnson was so honest in his defense of the common folks that the Democrats' own "Nashville clique" feared him as much as the Whigs did. But common folks rewarded Johnson with the governorship, and then in 1857 with a seat in the Senate, where he became the most stinging southern foe of secession. Johnson supported every Union war measure, served as military governor in Tennessee, and told Negroes he was their Moses. But his nationalism was that of an old Jacksonian, as were his resentment of wealthy planters, belief in states' rights, opposition to big capital and government, and ramrod determination to use presidential authority to check federal power. The Radical Republicans, hypnotized by their faith that a great civil war must work a great revolution, applauded when Johnson said on the day Lincoln died that he meant to make treason infamous. It did not occur to them Johnson wanted to lead the nation thirty years *backward*, not forward to a Yankee millennium. Nor was he a fool. The president made by John Wilkes Booth would give the Radicals as good as he got. Nor was he a drunk, his slurred speech at his inauguration as vice president notwithstanding. He was suffering the effects of typhoid fever and wine from the night before, and thought a stiff whiskey might see him through. What the incident really revealed was Johnson's poor judgment: something the nation could hardly afford after a war and presidential assassination.[9]

Still, it is hard to imagine even Lincoln persuading northern Americans to bear the burden of rebuilding the South after the suffering its rebellion had caused. The summer of 1865 found all southerners stone broke except for a few war profiteers. Many cities and rural counties were simply laid waste. Planters and farmers who managed to get seed into the ground wondered if labor would be available or whether they would still own their land come harvesttime. Real estate and commodities markets collapsed. Slave owners lost over $2 billion in capital to emancipation. The former slaves wandered dusty roads hunting for their families and for food. The tons of clothing, machinery, and building supplies that northern merchants shipped into southern cities gave the appearance of rapid recovery, but the South could buy little, since agricultural recovery was frustrated by a severe drought, the refusal of blacks to work for their former masters, the inability of impoverished whites to pay them, and the attraction of cities. Negroes seeking work or expecting a biblical jubilee clustered in wretched "dark towns" where malnutrition, smallpox, tuberculosis, and diseases thriving on filth carried off as much as one-third of the population.

Nor did violence cease. In the Mississippi Valley and Texas some Confederate veterans turned bandit and vigilante in defense of white rule. The Ku Klux Klan, founded by Nathan Bedford Forrest in 1866, spread from Tennessee across the Deep South. But in spite of all that, African-Americans, led by the free antebellum communities in New Orleans, Mobile, Charleston, and elsewhere, made a bold start at claiming the entitlements of American liberty. They formed Union Leagues, Equal Rights Leagues, schools, Masonic lodges, orphans' asylums, and churches. A flurry of matrimony regularized family relations. Blacks adopted respectable surnames and insisted on being called "Mr." and "Mrs." Some men wore suits and some women high-fashion dresses picked from the flotsam and jetsam of shops and plantation mansions. It drove many whites crazy to see Negroes putting on airs. In truth, former slaves were just grasping for dignity in the few ways they could. Most important, blacks learned that freedom meant little without civil rights, suffrage, and opportunity: "If I cannot do like a white man I am not free." But the only means of sustaining a modicum of freedom for blacks were guns in their own hands and in the hands of federal troops, or carrots and sticks sufficient to make whites acquiesce in the elevation of former slaves.[10]

Americans' civic religion and their language of liberty have rendered them almost incapable of imagining Reconstruction as anything but a morality play. People in the South and North, at the time and 140 years later, seethed, blushed, and above all stereotyped the characters in the drama most notably portrayed in a novel of the 1930s, *Gone with the Wind*. The first generation of northern historians lauded the Radical Republicans' struggle to bestow equal rights on the Negroes and educate them for citizenship. Their villain was a willful, bigoted Andrew Johnson, who usurped power in order to sacrifice racial equality on the altar of swift reunion. They mourned his success. By the turn of the twentieth century, however, the nation at large decided southern apologists had been right. Blacks could not be equal; hence Jim Crow segregation and discrimination were right and proper. So historians began blaming the Radical Republicans for the white backlash, violence, and delay in reconstructing the Union. Woodrow Wilson even draped Jim Crow in a Progressive mantle.[11] By the 1930s revisionists such as Howard K. Beale questioned the Radicals' motives, arguing that their fight for black suffrage concealed a desire to impose the yoke of big business on the South. Accordingly, revisionists mourned Andrew Johnson's failure. Finally, the post–1945 civil rights movement inspired another school to revise the revisionists. This school granted that the Radical Republicans had underestimated the effort required to uplift the freedmen, but otherwise mourned Johnson's success in postponing black people's civil rights for a century. The ongoing penchant for blaming (and mourning) is not surprising, given that racial antipathies are the most emotional, intractable, embarrassing scandal of the American civic religion. But blaming robs blacks and whites both of their full humanity. In truth, flawed human nature as revealed through personalities, politics, confusion, and clashes of interests within all camps after 1865 so muddled Americans' efforts to put Humpty-Dumpty together that one is left wishing Reconstruction *had* been a simple morality play.[12]

Imagine a sober analogy. Suppose that the U.S. Army has overthrown an oppressive regime in the name of expanding freedom. The victors expect to be cheered by all save the old regime's minions and to help the people realize democracy. Instead, a hard, lengthy occupation ensues because bad guys mount an insurgency and good guys bicker among themselves. Ethnic, regional, and political factions become violent, if only in self-defense, and all reassess monthly who is most likely to come out on top when the U.S. soldiers go home. Meanwhile, the clock is ticking as the Americans grow weary and wonder if the occupation is only making things worse. Of course, history knows no true analogies, but the parallels between Reconstruction and Operation Iraqi Freedom are worth pondering. And just as Americans had to puzzle out the Sunni, Shiite, and Kurdish factions in Iraq, so the way to grasp what happened from 1865 to 1877 is to review the American types (not stereotypes) that clashed over what to do with the South.

First, forget the canard that evil Yankees flocked south with all their possessions in carpetbags for the purpose of duping ignorant blacks and gouging desperate whites. Of course there were hustlers, bummers, whores, and hypocrites in the mix. But the northerners who played prominent roles in Reconstruction included philanthropists willing to risk themselves and their property for the chance to assist southern blacks and whites alike. Many were veterans beguiled during the war by their encounters with Negro music, worship, community life, and (in some cases) women. Others were simply investors looking, in good American fashion, to help themselves while they helped rebuild the South. Lincoln's secretary John Hay invested in orange groves, as did Harrison Reed, a native of Massachusetts who became the governor that helped Florida get back on its feet. A Union general, Willard Warner of Ohio, bought plantations in Alabama and served his adopted state in the Senate, where he bitterly protested the lack of federal assistance to the South. Albert T. Morgan of Wisconsin, a crippled war veteran, went broke trying to raise cotton in Mississippi, but as sheriff of Yazoo County he sheltered blacks from violence by whites. The Pennsylvanian cavalryman Powell Clayton got rich as a planter in Arkansas, but also labored as a governor and senator of Arkansas to build railroads and schools. Albion W. Tourgée of Ohio, a severely wounded veteran, failed as a land speculator in North Carolina, but fought as a judge for blacks' civil rights. Most typical were the hundreds of anonymous women who sacrificed middle-class comfort to teach barefoot black children their letters so that some, at least, might matriculate at Fisk, Alcorn, Tuskegee, and other Negro colleges founded after the war. The sociologist W. E. B. Du Bois would reminisce about the Yankee "school ma'ams" with obvious love in *The Souls of Black Folk*. Still, there were reasons "*carpetbagger* was only a smear word that stuck and continues to stick," the major one being that some southern whites, not just blacks, viewed the carpetbaggers as indispensable allies.[13]

The smear word for those southern whites was scalawag. An editor in Tuscaloosa spoke for all former secessionists when he called the scalawag "the local leper of the community. Unlike the carpetbagger, he is native, which is so much the worse. Once he was respected in his circle; his head was level, and he could look his neighbor in the face. Now possessed of the itch of office and the salt rheum of Radicalism, he is a mangy dog, slinking through the alleys, haunting the Governor's office." If this legend is true, then the leper colony numbered nearly one-fifth of all southern whites, and a majority in upland counties of Tennessee, North Carolina, and Georgia. Their votes would be needed to put Republicans over the top in every former Confederate state except the ones with an absolute Negro majority (South Carolina, Mississippi, and Louisiana). Doubtless the southern Republican parties attracted opportunists. But their rank and file consisted of former Whig and Unionist farmers, businessmen, and professionals who believed that the planter class had led the South to its ruin. Their leaders included such prominent antebellum politicians as the Charlestonian doctor and Masonic grand master Albert G. Mackay; the Mississippian levee commissioner James Lusk Alcorn; the North Carolinian journalist William Woods Holden; and the Texan judge Edmund J. Davis, a graduate of West Point who had tried to persuade Robert E. Lee to reject secession. Governor Joseph E. Brown of Georgia, a fierce critic of Jefferson Davis, turned Republican for a while after the war. The "Gray Ghost" John Mosby and generals George Pickett and James Longstreet embraced the Republican Party permanently, on the theory that the "arbitrament of war" had ended debate over the course of American institutions. Longstreet even served for a time as commander of Louisiana's Reconstruction militia while quietly funding Negro schools. Andrew

Johnson's perennial rival William Brownlow became the Radical governor of Tennessee who launched the first campaign to suppress the Ku Klux Klan. Washington Duke hiked home from Appomattox to North Carolina, meditating on the horrors wrought by secession. He turned Republican, rolled up his sleeves, and built a global cigarette empire that made his state prosper.

Nevertheless, Democrats claimed that the Republican regimes during Reconstruction were corrupt tyrannies imposed by Union bayonets and Negro ballots. They certainly did tax and spend to a degree appalling to any Jacksonian, and many of their ambitious railroad schemes and other projects failed the test of creative corruption. But they were no more corrupt than northern governments at the time or during the notorious 1850s. What galled former rebels was that Negroes, lettered or not, former slaves or not, now postured and hustled in state capitols. Therein lay the southern Republicans' real vulnerability. The very existence of their fragile coalition of carpetbaggers, scalawags, and former slaves stiffened an increasingly unified white Democratic resistance. If federal protection was withdrawn, just one lost election might shrivel the Republicans' power as blacks lost the vote, carpetbaggers left in disgust, and scalawags returned to their roots under pressure from their neighbors.[14]

The northern political spectrum was just as splintered after the Civil War, not least because of Lincoln's decision to recast the Republican Party as a Unionist omnibus in 1864. Radical Republicans, led in the Senate by the bombastic Benjamin Wade of Ohio and by the crusader Charles Sumner of Massachusetts, and in the House by the ideologue Thaddeus Stevens of Pennsylvania, believed that winning the peace meant punishing traitors and securing full civil rights for black men. But they were just a minority. So were the conservative Republicans who cared only for Union and were ready to endorse rule by southern whites if that facilitated Reconstruction. Copperhead Democrats were whipped by secession into minority status. As a result, none of the above could patch together a governing coalition without winning over the large moderate Republican faction led by senators Lyman Trumbull of Illinois, John Sherman (William Tecumseh's brother) of Ohio, and William Pitt Fessenden of Maine, plus congressmen James G. Blaine of Maine and James Garfield of Ohio. Moderates hoped for rapid Reconstruction so the nation could move on, and expected a free labor market and minimal legal protections would suffice to give blacks a fair shake. They did not necessarily favor suffrage for blacks and certainly did not favor redistribution of property. Who did in the North besides New England zealots? In sum, Republicans agreed on little except that secessionists must not regain power and the war must give birth to a more perfect Union—so long as it did not cost very much.

What about that Whiggish economic package passed during the war: the "second American Revolution" including high tariffs, railroads, homesteads, federal banks, and greenbacks? Historians on the left and the right once thought that this, at least, united Republicans after the Civil War, but it was not so. To be sure, most Radical Republicans from the Northeast favored protective tariffs against foreign imports and feared that the revival of representation of white southerners might threaten their high-tariff policy. Indeed, the abolition of slavery would increase southerners' power in Congress and the electoral college because it implied abrogating the constitutional clause counting slaves as "three-fifths" of a person. Given the stakes, it surely seemed plausible that Sumner and others of his persuasion itched to create millions of black Republican voters for reasons affecting their pocketbooks. But even if that could be proved (it can't), what then to make of praise in Radical Republican newspapers such as the *Chicago Tribune* for Senator Trumbull's campaign *against* high tariffs? The Midwest, the *Tribune*'s editor claimed, was "being consumed by the goods of New England and Pennsylvania. . . . The remedy for the evils of which I complain will not be found until the process of robbing by law becomes plain to the farmers whose money is now profusely poured into the capacious pockets of the manufacturers." Nor were Republicans divided only East versus West, because every American industry and firm calculated its own bottom line. Mature industries that did not need protection from imports rued the adverse impact of tariffs on exports. Highly competitive firms opposed protection if it helped domestic start-ups grab market share. Manufacturers dependent on imported raw materials favored free trade. Merchants and bankers who prospered on the volume, not balance, of trade also hated high tariffs. All that is needed to explode the idea of a Yankee capitalist conspiracy is to note that *Hunt's Merchants' Magazine* and the *Journal of Commerce*, the voices of Wall Street, challenged Republican orthodoxy while endorsing Andrew Johnson's policies for Reconstruction.

Most tellingly, business split over currency issues. The Civil War saddled the United States with a national debt that once had been thought impossible or else immoral. Surely the federal government ought to pay off the debt as rapidly as revenues would permit. But by the same token the war had obliged investors to adjust to a soft-money environment. Millions of greenbacks lubricated transactions; physical plants, raw materials, and inventories were purchased at inflated prices; loans had been made at high interest rates. Monetary contraction would spell disaster for many employers and their workers throughout the North. The gambling railroad entrepreneur Jay Cooke, once a sound money man, now saw virtue in a paper currency and a floating government debt. He pleaded his case in patriotic rhetoric: "Why should this Grand and Glorious Country be stunted and dwarfed—its activities chilled and its very life blood curdled by these miserable 'hard coin' theories—the musty theories of a by gone age?" Nevertheless, Hugh McCulloch, the secretary of the treasury appointed by Lincoln, experimented with contraction in 1866. He bought back some government bonds, pledged to redeem the greenbacks in time, and kicked off a deflation fatal to numerous firms. At that point Senator Sherman brokered a compromise. Greenbacks would remain in circulation for the time being, and redemption of bonds would be financed, not through higher taxes, but through reductions in government spending. So the federal budget, which naturally shrank from $1.3 billion to $520 million the year after Appomattox, shrank by more than a half again until bottoming out at $241 million in 1877. All this might amount to another boring lesson in public finance except that as a result, no funds were available to succor the impoverished South during the very years southern whites needed carrots as well as sticks to reconcile themselves to something like equality for the Negro.[15]

In sum, congressional Republicans could not agree about Reconstruction; wrestled with economic, political, and moral angels; and in any case were not even in session when President Johnson took the initiative to rebuild the South. All that the Republicans left behind on adjournment in 1865 was the Bureau of Refugees, Freedmen, and Abandoned Lands to relieve immediate suffering. The law provided for exactly one commissioner, ten clerks, a one-year mandate, and zero dollars. It was assumed that the War Department would staff this Freedmen's Bureau with personnel already on the payroll.

* * *

Believing, as Lincoln had believed, that the southern states never left the Union, Johnson launched Reconstruction, as Lincoln had launched the war, on his own authority. His objectives were to get southern state governments up and running; put them in the hands of white "plebeians and mechanics"; and build a base for a presidential run of his own among southern whites, northern Democrats, and conservative Republicans. Even his coolness toward suffrage for blacks came in part from a conviction that the Negro "would vote with his late masters, whom he does not hate, rather than with the non-slaveholding whites, whom he does hate." Accordingly, the president issued two proclamations on May 29, 1865, that appeared true to the legacy of his predecessor. At least Lincoln's holdover cabinet thought so, because it unanimously endorsed them. The first proclamation empowered provisional governors to hold state conventions required to repeal secession, repudiate the Confederate debt, and draft voting laws based on the registry of 1861 (i.e., no Negroes). States might then elect officials, ratify the Thirteenth Amendment, and send stalwart bottoms to fill their vacant seats on Capitol Hill. The second proclamation offered presidential pardons and security against confiscation of property to all former rebels in exchange for a loyalty oath, but excluded fourteen categories of persons (Lincoln had only six). Among those excluded was everyone owning $20,000 in property. Yet even rebel kingpins soon won reprieves because Johnson's will weakened (or his ambition strengthened). He eventually pardoned 13,000 Confederates: nearly everyone except the warden of Andersonville prison and those who had conspired with John Wilkes Booth.

Southern states wasted no time taking advantage of this presidential Reconstruction. One by one their conventions met Johnson's minimal demands. South Carolina and Mississippi did even less. None of the southern states even considered giving the vote to the blacks. When Carl Schurz toured the region that autumn, he was appalled by the wreckage, and even more by the abject condition of Negroes and the haughtiness displayed by whites. The Carolinian Christopher Memminger, former Confederate treasurer, explained to Schurz that when Johnson held out "the hope of a 'white man's government'" it was only "natural that we should yield to our old prejudices." Natural—and for many a psychological necessity, because to confess that blacks could function as responsible citizens would mock the whites' sacrificial war effort. Finally, the new southern regimes legislated "black codes" to quell widespread rumors of insurrection by Negroes and alleviate the planters' shortage of labor. While granting Negroes the right to marry, own property, and have standing in court, they also imposed labor contracts and licenses that made work in the fields or the trades little better than peonage. Was this a shocking attempt to resurrect slavery under another name? Not exactly: most of the provisions of these black codes were cut and pasted from northern state laws regulating vagrancy and apprenticeships. Johnson admired the ploy and urged his provisional governor in Mississippi to do likewise for suffrage. "If you could extend the elective franchise to all persons of color who can read the Constitution of the United States in English and write their names, and all persons of color who own real estate valued at not less than two hundred and fifty dollars, and pay taxes thereon, you could completely disarm the adversary [the Radicals] and set an example for other states to follow." In other words, just do what northern states do and you will have political cover to deny the vote to 95 percent of the Negroes. But southern states stupidly refused the strategem, and Johnson never raised it again. Most heartbreaking to former slaves was the president's retroactive cancellation of the distribution of confiscated land. That obliged O. O. Howard, known, because of his mercy, as the "Christian general," to summon 2,000 black men and women to Edisto Island, South Carolina, and tell them that the Sherman lands were no more. They must sign contracts to work for their former masters or vacate. A black committee filed a pitiful petition asking, "Are not our rights as a free people and good citizens of these United States to be considered before those who were found in rebellion?" Most shocking to Radicals was the president's toleration of southern militias to help federal troops preserve law and order.[16]

What is often forgotten, however, is that Johnson resisted the temptation to exploit the most powerful ploy on behalf of swift Reconstruction: a foreign war with the French over their ongoing occupation of Mexico. It would have been easy. Radical Republicans outraged by Napoléon III's imposition of "Emperor" Maximilian on the Mexicans believed that defeat of the South should free American arms to defend the Monroe Doctrine. Grant himself cried, "Now for Mexico!" on the morrow of Appomattox. In May 1865 he sent 52,000 troops to the Rio Grande and even slipped Major General John Schofield into northern Mexico to rally Confederate fugitives to the Stars and Stripes. The influential Montgomery Blair and Francis Preston Blair, Jr., begged for a declaration of war that would reunite Americans, restore the patriotic bona fides of Democrats, and clothe "Tennessee Johnson" in the mantle of Jackson and Polk. But the president deferred to his secretary of state, William Henry Seward, whose intelligence told him that Napoléon would soon repent of this foolhardy, costly, unpopular venture so long as Yankee belligerence did not offend French amour propre. Seward made sure that his "wait and see" policy would stick by recalling Schofield and sending polite but increasingly trenchant démarches to Paris while Grant, on Johnson's orders, demobilized the 1-million-man Union army. Most important, perhaps, Charles Sumner—chief Radical, chairman of the Senate Foreign Relations Committee, and former hawk on Mexico—swung behind Seward. He had come to realize that a foreign war would perversely distract the nation from the plight of the former slaves. Seward was gratified in April 1866 when the French minister confirmed that the evacuation of Mexico would proceed. Perhaps Johnson was just complacent in the belief that presidential Reconstruction would prevail without a foreign war. Still, credit him with nudging monarchists out of the New World with no blood spilled save that of the honest, gullible Maximilian himself.[17]

Indeed, presidential Reconstruction proved too successful too quickly. When Congress convened in December to find sixty-nine former Confederate officials claiming seats in the Capitol, the Republicans angrily revoked their credentials and launched a counterattack on the "imperial presidency" bequeathed by Lincoln. First, they created a Joint Committee on Reconstruction that mirrored Wade's Joint Committee on the Conduct of the War. Next, they passed a bill extending the life of the Freedmen's Bureau and granting it 3 million acres to dispense to former slaves. Johnson vetoed that on the argument that it was no longer a war measure, and hence was unconstitutional. Next, the Republicans passed a Civil Rights Act that struck down black codes by defining all native-born Americans (except untaxed Indians) as citizens equal before the law. Johnson vetoed this too, on the argument that the federal government had never presumed to usurp the right of states to define their own citizens. But that veto, even if technically right, was a political blunder. The civil rights measure fell far short of what Radical Republicans wanted and was instead a peace offering to the president from the moderate leader Trumbull. After all, it said nothing about Negroes' voting. But Johnson's furious Jacksonian veto forced the moderates to join forces with Radicals or else fold up their tents. Between April and July 1866, Congress marshaled two-thirds majorities to override the vetoes and pass to the states a proposed Fourteenth Amendment to preempt judicial challenges. It disqualified from office all persons who had practiced or aided insurrection, declared Confederate debts null and void, repealed the "three-fifths" compromise, defined as citizens all persons born or naturalized in the United States, and denied states the power to make or enforce laws depriving "any person of life, liberty, or property, without due process of law; nor deny to any person within its jurisdiction the equal protection of the laws." Even that did not go far enough to unfurrow the brow of Thaddeus Stevens, but he settled for it "because I live among men and not among angels."[18]

Thus were the battle lines drawn for the first postwar elections. Johnson surrendered all pretense of membership in Lincoln's party by summoning a National Union Convention to Philadelphia in mid-August. His alloy of states' righters, northern Democrats, and conservative Republicans never fused, because the unelected, increasingly isolated president had nothing to offer but patronage over which his erstwhile supporters predictably fought. But what really undercut Johnson's new party was the racial violence that climaxed in New Orleans just two weeks before Johnson's convention. The Unionist governor of Louisiana elected under Lincoln's "ten percent" dispensation wanted to call a convention to mandate suffrage for blacks. Former Confederate officials, including the mayor of New Orleans, won a court injunction against the governor and vowed to deploy the police and militiamen to break up the assembly. Major General Absalom Baird, commanding federal troops in Louisiana, wired Secretary of War Stanton an urgent request for guidance. Stanton sat on it. He figured the Republican cause would be served whether the convention met or was forcibly broken up. So the latter occurred on July 30, 1866, when 400 conventioneers, about 90 percent black, were beset by a mob led by policemen who shot, clubbed, or stabbed to death thirty-eight people and wounded 146.[19]

Northern newspapers screamed bloody murder: Who could deny that such barbarity was a direct consequence of Johnson's leniency? Nor did Republicans shy from trumpeting the presence at Johnson's convention in Philadelphia of former Confederates and copperheads led by Clement Vallandigham and Fernando Wood. Finally, Johnson killed his own party's dwindling hopes by stumping for candidates on a "swing around the circle." That was beneath the dignity of his office, and Johnson lowered it farther by delivering such harangues and insults that even his friends slunk away. One of them, Governor Jacob D. Cox of Ohio, observed: "He is obstinate without being firm . . . pugnacious without being courageous. He is always worse than you expect." Thomas Nast, the devastating cartoonist for *Harper's Weekly*, had a field day depicting the "drunken tailor." Republicans won in a landslide.[20]

The outgoing Congress took that as a mandate. In February and March 1867 the Republicans passed legislation that amounted to their own blueprint for a new American South. The First Military Reconstruction Act dissolved all the state governments set up in 1865 except for Unionist Tennessee and created five military districts. The generals in charge were to ensure that new constitutional conventions approved suffrage for Negroes and ratified the Fourteenth Amendment. Such imposition of martial law in time of peace was of dubious constitutionality. Indeed, it challenged *Ex Parte Milligan*, a Supreme Court decision, in December, that military courts were out of bounds wherever civil courts were open for business. But Republicans anticipated every conceivable dodge. A Command of the Army Act obligated generals appointed by Johnson to take orders directly from General Grant in contravention of the president's powers as commander in chief. A Tenure of Office Act forbade the president to dismiss his own cabinet members (i.e., Stanton) without the Senate's advice and consent. A Second Reconstruction Act empowered district commanders to enroll voters themselves lest southerners undermine the process through simple foot-dragging. Congress also legislated suffrage for blacks in the District of Columbia (against the will of 95.5 percent of its white residents). Finally, to prevent the executive branch from reasserting itself during the summer recess, the Republicans summoned the new Fortieth Congress into immediate session. Johnson vetoed the whole agenda, of course, but the Senate and House overrode him in every case except one: statehood for the violent, gold-greedy, and wholly unqualified territory of

Colorado. Nebraska, by contrast, gave the Republicans no pangs of conscience.

The Thirty-Seventh State: Nebraska, 1867

"I simply don't care a damn what happens in Nebraska no matter who writes about it," grumbled an editor in New York when asked to review Willa Cather's crackling novel *O Pioneers!* about land speculation on the Great Plains frontier. Nor was that just East Coast snobbery. John F. Kennedy's assistant Ted Sorensen, a Nebraskan, called his home state "a place to come from or a place to die." The comedian Johnny Carson said, "A lot of good people came out of Nebraska; and the better they were the faster they came!" The estimated 500,000 pioneers who trekked west across the territory from 1841 and 1856 felt their hearts leap when the first butte or mesa appeared, because it meant dreary Nebraska was behind them. To this day one can drive U.S. 20 for hundreds of miles across the Sand Hills south of the Badlands and see nothing but windmills pumping groundwater for invisible cattle. Temperatures soar to 100 degrees in summer and plunge below zero in winter. Nebraska broke the spirits of many a pioneer who tried to sink roots in its soil. Native Americans knew better. The grasslands watered by North America's largest aquifer and the river system they called Nibraskier ("flat water," hence the French "Platte") sustained an ocean of maize, beans, and squash plus tens of millions of bison.[21]

The first white settlements around Bellevue trading post emerged after 1833, when the Pawnee ceded land south of the Platte. But the Kansas-Nebraska Act of 1854 coincided with treaties in which the Omaha and Otoe tribes opened a 100-mile stretch west of the Missouri River. Hustlers from Council Bluffs, Iowa, immediately ferried across to the "Lone Tree" shading the pier of what they named Omaha City. They meant to make it the capital. But promoters of sixteen other towns pressed their own claims. When the governor appointed by President Pierce died two days after arriving, the decision belonged to Thomas B. Cuming, the slick, mustachioed twenty-five-year-old territorial secretary. A budding journalist with friends in Iowa, Cuming knew what to do. Despite being "plied, begged, pressed, entreated, assailed and even threatened" by settlers south of the Platte, he rigged the electoral districts to ensure the selection of Omaha. That wouldn't stick, but the regional rivalry in Nebraska politics survives to this day.

The second usual frontier hustle, land speculation, seduced and fleeced many of the 50,000 settlers who arrived in the ensuing twelve years. "There were but few Christians among that varied population," regretted a Baptist preacher. "The great mass seemed in a terrible hurry to build their houses, and push their various enterprises to success and wealth." As in Minnesota, success was delayed by the Panic of 1857, droughts, floods ("The Big Muddy is Mad and has gone out of its banks"), infestations, and so much Indian strife that even the Civil War seemed a distraction. When the Second Nebraska Cavalry mustered in 1862 it was to fight, not rebels, but Lakota Sioux who were terrifying the upper plains. The Sand Creek massacre in Colorado in 1864 inflamed the Cheyenne. Then the influx of whites on the Bozeman Trail to Montana obliged the Oglala statesman Red Cloud to wage war in defense of his hunting grounds. No sooner did he agree to cede northwestern Nebraska in 1868 than the Cheyenne erupted again in southern Nebraska, threatening settlers and the Transcontinental Railroad. Not until June 1869, when cavalrymen aided by the Pawnee and the scout William Cody routed Tall Bull's war party, did a nervous peace settle on Nebraska. The Indians departed with "unspeakable sadness," but there was no point in staying. Professional hunters had systematically exterminated the buffalo.[22]

At the time of the Civil War, Nebraska fell far short of the population required for statehood, but Congress habitually abandoned its rules in favor of political advantage. So President Lincoln urged his new territorial governor, Alvin Saunders, to press for statehood as soon as the Republicans could manufacture a majority. They figured they had it by 1864 because the Homestead and Pacific Railroad acts seemed to be the twin pillars of frontier prosperity. But the voters split fifty-fifty (if the fraudulent tallies are credible) and rejected the call for a constitutional convention. So Saunders pressured the territorial assembly to appoint a committee to draft a constitution. Three men did it in secret, slipped it through the committee by a margin of seven to six, and won the assembly's approval without even making the document public. When the petition for statehood reached Congress, Sumner insisted the clause restricting the vote to whites be expunged, but otherwise the Radical Republicans were eager to welcome Nebraska because it meant two more votes for their Reconstruction majority. On March 1, 1867, Nebraska became the only territory to achieve U.S. statehood over a presidential veto.

David Butler became the state's first governor in a similarly high-handed fashion. After the Panic of 1857 ruined the livestock trade of his family in Indiana, he had started over in Pawnee City. His ranch was more a home than a business, however, because the glib, ambitious Butler knew that the way to get rich quick in the West was through politics. His campaign consisted of calling his opponent "the worst copperhead and rebel" on the frontier; the Democrats retorted, "Every vote for Butler is a vote in favor of negro suffrage." The mudslinging contest was a virtual tie, but Butler was judged to have won by 109 votes because Republican officials threw out all the contested Democratic ballots and allowed Republican soldiers at Fort Kearny to vote as if they were residents. Yet Butler reigned as if his mandate were from on high. He pacified Democrats south of the Platte by bribing legislators to build a new state capital and a land-grant university in the middle of nowhere. Then he pleased the Republicans by fixing the name Lincoln on the town. Everyone applauded when Butler lobbied Washington to clear out the Indians and hasten the building of railroads. But he also made many enemies in the course of spending unappropriated funds, speculating in town lots in Lincoln, handing out contracts to friends, pocketing bribes, "borrowing" from the state treasury, and purchasing favors with public land. All that was business as usual in every new state: the *Omaha Herald* just called Butler "the boldest and most bungling" of Nebraska's "official highwaymen." But it sufficed to get him impeached in 1871.[23]

The Union Pacific laid its first rails in 1865 near the Lone Tree in Omaha. As construction teams inched across the state, they left behind brawling boomtowns with names like Fremont, Grant, and Kearny. Branch lines splayed off the Transcontinental like veins of a leaf, planting more towns and more farms. Federal land grants endowed the railroads with one-sixth of the state's acreage (by definition the most valuable sixth), so the railroads advertised far and wide the paradise that was Nebraska. When the Civil War was over, it was not hard to find takers. In just three years after the Union Pacific's completion, Nebraska grew 250 percent to 123,000, then 370 percent and 230 percent in the decades that followed, to surpass 1 million people by 1890. Most of these people came from the Middle Atlantic and Middle West, but Germans and Swedes also arrived in large numbers. They discovered the tricked-up constitution and the railroads, which gave life to the state, were also the chief impediments to their pursuit of happiness. Railroads monopolized markets and gouged freight customers; government barely functioned at all. Still, the voters rejected a new constitution in 1871, not because it regulated corporations, but because it also restricted rights-of-way (people wanted more railroads, but under control). It also contained a bigoted tax on large religious institutions (Catholic churches put others to shame). In the hard times following the Panic of 1873, however, a better draft won approval. It bent the state's twig toward a curious combination of conservatism and community action. Except for the populist era led by the favorite son William Jennings Bryan, Nebraskans voted Republican—yet they applauded state management of the economy—yet they rejected the liberalism of their neighbors in Kansas in such matters as women's suffrage and temperance. Look no farther than Omaha, whose name means "against the current." In 1869 *Harper's Magazine* teased the muddy urban pretender: "Hast ever been to Omaha, / Where rolls the dark Missouri down, / And four strong horses scarce can draw / An empty wagon through the town? . . . Where whisky-shops the livelong night / Are vending out their poison-juice; / Where men are often very tight. / And women deemed a trifle loose? . . . If not, take heed to what I say; / You'll find it just as I have found it; / And if it lies upon your way, / For God's sake, reader, *go around it*!" By 1900 Omaha was home to Creighton College, striking architecture on display for the Trans Mississippi Exhibition of 1898, giant grain elevators, and the choicest steaks in the United States. But it remained an overgrown railroad and cow town that smelled of manure and stockyards and supported more bordellos than churches, plus a prosperous gambling industry run by "Canada Bill." [24]

The bent twig is an apt metaphor because J. Sterling Morton, the man Butler diddled out of the governor's chair, persuaded Nebraska's board of agriculture to found Arbor Day in 1872. What few trees had clung to the banks of the Missouri River had been chopped down to provide ties and fuel for the railroads. Morton's campaign for Arbor Day caught the imagination of Nebraska's community-minded conservatives. By 1890 about 350 million trees, including the world's largest artificial forest, were planted in Nebraska, and every other state but two had proclaimed its own Arbor Day. [25] The bent twig sprouted again in the Great Depression. The Nebraska legislature and voters abolished their state senate in favor of the nation's only unicameral legislature (it was supposed to minimize lobbying and corruption). Next, they mandated publicly owned irrigation districts and electric power authorities, a socialist system culminating in the buyout of Omaha Power and Light in 1946. Finally, they put the finishing touch on Lincoln's rise to cultural prominence by completing a new state capitol in 1934. The architect Bertram Grosvenor Goodhue of New York won the competition with a stunning design for a neoclassical structure with no dome, but a soaring tower topped by a sculpture: *The Sower*, strewing seed on the prairie.

Two more examples convey Nebraska's lovable quirkiness. In 1917 the Irish-born Father Edward Flanagan took pity on waifs in wicked Omaha and founded Boys Town to teach conservative virtue and the radical value of loving one's neighbor. "He ain't heavy; he's my brother" was its motto. Flanagan died in 1948, just as the new U.S. Air Force was choosing a site for the global headquarters of its nuclear-armed Strategic Air Command. It settled on Offutt Air Base, a few miles southeast of Boys Town.[26]

An undeservedly obscure book on American debates over how to modernize South Vietnam called the two major schools of thought the "Whigs" and the "Tories." The former consisted of officials who argued that the first priority in nation building must be democracy and the rule of law, because once good government was in place dynamic economic and social progress would follow. The latter consisted of officials who argued that the first priority must be economic growth and technology transfer, because once a prosperous middle class was in place democracy would take care of itself.[27] In the case of congressional Reconstruction, America's first experiment in nation building, the only "Tories" were the blacks themselves plus a few Radical Republicans. They included politicians like Stevens, generals like Sherman, and observant carpetbaggers like Tourgée, who rendered a bitter judgment: "Republicans gave the ballot to men without homes, money, education, or security, and then told them to use it to protect themselves. . . . It was cheap patriotism, cheap philanthropy, cheap success!" Be that as it may, the authors of congressional Reconstruction were mostly dyed-in-the-wool "Whigs" who believed that a few years of education for Negroes in the ways of democracy would set all to rights. When Stevens revived the idea of land distribution in spring 1867, his own party shouted him down. To give "special favors to special classes of people," the *Nation* argued, would amount to a "war on property." Trust in black suffrage, counseled the *Chicago Tribune*, to "change the whole structure of society in the South, without confiscation, without vindictive measures of any kind." The revolution was wrought and nothing more was required, decided the *New York Tribune* after a southern tour by Greeley.

For a while it seemed they were right. African-Americans quickly proved as adept as Anglo-Saxons at forming parties, making alliances, distributing patronage, canvassing votes, politicking, and horse-trading. "You never saw a people more excited on the subject of politics than are the Negroes of the South," wrote a white Alabaman. "They are perfectly wild." Under the umbrella of federal military protection blacks and whites together elected conventions, drafted constitutions acceptable to Congress, and applied for readmission to the United States. By 1868 the Carolinas, Alabama, Arkansas, Louisiana, and Florida would join Tennessee as states in good standing. What is more, the Republican coalition of blacks, scalawags, and carpetbaggers won initial control in all of them, thanks to the disenfranchisement of former rebels, and vote fraud when necessary. Their opponents knew what was afoot. If ballot boxes policed by soldiers were to be the Republicans' agency for social revolution, then the Ku Klux Klan would see to it that blacks never got to the polls and Andrew Johnson would see to it that no Republican generals skewed the tallies. In August 1867 the president ordered Grant to reassign Radicals such as Sheridan from southern districts and ordered Secretary of War Stanton to clean out his desk. "The turning point has at last come," Johnson muttered; "the Rubicon is crossed.[28]

Republicans had whispered the word "impeachment" in Capitol cloakrooms as early as October 1865. Now, it appeared, Johnson was asking for it because firing the secretary of war seemed a brazen violation of the Tenure in Office Act. But it appeared to the president that he was dealing from a position of waxing authority. He expected that whites' resistance to Negroes' civil rights would spread in the North. He did not think the Tenure in Office Act would survive a court test, especially when the cabinet member in question had been named by his predecessor. He was confident that the Senate would not dare reject his choice of Ulysses Grant as Stanton's interim replacement. Johnson even scored a serendipitous triumph in foreign policy—indeed, a bold leap forward for manifest destiny—that senators grudgingly applauded in 1867. The czarist government apparently wanted to sell the unremunerative colony of Russian America rather than wait for the British to seize it in time of war or Yankee filibusters to overrun it in time of peace. When Russia's minister in Washington, Baron Édouard de Stoeckl, reported that his government was willing to part with the gigantic province for $7.2 million (less than two cents per acre) Seward kept him up all night drafting a treaty. The press spun the deal as a diversion and an extravagance, in light of the Civil War debt. Radical Republicans were loath to bless anything the Johnson administration proposed. But Seward's détente with Sumner, dating from the Mexican business the previous year, paid golden dividends. The chairman of the Senate Foreign Relations Committee put in a week of intensive research, then regaled his colleagues with a three-hour sales pitch for the empire known by the Aleuts as Alaska ("Great Land"). Sumner assured the Senate that Alaska's resources of fish, timber, furs, most likely gold, and (not to be sniffed at) ice, would cover the cost of purchase many times over. He recalled how the acquisitions of Louisiana, Florida, Texas, and California had all been scoffed at initially. He reminded the chamber of czarist Russia's support for the Union cause during the rebellion. He boasted of removing another monarchy from North America. He warned that the British would grab Alaska if the United States refrained. He conjured up Pacific empires and rich commerce with Asia. Most of all, he gave the Radicals political cover by associating the treaty with himself at least as much as with Johnson and Seward. The Senate ratified it on April 9, thirty-seven to two. Another fourteen months and copious bribes from Stoeckl would be needed to goose the House of Representatives into voting the cash. But Seward ensured the outcome by rushing the transfer ceremony at Alexander Baranov's old castle in Sitka and then *daring* Congress to strike Old Glory.[29]

Johnson's political comeback seemed undeniable when the elections of autumn 1867 lifted northern Democrats off the floor. Whether that represented a natural postwar pendulum swing or—as many surmised—a backlash against Negro suffrage, it made the president eager for a fight he could wage on his own terms. Johnson's annual message of December 3 conceded that "blacks in the South are entitled to be well and humanely governed," only to suggest that this might mean they needed to be governed by whites. "But under the circumstances this is only a speculative point. It is not proposed merely that they shall govern themselves, but that they shall rule the white race, make and administer State laws, elect Presidents and members of Congress, and shape to a greater or less extent the future of the whole country. Would such a trust and power be safe in such hands?" He went on to call congressional Reconstruction unconstitutional and ask northern voters to consult their fears. Sumner anathematized the message as "an incendiary document" designed "to provoke civil war." But the House disagreed. Four days later it voted almost two to one against a motion to impeach the president, and sixty-eight of the nays were Republican. Johnson charged ahead with an imperious lecture defending his dismissal of Stanton and the generals in southern commands.

What stalled Johnson's rebound was the ongoing battle over checks and balances. Republicans might waffle and differ on blacks' rights, Reconstruction, tariffs, and currency, but they had to rally in defense of Congress when the president refused to execute laws. In January the Senate censured Johnson and threatened Grant with prison for assuming "illegal" control of the War Department. The worried general promptly returned the office keys to Stanton. The livid president fired Stanton all over again, tried to entice Sherman, and settled on the adjutant general Lorenzo Thomas, the man who carried the keys. Needless to say the Senate refused to confirm Thomas, and the House reached for the ultimate weapon. On February 24, 1868, it voted to impeach the president before the charges had even been specified. At length, the House managers drafted eleven articles of impeachment, eight derived from the Tenure of Office Act. Johnson relished the fight. He suspected the Radicals had at last overreached themselves, and for once he guessed right, because moderate and conservative Republicans had their eyes fixed, not on the fate of blacks in the South, but on the coming elections. The unprecedented act of removing the president—the American equivalent of "killing the king"—might do the party terrible harm and could not possibly do any good.[30]

The impeachment of Andrew Johnson was an affair of high constitutional principle for Radicals and some moderates. To the rest of the Republicans and all Democrats, however, it seemed an act of raw partisanship, the outcome of which would hinge on individual calculations as to its ancillary effects. Some effects were imponderable. Would impeachment prolong military rule in the South, permitting Republicans to carry the region in November 1868? Or would northern and southern white voters recoil from impeachment and vote Democratic? Moderates and conservatives feared a vote to convict would brand them as Radicals in the minds of their constituents. Other ancillary effects were all too ponderable. If Johnson was ousted the presidency of the United States would devolve on the president pro tempore of the Senate, Benjamin Wade. An outspoken Radical in favor of high tariffs, soft money, and bullying tactics, Wade was the last person likely to unify the Republican Party. Indeed, it is likely that some moderates voted to make him president pro tempore because it would make the impeachment of Johnson more distasteful. By the same token some Democrats made a pretense of defending Johnson while secretly hoping he would lose. Everyone had to pretend that the trial was as lofty and solemn as *Julius Caesar* or *Macbeth*. So it dragged on for two months while Chief Justice Chase made tortured procedural rulings, the chief defense counsel William M. Evarts parsed the separation of powers, and the House manager Benjamin Butler pinned the "hopes of free institutions" on the outcome. Butler, incompetent during the war and now an ambitious demagogue, waved the first "bloody shirt" while saying that Johnson was "the elect of an assassin" implicitly in cahoots with Booth. No one bought that, but would enough people buy the charge of "high crimes and misdemeanors"? The Senate answered on a sunny spring afternoon while citizens in the gallery held their breath. All twelve Democrats plus seven moderate Republicans led by Trumbull and Fessenden chose to acquit Johnson of the charge deemed most likely to stick. The thirty-five Republicans voting to convict thus came up a vote shy. The date, May 16, was significant. Many Republicans were already aboard trains to Chicago, where their national convention was scheduled to open three days later.[31]

Was Andrew Johnson an abject failure? That depends on one's definition of success under the most trying conditions. He surely failed to build a new party and prolong his presidency (although he did make a comeback in Tennessee and returned briefly to the Senate, the only former president to do so). He also overestimated the powers inherited from Lincoln, and this mistake caused him to alienate moderates and invite the unfriendly takeover of Reconstruction by Congress. On the other hand, Johnson understood that sooner or later the freedmen would have to find their own modus vivendi with southern whites, a process that military rule would render more violent because it was bound to inflame the prejudice of the crackers he knew so well. So if one adopts a negative definition of success— frustration of Radical Reconstruction—it might be said that Johnson prevailed after all, because the Republicans who gathered in Chicago in 1868 went out of their way to shun their own Radical wing.

* * *

One night in April 1868 Julia Grant popped the question. "Ulys, do you want to be president?" General Grant honored his wife's frankness with some of his own. "No. But I don't see that I have anything to say about it. The national convention is about to assemble and, from all I hear, will nominate me, and I suppose if I am nominated, I will be elected." For months Republican heavyweights had probed Grant's politics and intentions. He just puffed on cigars beneath Lincoln's bust in his parlor in Washington, grunted, or changed the subject to horses. The good general had no use for politicians, parties, or an office whose torments he had witnessed firsthand. Then Wade queried Grant's brother-in-law in Kentucky and learned that the general considered himself a Republican sympathetic to freed slaves. It was all Wade needed to hear. The party nominated Grant on the first ballot and paired him with a moderate congressman, Schuyler Colfax of Indiana. The platform was more moderate still. It congratulated the American people "on the assured success of the reconstruction policy of the Congress," and pledged the federal government would prevent the South from reverting "into a state of anarchy or military rule." Those code words warned southern whites to refrain from violent abuse of civil rights, but otherwise hinted that the North was ready to declare victory and go home. Grant's acceptance letter gave the campaign its motto: "Let us have peace." Then he repaired to quiet Galena, Illinois, half hoping he would end up the loser.

Republicans had cause to be cautious. By fall 1868, reactionaries and Klansmen were already intimidating Republican candidates and black voters in some southern states, and the northern Democrats recaptured their base now that secession and war were history. Their presidential candidate, Horatio Seymour of New York, even broke with tradition by touring on his own behalf. But unlike Andrew Johnson, he delivered smooth speeches in staged settings that promised easy credit and an immediate end of Reconstruction. Seymour's 47.3 percent of the popular vote stunned Republican bosses. To be sure, Grant won the electoral college by a comfortable 214 to eighty. But he lost New York, New Jersey, Maryland, Delaware, Kentucky, and Oregon; barely carried Pennsylvania, Indiana, and California; and, most ominously, won five southern states thanks only to Negro votes.[32] The reciprocal fraud in the Deep South can be appreciated by comparing these numbers: Grant won South Carolina with 58 percent and Alabama with 51; Seymour swept Louisiana with 71 percent and Georgia with 64. The white vote nationwide split fifty-fifty. That first postbellum election made it crucial for Republicans to capture moderate northern white voters while somehow defending rights deemed radical for southern black voters. Congress tried to work that magic a month after the election by forwarding to the states a Fifteenth Amendment. It affirmed that the right of all citizens to vote "shall not be denied or abridged by the United States or by any State on account of race, color, or previous condition of servitude" and empowered Congress to enforce the provision by (the usual weasel word) "appropriate" legislation. The amendment sounded radical. But nothing in it precluded states in the South—or, for that matter, in the North—from imposing literacy or property requirements that in effect disqualified almost all blacks, Irish, Chinese, or Italians depending on the locale. In other words, it left the door open for states to pretend to uphold equal rights simply because their suffrage laws were not based on race. That was exactly what Andrew Johnson had recommended southern states do in 1865. Southern states under Republican control ratified the amendment at once, but so did states slipping back under Democratic control. The loopholes were too gaping to miss.[33]

Poor Grant. Democrats hungry for office branded his administration the most corrupt ever, though it was no worse than most and the president was above reproach. Republicans hungry for office turned against Grant because he chose his own cabinet and spurned the spoils system. Many easterners decried Grant's tight money policy. Westerners resented his pacific policies toward the Indians. What mattered most for Reconstruction was that southern whites hated Grant's efforts to uphold civil rights, and northern whites tired of the conflict those efforts engendered. Nor did Grant have the time or resources needed to plot a new course in the South. Congress kept slashing spending, with Grant's approbation because he pledged to make the dollar "good as gold" again. Since the most vulnerable budget line was that of the War Department, the 38,700 federal troops patrolling the South in 1866 were reduced to just 8,000 by 1870 and 3,000 by 1876: hardly enough to police Cobb County, Georgia. In any case Grant's replacement as commanding general did not want to patrol anything. William Tecumseh Sherman hated martial law and wanted to liquidate Reconstruction. What then of the Freedmen's Bureau, whose overburdened, underfunded officials had done much to promote education and welfare for blacks? Congress abolished the bureau in 1869 on the pretense that Reconstruction had "succeeded."[34]

Yet the same Congress pronounced itself shocked when sensational hearings exposed the terror visited on "uppity" blacks by the Ku Klux Klan, the Knights of the White Camelia in Louisiana, and desperadoes in Texas killing and robbing for sport. Contrary to the twentieth-century image of Klansmen as white-robed ghouls pretending that a lynching was a solemn judicial act, the original "night riders" performed their rituals as if they were flamboyant circus acts or minstrel shows. They donned all manner of weird costumes including animal skins and women's dresses, hooted and keened outside Negroes' shacks, and butchered or burned alive victims as if they were animal sacrifices. Frustrated by their failure to win a courageous war, Klansmen took out their frustration in cowardly persecution. Congress indulged its knack for unfunded mandates by passing four Enforcement Acts obliging President Grant to defend blacks' Fifteenth Amendment rights. By golly, he tried. Since violation of civil rights was now a federal offense, the president ordered the attorneys of the new Department of Justice to indict hundreds of suspected Klansmen and threaten them with black juries (usually enough to pry out a confession). That, plus a minimal show of force by the military, buried the Klan by 1872.[35]

Of course, Americans never met a statute they could not get around. Paramilitary Democrats simply began to march in broad daylight, without hoods (hence legally) as members of "clubs" like the South Carolina Red Shirts and Mississippi Rifles. Every election season their conspicuous presence in key precincts and county seats discouraged blacks from venturing to polling places and encouraged whites to venture there. But far more important than intimidation was the shift in southern Democrats' strategy. They simply reinvented themselves as redemptionists pledging to redeem their states in the name, not of race, but of home rule and good government. Their growing strength plus an evident loss of will among northern Republicans to "stay the course" restored southern whites' morale. Virginia fell under Democratic control the moment it regained full statehood in 1870. North Carolina flipped the same year and Georgia the next when Governor Brown bolted the Republican coalition and took most of north Georgia's dirt farmers with him. Texas returned to the party of Sam Houston in 1873. It seemed just a matter of time before Republican rule would vanish even in states with heavy concentrations of blacks—South Carolina, Alabama, Mississippi, and Louisiana—unless the federal government poured in money and put more boots on the ground. Political and economic trends in the North rendered that extremely unlikely. And this in turn gave America's fulcrum, the border states, the chance they had been denied in 1861 to strenghten the Union and define race relations for the foreseeable future.[36]

One mighty political trend was the literal and figurative death of the Radical Republican wing. Thaddeus Stevens died in 1868 boasting of his humility (he asked to be buried in a grave for indigent Negroes). Stanton died the following year and Sumner in 1874. Wade lost his seat when the Democrats took over Ohio in 1869. The only young crusader to emerge was George Frisbie Hoar. A Harvard graduate and cofounder of Worcester Polytechnic Institute, Hoar entered Congress in 1869. He vowed to push New England's economic agenda, but not at the expense of "the needs of the Freedmen" in the face of "Southern outrages." Hoar's panacea was public schools, and his motto was, "Send down Freedom and Education to the South." But his was a lonely voice because a second political trend was the drift of mainstream, or "stalwart" Republicans into the torpor and influence-peddling characteristic of parties too long in office. Let the lawyer Roscoe Conkling of Utica, New York, stand for them all. He was a big, bearded man with a spit curl; his preferred relaxation was boxing; and his sole occupation was garnering patronage for his own profit and power. Anyone not in his pocket was by definition a rival, including his fellow Republicans Schurz, Greeley, James G. Blaine, Rutherford B. Hayes, and Chester A. Arthur. Conkling even froze out his daughter for marrying the wrong man. Conkling's principal gift to the nation during twelve years in the Senate was the cynical lie that he had purposely worded the Fourteenth Amendment to protect, not Negroes' rights, but corporate property.[37] It did not take long before such leadership conjured an intraparty revolt among would-be reformers. They called themselves liberal Republicans. The "stalwarts" called them "half-breeds."

The revolt began, significantly, in Missouri, Tennessee, and West Virginia. A critical mass of whites in the border states held recalcitrant southern "bourbons" and corrupt northern capitalists in equal contempt. That is not to say that whites in the border states were enlightened. Federal indictments for lynching, arson at black churches and schools, and electoral and judicial fraud in Kentucky, Tennessee, and Missouri exceeded those in the Deep South. Tennessee, West Virginia, and Delaware pioneered poll taxes and segregation. But the border states also led in melding freedmen into the wage labor market and granting them education and limited legal protections. In short, they offered a compromise based on second-class citizenship for blacks that the whole nation might find it could live with. Most of all, liberal Republicans in the border states wanted to liquidate sectional conflict so that Americans could get back to building their glorious future. Accordingly, their platform stressed, not issues of race, but lower taxes, hard money, reform of the civil service, free trade, and free enterprise without the sweetheart deals embraced by corporate Republicans.[38] Since Grant presided, if only naively, over those evils, the "half-breeds" wanted him out.

The liberals made high-profile converts of the midwesterners Trumbull, Schurz, and Chase, then won over national figures such as Greeley, Gideon Welles, Charles Francis Adams, and William Cullen Bryant. When their convention in Cincinnati nominated Greeley and a running mate from Missouri in 1872, Grant faced an unpredictable three-way contest. When the Democrats, lacking star candidates of their own, cut a deal with Schurz and nominated Greeley as well, it seemed that the campaign's main issue would be Grant himself. Instead, the main issue was Greeley. The towering journalist had grown eccentric in his old age. He touted vegetarianism, women's suffrage, labor unions, and other causes still too outré for the mainstream. He also seemed to forget that he was the Democrats' candidate; he tossed gratuitous insults at their leaders both past and present. Stalwart Republicans, by contrast, smartly stole the insurgents' agenda by promising lower taxes, lower tariffs, and reform, then "waved the bloody shirt" by reminding voters who had crushed the rebellion (Grant) and who had advocated, not just pardons, but amnesty for all rebels (Greeley). Best of all, a war scare turned into a diplomatic triumph for Grant (his only one, really) when an international tribunal obliged Britain to pay the United States $15.5 million in damages for having failed to exercise "due diligence" with regard to the Confederate commerce raiders built in its shipyards.[39]

Consequently, Grant won reelection with the largest percentage of the popular vote (55.6 percent) garnered by any candidate between Andrew Jackson and Theodore Roosevelt. He even carried the Deep South except for Georgia and Texas, a sure sign that some blacks were still voting. But like so many failed insurgencies in American politics, the liberal Republicans' "moment" presaged new realities that even the winners had to acknowledge. The border states had declared themselves sick and tired of sectional conflict, and most northerners replied, "So are we." Grant heard vox populi when he lowered the military profile in the South. So did Hoar, whose commitment to the rights of blacks did not stop him from becoming a half-breed and endorsing Greeley. So did the Democrats, who demanded an end to Reconstruction and pledged not to overturn its amendments and laws. In other words, everyone fled for the middle.[40]

Then the world economy stumbled. Europeans would call it a great depression. Americans just referred to the Panic of 1873, when Jay Cooke's financial house of cards fell and Wall Street shut down for two weeks. Railroad construction ceased. Factories closed or reduced wages, provoking the largest industrial strikes in American history. Congress picked that very year to vote itself a retrospective pay raise, provoking a national outcry. Grant's treasury had to reverse course by issuing more greenbacks to forestall a severe deflation. Cotton prices fell by half anyway, forcing southern white farmers into sharecropping arrangements no better than what the blacks had. That made rural whites all the more determined to enforce a racial hierarchy. Most heartbreakingly, some 60,000 African-Americans who trusted the Republicans' free-market ethic lost their life savings when the Freedmen's Savings Bank went bust in 1874. Frederick Douglass poured his own money into the bank to no avail, and the government was not equipped to do anything absent a central bank. In the elections of fall 1874 the Republicans lost control of the House of Representatives; and two more southern states—Arkansas and Alabama—were "redeemed" by Democrats.

Even the Supreme Court piled on, if inadvertently. In 1873 it ruled in the famous Slaughterhouse cases that the state of Louisiana could charter a butchers' monopoly (for reasons of public health) inasmuch as the equal protection clause of the Fourteenth Amendment applied only to rights conferred on former slaves by the federal government. In all other matters states were still free to discriminate against certain groups.[41] That established the precedent for the decisions of 1876 in *United States v. Cruickshank* and *United States v. Reese*. In the first, the Court threw out murder convictions won by Grant's prosecutors in deaths resulting from a race riot, on the grounds that federal authorities had jurisdiction only in cases where states, not individuals, violated civil rights. In the second, the Court held

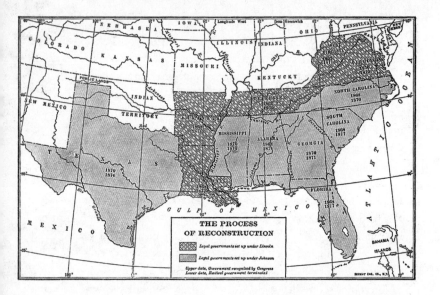

what everyone already guessed: the Fifteenth Amendment guaranteed no one the vote, but only proscribed denial of suffrage on the basis of race.[42] White rabbles celebrated by mobbing all the more. Frightened black communities formed militias in self-defense, but in vain. As with Indians, any act of violence by a Negro so enraged insecure whites in places like Memphis, Mobile, or Baton Rouge that they demanded all blacks be disarmed and repressed. Grant felt shamed by the violence, but he had not a hundredth of the army needed to keep peace in the South and anyway had begun to suspect that "the spirit of hatred and violence is stronger than law."

Early in 1875 the outgoing Republican Speaker of the House James G. Blaine rang the knell for Reconstruction when he rejected Grant's request for suspension of habeas corpus. "It could not have saved the South," said Blaine, but would have made defeat across the North "a foregone conclusion" for Republicans. One of his colleagues put it more colorfully: "Our people are tired out with this worn out cry of 'Southern outrages'!!! Hard times & heavy taxes make them wish the 'nigger,' 'everlasting nigger,' were in [hell] or in Africa." As state elections approached later that year, the president himself threw in the towel. Mississippi's carpetbag Governor Adelbert Ames wired an anxious plea for more soldiers, even though he knew it would burst "like an exploding shell to the political canvass at the North." Grant grumbled to his attorney general, "The whole public are tired out with these annual, autumnal outbreaks in the South, and there is so much unwholesome lying done by the press and people in regard to the cause and extent of these breaches of the peace." In other words, having done all he could for southern Republicans, Grant now pretended the threat was exaggerated to justify giving up. No troops were deployed. Democrats depressed the black turnout in enough key districts to take over the legislature. Ames was impeached; Mississippi was "redeemed." [43]

Only a farcical anticlimax remained. In 1876 Americans celebrated their centennial year by staging a thoroughly corrupt presidential election between two parties running against corruption. Democrats sensed that outrage over the "Grant" scandals, economic distress among northern workers and farmers, and the resurgence of white southerners' power added up to a grand realignment. Their platform called for more greenbacks, abolition of federal banks, reform of the civil service, and evacuation of the last paltry garrisons from the South. Their candidate, the former governor of New York, Samuel J. Tilden, was known as the man who had cleaned up Tammany Hall and Albany. He pledged to do likewise in Washington. Republicans countered by styling themselves the reformers, a pose Grant assisted by juggling his cabinet to woo back "half-breeds" and root out fraud in the revenue services. Their candidate, Governor Rutherford B. Hayes of Ohio, was squeaky-clean; their strategy was cleverly soiled. Republicans pledged both a sound dollar *and* moderate expansion of the money supply. They "waved the bloody shirt" even as they abandoned Reconstruction (Hayes endorsed "local self-government"). Most of all, they distracted northern voters from substantive issues by whipping up anti-Catholic hysteria. Nothing proved the Civil War era was ending more than this reprise of the Know-Nothings' agenda. Protestant leaders insisted that the pope, whom the Vatican council of 1871 had pronounced infallible on matters of faith and morals, meant to indoctrinate American children and subvert the republic. Grant himself launched the diversionary campaign in December 1875 when his annual message called for an amendment banning public money for parochial schools. The Republican platform gave it a prominent place. Neither party said much about Negroes.

It served them right that neither candidate won. Tilden seemed to win the popular vote by 51 to 48 percent, but ubiquitous fraud made even that dubious. Violence peaked, with Aiken County, South Carolina, hogging the headlines. Low-country blacks had been among the first liberated and were most militant in defense of their rights. In July a white mob fought a pitched battle against black militiamen in the town of Hamburg that ended when five freedmen were pulled from jail and shot in the back. At nearby Ellenton in September, paramilitary Red Shirts leveled cannons at black militiamen holed up in a church, killing 100. The governor insisted that the "rifle clubs" disband. They just renamed themselves the Mounted Baseball Club or the Baptist Church Sewing Circle. The state's Democratic leader even declared every white man "honor bound to control the vote of at least one Negro, by intimidation, purchase, keeping him away, or as each individual may determine." The election returns were risible. In states already "redeemed" such as Arkansas, Mississippi, and Alabama, precincts that voted for Grant in 1872 miraculously reported no Republican votes at all in 1876. In states still up for grabs, like Louisiana, Republican registrars tallied votes they said would have been cast by Negroes if not for the violence. The result in the electoral college was sillier still. Tilden's 184 votes would have put him over the top four years before. But they didn't in 1876, because the Republicans had slipped another eleventh-hour new state into the Union, leaving Tilden a vote shy of becoming president-elect. This meant the outcome hinged on twenty disputed votes: one from Oregon, where Republicans accidentally named an ineligible elector; and nineteen from South Carolina, Florida, and Louisiana, where biracial coalitions still hung by threads.

Lawyers from both parties flocked to those states hoping to execute slow-motion steals. "Money and intimidation can obtain the oath of white men as well as black to any required statement," wrote a Yankee observer of the recount in Florida. "A ton of affidavits could be carted into the state-house tomorrow, and not a word of truth in them." Local Republicans won the first round by certifying all the disputed votes for Hayes. But Democrats had a fire wall: their majority in the House of Representatives. They refused to convene in joint session unless the Republican president of the Senate recused himself from the job of counting the tally (the Twelfth Amendment just stated "the votes shall then be counted"). Should no victor be declared, the election would be thrown into the House. So the Republican "stalwarts" dug in their heels until it seemed that Americans might have no president on Inauguration Day. That prospect unsettled businessmen on Wall Street who prevailed on (who else?) Roscoe Conkling to break ranks and cut the first in a series of deals. On January 26, 1877, Congress established a bipartisan commission numbering ten of its own members plus five Supreme Court justices of whom the swing vote was David Davis, a liberal Republican. It is hard to imagine he would not find cause to award Tilden at least one of the twenty ambiguous votes, but nobody knows, because the Democrats of Illinois stupidly chose that moment to make Davis their senator. This forced the commission to choose a replacement from the remaining justices, all of whom were Republican stalwarts. The commission made a show of weighing the merits, then ruled eight to seven in every case in favor of Hayes.

A series of deals? Yes, because Republicans had to blunt a new threat by the Democrats to abort the inauguration by recessing Congress. Grant wielded the stick by refusing to withdraw U.S. Army protection for the rickety Republican governments in Columbia and Baton Rouge until Hayes was safely in office. The Hayes camp dangled the carrot by promising to keep hands off the South, name a southerner to the cabinet, and support the Texas Pacific Railroad. Nobody knows just how this Compromise of 1877 occurred, because it was done with secret handshakes, not signatures. But the upshot was that Rutherford B. Hayes was inaugurated on March 5 without incident. Having weathered secession, civil war, and impeachment, sensible Americans knew that pretending the president was legitimate was a small price to pay for tranquillity. Hayes did his part by promising "the permanent pacification of the country upon such principles and by such measures as will secure the complete protection of all its citizens in the free enjoyment of all their constitutional rights."[44]

Within a month those last two southern Republican governors were in exile.

The Thirty-Eighth State: Colorado, 1876

"Pikes Peak or bust" boasted westbound Fifty-Niners in their Conestoga wagons. "Busted by God!" grumbled eastbound "go-backers" who blamed the purveyors of picks, guidebooks, and town lots for spreading rumors of Rocky Mountain gold. One disconsolate party thought better of lynching a promoter they met on the trail, but planted a tombstone anyway: "Here Lies D. C. Oakes, Killed for Starting the Pikes Peak Hoax." It wasn't a hoax: the famous journalists Henry Villard and Horace Greeley confirmed "unparalleled riches" that made the western reaches of Kansas Territory a "new Eldorado." Nor did promoters start it: some Cherokee did. Having noticed gold flakes near the confluence of Cherry Creek and the South Platte River, they guided William Green Russell to the site in 1858. The poke of gold his prospectors panned that summer was enough to captivate midwesterners left high and dry by the Panic of 1857. Russell immediately planted a town named for his home in Georgia, Auraria. But competition arrived in the person of William H. Larimer, a Pennsylvanian politician and speculator. He founded a town across Cherry Creek, courted patronage by naming it after the former governor of Kansas James W. Denver, and shrewdly gave away lots to Freemasons, churches, families, and a stagecoach line. In 1860 Denver absorbed Auraria to form an "instant city" of 5,000 including "as sharp and sagacious merchants, as shrewd real estate speculators, as cunning and ambitious lawyers, as numerous doctors, as fine looking young men, and as handsome and stylish women" as any west of Saint Louis. They included the twenty-year-old feminist Julia Archibald Holmes, the first woman to scale Pikes Peak; and Ada Lamont, the nineteen-year-old wife of a preacher but now turned queen of a bawdy house. Larimer exulted, "We are bound to have a territory if not a state, and the capital will be Denver City." William N. Byers dragged a printing press across Nebraska to found the *Rocky Mountain News* "because we wished to help mold and organize the new population, and because we thought it would pay." The census of 1860 counted 34,277 people in the Pikes Peak district hungry for more government than the miners' "claims clubs" provided. Congress obliged in 1861. But the Committee on Territories rejected the settlers' name, Jefferson, in favor of Osage, rejected in turn by the residents. Others suggested Yampa (mountain bear), Idahoe (mountain gem), Lula (mountain fairy), Arapahoe (mountain Indian), and Tahosa (mountain dweller) until the miners' delegate worked the cloakroom to pass a bill creating the Territory of Colorado.[45]

The "spherical rectangle" traced by its arbitrary boundaries encompassed four topographical zones. Colorado's eastern third is sparsely populated grassland, "western Kansas" indeed. But the fifty-mile-wide Piedmont east of the mountains is hilly, cooler, and watered by snowmelt. That was the corridor destined for 80 percent of the population. The Piedmont is also a wind tunnel, especially when the dreaded winter chinook roars down from the third zone, the great southern Rockies. That youthful, jagged range on the Continental Divide has fifty-three peaks above the timberline (11,000 feet) and isolates the dry, forested Colorado plateau to the west. Cottonwood, oak, juniper, piñon pine, fir, and spruce thrive at their preferred elevations; and elk, deer, mountain sheep, and their predators visit them all as the seasons turn. Bracing air and wondrous scenery drew tourists well before the founding of Rocky Mountain National Park in 1915, and then Estes Park, Garden of the Gods, Royal Gorge, Mesa Verde, and Red Rocks parks. But Colorado was also bloodied, polluted, and scarred in the rush to extract its mineral wealth.

The 1860s were disastrous. William Gilpin, Lincoln's first governor, no sooner arrived than he faced an invasion by Confederate Texans. The First Colorado Infantry weathered that threat in March 1862 at the battle of Gloriéta, New Mexico, but it also propelled John Chivington to the leadership role he exploited to slaughter the Cheyenne at Sand Creek. Hailed by the *Rocky Mountain News* as one of "the brilliant feats of arms in Indian warfare," the massacre led to five more years of war with the Indians. Meanwhile, the easily refined surface gold played out and greenbacks played havoc with prices. The territory's metallic income fell from $7 million per year to just $2 million by 1868. Another wave of "go-backers" left ghost towns behind, and Denver shrank amid fire, flood, and bankruptcy. But boosters led by the second governor, John Evans, just would not give up. Evans—who was a physician, scientist, railroad man, speculator, and Republican stalwart—had helped to develop Chicago's suburb of Evanston and its Northwestern University. He applied all his talents to persuade the legislature to make Denver the permanent capital, design a government modeled on that of Illinois, lure railroads, and recruit the eastern capital and expertise needed to smelt "refractory" ore.[46]

It all turned around in 1870, when the silver spikes were driven to finish the Denver Pacific and Kansas Pacific, and a railroad boom followed. Soon more than 100 feeder lines crawled up the passes to mines and smelters, and another trunk line ran south to Pikes Peak, where Colorado City sprang up, and to the iron and coal deposits that made an industrial city of Pueblo. In Golden (where the Prussian immigrant Adolphe Coors founded his brewery in 1873) factories sprouted to service the mines and railroads. Only statehood aborted. Evans's "Denver crowd" put together a bare majority over resistance from a rival "Golden crowd" plus Mexican settlers cheated of their land grants and miners fearful of taxes. But Andrew Johnson vetoed statehood, and while congressional Radicals were in no mood to fight for greedy frontiersmen who massacred Indians and denied Negroes the vote. Statehood was postponed until the Republicans needed to increase their electoral count in 1876. The Coloradans were ready with a revealing state constitution. It promised strict regulation of corporations yet provided no enforcement mechanisms lest it scare off outside investors. Voters also rejected public funding for parochial schools, rejected women's suffrage despite (or because of) a campaign led by Susan B. Anthony, and affirmed their "reverence for the Supreme Ruler of the Universe." Three days before the nation's centennial they approved the draft, almost four to one, allowing President Grant to welcome the Centennial State on August 1.[47]

That same year, two miners gleaning an old goldfield near the source of the Arkansas River guessed that the despised "blue stuff" in the ground meant silver. A refining company in Saint Louis sent a metallurgist trained in Germany to examine the site. Sure enough, the carbonates were loaded with silver and lead, but chemical expertise on an industrial scale would be needed to tease out the metals. Within four years the gulch became Leadville, a city of 15,000 boomers perched 10,000 feet above sea level. Its thirty mines, fourteen smelters, and Alpine railroad offended the eye and nose, but its yield ($11.5 million in 1880 alone) was second only to that of the Comstock Lode and the mix of other metals such as lead, zinc, and copper helped Colorado survive a severe drop in the value of silver in 1893. Production soared from $35 million in the 1870s to $185 million in the 1880s. But Colorado needed the skills of people like Samuel F. Emmons of the new U.S. Geological Survey, and the capital of people like Andrew Carnegie. As romantic mining frontiers turned into corporate fiefdoms, the gold strike of 1891 on Cripple Creek proved to be the last hurrah for the individual prospector. By 1901 the Colorado School of Mines in Golden had acquired a national reputation and the American Smelting and Refining Company (with Rockefeller and Guggenheim money) had locked up the state's minerals in a gigantic trust.[48]

"There the great braggart city lay spread out . . . upon the brown and treeless plain, which seemed to nourish nothing but wormwood and Spanish bayonets." The treeless plain described by Isabella Bird in 1873 proved more important than mining to Colorado's future. The soil of the Piedmont just needed strong backs and water. Civil War veterans and their willing wives provided the former when Nathan Meeker of the *New York Tribune* formed Union Colony in a northern Colorado town that he named for his boss, Horace Greeley. An editor in Illinois scoffed: "Greeley is located . . . on a barren, sandy plain, part and parcel of the Great American Desert, midway between a poverty stricken ranch and a prairie-dog village." But the success of the cooperative inspired imitators from Chicago, Saint Louis, and elsewhere. Water was the concern of a young engineer from Ohio, Benjamin Eaton. When his teenage wife died in childbirth, Eaton tried farming in Iowa and prospecting and ranching in Colorado until the Union Colony's need for irrigation called him back to his first profession. Eaton dug a network of canals in the South Platte valley that made thousands of arid acres into gardens. He lobbied for Irrigation Acts in 1879 and 1881, served a term as governor, and then founded the Windsor Reservoir and Storage Company in 1890 to conserve each winter's snowmelt. Year-round water from reservoirs permitted farmers to plant a third crop of alfalfa, late potatoes, and sugar beets. When Eaton died in 1904, the *Denver Times* wrote that "thirty years ago the world was astonished at his vision of making fertile vast areas of barren waste . . . its agricultural yield now is many millions beyond the wealth of gold mines." [49]

The last chapter in the saga of development was, as usual, Indian removal. The Ute managed to share the Rocky Mountains with prospectors until gold was found on the western slope and whites streamed into the San Juan River country. In 1879, Meeker hoped to rescue the Ute by putting their hands to the plow in another agricultural colony. The Indians fled, Meeker summoned the army, and a road to hell was paved with good intentions. A band of distrustful Ute killed a dozen men, including Meeker himself, and kidnapped Meeker's wife and daughter. White Coloradans decided that the "Ute must go!" By 1881, when the last of them walked a trail of tears, another former governor of Kansas was already busy platting the town of Grand Junction on the western plateau. [50]

Thanks to mining, farming, industry, and land speculation, Denver, the "mile-high city," became the queen city of the plains after all. Denver numbered over 100,000 by 1890 and its initial gender gap (six males to one female) had almost disappeared. Eastern fashions and culture were on display in Victorian estates, theaters, and public buildings that received distinguished visitors including royalty, divas, and Theodore Roosevelt. Gone were the "go-backers," bushwhackers, swindlers like those who seeded mines in the "great diamond hoax," cardsharps like "Poker Alice" Stubbs, and confidence men like "Soapy" Smith. Politics took on a veneer of civilization by institutionalizing corruption, vice, and bosses. (One reform candidate blamed his loss on "15,000 gamblers and lewd women" who stormed city hall to protest his effort to clean up the police board in 1891.) For two decades Mayor Robert Walter Speer reigned over Denver, to the disgust of Populists and the Women's Christian Temperance Union. A plucky Pennsylvanian who usually wore a bowler hat, Speer went west in 1878 seeking relief from tuberculosis and soon built an urban Democratic machine in league with saloon keepers, ward heelers, and Republican businessmen happy to give him city hall in return for control of the state. When prohibition was passed by a Progressive-evangelical coalition in 1916, it actually enhanced the chances for underground profits, police payoffs, and political favors. But Colorado remained a "businessman's state" in part because labor unions, women's movements, Hispanic Catholics, moral crusaders, and muckraking journalists had competing agendas. Nor was rule by bosses necessarily bad. For example, Speer initiated Denver's "city beautiful" movement and spent lavishly to expand parks, infrastructure, and schools. He preached, "The time will come when men will be judged more by their disbursements than by their accumulations. Denver has been kind to most of us by giving to some health, to some wealth, to some happiness, and to some a combination of all. We can pay a part of this debt by making our city more attractive." [51]

If anything besides scenery pops into Americans' minds about Colorado, it is likely to be the Air Force Academy. That, too, was Denver's creation. Realizing that the Piedmont's climate and open expanses were ideal for aviation, the city fathers saw an opportunity in 1938, when General H. H. Arnold began preparing the Army Air Force for war. The city floated $750,000 in Depression-era bonds to purchase land for a base and bombing range and then donated them to the military. Lowry Air Base emerged, and then Buckley Field, to train pilots after Pearl Harbor. The investment paid for itself 100 times over, even though the Air Force Academy departed Denver for Colorado Springs in 1958. In the 1890s, when the silver collapse hit the western slope, there was talk in Grand Junction about secession from Colorado, if only in hopes that "in time Denver may learn that she is not the state."[52] More than a century later all one can say is: don't bet on it.

After the Compromise of 1877, the *Chicago Tribune* proclaimed, "The long controversy over the black man seems to have reached a finality." Most white Americans wanted that to be true; this is why they bestowed Henry Clay's title "the great pacificator" on Rutherford B. Hayes. Many teachers and students still want to believe, for simplicity's sake, that Reconstruction ended when the Democrats ousted the "black Republicans" and made the former Confederacy the Solid South. Reality, as always, confounds our wants.

First, the "black" Reconstruction regimes were primarily white everywhere but South Carolina and Louisiana. Over 600 African-Americans served in state legislatures during those years, but they were underrepresented relative to scalawags and carpetbaggers. No black served as a governor; just one served on a state supreme court; sixteen served in Congress; and two (the distinguished Mississippians Hiram Revels and Blanche K. Bruce) served in the Senate. The real damage to ordinary African-Americans' security, welfare, and dignity was done when white Democrats "redeemed" local law enforcement from hundreds of black sheriffs, judges, and justices of the peace.

Second, the curtain did not drop on blacks overnight. Southern states' voting laws remained in flux for the rest of the century as people of all classes, interests, and bloodlines negotiated what freedom meant in a new South struggling to find a niche in industrial America. Farmers of both races sometimes had a common cause in populist movements. Wage workers of both races joined the Greenback and Labor parties, sometimes in alliance with rump Republicans. In the 1880s biracial Readjuster parties briefly took back control of Tennessee and Virginia. Thus, no straight line led from the abandonment of Reconstruction to the Jim Crow era. Only gradually did Democratic elites—"the better class of whites"—learn how to exploit the conflicted loyalties of their electorates to gain unchallenged control. The race card was obviously trump. What did it mean to be free, white, or for that matter a man, now that blacks were no longer enslaved? Rather than define themselves as mudsills on a level with blacks, poor whites increasingly sought self-esteem in the notion that whiteness, not

legal rights or social status, conferred a superiority which only the Democrats could be trusted to maintain. But the social class card proved effective as well among middle-class whites who were no more willing to be
governed by "poor white trash" than by Negroes. Literacy and property
tests disenfranchised many whites as well as blacks, and rigged apportion-
ment of electoral districts, lynching, and fraud did the rest. ("It is the religious duty of Democrats to rob Populists and Republicans of their votes,"
insisted an editor in Louisiana. "Rob them! You bet! What are we here
for?") By the end of the century northern elites even handed southern
elites intellectual justifications for white "Bourbon" rule. So-called progressive trends in anthropology and a new discipline called political science blessed racial hierarchy, fostered distrust of democracy, and endorsed
government by experts—"the better class of men."[53]

What of the southern economy? Should Reconstruction be deemed a
failure because many former slaves just turned into peons for white landlords? Or was it a success because one-fifth of black farmers managed to
own some land by 1880, another 40,000 took up homesteads on the Great
Plains, and skilled former slaves were free to profit, however modestly, as
tradespeople or wage workers for railroads and industry? No generalizations are possible, since conditions in the upper South varied greatly from
those in the plantation "black belts," and the postwar markets for cotton,
sugar, rice, and tobacco showed unique patterns or no patterns at all. No
doubt many African-Americans, such as the elderly and disabled, found
themselves worse off than they had been under slavery. But then, everyone
in the South was worse off. Cotton prices were high just after the war, but
production was only 60 percent of what it had been in 1859. Once production recovered, the world market for cotton became so glutted that prices
tumbled from forty-three cents to just five cents a pound by the 1890s.
Then the boll weevil arrived to teach black sharecroppers that "all God's
dangers ain't a white man." All told, per capita income in the South
dropped 30 percent below its prewar (white) level and stuck there for the
rest of the century. Since money was scarce, most planters leased land to
black (and increasingly to white) farmers under the sharecropping system.
Farmers in turn had to pledge more of their harvests to merchants supplying tools, seed, and supplies. It was once thought that this crop lien system
played into the hands of factors who exploited black labor as surely as slave
masters had. Economic analyses now suggest the merchants' take averaged

a modest tenth of a farmer's income. But inasmuch as the rent demanded of sharecroppers rose and fell with the price of cotton, only a few were able to save enough cash to purchase their homesteads.

Given that emancipation elevated blacks' per capita income from nearly zero to 52 percent of whites' per capita income by 1880, one might claim that Reconstruction brought the most dramatic redistribution of wealth in American history. By the same token, however, the stagnation of the new South relative to the rest of the United States suggests Reconstruction freed rural blacks to live only as poorly as white subsistence farmers. Ignorance, illiteracy, hopelessness, laziness, envy, and fear held the South back. But most of all, it was starved for capital. The federal government, far from helping to rebuild the "eleven devastated provinces" that Winfield Scott had predicted, actually punished the South by withholding funds for internal improvements. Yankee capitalists found more lucrative places to invest once the carpetbaggers lost their enthusiasm. Jacksonian penury inhibited the local accumulation of capital. That made Booker T. Washington's passionate pleas for Negroes to win whites' respect through education, hard work, and thrift of circumscribed relevance. It didn't work anyway; a former black sheriff from Mississippi complained: "Education amounts to nothing, good behavior counts for nothing, even money cannot buy for a colored man or woman decent treatment and the comforts that white people claim and obtain." The idealistic carpetbagger Albion W. Tourgée's complaint was aptly summed up in his autobiographical novels—*A Fool's Errand* (1879) and *Bricks without Straw* (1880). "The South must take care of itself now," he wrote. "The nation had done its part; it had freed the slaves, given the ballot, opened the courts to them, and put them in the way of self-protection and self-assertion. The 'root-hog-or-die' policy . . . became generally prevalent. The nation heaved a sigh of relief." [54]

Politics and economics aside, the most damaging by-product of emancipation surely was segregation. Under slavery, racial hierarchy was mitigated and sometimes humanized by close interaction between the races. A potential for sympathy and even affection existed between masters or mistresses and domestic black servants; and even the field hands' shacks were usually near to the plantation big house. Under Jim Crow laws and customs, racial hierarchy was dehumanized by strict separation of the races in all things save debauchery. Du Bois felt this deeply. "In a world where it means so much to take a man by the hand and sit beside him, to look frankly into his eyes and feel his heart beating with red blood . . . one can imagine the consequences of the almost utter absence of such social amenities between estranged races. . . . Human advancement is not a mere question of almsgiving, but rather of sympathy and coöperation among classes who would scorn charity. And here is a land where, in the higher walks of life, in all the higher striving for the good and noble and true, the color-line comes to separate natural friends and co-workers; while at the bottom of the social group, in the saloon, the gambling-hell, and the brothel, that same line wavers and disappears."[55] After the Civil War, did any cultural force exist in the South that might have reconciled the races, equal or not, to mere coexistence—which might, over time, have melted antipathy? One might expect that force to have been religion, but alas, one would be wrong.

For a century black Christians prayed for deliverance in the belief slavery was the root of all evil. Northern abolitionists knew it wasn't that simple, but they hoped evangelization plus education could develop former slaves into citizens and melt white men's hearts. After all, most Americans of both races were Baptists, Methodists, or Presbyterians whose shared gospel and ethics were infinitely more important than their liturgical idiosyncracies. Since slavery had caused the antebellum Protestant schisms, it seemed natural that the end of slavery should repair them. Yet northern divines who went south after the war quickly learned that not only did southern whites spurn institutional unity; southern blacks did as well. The freedmen and freedwomen wanted no part of white pastoral care and even insisted on founding their own southern denominations in defiance of northern black churches. After initial dismay, southern whites were happy to see the Negroes abandon the balconies and rear pews of their chapels. "Disguise it as we may," said a North Carolinian pastor, "our colored brethren are disposed to independent action—they want preachers and churches of their own."

A few dynamic black pastors resisted the tendency of both races to segregate. Henry McNeal Turner, a freeborn African Methodist Episcopal preacher, attracted biracial congregations. Elected to Georgia's Reconstruction legislature, Turner promoted blacks' education and economic advance while doing everything possible "to please the white folks." But "redemption" in 1871 cost him his seat; resistance from northern Methodists, both white and black, undercut him; and Jim Crow finished him off. By the 1880s Turner was so embittered that he damned the American flag as a "rag of contempt." Isaac Lane and Lucius Holsey fared no better. Their Colored Methodist Episcopal Church, founded in 1870, won initial praise from southern whites for its "respectful attitude" and charitable works. Holsey, the child of a white master and an enslaved mother, dreamed of a new South based on biracial paternalism inspired by faith. But since he refused to take orders from northern Methodist and African Methodist Episcopal leaders, they denounced his movement as a "rebel church." Only around New Orleans did some integrated churches survive, but for social and political more than religious reasons. A few Catholics, such as Martin John Spalding, urged charity toward the freedmen in the belief that emancipation was "a golden opportunity for reaping a harvest of souls, which neglected may not return." But for the most part, the Roman church, too, surrendered to segregation.

Cultural styles had a great deal to do with it. Most African-Americans—90 percent of whom remained in the South—loved to be yanked to their feet by ecstatic, spontaneous preaching and loved to sway to spirituals and gospel chants. Respectable white folks spurned such "primitive" worship in favor of familiar hymns, communal prayers, and biblical exegesis. If a potential existed for biracial worship, it was represented by hillbillies, whose backwoods revival preaching and music was not so different from that of Negroes. Their cross-fertilization would eventually lure poor southern whites like Elvis Presley to black Baptist meetings. But a century passed before old-time religion began to transcend racial hostility and distrust. Did not the black clergy so conspicuous in the Union Leagues and Reconstruction regimes foster political evangelization and chastise parishioners who neglected their God-given duty to vote Republican? Did not white Redemptionists believe it their divine calling to stamp out the false "gospel according to Radicalism"? Freedmen testified how armed whites "used to come 'round" and listen for "the least thing nigger preacher say. . . . They whoop several."

Cerebral Presbyterians took up the work of codifying postwar white orthodoxy. Bitter-enders like Robert Lewis Dabney, a former adjutant for Stonewall Jackson, continued to pray that malicious Yankees might "be whirled aloft and plunged downward forever and ever in an endless retribution." More chastened theologians retained the notion of southern whites as a righteous remnant, but otherwise jettisoned the triumphalism they now regarded as the Confederacy's fatal flaw. Evidently the Lord did not bless secession or slavery. But scripture did mandate segregation and racial purity, which in turn made miscegenation an abominable sin. Southern Methodists and Baptists, who together made up 95 percent of white church members, embraced this "religion of the lost cause" that amounted, in effect, to a reformed southern civic religion. The influential theologian Benjamin Morgan Palmer took as his text the rebellion of Nimrod and the Tower of Babel. Clearly God had "divided the human race into several distinct groups, for the sake of keeping them apart." Since then all efforts to mix the races on the basis of "infidel humanitarianism" had been "providentially and signally rebuked." In 1874 Robert Lewis Dabney wrote that blacks used their new freedom to separate themselves in simple obedience to the "most controlling sentiment known to the human heart—*the instinct of race.*" Thomas Dixon, an archetypal child of Reconstruction, grew up on such theology in North Carolina. His father, a Baptist farmer and preacher ruined by the war, sought release from humiliation by joining the Klan. The son turned inveterate striver. He mastered ancient and modern languages, earned an advanced history degree from Johns Hopkins, and displayed a talent for journalism, acting, and politics. At length Dixon tapped all his talents as a phenomenal religious entrepreneur preaching, not the equality of all sinners in the eyes of heaven, but the inequality of Catholics and Negroes in the eyes of America. It was Dixon who wrote the pro–Ku Klux Klan novel *The Clansman*, which inspired the film *Birth of a Nation*, which in turn reinforced the solid South's version of Reconstruction in the minds of millions of Americans.[56]

How did segregationist theologians explain racial hierarchy? They could not argue (as many Yankees argued) that blacks had evolved from apes later than whites, because evangelicals rejected the theory of evolution published by Charles Darwin in 1859. So they clung to the biblical notion that blacks were the cursed children of Ham; or else they wandered off into polygenesis, holding that Negroes were not descended from Adam at all. It amounted, by 1890, to "a powerful theological imprimatur for American apartheid." Of course, segregation did not preclude benevolence. Joel Chandler Harris assuaged northerners' and southerners' conscience by sentimentalizing former slaves in the delightful *Uncle Remus* tales he began to publish in 1879. Moreover, the same churches that blessed segregation taught charity toward the downtrodden and needy. Northern philanthropists such as John D. Rockefeller and Andrew Carnegie donated millions to black education and welfare. Southern white charities chipped in as well. But as Du Bois would insist, human advancement is not just a matter of almsgiving, and whatever good may have come through emancipation, "the shadow of a deep disappointment rests upon the Negro people,—a disappointment all the more bitter because the unattained ideal was unbounded by the simple ignorance of a lowly people." Not for nothing did Du Bois begin and end his essay on Reconstruction with a prophecy: "The problem of the Twentieth Century is the problem of the color-line."[57]

The Civil War was not a tragedy in the classical sense—unless Americans' tragic flaw was their refusal to acknowledge tragedy at all. The Americans' civic religion simply did not permit them to imagine national failure, indelible sin, a pilgrim's regress. So even though millions of Caucasian, African, and Native American men and women emerged from the Civil War as losers in terms of pursuing their happiness, the dominant national memory of the war quickly became that of a glorious, victorious crusade. Even Memorial Day, first observed in 1865 and made a legal holiday by New York state in 1873, became an occasion for flowers and flags (Decoration Day) at cemeteries and ceremonies to honor the dead on both sides and the glorious chapter they added to the American epic. What, no soul-searching? How did that happen?

The most obvious answer is that history is written by the victors. Yankee intellectuals exploited their cultural dominance to profess that the war proved the moral and material superiority of northern civilization: survival of the fittest raised to the level of providence. There could be nothing tragic about a war that freed the slaves and saved the Union (ignoring the questions of why, how, and for what the war had been waged). Thus did Thomas Wentworth Higginson observe in 1870: "We are accustomed to say that the war and its results have made us a nation, subordinated local distinctions, cleared us of our chief shame, and given us the pride of a common career. This being the case, we may afford to treat ourselves to a little modest self-confidence." The key words are the first three: "We are accustomed." In truth, the Civil War caused a wound in American life that no stitches could mend, and so few cultural surgeons even tried to mend it. Except for the poems of Melville and Whitman, the war inspired almost no lasting literature. James Russell Lowell, Henry Wadsworth Longfellow, John G. Whittier, and William Dean Howells did not address the conflict. (Howells confessed that the war had "laid upon our literature a charge under which it has hitherto staggered very lamely.") Henry Adams spent the war in London. Henry James, Jr., ashamed to enjoy "almost ignobly safe stillness" while his younger brothers fought, expatriated to Europe in 1875. Ambrose Bierce and Oliver Wendell Holmes did witness the carnage, but emerged from it cynics, not tragedians. Mark Twain fled west (H. L. Mencken called him a draft dodger) and ignored the war's issues until he gently, belatedly touched on race in *Huckleberry Finn* (1884). Stephen Crane's naturalistic *The Red Badge of Courage* delved into the soldier's psyche, but he himself was not a veteran and did not publish until 1895. What Americans mostly read about their national passion were hundreds of repetitive, congratulatory memoirs suitably bowdlerized for women and children. Whitman expected "the real war will never get in the books," and critics a century later pronounced him correct. They told of "patriotic gore," "the unwritten war," the concoction of "incredible, self-serving myths," and "that spiritual censorship which strictly forbids the telling of truth about any American record until the material of such an essay is scattered and gone." One might add to that list of inhibitions the death of Romanticism (Romantics at least brooded on the mysteries of the human condition) and the birth of a literary realism focused on minutiae, manners, and muckraking. Realism, Bierce said, was "The art of depicting nature as it is seen by toads." To put it another way, Americans willingly chose not to see the forest for the trees because to do so might oblige them to admit that they had lost their way. Better to deem the war necessary, heroic, *over*, and move on.[58]

A second reason for the lack of soul-searching was the sheer exhaustion of the northern clergy. Having mortgaged Christian theology (or, in the case of Unitarians, whatever was left of it) to the Union's civic religion, they were hardly in a position to wring their hands over the war's results, cost, or attendant corruption. Before 1860, revivalist preachers and professors of divinity had captained the nation's conscience and had often driven its political discourse. After 1865, the clergy risked becoming irrelevant unless they trimmed their theology to suit political trends; and scientists and engineers assumed the leadership of universities devoted to material progress in the national interest. Lawrence Scientific School at Harvard predated the war, but it became a model for the whole Ivy League when the chemist Charles William Eliot became the president of Harvard in 1869. New institutions such as Cornell, Johns Hopkins, MIT, and the land-grant colleges modeled themselves on research-oriented German institutions, partnered with American industry, and followed the money. The remaining churchmen on faculties transformed themselves into philosophers, the better to weather storms of skepticism blown over from Europe. German and French "higher criticism" challenged the Bible's divine inspiration and inerrancy. Geology and paleontology challenged the Bible's accounts of creation. Darwin's *On the Origin of Species* (1859) and, even more shockingly, *The Descent of Man* (1871) challenged the Bible's teaching on the nature of humankind. So-called social Darwinists led by Herbert Spencer suggested that laissez-faire competition and conflict among races, nations, classes, industries, firms, and individuals were not only natural, but necessary for the improvement of the human race. The prominent publisher Henry Holt in New York said, "I got hold of a copy of Spencer's *First Principles*, and had my eyes opened to a new heaven and a new earth." A depiction of human life as dog-eat-dog competition struck everyone from businessmen to factory workers as far more credible than the "love thy neighbor" preachments of Christians or the utopian solidarity urged by socialists. What is more, many Americans were not overly troubled at first by the theory of evolution. On the contrary, it neatly meshed with their faith in unending human progress with America in the van. It also made the course and outcome of the Civil War seem a simple matter of natural selection rather than divine judgment on a nation's sins.[59]

The onslaught of scientific materialism threw religion on the defensive for the rest of the century. Fundamentalist Protestants would take refuge in an infallible Bible, Catholics in an infallible pope, Mormons in theocracy, and mystics in a crop of new cults. Charles Taze Russell founded Jehovah's Witnesses in 1872, Helena Blavatsky the Theosophical Society in 1875, and Mary Baker Eddy the first Church of Christ, Scientist in 1879. Mainline Protestant divines, however, invented one or another new, or liberal, theology in hopes of reconciling faith *with* science and preventing faith in science from sliding into atheism or, as was more likely, into pantheism. In practice, syncretic theology amounted to a surrender of Christian doctrines in the hope of salvaging Christian ethics. In other words, it reverted to the Puritans' old heresies of Arminianism, salvation by works rather than faith; and Pelagianism, the belief in man's power to perfect himself. Accordingly, many members of the Calvinist, Methodist, and Episcopal clergy (Baptists were immune) stopped pointing fingers toward heaven or at their jaded parishioners, and began pointing fingers outside their stained glass windows at the ills of society (i.e., other people's ills) that were frustrating America's quest for a heaven on earth.

In noxious industrial cities of the late nineteenth century they found ample squalor, poverty, ignorance, immigrants, socialism, booze, gambling, prostitution, and abuse of the Sabbath to fill the void left by the abolition of the nation's original sin, slavery. This social gospel movement, although cloaked in the rhetoric of Christian virtue, inveighed against Catholicism, often blessed segregation, and always extolled the same Anglo-Saxon mission that social Darwinists preached. To be sure, the social gospel systematized charity through organizations like the Women's Christian Temperance Union (1874) and the U.S. branch of the Salvation Army (1880). But a movement devoted to purging demons from the new urban America amounted to a confession that the millenarian interpretation of the Civil War had been wrong. Americans never doubted their calling to redeem the world, but henceforth secular intellectuals assumed the command posts while the clergy and laity peopled the trenches.[60]

To appreciate the demoralization of mainline Protestantism, look no farther than "Mr. Christianity," patriarch of America's first family of faith and pastor of its first megachurch. That would be Henry Ward Beecher. In 1861 the forty-eight-year-old stem-winding pastor of Brooklyn's huge Plymouth Church congregation had leaped to the van of cheerleaders for the Civil War. "We must not stop to measure costs," he would cry, but "put our honor and religion into this struggle." Equating the causes of church and state, Beecher helped recruit regiments, preached abolition, and pleaded the Union's cause in England. Lincoln gave him the honor of speaking at the flag-raising over Fort Sumter in 1865. Beecher emerged from the war the most famous American man of the cloth, by far the most richly paid, and the most influential, thanks to the *Independent*, a Christian journal bankrolled by Henry C. Bowen and edited by Theodore Tilton. The three men were devoted friends; mutual disciples; and apparent paragons of civil, religious, and family values. In truth, Bowen was a crooked customs official. The mother of his ten children, just before she died in 1863, confessed to an affair with Beecher—a bone Bowen swallowed in silence because Beecher's sermons kept his journal flush. Tilton kept quiet about ghostwriting Beecher's columns and having dalliances on the road while he peddled the paper. All three indulged in "free love" under the guise of supporting women's rights. Tilton was especially enamored of the fantastic spiritualist Victoria Woodhull, who insisted that marriage and Christianity must follow monarchy into America's trash can. Beecher cultivated discreet friendships with the feminists Elizabeth Cady Stanton and Lucy Stone. Where he stood on theology was anyone's guess because his theatrical sermons alternated between Calvinist granite and Universalist mush. "Orthodoxy is my doxy," he quipped, "and Heterodoxy is your doxy, that is if your doxy is not like my doxy."[61]

The Protestant mainstream did not know about Beecher's ménage or care about his theology so long as his politics and public demeanor met expectations. Politically, Beecher asked for trouble when he repented of his wartime crusade and endorsed Andrew Johnson's Reconstruction agenda. Of course, he insisted he was a good Republican (heaven forbid a Democrat in the pulpit!), but otherwise he warned against forcing blacks' civil rights on the South and cited the Klan's violence as proof that he was right. Beecher then mended his political fences by befriending Ulysses Grant. Tilton objected to such political trimming. Earnest about race issues at least, the editor anticipated Tourgée by insisting that the "Negro problem" was really white people's problem. "A freedman's bureau," Tilton wrote in the *Independent*, "is less needed than a rebel's bureau." Beecher's spiteful retort was to quit contributing to the journal. He also pocketed a $24,000 advance for a novel the publisher hoped would sell like his sister's *Uncle Tom's Cabin*. While writing it, Beecher received "encouragement" from Tilton's wife Elizabeth. The novel itself, a New England romance of flowers and bosomy sighs called *Norwood*, doubled as a seduction ploy tricked out as "new theology" that amounted to warmed-over Emerson. As Beecher explained, "I propose to delineate a high and noble man, trained to New England theology, but brought to excessive distress by speculations and new views. . . . The heroine is to be large of soul, a child of nature, and, although a Christian, yet in childlike sympathy with the truths of God in the natural world, instead of books." The characters would end up "nest hiding," which for Beecher meant that his "woman of nature and simple truth . . . is to triumph."

Guess what that meant. On July 3, 1870, the evening before President Grant attended Beecher's Independence Day gala, Elizabeth Tilton tearfully confessed to her husband that she had been nesting with their beloved pastor for eighteen months. Tilton repented at once of his advanced views on free love and fled for advice to his fellow cuckold, Bowen. They agreed that Beecher must resign or face public disgrace. But neither outcome occurred, because Bowen had already decided to fire the Radical Tilton and piggyback on the illustrious Beecher into Grant's inner circle. So he double-crossed Tilton by posing as a friendly go-between merely delivering, not endorsing, a letter of accusation. Beecher, unaware that Elizabeth had spilled the beans, played dumb. So Tilton had his wife write a confession and flung it in Beecher's face. The pastor, who liked to preach, "My friends, we are living an invisible life. There is a kingdom of God within us," now sobbed, "Theodore, I am in a dream; this is Dante's inferno." Tilton agreed for the moment to keep silent, in deference to his wife's reputation (not to mention his own), but the green-eyed monster clawed his guts all the more when Elizabeth turned out to be pregnant. She lost the baby on Christmas eve. What child is this? Tilton's or Beecher's? Only she knew, and she wasn't telling. How did the child die? By abortion or miscarriage? Only Elizabeth's mother knew, and she wasn't telling. It was then that the desperate Beecher sent Tilton a contrite note begging forgiveness, but confessing nothing.[62]

Three years passed, during which Beecher's stock soared. In 1872 Yale established the Lyman Beecher Lectures on Preaching and asked the honoree's son to deliver the first three. A rich disciple volunteered to launch a new journal for Beecher, in competition with Bowen's; it soon had 132,000 subscribers. A triumphal lecture tour gleaned $60,000. All the while Beecher reigned over genteel Brooklyn Heights and entertained overflow crowds with what *Vanity Fair* called "Sabbath harlequinades— motley in the pulpit." Tilton said the crowds included "forty of his m[istresses] each Sunday morning." Still, his masculine notions of honor required him to keep his mouth shut in the foolish belief that female's tongues would not wag. There is no way to sort out who told what to whom or what psychosexual political dynamics drove the contradictory rumors and schemes the gossip inspired. Family members took sides. Feminist confidantes calculated how a scandal might help or hurt their causes. But no one went public until Tilton's old flame Victoria Woodhull declared herself fed up with hypocrisy. "The Woodhull," as the formidable woman was known, had won Commodore Vanderbilt's patronage by feeding him stock tips from the spirit world. Vanderbilt repaid his "she broker" by subsidizing the incendiary *Woodhull and Claflin's Weekly*, in which she made good her threat: "Wherever I find a social carbuncle I shall plunge my surgical knife of reform into it up to the hilt!" In November 1872 she plunged her knife into America's Mr. Christian. She damned Beecher for covering up his dalliances because she knew very well that he also believed marriage to be "the grave of love." Hence "the fault with which I charge him is not infidelity to the *old* ideas, but unfaithfulness to the *new*." It was a sorry country that paid "forty thousand preachers to lie to it from Sunday to Sunday." That tore it. Now Beecher had to go public with his denials and Tilton had to go public with accusations that shocked the good people at Plymouth Church, and indeed the whole nation.[63]

Woodhull's accusations kicked off a "trial of the century," or more precisely four trials that drove Reconstruction off the front pages for two and a half years. What makes trials of the century is their raising of issues that puncture existential pretenses regardless of the actual parties' guilt or innocence. That's scary—and that is why sensational trials often end in farcical bargains (think of Scopes and O. J. Simpson). The first to be tried were Woodhull and her sister Tennie Claflin, for libel and obscenity. That suited them fine: jail time would enhance the publicity achieved by the sale of 100,000 copies of their broadside. Better yet, Benjamin Butler stepped forth as an amicus curiae and got them sprung on a technicality. The venue of the second trial was Plymouth Church, now forced to decide whether its pastor was a lecher or a victim. The decisive witness was Elizabeth Tilton, and unless she wanted to pin a scarlet A on her breast she had no choice but to recant. When she confessed only to having an inordinate affection for Beecher, the relieved elders voted 210 to thirteen to expel her husband for slander, i.e., for telling the truth.

Beecher gloated, but not for long. Tilton's life was so thoroughly wrecked that there was nothing left to be saved by silence. He sued Beecher for alienation of affection, dumped boxes of juicy letters on the Brooklyn City Court, and hired a crack legal team led by William A. Beach. Beecher countered with none other than William M. Evarts, the chief lawyer for Andrew Johnson in the impeachment trial. Dozens of witnesses offered 2 million words in litigation lasting six months—the most sensational "he said, she said" melodrama in American history. Scalpers hawked seats in the gallery. Reporters from as far away as San Francisco buttonholed witnesses on the street in hope of extracting a juicy exclusive. The summations alone took over a month because Beach meant to prove Beecher an adulterous apostate whose conviction would save American Christianity from modernist heresy, whereas Evarts meant to prove that Tilton was an adulterous liar bent on the ruin of his own wife and America's greatest preacher. They agreed that the nation stood at the fork of two roads, one leading to heaven and the other to hell; they just disagreed about which was which. So did the jurors, befuddled by witnesses who changed their stories so often that hardly a one had not been impugned. On July 2, 1875, after fifty-two polls over eight sweaty days, the jury stuck at three votes to convict and nine to acquit. Henry Ward Beecher was too dear an idol to smash. The mistrial was deemed a deliverance.

The fourth and last trial was Beecher's victory lap. He insisted the Congregational church convene a "scandal bureau" and exonerate him once and for all. The handpicked delegates needed little time to do that, because no accusers came forward. Elizabeth Cady Stanton called the affair a "holocaust of womanhood." She spat at a nation unable to "convict its fondest self-image." Precisely! Beecher returned to the pulpit, more famous than ever, to soothe Americans with his "new theology" positing a "watching, caring" Darwinian God "whose very life it is to *take care of life and bring it from stage to stage.*" When Beecher died at age seventy-three in 1887, the whole nation mourned. His tombstone read: "He Thinketh No Evil."[64]

The third and perhaps the principal reason Americans neglected to search their souls during Reconstruction was simple distraction. As George Santayana reflected in 1918, near the end of another war: "The American has never yet had to face the trials of Job. Great crises, like the Civil War, he has known how to surmount victoriously; and when he has surmounted the present crisis victoriously also, it is possible that he may relapse, as he did in the other case, into an apparently complete absorption in material enterprise and prosperity." Americans, however, would never consider that they had *relapsed*: only a foreign-born Catholic philosopher could have chosen such a gloomy, clinical verb. Americans considered it an almost orgasmic release of their pent-up ambition to pursue happiness through agriculture and industry, technical and managerial innovation, national marketing, exploitation of the Great Plains and mountain frontiers, foreign expansion and trade, and indeed manifest destiny in all its worldly forms. But in this one case the Civil War did provide a model for how to accelerate national growth once peace returned. The Union's industrial mobilization, logistical support of a 1-million-man army, and integration of transport and communications taught engineers, businessmen, financiers, and politicians a *systems approach* that made the whole world seem infinitely malleable by human hands. Senator Sherman summed it up in a letter to his brother William Tecumseh Sherman: "The truth is the close of this war with our resources unimpaired gives an elevation, a scope to the ideas of leading capitalists, far higher than anything ever undertaken in this country before. They talk of millions as confidently as formerly of thousands."[65]

Railroads still led, and the transcontinental project led the railroads. Congress granted the Union Pacific and Central Pacific railroads rights-of-way between Omaha and California; half the real estate lying within ten miles of the tracks; and $65 million in loans, which the managers adroitly leveraged to raise millions more through stock sales, revenue from portions already completed, California state bonds, and construction contracts with their own shadow subsidiaries. The saga of the Central Pacific in particular is often told as a gigantic hustle, because Collis P. Huntington, Leland Stanford, Charles Crocker, and Mark Hopkins, the "Four Associates," spent $50 million of other people's money to build the railroad over the High Sierra, but somehow came out of it with personal fortunes worth $200 million. True enough, but they also labored heroically and unstintingly, sometimes to the verge of nervous breakdowns, to realize what Greeley called "the grandest and noblest enterprise of our age." And they did this in just five years because, as everyone knows, the final spike was driven in Brigham Young's Utah in 1869. The Four Associates' systems approach to mobilizing an army of workers in the nation's most rugged terrain and keeping it supplied with food, water, wood, rails, and two newfangled technologies—dynamite and pneumatic drills—was neither more nor less than a peacetime application of Stanton's war economics. Suffice it to say that the triumphant telegram humming from Promontory Point to all corners of the United States called the last-spike ceremony the

"Appomattox" of the war between steam and distance. The Four Associates' profits, obscene or not, were the wages the American people paid for
an extraordinary public service, a point hammered home by a haggard,
impatient Huntington in congressional hearings many years later.[66]

After the Civil War every capitalist cried, "All aboard!" Settlement of
the plains, California's growth, Nevada's silver rush, and commerce with
Asia combined to lift railroad construction higher than Reconstruction on
the nation's list of priorities. Between 1865 and 1879 American railroads
laid 53,000 miles of new track. Government subsidies and corruption grew
proportionately, mocking anew the notion that the war had been purga-
tive. Congressmen happily pocketed bribes for handing out 100 million
acres of federal land. State legislators awarded exclusive charters, land
grants, public funds, rights-of-way, banking privileges, tax exemptions,
and bond issues, usually in exchange for a piece of the action. The Illinois
Central was given 2.5 million acres, which it sold for an average pure profit
of more than $10 per acre. The Union Pacific's reward for befriending
Congress was more than 12 million acres. By the end of the 1870s not one
but four transcontinental lines were under construction or being planned.
The Dakotas, supposedly Indian territory, were a fief of Jay Cooke's Northern Pacific Railroad. Its nonexistent metropolis was baptized Bismarck in
hopes of attracting German immigrants. In the East, any new railroad
security was gobbled up by investors, especially gullible Europeans, whose
share of U.S. railroad equity doubled to 20 percent by 1873. That flood of
cash enabled manipulators like Jay "the Hatchet Man" Gould, James "the
Spider" Fisk, and Daniel "the Undertaker" Drew to rig the market at will.
Drew, a former cowboy, knew all about feeding salt to cattle so that they
would bulk up on water before they were weighed. He introduced the
term "watered stock" to Wall Street. Fisk, who coined the phrase "Never
give a sucker an even break," watered and sold so much Erie stock that the
once prosperous railroad went bankrupt.

In truth, railroads rarely proved to be geese laying golden eggs, because the expansion of trunk and local lines made high profits and dividends the exception. Freight rates on the competing New York Central
and Pennsylvania systems dropped 25 percent by 1872 and another 50 percent by 1879. Marginal railroads went into receivership. In other words,
wherever competition existed the capitalists lost and the customers won.
The executives' answer was more systematization in the form of coopera-

tives or "pools" among railroads, express firms like Wells Fargo, and shippers of coal, grain, and other commodities. They meant to divide market share, fix rates, and thus lock in decent profits. In practice (confessed a veteran) the so-called Red and Blue Lines between the East Coast and Chicago, the southern Green Line, the Iowa Pool, and others "were generally made by the managers with the purpose of practicing deception upon each other." Indeed, a profit squeeze—not gouging—was the context for the misunderstood remark by the New York Central's Vanderbilt: "The public be damned." To be sure, homesteading farmers were damned wherever monopolistic railroads presided like feudal landlords. That kindled America's first "prairie fire," the Grange movement, among populist farmers demanding regulation of shipping rates. But throughout most of the eastern half of the nation the post–Civil War railroad boom resulted in service so uniform, efficient, and cheap that the average rate for hauling a ton of freight dropped from twenty cents per mile in 1865 to less than two cents by the end of the century. Add *safe* to that list of boons, because George Westinghouse's compressed air brakes, David Rousseau's automated railway signals, automatic couplers, the standard four- by eight-and-a-half-inch gauge, and scores of innovations in locomotive design made for fewer delays, breakdowns, and accidents. Despite the notorious stockjobbing, railroads became a national treasure and a source of pride, a fact expressed by the palatial downtown terminals (often named Union Station) erected by city fathers in cahoots with railroad executives. Indeed, if one date can stand for the triumph of the systems approach, let it be November 18, 1883, when every American railroad conductor reset his watch to the new "standard time" and bade his fellow citizens over four time zones do likewise.[67]

> *Law of thyself, complete, thine own track firmly holding*
> *(No sweetness debonair or tearful harp or glib piano thine)*
> *Thy trill of shrieks by rocks and hills returned,*
> *Launched o'er the prairie wild, across lakes*
> *To the free skies unpent and glad and strong.*
> —Walt Whitman, "To a Locomotive in Winter" (1876)

The telegraph accompanied railroads ocean to ocean. But no sooner did the Civil War end than Americans were thrilled to learn that the telegraph could span oceans, too. Before the war, a visionary, Cyrus W. Field, a retired paper supplier in New York, formed a consortium to lay cable beneath the Atlantic. Samuel F. B. Morse and such British luminaries as Michael Faraday and William Thompson (the future Lord Kelvin) helped design a copper cable wrapped in gutta-percha for insulation, tarred hemp for waterproofing, and iron strands for strength. But their first two cables snapped and the third quit just a month after Queen Victoria and President Buchanan exchanged greetings in Morse code in 1858. When the Civil War postponed the Atlantic project, two other Americans concocted a scheme worthy of Jules Verne. Perry M. Collins returned from service as U.S. consul on the rugged Amur River between China and Siberia with a plan for stringing telegraph wire from California north to Russian America, under the Bering Strait, and south to the trans-Siberian line being strung by the czarist government. Hiram Sibley, president of Western Union, bought the mad scheme, as did Seward, who helped win a congressional subsidy in 1863, and Czar Alexander II, who sold Collins a right-of-way for $100,000. The Collins Overland Telegraph teams (George Kennan, the diplomat's uncle, among them) spent $3 million in a futile struggle against Arctic fogs and desolation. They left behind great coils of wire and thousands of telegraph poles standing, like white men's totems, in a fairy ring around the Bering Sea. As soon as the war ended, the Atlantic consortium hired *Great Eastern*, the largest ironclad steamer on earth, to trail sturdier, better-insulated cables over the ocean floor. In July 1866 Europe and North America exchanged greetings again, this time having established a permanent connection. The same year, the three domestic telegraph companies that Stanton had merged for military purposes merged again for commercial purposes into an efficient national system. Those national rail and telegraphy networks also helped the original federal bureaucracy become the world's most efficient post office while reducing the cost of first class mail.[68]

Systematic production and marketing expanded with equal speed, at least in the North, Midwest, and West, as soon as the Civil War ended. Returning soldiers and a new tide of immigrants, which would peak at 460,000 in 1873, swelled the workforce, the consumer base, and the housing market. Thanks to postwar reconversion, breakneck rail construction, and feeding frenzies at the nations's rich energy pools, U.S. factory facilities grew 80 percent. They mass-produced flour (Pillsbury), packaged meats (Armour), tinned food (Heinz), beer (Schlitz), milk (Borden), firearms (Remington), lumber (Weyerhauser), sewing machines (Singer), pocket watches (Elgin), pharmaceuticals (Lilly and Squibb), clothing, accessories, household goods, and notions. Wholesalers tapped a national market while opening swanky urban flagship emporiums. Marshall Field, Montgomery Ward, Stewart's, Macy's, Carson, Pirie and Scott, Strawbridge and Clothier, and Jordan Marsh all opened their doors in the 1860s and 1870s. In 1869 the Great Atlantic and Pacific Tea Company (A&P) started the first national grocery chain. George Pullman, in 1858, had built his first "hotel car," which was five times more expensive than previous carriages; but not until after the Civil War did business travelers and vacationers demand elegant sleepers and dining cars on long-distance trains. Farm implements manufactured by McCormick and John Deere antedated the war as well, but factory runs redoubled during a postwar homesteading rush that Seward guessed was pushing back the frontier thirty miles every year.

The age of steel arrived just after the Civil War, when the Bessemer process was imported from England. Before then, steel could be made in bulk only by guessing when to stop the smelting of pig iron so as to leave the right proportion of carbon in the metal; or else by removing all the carbon from pig iron, cooling the wrought iron, then smelting it all over again in a charcoal furnace to reintroduce the required carbon. Henry Bessemer designed a converter for blast furnaces that did it all in a single smelting. As a result, far greater quantities of higher-quality steel could be forged at much lower unit costs. To appreciate the scale of this triumph, consider that a Bessemer mill could manufacture a ton of high-quality "crucible steel" at the expenditure of just 2.5 tons of fuel (metallurgical coke), whereas before it took seven tons of coal to make one ton of low-quality "blister steel." When the first American converters fired up in 1867, the *Pittsburgh Chronicle* exulted, "This country has thrown off another shackle that has hitherto bound it to England."

Cleveland, Bethlehem, Troy, Chicago, and Joliet, Illinois, were among the early steel towns, but Pittsburgh became "steel city" thanks to Henry Clay Frick's discovery that nearby Connellsville had rich deposits of "coking" coal, and thanks to the Scottish immigrant Andrew Carnegie's business acumen. Having already prospered in the linen trade, railroads, and telegraphs, Carnegie got into metallurgy through the invitation of friends seeking to drive a competitor out of the iron business. He moved on to steel, convinced that the way to corner an industry was to cut costs, improve quality, and buy out other firms when they could not compete. The way to control costs was through vertical integration, which meant owning not only the mills but the iron and coal mines that supplied them and the shipping and marketing firms that distributed their finished products. The way to enhance quality was to invest in the latest technology and hire the smartest chemists, metallurgists, and managers (e.g., Charles Schwab). Was Carnegie a ruthless and greedy monopolist? He certainly was ruthless to steelmakers less shrewd and industrious than himself. But largely under his aegis steel production rose from 20,000 tons in 1867 to more than 1 million tons by 1879 even as the price dropped from $166 to forty-five dollars per ton. Steel made railroad tracks stronger and cheaper. Steel (plus the hydraulic elevator invented in 1870) made possible skyscrapers that turned genteel Chicagoans and New Yorkers into cliff dwellers. Steel put cheap, superior tools and utensils into everyone's hands. Steel made Carnegie the richest man in America by the time he sold out to J. P. Morgan in 1901 for $480 million and began storing up treasure in heaven by donating 90 percent of his fortune to philanthropy. Carnegie made plenty of enemies and doubtless deserved them. But his "gospel of wealth" was mainstream theology after the Civil War, even more than it had been before the war; and the question remains whether any other capitalist, much less any socialist commissar, could have built the U.S. steel industry as quickly and well as Carnegie did.[69]

The age of petroleum arrived in 1859, when Edwin L. Drake applied drilling techniques used in salt wells to strike oil sixty-nine and a half feet under the ground at Titusville, Pennsylvania. Again, not much happened during the Civil War, but as soon as it ended prospectors drilled for oil as avidly as sourdoughs panning for gold. By 1870 small producers were pumping 25 million barrels per year, the first little pipeline designed by Samuel Van Sickle was in operation, and the first railroad tank cars and tanker ships were supplying fuel for kerosene lamps in cities across the North. Nobody needed expensive whale oil anymore, and this situation completed the work done by the CSS *Shenandoah* when its crew, unaware that the war was over, wiped out the Yankee whaling fleet in the North Pacific late in 1865. By 1873 more than 100 refineries had sprung up in western Pennsylvania and Ohio to provide products of uneven purity at unpredictable prices. The oil industry, like steel, needed the systematization that John D. Rockefeller was prepared to supply. A stern Baptist from New York state who drove himself harder than he drove anyone else, Rockefeller moved to Cleveland, borrowed $1,000 from his father, and entered the wholesale food business with Maurice B. Clark in 1859. Their timing was prefect. Profits from the Civil War allowed Rockefeller to buy out his partner for $72,500 in 1865, but not before Clark talked him into buying a lamp oil refinery. That was "the beginning of the modern oil industry." Within five years a series of deft partnerships, maneuvers to undercut and buy out small producers, and bargains with railroads for preferential shipping rates allowed Rockefeller to found Standard Oil.

Like Carnegie, he focused on quality standards, cost control, the best science and management, and vertical integration. Like Carnegie, he was a tidy man who thrived in free-for-all competition because he in fact hated it. Rockefeller's goal was to impose stability on the industry and get rich on volume, not price gouging. By 1879 he controlled most of the oil fields, 90 percent of the refineries, and all the pipelines in the East. Was Rockefeller a ruthless and greedy monopolist? He certainly was ruthless to refiners less shrewd and industrious than himself. But largely under his aegis oil production rose from 8,500 barrels of refined crude in 1859 to more than 26 million barrels by 1879, while prices declined from sixteen dollars per barrel in 1860 to less than one dollar by 1879 and remained there for the rest of the century. When at last the courts threatened his trust in 1911, Rockefeller broke up Standard Oil into thirty-eight profitable companies and retired with $900 million. He is lauded for philanthropic donations totaling $540 million. But his greatest gift was oil: for illumination, lubrication, paints, dyes, and all the products of organic chemistry from fertilizers to aspirin. As early as 1882—decades before the automobile made gasoline the national blood—a superintendent at the Baltimore and Ohio railroad proclaimed that without Rockefeller's petroleum "every wheel would grind to a stop within twenty-four hours."[70]

Lincoln had imagined a nation of personal strivers, and lots of them thrived in America after slavery had ended. But the real winners were impersonal systems, of which the most ubiquitous and most powerful was the corporation. In the 1870s just 5 percent of the 10,000 firms in Massachusetts were incorporated, but those few commanded 96 percent of the state's capital and employed 60 percent of its workers. Investment in new machinery and national marketing networks put a premium on capital that corporations alone could accumulate. Innovative, expansive corporations, in turn, put a premium on the pyramidal, departmentalized structures that the railroads had pioneered and that Frederick W. Taylor would call "scientific management" in 1895. Finally, corporations unable to establish monopolies had every incentive to secure market share and control prices and costs by forming trusts, rings, and cartels with their erstwhile competitors. Financial institutions rode the tide as national and local savings banks sprang up everywhere. Jay Cooke started the first underwriting syndicate in 1870. Anthony Drexel and J. P. Morgan teamed up in 1871. The Pennsylvania Railroad invented the holding company. All large corporations made strategic alliances with financial institutions. Downtown office space sold at a premium and set off construction booms. Everyone organized, from the Iron Founders Association and the American Medical Association to institutes of mining, mechanical, and electrical engineers. The peripatetic Thomas Alva Edison invented invention itself in his laboratory at Menlo Park, New Jersey. Within one generation planned research and development became imperative for every large corporation.

Even the prolonged business slump after the Panic of 1873 fed concentration, because some 10,000 insolvent firms got swallowed by bigger fish every year until 1878. Indeed, recessions and sluggish recoveries increased the comparative advantage of firms able to maintain profit margins through economies of scale. Not least, hard times depressed wages for industrial workers while crushing the National Labor Union promoted by William H. Sylvis in 1866 and the Knights of Labor founded in 1869. Between 2 million and 3 million people lost their jobs or knew that immigrants were waiting to take those jobs if they struck for more than two dollars per day. Union membership tumbled in New York City from 44,000 to just 5,000. Unions that dared to be militant, like the Irish anthracite miners (Molly Maguires) in Pennsylvania and the textile employees in New England, were beaten by lockouts. The era climaxed in 1877 in the war on the railroads, America's first national strike. When railroad executives announced a 10 percent wage cut, workers seized whole stretches of the Baltimore and Ohio, where Maryland militiamen and federal troops fought pitched battles against workers. The strike spread to the New York Central, Erie, and Pennsylvania railroad, whose rioting employees sacked Pittsburgh's Union

Station until soldiers shot fifty of them. But without strong unions workers had no strike funds, which meant no money to live on during a showdown. In just three weeks the men crept back to work for whatever management was willing to pay.

Who benefited from all these swift developments in technology, management, mass production, and labor conditions? Consumers, because the changes all fed a prolonged deflation that made everything cheaper. So even though the Goulds and Carnegies commanded headlines and most people still lived on farms, this was the era when the urban (and incipient suburban) middle class captured the heights of American culture.[71]

In the 1870s New York City plus Brooklyn reached a population of 1.8 million, Philadelphia 850,000, and Chicago 500,000. All relied on horse-drawn hackneys, omnibuses, and trolleys for transport, so their streets were coated with squashed manure. But New Yorkers saw the future in 1870 when the first locomotive chugged up Ninth Avenue to Thirtieth Street on an elevated rail line. Likewise, wooden buildings were still the rule, vulnerable to calamitous urban fires. But again it was 1870 when New Yorkers experimented with concrete, an import from Louis Napoléon's technocratic regime. That plus asbestos and steel skeletons made fireproof buildings the norm in the nation's downtowns. The French imperial style also inspired urban planners to create boulevards, parks, and architectural monuments such as Philadelphia's City Hall and the Old Executive Office Building next to the White House in Washington. Yet again in 1870 the first elegant apartment buildings known as "French flats" rose in lower Manhattan. They soon became the residence of choice for upwardly mobile urbanites decked out (thanks to sewing machines) in affordable copies of European high fashion. But the flow of goods and ideas was anything but one-way. In 1871, for the first time in history, American exports exceeded imports; and (except for a few fluky months in the 1880s) the U.S. balance of payments remained in the black until well after World War II. The annual volume of American trade tripled between 1865 and 1879 to more than $1.5 billion, thanks mostly to the bursting silos of the Midwest. In 1876 an event of global importance occurred when power to fix the wholesale price of cereals shifted from Danzig, Germany, to the "wheat pit" in Chicago's Board of Trade.

Corporate industry, high technology, and the agricultural breadbasket made America a cornucopia for the middle class. Pursuing their happiness, white-collar Americans also laid claim to leisure pursuits once confined to the rich, including vacations to Newport, the White Mountains, or the Jersey shore; grand tours of Europe or the mystical Holy Land (just opened to tourists by the Ottoman Turks); and day trips to Saratoga or the Jockey Club's new Jerome Park racetrack in Westchester County. Sports of all sorts became big in a hurry. In the 1850s, when avant-garde women in bloomers extolled the virtues of exercise, few paid any attention. No sooner did the Civil War end than sport turned into a system. The New York Athletic Club was founded in 1868 to promote the best English notions of amateur sportsmanship. Men's and women's clubs for gymnastics, equestrianism, tennis, cycling, and croquet proliferated. In 1876 the Amateur Athletic Association, Intercollegiate Association of Amateur Athletes, and Intercollegiate Football Association (Harvard, Yale, Princeton, Columbia) all sprang into life. The entrepreneurial mood ensured that professional spectator sports would evolve just as fast, starting with a game that soldiers from every state learned during the Civil War from Yankees and Yorkers: baseball. The Cincinnati Red Stockings toured to great acclaim in 1869. Jealous Chicagoans, led by the hotelier Potter Palmer, promptly incorporated the Chicago White Stockings and competed for talent in the Middle West. By spring 1876 the professional National League began to play.[72]

Northern, white, middle-class Americans were on the move, and proud of it, in 1876. Far from lamenting the Civil War or its outcome, they believed that their nation had finally "arrived" vis-à-vis Europe. E. L. Godkin, the *Nation*'s spark-spitting Anglo-Irish editor, wasn't sure. Americans after the Civil War were "far less raw and provincial than their fathers," he wrote, but theirs was a "chromo civilization" flashier on the outside than its innards warranted, almost as if it were gilded.

"Author Mark Twain names an era when he writes a novel called *The Gilded Age*, which describes the extravagance and corruption of post-Civil War life." So wrote the New York Public Library's reviewer in 1873. In fact the novel had a coauthor—Charles Dudley Warner—and ranks as one of Twain's weakest efforts. But it fixed on the era an image of plundering capitalists and crooked politicians, an image reinforced by the journalist Matthew Josephson's book *The Robber Barons* (1934). It hurled industrialists, railroad magnates, Wall Street speculators, venal congressmen, and big-city bosses into the same historical dungeon reserved for the American republic's most heinous villains: privileged aristocrats responsible for "the fearful sabotage practiced by capital" on honest mill workers and tillers of the soil. That was a useful past during the New Deal, when Franklin D. Roosevelt focused voters' wrath on "money-changers in the temple." But indiscriminate tags cease to be useful when they infiltrate textbooks. To understand what really was going on after the Civil War, one need only recall some familiar themes that date back to the colonial era. Americans are *hustlers*, in the good and pejorative senses of that word. Americans tolerate and even encourage corruption as long as it appears creative in the sense of evading artificial constraints, hastening development, and expanding opportunity for the many. Moreover, corruption is prevalent in all societies undergoing rapid technological, social, and political change. Since the United States has been the most dynamic nation on earth, thanks to its freedom and rich endowments, it is only to be expected that every age of American history is awash in old and new forms of corruption at every level of business and government.

The post–Civil War generation was not distinctive for being greedier or more lawless than others, or because government failed to regulate business in new ways. It was distinctive because the sudden spurts of industrialization, corporatism, urbanization, immigration, and nationalization of markets threw up new temptations for ambitious businessmen and politicians, while at the same time posing new challenges that American law and culture were not yet equipped to address. Carnegie broke no laws while building up U.S. Steel. Nor did any full-disclosure laws deter Jay Gould from spreading rumors over the telegraph, plundering companies' assets, or engaging in insider trading. So the question to ask is not whether this or that move in the nation's myriad, simultaneous Monopoly games was corrupt, but whether the alleged corruption was creative. The answer then becomes clear. During the so-called gilded age, manufacturers made the U.S. economy the largest and richest on earth. Financiers and railroad promoters were sometimes effective shepherds of capital, but too often wolves in sheep's clothing. Politicians achieved little besides making immigrants docile, blacks invisible, and democracy a bad joke.[73]

"There's an honest graft, and I'm an example of how it works," recollected George Washington Plunkitt. "I might sum the whole thing up by sayin':

'I seen my opportunities and I took 'em.' Just let me explain by examples. My party's in power in the city, and it's goin' to undertake a lot of public improvements. Well, I'm tipped off, say, that they're going to lay out a new park at a certain place. I see my opportunity and I take it. I go to that place and I buy up all the land I can in the neighborhood. Then the board of this or that makes its plan public, and there is a rush to get my land, which nobody cared particular for before. Ain't it perfectly honest to charge a good price and make a profit on my investment and foresight? Of course, it is. Well, that's honest graft." Plunkitt, a shanty Irishman from Nanny Goat Hill on New York's Upper West Side, quit school at age eleven thinking to learn the butcher's trade. But he did his real apprenticeship under the Tammany sachem and mayor Fernando Wood. So well did Plunkitt work the streets, rig elections, and dispense payoffs that he would rise from ward heeler to district captain, magistrate, alderman, supervisor, state senator, and second in command of a systematic machine built on 30,000 party hacks, 12,000 patronage jobs, and a payroll exceeding that of the New York Central. The machine was more powerful than the railroad, too, because its assets included the votes of tens of thousands of real people (and as many fake ones as might be needed) in the nation's leading state and metropolis.

Plunkitt on ending the spoils system: "This civil service law is the biggest fraud of the age. It is the curse of the nation. There can't be no real patriotism while it lasts. How are you goin' to interest our young men in their country if you have no offices to give them when they work for their party?" *Plunkitt on reformers*: "They were mornin' glories—lookin' lovely in the mornin' and withered up in a short time, while the regular machines went on flourishin' forever, like fine old oaks." *Plunkitt on national politics*: "The Democratic party ain't dead, though it's been givin' a lifelike imitation of a corpse for several years. It can't die while it's got Tammany for its backbone. The trouble is that the party's been chasin' after theories and stayin' up nights readin' books instead of studyin' human nature." *Plunkitt on Flatbush*: "I have made a careful study. . . . The Brooklynite is a natural-born hayseed, and can never become a real New Yorker. . . . There's no place in the world for him except Brooklyn." *Plunkitt's parting words*: "I see a vision. I see the civil service monster lyin' flat on the ground. I see the Democratic party standin' over it with a foot on its neck and wearin' the crown of victory. I see Thomas Jefferson lookin' out from a cloud and sayin': 'Give him another sockdologer; finish him.' And I see millions of men wavin' their hats and singin' 'Glory Hallelujah!'"[74]

Systemized corporations emerged in all industrial countries by the late nineteenth century. Only in America did machines systematize urban politics. The reasons are not far to seek. The extension of the franchise to almost all white males and the proliferation of elected offices—trends begun in the Jacksonian era—confronted voters with ballots as long as their arms listing scores of unfamiliar candidates for offices just as obscure. There was simply no point in voting unless someone told you how and then paid you to do it. (It helped that most primary balloting occurred in saloons.) That was especially the case among immigrants and their children, who numbered no less than three-fourths of New York City's electorate. Many foreign-born residents would not have been eligible to vote at all if not for machine judges who naturalized thousands before each Election Day (a process described in Rufus Shapley's hilarious satire *Solid for Mulhooly*). Explosive urban growth meant gigantic public works (e.g., Central Park and the Brooklyn Bridge) whose budgets begged for graft, kickbacks, sweetheart contracts, and unabashed skimming. Finally, machines prospered because "respectable" municipal leaders found it less bothersome to make their peace with the bosses, or else were distracted by their own bustling enterprise, or else had no stomach for the street-smart underworld. The muckraker Lincoln Steffens disapproved in *The Shame of the Cities*: "The typical businessman is a bad citizen; he is busy." Richard Croker of Tammany, known as Boss Croker, or "the easy boss," exulted: "All your high principles will not induce a mugwump to take more than a fitful interest in an occasional election. . . . They admit it themselves." Principles had nothing to do with machine politics, as was proved by the national debate over whether to issue silver coinage or stick with the gold standard. Businessmen cloaked their interests in complex economics that made no sense to workers. Croker said, "I'm in favor of all kinds of money—the more the better." Guess who won over the masses?[75]

Sociologists have argued that urban machines served a valuable function in the era before government agencies saw to the health, education, and welfare of the poor. One might even say that the otherwise crooked politicians performed better than federal and state agencies because they needed no paperwork, dispensed instant service, and treated their clients like human beings. A big American city was the heart of darkness for Irish,

Italians, and Poles with little English, little money, no job, and no understanding of the law. Who met them where they lived, gave them work on or off the municipal payroll, and procured needed licenses and leases? Who fixed their troubles with the police or the courts? Who arranged and even attended their weddings and funerals, and found medical care for families in need? Who scoffed at state law so that teachers in Catholic or Jewish neighborhoods could ignore the mandated curriculum? Who winked at liquor sales on Sunday? Who appropriated funds for a new park or sewer in the midst of the tenements? Who asked nothing in return save that immigrant males vote early and often? Let no one think precinct captains, ward heelers, and bosses did not earn their boodle. "A boss might as well give up the idea, for all time, of having fun," lectured Croker, "because there is no end of work in it."[76]

That said, urban machines raked in outrageous boodle, especially during the inglorious reign of William Marcy "Boss" Tweed. The gnarled, whiskered, 300-pound Falstaff immortalized by Thomas Nast's devastating cartoons wrested control of Tammany Hall from Fernando Wood during the Civil War. Serving simultaneously as a supervisor, fire commissioner, and Central Park commissioner, Tweed honed his skills selling $1,500 benches to the city for $169,000, and $3,080 in marble for $420,000. But the state constitution limited what a city machine could do unless it also controlled Albany. So Tweed got Tammany candidates elected to the statehouse and legislature in 1868 by bribing friendly judges to manufacture 60,000 new voters. The turnout in Manhattan that year was 115 percent. Himself a state senator, Tweed perfected the art of "ringing the bell," which meant proposing legislation harmful to some industry and then waiting for the bribes to roll in. But the real prize was New York City's budget, which Tweed pried loose from state supervision by greasing a Home Rule Bill through the Albany legislature. That permitted the Tweed Ring, described by his biographer as an engineering marvel, to rake off nearly two-thirds of all municipal spending. The loot was divided into five equal cuts: one for Boss Tweed, one for Mayor A. Oakley "O.K." Hall, one for Richard B. "Slippery Dick" Connolly, one for Peter "The Brains" Sweeney, and one for "the street" according to Tweed's motto "something for everyone." Catholic churches pocketed $1.5 million; this helps to explain Grant's campaign against funding parochial schools in 1872.

Accountants never were able to sort out the books, but the Tweed

Ring siphoned at least $75 million between 1869 and 1871, and perhaps as much as $200 million. How did its system work? Easily: appropriate funds for a new courthouse, contract with a Tammany firm to build it for $350,000, then charge the city $8 million. How did the Tweed Ring get away with it? Easily: control all city offices, make strategic alliances on Wall Street (Jay Gould was a big Tammany player), and buy up or gull the newspaper boys (even, for a time, Horace Greeley). How was it all financed? Easily: Tweed was too smart to raise taxes on the poor or the rich, so he tripled the city's debt from $30 million to $90 million in three years. That was Tweed's (though not the machine's) undoing. In 1871 suspicious domestic and foreign banks declared the city's bonds no longer welcome. That finally got the attention of New York's "wisest and best citizens." An investigatory Committee of Seventy headed by Henry G. Stebbins concluded, "There is not in the history of villainy a parallel for the gigantic crime against property conspired by the Tammany Ring." Republicans led by Samuel Tilden, plus Germans and other ethnic blocs outside the machine, pounded the streets during that autumn's campaign. But the decisive shift occurred when lesser Tammany bosses turned against Tweed to preserve the machine and leaked damaging documents to the *New York Times*. When reformers swept the elections, they buried Tweed in indictments. He beat the first rap ("No jury will convict me," he boasted) because the city's audits were so chaotic that the prosecution was at pains to prove anything. But a second trial, held during the Panic of 1873, landed Boss Tweed in prison (he later escaped). Tammany Hall itself lay low until the scandal blew over, then quietly resumed its control of the city under "Honest John" Kelly and Richard Croker. The "mornin' glory" reformers quickly lost interest in politics. The new sachems learned their lesson. Henceforth they would stick to "honest graft."[77]

Urban machines were not unique to New York or the Democrats. In Philadelphia "King James" McManes climbed the ladder as a ward heeler for Simon Cameron's Republican machine. His prize was control of a gas trust created in 1841 to insulate the powerful utility from politics. No sooner did McManes become a trustee in 1865 than he fired all 2,000 employees to make room for his vassals. Rake-offs, bribes, and a private militia of "plug-uglies" patrolling the polls soon made "King James" mighty enough to challenge Cameron's control of all Pennsylvania. As in New York, a party of reform-minded businessmen belatedly coalesced to topple the boss with

Cameron's assistance. As in New York, it made no difference over the long term. The silk-stocking goo-goos (as "good government" reformers were known) dropped out of politics as quickly as they had come in, allowing the next boss of the gas trust to revive the machine. In Pittsburgh, two businessmen—Christopher Lyman Magee and William Flinn—ran for office promising a "decent machine," then proceeded to add sixty relatives to the payroll and funnel all contracts to family members and friends. Mayor Albert Ames of Minneapolis, an amiable physician, won four races: two as a Democrat and two as a Republican. He did it by replacing the entire police force with mercenaries who gave gamblers, confidence men, and pimps the run of the town in exchange for payoffs used to buy votes. When at last he was arrested for bribery, Ames had the chutzpah to blame his poor judgment on a kidney ailment. Cleveland's rough-and-tumble ninth ward taught Mark Hanna the ropes in the 1870s. He would boss the Ohio machine that propelled William McKinley to the White House. Politics in San Francisco were a veritable perpetual-motion machine.[78]

Champions of decency included Henry Adams, who said that whoever betrayed the "sanctity of fiduciary relations," be it a railroad director or a member of Congress, was not just a thief but the "common enemy of every man, woman, and child who lives under representative government." Yet Adams had to admit that few bills stood a chance of passage unless financial persuasion opened legislators' eyes to their merits. Congressmen and state officials in the decades after the Civil War likewise took for granted the need to doctor the vote in precincts and counties they controlled, because the other party was bound to do so on its turf. "Why, dang it all," went the joke, "it costs too blamed much to carry an honest election in this country!" But even honest politicians such as President Grant could not help being tarred by corruption, because the Civil War had so greatly expanded the federal government's reach and largesse that it could scarcely police itself. Why did James Fisk and Jay Gould think they could corner the gold market in 1869 and then get out before the bubble burst? Because they assumed that the Treasury, desperate to redeem or prop up the greenbacks left over from the war, would not dump government gold to stabilize private markets. Just to make sure, they enlisted Grant's brother-in-law to slip letters of "advice" to the president. Grant woke up just in time to authorize the sale of $4 million in gold reserves. On September 24, Black Friday, telegraph wires literally melted from overload as panicky sellers tried to cash out. Many innocent people lost everything, and the president took the heat.

Why was the notorious Whiskey Ring, headquartered in Saint Louis, able to steal millions of dollars from the Treasury? Because the stiff excise taxes imposed during the war tempted manufacturers—in this case, distillers—to bribe revenue agents to fudge their numbers. Of course, that was no different from what merchants in the thirteen colonies had done with every British customs agent who turned up. But this time the king-pin was none other than John McDonald, a Civil War general who gave Grant an elegant brace of horses and bribed the president's secretary to quash the Treasury's inquiries. Again Grant was innocent, but again he took the heat. Why was the Union Pacific able to bilk taxpayers out of millions more dollars, make tycoons of Congressman Oakes Ames of Massachusetts and his brother, and bribe the vice president and half the members of Congress? Because the Pacific Railway Act of 1864 promised to reimburse the Union Pacific's construction costs. So its directors, like the Four Associates and for that matter Boss Tweed, formed a shadow subsidiary called Crédit Mobilier to dun the railroad, i.e, the government, with outrageous invoices. The executive branch was not involved at all, but once again Grant took the heat.[79]

Libertarians have a saying, which one need not be libertarian to af-firm: whenever government does *anything* it creates a privileged class and a criminal class. That is worthy of contemplation. The old muckraker and New Deal adage about business corrupting government is certainly valid, but the so-called Gilded Age, for the first time, exposed the power of gov-ernment to corrupt business.

The post–Civil War era was momentous because it proved a moment for so many things: a moment to forget the real past and fixate on the imag-ined future; a moment to welcome a "new theology" hardly distinguishable from the civic religion that had triumphed in 1865; a moment to thrill over new, amazing technology; a moment to systematize human life in imita-tion of nature; a moment to indulge new and old forms of corruption, creative or otherwise; and a moment for white Americans, at least, to dis-play as never before a resilience born of optimism, opportunity, pragmatism, and pretense. Those moments were captured colorfully and completely in a city that thrived on the Civil War and according to Lincoln did more than any but Boston to cause the Civil War. That was Chicago.[80]

An astute Chicagoan of that era observed, "The cities have not made the country; on the contrary, the country has compelled the cities." However much urban sophisticates and rural moralizers like to pass judgment on each other, they have been locked in unhappy marriages since ancient times. Primary producers need transporters, brokers, bankers, processors, wholesalers, and merchants to turn their commodities into consumer goods on the shelf. The middlemen, in turn, would not even exist without the farmers, breeders, and foresters who supply the cereals, vegetables, dairy products, livestock, and lumber. In other words, cities and their hinterlands form systems, and North America's greatest was bound to be the upper Midwest. Saint Louis looked likely to become the system's premier "capital" until 1861, when war on the Mississippi River and throughout divided Missouri chased business away and alienated the Lincoln administration. Congress made Chicago the jumping-off point for Omaha and the transcontinental route. The War Department made Chicago the principal warehouse for military supplies in the western theater. That made sense, given the city's Great Lakes port, canal and river links to the Mississippi, terminals for eastern rail networks, and Illinois Central trunk line to the federal bastion at Cairo. So the gloom of "secession winter," when eighty-nine of the 110 state banks in Illinois failed, turned to glee the moment Captain Joseph A. Potter opened his quartermaster office in Chicago and bought up $900,000 worth of supplies in just ninety days, then another $4.8 million in 1862 alone. Chicago lost its bid for the federal arsenal built to replace Harpers Ferry. That went to nearby Rock Island, Illinois. But no matter. Towering grain elevators stored the rich harvests shipped from Indiana, Illinois, and Iowa. Warehouses burst with freshly cut timber from Wisconsin and Michigan. The slaughterhouses of Armour, Swift, and Nelson Morris made Chicago the nation's beef capital and took from Cincinnati the title "Porkopolis." Cattle were so dense in the city that one herd collapsed a drawbridge on the Chicago River. By 1864 the French-born engineer Octave Chanute designed the great Union stockyards, whose supply and disposal systems for water, fodder, meat, hides, blood, and tallow were so elegant that the *Tribune* thought the "bovine city will be far ahead of the human city which it adjoins." The flow-through of goods caused an acute shortage of rolling stock, so iron and steel mills arose to support Pullman's railroad car industry. Marketing and manufacturing in turn required financial institutions as big and agile as those in the East. The Republicans' wartime system of federal banks made that possible. The First National Bank of Chicago emerged during the month of Gettysburg, and thirteen national banks with assets of $30 million sprang up by the war's end.[81]

So many people accompanied the cattle, hogs, grain, and wood to Chicago that the population in 1870 was triple what it had been when the Republicans met there to nominate Lincoln. The newcomers included eastern entrepreneurs, farm boys and farm girls seeking work and excitement, war veterans, Germans, Irish, freedmen, and gangsters, gamblers, and lowlifes from downriver towns such as Memphis and Vicksburg. Destined to become the nation's "second city," Chicago already had a reputation second to none for crime, vice, filth, corruption, and violence. It was known as "shock city" and the "wickedest city in America."[82] Prostitutes outnumbered policemen by almost five to one. Gambling dens outnumbered churches. Politicians ruled the wards with payoffs from the vice trade and swelled the voter registries with names copied off gravestones. Although no one-party machine ruled Chicago in the manner of Tammany Hall, city councilmen were in the same business: increase the municipal budget by a factor of six (during the 1860s), float bonds to support public works, then gorge at the pork barrel. Goo-goo committees of businessmen seconded by the clergy periodically cried for reform, as elsewhere. But Chicago's politicians and gangsters took some of their cues from business itself. Railroads threw off bribes as their locomotives threw off sparks. The Board of Trade Exchange was a gigantic casino where traders bilked each other—and defenseless farmers—by manipulating the price of futures contracts in collusion with the managers of grain elevators. Chicago's Irish underworld boss even modeled a huge horse betting pool on the Exchange. It was enough to make the *Tribune* wonder "what posterity must think of the moral character of our age. . . . The cynical must consider it as a biting satire on the degeneracy of our time."

Money roared in Chicago. Those with a pile of it had a big stake in the city's health and future. Accordingly, as in New York, civic boosters attempted to meet the challenges of growth through public works commissions insulated from politics. In 1865 the city floated bonds for an enormous new sewer system. In 1866 it purchased a steam engine sufficient to pump 18 million gallons of water per day from a tunnel extending two miles into Lake Michigan (little fish wriggled out of kitchen taps until adequate filters were installed). The city's most spectacular feat of engineering reversed the flow of the Chicago River so that its sewage no longer polluted the lake. That $3 million project was completed in July 1871.[83]

And then there came a day of fire, on the evening of Sunday, October 8. It began in Catherine O'Leary's barn on De Koven Street southwest of downtown, albeit without the aid of her cow. Drought, a stiff southwest breeze, and densely packed shanties, warehouses, and lumberyards turned the blaze into an inferno and then a firestorm as air rushed in to replace the superheated gases leaping skyward. The flames jumped the south branch of the Chicago River opposite what is now Grant Park, then marched north to consume even the stone and marble structures in what later would be known as the Loop. "There has never been a fire which so completely attended to its business," wrote Greeley's reporter. The main branch of the Chicago River seemed a barrier until the fire leaped over it, too, in the early hours on Monday. That gutted the city's final defense, the pumping station at the castle-like Water Tower, terrifying the residents of the densely packed, wooden North Side. Mostly Germans and Swedes, they stampeded northward or else waded into the chilly lake to watch all they had disappear. The fire ended after twenty-six hours, when it ran out of fuel above Fullerton Street (near present-day Lincoln Park Zoo). Some 18,000 buildings, 300 lives, and property worth millions of dollars had gone up in smoke. Survivors were quick to scapegoat Mrs. O'Leary, or Irish drunks said to have invaded her barn, or arsonists hoping for loot, or revolutionaries inspired by the uprising of the Paris Commune that year. Moralists pronounced the fire a divine judgment on a city of Ahabs and Jezebels. In *Chicago Burned*, a popular instant history, the Hoosier author wrote, "Politicians have worshiped their offices, and merchants their business. . . . How often have they seen their works perish!" Pastor Edward Payson Roe's best-selling novel *Barriers Burned Away* likened the pyre to the nation's purification in the war. Indeed, the first two media events experienced by all Americans in almost real time were the Civil War and the Great Chicago Fire. A poor girl provided the most illustrative anecdote. She had fled through the streets with her sole means of livelihood, a sewing machine, only to have it wrested from her by a thieving teamster. "Do you wonder," she asked, "that Chicago burned?"[84]

In the aftermath, Chicagoans displayed pluck in abundance. But barriers were by no means burned away. When General Sheridan (who had helped fight the initial blaze on the West Side) enforced law and order, opponents led by the governor cried tyranny. Imposition of martial law in peacetime reduced Illinois to the status of the occupied South. When the mayor gave the charitable donations pouring in to the evangelical Relief and Aid Society (he wasn't about to let the city council lay hands on the money), Jewish and Lutheran relief organizations objected. Bums (read: Irish), "never taught that any labor is honorable," were sure to exploit soup kitchens and clothing centers meant for genuine victims of the fire. When workers and tradesmen asked for wages commensurate with their hazardous overtime labor, construction firms and clients accused them of exploiting the misery. Most of all, Chicagoans split into the usual factions over how to rebuild bigger and better than ever.

The campaign to rebuild began while the ashes still glowed. William Bross, who was a former lieutenant governor and the city's number one booster, set out on a whistle-stop tour of the East. "Go to Chicago now!" he cried. "Young men, hurry there! Old men, send your sons! Women, send your husbands! You will never again have such a chance to make money!" All could "start even," said Bross, and "never was there such a field for employment since God said 'let there be light!'" That was like the American spirit New York had displayed after its horrible fire in 1835. But the 30,000 people who answered the call aggravated the desperate shortage of housing while driving down wages and intensifying the unions' militancy. Captains of business in Chicago likewise copied the New York elite that rebuilt lower Manhattan after 1835. Swearing off politics as usual, a bipartisan committee of "wise, prudent, and honest men" formed a Union Fireproof Party to compete in the municipal elections held just a month after the fire. After offering the top spot to a German banker, a gesture to win over that ethnic bloc, Joseph Medill, part owner of the *Tribune* and a free-labor disciple of Lincoln's, ran for mayor. He promised to make the "fire limits" mandating stone and brick building conterminous with the city limits. When the Union Fireproofers won by a landslide (in an understandably very light turnout), the benevolent, bearded Medill imagined the emergency sweeping away all "knavish combinations of unscrupulous partisans."

Every good deed gets punished, but Medill never got a chance to do his good deeds. No sooner did he introduce bills to extend the fire limits than Germans on the North Side cried, *Gott in Himmel!* They had turned the prairie into a city with their own hands and were willing to raise up homes, shops, and churches again. But to build in brick, stone, or iron was simply beyond their means. So Medill had to climb down just two months after taking office. The fire zone was extended to a strip along the northern lakefront (where Sheridan Road would later begin), but otherwise exempted working-class neighborhoods. That ruling enabled Chicagoans to rebuild with unbelievable zest. An average of 5,000 wagons dumped rubble into the lake every day for a year after the fire. From April to November 1872, masons laid 1 million bricks per day and construction teams topped off a new fireproof building between four and six stories in height every sixty minutes during working hours. These buildings included magnificent banks, office buildings, and hotels that cemented Chicago's well-deserved reputation for architectural brilliance (Louis Sullivan arrived in 1873). But for every landmark lovingly created, 1,000 cheap wooden structures got banged together. Medill despaired of fireproofing the city. He also despaired of ending partisanship. Labor and capital jostled and threatened each other. Established tradesmen made newcomers unwelcome. Newspapers boosted their circulation with screams about imminent general strikes. On May 15, 1872, the mayor had to stand in the rain and exhort 20,000 angry workers to "practice economy in personal expenditures; drink water instead of whiskey; keep out of debt; put your surplus earnings at interest . . . ; go with your wife to church on Sunday, and send your children to school." Medill's soft words turned away wrath: the crowd dispersed. But once Chicago's railroads and industry got back up to speed, the mayor himself began muttering about a new civil war, between labor and capital.

Finally, Medill despaired of purging Chicago's bad habits. More saloons were licensed in the year after the fire—2,218, or one per 150 residents—than any other category of business. A crime wave washed over the town, causing advocates of temperance to insist that the sale of booze be banned at least on the Sabbath. Germans and Irish pleaded that was their only day off from work, and chided a republic that would outlaw a pleasure even their European oppressors had permitted! Medill sympathized, but at last bowed to the goo-goos and evangelicals in early 1873. That ensured the landslide defeat of the reformers in the autumn election. City councilmen immediately repealed the Sunday liquor ban while the clergy and women prayed for their souls in the street. The incoming mayor, Harvey Colvin, even moved City Hall to a garish casino known as The Store, which Mike "Sure Thing" McDonald had built on the corner of Clark and Monroe. Medill was long gone. In August 1873 he resigned in disgust and left for a long vacation in Paris. Good riddance, according to the Germans' *Chicago Staats-Zeitung.* The mayor's only distinction was "to have been esteemed as highly, and later to have been esteemed as little, as no one else has previously." [85]

Chicago was back in business. New neighborhoods splayed north, west, and south as the city resumed its sprint toward 1 million people. Hustling of all sorts increased. Ethnic groups, sexes, and classes did battle. During the railroad strike in 1877 Chicago had to be placed under martial law again. That unnerved President Hayes. To be sure, there were no more army garrisons in the South, but the labor unrest forced him to peel off six companies of infantry from the Dakotas, where another American nation-building project had just gone awry.

In January 1869 president-elect Ulysses Grant met with some Quakers whose hearts were rent by the ongoing Indian wars. Morality and economy alike, they said, enjoined the government to practice Christian charity toward its aboriginal wards. "Gentlemen," replied Grant, "your advice is good. I accept it." Thus began Grant's celebrated "peace policy," to be carried out by a Board of Indian Commissioners composed of Protestant divines: call it a faith-based initiative. Rather than make more treaties with tribes, Congress agreed to adopt them until education and agriculture equipped them to enter the nation's mainstream. Since Ely Parker, that Seneca chief who rose to the rank of brigadier general, seemed the perfect role model, Grant put him in charge. Parker soon realized the assignment was hopeless. His budget, puny to start with, disappeared into the pockets of Indian agents far beyond the reach of the pious commissioners. The War Department and the Interior Department feuded over Indian policy as military officers damned corrupt, incompetent civilians, and civilians damned belligerent generals. Some tribes tried farming where once the buffalo roamed, but the soil on their reservations was awful. Anyway, young braves hated farming because, to Plains Indians, that was women's work. Nor did the peace policy end encroachments by whites. Horse thieves, buffalo hunters, traders peddling cheap whiskey, homesteaders, prospectors, and railroad planners constantly violated Indian Territory even as they protested the Indians' permissible annual hunts outside their reservations. Within twenty-four months of its inception, the whole Indian agency system became the subject of a congressional investigation. Ely Parker was cleared of wrongdoing, but he quit in 1871, declaring that it was "no longer a pleasure to discharge patriotic duties."

The peace policy began to unravel the following year, when the Northern Pacific Railroad undertook surveys from Bismarck to the Yellowstone River. The Sioux had every right to say no—and did—but surveyors proceeded under an escort provided by General Sherman. When the Panic of 1873 bankrupted the railroad, immediate war was averted. But the incident convinced Sherman that the army needed a garrison in the heart of the Sioux country. Grant approved, since only soldiers could deter white violators of Indians' rights as well as chase renegade Indians. But the man picked to reconnoiter, George Armstrong Custer, angered the Sioux by galloping all over their sacred Black Hills; then returned in September 1874 with the news of "gold in them thar hills." Miners streamed in the following spring. Red Cloud warned that even if he clung to peace the off-reservation braves led by young Sitting Bull and Crazy Horse surely would not. Thus Grant faced in the West the same contradiction he had faced in the South: the only way to defend the rights of the victimized was through a permanent military occupation. Nobody wanted that, so a blue-ribbon committee from Washington dispatched an offer to purchase the Black Hills. Sitting Bull threw dirt in the face of their emissary. Grant was out of ideas. When a war council at the White House decided in November 1875 to cease efforts to evict prospectors from the Black Hills and instead make the off-reservation Indians a pretext for war, the president sadly approved. All Sioux, Cheyenne, and northern Arapaho failing to turn themselves in by January 1876 would be considered hostile to the United States of America.

They failed to turn themselves in. Sheridan, knowing that Indians did not like traveling in winter, expected as much. So he ordered three converging attacks to corral the "hostiles" somewhere in eastern Montana Territory. He meant to squeeze the enemy into one mass concentration, so it was imperative that the three U.S. columns also bring their whole weight to bear in a coordinated final assault. But that did not happen in winter, because two of Sheridan's commanders dithered for months before marching; and the third, General George Crook, took a beating on the Powder River in March. Nor did it happen in spring, because Crazy Horse, crying "Come on, Lakota, it's a good day to die!" bloodied Crook's column again on the Rosebud River. Nor did it happen on June 25, 1876, because Custer grossly underestimated the strength of Sitting Bull's camp on the Little Bighorn, heedlessly galloped toward it under the constant gaze of Indian scouts, and then divided his troopers in hopes of pulling off an envelopment with just a third of the force required. As a result, nearly half of the Seventh Cavalry perished in Montana's greasy grass and choking dust. They included Custer, fifteen other officers, 242 troopers, and ten civilian scouts.

Americans pretended that Custer was a hero. Sitting Bull attested, "I tell no lies about dead men. Those men who came with 'Long Hair' were as good men as ever fought." Custer's wife Libby and Buffalo Bill Cody used their considerable public relations acumen to depict a botched offensive against a camp full of women and children as a courageous, defensive "last stand." Congress inquired, but saw the wisdom of considering Custer a martyr to the civic religion. Walt Whitman wrote a sonnet. But except insofar as Little Bighorn gave Native Americans a vengeful memory to savor, the battle accomplished nothing. Sheridan reinforced, vowed that the next campaign would unfold in winter, and got lucky when blizzards and icy winds lacerated the high plains in November. Somehow, Sitting Bull and a few survivors trekked to a Canadian refuge in February 1877. Crazy Horse surrendered his people in May but was assassinated at Fort Robinson, Nebraska. Chief Joseph of the Nez Percé led his women, children, old folks, and braves on a truly heroic march over the Idaho Rockies, hoping to reach Canada before being trapped. Snow and sniper fire stopped him in October. Promised that his people were free to go home, Chief Joseph spoke his famous vow, "From where the sun now stands, I will fight no more forever." The government's promise was not kept.[86]

Concentration—the reservation system—had to succeed sooner or later. But its purpose was to insulate Native Americans from violence and depredation while they learned white men's ways. In that respect the peace policy utterly failed. A military inspector blamed the Indian Bureau, a

"dumping ground for the sweepings of the political party that is in power."
Hayes's secretary of the interior, Carl Schurz, considered "a thoroughly
competent, honest, and devoted Indian agent" to be a "rare jewel." But an
Episcopal missionary, Bishop William Hare, insisted in 1877 that even
honorable agents were no answer: "Wish well to the Indians as we may,
and do for them what we will, the efforts of civil agents, teachers, and missionaries are like the struggles of drowning men weighted with lead, as
long as by the *absence of law* Indian society is left without a base." The rule
of law through Indian police and Indian courts became the new fixation
for liberal reformers. But that meant breaking down tribal authorities and
traditions, which in turn meant taking children away from their parents
and educating them as white men in the East. Richard Henry Pratt, who
had commanded black buffalo soldiers in Oklahoma and served as warden
to Indian prisoners in Saint Augustine, Florida, thought he had the answer. With support from Schurz and the War Department, Pratt opened
Pennsylvania's Carlisle Indian Industrial School in 1879. Its official motto
was "Kill the Indian and save the man."[87]

In 1866 Professor John L. Campbell of Wabash College, the main speaker
at the Smithsonian's tercentenary celebration of Galileo; and General
Charles B. Norton, U.S. commissioner to the Exposition Universelle of
Napoléon III, sent up similar trial balloons. Why not stage a world's fair
in America to celebrate the nation's centennial ten years hence? At first
everyone shot them down. Where would it be held? Who would pay for it?
More to the point, who would profit from it? But in 1870 the city fathers of
Philadelphia volunteered it as the obvious site and won over the state legislature. The following year Congress endorsed "an exhibition of American and foreign arts, products, and manufactures," but was careful not to
promise a dime. That left everything up to a voluntary, nonprofit Centennial Commission headed by a former governor of Connecticut, Joseph
Hawley, and including Campbell, Norton, and the Philadelphian merchant Joseph Welsh as chief of its financial board. But these hard-driving,
civic-minded Americans skilled in systematic administration were up to
the job and fully aware of their calling, which was to transcend Reconstruction and sell the whole world on the truth of the American dream.

"The masses of the American people," said Hawley, "desired to make long strides in the Centennial year toward perfect reconciliation. Divine Providence gave us a splendid opportunity to shake hands." He might have added "and pat ourselves on the back."

The commission chose for its motto "A Century of Progress." It attracted exhibits from most of the states (some in Dixie still pouted) and thirty-nine foreign countries (including monarchies, which scoffers said would not deign to congratulate a republic). It sold $10 million in small shares to the public, promising reimbursement from receipts. It chose a bucolic site in Fairmount Park, south of the Schuylkill River. Then, because time was short, the commissioners put the fair's preparations on a timetable more rigorous than any railroad's. In 1873 the tract was leveled, platted, and fitted with utilities. In 1874 blueprints were finalized for more than 200 structures. In 1875 an army of workers erected the buildings, including Main Hall (at twenty acres the largest enclosed space in the world), Machinery Hall, and Agricultural Hall. In January 1876 exhibits began to arrive. On May 10, against all predictions, the Centennial Exhibition opened on schedule. There had not been so much as a whiff of corruption. Nor were there heated disputes, thanks to a series of equable judgments. In deference to evangelicals the commissioners closed the Exhibition on the Sabbath. In deference to workers they set aside special weekdays when admission was cut in half. In deference to Catholics they permitted alcohol to be sold (albeit the Catholic Total Abstinence League erected a fountain gushing potable water). In deference to skeptics they banned religious displays. They prohibited smoking, but more to prevent fire than for the sake of morals: chewing tobacco (hence spitting) was allowed. Needless to say, the city at large hustled to welcome the world and profit thereby. Hotels prepared 150,000 beds. Two railroads ran tracks to the Exhibition's entrance on Belmont Street. Trolleys and riverboats cycled thousands of people per hour from Center City to the fairgrounds and back. The Exhibition had its own miniature railroad that looped around the exhibits to spare patrons sore feet. Outside the gates peddlers hawked snacks, souvenirs, and Barnumesque freak shows.[88]

On opening day an orchestra huge enough to gratify Johannes Brahms played fanfares and a Centennial March by the German composer Richard Wagner. (By all accounts the march confirmed Mark Twain's quip that

Wagner's music is better than it sounds.) Foreign dignitaries led the parade to the podium followed by Hawley and Welsh, then President Grant accompanying Emperor Dom Pedro of Brazil. Grant did the honors when church bells chimed noon, then strolled to Machinery Hall to pull a valve and summon a monster to life. Serving as power plant for all the other intriguing machines in the hall was a forty-foot-tall, 680-ton steam engine designed by George Corliss of Rhode Island. Its cylinders were forty-four inches wide. Its eight giant shafts generated 1,600 horsepower. Its cams, cranks, and pulleys transferring raw power like the tentacles of a giant squid symbolized, not just the industrial system, but a system of systems. The pavilion was packed with Goodyear rubber, Sharps rifles, Yale locks, Edison telegraphs, Westinghouse air brakes, Pratt and Whitney machine tools, Otis elevators, Pullman cars, arc lights, typewriters, food canning processes, refrigerators, steel cables for the Brooklyn Bridge, Krupp artillery, and a newly patented gadget that its inventor called a telephone. Dom Pedro dropped Alexander Graham Bell's device with a shocked "My God, it talks!" Mining technology, farm implements, watchmaking, and other industries had halls all to themselves. The United States pavilion (Congress belatedly kicked in $500,000) showed off the activities and equipment of the various departments of government. A Woman's Pavilion featured textiles, appliances, and medical devices designed not just for, but by, American women. That made William Dean Howells bristle. If women insisted on seceding from the human race, he asked, then why not have a Men's Pavilion? Susan B. Anthony bristled, too, because no females were put on the program and her petition to present a Declaration of Rights of Women was denied.[89]

Welsh, the Exhibition's bean counter, hoped for 10 million paid admissions over six months. But attendance fell from 76,172 on opening day to an average of just 25,000 per day over the following months. Welsh blamed the newspapers for insufficient publicity, the railroads for failing to offer special excursions, and Congress for quibbling about money. What really held down the crowds was Philadelphia's sweltering summer. The park's asphalt promenades, a marvel in May, became gooey, and the heat and noise inside the pavilions were insufferable. Even Centennial Day proved disappointing; Grant sent Vice President Thomas W. Ferry in his stead. ("The president's absence put the finishing stroke to the sum of his offenses," said the *Atlantic*.) Given the heat, the commission wisely staged the mass celebration of the Fourth of July at midnight, and not at the fairgrounds but outside Independence Hall. A new Liberty Bell rang thirteen times. The crowd sang "The Star-Spangled Banner," the Doxology ("Praise God from Whom all blessings flow"), and spontaneous patriotic or drinking songs throughout the early hours. Cannons boomed at first light and fireworks in the evening. But the red-white-and-blue mood turned black on July 6, when news of Custer's last stand buzzed over the wires; and again on July 9, when the massacre of black celebrants in South Carolina became known. Then on August 2 came the murder of "Wild Bill" Hickok, a man Custer thought "entirely free of bluster and bravado," in Deadwood, Dakota. On September 7 Jesse James's gang of angry former rebels shot up the First National Bank of Northfield, Minnesota. The whereabouts of the fugitive Boss Tweed competed for headlines throughout the year. Foreign correspondents, whose good opinion the commission especially valued, insisted on contrasting "the pretentious national Exhibition" with the poverty, filth, and labor strife in shantytowns just a few blocks away. Then at last, in the relative cool of autumn, the Centennial Exhibition picked up momentum. September 14, New York Day, attracted 118,000 people and September 28, Pennsylvania Day, 257,169. Welsh lamented the decision to close the Exhibition on November 10, because one more month might have allowed him to reach his goal. Or not—because three days before the closing date American voters failed to elect either Tilden or Hayes and careened into a constitutional crisis. Still, some 8 million paying customers, including one in every fifteen U.S. citizens, had made the pilgrimage to the Exhibition, and the business broke even with receipts of $3.8 million.[90] Systems worked. America worked. *Americans* worked.

But for what? That was the question posed by one of the foremost celebrities who steamed across the Atlantic to see for himself what the century-old United States had to offer. Thomas Henry Huxley, "Darwin's bulldog," was already famous and infamous for his stubborn, scathing, often sardonic advocacy of natural selection at the bar of British public opinion. He accepted an invitation to visit America from a committee of scientists including Daniel Coit Gilman, first president of newly chartered Johns Hopkins University. Huxley arrived with his wife on August 5 to evoke instant fascination and ridicule. Newspapers "monkeyed" with the evolutionist in words and cartoons, and Huxley counterpunched deftly. When asked to sign the guest book at Harvard, he instead penciled a sketch of a simian Eve offering a simian Adam the apple to the glee of a serpentine spectator. Huxley toured campuses; addressed a convention of the American Association for the Advancement of Science in Buffalo; and went down to Tennessee, where his sister lived with her American husband. He did not attend the Centennial Exhibition. But the first and last words Huxley pronounced on his trip captured America in 1876 better than anything uttered in Philadelphia. On catching his first glimpse of Manhattan from the deck of the *Germanic*, Huxley inquired about two towers he spied on the island. Told that those were the headquarters of Western Union and the *New York Tribune*, he replied, "Ah, that is interesting; this is America. In the Old World the first things you see as you approach a city are steeples; here you see, first, centers of intelligence." Nearing the wharf, Huxley was captivated by something else: the sheer hustle and bustle of everyone onshore or else darting about the bay in purposeful tugboats. "If I were not a man," he mused, "I think I should like to be a tug."

Huxley's far more pensive farewell remarks were just as incisive. "Size is not grandeur, and your territory does not make a nation," he admonished Americans. "The great issue, about which hangs a true sublimity, and the terror of overhanging fate, is what are you going to do with all these things? What is to be the end to which these are the means?" Huxley predicted that the United States would surpass 200 million by the time of its second centennial. But he wondered whether democracy must end in disguised despotism, whether centralization could be had without disguised monarchy, whether America's "shifting corruption" would prove less of a millstone than entrenched bureaucracy. "Truly America has a great future before her; great in toil, in care, and in responsibility; great in true glory if she be guided by wisdom and righteousness; great in shame if she fail." He considered America's only safeguard to be "the moral worth and intellectual clearness of the individual citizen," but warned that education alone could not impart such qualities. Huxley, chief exponent of the survival of the fittest, steamed away from New York uncertain how fit the Yanks really were.

Captains of American culture never doubted their nation's fitness. The month Huxley departed, the mass literary magazine *Scribner's* published encomiums to the Centennial Exhibition including Richard Henry

Stoddard's long ode "Hospes Civitatis Annus Mirabilis MDCCCLXXVI." Confusing America with the Almighty (notice the capitalization), Stoddard sang: "Henceforth, America, Man looks up to Thee. . . . Be humble, and be wise; / And let thy head be bowed / To the Unknown, Supreme One, who on high / Has willed thee not to die!"[91]

By God's will or otherwise, the United States did not die from the Civil War or the pretense of Reconstruction. On the contrary, Americans drew such self-confidence from their scientific, industrial, and presumed moral progress that in just twenty more years they would step forth as redemptive crusaders, certain they knew how to do for Cuba, the Philippines, Mexico, China, and the world itself what they were manifestly unable to do for their own conquered South.

Chapter 9

TRUTH-TELLERS?

Gimlet Eyes on a Republic of Pretense

Granted, Mark Twain turned misanthrope in old age, but he certainly knew his subject.

"In the course of his evolutionary promotions, his sublime march toward ultimate perfection, he has been a gambler, a low comedian, a dissolute priest, a fussy woman, a blackguard, a scoffer, a liar, a thief, a spy, and informer, a trading politician, a swindler, a professional hypocrite, a patriot for cash, a reformer, a lecturer, a lawyer, a conspirator, a rebel, a royalist, a democrat, a practicer and propagator of irreverence, a meddler, an intruder, a busybody, an infidel, and a wallower in sin for the mere love of it. The strange result, the incredible result, of this patient accumulation of all damnable traits is, that he does not know what care is, he does not know what sorrow is, he does not know what remorse is, his life is one long thundering ecstasy of happiness, and he will go to his death untroubled, knowing that he will soon turn up again as an author of something, and be even more intolerably capable and comfortable than ever he was before."

Of course, the irreverent humorist did not claim to be chiding his fellow Americans in the manner of Herman Melville's *The Confidence-Man*. Twain pretended that his subject was a sassy murder (flock) of crows he encountered in India while "following the equator."[1] But Twain rarely, if ever, picked up a pen except to tell truths about us. From "The Notorious

Jumping Frog of Calaveras County" (1865) to "The Facts Concerning the Recent Carnival of Crime in Connecticut" (1876) to "The Man That Corrupted Hadleyburg" (1899), Twain's modus operandi was to expose—in suitably palatable fashion—the lies Americans told themselves in order to get on with the business of realizing their personal and national destinies. Indeed, "Carnival of Crime" is a miniature *Moby Dick*. It tells of a respectable author in Hartford (Twain himself) whose conscience materializes in the form of an ugly, truth-telling imp. Their conversation grows ever more torturous until the author at last figures out how to kill him.

Remember all those European visitors in the 1830s and 1840s who suspected that the apparent success of "democracy in America" derived from powerful pretenses imposed by public opinion. Guess what—visitors in the 1860s and 1870s suspected the same. Fanny Trollope's more famous son Anthony tasted life in the United States in 1861–1862. His time was short and his travel restricted by the Civil War, so Trollope sought to learn Americans' ways by immersing himself in their public press. But he found the newspapers incredible except insofar as their very incredibility met the needs of the people: "All idea of truth has been thrown overboard. It seems to be admitted that the only object is to produce a sensation, and that it is admitted by both writer and reader that sensation and veracity are incompatible. Falsehood has become so much a matter of course with American newspapers that it has almost ceased to be falsehood." Nor had American bravado retreated an inch since his mother's time, despite the fact the republic was at war with itself. "I do not know," Trollope wrote, "that an American as an individual is more thin-skinned than an Englishman; but as the representative of a nation it may almost be said of him that he has no skin at all. Any touch comes at once upon the net-work of his nerves and puts in operation all his organs of feeling with the violence of a blow." The American would insist, "Do you like our institutions, sir? Do you find that philanthropy, religion, philosophy, and the social virtues are cultivated on a scale commensurate with the unequalled liberty and political advancement of the nation?" Yet Trollope's observation of poor immigrants taught him the value of an egalitarian pose. "The Irishman when he expatriates to one of those American States loses much of that affectionate, confiding, master-worshipping nature which makes him so good a fellow at home. But he has become more of a man. He assumes a dignity which he never has known before. He learns to regard his labour as his own property. . . . To me personally he has perhaps become less pleasant than he was. But to himself—! It seems to me that such a man must feel himself half a god, if he has the power of comparing what he is with what he was. . . . If this be so,—if it be acknowledged that it is so,—should not such knowledge in itself be sufficient testimony of the success of the country and her institutions?" [2]

William Hepworth Dixon arrived from England just after the Civil War. He marveled at the hash Americans made of Reconstruction. The president and Congress were at loggerheads while the South was starving and freedmen were abandoned! There were endless debates over how to repair a Constitution that the Yanks previously claimed had descended from heaven like the tablets of Moses! "While Americans are busy, unmaking and amending their Constitution, may they not fairly put to themselves the question, What is the use of this record? At best, when the letter of a constitution is true in every detail—true to the designs of God in His moral government of men, true to the life and hope of the people in whose name it is drawn up—it is only a definition of facts. . . . But the fact of defining is also one of narrowing, limiting, restricting. Why should the life of a great continent be narrowed down to a phrase. How can a progressive country *pretend* to limit its power of future growth?"

Dixon was especially perplexed about what made Americans tick after encountering two phenomena so protean he devoted half a volume to each. The first was Salt Lake City, where he heard the Mormon patriarch Brigham Young greet a new band of emigrant converts. "Brothers and sisters," cried Young, "you have been chosen from the world by God, and sent through His grace into this valley of the mountains, to help in building up His kingdom. You are faint and weary from your march. Rest, then, for a day, for a second day, should you need it; then rise up and see how you will live. Don't bother about your religious duties. . . . Your first duty is to learn how to grow a cabbage, and along with this cabbage an onion, a tomato, a sweet potato; then how to feed a pig, to build a house, to plant a garden, to rear cattle, and to bake bread; in one word, your first duty is to live. The next duty,—for those who, being Danes, French, and Swiss, cannot speak it now—is to learn English; the language of God, the language of the Book of Mormon, the language of these Latter Days. . . . [But] the first thought of a convert, the first counsel of an elder, is always, that the Saint shall look upon labour, labour of the hand and brain . . . as the appointed sacrifice through which, by God's own law, a man shall be purged from sin and shall attain everlasting peace."

The second phenomenon that enthralled Dixon was a spiritualist convention in Providence, where feminist mystics advocated the abolition of marriage, property, and Christianity in favor of a new spiritualist church. He later asked a New England Republican how such extremes could coexist. "Well, says the judge, while we are divided in opinion . . . we must have one law in this Republic. Union is our motto, equality our creed. Boston and Salt Lake City must be got to shake hands, as Boston and Charleston have already done. . . . And now pass the wine."[3]

Sir George Campbell, member of Parliament, toured the United States just as Americans were ending Reconstruction, on the assumption that it had succeeded. His memoir *White and Black* exposed that pretense. He readily saw how America's geographical, demographic, and industrial growth would soon elevate it to world power, but he predicted that its races would never mix easily or on terms of equality, any more than they could in the British Empire. Just as Benjamin Disraeli had written of

DISTRIBUTION OF POPULATION
1880

Britain's rich and poor, so would America's whites and blacks make up "Two nations; between whom there is no intercourse and no sympathy; who are as ignorant of each other's habits, thoughts, and feelings, as if they were dwellers in different zones, or inhabitants of different planets; who are formed by a different breeding, are fed by a different food, are ordered by different manners, and are not governed by the same laws." [4]

Did no American tell truths in the nineteenth century? On the contrary, most people did tell truths part of the time concerning some things. Human nature finds it difficult to live self-conscious lies; this is what makes pretense so fetching. Americans, being uniquely free to think, believe, and act as they wish, needed to believe that they clung to some truth, the better to feel good about doing well. But their very freedom to think, believe, and act however they wished meant that their versions of truth rested on no authority unless vox populi really was *vox Dei* and the American civic religion really did reflect transcendent reality. Fiercely honest Americans who could not swallow that—think of Melville, Lincoln, or Twain—sought refuge in prophecy, tragedy, and satire. Truth did exist—and truth could be told even if telling it did the republic no obvious good. One patriot, however, went farther. He was not of the stature of the aforementioned trio. But he sniffed out authenticity like a bloodhound. The man's best friend put it this way: "His predominant passion was love of truth. This was all his glory and all his trouble; all his quarrels, friendships, aversions, perplexities, triumphs, labors." His adult life span coincided exactly with American history from the rise of Jacksonian democracy to the demise of Reconstruction. His collected works fill twenty ponderous volumes, making him the nation's most prolific political philosopher. His spiritual and intellectual intensity, plus his sheer lust for life, exhausted allies and confounded foes. His name was Orestes Augustus Brownson. [5]

   In 1803 a poor farm wife of Puritan stock gave birth in Stockbridge, Vermont, to her fourth and fifth children, named Orestes and Daphne. When her husband soon died, the mother struggled but failed to hold the family together. At the age of six the twins fell to the care of foster parents, an elderly couple in Ballston, Vermont. Though not church members, the couple instilled strict Calvinist habits in their wards, including probity, hard work, and self-criticism. Orestes would later lament that "properly speaking, I had no childhood." The old couple evidently taught him to read, since he had no formal schooling, but the only books in the house were the Bible and a shelf of Congregationalist treatises. When not busy with chores the boy all but committed the books to memory, turning them over in his mind and examining his own soul. Jonathan Edwards's apparent denial of free will especially troubled him. Where did truth lie, or authority, or freedom? Joseph Smith, a neighbor in Ballston who was two years younger, would pose similar questions until paid a visit by the angel Moroni. Orestes sought no new revelation. He sought to make sense of the old revelations, especially since culture and society in his allegedly Christian country reflected the Bible's message so poorly. Other nine- or ten-year-old boys played Indians or fished. Orestes chewed over dogmatic theology. Other boys dreamed of fairies and candy. Orestes dreamed about heaven and hell. Other boys counted their pennies. Orestes counted his sins and realized that he could not live an hour without degrading his soul. Then, after attending a revival at age thirteen, he had a beatific vision in the dark of his bedroom. "All my guilt, all my grief, all my anguish were gone and I felt ushered into a new world, where all was bright and lovely, where the air was perfumed with sweet spices, where soft and thrilling music breathed from every dwelling and warbled from every grove. . . . I broke out so loud that I was heard all over the house: 'I have tasted heaven today, what more can I contain?' Thus was I born again."

The following year their mother reclaimed the twins and moved to upstate New York. Work as a printer's apprentice taught Brownson that authorship was one of his callings. Exposure to the competing Protestant sects in this "burned-over district" taught him his other calling was ministry. But ministry in which denomination? Brownson was baptized as a Presbyterian, but within two years he rejected a faith whose doctrines of depravity and predestination offended his reason and whose parishioners displayed a "singular mixture of bigotry, uncharitableness, apparent zeal for God's glory, and a shrewd regard for the interests of this world." In 1823 he took a job teaching in Michigan, caught the malaria prevalent on that boggy frontier, and spent his convalescence reading Universalist tracts. He came home convinced that a God of love had no choice but to bestow salvation on everyone. Hence the purpose of the Christian life was not to repent, but to reform. By 1826 he was ordained to preach and by 1827 he was married to Sally Healy, of Elbridge, New York, cousin of a law partner to Daniel Webster. A dutiful, loving wife, Sally would bear eight children and follow every geographical, financial, and philosophical twist and turn of her husband's dialectical quest for truth.[6]

Picture an oval face framed by short brown hair over a high forehead and a short brown beard under a jutting chin. Picture a clean-shaven lip, ruddy cheeks, and pale forehead. Picture dark, alert eyes behind small, wired-rimmed spectacles: eyes that could radiate warmth yet flash coldly at a suspicion of falsehood. It was a face that impressed everyone Brownson encountered during his years as a Universalist determined to reform the world by sundown. In 1828 he founded the *Gospel Advocate and Impartial Investigator* to promote every betterment project that came down the pike. The first was William Henry Seward's Anti-Masonic Party, but Brownson moved on to socialism and abolitionism after hearing Fanny Wright lecture in Utica. He quickly learned the only cause he could *not* promote was Universalism itself. Its God was passive, remote, and above all *constrained*; this struck him as a contradiction in terms. Moreover, if all human beings were destined for heaven, then virtue and vice were neither rewarded nor punished; reform had no ultimate purpose; and the Bible itself was a fraud because it repeatedly witnessed to death, judgment, heaven, and hell. On what authority, then, did Universalism rely? The sect's elders grew uncomfortable with this querulous pastor, then hostile when he decided that Robert Owen's utopian socialism accorded with Christian charity far better than did the wage labor system. When the Universalist general convention declared Brownson in "disfellowship" he could only smile at the sectarianism of the allegedly nonsectarian sect. In 1832 he disavowed all denominations, but answered a call from a Unitarian church in Walpole, New Hampshire. There Brownson completed his self-education through a prodigious curriculum including Latin, French, German, philosophy, and solitary chess and backgammon to hone his powers of logic. He frequently ventured to Boston, where he attended Unitarian lectures, toured the textile mills, and decided that industrialization was soiling American life. He deplored the materialism of the rich and the hardships of the poor. He became a fervent Jacksonian Democrat. Yet he doubted democracy could reform society without a genuine spiritual rebirth that Christian sects seemed unwilling or unable to foster.

There was nothing for it but to design a new "church of the future."

Brownson outlined his vision in "New Views of Religion, Society, and the Church" in 1834. Christianity was not the answer, because it excluded most of the human race, splintered its own followers, stressed only individual sanctification, and encouraged an otherworldly attitude. The true path toward peace, equality, and heaven on earth, he imagined, must be communal sanctification through a church and state not separate, but unified. Taking the Declaration of Independence as his holy text, he imagined that an overt civic religion could transform America and someday the world into a genuine promised land wherein "Man will be sacred in the eyes of man." When the new church triumphed, slavery, war, ignorance, and oppression would disappear and "the service of God and the service of man" would become identical.

In 1835 Brownson took a Unitarian post in Canton, Massachusetts, where he hired Henry David Thoreau as a teacher. Then he moved again, to Mount Billington, named for a family that had arrived on the *Mayflower*. The town's proximity to Boston allowed him to hobnob with Ralph Waldo Emerson, Amos Bronson Alcott, Theodore Parker, Margaret Fuller, and William Henry Channing. He found them somewhat lacking in "robustness, manliness, and practical aims. They are too vague, evanescent, aerial." Yet he did not doubt that these were "the men and women who are to shape our future." Together they founded the Transcendentalist Club. In 1836 Brownson assumed the editorship of the *Boston Reformer* to serve as a mouthpiece for the club's humanitarian agenda. Its motto: "We know no party but mankind."[7]

Brownson left Christianity behind, but was far too cerebral for Emerson's mysticism. Indeed, the Vermont farm boy's "cultivated bluster" soon earned him a reputation as the Transcendentalists' most contentious, aggressive, and frankly philosophical member. (Edgar Allan Poe noticed and modeled some of his hyperrational characters on Brownson.) Whereas Emerson wafted skyward on deft contradictions, Brownson insisted that words must have definitions, arguments syllogisms, and ideas consequences. He much preferred Channing, who planned a system of schools to elevate manual workers to intellectual, cultural, and eventual economic equality. Nor did Brownson seek to withdraw from the world in the manner of Thoreau. Rather, he meant to transform the world; that is why his journal condemned capitalism in ever more scathing terms. When the Panic of 1837 hit, Brownson exulted, "Babylon is falling," and predicted the demise of America's "universal fraud and injustice." At a mass Fourth of July meeting of Jacksonian Democrats on Bunker Hill, Brownson exhorted the party to thwart the capitalist Whigs and Biddle's Second Bank of the United States before they captured the federal government in their selfish interests. He even redeployed John C. Calhoun's doctrine of states' rights in defense of the northern Loco-Foco faction of Democrats. That won Brownson an appointment as steward of the Marine Hospital in Chelsea, Massachusetts, a valuable patronage post for the struggling philosopher. But Brownson was no hired pen. He truly believed that democracy would be no more than a pretense unless equality was extended to the possessions and power relationships of all citizens.

Brownson started his own journal, the *Boston Quarterly Review*, to preach the gospel of democracy. He wrote furiously about politics, welfare, money, religion, and Indian removal (he lamented its necessity, given the progress made by the Five Civilized Tribes). In almost every article he strove to expose the false premises, facts, or conclusions inhibiting public debate. He even wrote "A Discourse on Lying" as uncompromising as anything by Cotton Mather. All attempts to deceive, spoken or not, are lies, and all lies are sins. The worst he judged to be "those not usually christened with that name." For instance, many people carried pieces of paper called bank bills, which "high-minded and honorable gentlemen" promised to redeem for a given amount of hard money. In truth, state banks had ceased for years to make good on their paper while "men of the highest standing in society, and the loudest in their pretensions to decency, intelligence, virtue, religion, have greatly applauded the lie" and held "that the public good demands its continuance." What, asked Brownson, could be more corrupting to the morals of youth—unless it was the lies told by politicians? The Whig or Democrat is loud in his praise for fair dealing, confident of the good sense of the people, and appalled by the other's tricks, "while at the same he pays liberally for the publication of falsehood, holds his secret meetings, spares no efforts to flatter or coerce the people into the support of his party." Politics was simply "a tissue of falsehood" and governments "splendid lies." But worst of all were cowards who willingly gave their assent to things they knew to be false lest they risk the penalty for confronting public opinion. Brownson warned that a nation afraid to acknowledge the truth must sooner or later fall to the ground, like Judas Iscariot, and gush forth its bowels.[8]

Nothing happened. The nation limped through Martin Van Buren's four years as president without recovery or reform. So Brownson decided that the only cure for the ills of democracy was more democracy. In July 1840, just as the Whigs began their "log cabin and hard cider" campaign for the White House, Brownson's essay "The Laboring Classes" appeared in the *Boston Reformer*. On page after acerbic page Brownson exposed the illogic and injustice of wage labor, inherited wealth, and every law and institution frustrating social democracy. Mind you, his closely reasoned critique of capitalism predated *The Communist Manifesto* of Marx and Engels by eight years. Of course, it proved a disaster. Whigs gleefully reprinted the piece as proof that the Loco-Focos were radicals. Democrats blanched, then made Brownson a scapegoat for Van Buren's defeat in November. Brownson himself was appalled at how easily voters were duped by the Whigs' pretentious slogans, Barnumesque hoopla, torchlight parades, and free booze. He began to dread the prospect of civil war—not between slave states and free states, not yet; but between social classes—unless a disparate people could unite around their common citizenship and humanity. Americans needed more than rules on how to play the political game; they needed a communal spirit that committed them to keep playing. But democracy as presently practiced in the United States did not work. Nor did his "church of the future." So the dogged logician, who had already tried and rejected Congregationalism, Presbyterianism, Universalism, Unitarianism, Transcendentalism, "Nothing-arianism," democracy, and socialism by age thirty-seven, retreated into his study—and into his past—to figure out where he had strayed from the narrow paths leading to truth.

It took several years before Brownson burned all his bridges. He struck up a friendship with the Democratic historian George Bancroft. He visited the Transcendentalists' utopian Brook Farm and sent one of his sons there for a spell. (The farm's denizens soon decided Orestes was daft; he returned the sentiment by declaring them icy cold.) Brownson even tried merging his journal with John ("Manifest Destiny") O'Sullivan's *Democratic Review*. They agreed on America's mission to absorb the whole continent and redeem mankind by example. But they soon parted ways because O'Sullivan remained an unreconstructed Jacksonian whereas Brownson wrote inscrutable lectures and essays that questioned the bedrock principles of the United States. In "Democracy and Liberty" he suggested that the former, absent a higher authority, threatened the latter. In "The Democracy of Christ" he implied that liberty without social justice tended toward self-destruction. His "Reform Spirit of the Age," "American Independence," "Civilization and Human Progress," and "Origin and Ground of Government" questioned Americans' faith in progress and even the enlightened, Protestant roots of their institutions. In 1844 these thoughts and more began to appear in *Brownson's Quarterly Review*, a one-man publication that would pepper the nation's cultural heights with grapeshot for two decades. O'Sullivan suspected that the renegade from Ethan Allen's Green Mountains was drifting into conservatism. He was right. But this was not something Brownson willed. Like everything else in his life, it was something that happened to him when theory and history imposed themselves on his relentless brain.[9]

Orestes Brownson was forty years old in October 1843 when Sally gave birth to their seventh child. The household was strapped. He had quit his Unitarian posts to concentrate all his energies on politics and had then lost his sinecure in Chelsea when the Whigs triumphed in 1840. To make ends meet he returned to the pulpit, edited the *Boston Quarterly Review* in addition to his own journal, and wore himself out on the lecture circuit. Yet the indefatigable Brownson found time over those years to rethink the meaning of life.

The initial goad to his inquiries was a celebrated lecture by Theodore Parker on "natural religion." While dabbling in Transcendentalism Brownson also entertained the notion that an individual's perception of nature, not revelation, was the only reliable source of truth about God, man, and history. On hearing Parker anew, Brownson realized that this bogus assumption could only end in self-worship, the negation of true religion.

Second, while meditating on these mysteries Brownson turned his mind's eye—for the first time in years—from mankind to God and had an epiphany. God was free! God must possess free will to have fashioned creatures possessing free will. "To pretend, as some do, that God is tied up by the so-called laws of nature, or is bound in His free action by them, is to mistake entirely the relation of Creator and creature."[10] Ergo, if God was not a transcendent spirit, natural force, or principle of necessity, but a sovereign deity, then it followed logically that he was free to reveal himself to his children, issue commandments for their welfare, chastise and judge them for their own good, and reconcile a fallen race so they might commune in love with their maker and with each other.

Third, Brownson did extensive research on the place of the United States in world history before lecturing on the subject during the winter of 1842–1843. He was drawn steadily backward in time, to the thirteen colonies, to Tudor-Stuart England, and at last to the medieval era. There he unearthed, to his surprise and momentary confusion, the roots of American liberty. Common law, natural law, the separation of powers, civic virtue, the dignity of each human soul: all were products, not of the Reformation and Enlightenment, but of the maligned Middle Ages! (Had not John Adams reached the same puzzling conclusion?)

Fourth and finally (though all these influences overlapped), Brownson discovered a concept that helped him link his epiphany about the "freedom of God" to his quest for a principle of community and authority. That was the "life by communion" advanced by the French philosopher Pierre Leroux, another apostate utopian socialist. In his *Réfutation de L'Éclectisme* and *De L'Humanité* of 1839–1840 Leroux swept aside the false divisions among people caused by their love for power, prestige, and wealth. Instead he imagined a fraternal republic in which citizens communed with each other in daily life the same way worshippers humbly communed with each other and God in church on Sunday. Why not? Why not realize an authentic Christian republic founded on truth because its logic was informed by communion with the author of truth?

By 1844 Brownson confessed to himself that no humanistic, self-referential philosophy yielded a scintilla of truth. Protestant divines might apprehend partial truths (he found Horace Bushnell least objectionable).[11] But ceaseless sectarian quarrels over biblical interpretation rendered their vaunted principle *sola scriptura* as empty as *vox populi vox Dei*. The Anglican church seemed a possible refuge for a prodigal son of the Puritans, especially the High Church Oxford movement. Its embrace of Anglo-Catholic tradition, elegant melding of patristic orthodoxy and mysticism, and raw intellectual power in such persons as John Henry Newman, E. B. Pusey, and John Keble, tempted Brownson. But ever the truth-telling curmudgeon, he rejected the Anglicans' "middle way" as a trimmer's dodge, while suspecting that their "bells and smells" liturgy was an aesthetic conceit. In any case, most American Episcopalians were Low Churchmen who rejected the very qualities Brownson admired in the

Oxford movement. Anglicanism could never serve as the basis for a Christian republic.

At this point, 999 out of 1,000 Yankees would have decided they had reached a dead end. Brownson did not so conclude. Nor did he tear his hair, relinquish his quest, become a cynic, or cry out to God. He just followed his dialectics to their logical conclusion and made the most "politically incorrect" judgment imaginable: "If we cannot have a church unless we go to Rome, let us go to Rome." Brownson sought out Boston's Monseigneur Benedict Joseph Fenwick for instruction. In October 1844 he was rebaptized as a Roman Catholic, joining a religion he had previously decried as "ignorant, degraded, enslaved, cowardly, and imbecile."[12]

What happened next may amaze twenty-first-century readers still more. Sally and all of Brownson's children respectfully followed their father into the Catholic church. So did his friend Isaac Thomas Hecker, a German-American New Yorker who became Brownson's disciple during their Loco-Foco campaigns. Hecker had monitored Brownson's pilgrimage through a voluminous correspondence. Persuaded now that secular socialism, or any strictly human "betterment" movement, was an update of the ancient gnostic heresy (absent an external measure, how could humans define "better"?), Hecker entered the priesthood. In 1858 he would found the Society of Saint Paul, devoted to outreach among American Protestants. None of these converts suffered emotional crises or had to make wrenching adjustments. "When Brownson embraced Catholicism," wrote Arthur Schlesinger, Jr., "he abandoned not reason, but the pride which exalted individual reason above the accumulated experience of mankind and raised private intuitions above the study of objective reality. . . . He believed profoundly in God, in morality and in logic; and he was passionately eager to combine the three in one final whole. If God were more than man's fancy, He must be absolute reality. If morality were more than local custom, it must have absolute rewards and penalties. If logic were more than human illusion, it must refer to absolute truth." So gentle was Brownson's transition that he never stopped paying attention to current events. In the presidential contest of 1844 he hoped John C. Calhoun would run, but agreed to support James K. Polk when Calhoun dropped out.

Nonetheless, Brownson soon learned that his leap of faith entailed plenty of change. His journal lost most of its former subscribers overnight, but then it gained even more new ones, thanks to the patronage of Catholic bishops. He received a thorough grounding in theology from Bishop John B. Fitzpatrick of Boston, himself educated at Saint-Sulpice in Paris. "Bishop John," as his Irish flocks called him, was nine years Brownson's junior, but his instruction "revolutionized" the mind of the convert and allowed him to see with new eyes. Not least, Brownson rejected his radical desire to meld church and state and enthusiastically endorsed their separation. But he did it for the opposite reason from Protestants. His study of history taught him that separate spheres of temporal and spiritual authority were a Catholic invention, which ideally checked the pretensions of both and promoted virtue without squelching liberty. Of course, the decline of the papacy in the late Middle Ages permitted Europe's monarchs to reduce their national clergies to agents of state power while corrupting the church at large. The Protestants rightly protested, but went still farther in the wrong direction. They either made their princes the heads of their Lutheran or Anglican churches, or else endowed clerics with secular power as in Calvinist lands. The United States, thanks to its blessed First Amendment, had a chance to restore the proper balance between church and state, a balance whose purpose was not to denude the public square of religion, but rather to allow a sovereign God's "higher law" (is that where Seward got hold of the phrase?) to discipline and inspire public life. Unfortunately the United States could not take advantage of that opportunity, because Americans lacked a church. As a result, their liberty degenerated into license—corruption, hedonism, mobbing—that in turn conjured self-appointed, self-righteous reformers bent on imposing their morality through civil law: abolition, prohibition, women's suffrage. Brownson would devote the rest of his life to the quixotic task of persuading Americans that their republic was at risk unless a critical mass embraced what he called Catholicity (as opposed to Catholicism, which implied ideology). In other words, he saw beyond the surface religiosity of the disparate, devolved, deranged, or dismembered Protestant sects to spy the greatest pretense of all: to wit, that the United States was a Christian country.[13]

That raised hackles. Critics tried to shrug Brownson off as a sort of weather vane, if not a crank.[14] They failed because *Brownson's Quarterly Review* contained enough pointed truth to make everyone angry. Brownson damned the Mexican War as "uncalled-for, impolitic, and unjust." He identified slavery as the paramount threat to liberty and respect for authority in the United States. Yet he denounced the fanatical abolitionists with equal vehemence and parsed the law to prove that Congress could not restrict slavery in the territories. Brownson railed against worship of Mammon, damned corruption in business and government, and defended the laboring classes with more sincerity than before. Yet he now condemned socialism as incompatible with Christian charity and as an impossible dream unless imposed by force, thus extinguishing liberty. Brownson eviscerated the solipsism and sexual mores of Unitarians, Transcendentalists, and utopian cults. Yet he fired as much polemical ordnance at lukewarm Episcopalians, Methodists, and Presbyterians. Brownson called the Know-Nothings of the 1850s by their real name—bigots. Yet he scolded Irish immigrants for their ignorance, dissolution, and crime, and blamed Catholic bishops and schools for indulging ethnic enclaves. Immigrants, he insisted, ought to receive the best secular as well as Catholic training and be encouraged to assimilate and succeed in America. That latter charge stung Bishop Fitzpatrick; that is why Brownson moved his brood again, this time to New York, in 1855. But he soon fell out of favor with Archbishop Hughes, who suspected that Brownson wanted to Americanize the church more than he wanted to Catholicize America. (Brownson admitted that his "tenure with Catholicism was more like a bed of spikes than a bed of roses.") So he repaired in 1859 to New Jersey and found temporary cover, at least, under the miter of a fellow convert, Bishop James Roosevelt Bayley, a nephew of Saint Elizabeth Seton.[15]

Picture Brownson as described above, but now fifty-seven years old. The hair on top is thinner and the beard is full and dabbled with white. The cheeks are ruddier than ever because on his conversion, Brownson became an "incarnational" Christian. That is to say, he dispensed with the asceticism dear to Transcendentalists (vegetarianism, graham crackers, teetotaling) in favor of bourbon and beefsteak, except during fasts. If that verged on gluttony, Brownson paid for his sin: he ballooned to 250 pounds and suffered from gout. But the eyes behind the wire-rimmed glasses—twinkling like Santa's or flaring like John Brown's—never changed. Nor did Brownson's passion for work, except that now it was centered because his intellect was no longer self-centered. He meant to serve as a sort of philosophical mason called to lay down a "Christian foundation for politics."[16]

His labors began with four lengthy treatises, all of them characteristic reassessments of discarded ideas. The first, "Transcendentalism" (1845–1846), identified that movement as no philosophy at all, but rather a gnostic self-deification that differed from ancient gnosticism only in its democracy. According to Emerson and like-minded philosophers, everyone, not just a few initiates, could get in touch with the "oversoul" and experience divine insight into nature. Brownson had once entertained that idea himself, but now considered it humbug. Transcendentalism "represents the irrational as superior to the rational, reverses all our common notions of things, declares the imperfect more perfect than the perfect, the less of a man one is the more of a man he is, the less he knows the more he knows." Transcendentalists offered vague, flattering promises of transfiguration ("Ye shall be as gods") based on nothing except intuition. They shunned not only reason, but the empirical fact of humanity's humanness, which is to say, sin.

In "Channing on Social Reform" (1849) Brownson spun out the consequences of the Transcendentalist temptation that urged Americans to perfect themselves and society but left it up to each soul to decide what perfection entailed. The reformers worshipped "improvement," but what were their standards of good, better, and best? Brownson still blessed Channing's motives and critique of materialism. But he now warned his old friend that sentimental humanitarianism untamed by reason, love, and humility might easily degenerate into zeal, coercion, and violence. "You cannot go on, year after year, denouncing social order, denouncing society itself, denouncing every restraint of law, all faith, all piety, all conscience, everything the race has held sacred, and hope that the multitude, if they heed you, will remain quiet . . . or that they will not take up arms to realize the visions of Mahomet's paradise on earth, with which you have maddened their brains." The universal lust to reform society, to reform other people in a spirit of ideology rather than faith, must at the last come to this: "Love me as your brother, or I will cut your throat." Brownson had cause to sound that alarm. In 1848–1849 violent revolutions swept across western and central Europe, setting off civil war among the ethnic groups of the diverse Hapsburg monarchy and a bloody class war in Paris during the infamous "June days." Americans liked to think themselves immune from such civil strife, but their secular, ideological, and increasingly demagogic democracy all but invited worms of passion to devour the fruits of republican law and liberty.

In "Socialism and the Church" (1849) Brownson completed that line of thought. Civil and religious liberty was the font of America's glory. Love of money and power was America's shame. But Brownson echoed Edmund Burke and anticipated Friedrich Nietzsche when he warned that efforts to purge America of its shame through statutory law backed by force must extinguish its glory. The "pretended, peaceful reformers" who wanted to revolutionize American life might fancy themselves "pure-minded, moral, and heroic." But they could overturn private property, marital relations, and other traditions deemed wicked only by inflaming a "cruel and despotic public opinion" and making the nation "a prey to all the low, base, beastly, cruel, violent, wild, and destructive propensities and passions of fallen nature." Socialism, Brownson predicted, could never triumph except by stifling liberty—even the liberty of thought.

Finally, in *The Spirit-Rapper: An Autobiography* (1854), Brownson un-leashed his prophetic imagination by crafting semifictional dialogues of the sort he had heard at Brook Farm. His earnest reformers, speaking with candor among like-minded friends, insisted that belief in the false God of the Bible had duped Jews and Christians into calling evil good and good evil for millennia. But now that the fraud was exposed, a "Messiah of the nineteenth century" might at last free mankind from the twin tyrannies of church and state. One earnest seeker, Priscilla, says, "Satan is my hero," because he was the first to defy authority in the name of freedom. *The Spirit-Rapper* is hardly great fiction, but neither can it be dismissed as the dyspeptic fantasy of a Catholic convert. Brownson knew whereof he wrote, and his dark probe into the modern "transvaluation of values" predated Dostoyevsky's dialogue of the "Grand Inquisitor" by a quarter century.[17]

Brownson's Christian foundation for American politics was still un-der construction, but two principles were already settled. First, "the true founders of nations, promoters of social order, and reformers of society" were those who devoted themselves to the spiritual health of individuals, not those who tried to mold individuals into a healthy society. America's founders, in that respect, had been true republican builders, not demo-cratic wreckers. Second, everything the self-appointed humanitarian re-garded as progress "is but a forgetting, is but his departure from truth, and an unhappy fall into error." Brownson translated those principles into political science in a lecture series he gave at a provocative venue: the same Broadway Tabernacle built for Charles Grandison Finney's evangelical revivals. What, he asked, did Americans have against Catholics? They did not deny that the Roman church was a good vehicle for translating souls to heaven. Rather, they deemed Catholicity bad for this world in the belief that it encouraged superstition and sloth rather than progress and enterprise. Americans also feared that the Catholic hierarchy meant to subvert their republic. On the contrary, Brownson argued, not only was Catholicity compatible with republican government; it was downright necessary to its health.

His argument went like this. All well-constituted societies comprise three elements: authority, liberty, and religion. Societies were imperiled whenever civil authority verged on absolutism or individual liberty verged on anarchy. The American system of checks and balances resisted those perils, but could not defeat them "without the presence and energetic support of religion." The founding fathers acknowledged as much. But for religion to inculcate the love and self-discipline a republic required it must ipso facto "stand on a basis of its own, independent both of the individual and the state." Protestant sects could not do that. They were either creatures of the civil authority, as in northern Europe, or else creatures of public opinion, as in America. In short, Protestantism was already a slave to the state or the marketplace. Brownson was careful not to deny that evangelical sects could express religious truths and beauty; he was considering only their political worth, which he judged to be nil. Indeed, the only way American sects could influence society was by legislating morality. Catholics, by contrast, held drunkenness to be sinful, but would never imagine lobbying for a "Maine liquor law."[18]

Brownson, the scion of Puritans, took as given his nation's special providence. America was called to become a Christian republic and beacon to the world. But it could not fulfill that mission, or even adhere as a nation, unless religion mediated the tension between individuals and government and kept both anarchy and oppression at bay. That Protestantism was not up to this job became glaringly obvious when the main denominations split over slavery, while all manner of crazy cults multiplied. Accordingly, Brownson addressed himself to his fellow Catholics, who were just a minority but might serve as an example even as America as a whole served as an example. For nowhere else in the world—certainly not in the "Catholic" monarchies—was the Roman church freer to be itself than in the United States: free to witness with authority; to be in the world but not of the world; to speak truth in love to power both public and private. So Brownson concluded his "Mission of America" (1856):

> *This manifest destiny of our country, showing that Providence has great designs in our regard, that he has given us the most glorious mission ever given to any people, should attach us to our country, kindle in our hearts the fire of a true and holy patriotism, and make us proud to be Americans. Especially should it endear the country to every Catholic heart, and make every Catholic, whatever his race or native land, a genuine American patriot; for it is the realization of the Christian ideal of society, and the diffusion through all quarters of the globe, for all men, whatever their varieties of race and language, of that free, pure, lofty, and virile civilization which the church loves, always favors, and has from the first labored to introduce, establish, and extend, but which owing to the ignorance, barbarism, and superstitions retained, in spite of her most strenuous exertions, from pagan Rome and the barbarian invaders of the empire, she has never been able fully to realize in the Old World.*[19]

* * *

The Civil War tortured Brownson. He had railed against the Dred Scott decision for failing to acknowledge the humanity of the enslaved Negroes. He had railed against southern Fire-Eaters on the argument that natural law condemned slavery; but he railed against abolitionists on the argument that statutory law permitted slavery and could be changed only by legal means. Given his faith in America's destiny, he considered secession a despicable sin and supported the Union war effort. But its ineffectiveness frustrated him. Brownson admired Stanton, distrusted Seward, and called Lincoln "thick-headed; he is ignorant; he is tricky, somewhat astute in a small way, and stubborn as a mule." He wrote to Charles Sumner, "Is there no way of inducing him to resign, and allow Mr. Hamlin to take his place?" In November 1862 the aging, corpulent Brownson wore himself out running for Congress but lost his New Jersey district in a landslide (being Catholic did not help). In 1864 he refused to endorse Lincoln for reelection until Chase, Butler, and Frémont stepped aside. Brownson even thought that the Radicals had committed fraud when they took the title Republican because he sensed they were driven by a moralistic, big-government spirit as threatening to authentic liberty as the rebels' anarchic Jacksonian spirit.

Suffice it to say that he strongly affirmed Pope Pius IX's *Syllabus of Errors* (1864), which condemned capitalism and socialism alike as destructive of faith and community. The Union might be saved, but at what cost? The war's cost, for Orestes and Sally, included two of their sons.

Still, Brownson shouldered wartime duties no one else could have performed. When Catholics, with at least tacit support from their clergy, screamed against the Emancipation Proclamation in 1862, *Brownson's Quarterly Review* screamed back at them. No right-thinking American could deny Lincoln's war powers, and no right-thinking Catholic could, if forced to choose, prefer slavery to freedom. When the Irish rioted against the draft and lynched black New Yorkers in 1863, Brownson issued a stinging rebuke: "sneer not at the nigger, for today it is in him we must find our Lord, and in serving him that we are to serve the Church of God." He was not sure how best to serve them until blacks swelled the ranks of the Union army. This convinced him that Negroes could not be denied full civil rights. Finally, by 1864, Brownson felt called to perform an even more strenuous duty, not toward Catholics or Negroes, but toward the white Protestant majority that would soon be deciding how to repair the Union. He suspended publication of his beloved quarterly, dropped out of the public eye, and gave himself entirely over to reintroducing the United States of America to its citizens.[20]

No nation in history was more in need of the ancient wisdom "Know thyself" than America, yet its vision and judgment were still obscured by "the wild theories and fancies of its childhood." So began Brownson's *The American Republic*, first issued in 1865. All nations had spiritual and moral lives in addition to physical and material lives, and all had some providential purpose. But Brownson believed the U.S. republic unique, since nothing resembling its institutions could be found among Aristotle's political types. America's mission was nothing less than to realize popular sovereignty without despotism and liberty without anarchy, and so reconcile the rights of the whole and the rights of the individual. Hence, Americans erred in thinking government a necessary evil. On the contrary, government was the minister of wrath to wrongdoers, the protector of property and religious freedom, the promoter of science and art, the basis for civilization itself. "Next after religion, it is man's greatest good; and even religion without it can do only a small portion of her work." But citizens could pervert or destroy their own government if they harbored false notions of its origins and purpose as an agency of divine grace. Americans had just attempted one sort of political suicide and were sure to try other sorts if they persisted in resenting their government while deifying themselves.

Sovereignty, posited Brownson, cannot exist as a theory or a mere claim advanced by subject peoples. Sovereignty is a historical fact that emerges whenever a self-conscious, organic community takes control over a defined territory and wins recognition from existing sovereign states. Disparate peoples might form some voluntary association, but "whatever its terms or conditions, [it] is only an alliance, a league, or a confederation, which no one can pretend is a sovereign state, nation, or republic." The late Confederacy got that wrong, but America's founders did not. "We, the People" constituted the nation and defined it as the coeval *United* States and United *States*. That is: no Union absent the states, no states absent the Union, and no Union or states absent the sovereign indissoluble people residing within indestructible borders.[21]

Where did the United States come from? Brownson believed that Americans erred again when they sought their origins in the Constitution or the Declaration of Independence. The Constitution constituted the law of the preexisting land, and the Declaration declared a preexisting sovereign people. The real origin of the American republic was to be found in the constitution bestowed by Providence when it first dawned on colonial patriots that they were, in fact, sovereign. This "unwritten constitution" was begotten, not made, and embodied the "City of God," to which the "City of Man" could aspire only if it was grounded on truth that legal genius alone could not impart. To be sure, Brownson thought that the framers in Philadelphia did a miraculous job. But this was because the taproots of their Constitution drank more deeply than the Enlightenment (faith in which had aborted the French republic of the 1790s) or even the Reformation (faith in which had aborted the Cromwellian republic of the 1650s). Rather, the taproots of American notions of natural rights, ordered liberty, due process, and separation of church and state could all be traced to the medieval schoolmen and to English common law. Brownson even went so far as to conclude that the American republic was truer to Catholic principles of dignity, law, and justice than any European regimes, especially those purportedly Catholic. That was why America's founders not only rejected monarchy and its established church, but shunned democracy in favor of a constitutional republic. They believed the purpose of government was neither to tyrannize citizens nor to indulge their wandering will, but to cultivate liberty under law, forbearance, and virtue best fostered by free exercise of religion.[22]

Brownson thought the founders got all of that right. A sturdy republic must rest on three pillars: the people, the state, and the church. A clever constitution might check and balance powers so as to inhibit the people and state from trampling each other, but it could not do so forever without the mediating power of an independent, orthodox church. But America's blueprint had a flaw. The religion the founders counted on to keep Americans loyal to their two constitutions and discipline the people and state did not exist. Protestantism was no church, but 1,000 churches, which were themselves merely creatures of the same public opinion driving politics. American pastors exercised no authority and religion had no rights which the white man was bound to respect (to borrow a phrase from Chief Justice Taney). Instead, Americans bent on pursuing their happiness established a culture in which all individuals were effectively free to craft their own religion, as if they were gods. Worse still for the health of the public square, many Americans claimed the freedom to shove their disparate moral agendas down others' throats, again as if they were gods. Hence the two dangers to the republic: egotism, license, and anarchy in the manner of frontier Jacksonians and urban mobs; or intolerance, intimidation, and coercion in the manner of self-righteous reformers. Nor could anyone say that Americans were too prudent to toy with those dangers, after the Walpurgis Night they had just endured during the Civil War.

Brownson blamed that on a churchless democracy. "The most marked political tendency of the American people has been, since 1825, to interpret their government as a pure and simple democracy. . . . Their tendency had unconsciously, therefore, been to change their constitution from a republic to a despotic, or from a civilized to a barbaric, constitution." Stuffy Whigs would no doubt have concurred. But Brownson was no Whig wagging his finger at unwashed Democrats, because he saw in Whiggism an equal and opposite danger. In fact, he meant to distinguish among three sorts of democracy, only one of which was conducive to liberty. That was what Benjamin Disraeli called "territorial democracy." He meant by this simply a sovereign people inhabiting a fixed territory and subject to laws whose force was coterminous with the territory. That may sound unexceptional until one contrasts it with the other sorts of democracy. The second, which Brownson called Jeffersonian, interpreted government by consent of the governed to mean that individuals and states might scoff at unpopular laws inside the national territory or even secede from it. Needless to say, such anarchic democracy was the solvent of ordered liberty and Union itself. The third sort of democracy, which Brownson called humanitarian, "scorns all geographical lines, effaces all in individualities, and professes to plant itself on humanity." Thus he sensed as early as 1865 a fanatical, Jacobin urge among American reformers that might someday spill over their national boundaries in a crusade to reform the whole world.[23]

The Civil War was a free-for-all among these competing forms. The South fought to save Jeffersonian democracy. The North fought to save territorial democracy. But Yankee humanitarian democracy piggybacked on the war for the Union and abolition to emerge more self-confident than before, more determined (said Brownson) to efface individuals into prescriptive society even as states' rights had been effaced. In other words, the Civil War purged the threat posed by secessionists, anarchists, and libertines but magnified the threat posed by humanitarians, whose ambition would grow along with the nation's raw power. Brownson feared that future William Lloyd Garrisons in league with future Henry Ward Beechers would seduce or cajole the people to make war on property, privacy, marriage, and even foreign countries in the name of perfecting mankind according to their lights. The humanitarian sees only "humanity, superior to individuals, superior to states, governments, and laws, and holds that he may trample on them all or give them to the winds at the call of humanity or the 'higher law.'" But since that higher law was the humanitarian's own invention, entrusting the republic's health to him was like letting a patient tell the doctor what to prescribe. Such altruism amounted to pride, not humility; socialism, not charity; the will to power, not love. Taken to its logical conclusion, it must eventually reduce to Jean-Jacques Rousseau's calamitous formula whereby government forces men to be free and sacrifices the rights of men in the name of the rights of man.

Slipping out of political science and back into Catholicism, Brownson blamed Satan, whose "favorite guise in modern times is that of philanthropy. He is a genuine humanitarian, and aims to persuade the world that humanitarianism is Christianity, and that man is God." Christian doctrine taught that the melting of hearts, preaching of morals, and sanctifying of souls were the church's business, not the state's. But the lack of an American church tempted moralists to redefine indelible sins as erasable vices and make civil authority the eraser. Must the federal government therefore usurp religious authority and over time impose a tyranny akin to the Spanish Inquisition? On the contrary, Brownson believed to the last that Providence intended the United States to reconcile liberty and authority through the influence of an independent, respected church enjoining truth, justice, prudence, and charity. But since Protestantism could not supply such a church, it would have to be Catholic, just as the principles undergirding the Constitution were Catholic. Indeed, Brownson insisted that Americans needed Catholicity more than other people in history lest, in their liberty, they simply run wild.[24]

Brownson was under no illusions that Catholics, a despised minority, were likely to convert many Protestants and freethinkers. But he had faith that revealed truth would flourish, not die, under religious liberty. Nor was his vision that of a Romantic medievalist. He knew Americans would "always be progressive as well as conservative," and he imagined the church channeling both tendencies in healthy directions. Nor did Brownson's apprehension about humanitarian crusades cause him to doubt America's destiny. "The effect of this mission on our country fully realized, would be to harmonize church and state, religion and politics, not by absorbing either into the other . . . but by conforming both to the real or Divine order, which is supreme and immutable." Such an America would grow and prosper as none before. Hence, "the American people need not trouble themselves about their exterior expansion. That will come of itself as fast as desirable. Let them devote their attention to their internal destiny, to the realization of their mission within, and they will gradually see the whole continent coming under their system, forming one grand nation, a really catholic nation, great, glorious, and free."[25]

Learned American Catholics paid polite attention to *The American Republic*. European Catholics eyed it with a certain suspicion. Mainstream American Protestants just thought it a curiosity. Brownson was too much the Yankee for Catholics, said Van Wyck Brooks, and too much the Catholic for Yankees. Indeed, the swivel in Brownson's weather vane had seemingly rusted shut, so it now pointed permanently into the wind. His notion that America was destined to be Catholic seemed so preposterous that his colleagues and rivals alike ceased taking him seriously. His broadsides against congressional Reconstruction (he thought building a new South on the strength of field hands and a few carpetbaggers absurd) alienated his Republican friends. His ongoing quarrels with Catholic bishops forced his old friend Father Hecker to distance himself. Worst, his wife Sally died in 1872. But since her last wish was that Orestes revive *Brownson's Quarterly Review* he obliged her by writing more witty critiques. One skewered Henry Ward Beecher, another the false pretenses under which Darwinists and evangelicals alike fought over science, yet another the Protestants' manic fear of the pope.[26] But at times the fast-aging Brownson seemed to caricature himself. He became ornery, especially about feminists, lost interest in the nation's corrupt politics, and redoubled his consumption of rich food and spirits. When at last his health failed he joined his son's family in Detroit, where he had caught malaria and imbibed Universalism so many years before. Brownson died there, after receiving communion brought by a priest, on Easter Monday in the centennial year of 1876.

His message had no chance of a hearing over the din of Americans' post–Civil War pandemonium, and subsequent eras seemed to prove Brownson wrong. Thanks to immigrants and large families, Catholics would in time become the United States' largest denomination. But that was the point: Catholics, too, were just a denomination and hardly a national church. In any case, the Vatican grew alarmed as early as the 1890s over the prospect that assimilation was "Americanizing" the church. (Orthodox rabbis sounded the same alarm at the time over the Americanization of Jewish immigrants.) Brownson's dialectical mind also led him to underestimate the common sense of his countrymen. In the century to come Americans proved they had learned the lesson of their Civil War, which was never to carry their pride, conceits, and heresies to ultimate, destructive conclusions. To be sure, secular and evangelical reform movements, most prominently those of the Progressive Era and Prohibition, the New Deal, and the Great Society, would empower government in myriad ways, for better or worse. But America never came close to the anarchy, communism, or theocracy Brownson feared. At times the more traditional Protestant sects even played (if belatedly) the role of the nation's conscience that he thought only Catholics could play.

Still, Brownson's diagnosis of the body politic, if not his prognosis, rang true. Lacking a shared spirituality and devotion to a common Creator, Americans clung to a civic religion. They pretended, like Walt Whitman, that they were priests in a church of democracy, that their nation was a heaven under construction, that their progress was a sufficient measure of truth, and that their corruption was forgivable so long as it proved creative. Oliver Wendell Holmes wrote that into the nation's laws, John Dewey wrote it into the nation's curriculum, and William James wrote it into the nation's psyche. James, the brilliant Harvard philosopher, made America's relation to truth explicit in his lectures of 1907 on pragmatism: "Truth *happens* to an idea. It *becomes* true, is *made* true by events." If believing in something—anything—helps the people achieve what they desire, that alone makes it so. Thus, "if the hypothesis of God works satisfactorily in the widest sense of the word, it is true." [27]

Nor was Brownson alone in his diagnosis. Soon after his death even some devolved Protestants echoed his suspicion that democracy without truth is idolatry. In *The Degradation of Democratic Dogma*, Henry Adams described the crisis of faith his grandfather John Quincy Adams suffered on losing the presidency to Andrew Jackson. "Democracy had failed to justify itself. Man alone, unaided by a supernatural power, could not resist the pressure of self-interest and of greed. . . . Above this level of servitude to 'the flesh,' or competition, democracy could not rise. On the contrary democracy then *deified* competition, preaching it as the highest destiny and true duty of man. And Mr. Adams himself found to his horror that he, who had worshiped education and science, had unwittingly ministered to the demon." [28]

Another sworn foe of pretense made the point even more eloquently. In 1886 an African Methodist Episcopal pastor in Washington, D.C., asked a certain celebrity—known to be a religious skeptic—why he frequented services and generously fed the collection plate. The layman replied in a letter crafted with care. He acknowledged all manner of doubts regarding the authorship of the Bible, the means of interpreting it, and the confusion of claims among Christian sects. But he attended church anyway because, he confessed:

I hold that the pulpit is capable of being a powerful agent in the dissemination of truth, and I hold that truth is the power of God for the salvation of the world, and I do not limit truth to mere spiritual matters, but to man in all his relations in the family, in the church, in the government, and in the world.

Very truly yours, Frederick Douglass.[29]

ACKNOWLEDGMENTS

The roster of generous and loyal supporters who made possible this second volume is identical to that for the first, with one outstanding addition: Bruce Hooper, a trustee of Philadelphia's Foreign Policy Research Institute (FPRI) and a former U.S. Marine. Thanks to his "*Semper Fi*" spirit I occupied the FPRI's Hooper Chair for 2005, during which most of this book was written. Otherwise, I am once again indebted to all the supporters of FPRI's Center for the Study of America and the West, especially the Earhart Foundation, Lynde and Harry Bradley Foundation, John M. Olin Foundation, and Smith Richardson Foundation. Once again I thank FPRI's president, Harvey Sicherman; its vice president, Alan Luxenberg; and business manager, Harry Richlin, for hosting (and hiding) me while I toiled. Likewise the roster of people who helped me complete this second volume is identical to that of the first, with two outstanding additions: Rob Crawford, Hugh Van Dusen's gracious, efficient assistant editor at HarperCollins; and Susan Gamer, the patient, skilled copy editor who vetted the final manuscript. Otherwise, I thank once again my wife, Jonna, and children Angela and Christopher for tolerating my long hours absent from home and my obsessive-compulsive moods while at home. (In a metaphor any mother would understand I equate writing a book of this length to being pregnant for four years.) Once again I thank Frank Plantan and Donna Shuler for running the International Relations Program at Penn during my leave. Sicherman, James Kurth, and my late ninety-one-year-old father, Dugald S. McDougall, read and commented on the chapters with care and insight. Allen Guelzo, Kevin Starr, Forrest McDonald, Michael Federici, and Gregory Butler were inspirations from afar: credit them for whatever I get right about Lincoln, California, the presidency, and Orestes Brownson. I benefited throughout from the advice or scholarship of colleagues at Penn, including Richard Beeman, Robert Engs, Ann Greene, Sheldon Hackney, Steven Hahn, Bruce Kuklick, Walter Licht, Stephanie McCurry, Daniel Richter, Thomas Sugrue, Arthur Waldron, Beth Wenger, and Michael Zuckerman. Most of all, I once again thank Christopher M. Gray, my sole (and mostly pro bono) research assistant, for his countless tips on historiography, praise or criticism of every word in the manuscript, and indispensable pep talks. Last time I likened Gray to a wizened golf caddy. This time, to honor his bluegrass roots in Kentucky, I shall liken him to a thoroughbred trainer who gets the most out of a mount, even an aging nag like myself.

NOTES

Chapter 1

1. Allan Nevins, ed., *The Diary of Philip Hone 1828–1851*, 2 vols. (New York: Dodd, Mead, 1927), I: 185–190 (account of fire), I: 244 (quotation about Jackson). See also the abridged, illustrated edition with introduction by Louis Auchincloss, *The Hone and Strong Diaries of Old Manhattan* (New York: Abbeville, 1989).

2. Eliza Leslie, "The New York Fire," *Parley's Magazine* (January 1838): 30–33.

3. On the great fire of December 1835, see Edwin G. Burrows and Mike Wallace, *Gotham: A History of New York City to 1898* (New York: Oxford University, 1999), pp. 596–598 ("flagrant, providential warning," p. 601); Edward Robb Ellis, *The Epic of New York City: A Narrative History* (New York: Kodansha, 1997), pp. 240–242; and Terry Golway, *So Others Might Live: A History of New York's Bravest—The FDNY from 1700 to the Present* (New York: Basic Books, 2002), pp. 43–60. Nathaniel Currier was a great artist, but his partner, the native New Yorker James Merrill Ives, contributed the business smarts behind the firm.

4. Washington Irving, his brother William, and James Kirke Paulding (William's brother-in-law) collaborated on a book of essays they called *Salmagundi* after the disgusting hash of that name, thereby inventing the literary style of the "Knickerbocker wits." In one story they likened New York to the medieval English village of Gotham. According to legend King John determined to build an estate there, which would saddle the locals with no end of work, responsibility, and unwelcome royal authority. So they conspired to behave like fools and idiots until the king threw up his hands and went elsewhere.

5. Robert Greenhalgh Albion, *The Rise of New York Port* (New York: Scribner, 1939), is still a valuable classic. On New York's geographical advantages and sudden rise (he calls 1817 an annus mirabilis), see pp. 1–37; on the cotton trade, pp. 95–121.

6. Steve Fraser, *Every Man a Speculator: A History of Wall Street in American Life* (New York: HarperCollins, 2005), pp. 34–48 ("bear's skin," p. 46). Fraser's epic survey of this premier American metaphor is not to be missed.

7. Ibid., pp. 49–69 (quotations: Jackson, pp. 50–51; Beecher and Channing, p. 55; southerners, p. 58); Charles Dickens, *The Life and Adventures of Martin Chuzzlewit* (New York: Penguin, 1977, orig. 1843), pp. 336–337.

8. On the recovery from the fire, see Burrows and Wallace, *Gotham*, pp. 598–602 (quotation from Gallatin, p. 598); George J. Lankevich, *American Metropolis: A History*

of New York City (New York: New York University, 1998), pp. 81–82; "Smiling faces" cited by Cynthia Crossen, "New York City Rebuilt with More Splendor—It Happened Before," *Wall Street Journal* (September 4, 2002), p. B1; Hone quoted in Nevins, ed., *Diary of Philip Hone*, I: 202.

9. Burrows and Wallace, *Gotham*, pp. 625–628; Nevins, ed., *Diary of Philip Hone*, II: 624–625. On the Croton water project, see Neil Fitzsimmons, *The Reminiscences of John B. Jervis, Engineer of the Old Croton* (Syracuse, N.Y.: Syracuse University, 1971).

10. Herbert Asbury, *The Gangs of New York: An Informal History of the Underworld* (New York: Knopf, 1928) and Alvin F. Harlow, *Old Bowery Days: The Chronicles of a Famous Street* (New York: Appleton, 1931) are entertaining classics. Their anecdotes, however, are not always trustworthy. For an authoritative scholarly history, see Tyler Anbinder, *Five Points: The 19th Century New York City Neighborhood That Invented Tap Dance, Stole Elections, and Became the World's Most Notorious Slum* (New York: Free Press, 2001).

11. Anbinder, *Five Points*, pp. 1–37 ("more filthy," p. 35); Burrows and Wallace, *Gotham*, pp. 542–546. Five Points festered for sixty years before the Danish immigrant journalist Jacob Riis shamed the city fathers with his sensation of 1890, *How the Other Half Lives*. By 1897 the city's bulldozers and vice squads had turned the district into a still humble but pleasant immigrant community graced with a park. This "New York approach" became a model of early slum clearing and urban renewal (Anbinder, *Five Points*, pp. 424–441).

12. David Carlyon, *Dan Rice: The Most Famous Man You've Never Heard Of* (New York: Public Affairs, 2001), pp. 46–47. We shall have occasion to hear much more of Dan, who borrowed T. D. Rice's famous name, in due course. For a comprehensive treatment, see Robert C. Toll, *Blacking Up: The Minstrel Show in Nineteenth-Century America* (New York: Oxford University, 1974).

13. On sex, tourism, and "drummers," see Burrow and Wallace, *Gotham*, pp. 533–541; on lotteries and the policy racket, see Herbert Asbury, *Sucker's Progress: An Informal History of Gambling in America from the Colonies to Canfield* (New York: Dodd, Mead, 1938), pp. 72–101.

14. See Bertram Wyatt-Brown, *Lewis Tappan and the Evangelical War against Slavery* (Cleveland, Ohio: Case Western Reserve University, 1969). Curiously, there are no biographies of Arthur Tappan except Lewis Tappan's *The Life of Arthur Tappan* (Westport, Conn.: Negro Universities Press, 1970, orig. 1870). Child quotation, *Letters from New York* (1844), cited in Anbinder, *Five Points*, p. 34. The Tappans' grandfather taught them to revere Northampton's great revivalist and theologian Jonathan Edwards, who did as much as anyone in the eighteenth century to popularize the notion of America's millenarian destiny.

15. John McDowell threw all his youthful energy into the work of the Magdalen Society. He was one of the first graduates of Amherst College, founded in 1827 to provide an orthodox Congregationalist alternative to Unitarian Harvard, and he went on a one-man mission to shock New York's clergy. The trove of pornography he displayed and the list of bawdy houses he published had the desired effect, but also the unintended consequence of advertising the best spots in town to procure prurient pleasures. For a sober survey of this titillating subject, see Timothy J. Gilfoyle, *City of Eros: New York City, Prostitution, and the Commercialization of Sex, 1790–1920* (New York: Norton, 1992).

16. Wyatt-Brown, *Lewis Tappan*; and Burrows and Wallace, *Gotham*, pp. 529–541.

On Finney's early career, see Keith Hardmann, *Charles Grandison Finney, 1792–1875: Revivalist and Reformer* (Syracuse, N.Y.: Syracuse University, 1987); and Charles E. Hambrick-Stowe, *Charles G. Finney and the Spirit of American Evangelism* (Grand Rapids, Mich.: Eerdmans, 1996).

17. Paul A. Gilje, *The Road to Mobocracy: Popular Disorder in New York City, 1763–1834* (Chapel Hill: University of North Carolina, 1988); Burrows and Wallace, *Gotham*, pp. 542–552 (Webb quoted on p. 552); Tappan, *Life of Arthur Tappan*, pp. 194–195; Anbinder, *Five Points*, pp. 7–13; businessman quoted in Ellis, *Epic of New York*, p. 236; Hone quoted in Nevins, ed., *Diary of Philip Hone* I: 134.

18. David Grimsted, *American Mobbing, 1828–1861: Toward Civil War* (New York: Oxford University, 1998), pp. 3–32 ("crest of rioting," p. 4); Burrows and Wallace, *Gotham*, pp. 553–560 ("time has not yet come," p. 559); on the Herschel hoax, perpetrated in all likelihood by the English-educated reporter Richard Adams Locke, see Ormond Seavey, ed., *The Moon Hoax, Or, A Discovery That the Moon Has a Vast Population of Human Beings* (Boston, Mass.: Gregg, 1975). Herschel, who in fact spent 1835 in South Africa observing Halley's comet, refuted the stories. Curiously, in the 1780s John's father, Sir William Herschel, had claimed that his telescope revealed live volcanoes on the moon.

19. Burrows and Wallace, *Gotham*, pp. 609–611 (quotation, p. 611).

20. On the origins of the Panic of 1837, see Bray Hammond, *Banks and Politics in America from the Revolution to the Civil War* (Princeton, N.J.,: Princeton University, 1957), pp. 457–467; effects on New York in Burrows and Wallace, *Gotham*, pp. 611–618; Hone quoted in Nevins, ed., *Diary of Philip Hone*, I: 249, 259. On the powerful influence of European investment on the developing American economy, see Mira Wilkins, *The History of Foreign Investment in the United States to 1914* (Cambridge, Mass.: Harvard University, 1989).

21. See, for instance, Henry Steele Commager, *America in Perspective: The United States through Foreign Eyes* (New York: Mentor, 1947).

22. I plead guilty. See Walter A. McDougall, *Promised Land, Crusader State: The American Encounter with the World since 1776* (Boston, Mass.: Houghton Mifflin, 1997), p. 6.

23. George Wilson Pierson, *Tocqueville in America* (Baltimore, Md.: Johns Hopkins University, 1996), the classic demystification of *De la Démocratie en Amérique*, was first published by Oxford University Press in 1938.

24. Ibid., pp. 43–83 (quotations, pp. 49, 65, 68, 71, 73).

25. Ibid., quotations pp. 458, 476, 488, 509, 672, 667, 678. Pierson's long book is a delight. Those interested in a brief synopsis may turn instead to Garry Wills, "Did Tocqueville 'Get' America?" *New York Review of Books* (April 29, 2004): 52–56.

26. One may excuse Tocqueville for not being a prophet. But another Frenchman, Michel Chevalier, toured America just two years later and made railroads a centerpiece of his study *Lettres sur l'Amérique du Nord* (1836).

27. Encounter with Chase in Pierson, *Tocqueville in America*, pp. 555–558 (quotations, p. 557). Tocqueville quoted in Phillips Bradley, ed., *Democracy in America by Alexis de Tocqueville*, 2 vols. (New York: Vintage, 1945), I: 207–208. "Principal Causes Which Tend to Maintain the Democratic Republic" in *Democracy in America*, I: 298–342.

28. Conversation at Baltimore in Pierson, *Tocqueville in America*, pp. 500–501; Tocqueville's conclusions on religion in *Democracy in America*, I: 45, 310–322 (quotations, pp. 316, 318, 321).

29. J. P. Mayer, ed., *Journey to America by Alexis de Tocqueville* (London: Faber and Faber, 1959), p. 177.

30. Quotation from a letter to one of her sons, cited in Richard Mullen's introduction to Frances Trollope, *Domestic Manners of the Americans* (New York: Oxford University, 1984), p. xix.

31. The ubiquitous "Have you read Mrs. Trollope?" is from E. T. Coke, *A Subaltern's Furlough* (London, 1838), cited ibid., p. ix. Fanny Trollope went on to write more than forty books of fiction, travel, and poems. Appearances to the contrary notwithstanding, it is unlikely that anything "improper" occurred between her and the tutor, an erstwhile revolutionary named August Jean Hervieu. *Tant pis!* See Helen Heineman, *Mrs. Trollope: The Triumphant Feminine in the Nineteenth Century* (Athens: Ohio University, 1979).

32. Trollope, *Domestic Manners*, pp. 1–26 (quotations, pp. 2, 12, 24). Trollope later learned that Fanny Wright abandoned her humanitarian idyll, sailed to Haiti, and set her slaves free.

33. Ibid., pp. 35–59 (quotations, pp. 36, 56, 58). On the "wolf traps," see Asbury, *Sucker's Progress*, pp. 271–279. More ironically still, Cincinnati later became home to the United States Playing Card Company, purveyor of Bicycle, Aviator, Bee, and Hoyle decks to the nation since 1867.

34. Trollope, *Domestic Manners*, quotations from pp. 44, 126–128, 85–86. Tocqueville had a similar wry encounter in Philadelphia. When told that most crime in America was caused by the cheapness of alcohol, he asked why legislators did not place a tax on intoxicating beverages. Because, he was told, any politicians who did so were sure to lose their seats. "Whence I am to infer," he replied, "that drunkards are the majority in your country" (*Democracy in America* I: 239).

35. Trollope, *Domestic Manners*, pp. 59–67, 89–95 (quotations, pp. 60, 89, 90).

36. Ibid., quotations from pp. 186–187, 207, 154–155.

37. Ibid., pp. 213–220 (Baltimore), 293–360 (New York state); quotations, pp. 213–215, 297, 360.

38. See Valerie Kossew Pichanick, *Harriet Martineau: The Woman and Her Work, 1802–1876* (Ann Arbor: University of Michigan, 1980); R. K. Webb, *Harriet Martineau: A Radical Victorian* (New York: Columbia University, 1960); and Seymour Martin Lipset, introduction to Harriet Martineau, *Society in America* (Garden City, N.Y.: Doubleday, 1962), pp. 1–42. Quotation from Martineau, *Society in America*, p. 50.

39. Martineau, *Society in America*, pp. 60–73 (quotation, p. 73).

40. Ibid., pp. 94–106 (quotations, pp. 97–100, 103, 248).

41. Ibid., *Society in America*, pp. 129–245 (quotations, pp. 131, 198–201, 207, 245).

42. Ibid., pp. 291–306 (Philadelphia, p. 262; women, p. 305).

43. Ibid., pp. 332–357 (quotations, pp. 334, 347, 350, 357).

44. See Jerome Meckier, *Innocent Abroad: Charles Dickens's American Engagements* (Lexington: University of Kentucky, 1990); and Michael Slater, ed., *Dickens on America and the Americans* (Austin: University of Texas, 1978).

45. Charles Dickens, *American Notes for General Circulation* (New York: Penguin, 2000, orig. 1842), pp. 33–89 (quotations, pp. 35, 36, 89).

46. Ibid., pp. 90–108 (quotations, pp. 97–99, 102) and introduction by Patricia Ingham, pp. xi–xxxi. All Dickens achieved with his acerbic indictment was to turn "slumming" into a trendy day trip for ladies and gentlemen (and their bodyguards). They

affected horror and pity while inwardly leering at the hussies and drunks lolling half-naked in gutters.

47. Ibid., pp. 125–162 (quotations, pp. 134, 151).

48. Ibid., pp. 152–183, 220 (Niagara), 250–277 (quotations, pp. 265, 267).

49. Nevins, ed., *Diary of Philip Hone*, II: 673. Hone had no use for "the famous Mrs. Trollope" either. Her book "abounds in ridiculous remarks on the national character and political institutions of the United States, and represents highly exaggerated caricatures of our domestic manners. . . . I paid it the compliment of devoting two or three hours to it when I ought to have been in bed" (I: 88). Harriet Martineau, on the other hand, impressed Hone with her intelligence and sense of humor.

50. John R. G. Hassard, *Life of the Most Rev. John Hughes, D.D.: First Archbishop of New York* (New York: Arno, 1969); this biography first appeared in 1865, two years after Hughes's death. The only recent biography is Richard Shaw, *Dagger John: The Unquiet Life and Times of Archbishop John Hughes of New York* (New York: Paulist, 1977). Hughes received valuable encouragement in his educational campaigns from Saint Elizabeth Seton. Bishop Du Bois had studied at the College of Louis Le Grand, where his classmates included the future revolutionaries Robespierre and Desmoulins.

51. On the struggle over the schools and the election riots of 1842 see Burrow and Wallace, *Gotham*, pp. 629–631 (Whitman is quoted on p. 631); Ellis, *Epic of New York*, pp. 252–254 ("build your own schools," p. 254); Anbinder, *Five Points*, pp. 154–158 ("so beaten about the head," p. 156).

52. On the riots of 1844 in Philadelphia and the threat to New York, see Grimsted, *American Mobbing*, pp. 218–219; Samuel Eliot Morison, *The Oxford History of the American People* (New York: Oxford University, 1965), p. 482; Burrow and Wallace, *Gotham*, pp. 631–633 ("Are you afraid," p. 633); Hone quoted in Nevins, ed., *Diary of Philip Hone*, II: 660.

53. P. J. Hayes, "John Hughes," *The Catholic Encyclopedia*, 1910; Online Edition, 2003, p. 3.

54. George H. Shriver, *Philip Schaff: Christian Scholar and Ecumenical Prophet* (Macon, Ga.: Mercer, 1987). Schaff made many trips back to Europe, but the invitation to Berlin occasioned his first trip, so news of him must have reached the old country independently. Schaff presided over Mercersburg for thirty years. In 1863 Lee's Army of Northern Virginia raided the town and dragged the pastor's African-American housemaid into slavery. Two years later Schaff moved to New York, and in 1870 he joined the faculty of Union Theological Seminary, where he spent his last twenty-three fruitful years. Schaff also founded the American Society of Church History (1888). His dying act, at the 1893 World's Parliament of Religions in Chicago, was to plead for Christian unity.

55. Philip Schaff, *America, A Sketch of the Political, Social, and Religious Character of the United States of North America, in Two Lectures* (New York: Scribner, 1855).

56. Ibid., pp. viii–x, 33–34.

57. Ibid., pp. xi–xiv.

58. Ibid., pp. 35–52 (quotations, pp. 43, 44, 55, 62). The quacks in question, Dr. Jayne and Dr. Townshend, were pioneers in the marketing of patent medicines. In addition to sarsaparilla (billed as a feel-good tonic and aphrodisiac) they sold concoctions reputed to cure every complaint from epilepsy to hair loss.

59. Ibid., p. 85.

Chapter 2

1. Robert V. Remini, *Andrew Jackson and the Course of American Freedom, 1822–1832* (New York: Harper and Row, 1981), pp. 1–11 ("slime," p. 2); 148–155 (Rachel's death). On Jackson's early life and driven personality, see James C. Curtis, *Andrew Jackson and the Search for Vindication* (Boston: Little, Brown, 1976), or the summary in Walter A. McDougall, *Freedom Just Around the Corner: A New American History 1585–1828* (New York: HarperCollins, 2004), pp. 457–458.

2. On Jackson's cabinet selections, see Remini, *Course of American Freedom*, pp. 159–166 (McLane is quoted on p. 162). Margaret O'Neall Timberlake Eaton's reputation as "the American Pompadour" was fixed for posterity in Gaillard Hunt, ed., *The First Forty Years of Washington Society in the Family Letters of Margaret Bayard Smith* (New York: Scribner, 1906), pp. 252, 282–289; and Meade Minnigerode, *Some American Ladies* (New York: Putnam, 1926), pp. 241–287 ("astonishingly pretty," p. 254). Peggy's side of the story has not been adequately told, because her autobiography, written with the assistance of her pastor (!) in 1873, is scarcely reliable. Queena Pollack, *Peggy Eaton: Democracy's Mistress* (New York: Minton, Balch, 1931), and Leon Phillips, *That Eaton Woman* (Barre, Mass.: Barre, 1974) are undocumented.

3. For Van Buren's youth and early career, see Donald B. Cole, *Martin Van Buren and the American Political System* (Princeton, N.J.: Princeton University, 1984); John Niven, *Martin Van Buren: The Romantic Age of American Politics* (New York: Oxford University, 1983); Robert V. Remini, *Martin Van Buren and the Making of the Democratic Party* (New York: Columbia University, 1959); and of course *The Autobiography of Martin Van Buren* (New York: Da Capo, 1973, orig. printed in 1920 in *Annual Report of the American Historical Association for the Year 1918*). "Millennium of minnows," John Niven, *John C. Calhoun and the Price of Union: A Biography* (Baton Rouge: Louisiana State University, 1988), p. 167; "general expression," "grieved," and "Even Van Buren" from Cole, *Martin Van Buren*, pp. 180–181.

4. Van Buren quoted in Cole, *Martin Van Buren*, pp. 107–108. On the election of 1828, see Robert V. Remini, *The Election of Andrew Jackson* (Philadelphia, Pa.: Lippincott, 1963), Richard P. McCormick, *The Presidential Game: The Origins of American Presidential Politics* (New York: Oxford University, 1982); and the summary in McDougall, *Freedom Just Around the Corner*, pp. 494–497.

5. *Autobiography of Martin Van Buren*, pp. 228–232.

6. See John F. Marszalek, *The Petticoat Affair: Manners, Mutiny, and Sex in Andrew Jackson's White House* (New York: Free Press, 1997) for a recent, thorough treatment. The visit to Floride in Niven, *Calhoun and the Price of Union*, pp. 167–168; Jackson's behavior during the Eaton affair in Remini, *Course of American Freedom*, pp. 203–210 ("as chaste" and "I did not come here," pp. 204–205); Masonic argument quoted in Steven C. Bullock, *Revolutionary Brotherhood: Freemasonry and the Transformation of the American Social Order 1730–1840* (Chapel Hill: University of North Carolina, 1996), p. 229.

7. Remini, *Course of American Freedom*, pp. 211–216 ("frank, open," p. 214); Cole, *Martin Van Buren*, pp. 187–189, 203–207; *Autobiography of Martin Van Buren*, pp. 339–343; "Calhoun leads the moral party," in Charles Francis Adams, ed., *Memoirs of John Quincy Adams*, 12 vols. (Philadelphia: Lippincott, 1875–1877), VIII: 185; the term "Eaton Malaria" was coined by Van Buren. There is no evidence that Calhoun was behind the cabinet's revolt, but after Eaton resigned in 1831 he cheered "the great victory that has been achieved in favor of the morals of the country, by the high-minded independence and

virtue of the ladies of Washington": Irving H. Bartlett, *John C. Calhoun: A Biography* (New York: Norton, 1993), p. 165.

8. J. D. Richardson, ed., *Compilation of the Messages and Papers of the Presidents*, 20 vols. (Washington, D.C.: GPO, 1908), II: 1005–1025.

9. Construction of racial hierarchies as a root cause for removal of the Indians is a theme in Robert F. Berkhofer, *The White Man's Indian: Images of the American Indian from Columbus to the Present* (New York: Knopf, 1978); Ronald T. Takaki, *Iron Cages: Race and Culture in Nineteenth-Century America* (New York: Knopf, 1979); Alexander Saxton, *The Rise and Fall of the White Republic: Class Politics and Mass Culture in Nineteenth-Century America* (New York: Verso, 1990); and Richard Drinnen, *Facing West: The Metaphysics of Indian-Hating and Empire-Building* (Minneapolis: University of Minnesota, 1995). Michael Paul Rogin, *Fathers and Children: Andrew Jackson and the Subjugation of the American Indian* (New York: Knopf, 1975); and James C. Curtis, *Andrew Jackson and the Search for Vindication* (Boston, Mass.: Little, Brown, 1976) make psychological arguments. The revisionist case that Jackson meant to act paternalistically toward Indians is best made by Francis Paul Prucha, *The Great Father: The United States Government and the American Indians*, 2 vols. (Lincoln: University of Nebraska, 1984), I: 183–213. His arguments caused Robert V. Remini, in *Andrew Jackson and His Indian Wars* (New York: Viking, 2001), to acknowledge Jackson's effort to foster nationalism, states' rights, and opportunity for whites, while still saving the Indians. The debate is summarized in Ronald N. Satz, "Rhetoric versus Reality: The Indian Policy of Andrew Jackson," in William L. Anderson, ed., *Cherokee Removal Before and After* (Athens: University of Georgia, 1991), pp. 29–54. See also Ronald N. Satz, *American Indian Policy in the Jacksonian Era* (Lincoln: University of Nebraska, 1975).

10. Philip Weeks, *Farewell My Nation: The American Indian and the United States in the Nineteenth Century* (Wheeling, Ill.: Harlan Davidson, 2001), pp. 34–40 ("Two centuries," p. 35); Prucha, *The Great Father*, I: 183–191; Jackson quoted in Richardson, ed., *Messages and Papers of the Presidents*, III: 1084. For McKenney's important role, see Herman J. Viola, *Thomas L. McKenney: Architect of America's Early Indian Policy 1816–1830* (Chicago, Ill.: Swallow, 1974); and Francis Paul Prucha, *American Indian Policy in the Formative Years: The Indian Trade and Intercourse Acts, 1790–1834* (Cambridge, Mass.: Harvard University, 1962).

11. In Georgia's defense it must be said that the federal government created the dilemma by making contradictory commitments to the state and the Cherokee. In 1802 Georgia relinquished its claim, based on its colonial charter, to the future states of Alabama and Mississippi. In return the Jefferson administration promised to retire all Indian land claims in the state. The federal government reneged on this promise even as it turned over vast portions of Ohio, Indiana, Illinois, Kentucky, Tennessee, Alabama, and Mississippi to their state governments.

12. On Cass and the Indians, see Anthony F. C. Wallace, *The Long Bitter Trail: Andrew Jackson and the Indians* (New York: Hill and Wang, 1993), pp. 41–49 ("controversial," p. 47); Willard Carl Klunder, *Lewis Cass and the Politics of Moderation* (Kent, Ohio: Kent State University, 1993), pp. 49–57; Lewis Cass, "Removal of the Indians," *North American Review* 30 (January 1830): 62–121. Given the lead time involved, Cass must have written his article before the president's annual message; hence he was probably put up to it by the Jackson camp.

13. The Cherokee constitution of 1827 declared the tribal lands a sovereign, inviolate

nation. That was what justified retaliatory acts by the state government. On the Five Civilized Tribes, see William G. McLoughlin, *Cherokee Renaissance in the New Republic* (Princeton, N.J.: Princeton University, 1986) and William G. McLoughlin, ed., *The Cherokees and Christianity 1794–1870: Essays on Acculturation and Cultural Persistence* (Athens: University of Georgia, 1994); Douglas C. Wilms, "Cherokee Land Use in Georgia before Removal," and Theda Perdue, "The Conflict Within: Cherokees and Removal," in Anderson, ed., *Cherokee Removal*, pp. 1–28, 55–74; Weeks, *Farewell My Nation*, pp. 38–40; Evarts quoted in Prucha, *The Great Father*, p. 204. The wealthiest Cherokee must have been Joseph Vann, who owned a brick mansion on a 300-acre plantation worked by 110 slaves, as well as a lucrative ferry, mill, and roadhouse. The tasteful historical parks preserving Chief Vann's mansion, the log cabin capital at New Echota, and the Trail of Tears memorial are well worth the short drive up Interstate 75 from Atlanta.

14. The governor and Eaton quoted in Perdue, "The Conflict Within," pp. 55–56. Jackson's promises (made initially to the Choctaw) in Prucha, *The Great Father*, p. 218; and Wallace, *Long Bitter Trail*, p. 74. Herman Melville used dark humor to explain "The Metaphysics of Indian-Hating" in his novel *The Confidence-Man: His Masquerade* (Indianapolis: Bobbs-Merrill, 1857), pp. 203–214. Indians who clung to their folkways proved the frontiersman's image of them as savage; Indians who embraced white culture admitted, by that very act, the depravity of their traditional ways. Either way, they couldn't win.

15. Prucha, *The Great Father*, pp. 214–229; Remini, *Course of American Freedom*, pp. 272–279 (on Greeley and Jackson's alleged remark, pp. 276–277); Russell Thornton, "The Demography of the Trail of Tears Period: A New Estimate of Cherokee Population Losses, in Anderson, ed., *Cherokee Removal*, pp. 75–95 ("cruelest work," p. 79); Weeks, *Farewell My Nation*, pp. 46–50 ("midst of death," p. 49); John S. D. Eisenhower, *Agent of Destiny: The Life and Times of General Winfield Scott* (New York: Free Press, 1997), pp. 184–194.

16. Francis Paul Prucha, *The Sword of the Republic: The United States Army on the Frontier, 1783–1846* (New York: Macmillan, 1968), pp. 273–300; John K. Mahon, *History of the Second Seminole War, 1835–1842* (Gainesville: University of Florida, 1967); Patricia Riles Wickman, *Osceola's Legacy* (Tuscaloosa: University of Alabama, 1991); and the good summary in Norman K. Risjord, *Representative Americans: The Romantics* (Lanham, Md.: Rowman and Littlefield, 2001), pp. 309–344.

17. Prucha, *The Great Father*, pp. 243–269, describes the eviction of the northern tribes. The defiant Sauk warrior with the "Mohawk" haircut lives on in the nickname of the National Hockey League's Chicago Black Hawks.

18. H. L. Mencken, *The American Language: An Inquiry into the Development of English in the United States*, 4th ed. (New York: Knopf, 1936), pp. 138–150 (quotations, pp. 138–139).

19. On the "golden age" of American oratory, see Russel Blaine Nye, *Society and Culture in America 1830–1860* (New York: Harper and Row, 1974), pp. 136–146. I am indebted to David Eisenhower for steering me, at this late date in life, to Aristotle.

20. Herman Belz, ed., *The Webster-Hayne Debate on the Nature of the Union: Selected Documents* (Indianapolis, Ind.: Liberty Fund, 2000), pp. vii–xv, 3–13 ("cause for complaint," p. 5).

21. Maurice G. Baxter, *One and Inseparable: Daniel Webster and the Union* (Cam-

bridge, Mass.: Harvard University, 1984); and Robert V. Remini, *Daniel Webster: The Man and His Time* (New York: Norton, 1997) are the definitive biographies. Merrill D. Peterson, *The Great Triumvirate: Webster, Clay, and Calhoun* (New York: Oxford University, 1987), is a masterful group portrait ("True eloquence," p. 111). But the best quick take on the man remains Stephen Vincent Benét's humorous tale (later a play), "The Devil and Daniel Webster," in which the great lawyer contests old Scratch himself for the soul of a humble New Hampshire man. This passage speaks volumes: "Dan'l Webster's brow looked dark as a thundercloud. 'Pressed or not, you shall not have this man!' he thundered. 'Mr. Stone is an American citizen, and no American citizen may be forced into the service of a foreign prince. . . .' 'Foreign?' said the stranger [the Devil]. 'And who call me a foreigner?' 'Well, I never yet heard of the dev—of your claiming American citizenship,' said Dan'l Webster with surprise. 'And who with better right?' said the stranger, with one of his terrible smiles. 'When the first wrong was done to the first Indian, I was there. When the first slaver put out for the Congo, I stood on her deck. Am I not in your books and stories and beliefs, from the first settlements on? Am I not spoken of, still, in every church in New England? Tis true the North claims me for a Southerner and the South for a Northerner, but I am neither. I am merely an honest American like yourself— and of the best descent—for, to tell the truth, Mr. Webster, though I don't like to boast of it, my name is older in this country than yours.' 'Aha!' said Dan'l Webster, with the veins standing out in his forehead. 'Then I stand on the Constitution! I demand a trial for my client!'"

22. Speech of January 20, 1830, in Belz, ed., *Webster-Hayne Debate*, pp. 15–33.

23. Speech of January 25, 1830, ibid., pp. 35–80 (quotations, pp. 45–51, 55, 80).

24. Speech of January 26–27, 1830, ibid., pp. 81–144. Webster stretched the truth when he implied that the framers of the Constitution meant to endow the Supreme Court with the power of judicial review. In fact, John Marshall's Court arrogated that power beginning in 1803.

25. Remini, *Course of American Freedom*, pp. 234–236; Niven, *Calhoun and the Price of Union*, pp. 171–173.

26. This is the metaphor of Louis P. Masur, *1831: Year of Eclipse* (New York: Hill and Wang, 2001), which elegantly describes the jarring developments occurring simultaneously in American politics, religion, and technology.

27. Remini, *Course of American Freedom*, pp. 306–309 ("dead man," p. 309), calls Calhoun's publication of the correspondence "sheer, inexplicable folly." Bartlett, *John C. Calhoun*, p. 174, argues that breaking with Jackson was the only way Calhoun might escape vice-presidential obscurity and contest Van Buren for party leadership, but he adds that personal honor was also at stake. Cole, *Martin Van Buren*, p. 217, insists that his subject "remained quietly on the sidelines." This rings true. The "little magician" was too smart to violate Napoléon's dictum: never interfere with your enemy when he's in the process of destroying himself.

28. "It will kill him" in Thomas Hart Benton, *Thirty Years' View, Or a History of the Workings of the American Government*, 2 vols. (New York: Appleton, 1856–57), I: 215; Masur, *Year of Eclipse*, pp. 107–114 ("house divided," p. 109). What became of the indomitable Peggy? She quickly befriended Queen Christina and was as much an ambassador as her husband in 1836–1840. On her return to the United States she succeeded in matching one of her daughters with a French duke and another with a U.S. Navy officer. When John Eaton died in 1856, Peggy was not only a free spirit again but well endowed with

money and land. In 1859 she married an Italian dancer one-third her age (he was twenty, she sixty) and moved to New York. But finally Peggy was scammed in turn. After talking her into signing over most of the estate, "Antonio" bolted for Canada with her fortune and one of her granddaughters. Peggy returned to Washington, D.C., where old-timers remembered her fondly, and died there in 1879.

29. Masur, *Year of Eclipse*, pp. 105–106 (Adams), 161–162 (Hayne).

30. Niven, *Calhoun and the Price of Union*, pp. 180–182 (quotation, p. 182).

31. Kenneth S. Greenberg, "Name, Face, Body"; David F. Almendinger, Jr., "The Construction of *The Confessions of Nat Turner*"; Herbert Aptheker, "The Event"; Thomas C. Parramore, "Covenant in Jerusalem"; and Vincent Harding, "Symptoms of Liberty and Blackhead Signposts," in Kenneth S. Greenberg, ed., *Nat Turner: A Slave Rebellion in History and Memory* (New York: Oxford University, 2003), pp. 3–102 ("terror and devastation," p. 41; "blood of Christ" and "first should be last," pp. 83–84). *The Confessions of Nat Turner, Leader of the Late Insurrection in Southampton, Virginia, as Fully Made to Thomas Gray . . . in 1831,* and the contemporary accounts compiled by Henry Irving Tragle, *The Southampton Slave Revolt of 1831: A Compilation of Source Material* (Amherst: University of Massachusetts, 1971), are summarized in Masur, *Year of Eclipse*, pp. 9–21 ("kind and indulgent," p. 10; "bleeding heap," p. 11; "blood-stained," p. 17; "gloomy fanatic," p. 19; "negroes found fault," p. 20. The rioting outside Virginia is described in David Grimsted, *American Mobbing 1828–1861: Toward Civil War* (New York: Oxford University, 1998), pp. 138–141.

32. James Sidbury, "Reading, Revelation, and Rebellion," in Greenberg, ed., *Nat Turner*, pp. 119–133; "smothered volcano" in Samuel Eliot Morison, *The Oxford History of the American People* (New York: Oxford University, 1965), pp. 508–509.

33. Elizabeth R. Varon, *We Mean to Be Counted: White Women and Politics in Antebellum Virginia* (Chapel Hill: University of North Carolina, 1998), pp. 41–70 (quotations, pp. 43, 49, 50).

34. Masur, "Nat Turner and Sectional Crisis," in Greenberg, ed., *Nat Turner*, pp. 148–161 ("I will not rest," p. 154; "as happy," 161). For the full story, see Alison Goodyear Freehling, *Drift toward Dissolution: The Virginia Slavery Debate of 1831–1832* (Baton Rouge: Louisiana State University, 1982).

35. Sometimes called the most prominent American politician never to win the presidency (my own choice is Charles Evans Hughes), Clay has a vast literature. The best and most readable works include Robert V. Remini, *Henry Clay: Statesman for the Union* (New York: Norton, 1991), although it is somewhat acerbic regarding Clay's ambition ("I am determined," p. 326; "wicked, passionate," p. 367); the Whigs' admirer Glyndon G. Van Deusen, *The Life of Henry Clay* (Boston, Mass.: Little, Brown, 1937); and Peterson, *Great Triumvirate*.

36. Excellent summations of Clay's American System include the chapter "Henry Clay, Ideologue of the Center," in Daniel Walker Howe, *The Political Culture of American Whigs* (Chicago: University of Chicago, 1979), pp. 123–149; and Peterson, *Great Triumvirate*, pp. 68–84; "defy the South" in C. F. Adams, ed., *Memoirs of John Quincy Adams*, VIII: 446.

37. The importance of Freemasonry and the origins of Anti-Masonry in the "William Morgan affair" are described in McDougall, *Freedom Just Around the Corner*, pp. 328–33, 494–96. Anti-Masonry as a cultural phenomenon will be described later. Clay's statement on Masonry is quoted in full by Remini, *Henry Clay*, p. 360. On Anti-Masonry

see Paul Goodman, *Towards a Christian Republic: Antimasonry and the Great Transition in New England, 1826–1836* (New York: Oxford, 1988).

38. On the soundness of the dollar under Biddle, see the classic by Bray Hammond, *Banks and Politics in America from the Revolution to the Civil War* (Princeton, N.J.: Princeton University, 1957), pp. 323–325; and Walter B. Smith, *Economic Aspects of the Second Bank of the United States* (Cambridge, Mass.: Harvard University, 1953), p. 76.

39. Hammond, *Banks and Politics*, pp. 286–305 ("clumsy," p. 294; "frankness," p. 290; "gently," p. 304). The best biography is Thomas P. Govan, *Nicholas Biddle: Nationalist and Public Banker, 1786–1844* (Chicago, Ill.: University of Chicago, 1959). Jean A. Wilburn, *Biddle's Bank: The Crucial Years* (New York: Columbia University, 1967), argues the case for the bank's popularity. Peter Temin, *The Jacksonian Economy* (New York: Norton, 1969), argues that Hammond exaggerates Biddle's role as a steward of the economy as a whole.

40. Robert V. Remini, in *Andrew Jackson and the Bank War: A Study in the Growth of Presidential Power* (New York: Norton, 1967), pp. 15–48, describes the apparent power of the BUS and the sources of Jackson's hostility. Remini makes a strong case that Jackson's motives were political, not economic. Temin, *Jacksonian Economy*, pp. 44–58, exposes the illogic of Jackson's economic arguments against the BUS (suggesting political motives) and explains the limits of Biddle's power to regulate the economy in the manner of a twentieth-century central banker. Hammond, *Banks and Politics*, pp. 326–361, explains the self-interestedness of many lieutenants in Jackson's Bank War ("The world is governed," p. 333).

41. Remini, *Jackson and the Bank War*, pp. 44–45; "I will kill it" in *Autobiography of Martin Van Buren*, p. 625.

42. Remini, *Henry Clay*, pp. 379–380, suspects that Clay's push for recharter was "a blatant act of political self-interest." That is, he was disingenuous when he told Biddle that Jackson was more likely to accept recharter before rather than after his reelection (assuming he won), and instead hoped to provoke a veto that would make the BUS a hot campaign issue. Peterson, *The Great Triumvirate*, pp. 206–208, thinks Clay was sincere in thinking he was serving the best interests of the BUS, noting that its counsel Daniel Webster also supported early recharter.

43. Jackson's veto message is in Richardson, ed., *Messages and Papers of the Presidents*, II: 1140–1153; and online at www.yale.edu/lawwab/avalon/presiden/veto/ajveto01.htm. On Kendall's role, see Lynn L. Marshall, "The Authorship of Jackson's Bank Veto Message," *Journal of American History* 50 (December 1963); and Donald B. Cole, *A Jackson Man: Amos Kendall and the Rise of American Democracy* (Baton Rouge: Louisiana State University, 2004); "to preserve the republic" in Remini, *Jackson and the Bank War*, p. 81.

44. Jackson's victory might seem unimpressive because he is the only president reelected with a smaller percentage of the vote than he received the first time around. In fact, he polled 80,000 more votes than in 1828 (and possibly even more, since Missouri did not report popular totals) in a field that included three other candidates. See Samuel R. Gammon, *The Presidential Campaign of 1832* (Baltimore, Md.: Johns Hopkins University, 1922), passim ("cause all the Election," p. 134); "Bank of the United States," *North American Review* 35 (October 1832): 485–519 ("wild and reckless," p. 517); Remini, *Jackson and the Bank War*, pp. 88–108 ("Golden vaults," p. 99; "he may be," p. 107); Allan Nevins, ed., *The Diary of Philip Hone 1828–1851*, 2 vols. (New York: Dodd-Mead, 1927), II: 96–97.

45. Jackson's ultimatum (Richardson, ed., *Messages and Papers of the Presidents*, II:

1217) won him rare plaudits from Philip Hone: "I say, Hurrah for Jackson! and so I am willing to say at all times when he does his duty" (*Diary of Philip Hone* II: 84.) On the origins of the nullification doctrine, see William W. Freehling, *Prelude to Civil War: The Nullification Controversy in South Carolina, 1816–1836* (New York: Harper and Row, 1966). Calhoun's strategy and fear of arrest in Merrill D. Peterson, *Olive Branch and Sword: The Compromise of 1833* (Baton Rouge: Louisiana State University, 1982), pp. 32–53, 82–84; and Niven, *Calhoun and the Price of Union*, pp. 179–199. "I can tell you" quoted in James Parton, *Life of Andrew Jackson*, 3 vols. (New York: Mason Brothers, 1863), III: 669–670.

46. Hammond, *Banks and Politics*, pp. 412–429; "ties of party" in Remini, *Jackson and the Bank War*, pp. 16–27; "shan't have" in Parton, *Life of Jackson*, III: 500. A Pennsylvanian critic of Jackson waxed sarcastic about the distribution of federal assets to favored banks. "Heaven forfend that any son of mine should be a Democrat in principle—being a good Democrat by trade, he got a snug slice of the federal deposits": Louis Hartz, *The Liberal Tradition in America* (New York: Harcourt, Brace, and World, 1955), p. 137.

47. Jackson lived true to his hard-money ethic (and incidentally courted popularity) by paying all his expenses from a pouch of gold coins. Just imagine how Old Hickory must spin in his grave every time we admire his portrait on a twenty-dollar bill! On Gouge and Benton, see Arthur M. Schlesinger, Jr., *The Age of Jackson* (Boston, Mass: Little, Brown, 1945), pp. 115–131 ("Banks on Gouging," p. 118; "Until the nature," p. 120; "PEOPLE, or PROPERTY," p. 125). Schlesinger got the economics backward, as noted below, but his account of the hard-money movement is excellent.

48. Temin, *Jacksonian Economy*, pp. 59–112.

49. Biddle kept the bank in existence by paying $6 million for a Pennsylvania charter, but it foundered. Biddle resigned, fought off lawsuits from angry stockholders, beat trumped-up charges of criminal conspiracy, and died of bronchitis in 1844, no doubt wishing he had devoted his life to literature.

50. Edward Pessen, "The Egalitarian Myth and the American Social Reality: Wealth, Mobility, and Equality in the 'Era of the Common Man,'" in Edward Pessen, ed., *The Many-Faceted Jacksonian Era: New Interpretations* (Westport, Conn.: Greenwood, 1977), pp. 7–46; Gallatin quoted in Temin, *Jacksonian Economy*, p. 177. On the politics of banking at the state level, see Howard Bodenhorn, *State Banking in Early America: A New Economic History* (New York: Oxford University, 2003); Bodenhorn argues that any growing economy needs banks to facilitate saving, and even corruption can be benign. See also James Roger Sharp, *The Jacksonians versus the Banks: Politics in the States after the Panic of 1837* (New York: Columbia University, 1970). Sharp's frontispiece (p. xv) quotes a Democratic campaign song of 1839: "Hard times, hard times is all the cry / The country's in confusion / The Banks have stop'd, but still they try / To mistify delusion / They give us trash and keep our cash / To send across the waters / To pay for things they bought of Kings / And gull our sons and daughters."

51. James L. Haley, *Sam Houston* (Norman: University of Oklahoma, 2002), pp. 3–17; Randolph B. Campbell, *Sam Houston and the American Southwest* (New York: Longman, 2002), pp. 1–12 ("wild liberty," p. 3). Houston's debts were not induced by whiskey, since moonshine was available for a few pennies or less. Houston went into debt buying shot, powder, blankets, and tools for his Indian friends.

52. Many years later a friend of Eliza claimed she had told of a madly jealous, suspicious Houston who locked her up for days without food or water. Another story was that Houston's bride was repelled by an oozing battle wound in his groin. Still another sug-

gested that he found her unchaste inasmuch as her doctor was later asked for testimony. Nobody knows the truth, but Houston's most thorough and recent biographer suspects that Eliza rebelled in the only permissible way against an arranged marriage to an older man. She could not buck her own father's will without dishonoring him. She *could* buck her new husband, however, because no southern gentleman could honorably impugn a damsel in distress (even if feigned). See Haley, *Sam Houston*, pp. 45–61.

53. Campbell, *Houston and the American Southwest*, pp. 27–41 ("Take my laurels," p. 37; "repair to TEXAS," p. 39); Haley, *Sam Houston*, pp. 62–89 ("I was dying out," p. 85). Houston's roof-raising summation sounds like a parody of Daniel Webster (perhaps it was): "So long as that flag shall bear aloft its glittering stars—bearing them amidst the din of battle, and waving them triumphantly above the storms of the ocean, so long, I trust, shall the rights of American citizens be preserved . . . till discord shall wreck the spheres—the grand march of time shall cease—and not one fragment of all creation be left to chafe on the bosom of eternity's waves."

54. William C. Davis, *Lone Star Rising: The Revolutionary Birth of the Texas Republic* (New York: Free Press, 2004), pp. 1–53, includes a detailed and colorful account of the early filibusters.

55. The careers of Moses Austin and that outlandish killer John Smith T are described in McDougall, *Freedom Just Around the Corner*, pp. 479–483. On the origins of the Texas project, see Gregg Cantrell, *Stephen F. Austin, Empresario of Texas* (New Haven, Conn.: Yale University, 1999), pp. 43–88; and the full-length biography, David B. Gracy II, *Moses Austin: His Life* (San Antonio, Tex.: Trinity University, 1987). Austin and his enslaved companion Richmond lived off berries and roots during their trek back to Arkansas, and for good measure fought off a panther attack.

56. Cantrell, *Stephen F. Austin*, pp. 132–201 ("fanaticism," p. 115; "perfect and complete," p. 134; "not our wish," p. 138); Davis, *Lone Star Rising*, pp. 54–76 ("I obey," p. 63; "resistance and obstinacy," p. 69). Numerous settlers accused Austin of cheating or showing favoritism toward large planters (shades of the thirteen colonies). Others displayed racial bigotry of the sort Austin despised (shades of Penn's colony). In December 1826 a band of "infatuated madmen" at Nacogdoches declared the independent republic of Fredonia (shades of the abortive state of Franklin carved out of Tennessee).

57. Randolph B. Campbell, *Gone to Texas: A History of the Lone Star State* (New York: Oxford University, 2003); T. R. Fehrenbach, *Lone Star: A History of Texas and the Texans* (New York: Da Capo, 2000, orig. 1968), pp. 152–189; Teresa Palomo Acosta and Ruth Winegarten, *Las Tejanas: 300 Years of History* (Austin: University of Texas, 2003), pp. 38–44; Davis, *Lone Star Rising*, pp. 77–145 ("no more doubts," pp. 133–134; "go for *Independence*," p. 79; "Tories," p. 133); "character of a religion" in Cantrell, *Stephen F. Austin*, p. 221.

58. See David M. Pletcher, *The Diplomacy of Annexation: Texas, Oregon, and the Mexican War* (Columbia: University of Missouri, 1973). So far as I know, the terms "cheap hawk" and "cheap hawkery" were coined by Harvey Sicherman. They are apt terms for this perennial American predilection.

59. Haley, *Sam Houston*, pp. 128–165, reviews the battle and the historical debate about the quality of Houston's command. Haley grants that the critics are right about Houston's failure to command the respect and obedience of numerous officers and men, but turns that evidence against them. So undisciplined was the Texas army that only a leader with Houston's prestige, grit, and willpower could have carried it through a long retreat and a pitched battle against a superior, professional foe.

60. Campbell, *Houston and the American Southwest*, pp. 89–99; Cantrell, *Stephen F. Austin*, pp. 344–364 ("war of extermination," p. 344; "independence of Texas," p. 364); Remini, *Course of American Freedom*, pp. 311–314 ("rash and premature," pp. 311–312; "Gentlemen, the Republic," in Fehrenbach, *Lone Star*, pp. 250–351. The bill passed the House easily, in part because members lukewarm about Texas expected the Senate to reject it anyway. But Senator Robert Walker of Mississippi cleverly attached the Texas provision to a pork-laden appropriations bill in which many senators had a stake. A motion to strike the rider failed by one vote.

61. As in the first volume of this new American history, I mean to ensure that every region and state receives due attention. Hence, each state admitted to the Union is presented in this form within the text. Readers eager to press on with the main narrative may skip the state profiles or return to them at leisure. But these historical travelogues do illustrate general themes in addition to providing rich, often humorous, local color.

62. "State of Arkansas" © Sanga Music, 1959; "unvaried sterility" in Brooks Blevins, *Hill Folks: A History of Arkansas Ozarkers and Their Image* (Chapel Hill: University of North Carolina, 2002), pp. 11–12; "if they work," "abundantly dirty," and "wilderness of sorrows" in S. Charles Bolton, *Arkansas 1800–1860: Remote and Restless* (Fayetteville: University of Arkansas, 1998), pp. xi, 6, 120; "tolerant state" and Washburn in Charles Bolton, *Territorial Ambition: Land and Society in Arkansas 1800–1840* (Fayetteville: University of Arkansas, 1993), pp. 1–2.

63. George R. Stewart, *Names on the Land* (Boston, Mass.: Houghton Mifflin, 1967), pp. 90, 137, 228, 336–338. English-speakers did better with the French shorthand for Arkansas's highlands: *aux Arks* became Ozarks.

64. Bolton, *Territorial Ambition*, pp. 38–56.

65. Bolton, *Arkansas 1800–1860*, pp. 23–43. Jim Bowie, the most fearsome brawler on the river, filed gigantic claims based on the forged signatures of fictitious Spaniards and sustained them for years because no one in the territory dared make him angry. It took the U.S. Supreme Court to throw out his claims. Nor was the "common man" above (or below) beating the system. When Congress provided generous compensation for 300 white families displaced by an Indian reservation, nearly 600 claims were filed.

66. Bolton, *Territorial Ambition*, pp. 90–102 ("turning loose," p. 26); Bolton, *Arkansas 1800–1860*, pp. 53–66 ("Arkansas is free," p. 58); the Smithson bequest and the fraud involving it are documented in William J. Rhees, *The Smithsonian Institution: Documents Relative to Its Origin and History, 1835–1899* (Washington, D.C.: GPO, 1901), and retold in Nina Burleigh, *The Stranger and the Statesman: James Smithson, John Quincy Adams, and the Making of America's Greatest Museum* (New York: HarperCollins, 2003), pp. 206–207.

67. *Diary of Philip Hone*, I: 121–126 (the winning candidate for mayor, Cornelius W. Lawrence, would preside over the Wall Street fire in December); Remini, *Henry Clay*, pp. 458–470 (Clay and Blair quoted on pp. 458–461); E. Malcolm Carroll, *Origins of the Whig Party* (Durham, N.C.: Duke University, 1925).

68. Cole, *Martin Van Buren*, pp. 256–281 ("pernicious," p. 272); David Crockett, *The Life of Martin Van Buren* (Philadelphia: R. Wright, 1835), p. 81 (the reference to red and gray whiskers connotes a fox); Harry L. Watson, *Liberty and Power: The Politics of Jacksonian America* (New York: Hill and Wang, 1990), pp. 198–253; and especially Richard P. McCormick, *The Second American Party System: Party Formation in the Jacksonian Era* (Chapel Hill: University of North Carolina, 1966). New York's radical Democrats were

called Loco-Focos after a blowup at Tammany Hall on October 29, 1835. As conservatives walked out in protest they turned off the gas to extinguish the lamps. Radicals lit candles with patented Loco-Foco friction matches and continued the meeting.

69. Watson, *Liberty and Power*, pp. 198–205; "let him say nothing" in Reginald C. McGrane, ed., *The Correspondence of Nicholas Biddle* (Boston, Mass.: Houghton Mifflin, 1919), p. 256; "humane, patriotic" in *The Presidents . . . Their Inaugural Addresses* (Chicago, Ill.: Whitehall, 1968), p. 66; "the *rising*" in Benton, *Thirty Years' View*, I: 735.

70. George Cornell, "Unconquered Nations: The Native Peoples of Michigan," pp. 25–42; and John Cumming, "Revolution in the Wilderness: Michigan as Colony and Territory," pp. 59–78, in Richard J. Hathaway, ed., *Michigan: Visions of Our Past* (East Lansing: Michigan State University, 1989). Wills Frederick Dunbar, *Michigan: A History of the Wolverine State* (Grand Rapids, Mich.: Eerdmans, 1965), pp. 121–223; "consummate politician" in Willard Carl Klunder, *Lewis Cass and the Politics of Moderation* (Kent, Ohio: Kent State University, 1996), p. 43. That was grudging praise from William Woodbridge, leader of the National Republican junto that opposed Cass's Jacksonian Democratic regime in Detroit. Cass's great seal shows an elk and a moose holding a banner that depicts a buckskin-clad pioneer on the shores of a lake at sunset. His pledge "TUEBOR" (I shall defend) invokes the territory's status as a march land against British Canada. The motto "*Si quaeris . . .*" paraphrases the memorial to Christopher Wren in St. Paul's.

71. At Fort Dearborn (Chicago) Harriet Martineau fearlessly chose to return to Detroit in a schooner because she wanted to traverse the entire lengths of Lake Michigan and Lake Huron and visit fabled Mackinac Island. Her chapter on "the delights of Mackinaw" [sic] is reprinted in Justin L. Kestenbaum, ed., *The Making of Michigan 1820–1860: A Pioneer Anthology* (Detroit, Mich.: Wayne State University, 1990), pp. 58–71 ("nine months winter," p. 66).

72. Klunder, *Lewis Cass*, pp. 17–57 ("influence of capital," p. 33); Dunbar, *Michigan: A History*, pp. 225–299 ("Don't go to Michigan," p. 242; "Come all you Yankee," p. 249). The popular Father Richard cut a strange figure on Capitol Hill with his long black skirts, French accent, and snuff habit. He returned to Detroit in triumph only to learn that a divorced parishioner whom he had excommunicated was suing for damages. Richard pleaded separation of church and state, but was jailed for contempt of court. A letter from Henry Clay sprang him: his standing in Congress gave him immunity. As for the absurd name of the college and its faculty of "didactors," these were Judge Augustus Woodward's brainstorm; Cass called the school the "Cathole-what's-its-name."

73. John M. Gordon, "A Speculator's Diary," in Hathaway, ed., *Michigan: Visions*, pp. 115–158 ("ripens into maturity," p. 134); Dunbar, *Michigan: A History*, pp. 249–269. The Michigan Dutch, noted today for their famous spring tulip festival, fled from potato blight and theological controversy after 1845. Not unlike John Winthrop two centuries before, the Dutch Reformed pastor Albertus van Raalte trusted that God would bless his efforts to plant a Calvinist city on a hill in the New World. The first pioneers suffered cruelly in the snow belt that is western Michigan, but the unity, charity, and education preached by van Raalte ("Don't hold a calf in higher esteem than your own child") helped the community prosper. See the touching memoir of Engbertus van der Veen, "Hollanders in Michigan," in Kestenbaum, ed., *Making of Michigan*, pp. 181–205 ("don't hold," p. 193).

74. Alec R. Gilpin, *The Territory of Michigan 1805–1837* (East Lansing: Michigan State University, 1970), pp. 139–182 (of the eighty petitioners Mason was able to identify,

forty-six turned out to be Clay men, eleven Anti-Masons, and twenty-two transients bribed to sign the document); Dunbar, *Michigan: A History*, pp. 276–312. The only casualty of the Toledo war was an Ohioan named Two Stickney (his parents named their sons "One," "Two," etc.), who received a minor knife wound in a tavern brawl.

75. Like the founding fathers of Indiana, Texas, and other states, Stevens T. Mason came to a sad, early end. A victim of the Panic of 1837, he was voted out, moved to New York and got married, but he died at the age of just thirty in 1843. His remains were repatriated to Detroit in 1905. Lewis Cass, of course, went on to a sterling career in the Senate and ran for president in 1848.

76. On Houghton's tremendous importance to American geology and mineralogy, see David J. Krause, *The Making of a Mining District: Keweenaw Native Copper 1500–1870* (Detroit, Mich.: Wayne State University, 1992). The Methodist preacher John H. Pitzel wrote a chilling description of the hard life of the Irish, Welsh, and German miners in the copper shafts. His *Lights and Shades of Missionary Life* (1857) is excerpted in Kestenbaum, ed., *Making of Michigan*, pp. 208–234.

77. Major L. Wilson, *The Presidency of Martin Van Buren* (Lawrence: University of Kansas, 1984); Hammond, *Banks and Politics*, pp. 451–499; Cole, *Martin Van Buren*, pp. 342–378; Howard Jones, *Mutiny on the Amistad: The Saga of a Slave Revolt and Its Impact on American Abolition, Law, and Diplomacy* (New York: Oxford University, 1986); Leland W. Meyer, *The Life and Times of Colonel Richard M. Johnson of Kentucky* (New York: AMS, 1967, orig. 1936). John Quincy Adams defended the fifty *Amistad* mutineers by arguing that the slave trade was illegal under U.S. and Spanish law alike. In 1841 Chief Justice Taney's pro-slavery Court found this persuasive. The refugees were shipped home to West Africa, whereupon the man who had led the mutiny, Cinqué, began trading in slaves himself.

78. On the race of 1840, see Robert G. Gunderson, *The Log-Cabin Campaign* (Lexington: University of Kentucky, 1957); Clay's maneuvering in Remini, *Henry Clay* ("mad and fatal," p. 525; "I had rather," p. 527). Though sometimes rendered "Sir, I would rather be right . . . ," the words above are correct. See also Peterson, *Great Triumvirate*, p. 289.

79. Remini, *Henry Clay*, pp. 545–560; Schlesinger, *Age of Jackson*, p. 287 ("I have found"); Watson, *Liberty and Power*, pp. 198–216 ("the necessity," p. 216; "have no faith," p. 212).

80. The most recent biographer of the ninth president is James A. Green, *William Henry Harrison: His Life and Times* (Richmond, Va.: Garrett and Massie, 1939), but see Norma Lois Peterson, *The Presidencies of William Henry Harrison and John Tyler* (Lawrence: University of Kansas, 1989). The significance of Harrison's "omnibus candidacy" for the Whigs is explained by Fischer, *Albion's Seed*, pp. 850–855. Harrison quote is in Watson, *Liberty and Power*, p. 212. The notion that Harrison "transformed" egalitarianism by reconciling Hamiltonian and Jeffersonian strains is from Hartz, *Liberal Tradition*, pp. 111–112.

81. *Diary of Philip Hone*, II: 553 ("There was rhyme"); I: 486 ("capital speeches"). "Give [Harrison] a barrel," in *Baltimore Republican* (December 11, 1839), cited by Remini, *Henry Clay*, p. 562. On women in the Harrison campaign, see Gunderson, *Log-Cabin Campaign*, pp. 127–129; on clergy, Richard J. Carwardine, *Evangelicals and Politics in Antebellum America* (New Haven, Conn.: Yale University, 1993), pp. 50–70.

82. The rowdiness and humorous anecdotes of the campaign of 1840, including the

origins of "booze" and "OK," are compiled by Paul F. Boller, Jr., *Presidential Campaigns* (New York: Oxford University, 1985), pp. 65–77; "We have taught them," *Democratic Review* VIII (September 1840), p. 198; "fraud, hallucination," in Cole, *Martin Van Buren*, p. 373.

83. See inter alia Schlesinger, *Age of Jackson*; Charles G. Sellers, *The Market Revolution: Jacksonian America 1815–1846* (New York: Oxford University, 1991) and the critique by William E. Gienapp, "The Myth of Class in Jacksonian America," *Journal of Policy History* 6 (1994): 232–259; Morton Horwitz, *The Transformation of American Law* (Cambridge, Mass.: Harvard University, 1977); Charles A. Beard and Mary R. Beard, *A Basic History of the United States* (New York: Doubleday, Doran, 1951); Sean Wilentz, *Chants Democratic: New York City and the Rise of the American Working Class 1788–1850* (New York: Oxford University, 1984); Lawrence H. White, *Democratick Editorials: Essays in Jacksonian Political Economy by William Leggett* (Indianapolis, Ind.: Liberty, 1984); Richard Hofstadter, *The American Political Tradition and the Men Who Made It* (New York: Knopf, 1948), pp. 44–66; Edward Pessen, *Riches, Class, and Power before the Civil War* (Lexington: University of Kentucky, 1973); Marvin Meyers, *The Jacksonian Persuasion* (Stanford, Calif.: Stanford University, 1957); Lee Benson, *The Concept of Jacksonian Democracy: New York as a Test Case* (Princeton, N.J.: Princeton University, 1961); Ronald P. Formisano, *The Birth of Mass Political Parties: Michigan 1827–1861* (Princeton, N.J.: Princeton University, 1971) and *The Transformation of Political Culture: Massachusetts Parties 1790s–1840s* (New York: Oxford University, 1983); Richard P. McCormick, "New Perspectives on Jacksonian Politics," *American Historical Review* 65 (1960): 288–301; Richard L. McCormick, "Ethno-Cultural Interpretations of Nineteenth-Century American Voting Behavior," *Political Science Quarterly* 89 (1974): 351–377; David Hackett Fischer, *Albion's Seed: Four British Folkways in America* (New York: Oxford University, 1989), pp. 847–852.

84. Don't take my word for this. Read the powerful argument of Lawrence Frederick Kohl, *The Politics of Individualism: Parties and the American Character in the Jacksonian Era* (New York: Oxford University, 1989), pp. 3–62. On the Jacksonians' appropriation of the "natural man," see John William Ward, *Andrew Jackson: Symbol for an Age* (New York: Oxford University, 1955).

85. Kohl, in *Politics of Individualism*, pp. 63–99, contrasts the "inner directed" Whig individualists with the "outer directed" Democratic individualists ("sobriety, industry," pp. 73–74; "sage Doctors," a quotation from Orestes Brownson, p. 29; "gospel of commerce" described, pp. 87–88). Howe, in *Political Culture of the American Whigs*, pp. 1–41, debunks the old myths about an ephemeral, insubstantial Whig Party ("economically diverse," p. 20); "the poor," a quotation from Edward Everett, cited in Hartz, *Liberal Tradition*, pp. 111–112. See also Major L. Wilson, "What Whigs and Democrats Meant by Freedom," in Pessen, ed., *Many-Faceted Jacksonian Era*, pp. 192–211.

86. Kohl, *Politics of Individualism*, pp. 52, 47, 38, 105, 95.

Chapter 3

1. The census commanded by David, objections raised by his military chief Joab, and the resulting anger of the Lord are recounted in 2 Samuel 24 and 1 Chronicles 21. There is no evidence that any member of the Constitutional Convention raised theological objections against the census.

2. This and subsequent paragraphs are based on an excellent monograph by Patricia Cline Cohen, *A Calculating People: The Spread of Numeracy in Early America* (Chicago, Ill.: University of Chicago, 1982). On the arithmetical revolution and change in female roles, see pp. 116–149, as well as Gerald Lee Gulick, *Pestalozzi and Education* (New York: Random House, 1968); Mary Beth Norton, *Liberty's Daughters: The Revolutionary Experience of American Women, 1750–1800* (Boston, Mass.: Little, Brown, 1980); and Barbara Welter, "The Cult of True Womanhood, 1820–1860," *American Quarterly* 18 (1966): 151–174.

3. John Cummings, "Statistical Work of the Federal Government of the United States," in John Koren, ed., *The History of Statistics* (New York: Macmillan, 1918), pp. 571ff.; Jacob E. Cooke, *Tench Coxe and the Early Republic* (Chapel Hill: University of North Carolina, 1978); Cohen, *A Calculating People*, pp. 150–175 ("marking the progress," p. 162; "reckoning, expecting," p. 175; "How many tears," p. 210). The famous cartographer John Melish published the ambitious *A Statistical View of the United States* in 1825.

4. Cohen, *A Calculating People*, pp. 175–226 ("innocence lost," p. 204; "unique, mysterious," p. 206). Population growth plus increased complexity pushed the cost of the census of 1850 to $1,318,027, which was four times the expense of the 1830 census and thirty times that of the original 1790 census.

5. *Abstract of the Returns of the Fifth Census* (Washington, D.C.: Duff Green, 1832); *Accurate Synopsis of the Sixth Census of the United States* (Philadelphia, Pa.: Augustus Mitchell, 1843); *The Seventh Census of the United States* (Washington, D.C.: Robert Armstrong, 1853); regional proportions in Douglass C. North, *The Economic Growth of the United States 1790–1860* (New York: Norton, 1966), p. 121. An excellent Web site for historical census data is http://fisher.lib.virginia.edu/collections/stats/histcensus. A lovingly crafted case study of the southern decline is David Hackett Fischer and James C. Kelly, *Bound Away: Virginia and the Westward Movement* (Charlottesville: University of Virginia, 2000).

6. The ensuing passage on Irish immigration is based largely on Kerby A. Miller, *Emigrants and Exiles: Ireland and the Irish Exodus to North America* (New York: Oxford University, 1985), whose exhaustive survey of emigrants' letters amply documents their self-identification as exiles; Kevin Kenny, *The American Irish: A History* (New York: Longman, 2000), pp. 1–130 ("The poor Irishman," p. 62; "niggers are worth," p. 63); Lawrence J. McCaffrey, *The Irish Catholic Diaspora in America* (Washington, D.C.: Catholic University, 1997); Peter Gray, *The Irish Famine* (New York: Abrams, 1995); K. Theodore Hoppen, *Ireland since 1800: Conflict and Conformity* (New York: Longman, 1989); Jay P. Dolan, *The Immigrant Church: New York's German and Irish Catholics, 1815 to 1865* (Baltimore, Md.: Johns Hopkins University, 1987); Dale T. Knobel, *Paddy and the Republic: Ethnicity and Nationality in Antebellum America* (Middletown, Conn.: Wesleyan University, 1986); William V. Shannon, *The American Irish* (New York: Macmillan, 1974); Carl Wittke, *The Irish in America* (New York: Russell and Russell, 1970); and David Gleeson, *The Irish in the South, 1815–1877* (Chapel Hill: University of North Carolina, 2004).

7. Miller, *Emigrants and Exiles*, pp. 26–130 (factional fights, pp. 60–69; beginnings of nationalist movement, pp. 88–101; Gaelic, pp. 119–120); McCaffrey, *Irish Catholic Diaspora*, pp. 1–43 ("It is vain," p. 43). The Irish Emancipation Act was passed in the teeth of opposition from Sir Robert Peel, the duke of Wellington, and King William IV himself.

8. Miller, *Emigrants and Exiles*, pp. 305–311; Kenny, *American Irish*, p. 95; McCaffrey, *Irish Catholic Diaspora*, p. 60. Many Irish reflexively interpreted the potato blight as God's

judgment on their wickedness. But its persistence after waves of repentance and prayer plus the absence of English pity made it easy for Irish to conclude that the plague on their people was an unalloyed evil associated with the Protestant ascendancy.

9. "The people" and "land-sharks" quoted in Kenny, *American Irish*, pp. 101–102; "Yankee tricksters" and "in heart and soul" quoted in Miller, *Emigrants and Exiles*, pp. 263, 323.

10. Dale B. Light, *Rome and the New Republic: Conflict and Community in Philadelphia Catholicism between the Revolution and Civil War* (Notre Dame, Ind.: Notre Dame University, 1996), pp. 247–258; Dolan, *Immigrant Church* ("The task," p. 57); Kenny, *American Irish*, pp. 77–80 ("pious lie," p. 80); Miller, *Emigrants and Exiles*, pp. 312–344. The sad Irish belief that unlike other Americans they were denied social mobility is evident in one of their famous work songs: "In Eighteen Hundred and Forty Three twas then I met sweet Finny McGee / An elegant wife she's been to me while workin' on the railway / Fill-a-mee-yoo-ree-yoo-ree-ay while workin' on the railway. . . . In Eighteen Hundred and Fifty Seven sweet Finny McGee she went to heaven / If she left one son she left *eleven* to work upon the railway."

11. The subsequent discussion of German immigration is based mostly on Don Heinrich Tolzmann, *The German-American Experience* (New York: Humanity, 2000); Frank Trommler and Joseph McVeigh, eds., *America and the Germans: An Assessment of a Three-Hundred-Year History*, Vol. 1, *Immigration, Language, Ethnicity* (Philadelphia: University of Pennsylvania, 1985); A. E. Zucker, ed., *The Forty-Eighters: Political Refugees of the German Revolution of 1848* (New York: Russell and Russell, 1950); Charlotte L. Brancaforte, *The German Forty-Eighters in the United States* (New York: Peter Lang, 1989); the classic by Albert B. Faust, *The German Element in the United States* (New York: Steuben Society, 1927); and of course Carl Schurz, *Reminiscences*, 3 vols. (New York: McClure, 1907–1908).

12. See Oscar Handlin, *The Uprooted: The Epic Story of the Great Migrations That Made the American People* (New York: Little, Brown, 1951); and the excellent corrective by John Bodnar, *The Transplanted: A History of Immigrants in Urban America* (Bloomington: Indiana University, 1985). Bodnar rightly argues that all people, not just immigrants, must adjust when changes in technology or market conditions reduce demand for their skills, and that most Europeans experienced capitalism before coming to America (indeed, displacement by market forces was a frequent cause of their flight). Far from being helpless, alienated victims in the New World, most immigrants maintained ethnic community ties and mutual support groups through extended families, civic and charitable organizations, and especially churches. Of course many suffered or were stuck for life in low-paying, insecure jobs. But inasmuch as immigrants came to the United States in search of social mobility, most "became staunch advocates of the rewards of capitalism and all that it involved. Thus, religious traditions and ethnic solidarity could be stressed, but not to the exclusion of Americanization, hard work, thrift, and free enterprise" (p. 143).

13. On settlement patterns, see Tolzmann, *German-American Experience*, pp. 124–150 ("remarkable," p. 131; "This costs," p. 147); Zucker, ed., *Forty-Eighters*, pp. 43–78 ("Our life here," p. 48).

14. Tolzmann, *German-American Experience*, pp. 187–193; Augustus J. Prahl, "The Turner," in Zucker, ed., *Forty-Eighters*, pp. 79–110; Kathleen Neils Conzen, "German-Americans and the Invention of Ethnicity," in Trommler and McVeigh, eds., *America*

and the Germans, pp. 131–147; Susan K. Appel, "The German Impact on Nineteenth-Century Brewery Architecture in Cincinnati and St. Louis," in Brancaforte, ed., *German Forty-Eighters*, pp. 243–256.

15. See the summary of German-American contributions to society in Tolzmann, *German-American Experience*, pp. 373–408. Roebling's extraordinary career is described in Brooke Hindle and Steven Lubar, *Engines of Change: The American Industrial Revolution, 1790–1860* (Washington, D.C.: Smithsonian Institution, 1986), pp. 240–248. On pianoforte manufacture, see Ronald Ratcliff, *Steinway and Sons* (San Francisco, Calif.: Chronicle, 2002). The famous Baltimore academy counted among its later graduates none other than H. L. Mencken.

16. As a graduate of New Trier Township High School in Winnetka, Illinois, I have wondered all my life why its athletic teams are bedecked in "the green and the gray." I may at last have stumbled on the answer. For German contributions to colonial America, see Walter A. McDougall, *Freedom Just Around the Corner: A New American History 1585–1828* (New York: HarperCollins, 2004), pp. 136–142.

17. Lawrence S. Thompson and Frank X. Braun, "The Forty-Eighters in Politics"; and Eitel W. Dobert, "The Radicals," in Zucker, ed., *Forty-Eighters*, pp. 111–181; "My country" quoted in Tolzmann, *German-American Experience*, p. 157; Philip Gleason, *The Conservative Reformers: German-American Catholics and the Social Order* (South Bend, Ind.: Notre Dame University, 1968); Light, *Rome and the New Republic*, pp. 294–304 (Kenrick's handpicked successor as bishop of Philadelphia was a Bohemian German, John Neumann).

18. Tolzmann, *German-American Experience*, pp. 164–166, 183–184; Annette P. Bus, "Mathilde Anneke and the Suffrage Movement," in Brancaforte, ed., *German Forty-Eighters*, pp. 79–92. Two male literary figures—Friedrich Gierstäcker and Otto Ruppius—are of particular interest for having written dozens of rip-roaring books about the American West. German readers ever since have been curiously addicted to the "cowboys and Indians" genre.

19. The president's address at a memorial service for Carl Schurz is cited in Zucker, ed., *Forty-Eighters*, p. 250. The following poem may be kitsch, but it earnestly expresses the cosmopolitan spirit of Schiller and Beethoven carried by Germans to the United States. It is cited by Frank Trommler, "The Use of History in German-American Politics," in Brancaforte, ed., *German Forty-Eighters*, p. 282:

> *Alles, was wir lernten lieben, Freiheit, Tugend, Licht und Recht*
> *Alles wollen frei wir üben, wie ein neues Gottgeschlecht,*
> *Und wenn dann die Stürme brausen, beugen sie nicht unsern Mut,*
> *Freudig stehen wir im Kampfe für ein klar erkanntes Gut.*

> (Everything we learned to love, Freedom, Virtue, Light, and Right
> All that we wish to practice freely, like a new-born race of gods,
> And if the storms should rage about us, they deflect our courage not,
> Joyful stand we firm in battle for a clear acknowledged good.)

20. Kohn quoted in Hasia R. Diner, *A New Promised Land: A History of Jews in America* (New York: Oxford University, 2003), pp. 28–29. The following passage is based on Jonathan Sarna, *American Judaism: A History* (New Haven, Conn.: Yale University, 2004);

Henry L. Feingold, *Zion in America: The Jewish Experience from Colonial Times to the Present*, rev. ed. (New York: Hippocrene, 1981); Hasia R. Diner, *The Jews of the United States 1654 to 2000* (Berkeley: University of California, 2004), and *A Time for Gathering: The Second Migration, 1820–1880* (Baltimore, Md.: Johns Hopkins University, 1992); Gerald Sorin, *Tradition Transformed: The Jewish Experience in America* (Baltimore, Md.: Johns Hopkins University, 1997); Michael A. Meyer, *Response to Modernity: A History of the Reform Movement in Judaism* (New York: Oxford University, 1988); and a dip into Jacob Rader Marcus, *United States Jewry, 1776–1985*, 4 vols. (Detroit, Mich.: Wayne State University, 1989–1993).

21. Feingold, *Zion in America*, pp. 52–67 ("thee looks," p. 59; "get money," p. 66; "have nothing," 71; "A German," p. 70); Diner, *Jews of the United States*, pp. 71–88 ("the desire," p. 85).

22. Feingold, *Zion in America*, pp. 68–81 ("gains in influence," p. 73); Diner, *Jews of the United States*, pp. 99–107.

23. The dean of Jewish American historians, Jacob Rader Marcus, quipped that there were as many American Judaisms as they were American Jews—and all were valid, in his eyes. But the only Jewish organization to which he belonged was B'nai Brith, because "They never tell me what to do except write a check. . . . And I write the check. They do good work and they're innocuous": Gary Phillip Zola, ed., *The Dynamics of American Jewish History* (Hanover, N.H.: Brandeis University and University Press of New England, 2004), p. 22.

24. Sorin, *Tradition Transformed*; Meyer, *Response to Modernity;* Feingold, *Zion in America* ("This country," p. 99; "law of the land," p. 89). Abolitionists and advocates of slavery alike appealed to the Jewish authorities in their midst to render judgments about what the Bible taught in the matter of slavery. Rabbis published learned essays rendering a Solomon's judgment to the effect that whether or not the Lord condoned slavery, the institution as practiced in ancient Judah and Israel differed so much from slavery in contemporary America that no lessons could be drawn.

25. See, above all, the classic by Stewart H. Holbrook, *The Yankee Exodus: An Account of Migration from New England* (New York: Macmillan, 1950); "Fifty years ago" and "I am thinking," pp. 2–3. Holbrook's genius for conveying encyclopedic information in a loving narrative recalls that of George Rippey Stewart, whose *Names on the Land: A Historical Account of Place-Naming in the United States*, rev. ed. (Boston, Mass.: Houghton Mifflin, 1958), also documents the swath of Yankee influence across America's upper tier from New York to Oregon and beyond.

26. Holbrook, *Yankee Exodus*, pp. 25–199 ("schoolhouse to be built," p. 64; "I was the young man," p. 138; "seemed to be agreed," p. 115; "profane, scandalous," p. 132).

27. The character and persistent influence of America's four "cradle cultures" are described in David Hackett Fischer, *Albion's Seed: Four British Folkways in America* (New York: Oxford University, 1989), and summarized in McDougall, *Freedom Just Around the Corner*, pp. 136–155.

28. Malcolm J. Rohrbough, *The Land Office Business: The Settlement and Administration of American Public Lands, 1789–1837* (New York: Oxford University, 1968), pp. 221–249 ("never saw" and "germ of an immense city," p. 240); A. M. Sakolski, *The Great American Land Bubble* (New York: Harper, 1932), pp. 232–254 ("predicted that the place," p. 250); Bessie Louise Pierce, *A History of Chicago*, 2 vols. (New York: Knopf, 1937), vol 1: *The Beginning of a City 1673–1848*, pp. 43–74 ("chaos of mud," p. 49; Chicago American, p. 62); William Cronon, *Nature's Metropolis: Chicago and the Great West* (New York: Norton,

1991), pp. 23–54 ("Their Great Father" and "more pagan," p. 28); James L. Meeriner, *Grafters and Goo Goos: Corruption and Reform in Chicago, 1833–2003* (Carbondale: Southern Illinois University, 2004), pp. 10–35; "achievement of government" a quotation from Jon C. Teaford in Alberta M. Sbragia, *Debt Wish: Entrepreneurial Cities, U.S. Federalism, and Economic Development* (Pittsburgh, Pa.: University of Pittsburgh, 1996), p. 65.

29. Peter Temin, *Causal Factors in American Economic Growth in the Nineteenth Century* (New York: Macmillan, 1975), pp. 21–30; North, *Economic Growth*, p. 101–121, 135–155.

30. Jeremy Atack and Fred Bateman, *To Their Own Soil: Agriculture in the Antebellum North* (Ames: University of Iowa, 1987); Clarence H. Danhoff, *Change in Agriculture: The Northern United States, 1820–1870* (Cambridge, Mass.: Harvard University, 1969); Paul W. Gates, *The Farmer's Age: Agriculture 1815–1860* (New York: Holt, Rinehart, and Winston, 1960), "nothing but mend and patch," p. 349; R. Douglas Hurt, *American Agriculture: A Brief History* (Ames: Iowa State University, 1994); Percy Wells Bidwell and John I. Falconer, *History of Agriculture in the Northern United States 1620–1860* (Washington, D.C.: Carnegie Institution, 1925), "All hail the inventor," p. 273. In the 1840s and 1850s Maryland increased its tobacco production 54 percent; and Fairfax County, Virginia, alone imported 1,000 tons of guano per year. Kentucky's other big cash crop was hemp, a sturdy plant whose fibers supplied rope and burlap sacks for the growing economy. One crop that tailed off in the upper South was flax, since farmers' wives substituted factory-made clothing for homespun. Only the demand for linseed oil kept flax in cultivation at all.

31. Hurt, *American Agriculture*, pp. 133–147; Bidwell and Falconer, *History of Agriculture*, pp. 281–305; Gates, *Farmer's Age*, pp. 271–194. See also Wayne G. Broehl, *John Deere's Company: A History of Deere and Company and Its Time* (New York: Doubleday, 1984).

32. Albert Lowther Demaree, *The American Agricultural Press, 1819–1860* (New York: Columbia University, 1941), pp. 1–18. The gradual "emancipation" of farm women back east, recounted in Joan M. Jensen, *Loosening the Bonds: Mid-Atlantic Farm Women, 1750–1850* (New Haven, Conn.: Yale University, 1986), lagged on the frontier.

33. Gates, *Farmer's Age*, pp. 358–382. On the early history of the Smithsonian Institution, see Albert E. Moyer, *Joseph Henry: The Rise of an American Scientist* (Washington, D.C.: Smithsonian Institution, 1997). On the rise of rural capitalism in early modern England, see McDougall, *Freedom Just Around the Corner*, pp. 17–24.

34. This Arnold Toynbee is not to be confused with his nephew, the historian Arnold J. Toynbee. The elder Toynbee's Oxford lectures were first published in 1884 and reprinted as *The Industrial Revolution* (Boston, Mass.: Beacon, 1956). The term "industrial takeoff," first popularized in Walt W. Rostow, *The Stages of Economic Growth: A Non-Communist Manifesto* (Cambridge: Cambridge University, 1960), referred to the stage at which a nation's net investment reached the critical mass needed to spark sustained growth. Rostow believed that the United States reached this point in the 1840s, thanks to the railroad boom, but was careful to stress that necessary, if not sufficient, conditions for a nation's takeoff included competitive markets and a secure rule of law protecting property rights. Charles G. Sellers, *The Market Revolution: Jacksonian America, 1815–1846* (New York: Oxford University, 1991), focused on those preconditions, arguing that capitalists and their lawyers destroyed an essentially pre-capitalist American society during the very years in which Jackson inveighed against entrepreneurial behavior. Sellers was right about the mighty engines of change, but

wrong to imagine a pristine nation of subsistence farmers hostile to progress. Perry Miller, *The Life of the Mind in America: From the Revolution to the Civil War* (New York: Harcourt, Brace, and World, 1965), p. 289, deems Bigelow's *Elements of Technology* a milestone in U.S. intellectual development. The paragraph's final image is that of Leo Marx, *The Machine in the Garden: Technology and the Pastoral Ideal in America* (New York: Oxford University, 1964).

35. The Charles River Bridge Company had long enjoyed a municipal monopoly over transport between Boston and Cambridge. The state legislature heard the plaints of citizens tired of paying artificially high tolls. So it chartered the Warren Bridge Company in 1837 on condition that its builders turn the bridge over to the public after its costs and a modest profit were covered. The Charles Company retained Daniel Webster and sued. Here was a clear conflict between the sanctity of contracts and the principle of free competition. The majority opinion of the Taney Court bowed to "the rights of private property," but added, "we must not forget that the community also have rights." Was this judgment antibusiness? On the contrary, it blessed capitalist competition as a positive good serving the larger rights of the people. From the Camden and Amboy Railroad company in the 1830s to AT&T and Microsoft every firm vulnerable to an antitrust case has been obliged to build its case on the public interest it allegedly serves.

36. Edward C. Walterscheid, *To Promote the Progress of Useful Arts: American Patent Law and Administration 1787–1836* (Littleton, Colo.: Rothman, 1998); "to promote the progress," p. 18; "the whole," p. 355; "discovery must not" (a case of 1816), p. 374; "the first inventor" (a case of 1829), p. 396; "promotion," p. 423; "put an end," p. 425. Statistics on patent registrations in *Historical Statistics of the United States: Colonial Times to 1970* (Washington, D.C.: GPO, 1976), pp. 958–959. Steven Lubar, "The Transformation of American Patent Law," *Technology and Culture* 32 (1991): 941–942, interprets the reform of 1836 as an expression of the Jacksonians' hostility to monopolies. If so, then Senator Ruggles hornswoggled them. On woodworking technology, see Carolyn C. Cooper, "A Patent Transformation: Woodworking Mechanization in Philadelphia, 1830–1856," in Judith A. McGaw, ed., *Early American Technology: Making and Doing Things from the Colonial Era to 1850* (Chapel Hill: University of North Carolina, 1994), pp. 278–327. *Scientific American* estimated in 1852 that Woodworth and his assignees grossed more than $15 million on an invention that was largely based on foreign technologies in the first place.

37. Alan I. Marcus and Howard P. Segal, *Technology in America: A Brief History* (New York: Harcourt, Brace, Jovanovich, 1989), pp. 88–129; Oscar Schisgall, *Eyes on Tomorrow: The Evolution of Procter and Gamble* (Chicago, Ill.: Ferguson, 1981); Reese V. Jenkins, *Images and Enterprise: Technology and the American Photographic Industry, 1839–1925* (Baltimore, Md.: Johns Hopkins University, 1975).

38. U.S. House of Representatives Report quoted by Marcus and Segal, *Technology in America*, p. 100. On telegraphy, see Carleton Mabee, *The American Leonardo: A Life of Samuel F. B. Morse* (New York: Knopf, 1944); and Robert Luther Thompson, *Wiring a Continent: The History of the Telegraphy Industry in the United States* (Princeton, N.J.: Princeton University, 1947).

39. William Hosley, *Colt: The Making of an American Legend* (Amherst: University of Massachusetts, 1996).

40. The proportion of students in the total population in 1850 was 25.7 percent in

New England, as compared with 21.7 percent in Denmark, 20.4 percent in the United States as a whole, and just 11.7 percent in Britain: Douglass C. North, *Growth and Welfare in the American Past* (Englewood Cliffs, N.J.: Prentice Hall, 1974), p. 83. Peter Temin, "The Industrialization of New England, 1830–1880," in Peter Temin, ed., *Engines of Enterprise: An Economic History of New England* (Cambridge, Mass.: Harvard University, 2000), pp. 109–152, argues, "The special quality of New England had to do with the high proportion of women employed in the cotton industry" (p. 115). The innovations that made the sewing machine commercial included the grooved, eye-pointed needle and lockstitching shuttle, patented by Elias Howe in 1846. Howe made his fortune collecting $5 in royalties on every machine plus some hefty infringement judgments won in court. Singer, whose firm opened on Broadway in 1857, eventually dominated the market, thanks to aggressive advertising. His factory in Elizabethport, New Jersey, built in 1873, was for a time the largest single-product plant in the United States. For the whole story, see David A. Hounshell, *From the American System to Mass Production* (Baltimore, Md.: Johns Hopkins University, 1984), pp. 67–123.

41. This discussion is based on four landmarks of social history: Anthony F. C. Wallace, *Rockdale: The Growth of an American Village in the Early Industrial Revolution* (New York: Norton, 1972); Jonathan Prude, *The Coming of Industrial Order: Town and Factory Life in Rural Massachusetts, 1810–1860* (New York: Cambridge University, 1983); Cynthia J. Shelton, *The Mills of Manayunk: Industrialization and Social Conflict in the Philadelphia Region, 1787–1837* (Baltimore, Md.: Johns Hopkins University, 1986); and Paul E. Johnson, *A Shopkeeper's Millennium: Society and Revivals in Rochester, New York, 1815–1837* (New York: Hill and Wang, 1978). The reference to the "making of the American working class" invokes E. P. Thompson's classic, *The Making of the English Working Class* (London: Gollancz, 1963). Everett is quoted in John F. Kasson, *Civilizing the Machine: Technology and Republican Values in America, 1776–1900* (New York: Penguin, 1976), p. 44. Mann's *Portrait of a Factory Village* (1833) continued, in Jeffersonian fashion, to damn factories as un-American: "The blood runs chilly from my heart, To see fair Liberty depart; And leave the wretches in their chains, To feed a vampyre from their veins. Great Britain's curse is now our own; Enough to damn a King and Throne."

42. On the "factory as republican community," see Kasson, *Civilizing the Machine*, pp. 53–106. Alonzo Potter elaborated his egalitarian views in *The Principles of Science, Applied to the Domestic and Mechanical Arts, and to Manufacture and Agriculture* (1841). The quotations from Potter are in Daniel Hovey Calhoun, *The American Civil Engineer: Origins and Conflict* (Cambridge, Mass.: MIT, 1960), pp. 197–199; the other clergymen and Perry Miller are quoted in Thomas C. Cochran, *Business in American Life: A History* (New York: McGraw Hill, 1972), pp. 88–110.

43. George Rogers Taylor, *The Transportation Revolution 1815–1860* (New York: Holt, Rinehart, and Winston, 1964), pp. 270–300; Wallace, *Rockdale*, pp. 243–397 ("My mind," p. 114; Carey quotations, pp. 394–397); Prude, *Coming of Industrial Order*, pp. 100–180 ("reclaimed," p. 112). Walter Licht, *Industrializing America: The Nineteenth Century* (Baltimore, Md.: Johns Hopkins University, 1995), pp. 46–78, synthesizes a great deal of material on the ineffectual workers' movements prior to 1860. It must be said that paternalism came naturally to wealthy New Englanders for whom public service and charity were traditions two centuries old. Clara Barton, founder of the American Red

Cross, was herself the spinster daughter of a mill owner in Oxford, Massachusetts, Stephen Barton.

44. Frederick Moore Binder, *Coal Age Empire: Pennsylvania Coal and Its Utilization to 1860* (Harrisburg: Pennsylvania Historical and Museum Commission, 1974), pp. 1–84 ("Commerce is President," p. 21); Peter Temin, *Iron and Steel in Nineteenth-Century America: An Economic Inquiry* (Cambridge, Mass.: MIT, 1964), pp. 13–124; Alfred D. Chandler, Jr., "Anthracite Coal and the Beginnings of the Industrial Revolution in the United States," in Thomas K. McCraw, ed., *The Essential Alfred Chandler* (Cambridge, Mass.: Harvard Business School, 1988): 307–342; Robert B. Gordon, "Custom and Consequence: Early Nineteenth-Century Origins of the Environmental and Social Costs of Mining Anthracite," in McGaw, *Early American Technology*, pp. 240–277 ("picturesque," p. 270).

45. Carroll W. Pursell, Jr., *Early Stationary Steam Engines in America: A Study in the Migration of Technology* (Washington, D.C.: Smithsonian Institution, 1969). Most early steam engines were custom-built for their intended use. This meant that the chief boilermaker (be he a hired mechanic, an engineer, or the owner himself) had to be "a walking encyclopedia of his own and others' experience. He could remember all sorts of devises contrived by this or that man, and was full of anecdotes to illustrate his points. His drawing board was any plank at his command and his instruments a bit of chalk. He knew where every pattern was placed in the loft and could combine them to meet any emergencies" (p. 107). See also John Alfred Heitmann, *The Modernization of the Louisiana Sugar Industry, 1830–1910* (Baton Rouge: Louisiana State University, 1987).

46. Marx, *Machine in the Garden*, p. 13.

47. Richard Reinhardt, ed., *Workin' on the Railroad: Reminiscences from the Age of Steam* (Palo Alto, Calif.: American West, 1970), p. 24.

48. Hindle and Lubar, *Engines of Change*, pp. 125–150; Taylor, *Transportation Revolution*, pp. 74–103; John H. White, Jr., *American Locomotives: An Engineering History, 1830–1880* (Baltimore, Md.: Johns Hopkins University, 1968) and *The John Bull: 150 Years a Locomotive* (Washington, D.C.: Smithsonian Institution, 1981). On the Camden and Amboy, Wheaton J. Lane, *Indian Trail to Iron Horse: A History of Transportation in New Jersey* (Princeton, N.J.: Princeton University, 1939); on railroads' promotion of nonexistent towns to boost real estate values, Paul W. Gates, *The Illinois Central Railroad and Its Colonization Work* (Cambridge, Mass.: Harvard University, 1934).

49. On the debate over the importance of railroads to industrialization, see, inter alia, Rostow, *Stages of Economic Growth*; Robert W. Fogel, *Railroads and American Economic Growth: Essays in Econometric History* (Baltimore, Md.: Johns Hopkins University, 1964); Albert Fishlow, *American Railroads and the Transformation of the Ante-Bellum Economy* (Cambridge: Harvard University, 1965); and the summary in Marcus and Segal, *Technology in America*, pp. 96ff. The notion of railroads as a "social invention" is argued by Bruce Mazlish, *The Railroad and the Space Program: An Exploration in Historical Analogy* (Cambridge, Mass.: MIT, 1965). The contrast between American and European railroads and the argument for creative corruption are discussed in Frank Dobbin, *Forging Industrial Policy: The United States, Britain, and France in the Railway Age* (New York: Cambridge University, 1994); and Charles Perrow, *Organizing America: Wealth, Power, and the Origins of Corporate Capitalism* (Princeton, N.J.: Princeton University, 2002). The argument for "mercantilism" is made by Taylor, *Transportation Revolution*, pp. 97–103.

50. Yes, the railroading Latrobe was the son of the great architect. Poor, the business journalist, was also a founder of Standard and Poor's. See Alfred D. Chandler, Jr., *Henry Varnum Poor* (Cambridge, Mass.: Harvard University, 1955).

51. The classic account of railroad management is Alfred D. Chandler, Jr., *The Visible Hand: The Managerial Revolution in American Business* (Cambridge, Mass.: Harvard University, 1977), especially pp. 79–121. Perrow, in *Organizing America*, faults Chandler for focusing too narrowly on rational economic causes of the rise of big corporations (and for dismissing the priority of the textile firms). Perrow, a sociologist, believes the subjection of American society to large, impersonal corporations was neither inevitable nor in some cases rational. He makes a good point. But the nature of the technology, the demands of market discipline (where competition existed), and the pressure from stockholders did behoove railroads to maximize efficiency in the ways Chandler describes.

52. Walter Licht, *Working for the Railroad: The Organization of Work in the Nineteenth Century* (Princeton, N.J.: Princeton University, 1983), is an elegant book, and all the more authoritative given that its author is not a railroad buff. The biggest wave of strikes drove the proud Erie into receivership, forcing McCallum to resign. It began in 1854 when McCallum imposed a new, rigid set of work rules including "the infamous Rule 6" holding engineers liable for running off an incorrectly set switch. The proud McCallum dug in his heels. Pride goeth before a fall.

53. Ann Norton Greene, "Harnessing Power: Industrializing the Horse in Nineteenth-Century America," PhD dissertation, University of Pennsylvania (2004). Greene constructs this important history almost ex nihilo, given the failure of economic and technological historians to appreciate the symbiosis between steam power and horse power ("What! the ladies," p. 61; "nature of the machine," p. 71; statistics, pp. 65, 164). Robert West Howard, *The Horse in America* (Chicago, Ill.: Follett, 1965), is a rich source of insight and anecdotes about horses in all "walks of life" (draft horses, pp. 168–180).

54. Hindle and Lubar, *Engines of Change*, pp. 249–268 (Crystal Palace); 76–77 (*Men of Science*); quotations on technology in Russell Blaine Nye, *Society and Culture in America 1830–1860* (New York: Harper and Row, 1974), pp. 277–282.

55. This was the era in which "rags to riches" novels first caught Americans' imagination. According to Kenneth S. Lynn, *The Dream of Success* (Boston, Mass.: Little, Brown, 1955), such books conflated "belief in the potential greatness of the common man, the glorification of individual efforts and accomplishment, the equation of the pursuit of money with the pursuit of happiness and of business success with spiritual grace" (p. 7). On economic growth estimates, see Temin, *Causal Factors in American Economic Growth*, pp. 12–20; Moses Abramovitz and Paul A. David, "Reinterpreting Economic Growth: Parables and Realities," *American Economic Review* 63 (1973): 428–439; Douglass C. North, *The Economic Growth of the United States* (New York: Norton, 1966), pp. 189–215; Taylor, *Transportation Revolution*, pp. 176–178, 246–249, 384–398 ("stood a better chance," p. 395). On Americans' faith in upward mobility before 1860, see Jim Cullen, *The American Dream: A Short History of an Idea That Shaped a Nation* (New York: Oxford University, 2003), pp. 5–73; David M. Potter, *People of Plenty: Economic Abundance and the American Character* (Chicago, Ill.: University of Chicago, 1954), pp. 95–97 ("immense moral power," p. 97).

56. Frank Luther Mott, *American Journalism* (New York: Macmillan, 1947), pp.

201–302; Allan Nevins, *Ordeal of the Union*, 3 vols. (New York: Scribner, 1947), III: 80ff; Nye, *Society and Culture 1830–1860*, pp. 366–372; Taylor, *Transportation Revolution*, pp. 149–152. On *Godey's Lady's Book*, see especially Karen Halttunen, *Confidence Men and Painted Women: A Study of Middle-Class Culture in America, 1830–1870* (New Haven, Conn.: Yale University, 1982), pp. 56–91 ("the conflict," p. 65; "shame," p. 64).

57. Jackson Lears, *Fables of Abundance: A Cultural History of Advertising in America* (New York: Basic Books, 1994); quotations from pp. 88–94.

58. George L. Miller, "Changing Consumption Patterns: English Ceramics and the American Market from 1770 to 1840"; Diana diZerega Wall, "Family Dinners and Social Teas: Ceramics and Domestic Rituals"; and Jack Larkin, "From 'Country Mediocrity' to 'Rural Improvement': Transforming the Slovenly Countryside in Central Massachusetts, 1775–1840," in Catherine E. Hutchins, ed., *Everyday Life in the Early Republic* (Winterthur, Del.: Henry Francis du Pont Winterthur Museum, 1994), pp. 175–200, 219–283 ("houses too large" and "our villages," pp. 193–194). On religious merchandising, see Colleen McDannell, *Material Christianity: Religion and Popular Culture in America* (New Haven, Conn.: Yale University, 1995).

59. David Carlyon, *Dan Rice: The Most Famous Man You've Never Heard Of* (New York: Public Affairs, 2001), is a delightful and erudite tour of nineteenth-century American popular culture. The brief summation that follows does not begin to do it justice. See also George L. Chindahl, *A History of the Circus in America* (Caldwell, Id.: Caxton, 1959).

60. Carlyon, *Dan Rice*, pp. 1–283 ("shouting Methodist," p. 12; "extravagantly fond," p. 30; "nigero," p. 47; "DAN RICE CIRCUS TRIUMPHANT!" p. 123; "Wit so Whimsical!" p. 222; "splendor, refinement," p. 225; "grammatical assassin," p. 256; "REFINED AND INTELLECTUAL," p. 257; "more brilliant, except in one instance," p. 195). Devoted as he was to the Union and the national audience it made possible, Rice showed signs of sympathy toward the South in the early stages of the Civil War but soon joined the eventual winning side. After the war he continued his career as a circus impresario, ran for office as a Democrat, and became an author. He also continued to travel throughout the country, finding his third wife in Texas. As the shadows lengthened, so did the legends swirling about Rice (thanks in good part to himself). He lived until 1900, revered by a younger generation of comedians and promoters as "the greatest and the most original clown this or any other country ever produced" (p. 416). Then he was forgotten, buried without a trace by the highbrow defenders of "authentic" culture.

61. Barnum even revised his own autobiography. Subsequent editions published in 1869 and 1889 altered or deleted "facts" revealed in the original 1855 edition. See the introduction by George S. Bryan to *Struggles and Triumphs: Or, The Life of P. T. Barnum, Written by Himself,* 2 vols. (New York: Knopf, 1927). Recent biographies are Philip B. Kunhardt, Jr., et al., *P. T. Barnum: America's Greatest Showman* (New York: Knopf, 1995); and A. H. Saxon, *P. T. Barnum: The Legend and the Man* (New York: Columbia University, 1989). The claim made for Mike McDonald, the "prince of sharpers" and friend to mayors of Chicago and governors of Illinois, is in Herbert Asbury, *Sucker's Progress: An Informal History of Gambling in America from the Colonies to Canfield* (New York: Dodd, Mead, 1938), p. 295.

62. Barnum's account of the Heth affair is in *Struggles and Triumphs*, I: 95–135. Benjamin Reiss, *The Showman and the Slave: Race, Death, and Memory in Barnum's America* (Cambridge, Mass.: Harvard University, 2001), is an outstanding history and cultural

analysis that includes painstaking research into the true biography of Joice Heth ("mere skeleton," p. 2; "The GREATEST," p. 34).

63. Barnum, *Struggles and Triumphs*, I: 136–233 ("This is a trading world," p. 102; "exhibition of stuffed monkey," p. 190; "educated dogs," p. 195; "The titles of 'humbug,'" p. 199).

64. Barnum, *Struggles and Triumphs*, II: 416–600 ("takes no interest," p. 571; "if I might provide," p. 814). Barnum died in Bridgeport, at age eighty, in 1891, after signing off on his obituary.

65. This is the judgment of Nye, in *Society and Culture in America*, pp. 146–157, and I dare say he is right.

66. On the nineteenth-century war over theater and art in general, see Lawrence W. Levine, *Highbrow/Lowbrow: The Emergence of Cultural Hierarchy in America* (Cambridge, Mass.: Harvard University, 1988), pp. 13–81 ("his pockets," p. 17; "When was Desdemona," p. 16; "because the audiences," p. 57). On "rube stories," see David Carlyon, "'Blow Your Nose with Your Fingers': The Rube Story as Crowd Control," *New England Theatre Journal* 7 (1996): 1–22.

67. Herman Melville, *The Confidence-Man: His Masquerade* (San Francisco, Calif.: Chandler, 1968, orig. 1857), p. 286.

68. On the Astor Place riot, see Edwin G. Burrows and Mike Wallace, *Gotham: A History of New York City to 1898* (New York: Oxford University, 1999), pp. 761–766; Allan Nevins, ed., *The Diary of Philip Hone 1828–1851*, 2 vols. (New York: Dodd, Mead, 1927), II: 867.

Chapter 4

1. "On the Discrimination of Romanticisms" and "Coleridge and Kant's Two Worlds," in Arthur O. Lovejoy, *Essays in the History of Ideas* (Baltimore, Md.: Johns Hopkins University, 1948), pp. 228–276 ("renascence," p. 276); David Morse, *Perspectives on Romanticism: A Transformational Analysis* (Totawa, N.J.: Barnes and Noble, 1981).

2. David Morse, *American Romanticism*, 2 vols. (Totawa, N.J.: Barnes and Noble, 1987), I: 1–29 ("burdened," p. 1). Granted, this scholar is British and is none too impressed with the "flowering" of American culture in the Romantic era. Nonetheless, evidence of the excessiveness he identifies will accumulate.

3. The Shakers got their name from the rhythmic dances they performed as a community to the glory of God. Needless to say, they could propagate themselves only through new conversions and the adoption of children, and they all but died out in the late nineteenth century. See Stephen J. Stein, *The Shaker Experience in America* (New Haven, Conn.: Yale University, 1994). The Oneida Community, which bears some comparison to the Mormons (to be discussed below, at some length), collectively supervised the sexual activity of its members, used methods of birth control, and even anticipated the selective breeding of human beings associated with the twentieth-century eugenics movement. The Oneida enterprise (it was run like a corporation) died with its founder in 1886. See Karen Lockwood Carden, *Oneida: Utopian Community to Modern Corporation* (Baltimore, Md.: Johns Hopkins University, 1969). The Amana Colony founded in 1844 by Christian Metz and planted in Iowa in 1855 ought not to be grouped with the homegrown American utopian sects because it traced its roots to eighteenth-century German Pietism. On the Millerites,

see Ruth Aldon Doan, *The Miller Heresy, Millennialism, and American Culture* (Philadelphia, Pa.: Temple University, 1987). Miller, a simple farmer and evidently no fraud, apologized to his followers and pronounced himself genuinely perplexed by his miscalculations.

4. On millennialism as a theme in colonial and early national American history, see Walter A. McDougall, *Freedom Just Around the Corner: A New American History 1585–1828* (New York: HarperCollins, 2004), and its principal sources for the merger of republican and evangelical expectations: J. G. A. Pocock, *The Machiavellian Moment: Florentine Political Thought and the Atlantic Republican Tradition* (Princeton, N.J.: Princeton University, 1975); Jonathan C. D. Clark, *The Language of Liberty, 1660–1832: Political Discourse and Social Dynamics in the Anglo-American World* (New York: Cambridge University, 1994); Ernest Lee Tuveson, *Redeemer Nation: The Idea of America's Millennial Role* (Chicago, Ill.: University of Chicago, 1968); Melvin B. Endy, Jr., "Just War, Holy War, and Millenarianism in Revolutionary America," *William and Mary Quarterly* 42 (1985): 3–25; Gordon S. Wood, *The Rising Glory of America, 1760–1820* (New York: Braziller, 1971); Conrad Cherry, ed., *God's New Israel: Religious Interpretations of American Destiny*, rev. ed. (Chapel Hill: University of North Carolina, 1998); Harold Bloom, *The American Religion: The Emergence of the Post-Christian Nation* (New York: Simon and Schuster, 1992).

5. On the effects of the "free exercise of religion" and the marketplace in spirituality it created, see inter alia: R. Laurence Moore, *Selling God: American Religion in the Marketplace of Culture* (New York: Oxford University, 1994); Nathan O. Hatch, *The Democratization of American Christianity* (New Haven, Conn.: Yale University, 1989); Joyce Appleby, *Inheriting the Revolution: The First Generation of Americans* (Cambridge, Mass.: Harvard University, 2000), "looking for God's will," pp. 250–251; John Thomas Noonan, *The Lustre of Our Country: The American Experience of Religious Freedom* (Berkeley: University of California, 1998).

6. The revivals collectively known as the Second Great Awakening began in Virginia in the 1790s, swept the frontier following the tent meeting of 1801 at Cane Ridge, Kentucky, and exploded onto the national scene through Finney's urban revivals of the 1820s and 1830s. For good summaries of their origins, methods, and effects, see Mark Noll, *America's God: From Jonathan Edwards to Abraham Lincoln* (New York: Oxford University, 2002), pp. 159–252; Robert William Fogel, *The Fourth Great Awakening and the Future of Egalitarianism* (Chicago, Ill.: University of Chicago, 2002), pp. 202–222, 68–107; Edwin S. Gaustad, *A Religious History of America*, rev. ed. (San Francisco, Calif.: Harper and Row, 1990); Charles E. Hambrick-Stowe, *Charles G. Finney and the Spirit of American Evangelism* (Grand Rapids, Mich.: Eerdmans, 1996). Sydney E. Ahlstrom, *A Religious History of the American People*, 2nd ed. (New Haven, Conn.: Yale University, 2004), pp. 583–614, has an especially good summary of the nexus between Romanticism and religion.

7. Daniel Walker Howe, *The Unitarian Conscience: Harvard Moral Philosophy, 1805–1861* (Cambridge, Mass.: Harvard University, 1970), pp. 137–138, aptly characterizes the Unitarians as Puritans without Calvinism. On Channing, see Andrew Delbanco, *William Ellery Channing* (Cambridge, Mass.: Harvard University, 1981). The split Unitarians caused in the Congregationalist church can be appreciated through a brief (but steep) stroll up the street from Plymouth Rock. At the top of the hill a Unitarian church

enjoys pride of place. Off to the side, in humbler surroundings, sits the church of the remnant that had to start over in hopes of preserving the faith of the Pilgrim fathers.

8. Finney's background and preaching are described in McDougall, *Freedom Just Around the Corner*, pp. 509–513. Excellent summaries of the debates among various Congregationalists, Presbyterians, and Unitarians are provided by Noll, *America's God*, pp. 269–299 (Taylor quotations, p. 298; Finney quotations, p. 308; "I'll meet you," p. 296); and the classic by Perry Miller, *The Life of the Mind in America: From the Revolution to the Civil War* (New York: Harcourt, Brace, and World, 1965), pp. 3–95 ("no safety," p. 36). Noll provides a useful table of the various Calvinist schools and leaders. Unitarians traced their lineage to Jonathan Mayhew; the "new divinity" men to Edwards, Dwight, Samuel Hopkins, and Nathaniel Emmons; the "new school" and "old school" Presbyterians to Witherspoon; the Finneyites to Edwards (I would add the preacher James McGready of Cane Ridge); and Romantic Christians such as Horace Bushnell and Lyman Beecher's extraordinary children Henry Ward Beecher and Harriet Beecher Stowe to Samuel Coleridge. On Unitarians and Universalists, see E. Brooks Holifield, *Theology in America: Christian Thought from the Age of the Puritans to the Civil War* (New Haven, Conn.: Yale University, 2003), pp. 197–233. On Beecher as a vital transitional figure, see Vincent Harding, *A Certain Magnificence: Lyman Beecher and the Transformation of American Protestantism, 1775–1863* (Brooklyn, N.Y.: Carlson, 1991).

9. Noll, *America's God*, pp. 300–364 ("than *in any sense*," p. 303; "merciless tyrant," p. 352; "my heart was emptied," p. 359). On "Methodist perfection," see Holifield, *Theology in America*, pp. 256–272. On Episcopalians, see Robert Bruce Mullin, *Episcopal Vision / American Reality: High Church Theology and Social Thought in Evangelical America* (New Haven, Conn.: Yale University, 1986); and Clarence E. Walworth, *The Oxford Movement in America* (New York: Catholic Historical Society, 1974, orig. 1895). Anglo-Catholicism made little headway, given the evangelical, low-church preferences of most American Episcopalians, especially in the South. But the mere toleration accorded parishes preferring "Romish" practices was enough to cause irate low churchmen to form the independent Reformed Episcopal Church in 1873.

10. Howard A. Barnes, *Horace Bushnell and the Virtuous Republic* (Metuchen, N.J.: American Theological Library Association, 1991); "Let's sin," p. 13; "kind of divine," p. 7; "excessive honesty," p. 13; "hypocrisy, flattery," p. 22; "mantle of sweetness" and "God made," pp. 37–38; "ante-natal," p. 48; "I see the nation," p. 18. Collections of Bushnell's works include David L. Smith, ed., *Horace Bushnell: Selected Writings on Language, Religion, and American Culture* (Chico, Calif.: Scholars, 1984); Horace Bushnell, *Work and Play* (New York: Scribner, 1903); Conrad Cherry, ed., *Howard Bushnell: Sermons* (New York: Paulist, 1985). See also Barbara Cross, *Horace Bushnell* (Chicago, Ill.: University of Chicago, 1958); and Robert L. Edwards, *Of Singular Genius, of Singular Grace: A Biography of Horace Bushnell* (Cleveland, Ohio: Pilgrim, 1992).

11. Barnes, *Bushnell and the Virtuous Republic*; "The gospel," "What form," "Sin takes," and "clear image," pp. 79–80; "the key to Horace," p. 82; "sacrifices and other Jewish," p. 90. Strangely, the grand U.S. histories by Schlesinger, Morison, Tindall, and others mention Bushnell only in passing or not at all. Historians of religion, however, appreciate his importance. Sydney Ahlstrom, *Theology in America* (Indianapolis, Ind.: Bobbs-Merrill, 1967), p. 62, put Bushnell "at the head of the main current of American theological liberalism." Bruce Kuklick, *Churchmen and Philosophers from Jonathan*

Edwards to John Dewey (New Haven, Conn.: Yale University, 1985), pp. 161–170; Gary Dorrien, *The Making of American Liberal Theology: Imagining Progressive Religion 1805–1900* (Louisville, Ky.: Westminster John Knox, 2001), pp. 111–178 ("Triunity," p. 155ff); and Holifield, *Theology in America*, pp. 452–466, devote whole chapters to Bushnell's Romantic imagination. On the conflation of civil and sacred religion in this era, see Paul C. Nagel, *One Nation Indivisible: The Union in American Thought* (New York: Oxford University, 1964). Among his other accomplishments, Bushnell chose the site (and was offered the presidency) of the new College of California at Berkeley while traveling there for his health in 1856. He preached Union until 1861, then became one of the loudest of preachers equating the federal cause with God's will. By his death in 1876 he was renowned as a pioneer of the notion that Christianity was a way of life, not a dogmatic creed, and hence as a progenitor of the Social Gospel.

12. No less a cultural critic than Harold Bloom agrees with Tolstoy. Bloom devotes an entire chapter of *The American Religion: The Emergence of the Post-Christian Nation* (New York: 1992), pp. 77–128, to Mormonism and Joseph Smith, whom he considers "an extraordinary religious genius" and the "most authentic of American prophets, seers, and revelators." Bloom's book argues that the United States is a gnostic nation tempted by self-worship and expert at bending traditional creeds to suit itself. A recent survey by Richard N. Ostling and Joan K. Ostling, *Mormon America: The Power and the Promise* (San Francisco, Calif.: HarperCollins, 1999), describes Smith's message as not just an American but "a *very* American gospel." Nathan O. Hatch, Jan Shipps, James Charlesworth, and Jacob Neusner are other non-Mormon scholars who judge the *Book of Mormon* to be of major importance to American cultural history.

13. See the preface to John L. Brooke, *The Refiner's Fire: The Making of Mormon Cosmology, 1644–1844* (New York: Cambridge University, 1994), for a survey of the scholarly literature on Mormon origins. Brooke explains that whereas Whitney Cross and David Brion Davis once traced Mormonism back to the culture of Puritan New England, more recent scholars have adopted a functionalist approach linking Mormonism (and other revivals and cults) to the social stress and anxiety of the Jacksonian era. See, for instance, Whitney R. Cross, *The Burned-Over District: The Social and Intellectual History of Enthusiastic Religion in Western New York, 1800–1850* (Ithaca, N.Y.: Cornell University, 1965); Gordon S. Wood, "Evangelical America and Early Mormonism," *New York History* 61 (1980): 359–386; Marvin S. Hill, *Quest for Refuge: The Mormon Flight from American Pluralism* (Salt Lake City, Utah: Signature, 1989); and Kenneth H. Winn, *Exiles in a Land of Liberty: Mormons in America, 1830–1846* (Chapel Hill: University of North Carolina, 1989). Still another school—including D. Michael Quinn, *Early Mormonism and the Magic World View* (Salt Lake City, Utah: Signature Books, 1987); Richard Bushman, *Joseph Smith and the Beginnings of Mormonism* (Urbana: University of Illinois, 1984); and Brooke himself—situates Mormonism in the tradition of the most radical wing of the Protestant Reformation.

14. The accounts of Smith's visions, discovery of the tablets, and translation can only be taken on faith or dismissed. Until recently, therefore, biographies of Joseph Smith ranged from the devotional to the vicious. But non-Mormon scholars now seek to explain rather than refute or ridicule Smith's prophetic claims. B. H. Roberts, *A Comprehensive History of the Church of Jesus Christ of Latter-Day Saints*, 6 vols. (Salt Lake City, Utah: Church of JCLDS, 1930) is the classic official account. It draws heavily on the

autobiographical work by Joseph Smith, *A History of the Church of Jesus Christ of Latter-Day Saints*, 7 vols., 2nd rev. ed. (Salt Lake City, Utah: Church of JCLDS, 1932–1951). Two useful missionary tracts (for which I am grateful to Cordell Clinger) are LeGrand Richards, *A Marvelous Work and a Wonder* (Salt Lake City, Utah: Deseret, 1988, orig. 1950); and *Truth Restored: A Short History of the Church of Jesus Christ of Latter-Day Saints* (Salt Lake City, Utah: Intellectual Reserve, 1979); "fell into many," p. 7; "He said there was a book," p. 8. The classic biography by a former Mormon disillusioned with Smith is Fawn Brodie, *No Man Knows My History: The Life of Joseph Smith the Mormon Prophet* (New York: Knopf, 1972); "Those days were never" and "Joseph Smith would put," pp. 60–61. Donna Hill, *Joseph Smith: The First Mormon* (Garden City, N.Y.: Doubleday, 1977); and Terryl L. Givens, *By the Hand of Mormon: The American Scripture That Launched a New World Religion* (New York: Oxford University, 2002), are thorough and objective analyses of Mormon origins.

15. *The Book of Mormon: An Account Written by the Hand of Mormon upon Plates Taken from the Plates of Nephi*, trans. Joseph Smith, Jun. (Salt Lake City, Utah: Church of JCLDS, 1981, orig. 1830). Beginning in 1982 the church decreed that every copy include the subtitle *Another Testament of Jesus Christ*. It also notes proudly that the Church of JCLDS is the only denomination to name itself after Jesus Christ rather than some peculiarity of doctrine or church governance. Yes, the *Book of Mormon* is repetitive, but deserving of attention from scoffers, seekers, and the merely curious. That Smith had a poetic (i.e., Romantic) soul cannot be doubted. If the book did emerge unaided from the head of this young, untutored man, he must have soaked up influences both ancient and modern like a sponge, and must have known intuitively how to refine them into what Nathan O. Hatch considers an extraordinary work of popular imagination.

16. Givens, *By the Hand of Mormon*, pp. 8–61, 155–175 ("greatest piece of superstition," p. 52; "every error and almost every truth," p. 163); Brodie, *No Man Knows My History*, pp. 50–82 ("Chloroform in print," p. 63). Mark Twain encountered the Mormons in Utah on his way to Nevada in 1861 and recorded his crack in *Roughing It*. Assuming that the *Book of Mormon* was not miraculously translated from golden plates, what did it reflect? The interest shown during the 1820s in the origins of the Indians and their mysterious mounds would seem one obvious source. But ironically, the evidence of pre-Columbian civilizations dug up in that era by Alexander von Humboldt, John Lloyd Stephens, and Frederick Catherwood and described in Joseph Priest's books lent verisimilitude to the account in the *Book of Mormon*. So there had been great cities, pyramids, and burial grounds in ancient America! The charge of plagiarism was itself a fraud perpetrated by Philastus Hurlbut, a disaffected Mormon disciple. He said that Smith poached from an unpublished novel by the Reverend Solomon Spaulding. In fact that manuscript described an odyssey of ancient Romans to Britain, not Jews to America, and contained no parallels to the Mormon account. That didn't stop Eber Howe from paying Hurlbut for his alleged evidence and throwing *Mormonism Unveiled* onto the market in 1834. Recent scholars have suggested that Smith wrote the "golden Bible" in a trance, either self-induced or brought on by epileptic fits (for which there is no evidence). Brodie described the book as an "obscure compound of folklore, moral platitude, mysticism, and millennialism" that expressed Smith's identity crises. But she also suggested that he may have believed in it, given his overpowering self-confidence and the absence of any note of cynicism in his voluminous private papers. Brooke, *Refiner's Fire*, does a brilliant job of

tracing Smith's themes of magic, Masonry, hidden wisdom, buried treasure, occult forces, and millennialism to seventeenth-century hermetic cults such as the Rosicrucians, Fifth Monarchy Men, Muggletonians, and Diggers. But Brooke is careful to add it is not at all clear how hermeticism might have been transplanted in nineteenth-century frontier New York. Most problematically, computer analyses of the vocabulary, syntax, and rhythms in the *Book of Mormon* reveal the fingerprints of as many as twenty-four different authors. Hence, a study done at the University of California, Berkeley, concluded that it was "statistically indefensible to propose Joseph Smith or Oliver Cowdery or Solomon Spaulding as the author of the 30,000 words . . . attributed to Nephi and Alma" (Givens, *By the Hand of Mormon*, p. 156).

17. Brigham Young's history illustrates the appeal of Mormonism to poor, uneducated children of the Yankee exodus. His great-grandfather was a shoemaker from the wilds of New Hampshire, his grandfather a surgeon in the French and Indian War, his father a veteran of Washington's army. His mother, Nabby Howe, traced her American roots to 1640 and was related to Elias and Samuel Gridley Howe. Brigham's father was a devout Methodist and rugged pioneer in Vermont, but rocky soil and crooked merchants ruined him. He moved to New York, then had to scatter his eleven children when Nabby died in 1815. Three of the boys became Methodist preachers. Brigham, the youngest but one, was apprenticed to a carpenter in Auburn, where he worked on a home later bought by the father-in-law of William Henry Seward. Just then the upstate revivals filled the air with anticipation. Two of Brigham's sisters and a friend, Heber Chase Kimball, experienced visions about great things soon to unfold. Then a stranger approached Phineas Young with "a book, sir, I wish you to read. The Book of Mormon, or, as it is called by some, the Golden Bible." The family was merely curious until, a year later, Kimball, Brigham and Joseph Young, and their father were out hewing trees and debating religion. When the subject of Mormonism arose they found themselves running in circles shouting "Hosannah." In January 1832 they rode Kimball's sleigh through the snow to visit a Mormon community and came away convinced that it was a genuine restoration of the Christian churches described in the Acts of the Apostles. A Methodist warned them that the Mormon leader was a fraud and a horse thief. Young replied, "Joseph Smith I never saw. . . . I went to my Father in heaven and asked with regard to the truth of the doctrines taught by Joseph Smith . . . and though Joseph Smith should steal horses every day, or gamble every night . . . I know that the doctrine he preaches is the power of God to my salvation." When Young journeyed to Kirtland, Ohio, and found the rangy prophet humbly wielding an ax, he felt that his soul's quest was fulfilled. Kimball also decided to stay. His alternative was debtors' prison in New York. See Stanley P. Hirshson, *The Lion of the Lord: A Biography of Brigham Young* (New York: Knopf, 1973), pp. 3–12.

18. All of Smith's major revelations and translations, most of which date from the years in Kirtland, appear in *The Doctrine and Covenants of the Church of Jesus Christ of Latter-Day Saints*, subtitled *The Pearl of Great Price* (Salt Lake City, Utah: Church of JCLDS, 1982); "glory of God is intelligence," 93: 36. Brooke, *Refiner's Fire*, pp. 209–234; and Brodie, *No Man Knows My History*, pp. 114–207 ("Nits will make lice," p. 237), narrate the church's Ohio phase. Brodie argues that Smith and his lieutenants were caught up in the same speculative frenzy as the rest of the nation before the Panic of 1837. Because of their shortage of capital, they engaged in ever more reckless financial transactions. The ensuing bankruptcies, lawsuits, and threats of criminal proceedings caused

the first great rift in the movement. It must be said that the Missourian's threat of a "war of extermination" *(Truth Restored,* p. 57) was thrown back by Rigdon, who warned that if the mob attacked "it shall be between us and them a war of extermination; for we will follow them till the last drop of their blood is spilled" (Brodie, *No Man Knows My History,* p. 223).

19. Smith played politics quite well for a few years, aided by the fact that the "Suckers" of Illinois and the "Pukes" of Missouri across the river despised each other. Thus authorities in Illinois sheltered the Mormon refugees from the writs and warrants issued by the governor of Missouri. Likewise, when Whig lawyers rushed to defend Smith against Missouri's charges, the Democrat Stephen A. Douglas tossed out the writs on a technicality. Both parties courted the Mormon vote until Smith overplayed his hand and made powerful enemies in both camps (Brodie, *No Man Knows,* pp. 266–274, 344–366).

20. Heber Kimball wrote a letter in June 1842 reporting the founding of a Masonic Lodge in Nauvoo: "Br. Joseph and Sidny was the first that was received into the Lodg[e]. . . . Thare is a similarity of preas hood [priesthood] in Masonry. Bro. Joseph ses Masonry was taken from preas hood but has become degenerated. But menny things are perfect" (Brooke, *Refiner's Fire,* p. 91). In one sense Smith was coming full circle. The men of his father's and mother's families in New England had strong Masonic affiliations, as did Oliver Cowdery and other early Mormon leaders. Of course, Smith insisted that all males join the Nauvoo lodge, with the result that its membership soon surpassed that of all other Illinois lodges combined. The state's grand master in Springfield was not pleased, and later revoked Nauvoo's charter.

21. On the Nauvoo period, see Brooke, *Refiner's Fire,* pp. 235–277, and Brodie, *No Man Knows My History,* pp. 256–379 ("seems to have hit," p. 270; "literal gathering," p. 278). Smith's revelation concerning eternal marriage and the plurality of wives was decreed July 12, 1843. It is styled a commandment from the Lord and includes this shocking verse concerning Smith's original wedded wife of sixteen years: "And I command mine handmaid, Emma Smith, to abide and cleave unto my servant Joseph, and to none else. But if she will not abide this commandment she shall be destroyed" (*Doctrines and Covenants* 132: 54). Ostling and Ostling, in *Mormon America,* pp. 56–75, record estimates of Joseph Smith's plural wives ranging from a low of "at least" twenty-eight (*Encyclopedia of Mormonism*) to a high of forty-eight (Brodie), but if ritual "sealings" of women to the prophet for their benefit in the hereafter are counted, the number may approach eighty-four.

22. The conversation in which Smith prophesies "we shall be massacred" is recorded in Roberts, *Comprehensive History of the Church,* 6: 545–546. Ostling and Ostling, in *Mormon America,* pp. 56–75, recount the Smith brothers' murders ("Oh Lord, my God," p. 17) and reproduce Smith's eulogy of April 7, 1844, for the martyred King Follett in its entirety, pp. 387–394.

23. Hirshson, *Lion of the Lord,* pp. 16–21 ("You know nothing," p. 18; "nothing could stop it," p. 20). Young took command of the largest contingent of Mormons, but the second largest war led—for fifty-four years—by the martyred prophet's son. See Roger D. Lannius, *Joseph Smith III: Pragmatic Prophet* (Urbana: University of Illinois, 1988).

24. Good summaries of the Romantic reform movements include Miller, *Life of the Mind,* pp. 3–95 ("every lover" and "haste to be rich," p. 74; "genius," "there is no other power," and "make public opinion," p. 68); Russel B. Nye, *Society and Culture in America, 1830–1860* (New York: Harper and Row, 1974), pp. 32–70; Ahlstrom, *Religious History of*

the American People, pp. 637–647; John L. Thomas, "Romantic Reform in America 1815–1865," in Lawrence W. Levine and Robert Middlekauff, eds., *The National Temper: Readings in American History* (New York: Harcourt, Brace, and World, 1968), pp. 166–189 ("religious party," p. 167); and a classic by Alice Felt Tyler, *Freedom's Ferment: Phases of American Social History to 1860* (Minneapolis: University of Minnesota, 1944). The origins of the reform movements in revivalist religion are examined by John R. Bodo, *The Protestant Clergy and Public Issues, 1812–1848* (Princeton, N.J.: 1954); Clifford Stephen Griffin, *Their Brothers' Keepers: Moral Stewardship in the United States, 1800–1865* (New Brunswick, N.J.: Rutgers University, 1960); Whitney R. Cross, *The Burned-Over District: The Social and Intellectual History of Enthusiastic Religion in Western New York, 1800–1850* (Ithaca, N.Y.: Cornell University, 1950); and Richard Carwardine, *Evangelicals and Politics in Antebellum America* (New Haven, Conn.: Yale University, 1993). Reinhold Niebuhr, in *A Nation So Conceived: Reflections on the History of America from Its Early Visions to Its Present Power* (New York: Scribner, 1963), p. 49, put his finger on the distinctively American (I would say Anglo-American) feature in Romantic reform: "Thus there grew on American soil a religious and evangelical version of the spirit of the secular enlightenment in France—a form of utopianism which regarded 'liberty, equality, and fraternity' as simple historical possibilities."

25. Paul Goodman, *Towards a Christian Republic: Anti-Masonry and the Great Transition in New England, 1826–1836* (New York: Oxford University, 1988), pp. 3–119, 234–245 ("darkest and deepest," p. 57; "average family," p. 102; "struggle of republican," p. 25; "put down Masonry," p. 3; "blackcoats," p. 101; "faction, corruption," p. 31). See also William Preston Vaughan, *The Anti-Masonic Party in the United States, 1826–1843* (Lexington: University of Kentucky, 1983); David Brion Davis, "Some Themes of Counter-Subversion: An Analysis of Anti-Masonic, Anti-Catholic, and Anti-Mormon Literature," in Levine and Middlekauff, eds., *The National Temper*, pp. 149–165.

26. American college fraternities trace their origins to Phi Beta Kappa, the originally secret honor society founded at the College of William and Mary in December 1776. After 1815 the romantic *Burschenschaften* founded by students in German universities served as models for English and American campuses. My own fraternity, Delta Kappa Epsilon, was founded at Yale in 1844. Within three years chapters emerged at Bowdoin, Colby, Princeton, and Amherst, my alma mater.

27. W. J. Rorabaugh, *The Alcoholic Republic: An American Tradition* (New York: Oxford University, 1979), pp. 3–21, 93–222 ("we may set," p. 209; "Here the disgusting," pp. 199–200); Eric Burns, *The Spirits of America: A Social History of Alcohol* (Philadelphia, Pa.: Temple University, 2004), pp. 1–81 ("Prince Alcohol," p. 81; "heavy perfume," pp. 69–70; "thing more and more," p. 69). On gambling, see Ann Vincent Fabian, *Card Sharps, Dream Books, and Bucket Shops* (Ithaca, N.Y.: Cornell University, 1990); Herbert Asbury, *Sucker's Progress: An Informal History of Gambling in America* (New York: Dodd, Mead, 1938), Greeley quoted on pp. 163–164; Karen Halttunen, *Confidence Men and Painted Women: A Study of Middle Class Culture in America, 1830–1870* (New Haven, Conn.: Yale University, 1982); and Jackson Lears, "Playing with Money," *Wilson Quarterly* 19 (Fall 1995): 7–23. As for the clergy, Methodists had long preached abstinence and most Baptists and evangelical Calvinists (Presbyterians and Congregationalists) now did so as well, citing numerous biblical passages inveighing against drunkenness. The so-called Hard-Shell or Forty-Gallon Baptists, however, clung to the Protestant doctrine of salvation by faith, not works. In any case, had not Jesus turned water to wine and

Paul advised Timothy to take a little wine for his health? The Episcopal bishop of Vermont, John Henry Hopkins, made a suitably high-minded rebuttal to teetotalers: their focus on a particular vice rather than the sanctification of souls was an unbiblical invitation to spiritual pride

28. Burns, *Spirits of America*, pp. 82–89. From his autobiography, *The Reminiscences of Neal Dow: Recollection of Eighty Years* (Portland, Me.: Express, 1898), one would think that he never made a misjudgment or suffered a setback.

29. On Graham and his imitators, see Stephen Nissenbaum, *Sex, Diet, and Debility in Jacksonian America: Sylvester Graham and Health Reform* (Westport, Conn.: Greenwood, 1980). Dr. Alcott, a cousin of Bronson Alcott, is quoted in Russel Blaine Nye, *Society and Culture in America, 1830–1860* (New York: Harper and Row, 1974), p. 350.

30. Unless I missed one, Jonathan Messerli, *Horace Mann: A Biography* (New York: Knopf, 1971) is still the most recent life. Risjord, in *The Romantics*, pp. 153–178, provides an excellent pen portrait ("I believe," p. 153; "most common thing," p. 157; "spirit of a martyr," p. 167; "Let the next," p. 168; "Education then" and "in a republic," p. 153). See also Nye, *Society and Culture*, pp. 376–380 ("will of God," "to such a degree," and "The very taxes," pp. 377–379); and Lawrence Arthur Cremin, ed., *The Republic and the School: Horace Mann and the Education of Free Men* (New York: Columbia University, 1966).

31. Only Oberlin was coeducational, but Mary Lyon's Mount Holyoke "seminary" in South Hadley, Massachusetts, and the Georgia Female College, both founded in 1836, pointed the way to higher education for women.

32. The sharp disparity between educational patterns in the North and those in the South was a later phenomenon. In 1860 one-seventh of southern white children were enrolled as compared with one-sixth in the North; but a slightly higher percentage of southern white males attended college. Literacy rates, however, clearly favored the North, suggesting that home and church schooling was far more prevalent there than in the southern backcountry. See Lawrence Arthur Cremin, *American Education: The National Experience, 1783–1876* (New York: Harper and Row, 1980); Nye, *Society and Culture*, pp. 372–399.

33. Elliott J. Gorn, ed., *The McGuffey Readers* (Boston, Mass.: Bedford/St. Martin's, 1998), pp. 1–36; "beautiful coins," p. 28; Richard D. Mosier, *Making the American Mind: Social and Moral Ideas in the McGuffey Readers* (New York: Russell and Russell, 1965); Dolores P. Sullivan, *William Holmes McGuffey, Schoolmaster to the Nation* (Rutherford, N.J.: Fairleigh Dickinson University, 1994); Charles H. Carpenter, *A History of American Schoolbooks* (Philadelphia: University of Pennsylvania, 1963); Joseph J. Ellis, *After the Revolution: Profiles of Early American Culture* (New York: Norton, 1979), pp. 161–212, "for the morals," p. 179; Ruth Miller Elson, "American Schoolbooks and 'Culture' in the Nineteenth Century," in Levine and Middlekauff, eds., *The National Temper*, pp. 113–134 ("Next to the fear," p 123).

34. Ruth Miller Elson, *Guardians of Tradition: American Schoolbooks in the Nineteenth Century* (Lincoln: University of Nebraska, 1964); "the only basis," p. 46; "Franklin is," p. 186; "the marks of divine," p. 61.

35. Ann Douglas, *The Feminization of American Culture* (New York: Knopf, 1977), pp. 169–180 ("I now set myself," p. 174); Nye, *Society and Culture*, pp. 101–124; George Calcott, *History in the United States, 1800–1860: Its Practice and Purpose* (Baltimore, Md.: Johns Hopkins University, 1970).

36. See Adam Hochschild, *Bury the Chains: Prophets and Rebels in the Fight to Free an Empire's Slaves* (Boston, Mass.: Houghton Mifflin, 2005), "No watchman now," p. 339; Edith F. Hurwitz, *Politics and the Public Conscience: Slave Emancipation and the Abolitionist Movement in Britain* (London: Allen and Unwin, 1973), "a national crime," p. 40; David Brion Davis, *The Problem of Slavery in Western Culture* (Ithaca, N.Y.: Cornell University, 1966). The Abolition of Slavery Act of 1833 passed in spite of the opposition of King William IV, the Tory leadership, and a portion of the Anglican bishops. It received nearly unanimous support from Nonconformists (non-Anglican Protestants), Whigs, laborers, manufacturers, and commercial interests. Indeed, the bankers in the City of London received a hefty portion of the compensatory payment because most of the West Indian plantations were mortgaged. The £20 million represented a huge sum equivalent to 40 percent of the realm's annual budget. William Wilberforce lived, barely, to witness the final triumph of a movement he had begun nearly fifty years before.

37. See James Brewer Stewart, *William Lloyd Garrison and the Challenge of Emancipation* (Arlington Heights, Ill.: Harlan Davidson, 1992); Henry Mayer, *All on Fire: William Lloyd Garrison and the Abolition of Slavery* (New York: St. Martin's, 1998); William E. Cain, ed., *William Lloyd Garrison and the Fight against Slavery: Selections from the Liberator* (Boston, Mass.: St. Martin's, 1995), pp. 61–94 ("I will not equivocate," p. 72; "unequalled by any other," p. 91). See also, generally, James B. Stewart, *Holy Warriors: The Abolitionists and American Slavery* (New York: Hill and Wang, 1976). Radicals, not least John Brown himself, considered themselves disciples of America's greatest theologian, the eighteenth-century revivalist Jonathan Edwards. The fact that gradualists preaching limited toleration of slavery also invoked Edwards's authority, however, proves how difficult it was even for northern divines to tease out a political position from biblical exegesis. See Kenneth P. Minkema and Harry S. Stout, "The Edwardsean Tradition and the Antislavery Debate, 1740–1865," *Journal of American History* 92, no. 1 (2005): 47–74.

38. Robert H. Abzug, *Passionate Liberator: Theodore Dwight Weld and the Dilemma of Reform* (New York: Oxford University, 1980); Norman K. Risjord, "Theodore and Angelina Grimké Weld: The Antislavery Dilemma," in *Representative Americans: The Romantics* (Lanham, Md.: Rowman and Littlefield, 2001), pp. 223–248 ("Its severity," p. 224; "preference for self-interest," p. 226); Benjamin P. Thomas, *Theodore Weld: Crusader for Freedom* (New Brunswick, N.J.: Rutgers University, 1950).

39. Risjord, *Representative Americans*, pp. 232–243; Gerda Lerner, *The Grimké Sisters of South Carolina: Pioneers for Woman's Rights and Abolition* (Boston, Mass.: Houghton Mifflin, 1967). Stephen Howard Browne, *Angelina Grimké: Rhetoric, Identity, and the Radical Imagination* (East Lansing: Michigan State University, 1999), is a penetrating analysis of her rhetorical strategies and the psychology they reveal ("I will lift up" and "Episcopalian-Presbyterian," p. 1; "Men, brothers" and "They knew not," pp. 152–153).

40. Anna M. Speicher, *The Religious World of Anti-Slavery Women: Spirituality in the Lives of Five Abolitionist Lecturers* (Syracuse, N.Y.: Syracuse University, 2000), pp. 1–108 ("pure" and "God designs," p. 78; "be not troubled," p. 62; "the powerful effort," p. 66; "abolitionism is Christianity," p. 89); Browne, *Angelina Grimké*, pp. 57–82 ("God designs," p. 78). Speicher includes a good summary of the literature on the religious roots of abolitionism, pp. 1–12. Anne Braude, *Radical Spirits: Spiritualism and Women's Rights in Nineteenth Century America* (Boston, Mass.: Beacon, 1989), challenged the view of Douglas, *Feminization of American Culture*, by arguing that spiritual piety did not reinforce

domesticity or the "women's sphere" but in fact emboldened women to speak and act in public. Robert H. Abzug, *Cosmos Crumbling: American Reform and the Religious Imagination* (New York: Oxford University, 1994), found religious conviction to be a consistent motivation for male and female reformers determined to rend the curtain separating the secular and the sacred in the United States. Perhaps the most striking example of both the importance of faith and the abolitionists' fondness for odd causes is the enslaved New York woman Isabella Van Wagenen. Freed by law in 1827, she deserted her husband and five children after God revealed himself to her in a bolt of lightning. She became a disciple of the Methodist evangelist John Newland Moffitt, who served as chaplain to Congress but ruined his career with women and drink. Next Isabella followed "Matthias the Prophet," who abused the women he called "match spirits." In 1843, after another epiphany, she reinvented herself as Sojourner Truth and followed the Millerites' "end of the world" craze. At last, in a commune in Northampton, Massachusetts, she found her true calling as a powerful abolitionist and feminist witness. See Nell Painter, *Sojourner Truth: A Life, A Symbol* (New York: Norton, 1996).

41. Lewis Perry, *Radical Abolitionism: Anarchy and the Government of God in Antislavery Thought* (Ithaca, N.Y.: Cornell University, 1973), pp. 1–91 ("fertile forms," "tendency of every reform," and "representative army," pp. 14–16; "slaves are men," p. 49; "God has a Government," p. 52; "total abstinence," "Every person," and "hope of the millennium," pp. 67–68.

42. The utilitarian philosopher and historian Richard Hildreth dissected the self-destructive characteristic of the radical abolitionists: "They denounce slavery not as a wrong, a crime, a delusion, a blunder, a folly—but as a sin against God, to be immediately repented of and abandoned. Thus they set themselves up not as expositors of mere human science, not as teachers of morals, and politics, and political economy, but as expositors of the will of God; and according to the usual course of things, from being expositors, they have proceeded to act as God's vicegerents, judges, and executors . . . [with] the claim of infallibility, and the domineering, denouncing, excommunicating spirit which appertains to all priesthoods" (Perry, *Radical Abolitionism*, p. 186).

43. James Kirke Spaulding, *Slavery in the United States* (New York: Harper, 1836), pp. 281–282.

44. Virginia Bernhard and Elizabeth Fox-Genovese, eds., *The Birth of American Feminism: The Seneca Falls Woman's Convention of 1848* (Saint James, N.Y.: Brandywine, 1995). Garrison was so foolish as to gloat that the split of the society in 1840 revealed a lack of commitment on the part of his opponents. Thus, he asked in *The Liberator* in August 1842: "Where is Theodore D. Weld and his wife?" In fact, Weld was ensconced at a desk in the Library of Congress gathering information on the evils of slavery. He lived across the street from the Capitol at Mrs. Spriggs's boardinghouse, where he supped on vegetables, graham crackers, and milk while the resident congressmen dug into succulent hams and roasts (Risjord, *Representative Americans*, p. 246).

45. Benjamin Quarles, *Black Abolitionists* (New York: Oxford University, 1969), pp. 42–167 ("solemn crisis," p. 45; "man who wears," p. 47; "Church is the Alpha," p. 69; "my friends," p. 53); William Still, *Underground Railroad Records* (Philadelphia, Pa.: W. Still, 1872); Catherine Clinton, *Harriet Tubman: The Road to Freedom* (New York: Little, Brown, 2004); Richard Newman, ed., *Narrative of the Life of Henry Box Brown Written by Himself* (New York: Oxford University, 2002, orig. 1849). In New

York many blacks celebrated July 5, the day that this state's law abolishing slavery went into effect.

46. The following passage adumbrates the chilling tale told by John Stauffer, *The Black Hearts of Men: Radical Abolitionists and the Transformation of Race* (Cambridge, Mass.: Harvard University, 2002).

47. Stauffer, *Black Hearts of Men*, pp. 1–7, 45–133 ("The heart of the whites," p. 1; "ransom", p. 143).

48. William L. Andrews, *The Oxford Frederick Douglass Reader* (New York: Oxford University, 1996), pp. 3–19 ("Representative American," p. 1); Quarles, *Black Abolitionists*, pp. 63–67 ("I appear before," p. 63); Stauffer, *Black Hearts of Men*, pp. 82–86, 110–114, 155–168 ("were sinners," p. 85; "In this your new home," p. 155). See also David W. Blight, ed., *Narrative of the Life of Frederick Douglass* (Boston: St. Martin's, 1993, orig. 1845); John W. Blassingame, ed., *The Frederick Douglass Papers: Speeches, Debates, and Interviews*, 5 vols. (New Haven, Conn.: Yale University, 1979–1992); William S. McFeely, *Frederick Douglass* (New York: Norton, 1991).

49. Stauffer, *Black Hearts of Men*, pp. 88–92, 118–123, 168–174 ("*very short*," p. 123; "The reforms which have swept," p. 236). "The true object" quoted in Charles L. Blockson, *The Underground Railroad* (New York: Berkley, 1989), p. 108.

50. Frank Moore, ed., *Speeches of Andrew Johnson* (Boston, Mass.: Little, Brown, 1866), p. 56.

51. See McDougall, *Freedom Just Around the Corner*, pp. 499–502.

52. The extraordinary coteries of New York in this era are the subject of James T. Callow, *Kindred Spirits: Knickerbocker Writers and American Artists, 1807–1855* (Chapel Hill: University of North Carolina, 1967).

53. See the classic by Barbara Novak, *Nature and Culture: American Landscape and Painting, 1825–1875* (New York: Oxford University, 1980), especially pp. 3–46, 203–274 ("God has promised," p. 15; "great in proportion," p. 9; "Art is in fact," p. 10; "The opportunity," p. 5; "infinitely malleable," p. 17; "something distinctive," p. 273); and Callow, *Kindred Spirits*, pp. 117–143 ("THERE IS A GOD," p. 120; "keep that earlier," p. 65). The themes of America as a Garden of Eden and nature as the face of God were explored earlier by Henry Nash Smith, *Virgin Land: The American West as Symbol and Myth* (Cambridge, Mass.: Harvard University, 1950); Richard W. B. Lewis, *The American Adam: Innocence, Tragedy, and Tradition in the Nineteenth Century* (Chicago, Ill.: University of Chicago, 1961); and Leo Marx, *The Machine in the Garden: Technology and the Pastoral Ideal in America* (New York: Oxford University, 1964).

54. Novak, *Nature and Culture*, pp. 5–7 ("The axe of civilization," p. 5). A fine biography, is Ellwood C. Parry III, *The Art of Thomas Cole* (Newark: University of Delaware, 1988), and a beautiful book of prints with fine commentary is Elizabeth Mankin Kornhauser, *Hudson River School: Masterworks from the Wadsworth Atheneum Museum of Art* (New Haven, Conn.: Yale University, 2003).

55. F. O. Matthiessen, in *American Renaissance: Art and Expression in the Age of Emerson and Whitman* (New York: Oxford University, 1941), says that "one might search all the rest of American literature without being able to collect a group of books equal to these in imaginative vitality" (p. vii); "devotion to the possibilities," p. ix. Marcus Cunliffe, in *The Literature of the United States*, 3rd ed. (Baltimore, Md.: Penguin, 1967), calls it New England's Day. George Brown Tindall, in *America: A Narrative History*,

2nd ed. (New York: Norton, 1988), writes of the American renaissance and the flowering of American literature. As noted at the beginning of this chapter, Morse, in *American
Romanticism*, argues that the defining feature of American culture was its excess.

56. Terence Whalen, *Edgar Allan Poe and the Masses: The Political Economy of Literature in Antebellum America* (Princeton, N.J.: Princeton University, 1999); "whole, energetic,"
"literary commodities," and "saleableness," p. 7; "truth and honor," p. 49; "Thirteen
Egyptian," p. 29. Did Poe fear that his own identity was being sucked out of him by the
need to cater to mass audiences? His tales "The Man of the Crowd" and "William Wilson" suggest so.

57. "Emerson the Man" in Kenneth S. Lynn, *The Air-Line to Seattle: Studies in Literary and Historical Writing about America* (Chicago, Ill.: University of Chicago, 1983), pp.
23–32 ("I have not the affections," p. 26; "ungenerous and selfish," p. 27). Lynn speculated that prior biographers and commentators on Emerson dismissed the evidence of a
cynical marriage because they wanted to see their hero Emerson and by extension themselves as "morally superior beings" (p. 32). He wrote this essay in response to Joel Porte,
Representative Man: Ralph Waldo Emerson in His Time (New York: Oxford University,
1979); and Gay Wilson Allen, *Waldo Emerson* (New York: Viking, 1980).

58. Richard Gravil, *Romantic Dialogues: Anglo-American Continuities, 1776–1862*
(New York: St. Martin's, 2000), pp. 23–68 ("comprehensive raid," p. 59).

59. Emerson will always have devoted admirers. These include Lawrence Buell,
Emerson (Cambridge, Mass.: Harvard University, 2003); and Kenneth S. Sacks, *Under-
standing Emerson:"The American Scholar" and His Struggle for Self-Reliance* (Princeton,
N.J.: Princeton University, 2003). Nor, whatever one thinks of the man or his ideas, can
Emerson's literary influence be denied. By the same token he ceased, decades ago, to be
an object of obeisance. In addition to the critiques by Lynn and Gravil cited above,
Cunliffe, in *Literature of the United States*, pp. 88–97 ("Emerson's oration," p. 92), wrote,
"The case against Emerson may seem overwhelming" and "The case for Emerson is less
easy to state," pp. 94–95. Morse, *American Romanticism*, I: 119–148, described Transcendentalism as "a bundle of attitudes" and "indignant rejection of any kind of restraint"
for which "thought itself is a work of demolition" (p. 119). Robert E. Spiller and Alfred
R. Ferguson, eds., *The Collected Works of Ralph Waldo Emerson*, 2 vols. (Cambridge,
Mass.: Harvard University, 1971), grants the derivative nature of much of Emerson's
thought ("misty, dreamy," I: 50). The Transcendentalist Henry Hedge coined the
phrase "dots of thought" for Emerson's style. The Club also included such notable intellectuals as Theodore Parker, George Ripley, Bronson Alcott, Henry David Thoreau,
Margaret Fuller, Nathaniel Hawthorne, and Orestes Brownson. Hawthorne eventually
became fed up with Transcendentalism. Brownson converted to Roman Catholicism.

60. Fuller edited *The Dial* from 1840 to 1842 but then quit because the Club failed
to come up with her promised salary. Emerson took on the position until 1844, then
folded the journal. Evidently, radical egoism is poor stuff on which to base a club. The
damning insight that Americans believe, or want to believe, they are "masterless" (an
ironic twist implying that they are *bereft* of a master) is elaborated in Wilfred M. Mc-
Clay, *The Masterless: Self and Society in Modern America* (Chapel Hill: University of
North Carolina, 1994). The quotations from Emerson's speeches and essays are in Spiller
and Ferguson, *Collected Works of Ralph Waldo Emerson*: "passionate love," I: xx; "Our age
is retrospective," I: 7; "the palace of eternity," I: 38; "man is a god" and "dwarf of him-

self," I: 42; "kingdom of man," I: 45; "the world is nothing," I: 67; "nation of men," I: 70; "miracle," "monster," I: 81; "moral nature of man," I: 87; "convert life into truth," II: 82; "Faith makes us," I: 92; "We know truth," II: 166; "Beware," II:183. Incidentally, Emerson's essay "Circles" describing the growth of man's knowledge as an ever-increasing array of circles, beginning with the self, was also borrowed from the radical European culture of the day. Utopian socialists, mystical technocrats, and underground revolutionary movements in the 1830s and 1840s formed conspiratorial "circles" (the circle being considered the perfect form) and convened at "centers." Hence the usage so common today, as in "Center for the Study of————." See James H. Billington, *Fire in the Minds of Men: Origins of the Revolutionary Tradition* (New York: Basic Books, 1980).

61. Scholars who know Thoreau far better than I and regard *Walden* as a masterpiece include Walter Harding, *The Days of Henry Thoreau*, rev. ed. (Princeton, N.J.: Princeton University, 1982); and Stanley Cavell, *The Senses of Walden* (San Francisco, Calif.: North Point, 1981). By contrast, Cunliffe, in *Literature of the United States*, pp. 97–103, considers Thoreau's feints into the forest memorable, but absurd; and Morse, in *American Romanticism*, I: 14–59 ("I went to the woods," pp. 154–155) considers Thoreau excessive and—what else?—pretentious.

62. Douglas L. Wilson, *The Genteel Tradition: Nine Essays by George Santayana* (Lincoln: University of Nebraska, 1998), p. 47.

63. Biographies illuminating the influences on Hawthorne's life and literature include Edwin H. Miller, *Salem Is My Dwelling Place: A Life of Nathaniel Hawthorne* (Iowa City: University of Iowa, 1991); T. Walter Herbert, *Dearest Beloved: The Hawthornes and the Making of the Middle Class Family* (Berkeley: University of California, 1993); James R. Mellow, *Nathaniel Hawthorne in His Times* (Boston, Mass.: Houghton Mifflin, 1980); and Rita K. Gollin, *Nathaniel Hawthorne and the Truth of Dreams* (Baton Rouge: Louisiana State University, 1979). See also Frederick C. Crews, *The Sins of the Fathers: Hawthorne's Psychological Themes* (New York: Oxford University, 1966).

64. Morse, *American Romanticism*, I: 169–220 ("Bewitching," p. 197); Newton Arvin, ed., *Hawthorne's Short Stories* (New York: Vintage, 1946), pp. v–xvii.

65. Arvin, ed., *Hawthorne's Short Stories*, pp. 44–59, 210–228; 165–179 ("Welcome, my children," pp. 176–177).

66. The literature on Melville is vast, but the places to start are the exhaustive chronology by Jay Leyda, *The Melville Log: A Documentary Life of Herman Melville, 1819–1891* (New York: Gordian, 1969); and the biography by Hershel Parker, *Herman Melville: A Biography*, 2 vols. (Baltimore, Md.: Johns Hopkins University, 1996).

67. Herman Melville, *Typee: A Peep at Polynesian Life* (New York: Modern Library, 2001); "anxious desire," p. xviii; "picturesque band," "unholy passions," and "grossest licentiousness," p. 15; "insensible," p. 29.

68. Parker, *Herman Melville*, I: 531–533; "sinful propensities," pp. 212–113. Parker notes that original sin was no outmoded theological conceit in Maria Melville's household. Whatever their religious choices later in life, Herman and his siblings learned to take sin seriously.

69. Cunliffe, *Literature of the United States*, pp. 115–128 ("that which is beneath the seeming" and "ontological heroics," p. 121; Morse, *American Romanticism*, II: 11–79; "broiled in hellfire" quoted in Nye, *Society and Culture*, p. 97. To Melville's gratification, *Moby-Dick* was first published in America by Harper Brothers.

70. Matthiessen, *American Renaissance*, p. 656; Ann Douglas, "Herman Melville and the Revolt against the Reader," in *Feminization of American Culture*, pp. 304–326; Lewis Mumford, *Herman Melville* (New York: Harcourt, Brace, 1929), p. 254. Pursuing her theme, Douglas argues that Melville's revolt was against the feminized culture of his time, which repressed true masculinity. He even urged women to shun *Moby-Dick*: "Don't you buy it—don't you read it" (p. 304). Hawthorne was similarly frustrated to note how his book sales paled beside those of sentimental novels written by "scribbling women."

71. Stauffer, *Black Hearts of Men*, p. 66. Carolyn L. Karcher, in *Shadow over the Promised Land: Slavery, Race, and Violence in Melville's America* (Baton Rouge: Louisiana State University, 1980), brilliantly buttresses this interpretation of Melville, not least through her analysis of Melville's novel of 1857, *The Confidence-Man: His Masquerade*, which indicted "a nation that was bringing God's judgment on itself by enslaving and massacring its nonwhite citizens while posing as a religious and political haven" (p. 257). But one need look no farther than Melville's spine-tingling short story "Benito Cereno" (1855). Based on a true incident, it described the encounter of Captain Amasa Delano of Duxbury, Massachusetts, with a derelict coastal slaver off the Chilean coast. Delano (an ancestor of Franklin Delano Roosevelt) boards the ship to discover that Don Benito, the Spanish captain, is somehow presiding alone over an entire cargo of obedient, abject, half-starved slaves. The good Yankee wants to believe the captain's mournful tale of uncanny calms, scurvy, and storms, but can't shake the suspicion that truth lies somewhere else, somewhere eerie and dangerous. I won't give away the ending, but it should already be clear that Delano's ship is the North, and Cereno's the South.

Chapter 5

1. John L. O'Sullivan, "The Great Nation of Futurity," *Democratic Review* 6 (November 1839): 426–430.

2. John L. O'Sullivan, "Annexation," *Democratic Review* 17 (July–August 1845): 5–10 (quotation, p. 5). O'Sullivan's life was as romantic as his prose. Descended from a line of Irish adventurers, he was born on a British warship off Gibraltar because his father—a sea captain—and mother were taken prisoner in the War of 1812. When his father died in service in 1825, the family won a judgment against the U.S. government that helped the young O'Sullivan, by then a graduate of Columbia University, to start his *Democratic Review*. O'Sullivan made his mark promoting the Van Buren wing of the Democratic Party and publishing Hawthorne, Emerson, Thoreau, Bryant, and Walt Whitman . He also served in the New York assembly, where he crusaded for abolition of the death penalty and founded the *New York Morning News* to promote Polk's campaign for president. But his poor management caused him to be dismissed from the *Democratic Review* and the *News* in 1846, and his poor business sense repeatedly bankrupted him. Like Hawthorne, he was able to get by with political appointments, serving as U.S. consul in Portugal from 1854 to 1858, then as a Confederate agent. He returned to America in the 1870s as a peddler of spiritualism, suffered a stroke in 1889, and died an invalid in 1895. He had no biography until the sympathetic treatment by Robert D. Sampson, *John L. O'Sullivan and His Times* (Kent, Ohio.: Kent State University, 2003).

3. Jefferson quoted in Bradford Perkins, *The Cambridge History of American Foreign Relations*, vol. 1, *The Creation of a Republican Empire, 1776–1865* (Cambridge: Cambridge

University, 1993), p. 170; Adams's letter of 1811 to his father in Worthington C. Ford, ed., *The Writings of John Quincy Adams*, 7 vols. (New York: Macmillan, 1913–1917), IV: 209; Douglas quoted in Harry Jaffa, *Crisis of the House Divided* (Seattle: University of Washington, 1973), p. 406; Lincoln's speech in Roy P. Basler, ed., *The Collected Works of Abraham Lincoln*, 9 vols. (New Brunswick, N.J.: Rutgers University, 1953), I: 109. This section adumbrates the discussion of Manifest Destiny in Walter A. McDougall, *Promised Land, Crusader State: The American Encounter with the World since 1776* (Boston, Mass.: Houghton Mifflin, 1997), pp. 76–90.

4. Ideologies of expansion are analyzed at length in Albert K. Weinberg, *Manifest Destiny: A Study of Nationalist Expansion in American History* (Baltimore, Md.: Johns Hopkins University, 1935); and Frederick Merk, *Manifest Destiny and Mission in American History* (New York: Vintage, 1966). The critique of expansionism as racist is made by Richard Drinnon, *The Metaphysics of Indian-Hating and Empire-Building* (Minneapolis: University of Minnesota, 1980); and Reginald Horsman, *Race and Manifest Destiny: The Origins of American Racial Anglo-Saxonism* (Cambridge, Mass.: Harvard University, 1981). Phrenology, the study of skull sizes and shapes in search of clues to racial characteristics, was pioneered in America by George Calvert's treatise of 1832 and culminated in Samuel George Morton, *Crania Americana* (Philadelphia, Pa.: J. Dobson, 1839); and Josiah C. Nott and George R. Glidden, *Types of Mankind* (Philadelphia, Pa.: Lippincott, 1854). Morton concluded that the intellectual faculties of American Indians were of "a decidedly inferior cast when compared with types of the Caucasian and Mongolian races" (Horsman, *Race and Manifest Destiny*, p. 127).

5. Thomas R. Hietala, in *Manifest Design: Anxious Aggrandizement in Late Jacksonian America* (Ithaca, N.Y.: Cornell University, 1985), challenged the myth that the spontaneous spread of American pioneers dragged the government into expansionism. But that came as no surprise to students of the Polk administration or to diplomatic historians in the realist tradition such as Norman A. Graebner, *Foundations of American Foreign Policy: A Realist Appraisal from Franklin to McKinley* (Wilmington, Del.: Scholarly Resources, 1985). Quotation from Benton is from his *Thirty Years' View; or, a History of the Working of the American Government for Thirty Years, from 1820 to 1850*, 2 vols. (New York: Appleton, 1854–1856), I: 468–469.

6. William B. Skelton, "The Army in the Age of the Common Man, 1815–1845, in Kenneth J. Hagan and William R. Roberts, eds. *Against All Enemies: Interpretations of American Military History from Colonial Times to the Present* (Westport, Conn.: Greenwood 1986), pp. 91–112 ("an art," p. 92); William B. Skelton, *An American Profession of Arms: The Army Officer Corps, 1784–1861* (Lawrence: University of Kansas, 1992); Charles M. Wiltse, *John C. Calhoun, Nationalist, 1782–1828* (Indianapolis, Ind.: Bobbs-Merrill, 1944). The origins of the American system of manufactures are summarized in Walter A. McDougall, *Freedom Just Around the Corner: A New American History 1585–1828* (New York: HarperCollins, 2004), pp. 488–489.

7. John S. D. Eisenhower, *Agent of Destiny: The Life and Times of General Winfield Scott* (New York: Free Press, 1997); Timothy D. Johnson, *Winfield Scott: The Quest for Military Glory* (Lawrence: University of Kansas, 1998); Allan Peskin, *Winfield Scott and the Profession of Arms* (Kent, Ohio.: Kent State University, 2003), "cursed," p. 67.

8. Forrest G. Hill, in *Roads, Rails, and Waterways: The Army Engineers and Early Transportation* (Norman: University of Oklahoma, 1957), pp. 209–210, calculated that from its inception to 1867, West Point graduated 2,218 men. The 139 who made careers in

education included twenty-six college presidents and eighty-five professors. The 334 who pursued engineering and technical careers included thirty-five railroad presidents and 155 prominent civil engineers. Merritt Roe Smith, "New England Industry and the Federal Government," in Peter Temin, ed., *Engines of Enterprise: An Economic History of New England* (Cambridge, Mass.: Harvard University, 2000), p. 168, concluded: "Given these impressive numbers, it is unlikely that any other entity in America could claim to have made a better investment in human capital than the federal government did with the U.S. military academy at West Point."

9. See Theodore J. Crackel, *West Point: A Bicentennial History* (Lawrence: University of Kansas, 2002); George S. Pappas, *To the Point: The United States Military Academy, 1802–1902* (Westport, Conn.: Praeger, 1993); James L. Morrison, Jr., *The Best School in the World: West Point, the Pre–Civil War Years* (Kent, Ohio.: Kent State University, 1986); Joseph Ellis and Robert Moore, *School for Soldiers: West Point and the Profession of Arms* (New York: Oxford University, 1974). On the officer corps generally, see Marcus Cunliffe, *Soldiers and Civilians: The Martial Spirit in America, 1775–1865* (Boston, Mass.: Little, Brown, 1968), "best school," p. 151; Samuel P. Huntington, *The Soldier and the State: The Theory and Politics of Civil-Military Relations* (Cambridge, Mass.: Harvard University, 1957); Morris Janowitz, *The Professional Soldier: A Social and Political Portrait* (Glencoe, Ill.: Free Press, 1960); Paul A. C. Koistenen, *Beating Plowshares into Swords: The Political Economy of American Warfare, 1606–1865* (Lawrence: University of Kansas, 1996); Russell F. Weigley, *The American Way of War: A History of United States Military Strategy and Policy* (New York: Macmillan, 1973); Allan R. Millett and Peter Maslowski, *For the Common Defense: A Military History of the United States of America* (New York: Free Press, 1984); and Skelton, "The Army in the Age of the Common Man," ("insane attachment" and "I verily believe," p. 98). Edward M. Coffman, in *The Old Army: A Portrait of the American Army in Peacetime, 1784–1898* (New York: Oxford University, 1986), eloquently describes the suffering and frustration of Army wives.

10. On the frontier army, see the following. Francis Paul Prucha, *Broadax and Bayonet: The Role of the United States Army in the Development of the Northwest, 1815–1860* (Lincoln: University of Nebraska, 1953). Francis Paul Prucha, *The Sword of the Republic: The United States Army on the Frontier, 1783–1846* (New York: Macmillan, 1969), "like productive corn," p. 332; "whose imaginations," p. 246). William H. Goetzmann, *Army Exploration in the American West, 1803–1863* (New Haven, Conn.: Yale University, 1959). Michael L. Tate, *The Frontier Army in the West* (Norman: University of Oklahoma, 1999). Biographies include James W. Silver, *Edmund Pendleton Gaines: Frontier General* (Baton Rouge, Louisiana State University, 1949); and Dwight L. Clarke, *Stephen Watts Kearny, Soldier of the West* (Norman: University of Oklahoma, 1961).

11. Goetzmann, *Army Exploration* and *Exploration and Empire: The Explorer and the Scientist in the Winning of the American West* (New York: Knopf, 1967). Army explorers were ably seconded by civilian naturalists, the greatest of whom was surely George Catlin. Born in Wilkes-Barre in 1796, he spent his youth "with books reluctantly held in one hand and a rifle or fishing pole firmly and affectionately in the other." Catlin attended Tapping Reeve's law school but gave up his practice to paint. When a delegation of Indian chiefs passed through Philadelphia in "silent and stoic dignity" he decided that his calling was to capture the Indians on canvas as they were "rapidly passing away." He went west in 1830 and after nine years recorded: "I have visited forty-eight different

tribes . . . containing in all 400,000 souls. I have brought home safe, and in good order, 310 portraits in oil, all painted in their native dress, and in their own wigwams; and also 200 other paintings in oil, containing views of their villages . . . and the landscapes of the country they live in." Catlin was delighted to "have viewed man in the innocent simplicity of nature," but he lamented to observe the Indian "shrinking from civilized approach, which came with all its vices, like the dead of night, upon him." Catlin traveled to Europe in 1839 and remained abroad thirty-one years. At the end he took up residence at the Smithsonian, where his collections reside. He died in 1872 (*Exploration and Empire*, pp. 185–189).

12. On Frémont see, in addition to Goetzmann, *Army Exploration*, pp. 65–108, the classic biography by Allan Nevins, *Frémont: Pathmarker of the West* (New York: Longman, 1955); the biography by Tom Chaffin, *John Charles Frémont and the Course of American Empire* (New York: Hill and Wang, 2002); and John Charles Frémont, *Memoirs of My Life* (Chicago; Ill.: Belford, Clarke, 1887) ("my mind had been quick," p. 65).

13. David Dary, *Entrepreneurs of the Old West* (New York: Knopf, 1986), pp. 16–41, 68–88 ("grateful evidence," p. 18); Larry M. Beachum, *William Becknell: Father of the Santa Fe Trade* (El Paso: Texas Western University, 1982); David Lavender, *Bent's Fort* (Garden City, N.Y.: Doubleday, 1954). Josiah Gregg's book *Commerce of the Prairies* (New York: Laug Toy, 1844) documented the Santa Fe trade up to 1843, when deterioration of relations between the United States and Mexico caused Mexican authorities to close their borders. The expanding U.S. military presence also spoiled Bent's business by provoking resistance from the Indians. He allegedly offered to sell his fort to the Army for $15,000 and angrily burned it down when the government offered him just $12,000. By then his firm had removed its main fur trade operations to Fort Saint Vrain on the South Platte River.

14. Dary, *Entrepreneurs of the Old West*, pp. 52–67 ("Come. We are done," pp. 65–66); Richard M. Clokey, *William H. Ashley: Enterprise and Politics in the Trans-Mississippi West* (Norman: University of Oklahoma, 1980); and the classic by Hiram M. Chittenden, *The American Fur Trade of the Far West* (New York: Barnes and Noble, 1935, orig. 1902). William H. Goetzmann, *Exploration and Empire: The Explorer and the Scientist in the Winning of the American West* (New York: Knopf, 1967), pp. 105–145, tells lovingly of the mountain men.

15. On the diplomacy of the Pacific Northwest, see Warren L. Cook, *Flood Tide of Empire: Spain and the Pacific Northwest 1543–1819* (New Haven, Conn.: Yale University, 1973); Howard I. Kushner, *Conflict on the Northwest Coast: American-Russian Rivalry in the Pacific Northwest, 1790–1867* (Westport, Conn.: Greenwood, 1975); Walter LaFeber, *John Quincy Adams and American Continental Empire* (Chicago, Ill.: University of Chicago, 1965); or the relevant chapters of Walter A. McDougall, *Let the Sea Make a Noise: A History of the North Pacific from Magellan to MacArthur* (New York: HarperCollins, 2004, orig. 1993). On the British presence in the Northwest, see E. E. Rich, *Hudson's Bay Company, 1670–1870*, 2 vols. (London: Hudson's Bay Record Society, 1959), "such a figure," 2: 577; G. P. V. Akrigg and Helen Akrigg, *British Columbia Chronicle, 1778–1846: Adventures by Sea and Land* (Vancouver: Discovery, 1975), "Here Before Christ," p. 180; Dary, *Entrepreneurs of the Old West*, pp. 89–109 ("poor blind people," p. 90).

16. Descriptions of Carson in Frémont, *Memoirs*, pp. 74, 82; "disposed to cowardice"

in Frémont, *Narratives of Exploration and Adventure* (New York: Longmans, Green, 1956), p. 61; "I was no longer" in Joaquin Miller, *Overland in a Covered Wagon* (1930), pp. 42–43, cited by Goetzmann, *Army Exploration*, p. 84; "were plundered" in Mason Wade, ed., *The Journals of Francis Parkman* (New York: Harper, 1947), p. 440; "Let the emigrants" cited in McDougall, *Let the Sea Make a Noise*, p. 224.

17. Needless to say, the Donner party, which attempted to reach California three years later, suffered a different fate. Misled by Lansford Hastings's *The Emigrant's Guide to Oregon and California* (1845), the pioneers followed "Hastings' Cutoff" south of Great Salt Lake, thereby losing precious weeks. They attempted the High Sierra late in the season and were trapped by a series of blizzards. Of the eighty-seven men, women, and children, just forty survived.

18. Frémont's status as a national hero suffered in the twentieth century. Historians could find little evidence that the intelligence he gathered on his three expeditions directly influenced U.S. foreign policy, although it certainly made a big impact on public opinion. His scientific and cartographic findings were sometimes inaccurate and rarely original. Needless to say, the romantic manifest destiny he personified has been roundly disparaged by historians sympathetic to the victims of U.S. expansion, including Indians, Mexicans, and the environment itself. Goetzmann concludes nonetheless that the sheer mass of Frémont's geographical, botanical, meteorological, and ethnographic data "added a cubit to the national stature" (*Army Exploration*, p. 108), and that his journals and maps helped countless pioneers reach the Pacific coast safely.

19. The fate of Native Americans under Spanish rule, particularly in the Franciscan missions, became a controversial subject in the late twentieth century, when the Catholic church considered the canonization of Father Serra. A thorough and scholarly treatment is Steven W. Hackel, *Children of Coyote, Missionaries of Saint Francis: Indian-Spanish Relations in Colonial California, 1769–1850* (Chapel Hill: University of North Carolina, 2005).

20. Larkin was a real piece of work, according to Harlan Hague and David J. Langum, *Thomas O. Larkin: A Life of Patriotism and Profit in Old California* (Norman: University of Oklahoma, 1990). Born on the Charles River in 1802, he grew into a meticulous, driven young man (we would say "anal") who saved his every receipt and recorded his every egocentric desire. Spurning religion, Larkin lived only to chase money and exploit the opposite sex. But his business ventures failed and his dalliances with married women forced him to start over in North Carolina and then, in 1831, California. "All love and no capital will never do for me," he confessed. "If I can make a fortune by marriage, I will" (p. 32). Instead, Larkin impregnated Rachel Holmes, a Massachusetts captain's wife, and married her after the husband conveniently died. (Rachel was the first American female resident in California.) But unfortunately she was not rich, so Larkin threw himself single-mindedly into business.

21. Dary, *Entrepreneurs of the Old West*, pp. 110–119; Richard Henry Dana, *Two Years Before the Mast* (New York: Collier, 1909, orig. 1841), pp. 81–88; Reuben L. Underhill, *From Cowhides to Golden Fleece* (Stanford, Calif.: Stanford University, 1939), pp. 19–28; Robert Glass Cleland, *From Wilderness to Empire: A History of California, 1542–1900* (New York: Knopf, 1944); David J. Weber, *The Mexican Frontier, 1821–1846: The American Southwest under Mexico* (Albuquerque: University of New Mexico, 1982). "Is there nothing but Yankees here" from G. P. Hammond, ed., *The Larkin Papers*, 10 vols.

(Berkeley: University of California, 1951–64), 4: 369, cited in John A. Hawgood, "The Pattern of Yankee Infiltration in Mexican Alta California, 1821–1846," *Pacific Historical Review* 27 (February 1958): 27.

22. John H. Schroeder, *Shaping a Maritime Empire: The Commercial and Diplomatic Role of the American Navy, 1829–1861* (Westport, Conn.: Greenwood, 1985), pp. 3–82 ("usually selected," p. 4; "most rickety," a quotation from Commander Samuel Francis Du Pont, p. 12); Harold and Margaret Sprout, *The Rise of American Naval Power 1776–1918* (Annapolis, Md.: Naval Institute, 1966), pp. 119–158; James E. Valle, *Rocks and Shoals: Order and Discipline in the Old Navy 1800–1861* (Annapolis, Md.: Naval Institute, 1980), pp. 1–101. Tocqueville quotation from *Democracy in America*, 2 vols. (New York: Vintage, 1945), I: 447. Paul A. Gilje, *Liberty on the Waterfront: American Maritime Culture in the Age of Revolution* (Philadelphia: University of Pennsylvania, 2004), describes the ugly lives of sailors during the years when the nation's image of Jack Tar was shaped by chanteys like "What Do You Do with a Drunken Sailor" or "O Whiskey Is the Life of Man," and Dana's aphorism: "A sailor's liberty is but for a day; yet while it lasts it is perfect" (p. 5).

23. The classic Hawaiian history is Ralph S. Kuykendall, *The Hawaiian Kingdom*, 3 vols. (Honolulu: University of Hawaii, 1947). A very readable scholarly account is Gavan Daws, *Shoal of Time: A History of the Hawaiian Islands* (New York: Macmillan, 1968). See also Harold Whitman Bradley, *The American Frontier in Hawaii: The Pioneers, 1789–1843* (Stanford, Calif.: Stanford University, 1942); and McDougall, *Let the Sea Make a Noise*, pp. 82–89, 103–111, 152–159, 211–218, for the years up to 1850. Hiram Bingham, *A Residence of Twenty-One Years in the Sandwich Islands* (Hartford, Conn.: Hezekiah Huntington, 1847), reveals in rich detail the mentality of nineteenth-century American missionaries. His devotion, humanitarianism, cultural arrogance, and powers of self-deception have characterized American state- and nation-building abroad ever since.

24. The origin of precious gray ambergris was not aromatic at all. Found in the intestines and rectums of sperm whales, it was thought at first to be dried feces. But whalers noticed that beasts whose oil yields were poor often had clusters of giant squid beaks in their guts. The indigestible beaks blocked digestion, limiting the intake of food. Naturalists deduced that the ambergris was a secretion that protected the whale's intestines from abrasion. So precious was ambergris that it sold for five to ten pounds sterling per ounce by the late 1800s.

25. Granville Allen Mawer, *Ahab's Trade: The Saga of South Sea Whaling* (New York: St. Martin's, 1997), pp. 1–35, 77–178 ("New England enterprise," p. 258; "either from shame," p. 110); Edouard A. Stackpole, *Whales and Destiny: The Rivalry between America, France, and Britain for Control of the Southern Whale Fishery, 1785–1825* (Amherst: University of Massachusetts, 1972); K. Jack Bauer, *A Maritime History of the United States* (Columbia: University of South Carolina, 1988), pp. 228–238. An international moratorium halted pelagic whaling in 1986 out of humanitarian concerns and fear that sperm whales had become an endangered species. However, Mawer, in *Ahab's Trade*, pp. 341–347, cites statistics showing that even at its peak the whaling industry reduced global sperm whale herds by no more than one-fifth of 1 percent, and that whales are as numerous now as they ever have been.

26. Schroeder, *Shaping a Maritime Empire*, pp. 1–55 ("haphazard", p. 7; "our best security," p. 35; "all our misfortunes," p. 43). Biographies of the dynamic young officers

include Samuel Eliot Morison, *"Old Bruin": Commodore Matthew C. Perry, 1794–1858* (Boston, Mass.: Little, Brown, 1967); John H. Schroeder, *Matthew Calbraith Perry: Antebellum Sailor and Diplomat* (Annapolis, Md.: Naval Institute, 2001); Gene A. Smith, *Thomas ap Catesby Jones: Commodore of Manifest Destiny* (Annapolis, Md.: Naval Institute, 2000); Frances Leigh Williams, *Matthew Fontaine Maury: Scientist of the Sea* (New Brunswick, N.J.: Rutgers University, 1963). On Jackson's "gunboat diplomacy," see John M. Belohlavek, *Let the Eagle Soar! The Foreign Policy of Andrew Jackson* (Lincoln: University of Nebraska, 1985); and Robert Erwin Johnson, *Far China Station: The United States Navy in Asian Waters, 1800–1898* (Annapolis, Md.: Naval Institute, 1979). The *Potomac's* Captain John Downes ordered a landing party of 250 marines and sailors to assault Quallah Batoo in Sumatra when the pirates refused to make good the $41,054 in cargo they had stolen from the American merchantman. After a morning of intense combat the Americans took the fort, burned the town, and killed over 100 Malays at the cost of just two of their own. As for reparations and indemnities, they came home with none. That, too, has been a characteristic of U.S. armed intervention abroad ever since.

27. Wilkes was born into a well-to-do merchant family in New York in 1798. His mother died when he was two years old, so he was put in the care of the humanitarian Elizabeth Ann Seton. Herself the widow of a bankrupt merchant, she converted to Roman Catholicism and founded nurseries, schools, and orphanages for children in need. In 1975 Seton was canonized the first saint born in the United States. Wilkes grew up with dreams of becoming America's Captain Cook. He learned his trade on merchant vessels, won appointment as a midshipman under Captain William Bainbridge of the USS *Independence*, and realized his dreams when Jones laid down command of the "Ex. Ex."

28. Schroeder, *Shaping a Maritime Empire*, pp. 37–78 ("I am steamed," p. 42; quotations from southerners, pp. 57–64). On the dynamic secretary of the navy, see Claude H. Hall, *Abel Parker Upshur: Conservative Virginian, 1790–1844* (Madison: Wisconsin Historical Society, 1964).

29. Wilkes required his officers to keep daily journals. Needless to say, they were not candid. But William Reynolds also wrote a private diary that reached 250,000 words. Two juxtaposed passages convey the schizophrenic character of the "Ex. Ex." After calling at Sydney in New South Wales, Reynolds recorded: "There was much difficulty between Captain Wilkes & the Officers . . . [He] made himself despised and contemptible; finally said he was glad the difficulty had occurred, for he was not dependent on us, &c, &c, &c. The foul scamp—he should be hung. I have not patience to speak of him with decency, neither does he deserve I should. Mercy, how he is hated. *Four sycophants cling to him*—the rest abhor him!" Immediately following this is a sublime account of the flotilla's search for an Antarctic continent: "I shall never forget the first Ice berg we met with & passed close by. There is no such thing as describing its appearance. I cannot tell either of the feelings excited or of the wondrous beauty of the floating mass—of the wildest forms, & weathered into the most fanciful shapes—& cold & Icy as they were, glowing with the most vivid & brilliant hues; blue as azure, green as emerald, and oh!, the contrast, whiteness like unto the raiment of an Angel. We all came on deck, and we all gazed until our very vision ached": Nathaniel and Thomas Philbrick, eds., *The Private Journal of William Reynolds* (New York: Penguin, 2004), pp. 124–125.

30. On the "Ex. Ex.," see Nathaniel Philbrick, *Sea of Glory: America's Voyage of Discovery, the U.S. Exploring Expedition, 1838–1842* (New York: Viking, 2003); Reynolds, *Private Journal*; William Stanton, *The Great United States Exploring Expedition of 1838–1842* (Berkeley: University of California, 1975); and Charles Wilkes, *Narrative of the United States Exploring Expedition*, 5 vols. (Philadelphia, Pa.: Lea and Blanchard, 1845).

31. Dan Monroe, *The Republican Vision of John Tyler* (College Station: Texas A&M University, 2003), "To my latest breath," p. 47; Drew McCoy, *The Elusive Republic: Political Economy in Jeffersonian America* (Chapel Hill: University of North Carolina, 1980); Oliver P. Chitwood, *John Tyler: Champion of the Old South* (New York: Appleton-Century, 1939); Robert Seager, *"And Tyler Too": A Biography of John and Julia Gardiner Tyler* (New York: McGraw-Hill, 1963); Norma Lois Peterson, *The Presidencies of William Henry Harrison and John Tyler* (Lawrence: University of Kansas, 1989); Robert J. Morgan, *A Whig Embattled: The Presidency under John Tyler* (Lincoln: University of Nebraska, 1954). Jackson also warmed to Tyler when Letitia Christian Tyler, in 1842, became the first First Lady to die in the White House. Two years later the fifty-four-year-old Tyler became the first sitting president to marry in office. His bride, a lovely New York socialite named Julia Gardiner, was twenty years younger, a fact that inspired many a chortle about Tyler's stamina. In fact, John and Julia had five children to go along with the two from his previous marriage. No president has sired a larger brood. The Tylers were also renowned for lavish White House entertainments and for popularizing the waltz in America.

32. Francis M. Carroll, *A Good and Wise Measure: The Search for the Canadian-American Boundary, 1783–1842* (Toronto: University of Toronto, 2001), illuminates the history of the exploration and botched surveys as well as the diplomacy. See also John F. Sprague, *The Northeastern Boundary Controversy and the Aroostook War* (Dover, Me.: Observer, 1910). "Britannia shall not rule the Maine," sang the loggers, "nor shall she rule the water; they've sung that song full long enough, much longer than they oughter" (pp. 110–111). "Mr. President" quoted from Eisenhower, *Agent of Destiny*, p. 197.

33. Carroll, *A Good and Wise Measure*, pp. 243–306; Peterson, *Presidencies of Harrison and Tyler*, pp. 113–131 ("new mode," p. 119). The federal treasury compensated Maine and Massachusetts with $150,000 each for their losses of territory and private property. Benton, an advocate of states' rights in other matters, insisted that those states' endorsements of the treaty were irrelevant, since international boundaries were a federal affair. The mineral deposits of Minnesota's Wesabi Range had been known since the original French explorations of the *pays d'en haut*, but nobody guessed their extent. The Webster-Ashburton Treaty also liquidated claims arising from the *Caroline* affair and subsequent *Creole* affair in which raiders from the American side destroyed a Canadian steamer.

34. Tyler's annual message of 1842 is in James D. Richardson, ed., *A Compilation of the Messages and Papers of the Presidents, 1789–1902*, 10 vols. (Washington, D.C.: Bureau of National Literature, 1903), 4: 211–214. There is much more to say about the Hawaiian affair. When Kamehameha III ascended the throne, he attempted a pagan revival, only to be reined in by his own chiefs, because of either their Christian devotion or their resentment of royal prerogative. Many Hawaiians returned to the old gods, however, until a new revival was initiated in 1837 by Titus Coan, a disciple of Charles Finney. His emphasis on the Holy Spirit instead of the dry catechism pressed by Bingham won thousands of Hawaiians. Coan baptized 3,200 himself and was for a time pastor of the largest

Protestant church in the world. The bizarre story of Richard Charlton, the British consul, is told from the documents of Britain's Public Record Office in McDougall, *Let the Sea Make a Noise*, pp. 211–218. Melville was an eyewitness to Lord George Paulet's takeover. In an appendix to *Typee* he defended the British captain's "by the book" rule of law as a civilized respite from the priggishness and caprice of the missionary-influenced native regime.

35. In international law most-favored-nation status means any privileges granted to some other nation must also be granted to you; hence U.S. merchants operated in China on the same footing as Britain's. "Exterritoriality" meant privileged foreigners were not subject to Chinese law. If accused of malfeasance they were remanded to their own nation's authorities for trial. Europeans and Americans demanded this privilege because the Chinese practiced torture and recognized no legal rights such as habeas corpus or trial by jury. See Carroll Storrs Alden, *Lawrence Kearny, Sailor Diplomat* (Princeton, N.J.: Princeton University, 1936); Claude Moore Fuess, *The Life of Caleb Cushing*, 2 vols. (New York: Harcourt, Brace, 1923); John King Fairbank, *Trade and Diplomacy on the China Coast: The Opening of the Treaty Ports, 1842–1854* (Cambridge, Mass.: Harvard University, 1953); and Frederic Wakeman, Jr., "The Canton Trade and the Opium War," *The Cambridge History of China* (Cambridge: Cambridge University, 1978).

36. The classic monographs on war and diplomacy in the Republic of Texas are Joseph William Schmitz, *Texan Statecraft, 1836–1845* (San Antonio, Tex.: Naylor, 1941); Stanley Siegel, *A Political History of the Texas Republic* (Austin: University of Texas, 1956) and *The Poet President of Texas: The Life of Mirabeau B. Lamar* (Austin, Tex.: Pemberton, 1977); William C. Binkley, *The Expansionist Movement in Texas, 1836–1850* (Berkeley: University of California, 1925); Ephraim Douglass Adams, *British Interests and Activities in Texas, 1838–1846* (Baltimore, Md.: Johns Hopkins University, 1910); and biographies of Houston including James L. Haley, *Sam Houston* (Norman: University of Oklahoma, 2002), pp. 209–291 ("Talleyrand of the Brazos," p. 264), and J. Ralph B. Campbell, *Sam Houston and the American Southwest* (New York: Addison Wesley, 2002), pp. 107–128 ("the only man," p. 115, was the judgment of James Morgan: "Old Sam H. with all his faults appears to be the only man for Texas—He is still unsteady, intemperate, but drunk in a ditch he is worth a thousand of Lamar and Burnet").

37. Peterson, in *Presidencies of Harrison and Tyler*, pp. 185–259, sorts out the complex correspondence ("depended upon," p. 187). Aberdeen was obliged to answer an interpellation by Henry Peter, Lord Brougham, the fiery (some said eccentric) architect of the nineteenth-century Liberal Party. Brougham believed that abolition in Texas would lead to abolition across the American South, something Aberdeen dared not endorse.

38. Monroe, *Republican Vision*, pp. 156–179 ("the great measure," p. 169). Tyler later confessed that he "regretted" Calhoun's linkage of Texas to slavery. He also later confessed that one of his major motives for wanting Texas was to preserve the near monopoly the United States enjoyed over production of cotton. It was, he thought, a more powerful source of leverage against Britain than armies and fleets. As for Calhoun, contemporaries and historians speculated at length why he damaged, perhaps sabotaged, a treaty whose purpose he surely avowed. Perhaps he hoped to force a breakup of the Union over the issue; to force Van Buren to take an unpopular stance against Texas and thus promote his own bid for the Democratic nomination; to force southern Whigs to abandon their party and join a unified sectional party led by himself; or just to force northerners and the British alike to accept once and for all the permanence of slavery. As was so often

the case with Calhoun, he achieved only negative effects. See the discussions in John Niven, *John C. Calhoun and the Price of Union* (Baton Rouge: Louisiana State University, 1988), pp. 264–282; Irving H. Bartlett, *John C. Calhoun* (New York: Norton, 1993), pp. 306–318; Peterson, *Presidencies of Harrison and Tyler*, pp. 211–228; Frederick Merk, *Slavery and the Annexation of Texas* (New York: Knopf, 1972), pp. 68–69.

39. Thomas M. Leonard, *James K. Polk: A Clear and Unquestionable Destiny* (Wilmington, Del.: Scholarly Resources, 2001), pp. 23–41 ("the most available," p. 37; "reoccupation," p. 38). There is evidence that Clay and Van Buren had a gentlemen's agreement as early as 1842 not to make a campaign issue of Texas: James C. N. Paul, *Rift in the Democracy* (Philadelphia: University of Pennsylvania, 1951), pp. 87–88.

40. A detailed and sensitive treatment of Polk's prepresidential biography is Charles Grier Sellers, Jr., *James K. Polk: Jacksonian 1795–1843* (Princeton, N.J.: Princeton University, 1957). See also Leonard, *James K. Polk*, pp. 1–85; and Paul H. Bergeron, *The Presidency of James K. Polk* (Lawrence: University of Kansas, 1987), pp. 1–64.

41. After defeat of the treaty Houston loudly announced that the Republic of Texas was "free of all involvements and pledges" and would pursue its national interest as it saw fit: Campbell, *Sam Houston*, p. 135; "Mr. Tyler's abominable" in Peterson, *Presidencies of Harrison and Tyler*, p. 228. The motives of senators opposed to the treaty were varied. That annexation meant an extension of slavery was enough to repel a number of northerners. The fear of war against Mexico (possibly backed by Britain), opposition to Tyler or Calhoun, party discipline, and posturing for the upcoming election swayed other senators. In sum, it was not what Congressmen call an "easy vote."

42. Glyndon G. Van Deusen, in *The Life of Henry Clay* (Boston, Mass.: Little, Brown, 1937), pp. 374–375, wryly described Clay's position as "consistent throughout—a consistent straddle." Depictions of Polk as an "ultra slaveholder" and Clay as a reprobate quoted in Charles Grier Sellers, Jr., "Election of 1844," in Arthur M. Schlesinger, Jr., ed., *History of American Presidential Elections, 1789–1968*, 4 vols. (New York: Chelsea House, 1971), I: 1785–1791.

43. Doggett told of an old sow who went to sleep on a few grains of corn: "before morning the corn shot up, and the percussion killed her dead." Melville would write a serious allegory about Ahab's obsession with a huge white whale. Thorpe anticipated him with a hilarious allegory about Doggett's obsession with a huge black b'ar. At last Doggett shoots the critter, but the bear "gave a yell, and walked through the fence like a falling tree would through a cobweb. I started after, but was tripped up by my inexpressibles, which either from habit, or the excitement of the moment, were about my heels, and before I had really gathered myself up, I heard the old varmint groaning in a thicket nearby, like a thousand sinners, and by the time I reached him he was a corpse." See the wonderful study by Kenneth S. Lynn, *Mark Twain and Southwestern Humor* (Boston, Mass.: Little, Brown, 1959), pp. 3–139 ("corruption and filth," p. 62; "was just five feet," p. 70; "Big Bear" story, pp. 93–96).

44. Stephen John Hartnett, in *Democratic Dissent and the Cultural Fictions of Antebellum America* (Urbana: University of Illinois, 2002), pp. 93–131, analyzes the subtle rhetorical strategies of "The South in Danger." Needless to say, they involved brazen pretense.

45. Arthur Charles Cole, *The Whig Party in the South* (Washington, D.C.: American Historical Association, 1913), pp. 64–118 ("aversion," p. 40; "For the present," p. 115); Daniel Walker Howe, *The Political Culture of the American Whigs* (Chicago, Ill.: University of Chicago, 1979), pp. 238–262 ("The annexation project," p. 242); Robert V. Remini,

Henry Clay: Statesman for the Union (New York: Norton, 1991), pp. 600–667 ("ugly" and "He wires in," p. 661). Martha McBridge Morrell, in *"Young Hickory": The Life and Times of President James K. Polk* (New York: Dutton, 1949), p. 211, tells a good story. When Clay's son informed him at his Ashland estate of Polk's nomination, Clay reached for the whiskey and sighed, "Beat again, by God!" Remini, in *Henry Clay*, p. 647, says that this story is not only apocryphal but bowdlerized, since it should have read, "Beat again, by hell!"

46. The ominous implications of Tyler's congressional coup have been forcefully and most recently emphasized by Joel H. Silbey, *Storm over Texas: The Annexation Controversy and the Road to Civil War* (New York: Oxford University, 2005). The three Whig senators who sustained the Walker amendment were W. D. Merrick of Maryland, John Henderson of Mississippi, and Henry Johnson of Louisiana. On the origins of the joint resolution, see Justin Harvey Smith, *The Annexation of Texas* (New York: Baker and Taylor, 1911), pp. 281–288; David M. Pletcher, *The Diplomacy of Annexation: Texas, Oregon, and the Mexican War* (Columbia: University of Missouri, 1973), pp. 144–185; and Charles Grier Sellers, Jr., *James K. Polk: Continentalist, 1843–1846* (Princeton, N.J.: Princeton University, 1966), pp. 205–220 ("war of desolation," p. 179). Adams's "apoplexy" cited by Peterson, *Presidencies of Harrison and Tyler*, p. 257. John Tyler left the White House satisfied and content. He returned to the Democratic Party in Virginia and voted for secession in 1861. His son, Lyon Gardiner Tyler, became president of the College of William and Mary and a vitriolic historical critic of Abraham Lincoln.

47. Charleton W. Tebeau, *A History of Florida* (Coral Gables: University of Miami, 1971), pp. 19–115. Since Ponce de Léon was drawn in part by rumors of a fountain of youth, it was fitting that his arrival in 1513 fell during the feast of the Resurrection. *La Pascua Florida* means flowery Easter. An estimated 25,000 Native Americans lived all over the peninsula around 1500, and the missionaries claimed over 13,000 converts: Michael V. Gannon, *The Cross in the Sand* (Gainesville: University of Florida, 1965). But the usual die-offs rendered the indigenous tribes so feeble that the Seminole invaders destroyed or absorbed them. A legend traces *cracker*, the disparaging term for poor whites in Florida and neighboring states, to *cuáqueso*, Spanish for Quaker and thus a pejorative word to Roman Catholics. More likely, it had to do with whip-cracking herders, or with cracking kernels for corn bread ("Jimmy crack corn"), or with cracking jokes or boasting. See the classic work on Florida's popular culture by Stetson Kennedy, *Palmetto Country* (Tallahassee: Florida State University, 1989, orig. 1942), p. 59; and a definitive scholarly work by Grady Whiney, *Cracker Culture: Celtic Ways in the Old South* (Tuscaloosa: University of Alabama, 1988). On Florida's lawlessness under American rule, see James M. Denham, *"A Rogue's Paradise": Crime and Punishment in Antebellum Florida, 1821–1861* (Tuscaloosa: University of Alabama, 1997 ("ingenious rascality," p. 1).

48. On the Adams-Onís (or Transcontinental) Treaty of 1819, see McDougall, *Freedom Just Around the Corner*, pp. 458–459; "command of the gulph" in *Niles' Weekly Register* (March 3, 1819), p. 44.

49. Sidney Walter Martin, *Florida during the Territorial Days* (Athens: University of Georgia, 1944), pp. 53–197 ("create petty," p. 156; "full of filth," p. 64); Tebeau, *History of Florida*, pp. 117–150; "grotesque place" cited by Kennedy, *Palmetto Country*, p. 63. On early cotton culture in Florida, see Clifton Paisley, *The Red Hills of Florida, 1528–1865* (Tuscaloosa: University of Alabama, 1989); and Lynn Willoughby, *Fair to Middlin': The*

Antebellum Cotton Trade of the Apalachicola River Valley (Tuscaloosa: University of Alabama, 1993).

50. Martin, *Florida during the Territorial Days*, pp. 258–277 ('little Jew," p. 256, was a slur by the Whig opponent, David Putnam). On Branch, see Marshall D. Haywood, *John Branch, 1782–1863* (Raleigh, N.C.: Commercial, 1915). Levy's career was only beginning. In 1846 he married a gentile, broke his father's heart by converting to Christianity, and legally changed his name to David L. Yulee (inspired by his Sephardic ancestry in Morocco). A prosperous sugar planter with eighty slaves, Senator Yulee became a staunch ally of John C. Calhoun, but shifted his position after chartering the Florida Railroad in 1853. That ambitious project for a cross-peninsula line to funnel cargoes from Gulf ports to the Atlantic required heavy financing from Yankee businessmen. Hence Yulee became a moderate on sectional issues, waffled over Florida's secession in 1861, and did little to assist the Confederate war effort. After 1865 he finagled to retain control of his railroad and real estate empire and lived rich and content until his death in 1886.

51. *Jook* or *juke* is thought to derive from the West African *dzug* connoting a wild life. The jitterbug evolved from the jubilee's Walkin' Jawbone and Jumpin' Jim Josey steps. On slave codes in Florida see Larry Eugene Rivers, *Slavery in Florida: Territorial Days to Emancipation* (Gainesville: University of Florida, 2000). An early example of the "education" Yankees received appears in the letters of Ellen and Corinna Brown of Portsmouth, New Hampshire. Drawn to Florida by the Second Seminole War, they soon sniffed at crackers and decided Negroes were shiftless unless threatened; hence there was no point freeing them. Corinna Brown thought John Quincy Adams a crazy "old ninny" who deserved expulsion from Congress for introducing abolitionist petitions. See James M. Denham and Keith L. Huneycutt, eds., *Echoes from a Distant Frontier: The Brown Sisters' Correspondence from Antebellum Florida* (Columbia: University of South Carolina, 2004); "old ninny," p. 163.

52. Pedro Santoni, *Mexicans at Arms: Puro Federalists and the Politics of War, 1845–1848* (Fort Worth: Texas Christian University, 1996), pp. 1–67; Pletcher, *Diplomacy of Annexation*, pp. 31–63. See also David Brading, *The First America: The Spanish Monarchy, Creole Patriots, and the Liberal State, 1492–1867* (New York: Cambridge University, 1991); and Donald F. Stevens, *Origins of Instability in Early Republican Mexico* (Durham, N.C.: Duke University, 1991). Santa Anna's carefully crafted image came at a price. In 1838 he lost a leg while defending Veracruz against a French punitive expedition (Mexico was in flagrant default on its loans). That restored his reputation for patriotic charisma. He later staged a grand public ceremony to give his leg a Christian burial.

53. Aberdeen rendered a sour judgment on the Mexicans' nine years of bungling with regard to Texas: "Following the good old Spanish customs which you have inherited, you do everything too late" (Adams, *British Interests in Texas*, p. 223). On the revolt of June 7, see Santoni, *Mexicans at Arms*, pp. 68–87; on the Anglo-French mediation, see Pletcher, *Diplomacy of Annexation*, pp. 185–207; on Polk and the Texans, see Bergeron, *Presidency of Polk*, pp. 51–64 ("prompt & energetic," p. 62; "prudence will dictate," p. 63). In his memoirs written several years later Anson Jones accused Polk of wanting to foment a U.S.-Mexican war in 1845, as proved by Commodore Stockton's drumbeating. See Glenn W. Price, *Origins of the War with Mexico: The Polk-Stockton Intrigue* (Austin: University of Texas, 1967). Since no documentary evidence of this exists, Bergeron sensibly concludes that the urgent but prudent position consistently taken by Donelson, the

official U.S. representative, accurately reflected Polk's intentions. Yes, he wanted to frighten the Texans if that sped up the process of annexation; but, no, Polk did not want a war then, and perhaps not at all.

54. T. R. Fehrenbach, *Lone Star: A History of Texas and the Texans* (New York: Macmillan, 1968), "great country," p. 279. Fehrenbach's work was a classic, but is now superseded by Randolph B. Campbell, *Gone to Texas: A History of the Lone Star State* (New York: Oxford University, 2003). See also C. Allan Jones, *Texas Roots: Agriculture and Rural Life before the Civil War* (College Station: Texas A&M University, 2005), "The United States as we understand," p. 103. To illustrate the range of climate: Brownsville has not recorded a snowfall in over a century, whereas Lubbock averages almost one foot of snow per winter.

55. Fehrenbach, *Lone Star*, pp. 247–267 ("each step up," p. 256; "blood memory"— Katherine Anne Porter's phrase—p. 257; "the Republic," p. 247). The Texans' complaint that the U.S. Army did not provide protection against Indians was another canard. The army built no fewer than twenty-six forts in central and western Texas between 1846 and 1858.

56. Teresa Palomo Acosta and Ruthe Winegarten, *Las Tejanas: 300 Years of History* (Austin: University of Texas, 2003), pp. 46–69. One glaring example was an effort by Charles Stillman to steal from the heirs of a Mexican widow the Spanish land grant on which stood the town of Brownsville. When Stillman's bogus deed was thrown out of court, his lawyers "persuaded" the Mexicans to sell the vast tract for one-sixth of its value.

57. Walter Struve, *Germans and Texans: Commerce, Migration, and Culture in the Days of the Lone Star Republic* (Austin: University of Texas, 1996). The survivors gradually set themselves up as artisans and merchants in San Antonio, Galveston, and New Braunfels, or as farmers along the Pedernales, and placed a German stamp on parts of Texas that is still discernible today. Admiral Chester W. Nimitz was from Fredericksburg, Texas.

58. Robert A. Calvert and Arnoldo De León, *The History of Texas* (Arlington Heights, Ill.: H. Davidson, 1990), pp. 99–125; Fehrenbach, *Lone Star*, pp. 268–324. So weak was the influence of churches that the temperance movement sweeping the nation counted a mere 3,000 adherents in Texas. Nevertheless, almost all schools in Texas were denominational before the modest beginnings of public education and the University of Texas made in 1858. Baptists founded Baylor College (1845) and Catholics opened schools at Galveston (1847) and San Antonio (1852).

59. On the Rangers, see Robert M. Utley, *Lone Star Justice: The First Century of the Texas Rangers* (New York: Oxford University, 2002); and the classic by Walter Prescott Webb, *The Texas Rangers: A Century of Frontier Defense* (Boston, Mass.: Houghton Mifflin, 1935); on the Cortina War, see Jerry D. Thompson, ed., *Juan Cortina and the Texas-Mexico Frontier, 1859–1877* (El Paso: University of Texas at El Paso, 1994).

60. On Texan politics in the 1850s, see Walter L. Buenger, *Secession and the Union in Texas* (Austin: University of Texas, 1984). The secession of Texas would be worth far more than a footnote had Colonel Robert E. Lee not stepped down as commander of the U.S. Army's Texas District a few months before. Riding east in February 1861, he arrived at his former headquarters in San Antonio just a few days after his successor, seventy-one-year-old General David Emmanuel Twiggs, had reluctantly surrendered to Texas all

federal property. Lee swore that if he were still in command he would have resisted. If so, the war might have begun two months before Fort Sumter, and Lee might have wound up on the side of the Union. See Jerry Thompson, ed., *Texas and New Mexico on the Eve of the Civil War* (Albuquerque: University of New Mexico, 2001), p. 182. Thompson's book also contains a fascinating description by the future Confederate general Joseph E. Johnston of the Army's West Texas forts in 1860.

61. Frederick Law Olmsted, *A Journey through Texas, or, A Saddle-Trip on the Southwestern Frontier* (New York: Dix, Edwards, 1857), pp. 136–137. Olmsted went on to make keen observations about the conflicts among races. "There is, besides, between our Southern American and the Mexican, an unconquerable antagonism of character, which will prevent any condition of order where the two come together. . . . People commonly go into Mexico from Texas as if into a country in revolt against them, and return to boast of the insolence with which they have constantly treated the religious and social customs, and the personal self-respect of the inhabitants. . . . The mingled Puritanism and brigandism, which distinguishes the vulgar mind of the South, peculiarly unfits it to harmoniously associate with the bigoted, childish, and passionate Mexicans. They are considered to be heathen; not acknowledged as 'white folks.' Inevitably they are dealt with insolently and unjustly. They fear and hate the ascendant race, and involuntarily associate and sympathize with the negroes" (pp. 455–456).

62. Calvert and De León, *History of Texas*, pp.153–178; Fehrenbach, *Lone Star*, pp. 552–568; Palomo Acosta and Winegarten, *Las Tejanas*, pp. 46–69. A few cultured *tejanas* exploited Anglos rather than vice versa. Doña Petra Vela de Vidal, daughter of a Spanish governor, survived an ordeal as a captive of the Comanche to marry a Mexican officer, bear six children, and inherit his fortune when he died. When Mifflin Kenedy, a Pennsylvania Quaker running steamboats on the Rio Grande, agreed to become a Catholic in exchange for her hand, the beautiful heiress remarried and bore five more children on their 172,000-acre ranch. Though proud of her Spanish heritage, Doña Petra advised her daughters to marry Anglos.

63. Webb quoted in Calvert and De León, *History of Texas*, p. 164. The range wars peaked in 1883, a year of drought when cowboys were desperate to find water and forage for their cattle. John Wesley Hardin hired himself out as a mercenary until he succumbed, like Billy the Kid, to a fellow gunslinger turned lawman. John Selman killed Hardin in 1895. Judge Roy Bean epitomized hustle and fraud on the frontier. Born in Kentucky in the mid-1820s, he ran off with two brothers to Sante Fe and then opened a trading post in Chihuahua. After killing a man in a gunfight he moved to San Diego, where his brother was mayor. When the mayor was shot over a woman, Roy left behind a reputation as gambler, duelist, boozer, and braggart, and headed back to New Mexico, where he smuggled guns for the Confederacy. After the war he settled in San Antonio with his Mexican wife and five children to make a living from stolen firewood and watered-down milk. Not until 1882 did he head to West Texas with the railroad construction crews and begin his two-decade career as a duly elected judge seated on the porch of his ramshackle saloon, rifle and jug close at hand. Bean's tough talk led people to confuse him with Isaac Parker of Fort Smith, Arkansas, the real "hanging judge" who sent eighty-eight men to the gallows. Bean did hand down a few capital sentences, but the miscreants managed to escape (no doubt after slipping the judge a share of their loot).

64. Laura Lyons McLemore, *Inventing Texas: Early Historians of the Lone Star State* (College Station: Texas A&M University, 2004), is an elegant essay describing the origins of Texans' self-image and arguing that the post–Civil War era is the place to look in order to understand the longevity of Texan mythology.

65. Hietala, *Manifest Design*; and Norman A. Graebner, *Empire on the Pacific: A Study in American Continental Expansion* (New York: Ronald, 1955), make the case for a grand strategy. Pletcher, *Diplomacy of Annexation*; Sellers, *James K. Polk: Continentalist*; and William H. Goetzmann, *When the Eagle Screamed: The Romantic Horizon in American Diplomacy, 1800–1860* (New York: Wiley, 1966), pp. 38–73 ("There are four great measures," p. 39), consider Polk an improviser who never desired war, but who got one and risked another because he misunderstood domestic politics in Britain and Mexico. For assessments of Polk's diplomacy over time, see Jerald D. Combs, *American Diplomacy: Two Centuries of Changing Interpretations* (Berkeley: University of California, 1983), pp. 24–32, 57–62, 105–110, 177–180, 293–295, 362–364.

66. For an excellent summary of Polk's presidential style, see Bergeron, *Presidency of Polk*, pp. 171–246.

67. The following paragraphs on Polk's policies regarding Oregon and Mexico are based primarily on Bergeron, *Presidency of Polk*, pp. 65–77, 113–135; Pletcher, *Diplomacy of Annexation*, pp. 225–417; Sellers, *Polk: Continentalist*, pp. 236–414; Merk, *The Oregon Question: Essays in Anglo-American Diplomacy and Politics* (Cambridge, Mass.: Harvard University, 1967); Frederick Moore Binder, *James Buchanan and the American Empire* (Selinsgrove, Pa.: Susquehanna University, 1994); and McDougall, *Promised Land, Crusader State*, pp. 90–98.

68. Pletcher, *Diplomacy of Annexation*, pp. 229–253 ("profound blunder," p. 241); "The only way to treat John Bull," in Milo M. Quaife, ed., *The Diary of James K. Polk during His Presidency, 1845 to 1849*, 4 vols. (Chicago, Ill.: A.C. McClurg, 1910), I: 155; "bold and undaunted course" and "England with all her boast" in Robert V. Remini, *Andrew Jackson and the Course of American Democracy, 1833–1845* (New York: Harper and Row, 1984), p. 513; "Fifty-Four Forty or Fight!" in Thomas M. Leonard, *James K. Polk: A Clear and Unquestionable Destiny* (Wilmington, Del.: Scholarly Resources, 2001), p. 101. Early in May 1845 Sam Houston also felt a need to consult Old Hickory. He rode from Texas to Tennessee but arrived at the Hermitage a few days after his old friend gave up the ghost. Jackson's family and friends, and even his loyal slaves, were still weeping.

69. Sources of quotations: "sleeps in the Chamber," Pletcher, *Diplomacy of Annexation*, p. 274; "over a breakfast" and "anxious to preserve" cited by Bergeron, *Presidency of Polk*, p. 69; "tyranny of the demagogues" and "a national spirit," Santoni, *Mexicans at Arms*, pp. 96, 232.

70. "Take the remedies" in Quaife, ed., *Diary of Polk*, I: 319.

71. "Hostilities may now be considered" in Pletcher, *Diplomacy of Annexation*, p. 377; "It was a day," in Quaife, ed., *Diary of Polk*, I: 386–390; Polk's war message in James D. Richardson, ed., *A Compilation of the Messages and Papers of the Presidents, 1789–1897*, 10 vols. (Washington, D.C.: GPO, 1896–1899), IV: 437–443.

72. Bergeron, *Presidency of Polk*, p. 77.

73. Curiously, the only member of Polk's cabinet to drag his feet at the end was the one who previously had been most in favor of compromise: Buchanan. Evidently he was looking to future presidential sweepstakes and did not want to be known as the man who surrendered fifty-four-forty. When Polk asked his secretary of state to draft the cover

letter for the Senate, "a very painful and unpleasant" conversation ensued (Quaife, ed., *Diary of Polk*, I: 456–462).

74. But wait! The Mexicans could never defend Alta California; hence going to war amounted to throwing that rich province away. That is precisely the point. The Mexicans could not even assert their authority over their own citizens there, much less resist a naval incursion. In fact, that happened in 1842 when Captain Thomas ap Catesby Jones briefly seized Monterey on false intelligence of a war. The only real hope to keep California from the Yankees was that war, not peace, might trigger a British intervention. On other Mexican calculations, see David Lavender, *Climax at Buena Vista: The Decisive Battle of the Mexican-American War* (Philadelphia: University of Pennsylvania, 2003), pp. 39–58. General histories of the war include K. Jack Bauer, *The Mexican War, 1846–1848* (New York: Macmillan, 1974); John S. D. Eisenhower, *So Far from God: the U.S. War with Mexico, 1846–1848* (New York: Random House, 1989); Seymour B. Connor and Odie B. Faulk, *North America Divided: The Mexican War, 1846–1848* (New York: Oxford University, 1971); and Otis A. Singletary, *The Mexican War* (Chicago, Ill.: University of Chicago, 1960). A thorough, if dated, narrative is Justin H. Smith, *The War with Mexico*, 2 vols. (New York: Macmillan, 1919).

75. Grant quoted in Lavender, *Climax at Buena Vista*, p. 76. Resaca de la Palma (called Resaca de la Guerrero by the Mexicans) was a dry riverbed 200 feet wide with banks four feet tall. Arista occupied the ravine with infantry and placed his artillery behind them on the southern bank. But the Americans succeeded, after two frustrating frontal assaults, in turning the Mexicans' flank and causing a panic. Captain Ringgold was the commander of an artillery battery that fended off a counterattack. He and Davy Branch, his handsome white thoroughbred, were killed by the same Mexican shell and were immortalized on canvas.

76. The U.S. Army in 1846 was a skeleton force with a Congressional ceiling of 8,613 men in fourteen regiments (eight infantry, four artillery, two dragoon). Even those regiments were dispersed among frontier forts, and they were 40 percent understrength, owing to disease, desertion, and recruitment shortfalls. That is why Lieutenant George Meade, thirty-one years old, wrote, "Well may we be grateful that we are at war with Mexico! Were it any other power our gross follies would surely have been punished before now." See Richard Bruce Winders, *Mr. Polk's Army: The American Military Experience in the Mexican War* (College Station: Texas A&M University, 1997); Meade quoted on p. 14.

77. Bauer, *Mexican War*, pp. 66–80; Bergeron, *Presidency of Polk*, pp. 78–81; Peskin, *Winfield Scott*, pp. 138–141. It must be said Scott invited his fate by writing to the president that he was not about to face Mexican fire while being fired at from Washington City in his rear. The truth was that Scott and Polk utterly distrusted each other, but Scott was bound by a code of honor whereas Polk made an equation of patriotism and partisan politics.

78. Historians have found no evidence suggesting the Mexican War was a slaveholders' plot, but Polk's avid commitment to the peculiar institution was all suspicious northerners needed to know. On Polk's private life, see William Dusinberre, *Slavemaster President: The Double Career of James Polk* (New York: Oxford University, 2003). Ironically, the passage of the Wilmot Proviso by the House persuaded Calhoun to turn against the Mexican War. He now feared that territories annexed from Mexico would become free states, not slave states (see John Niven, *John C. Calhoun and the Price of*

Union: A Biography (Baton Rouge: Louisiana State University), pp. 303–221). For the debates on the origins and rectitude of the war, see John H. Schroeder, *Mr. Polk's War: American Opposition and Dissent, 1846–1848* (Madison: University of Wisconsin, 1973), pp. 3–62 ("all our difficulties," p. 43); "mischievous," in Bergeron, *Presidency of Polk*, p. 85; "Upper California, New Mexico," in Quaife, ed., *Diary of Polk*, II: 76–77.

79. Schroeder, *Mr. Polk's War*, pp. 3–50 ("aggressive, unholy," p. 31); James Russell Lowell, *The Biglow Papers* (Cambridge, Mass.: G. Nichols, 1848), pp. 6–7; Charles Ellsworth, "The American Churches and the Mexican War," *American Historical Review* 45 (1940): 301–326.

80. This story has been told in depth many times since Hubert Howe Bancroft first pieced it together. See, for example, Dale L. Walker, *Bear Flag Rising: The Conquest of California, 1846* (New York: Tom Doherty Associates, 1999).

81. Bauer, *Mexican War*, pp. 127–144, 183–200. On Magoffin's probable activities in New Mexico, see Elbert B. Smith, *Magnificent Missourian: The Life of Thomas Hart Benton* (Westport, Conn.: Greenwood, 1973), pp. 212–213. On the roles of the Navy and Marine Corps in California, see K. Jack Bauer, *Surfboats and Horse Marines: U.S. Naval Operations in the Mexican War, 1846–1848* (Annapolis, Md.: U.S. Naval Institute, 1969), pp. 149–204. On the exploits of Colonel Doniphan's regiment, see Joseph G. Dawson, *Doniphan's Epic March: The 1st Missouri Volunteers in the Mexican War* (Lawrence: University of Kansas, 1999). After serving as interim governor in Santa Fe, Doniphan guided his 1,000 troopers through New Mexico's bleak Jornada del Muerto and invaded Chihuahua, where they helped conquer the large province with minimal losses. His men then trekked all the way to New Orleans, where the most accomplished volunteer regiment of the war mustered out. An epic march, indeed.

82. On the coup, see Santoni, *Mexicans at Arms*, pp. 101–128; and Pletcher, *Diplomacy of Annexation*, pp. 443–448. "Every day that passes" quoted in Lavender, *Climax at Buena Vista*, p. 121.

83. Bergeron, *Presidency of Polk*, pp. 86–95; Lavender, *James K. Polk*, pp. 122–144 ("Ours is a go-ahead," "bold blow," and "rapid, crushing," p. 140); Elbert B. Smith, *Magnificent Missourian: The Life of Thomas Hart Benton* (Westport, Conn.: Greenwood, 1973), pp. 216–224; Eisenhower, *Agent of Destiny*, pp. 229–230.

84. Prucha, *Sword of the Republic*, pp. 370–371, 384–388; Leland Sage, *A History of Iowa* (Ames: Iowa State University, 1974), pp. 56–57. Does the name Albert Lea sound familiar? The little crossroads in southern Minnesota named after the lieutenant is now a bustling town of motels, gas stations, and malls at the intersection of interstate highways 90 and 35. The name Des Moines is a mistake. French explorers heard from the "Peouarea" tribe (hence Peoria, Illinois) of a river that sounded like Moingouena, which they shortened to Rivière des Moings. Since that word meant nothing in French, subsequent mapmakers decided it was supposed to have been Rivière des Moines (River of the Monks). See George R. Stewart, *Names on the Land: A Historical Account of Place-Naming in the United States* (Boston, Mass.: Houghton Mifflin, 1967), pp. 89–91.

85. That incident, taught to every Iowan child as a tragic exception, was the Spirit Lake Massacre of 1857. A small group of families had ignored Iowans' usual caution and squatted near the Minnesota border far north of Fort Dodge. The pioneers suffered during the frigid winter of 1856–1857. So did some renegade Wahpeton Sioux led by the chieftain Inkpaduta. They came to beg food from the family of Rowland Gard-

ner, but evidently decided there would be more to go around if the Gardners and the other whites were all killed. Just two of the thirty-six whites on Spirit Lake survived. The renegades then escaped to southern Minnesota, where a new round of massacres ensued in 1862. That obliged the U.S. Army to send troops even though it had more urgent business in the year of Antietam. See Sage, *A History of Iowa*, pp. 107–108.

86. Dorothy Schwieder, *Iowa: The Middle Land* (Ames: Iowa State University, 1996), pp. 3–34 (Dubuque tale, pp. 23–24; "a fair price," p. 33); Martha Royce Blaine, *The Ioway Indians* (Norman: University of Oklahoma, 1979). Iowans also disputed their boundary with Missouri. Thanks to an inaccurate survey, Missouri claimed sovereignty over a sliver of land that should have been part of Iowa. The arrival of Missourian tax collectors made Iowans reach for their muskets. Instead, the case went into litigation that dragged on until 1857, when the Supreme Court decided the incorrect survey must stand, since it had also been used for Indian treaties.

87. "No one who," written by James Hearst, a professor of English at the University of Northern Iowa, cited in Sage, *A History of Iowa*, p. 3; Schwieder, *Iowa: The Middle Land*, pp. 35–65 ("I shall never forget," p. 39; "There are no trees," p. 38); Davidson's diary in Glenda Riley, ed., *Prairie Voices: Iowa's Pioneering Women* (Ames: Iowa State University, 1996), pp. 23–40; economic growth described by Morton M. Rosenberg, *Iowa on the Eve of the Civil War: A Decade of Frontier Politics* (Norman: University of Oklahoma, 1972), pp. 6–34.

88. Schwieder, *Iowa: The Middle Land,* pp. 35–65 ("railroads were literally," p. 59); Rosenberg, *Iowa on the Eve of the Civil War*, pp. 55–78 ("The railroad mania," p. 78).

89. On the politics of the 1850s, see Rosenberg, *Iowa on the Eve of the Civil War*, pp. 146–161; on Iowa's striking turnaround in matters of racial discrimination, see Robert R. Dykstra, *Bright Radical Star: Black Freedom and White Supremacy on the Hawkeye Frontier* (Cambridge, Mass.: Harvard University, 1993).

90. Skeptical Iowa spawned, among others, the showman "Buffalo Bill" Cody, the jazz musicians Bix Beiderbecke and Glenn Miller, and the frontier populist Henry Wallace.

91. Lavender, *Climax at Buena Vista*, pp. 176–213 ("You are surrounded," p. 179; "A little more grape," p. 210); Bauer, *Mexican War*, pp. 201–231. Lavender suspects that Taylor's Whig handlers deleted a few expletives from his remark to Braxton Bragg during the presidential campaign of 1848. Bauer records Taylor as saying, "Well, double-shot your guns and give 'em hell, Bragg" (p. 216). After the battle a volunteer from Illinois wrote home, "They had us nearly whipped if they had known it." But the brave Mexicans had already suffered so many casualties that a Pyrrhic victory would not have availed. Santa Anna needed to invade U.S. territory in force in order to preempt the U.S. invasion of central Mexico.

92. Bauer, *Surfboats and Horse Marines*, pp. 75–97 ("We could not," p. 79; We, of course," p. 88); Bauer, *Mexican War*, pp. 232–258; Eisenhower, *Agent of Destiny*, pp. 233–244. Estimates of Mexican dead ranged wildly, but the British observer thought at least 100 civilians were killed in the barrage.

93. "Mexico has no longer" and "For God's sake" cited in Peskin, *Winfield Scott*, p. 169; "greatest living," p. 191.

94. Scott and Trist covered themselves politically by gaining General Gideon J. Pillow's consent to the bribe. Pillow was known to be a good friend of Polk.

95. Bauer, *Mexican War*, pp. 279–325 ("Too much blood," p. 307; "God is a Yankee," p. 318). Scott acted at once to ensure real pacification. To stop sporadic shooting at U.S. troops he informed the city council that Mexico City would be destroyed if guerrilla activity did not cease. It did in twenty-four hours. When U.S. troops engaged in sporadic looting and violence against Mexicans, he imposed draconian martial law: "No officer and no man, under my orders, shall be allowed to dishonor me, the army, or the U. States with impunity" (Peskin, *Winfield Scott*, p. 194).

96. Pletcher, *Diplomacy of Annexation*, pp. 530–563 ("acted worse," p. 558). Partisan activity, especially against supply lines, forced Scott to deploy 25 percent of his manpower to deal with guerrillas. See Irving W. Levinson, *Wars within War: Mexican Guerrillas, Domestic Elites, and the United States of America, 1846–1848* (Fort Worth: Texas Christian University, 2005).

97. Koistenen, *Beating Plowshares into Swords*, pp. 89–98; James L. Morrison, Jr., "Military Education and Strategic Thought, 1846–1861," in Hagan and Roberts, eds., *Against All Enemies*, pp. 113–131 ("I give it," p. 116). See also Winfield Scott, *Memoirs of Lieutenant-General Scott*, 2 vols. (Freeport, N.Y.: Books for Libraries, 1864). The power of American manufacturing and the efficiency of the ordnance and quartermaster corps are illustrated by the fact that Scott sent from Veracruz an urgent request for forty-nine massive mortars plus 50,000 ten-inch-thick shells and received them in just four months. Of course, the usual profiteering and corruption sullied war mobilization. Thanks to the patronage of Secretary of the Treasury Walker, Corcoran and Riggs, the Democrats' favorite brokers, diverted $2 million of army funds into stock speculation, while making fat profits from commissions on the sale of government bonds.

98. Winders, *Mr. Polk's Army*, pp. 3–87 ("An individual," p. 63; "hardy pioneers," p. 64).

99. Robert W. Johannsen, *To the Halls of Montezuma: The Mexican War in the American Imagination* (New York: Oxford University, 1985), pp. 1–112, 186–200 ("in a state," p. 10; "knights of old" and "war of reconciliation," p. 32; "Cold must be," p. 51; "fabled personage," p. 115; "pride and sorrow," p. 125; "the great prompter," pp. 199–200; "The Spanish maid," p. 300). The importance of the Mexican War to the emerging American cults of the flag and martyred patriots is well argued by Cecilia Elizabeth O'Leary, *To Die For: The Paradox of American Patriotism* (Princeton, N.J.: Princeton University, 1999). On the "All-Mexico" movement, see John D. P. Fuller, *The Movement for the Acquisition of All Mexico, 1846–1848* (Baltimore, Md.: Johns Hopkins University, 1936).

100. Many of the Irish-American "volunteers" were destitute immigrants lured by rum and the promise of loot. At least that it what Bostonian Protestants said. They lined the streets to hurl insults and threats when Caleb Cushing's recruits (one-third of whom were Irish) marched to the pier. American bishops, by contrast, blessed the war and Irish recruitment as a means of demonstrating Catholics' patriotism. That caused the intellectual Catholic convert Orestes Brownson to quit editing the *Catholic Observer* and start a dissenting Catholic review. In Matamoros Cushing's regiment was riddled with disciplinary problems and desertions to the enemy. But the famous San Patricio Battalion was by no means made up exclusively of American Irish. Robert R. Miller, in *Shamrock and Sword: The Saint Patrick's Battalion in the U.S.-Mexican War* (Norman: University of Oklahoma, 1989), established that roughly 40 percent were native Irishmen, and that

deserters from the U.S. Army included German Catholics as well. The highest-ranking deserter, John Reilly, was an Irishman who had previously deserted from British service. Bauer, in *Mexican War*, pp. 41–42, notes that the Mexicans themselves referred to the San Patricios as the "foreign legion" because of their diverse composition. The battalion fought courageously in several battles, specializing in just what the Mexicans needed: artillery.

101. Paul Foos, in *A Short, Offhand, Killing Affair: Soldiers and Social Conflict during the Mexican-American War* (Chapel Hill: University of North Carolina, 2002), pp. 83–137, documents the desertions and atrocities ("God damned set," p. 123; "I wish I had," p. 128). American soldiers wrote fondly of Mexican food once they got used to it. Tortillas; hot chili peppers; hot chocolate; and alcoholic *pulque, mescal,* and *aguardiente* were favorites (tequila as we know it was not yet being made). Soldiers' letters expressed contempt for the Roman church, the apparent laziness of Mexican men, and the backward state of Mexican agriculture. Not surprisingly, they were bewitched by the black-haired, olive-skinned Mexican women dancing the fandango. The men who probably benefited most from the cultural exchange were the American officers who socialized with upper-class Mexicans, often in their respective Masonic lodges. See Johannsen, *To The Halls of Montezuma*, pp. 113–185. Dapper U.S. veterans also brought home from Mexico a major male fashion statement: the mustache.

102. *American Review* cited in Leonard, *James K. Polk*, p. 172; Gallatin cited in Reginald Horsman, "Scientific Racism and the American Indian in the Mid-Nineteenth Century," *American Quarterly* 27 (1975): 168; Polk's speech in Richardson, ed., *Messages of the Presidents*, 4: 632.

103. Richard Nelson Current, *Wisconsin: A History* (Urbana: University of Illinois, 2001), pp. 3–33 ("The Ouisconsin," p. 8; "the finest portion," p. 9).

104. Robert C. Nesbit, *Wisconsin: A History*, 2nd ed. (Madison: University of Wisconsin, 1989), pp. 76–117; Nancy O. Lurie, *Wisconsin Indians*, 2nd ed. (Madison: University of Wisconsin, 1980). The missionary was Florimond Bonduel, who founded a school, a church, and an agricultural community among the Menominee. The whole project was ruined when whiskey sellers and loggers incited unconverted Indians against Bonduel's converts.

105. Nesbit, *Wisconsin*, pp. 119–132; Alice E. Smith, *The History of Wisconsin from Exploration to Statehood* (Madison, Wis.: State Historical Society, 1985), pp. 249–259. The honorable Dodge was disgusted by the competitive influence peddling engaged in by the territorial assemblymen. Some of his own supporters sold out to promoters. Doty feared that Dodge's reports of malfeasance might spark resistance in Congress to the choice of Madison as the capital. So Doty himself affected outrage, telling the innocent Wisconsin delegate to Congress: "Speculation and a thirst for gain appear to run everything here. Patriotism and duty is [sic] apparently lost sight of. I am to have all the rascally Speculators arrayed against me" (p. 258). If you are interested in the Badger State, get the whole story from Alice E. Smith, *James Duane Doty: Frontier Promoter* (Madison: University of Wisconsin, 1954). Not surprisingly, Doty was up to his ears in the wildcat banks chartered by the territory. After the Green Bay corporation defaulted in 1837 an audit revealed that it had issued $100,000 in banknotes against assets consisting of an $18,000 promissory note from Doty and a nail keg containing coins worth $86.20. Doty began a second career in 1861, when President Lincoln appointed him Indian agent and

then governor in an even stormier political venue: Mormon Utah. Doty died in 1865 and is buried in Salt Lake City.

106. Nesbit, *Wisconsin*, pp. 209–225, 416–434 ("We are overrun," p. 419); Kathleen Neils Conzen, *Immigrant Milwaukee: Accommodation and Community in a Frontier City* (Cambridge, Mass.: Harvard University, 1976), pp. 10–43 ("Land speculators," p. 12). Nesbit, dean of the state's historians, did not mince words: "Wisconsin's founding fathers of 1846 did not remotely resemble the men of Philadelphia in the summer of 1787" (p. 209).

107. Conzen, in *Immigrant Milwaukee*, p. 10, quotes Anthony Trollope, who remarked in 1860: "Milwaukee is a pleasant town, a very pleasant town. Why it should be so, and why Detroit should be the contrary, I can hardly tell." A *Frau* from Saint Louis agreed. "Ach, Herr Jesu, so schön ist es doch nergends wie in Milwaukee" (Lord Jesus, truly nowhere is it so lovely as in Milwaukee). See also Nesbit, *Wisconsin*, pp. 226–241, 416–434, 585–595 ("Resolved that all property," p. 229); Current, *Wisconsin: A History*, pp. 34–66 ("the significant transition," p. 47; "strong, blooming," p. 63); and Richard H. Zeitlin, *Germans in Wisconsin*, rev. ed. (Madison, Wis.: State Historical Society, 2000). On Catholic parishes and schools for the Germans and Irish, see Peter Leo Johnson, *Crosier on the Frontier: A Life of John Martin Henni* (Madison: University of Wisconsin, 1959). The Episcopal bishop Jackson Kemper and Father Richard F. Cadle founded the Nashotah House seminary and twenty-five Wisconsin parishes strongly influenced by the Oxford movement. It has been known ever since as the "biretta belt." A must-see in Fond du Lac is the Anglo-Catholic Cathedral of Saint Paul's, whose interior has some of the most beautiful wood and stone carving in North America. Wisconsin's Protestant colleges predating the Civil War include Carroll, Lawrence, and Beloit.

108. Current, *Wisconsin: A History*, p. 174–215 ("paradise of folly," p. 175). On the upstanding Irish barrister, see Alfons J. Beitzinger, *Edward G. Ryan: Lion of the Law* (Madison: University of Wisconsin, 1960). When the vote fraud was exposed the governor tried to save the state house for the Democrats by resigning in favor of his lieutenant governor. His name was Arthur McArthur, grandfather of General Douglas MacArthur (who changed it to "Mac").

109. Current, *Wisconsin: A History*, pp. 67–94 ("Speak to a cow," p. 74); and Eric E. Lampard, *The Rise of the Dairy Industry in Wisconsin: A Study in Agricultural Change* (Madison, Wis.: State Historical Society). Over these same decades two journalists in Wisconsin gave America and the world a different sort of gift: the typewriter. In 1867 Christopher Latham Sholes and James Densmore began tinkering. First came the hammers and ink; then the paper roller, the "return" handle, and the keyboard arrangement still standard today; and at last the shift key. But their modest plant in Milwaukee was inadequate to meet the growing demand. So Densmore took the advice offered by a friend back east: try the Remington arms factory in Ilion, New York. It had the capacity and expertise to mass-produce reliable machines. By the time of his death in 1889, Densmore had grown rich on royalties from Remington typewriters. Sholes had sold out early, but he took solace in having helped to create "a blessing to mankind, and especially to womankind. I am very glad I had something to do with it." See Richard N. Current, *The Typewriter and the Men Who Made It* (Urbana: University of Illinois, 1954); "a blessing," pp. 126–127.

110. So fractious had New York Democrats become that they sent two delegations to Baltimore: pragmatic "Hunkers," who wanted to win the election because (it was said)

they hankered after political appointments; and antislavery "barn burners", who were prepared (it was said) to burn down the barn to get rid of the rats. The convention tried hard to mend the rift, even offering to seat both delegations. Since both refused, New York was denied a voice in the nomination and the platform. "Mr. Van Buren is the most fallen man" in Quaife, ed., *Diary of Polk*, 4:67.

111. Democrats mocked Taylor's lack of experience and opinions, but Congressman Abraham Lincoln suggested in a hilarious speech that a candidate without a platform was ideal: Taylor would simply carry out the will of the people whatever the issue. See Joseph G. Rayback, *Free Soil: The Election of 1848* (Lexington: University of Kentucky, 1970); Frederick J. Blue, *The Free Soilers: Third Party Politics, 1848–1854* (Urbana: University of Illinois, 1973); and Willard Carl Klunder, *Lewis Cass and the Politics of Moderation* (Kent, Ohio: Kent State University, 1996), pp. 175–234.

112. Polk's fourth annual message in Richardson, ed., *Messages and Papers of the Presidents*, 4: 636–642; "a clean table" in Bergeron, *Presidency of Polk*, p. 256.

Chapter 6

1. A spate of scholarship on civil religion was inspired by Robert N. Bellah's consciousness-raising article in *Daedalus* (winter 1967). It was reprinted in Russell E. Richey and Donald G. Jones, eds., *American Civil Religion* (New York: Harper and Row, 1974), pp. 21–44. For evidence of the reality and origins of the phenomenon, plus a variety of interpretations, see Catherine L. Albanese, *Sons of the Fathers: The Civil Religion of the American Revolution* (Philadelphia, Pa.: Temple University, 1976); Robert Bellah and Phillip E . Hammond, *Varieties of Civil Religion* (San Francisco, Calif.: Harper and Row, 1980); Ruth Block, *Visionary Republic: Millennial Themes in American Thought, 1756–1800* (Cambridge: Cambridge University, 1985); Harold Bloom, *The American Religion* (New York: Simon and Schuster, 1992); Conrad Cherry, *God's New Israel: Religious Interpretations of American Destiny* (Englewood Cliffs, N.J.: Prentice-Hall, 1971); Jim Cullen, *The American Dream: A Short History of an Idea That Shaped a Nation* (New York: Oxford University, 2003); Richard T. Hughes, *Myths Americans Live By* (Urbana: University of Illinois, 2003); Pauline Maier, *American Scripture: Making the Declaration of Independence* (New York: Knopf, 1997); Carolyn Marvin, *Blood Sacrifice and the Nation: Myth, Ritual, and the American Flag* (New York: Cambridge University, 1999); Elwyn A. Smith, ed., *The Religion of the Republic* (Philadelphia, Pa.: Fortress, 1971); Ernest Lee Tuveson, *Redeemer Nation: The Idea of America's Millennial Role* (Chicago, Ill.: University of Chicago, 1968). On the sacred Union, see Paul C. Nagel, *One Nation Indivisible: The Union in American Thought, 1776–1861* (New York: Oxford University, 1964); "overturning," p. 165, is a quotation from Congressman Samuel Gordon of New York in 1847. Chesterton coined his famous phrase in the book *What I Saw in America* (New York: Dodd, Mead, 1922).

2. Needless to say, the miracles are disputed. According to legend a freakish plague of locusts threatened to devour the Mormons' entire crop until they gathered as one to pray for deliverance. Behold! On June 9 a great flock of gulls never seen since in the valley wafted down from the sky to eat the locusts and save the colony. But why do some letters and diaries written at the time refer to the gulls whereas others say the pioneers saved much of their crop by flooding ditches and forming lines to thrash the crickets? Perhaps the bugs or birds were only a local phenomenon. In any event, the "Mormon

crickets," a variety of Rocky Mountain katydids that swarm naturally every seven years, would be expected to attack the rich fields instead of fanning out over the scrub, whereas the gulls (according to ornithologists) were a species native to the Great Basin and known to reconnoiter for the pools that form after heavy rains. But the existence of a plausible natural theory for something cannot disprove divine intervention any more than the existence of faith can prove it. The only thing certain about the moving tale is that a people who had come through enormous hardship trusted their lives to a single year's crop in the desert and believed that their trust was rewarded.

3. Stanley P. Hirshson, *The Lion of the Lord: A Biography of Brigham Young* (New York: Knopf, 1973), pp. 70–115 ("This is the place," p. 85; "industrious," p. 90). Recipients of land in Utah paid just one dollar for survey costs and fifty cents to record the deed. But all land remained property of the Church of JCLDS, which decided how much each family received and evicted tenants who did not sufficiently improve their plots. The Mormons' elegant plans to make Salt Lake City a New Jerusalem on American soil are strikingly contrasted with the Spanish pueblo of Santa Fe and the gold rush towns of California and Colorado in Gunther Paul Barth, *Instant Cities: Urbanization and the Rise of San Francisco and Denver* (New York: Oxford University, 1975). On the cold war between the Utah territory and the federal government see inter alia Sarah Barringer Gordon, *The Mormon Question: Polygamy and Constitutional Conflict in Nineteenth-Century America* (Chapel Hill: University of North Carolina, 2002). To appreciate the genius of the Mormons' irrigation, see Donald W. Meinig, *The Shaping of America: A Geographical Perspective on 500 Years of History:* Vol. 3, *Transcontinental America 1850–1915* (New Haven, Conn.: Yale University, 1998), pp. 89–113 ("one of the most remarkable," p. 89). Deseret was the name of a honeybee in the Book of Mormon.

4. I describe the gold rush in more detail in *Let the Sea Make a Noise: A History of the North Pacific from Magellan to MacArthur* (New York: Basic Books, 1993), pp. 239–248. The inimitable Brannan's role is described in James J. Rawls and Walton Bean, *California: An Interpretive History*, 6th ed. (New York: McGraw-Hill, 1993), pp. 82–95 ("the Lord's money" and "Gold! Gold!" p. 84); and Paul Dayton Bailey, *Sam Brannan and the California Mormons* (Los Angeles, Calif.: Westernlore, 1959). Good accounts of the gold rush include H. W. Brands, *The Age of Gold: The California Gold Rush and the New American Dream* (New York: Doubleday, 2002); Donald Dale Jackson, *Gold Dust* (New York: Knopf, 1980); and J. S. Holliday, *The World Rushed In: The California Gold Rush Experience* (New York: Simon and Schuster, 1981).

5. In addition to the books noted above about the gold rush, see Mary McDougall Gordon, "Overland to California in 1849: A Neglected Commercial Enterprise," *Pacific Historical Review* 52 (1983): 17–36; David Dary, *Entrepreneurs of the Old West* (New York: Knopf, 1986), pp. 89–109; Raymond A. Rydell, "The Cape Horn Route to California, 1849," *Pacific Historical Review* 17 (1948): 149–163; Raymond A. Rydell, "The California Clippers," *Pacific Historical Review* 18 (1949): 70–83; John Haskell Kemble, *The Panama Route, 1848–1869* (Berkeley: University of California, 1943). On clipper ships, see the illustrated scholarly volumes by Carl C. Cutler, *Greyhounds of the Sea* (Annapolis, Md.: Naval Institute, 1984, orig. 1930); David R. MacGregor, *Fast Sailing Ships: Their Design and Construction, 1775–1875* (Annapolis, Md.: Naval Institute, 1988, orig. 1973); and James P. Delgado, *To California by Sea: A Maritime History of the California Gold Rush* (Columbia: University of South Carolina, 1990).

6. On Clayton-Bulwer, see Kenneth Bourne, *Britain and the Balance of Power in North America, 1815–1908* (London: Longmans, Green, 1967). The Americans' failure to face down the British when the latter strengthened their grip on Belize (British Honduras) stemmed partly from Clayton's Whiggish commitment to Anglo-American cooperation and partly from the firmness of Palmerston, who was once again foreign secretary.

7. Thomas Hart Benton, *Thirty Years View, or A History of the Working of American Government from 1820 to 1850*, 2 vols. (New York: Appleton, 1854–1856), II: 695–696.

8. Polk was in Buchanan's camp. He willingly signed the Oregon bill as consistent with his endorsement of the Missouri Compromise line. Had New Mexico, for instance, been up for territorial status, Polk would have insisted that slavery be permitted there. See David M. Potter, *The Impending Crisis 1848–1861* (New York: Harper and Row, 1976), pp. 51–89.

9. All these statesmen deserve far more of an introduction than is possible in a synthetic history. For starters see the following excellent biographies. Robert Walter Johannsen, *Stephen A. Douglas* (New York: Oxford University, 1973); William J. Cooper, *Jefferson Davis, American* (New York: Knopf, 2000); Glyndon G. Van Deusen, *William Henry Seward* (New York: Oxford University, 1967); John Niven, *Salmon P. Chase: A Biography* (New York: Oxford University, 1995).

10. Holman Hamilton, *Prologue to Conflict: The Crisis and Compromise of 1850* (New York: Norton, 1964), pp. 1–42; John C. Waugh, *On the Brink of Civil War: The Compromise of 1850 and How It Changed the Course of American History* (Wilmington, Del.: Scholarly Resources, 2003), pp. 33–62 ("politics and his patriotism" and "This Union is the rock," p. 47; "I will hang them," p. 59). Waugh's book is outstanding, but it seems to me that his own thesis belies his subtitle. The Compromise of 1850 did not change the course on which the republic had embarked. "I am for disunion" in Potter, *Impending Crisis*, p. 94. The quotations from Taylor are from his inaugural address in Fred L. Israel, ed., *I Do Solemnly Swear: The Inaugural Addresses of the Presidents of the United States* (Philadelphia, Pa.: Chelsea House, 2001), p. 124.

11. Robert V. Remini, *Henry Clay: Statesman for the Union* (New York: Norton, 1991), pp. 730–761; John Niven, *John C. Calhoun and the Price of Union* (Baton Rouge: Louisiana State University, 1988), pp. 322–345; Claude M. Fuess, *Daniel Webster*, 2 vols. (Boston, Mass.: Little, Brown, 1930), II: 198–227; Van Deusen, *Seward*, pp. 122–128; Potter, *Impending Crisis*, pp. 90–105; Hamilton, *Prologue to Conflict*, pp. 43–101; Waugh, *On the Brink of Civil War*, pp. 63–123 ("one of the high priests"—the words of Senator Andrew P. Butler of South Carolina—p. 73; "great scheme," p. 75; "aggressive majority," p. 81; "the South is right" and "let us come out," p. 104; "radically wrong," "unjust," and "there is a higher law," p. 113).

12. Was Zachary Taylor poisoned? According to the official story he attended the laying of the cornerstone for the Washington Monument on a very hot Fourth of July, then returned to the White House to gorge himself on cherries and cream. He came down with severe stomach pains and died on July 9. Given the suddenness of the illness, the vagueness of its symptoms, and his tough, unpopular stance on the issues of the day, it was later rumored that he might have been poisoned (presumably by southern extremists). His remains were exhumed in the 1990s and tested for poison. Sorry: the results were negative. The pathology suggests he probably ate fruit contaminated with typhoid or cholera.

13. Potter, *Impending Crisis*, pp. 106–121 ("The question of slavery," p. 116; "final and irrevocable"—in Fillmore's annual message of December 1850—p. 121) coined the term "armistice of 1850." Waugh, *On the Brink of Civil War*, pp. 163–194 ("What will become," p. 128; "holy citadel," p. 187), concludes that the ultimate significance of the Compromise of 1850 was that it rendered impossible further compromise. Hamilton, in *Prologue to Conflict*, describes the influential Texas bond lobby centered on the Philadelphia stock exchange, and appends a chart of every congressional vote on the various measures. In Hamilton's judgment, Zachary Taylor's realism and toughness might have spared the nation a sterile "compromise" that permitted it to indulge its preference for procrastination and distortion of truth (pp. 182–190). All that can be said for the alleged bargain is that it gave the North ten additional years to strengthen itself prior to civil war.

14. Rawls and Bean, *California: An Interpretive History*, pp. 82–95 ("I looked on," p. 85); Herbert Asbury, *The Barbary Coast: An Informal History of the San Francisco Underworld* (Garden City, N.Y.: Garden City, 1933), pp. 3–48 ("the people," p. 31). After a miners' debauch gleaners would sweep the floors to gather up ounces of gold dust from the detritus. The paucity of women gave rise to a curious gallantry even toward harlots, more than 2,000 of whom sailed in from Mexico, Peru, New Orleans, New York, and Europe in the first six months of the gold rush. Their brothels lined Pacific Avenue inland from the Embarcadero, and presided on Telegraph Hill.

15. The military governors Sloat, Stockton, Frémont, Kearny, and Mason, in addition to presidents Polk, Taylor, and Fillmore, all had prominent streets in San Francisco named for them.

16. There are two touches of irony in the state's ursine seal and flag. First, General Vallejo objected to a reminder that the Bear Flag Revolt had brought an undignified end to Mexican rule. He wanted the bear to be depicted as lassoed by a vaquero but relented when he was assured no ethnic insult was intended. Second, the California grizzly or "golden" bear, which still serves as the totem for the University of California and UCLA, was promptly hunted to extinction.

17. Kevin Starr's series of monographs represents, in any judgment, the gold standard of California history. On the nineteenth century, see his *Americans and the California Dream, 1850–1915* and *Inventing the Dream: California through the Progressive Era* (New York: Oxford University, 1973 and 1985). On the origins of state government, see also Theodore Grivas, *Military Governments in California, 1846–1850* (Glendale, Calif.: Clark, 1963); and William Henry Ellison, *A Self-Governing Dominion: California, 1849–1860* (Berkeley: University of California, 1950).

18. The great early historian of the West, Hubert Howe Bancroft, praised the vigilantes in *Popular Tribunals* (San Francisco, Cal.: History Company, 1887), but recent studies document prejudice and class warfare in the vigilance committees. See Richard M. Brown, *Strain of Violence: Historical Studies of American Violence and Vigilantes* (New York: Oxford University, 1975); Peter R. Decker, *Fortunes and Failures: White Collar Mobility in Nineteenth Century San Francisco* (Cambridge, Mass.: Harvard University, 1978); and Roger D. McGrath, *Gunfighters, Highwaymen, and Vigilantes: Violence on the Frontier* (Berkeley: University of California, 1984).

19. Paul Wallace Gates, in "The California Land Act of 1851," *California Historical Quarterly* 50 (1971): 395–405, argued that this act has been unfairly vilified. Having learned from the cases of Louisiana, Missouri, and Florida how difficult foreign land

grants were to adjudicate, Congress tried hard to be fair and accorded losers the right to appeal the commission's decisions in federal courts. But the situation was rendered hopeless by the vagueness of rancho titles, 288 of which dated from the last three years of Mexican rule and none of which was surveyed. Moreover, the commission applied the same rules to white grantees such as Sutter, who lost his vast empire around Sacramento; and squatters, who fought pitched battles in defense of their plots in Sacramento and elsewhere. On the other hand, Douglas Monroy, in *Thrown among Strangers: The Making of Mexican Culture in Frontier California* (Berkeley: University of California, 1990), revives H. H. Bancroft's earlier vilification of the Act of 1851 on the grounds that Californios were just defenseless against Yankee law and lawyers. On the various ethnic minorities, see inter alia Leonard M. Pitt, *The Decline of the Californios: A Social History of the Spanish-Speaking Californians, 1846–1890* (Berkeley: University of California, 1966); Gunther P. Barth, *Bitter Strength: A History of the Chinese in the United States, 1850–1870* (Cambridge, Mass.: Harvard University, 1964); and James J. Rawls, *Indians of California: The Changing Image* (Norman: University of Oklahoma, 1984).

20. Dary, *Entrepreneurs of the Old West*, pp. 123–128 ("the omnipresent," p. 124). The quoted journalist from Massachusetts went on to give his readers a warning: "People, who know they are smart in the East, and come out here thinking to find it easy woolgathering, are generally apt to go home shorn. Wall Street can teach Montgomery Street nothing in the way of 'bulling' and 'bearing,' and the 'corners' made here require quick and long breath to turn without faltering. . . . [The California businessman displays] a wide practical reach, a boldness, a sagacity, a vim, that I do not believe can be matched anywhere in the world" (125–126). Spending a day in the wine country? Forget Napa: buy cheese and bread in the town of Sonoma; picnic at Haraszthy's quaint Buena Vista estate; then sample more wines along the road north to Alexander Valley.

21. Rawls and Bean, *California: An Interpretive History*, pp. 165–199; Meinig, *Transcontinental America 1850–1915*, pp. 36–69; Robert M. Fogelson, *The Fragmented Metropolis: Los Angeles, 1850–1930* (Cambridge, Mass.: Harvard University, 1967), pp. 1–63 ("no happier paradise," p. 63).

22. See Donald Worster, *Rivers of Empire: Water, Aridity, and the Growth of the American West* (New York: Pantheon, 1985); and Norris Hundley, Jr., *The Great Thirst: Californians and Water, 1770s–1990s* (Berkeley: University of California, 1992). Jack Nicholson's film *Chinatown* is based on the dirty tactics applied on behalf of the Owens River Aqueduct. Some vigilantes in Owens valley remained so angry that as late as the 1920s they repeatedly sabotaged the aqueduct.

23. On *Ramona*, see Starr, *Inventing the Dream*, pp. 54–63; "a commodity" in Meinig, *Transcontinental America, 1850–1915*, p. 64; "linked imaginatively" in Starr, *Americans and the California Dream*, p. 68. Helen Hunt Jackson's life adumbrates much of the nation's experience in the nineteenth century. Born in Amherst in 1830, she lost both her father and his New England faith while she was a teenager. She traveled widely in the company of her first husband, a soldier, until he was killed in an ordnance explosion during the Civil War. That plus the death of her two sons plunged her into a depression that dabbling in spiritualism did not cure. So she became a travel writer, like the widow Anne Newport Royall a generation before. In 1875 Helen Hunt remarried and moved to Colorado, but kept on the road. In 1881 *Century* magazine dispatched her to California, where Bishop Francisco Mora and the rancher Don Antonio de Coronel taught her to love the

land, its people, and its history. She wrote an essay eulogizing Father Serra, then undertook *Ramona*. She died in San Francisco in 1885.

24. To his credit Fillmore remained in the arena. The American Party ("Know-Nothings") nominated him for president in 1856. He was not especially xenophobic, but he accepted the call because he firmly (and rightly) believed that only a conservative party with national appeal could forestall secession and civil war. His campaign is characterized as disastrous (the way he himself saw it), but in fact he won 22 percent of the vote and a plurality in Maryland. See Robert J. Rayback, *Millard Fillmore: Biography of a President* (Buffalo, N.Y.: Buffalo Historical Society, 1959); and Elbert B. Smith, *The Presidencies of Zachary Taylor and Millard Fillmore* (Lawrence: University of Kansas, 1988); "till we can get rid" in Frank H. Severance, ed., *Millard Fillmore Papers*, 2 vols. (Buffalo, N.Y.: Buffalo Historical Society, 1959), II: 173–174. On the contest of 1852 between Pierce and Scott, see Larry Gara, *The Presidency of Franklin Pierce* (Lawrence: University of Kansas, 1991), pp. 17–41; John S. D. Eisenhower, *Agent of Destiny: The Life and Times of General Winfield Scott* (New York: Free Press, 1997), pp. 321–330; Arthur Charles Cole, *The Whig Party in the South* (Washington, D.C.: American Historical Association, 1913), pp. 212–276.

25. See Robert Manson Myers, ed., *The Children of Pride: A True Story of Georgia and the Civil War* (New Haven, Conn.: Yale University, 1972), pp. 7–31. Midway Church in Liberty County was founded by descendants of the Puritan migration who moved to South Carolina in 1695, then to Georgia in 1752. The small congregation counted among its members two signers of the Declaration of Independence, a Revolutionary War general who became a great-grandfather of Theodore Roosevelt, a grandfather of Woodrow Wilson's wife Edith, the father of Samuel F. B. Morse, the father of the scientist Joseph LeConte of the University of California, and the father of Justice Oliver Wendell Holmes. The thousands of letters exchanged among members and friends of the Jones family between 1854 and 1868 are the most complete firsthand account of the period in which the novel *Gone with the Wind* is set. Dr. Jones's weak chest resulted from a childhood accident in which he fell on a pointed stick, piercing a lung. On his efforts to prevent the breakup of slave families, see inter alia his letters of January 1856 and March 1857 (pp. 183–184, 309–310).

26. Ibid., pp. 35–38 (letters dated May 22 and May 30, 1854).

27. Gara, *Presidency of Pierce*, pp. 105–111 ("Your conduct," p. 107); Richard H. Abbott, *Cotton and Capital: Boston Businessmen and Antislavery Reform, 1854–1868* (Amherst: University of Massachusetts, 1991), cited in Allen C. Guelzo, *The Crisis of the American Republic: A History of the Civil War and Reconstruction Era* (New York: St. Martin's, 1995), pp. 51–53 ("We went to bed," p. 53); Potter, *Impending Crisis*, pp. 130–139. On the victimized fugitive himself, see Albert J. von Frank, *The Trials of Anthony Burns* (Cambridge, Mass.: Harvard University, 1998).

28. Wise counted twenty-two American ships engaged in the slave trade. Four sailed out of Baltimore and all the others from northern ports. On the illegal slave trade abroad, see Warren S. Howard, *American Slavers and the Federal Law 1837–1862* (Westport, Conn.: Greenwood, 1976), pp. 28–69; and Craig M. Simpson, *A Good Southerner: The Life of Henry A. Wise of Virginia* (Chapel Hill: University of North Carolina, 1985), pp. 61–77. On the controversy over fugitive slaves at home, see Paul Finkelman, *An Imperfect Union: Slavery, Federalism, and Comity* (Chapel Hill: University of North Caro-

lina, 1981); Stanley W. Campbell, *The Slave Catchers: Enforcement of the Fugitive Slave Law, 1850–1860* (Chapel Hill: University of North Carolina, 1968); and Harold M. Hyman and William M. Wiecek, *Equal Justice under Law: Constitutional Development 1835–1875* (New York: Harper and Row, 1982). Hyman and Wiecek describe five cases of capture or rescue that obliged federal courts to render significant constitutional judgments from 1851 to 1854. Definitions of states' rights, habeas corpus, and (in the Christiana case) treason were among the issues raised. But Taney's Supreme Court dithered until 1859 before ruling on the constitutionality of the Fugitive Slave Act itself, by which time seven northern states had passed laws effectively nullifying it. On Emerson's indictment of the act, see Ralph Waldo Emerson, "American Slavery" (1855), in David M. Robinson, ed., *The Political Emerson: Essential Writings on Politics and Social Reform* (Boston, Mass.: Beacon, 2004), pp. 120–137; "this filthy enactment," from his journal, quoted in George Brown Tindall, *America: A Narrative History*, 2 vols. (New York: Norton, 1988), I: 612.

29. Joan D. Headrick, ed., *The Oxford Harriet Beecher Stowe Reader* (New York: Oxford University, 1999); "Do you say," p. 402. Assailed for peddling a slanderous fable, Stowe published *A Key to Uncle Tom's Cabin* in 1853; it included documentation from life of all the characters and events in the novel. "An Appeal to the Women of the Free States" followed in 1854.

30. Headrick, ed., *Oxford Harriet Beecher Stowe Reader*, pp. 1–19 ("1st. To Conceal Ideas," p. 3; "great American novel," p. 8). John William DeForest invented the phrase in *The Nation* in 1868, judging *Uncle Tom's Cabin* the only tale "which paints American life so broadly, truly, and sympathetically." Henry James and others later praised Stowe for pioneering the realistic novel. See Kenneth S. Lynn, *Visions of America* (Westport, Conn.: Greenwood, 1973), pp. 27–48; and Ellen Moers, *Harriet Beecher Stowe and American Literature* (Hartford, Conn.: Stowe-Day, 1978). On the book's influence, see Edmund Wilson, *Patriotic Gore: Studies in the Literature of the American Civil War* (New York: Oxford University, 1962), pp. 3–58 ("make this whole nation," pp. 31–32); and Guelzo, *Crisis of the American Republic*, pp. 53–54. To be sure, most critics ritually denounce *Uncle Tom's Cabin* for its sentimentality, stereotypes, unbelievable dialogue, and coincidences. After learning of Stowe's background and her other achievements I am not inclined to do that. Or maybe I'm less of a cynic than some of my critics believe.

31. Johannsen, *Stephen A. Douglas*, pp. 375–410 ("saddle to the other horse," p. 410); Potter, *Impending Crisis*, pp. 145–160 ("There is a power," p. 152). On Asa Whitney, the early transcontinental railroad plans, and the Gadsden Purchase, see Robert Royal Russel, *Improvement of Communication with the Pacific Coast as an Issue in American Politics, 1783–1864* (Cedar Rapids, Iowa: Torch, 1948); and Paul Neff Garber, *The Gadsden Treaty* (Philadelphia: University of Pennsylvania, 1923). The F Street Mess was the capital's little island of plantation culture where Negro slaves in starched livery served juleps while senators caucused and plotted as if they were ambassadors in a foreign capital. See William W. Freehling, *The Road to Disunion: Secessionists at Bay 1776–1854* (New York: Oxford University, 1990), pp. 550–552.

32. Johannsen, *Stephen A. Douglas*, pp. 411–434 ("to take and hold their slaves," p. 411; "superceded by the principles," p. 415; "I had the authority," p. 434); Potter, *Impending Crisis*, pp. 161–176; Michael A. Morrison, *Slavery and the American West: The Eclipse of Manifest Destiny and the Coming of the Civil War* (Chapel Hill: University of North

Carolina, 1997), pp. 142–156; Roy F. Nichols, "The Kansas-Nebraska Act: A Century of Historiography," *Mississippi Valley Historical Review* 43 (1956): 187–212; "to guard the domain" in Thurlow Weed Barnes, ed., *Memoir of Thurlow Weed* (Boston, Mass.: Houghton Mifflin, 1884), p. 221.

33. Gara, *Presidency of Pierce*, pp. 111–123 ("Kansas has been invaded," p. 114); Potter, *Impending Crisis*, pp. 199–217 ("I kem to Kansas," p. 203); "the harlot, Slavery," quoted in David Donald, *Charles Sumner and the Coming of Civil War* (New York: Knopf, 1960), pp. 286; David Grimsted, *American Mobbing, 1828–1861: Toward Civil War* (New York: Oxford University, 1998), pp. 270–275 ("God's chosen instrument," p. 271); Oskar Garrison Villard, *John Brown, 1800–1859: A Biography Fifty Years After* (Boston, Mass.: Houghton Mifflin, 1910), pp. 165–188. "Bully" Brooks would die of a cold the subsequent winter, but for the moment he escaped with a censure and was promptly reelected to Congress. So was Sumner, even though his wounds (or shock) kept him from his duties for more than two years (southerners insisted that he was faking). John Brown's vengeance was also hyperbolic: just one man had died in the sack of Lawrence, and that was because a burning eave of the Free State Hotel happened to fall on his head. The butchery at Pottawatomie was performed by Brown's men (two of his sons soon repented); Brown himself stood aloof except to claim symbolic responsibility by firing a bullet into the head of a man already dead. But ultimate responsibility for "Bleeding Kansas" must fall on President Pierce. Not only did he support Douglas's plan to make the territory a free-for-all favoring whichever side got there "fustest with the mostest"; his appointment of Reeder was folly. A Pennsylvanian Democrat with southern sympathies and personal ambitions, Reeder had never held public office before or experienced conflict more violent than that of an orphans' court. The border ruffians' outrages eventually inclined Reeder to favor the Free-Soilers, whereupon Atchison and Jefferson Davis prevailed on Pierce to dismiss him. Ironically, the perfect choice for territorial governor of Kansas would have been Winfield Scott, the man Pierce defeated for president.

34. Alice Nichols, *Bleeding Kansas* (New York: Oxford University, 1954), pp. 265–296; James Claude Malin, *John Brown and the Legend of Fifty-Six* (Philadelphia, Pa.: American Philosophical Society, 1942), pp. 92–94; "Glorious Triumph" cited in Gara, *Presidency of Pierce*, p. 120.

35. George Fitzhugh, "Southern Thought," *De Bow's Review* 23 (1857): 338–350, reprinted in Drew Gilpin Faust, ed., *The Ideology of Slavery: Proslavery Thought in the Antebellum South, 1830–1860* (Baton Rouge: Louisiana State University, 1981), pp. 272–299 ("Twenty years ago," pp. 274–275; "wise and prudent," pp. 298–299). Not surprisingly, southerners ran from the argument for white slavery in Fitzhugh's *Sociology for the South* (1854): it fed northern paranoia while undercutting their racial arguments for African slavery.

36. The nineteenth-century evolution of the four "cradle cultures" dating from colonial times and their role in exacerbating sectional conflict are elegantly summarized in the concluding chapters of David Hackett Fischer, *Albion's Seed: Four British Cultures in America* (New York: Oxford University, 1989). See also Walter A. McDougall, *Freedom Just Around the Corner: A New American History 1585–1828* (New York: HarperCollins, 2004), pp. 136–167. On the Methodist convention, see Randy J. Sparks, "'To Rend the Body of Christ': Proslavery Ideology and Religious Schism from a Mississippi Perspective," in John R. McKivigan and Mitchell Snay, eds., *Religion and the Antebellum Debate over Slavery* (Athens: University of Georgia, 1998), p. 273–295.

37. See Jesse T. Carpenter, *The South as a Conscious Minority, 1789–1861* (Columbia: University of South Carolina, 1990), quotation from Stephens, p. 148; Forrest McDonald, *States' Rights and the Union: Imperium in Imperio, 1776–1876* (Lawrence: University of Kansas, 2000); Bertram Wyatt-Brown, *The Shaping of Southern Culture: Honor, Grace, and War, 1760s–1880s* (Chapel Hill: University of North Carolina, 2001); Eugene D. Genovese, *The Slaveholders' Dilemma: Freedom and Progress in Southern Conservative Thought, 1820–1860* (Columbia: University of South Carolina. 1992); Eugene D. Genovese, *The Southern Tradition: The Achievement and Limitations of an American Conservatism* (Cambridge, Mass.: Harvard University, 1994), "will not always," pp. 53–54. Joseph G. Baldwin was no Pollyanna. He exposed the hustling and violence of southern life in *The Flush Times of Alabama and Mississippi* (New York: Appleton, 1853). Yet his constitutional arguments in *Party Leaders* (New York: Appleton, 1855) were as subtle as John C. Calhoun's.

38. On southern intellectuals, see Adam L. Tate, *Conservatism and Southern Intellectuals 1789–1861* (Columbia: University of Missouri, 2005), especially good on Baldwin, Hooper, Tucker, and Simms ("father of southern literature," p. 138); Michael O'Brien, *Conjectures of Order: Intellectual Life and the American South, 1810–1860*, 2 vols. (Chapel Hill: University of North Carolina, 2004); Drew Gilpin Faust, *A Sacred Circle: The Dilemma of the Intellectual in the Old South, 1840–1860* (Baltimore, Md.: Johns Hopkins University, 1977), especially pp. 112–131; Drew Gilpin Faust, *James Henry Hammond and the Old South: A Design for Mastery* (Baton Rouge: Louisiana State University, 1982); Faust, ed., *Ideology of Slavery*, pp. 1–20 ("sober and cautious," p. 4); Eugene D. Genovese, *The World the Slaveholders Made: Two Essays in Interpretation* (New York: Pantheon, 1969), and *The Slaveholders' Dilemma*; Michael O'Brien, *Rethinking the South: Intellectual Life in Antebellum Charleston* (Baltimore, Md.: Johns Hopkins University, 1982); Charles S. Sydnor, *The Development of Southern Sectionalism, 1819–1848* (Baton Rouge: Louisiana State University, 1968), "misguided spirit," "cornerstone," and "no evil," pp. 331–339. It was David Donald who was "astonished" in "The Proslavery Argument Reconsidered," *Journal of Southern History* 37 (1971): 3–18 (p. 4). Louis Hartz imagined an "alien child" in *The Liberal Tradition in America: An Interpretation of American Political Thought since the Revolution* (New York: Harcourt, Brace, 1955), p. 8.

39. William J. Grayson, "The Hireling and the Slave," reprinted in Paul Finkelman, *Defending Slavery: Proslavery Thought in the Old South: A Brief History with Documents* (Boston, Mass.: Bedford/St. Martin's, 2003), pp. 173–187. Grayson's delicious last stanza contrasts the irenic South and anarchic North:

> "And yet the master's lighter rule insures / More order than the sternest code secures;
> No mobs of factious workmen gather here, / No strikes we dread, no lawless riots fear;
> Nuns, from their convent driven, at midnight fly, / Churches, in flames, ask vengeance
>   from the sky,
> Seditious schemes in bloody tumults end, / Parsons incite, and senators defend,
> But not where slaves their easy labors ply, / Safe from the snare, beneath a master's eye;
> In useful tasks engaged, employed their time, / Untempted by the demagogue to crime,
> Secure they toil, uncursed their peaceful life, / With labor's hungry broils and
>   wasteful strife,
> No want to goad, no faction to deplore, / The slave escapes the perils of the poor."

40. That a Christian theology justifying slavery had "official preeminence" was the judgment of Wyatt-Brown, in *Shaping of Southern Culture*, p. 101, and has now been argued at great length by Elizabeth Fox-Genovese and Eugene D. Genovese, *The Mind of the Master Class: History and Faith in the Southern Slaveholders' Worldview* (New York: Cambridge University, 2005). See also Thornton Stringfellow, "A Brief Examination of Scripture Testimony on the Institution of Slavery" (1841) in Faust, ed., *Ideology of Slavery*; Albert D. Kirwan, ed., *The Civilization of the Old South: Writings of Clement Eaton* (Lexington: University of Kentucky, 1968), pp. 181–207; Edward R. Crowther, "'Religion Has Something . . . to Do with Politics': Southern Evangelicals and the North, 1845–1860," in McKivigan and Snay, eds., *Religion and the Antebellum Debate over Slavery*, pp. 317–342. Frederick A. Ross, a Presbyterian pastor in Huntsville, was even shriller in his denunciation of the North (p. 336): "Ye men of Boston, New York, London, Paris— Ye hypocrites—Ye brand me a pirate, a kidnapper, a murderer, a demon fit only for hell, and yet ye buy my cotton. Why don't you throw the cotton into the sea, as your fathers did the tea? . . . Ye New England hypocrites—ye Old England hypocrites—ye French Revolution hypocrites—Ye Uncle Tom's Cabin hypocrites. . . . Oh, your holy twaddle stinks in the nostrils of God, and He commands me to lash you with my scorn, and His scorn, so long as you gabble about the sin of slavery . . . [yet] buy and spin cotton."

41. See Josiah C. Nott, *Two Lectures on the Natural History of the Caucasian and Negro Races* (Mobile, Ala.: Dade and Thompson, 1844); James Henry Hammond, *Two Letters on Slavery in the United States, Addressed to Thomas Clarkson, Esq.* (Columbia, S.C.: Allen, McCarter, 1845); William Harper, *Memoir on Slavery: The Pro-Slavery Argument as Maintained by the Most Distinguished Writers of the Southern States* (Charleston, S.C.: Walker, Richards, 1852); George Fitzhugh, *Sociology for the South; or the Failure of Free Society* (Richmond, Va.: A. Morris, 1854) and "Southern Thought," in Faust, ed., *Ideology of Slavery*, pp. 272–299 ("Now, is it," p. 298). On anthropological theories of racial inequality, see George M. Fredrickson, *The Black Image in the White Mind: The Debate on Afro-American Character and Destiny, 1817–1914* (New York, Harper and Row, 1971); and William Stanton, *The Leopard's Spots: Scientific Attitudes toward Race in America, 1815–1859* (Chicago, Ill.: University of Chicago, 1960).

42. On southern critiques of northern progress, see Genovese, *The Slaveholders' Dilemma* and Eugene D. Genovese, *The Southern Tradition: The Achievement and Limitations of an American Conservatism* (Cambridge, Mass.: Harvard University, 1994); Kirwan, ed., *Civilization of the Old South*, pp. 202–226 ("conservative portion" and "the breakwater," pp. 224–225); Sydnor, *Development of Southern Sectionalism*, pp. 331–339 (Millerites, Mormons, etc., p. 336); and Tate, *Conservatism and Southern Intellectuals*, pp. 175–230 ("All the steam power," p. 199; "I have no hope," p. 186). On Paulding's "conversion" see William R. Taylor, *Cavalier and Yankee: The Old South and American National Character* (Cambridge, Mass.: Harvard University, 1979), pp 225–259 ("When the love of self," p. 259). On southern appropriation of the Cavalier ethic (minus monarchism, of course), see Michael Johnson, *Toward a Patriarchal Republic: The Secession of Georgia* (Baton Rouge: Louisiana State University, 1977); and Anne Norton, *Alternative Americas: A Reading of Antebellum Political Culture* (Chicago, Ill.: University of Chicago, 1986), pp. 116–131.

43. On the filibustering expeditions of the 1850s, see Charles H. Brown, *Agents of Manifest Destiny: The Lives and Times of the Filibusters* (Chapel Hill: University of North

Carolina, 1980); Robert E. May, *The Southern Dream of a Caribbean Empire, 1854–1861* (Baton Rouge: Louisiana State University, 1973); William Earl Weeks, *Building the Continental Empire: American Expansion from the Revolution to the Civil War* (Chicago, Ill.: Ivan R. Dee, 1996); and William H. Goetzmann, *When the Eagle Screamed: The Romantic Horizon in American Diplomacy, 1800–1860* (New York: Wiley, 1966), pp. 74–91 ("Infiltration and internal subversion," p. xvi).

44. Weeks, *Building the Continental Empire*, pp. 140–164 ("If the slave," p. 145; "Never has there been," p. 153).

45. Brown, *Agents of Manifest Destiny*, pp. 174–358, 433–457 ("grey-eyed man," p. 176; "A dash of sadness," p. 177; "I hold that the government," pp. 340–341; "He never displayed," p. 456). Walker's epithet derived from a Central American Indian legend about a gray-eyed champion who would come to expel the oppressive Spaniards. Walker's greatest asset, besides southern support, was Americans' hostility to Britain, the major rival for influence in Central America and a threat to the Monroe Doctrine. His greatest blunder was the imperious attempt he made as president of Nicaragua to seize control of Vanderbilt's Accessory Transit Company. That turned the powerful shipping magnate Vanderbilt into an enemy and deprived Walker of his "navy." But what surely doomed him was the reintroduction of slavery in Nicaragua, since that measure, the payoff so far as his American supporters were concerned, was odious to the local Indian and Hispanic population.

46. An excellent survey of the historiography opens the article by Gary J. Kornblith, "Rethinking the Coming of the Civil War: A Counterfactual Exercise," *Journal of American History* 90, no. 1 (2003): 76–105. Arguments for an "irrepressible conflict" between capitalism and southern agrarianism were made originally by Charles A. Beard and Mary R. Beard, *The Rise of American Civilization*, 2 vols. (New York, Macmillan, 1927); and Arthur C. Cole, *The Irrepressible Conflict* (New York, Macmillan, 1934); and again by Eugene D. Genovese in his Marxist phase in *The Political Economy of Slavery: Studies in the Economy and Society of the Slave South* (New York: Vintage, 1967). The case for the "blundering generation" was made by Avery O. Craven, *The Repressible Conflict 1830–1861* (Baton Rouge: Louisiana State University, 1939); and James G. Randall, *Lincoln the President: Springfield to Gettysburg*, 4 vols. (New York: Dodd, Mead, 1945–1955). That a showdown over slavery was inevitable for moral and political reasons was first popularized by Allan Nevins, in *The Ordeal of the Union* (New York: Scribner, 1947) and *The Emergence of Lincoln*, 2 vols. (New York, Scribner, 1950); and reinforced by James M. McPherson, *Ordeal by Fire: The Civil War and Reconstruction* (New York: Knopf, 1982). Outstanding subsequent political historians have in turn stressed the play of contingencies through exhaustive reconstruction of the narrative of the 1850s. They include Michael F. Holt, *The Political Crisis of the 1850s* (New York: Wiley, 1978); Joel Silbey, *The Partisan Imperative: The Dynamics of American Politics before the Civil War* (New York: Oxford University, 1985); and William E. Gienapp, *The Origins of the Republican Party, 1852–1856* (New York: Oxford University, 1987). For summaries of the debate besides Kornblith's, see Thomas Pressly, *Americans Interpret Their Civil War*, 2nd ed. (New York: Free Press, 1965); Eric Foner, *Politics and Ideology in the Age of the Civil War* (New York: Oxford University, 1980); Kenneth M. Stampp, ed., *The Causes of the Civil War*, rev. ed. (Englewood Cliffs, N.J.: Prentice-Hall, 1991); and Gabor S. Boritt, ed., *Why the Civil War Came* (New York: Oxford University, 1996).

47. Clearly, I come down on the side of Eric Foner, *Free Soil, Free Labor, Free Men: The Ideology of the Republican Party before the Civil War* (New York: Oxford University, 1970), pp. 1–10; Potter, *Impending Crisis*, pp. 18–50; and Kornblith, "Rethinking the Coming of the Civil War," with regard to an intractable conflict over extension of slavery following the Mexican War. But I also bow to Holt and Gienapp with regard to the contingent events that defined how and when the conflict erupted. In their own ways they all refuse to see economic, cultural, and moral factors as mutually exclusive. They were all of a piece or, as Foner puts it, a gestalt.

48. Foner, *Free Soil, Free Labor, Free Men*, pp. 11–72 ("morally unjust," p. 40; "Enslave a man," p. 46; "to destroy slavery" from George Weston, *The Progress of Slavery in the United States*, 1857, p. 57). Daniel Walker Howe, *The Political Culture of the American Whigs* (Chicago, Ill.: University of Chicago, 1979), pp. 181–209, 263–269 ("You are destitute," p. 269; "a blessed boon," p. 189); Robert W. Fogel, *Without Consent or Contract: The Rise and Fall of American Slavery* (New York: Norton, 1989), pp. 320–344 ("pistols, dirks" and "erotic society," p. 327. Emerson ("more civilized") is quoted in Lewis O. Saum, *The Popular Mood of Pre–Civil War America* (Westport, Conn.: Greenwood, 1980), p. 170. On southern illness, see Todd L. Savitt and James Harvey Young, eds., *Disease and Distinctiveness in the American South* (Knoxville: University of Tennessee, 1988). On "female afflictions," see Catharine Clinton, *The Plantation Mistress: Woman's World in the Old South* (New York: Pantheon, 1982); and Deborah Gray White, *Ar'n't I a Woman? Female Slaves in the Plantation South* (New York: Norton, 1985). For an idiosyncratic interpretation of southern sexuality, see Earl E. Thorpe, *Eros and Freedom in Southern Life and Thought* (Westport, Conn.: Greenwood, 1967). He insists on the "mutual respect, harmony, and affection" prevailing between the races (p. 3), citing as proof the mixed offspring. He means to say that some white swains, at least, must have felt tenderness toward their intimate partners and mulatto children. As for the overall charge of southern debauchery, Thorpe writes (p. 96): "Historians must accept the fact that, at one pole of thought or behavior, the Old South was primitive, barbaric, wild, sexually uninhibited, hedonistic. So was the North." Touché.

49. Foner, *Free Soil, Free Labor, Free Men*, pp. 73–102 (Chase quoted on p. 87); Leonard L. Richards, *The Slave Power: The Free North and Southern Domination 1780–1860* (Baton Rouge: Louisiana State University, 2000), pp. 1–27; David Brion Davis, *The Slave Power Conspiracy and the Paranoid Style* (East Lansing: Michigan State University, 1969); Russell Nye, *Fettered Freedom* (East Lansing: Michigan State University, 1963), pp. 282–315. Belief in a slave-power conspiracy was such a potent political weapon in the 1850s and such a moral balm during and after the Civil War that not until 1921 was it seriously challenged by historians. See Chauncey S. Boucher, "In Re That Aggressive Slaveocracy," *Mississippi Valley Historical Review* 8 (1921): 13–80. Boucher was the whistleblower who showed that southern politicians, although powerful, were divided among themselves and incapable of mounting a grand conspiracy. They couldn't even form a united front under the Confederacy.

50. On Seward's commercial expansionism, see Ernest N. Paolino, *The Foundations of the American Empire: William Henry Seward and U.S. Foreign Policy* (Ithaca, N.Y.: Cornell University, 1973); Glyndon Van Deusen, *William Henry Seward* (New York: Oxford University, 1967), "one nation, race" and "take up the cross," pp. 206–207; and John H. Schroeder, *Shaping a Maritime Empire: The Commercial and Diplomatic Role of the Ameri-*

can Navy, 1829–1861 (Westport, Conn.: Greenwood, 1985), pp. 79–99 ("gospel of commerce" and "Commerce is now," p. 82; "no seat of empire," p. 91).

51. Emerson popularized the phrase "Young America" in an essay of 1844 in *The Dial,* but as usual he was borrowing a concept well established in Europe. The movement grew on the strength of Manifest Destiny and the European revolutions of 1848. Young Americans especially cheered Louis Kossuth when he toured the United States in hopes of winning recognition and help for the Hungarian rebellion against the Hapsburg empire. Their most prominent political patrons were Douglas and intellectuals and businessmen in New York hoping to transcend sectional strife in pursuit of a "higher" American nationalism. But their program endorsing U.S. expansion in the Caribbean and elimination of protective tariffs struck northern Free-Soilers and manufacturers as a facade for southern interests. See Johannsen, *Stephen Douglas,* pp. 339–373; Donald S. Spencer, *Louis Kossuth and Young America: A Study of Sectionalism and Foreign Policy, 1848–1852* (Columbia: University of Missouri, 1977); Edward L. Widmer, *Young America: The Flowering of Democracy in New York City* (New York: Oxford University, 1999).

52. On the origins of Perry's and Rodgers's expeditions to Japan, see Schroeder, *Shaping a Maritime Empire,* pp. 139–164. A comprehensive narrative of the mission is Peter Booth Wiley, *Yankees in the Land of the Gods: Commodore Perry and the Opening of Japan* (New York: Viking, 1990). The complicated diplomacy surrounding the mission, competition from the British and Russians, and internal debates in the *bakufu* are described in McDougall, *Let the Sea Make a Noise,* pp. 269–276.

53. John H. Schroeder, *Matthew Calbraith Perry: Antebellum Sailor and Diplomat* (Annapolis, Md.: Naval Institute, 2001), pp. 154–248 ("Columbus, Da Gama," pp. 245–246); Lewis Bush, *77 Samurai: Japan's First Embassy to America* (Tokyo: Kodansha International, 1968); "a gift of Providence" in *The Writings and Speeches of Daniel Webster,* 14 vols. (Boston, Mass.: Little, Brown, 1903), 14: 427–429. Secretary of the Navy William Graham first appointed Commodore John H. Aulick to command the mission, but he never got past Brazil after undiplomatic run-ins with his hosts and the American consul. Perry himself hoped for command of the Mediterranean squadron, accepting the Japanese mission only after the Navy Department granted him a sufficiently intimidating fleet and plenipotentiary powers. Perry had in mind the humiliation suffered by Commodore James Biddle in 1846. Biddle had sailed into Edo Bay with just two ships and no interpreter, allowed his craft to be boarded by crews from Japanese gunboats, then obeyed a minor official's order to be gone. The city and bay of Edo were renamed Tokyo after the Meiji restoration toppled the Tokugawa shogunate in 1868.

54. The full story of the Jones family plantations in Liberty County, Georgia, has been brilliantly reconstructed by Erskine Clarke, *Dwelling Place: A Plantation Epic* (New Haven, Conn.: Yale University, 2005).

55. For descriptions of slavery in the colonial and early national periods see McDougall, *Freedom Just Around the Corner,* pp. 155–164, 445–448, et passim. These generalizations about the lives of the enslaved in the mid-nineteenth century are based on Fogel, *Without Consent or Contract;* Robert William Fogel and Stanley L. Engerman, *Time on the Cross: The Economics of American Negro Slavery* (Boston, Mass.: Little, Brown, 1974); Eugene D. Genovese, *Roll, Jordan, Roll: The World the Slaves Made* (New York: Vintage, 1976), "white trash," p. 72; Larry E. Hudson, Jr., ed.,

Working toward Freedom: Slave Society and Domestic Economy in the American South (Rochester, N.Y.: University of Rochester, 1994); Walter Johnson, *Soul by Soul: Life inside the Antebellum Slave Market* (Cambridge, Mass.: Harvard University, 1999), "likely," p. 138; Albert J. Raboteau, *Slave Religion: The "Invisible Institution" in the Antebellum South* (New York: Oxford University, 2004); Timothy E. Fulop and Albert J. Raboteau, eds., *African-American Religion: Interpretive Essays in History and Culture* (New York: Routledge, 1997), "The ante-bellum Negro," p. 73; Ira Berlin, *Generations of Captivity: A History of African-American Slaves* (Cambridge, Mass.: Harvard University, 2003), "has been carried," p. 244; Eddie S. Glaude, Jr., *Exodus! Religion, Race, and Nation in Early Nineteenth-Century Black America* (Chicago, Ill.: University of Chicago, 2000), "Gwine to write," p. 4; Lawrence W. Levine, *Black Culture and Black Consciousness: Afro-American Thought from Slavery to Freedom* (New York: Oxford University, 1977).

56. Taylor, *Cavalier and Yankee*, pp 148–176 ("new-rich swells," p. 152; "woman wields a power," pp. 162–163), analyzes plantation novels; Drew Gilpin Faust, *Mothers of Invention: Women of the Slaveholding South in the American Civil War* (Chapel Hill: University of North Carolina, 1996), describes the self-reliance of plantation women; Stephanie McCurry, *Masters of Small Worlds: Yeoman Households, Gender Relations, and the Political Culture of the Antebellum South Carolina Low Country* (New York: Oxford University Press, 1995), argues that defense of male prerogatives helps to account for the support of planters by non-slaveholding men; Chesnut quoted in George Tindall Brown, *America: A Narrative History*, 2 vols., 2nd ed. (New York: Norton, 1988), 1: 573.

57. I rely for this passage on the learned, delightful monograph by Mark W. Summers, *The Plundering Generation: Corruption and the Crisis of the Union, 1849–1861* (New York: Oxford University, 1987); quotations from pp. 7, 283.

58. The place of "hustlers" and "creative corruption" in the origins of the United States is a principal (but not the only) theme of McDougall, *Freedom Just Around the Corner*. See especially pp. 1–16.

59. "Men's convictions as to the Truth, or what they receive as the truth . . . depend entirely upon their understanding of the facts. Convictions are always sincere." So wrote the Georgian Whig Alexander Stephens in *A Constitutional View of the Late War between the States* (1865). His imaginary colloquy purported to prove through relentless logic that northerners' claims of a conspiratorial slavocracy were bogus, whereas southerners' claims of an abolitionist conspiracy were true. See "The Conspiracy Theory Joined and Defined" in Michael A. Morrison, *Slavery and the American West* (Chapel Hill: University of North Carolina, 1997), pp. 157–187 ("Men's convictions," p. 157).

60. Summers, *Plundering Generation*, especially pp. 23–182, 281–296 ("They will steal," p. 143; "were no longer," p. 111; "My God, we have voted," p. 52; "old fogies," p. 179; "arrogant, aggressive," p. 290). The most notorious lobbying scandal of the era occurred in 1854, when Samuel Colt petitioned for an extension of the patent on his revolver. The "borers" hired by his agents seduced the House of Representatives with gifts of wine, women, pistols, and up to $60,000 in cash. But the business raised such a stink that the bill died in the Senate (pp. 202–203). Northern corruption as a motive for southern secessionism is argued by Kenneth Greenberg, *Masters and Statesmen: The Political Culture of American Slavery* (Baltimore, Md.: Johns Hopkins University, 1985). On nineteenth-century corruption generally, see also George C. S. Benson, *Political Corrup-*

tion in America (Lexington, Mass.: Heath, 1978); David Loth, *Public Plunder: A History of Graft in America* (New York: Carrick and Evans, 1936); Culver H. Smith, *The Press, Politics, and Patronage: The American Government's Use of Newspapers, 1789–1875* (Athens: University of Georgia, 1977); Joel H. Silbey, *The Partisan Imperative: The Dynamics of American Politics before the Civil War* (New York: Oxford University, 1985); and Henry Cohen, *Business and Politics in America from the Age of Jackson to the Civil War* (Westport, Conn.: Greenwood, 1971).

61. The snippets from "On Blue Ontario's Shore" (1859 edition) are taken from Michael Moon, ed., *Walt Whitman: Leaves of Grass and Other Writings* (New York: Norton, 2002), pp. 286–299. Of the many biographies of Whitman, Gay Wilson Allen, *The Solitary Singer: A Critical Biography of Walt Whitman*, 2nd ed. (New York: New York University, 1967), is especially complete. See also Philip Callow, *From Noon to Starry Night: A Life of Walt Whitman* (Chicago: Dee, 1992); "simmering" and "to a boil" are cited in Jerome Loving, "Walt Whitman," *American National Biography Online* (American Council of Learned Societies and Oxford University Press, 2000). The seemingly homoerotic themes in Whitman's poems around 1859–1860 raised allegations, which he hotly refuted by claiming to have fathered numerous illegitimate children. Indeed, *Leaves of Grass* became a strong seller only after 1882, when it was banned in Boston because of its steamy heterosexual content. Like all great poetry, however, Whitman's speaks for itself, and speaks volumes. *Drum-Taps* (1865) secured his place as America's psalmist by depicting Union soldiers and at last Lincoln himself as Christlike figures redeeming the nation. After the war Whitman, who never married, lived with his brother in Camden, New Jersey, and died there in 1892.

62. Gienapp, *Origins of the Republican Party*, pp. 3–67, draws a sharp distinction between the processes of party collapse and party construction, then backs it up with a mass of local evidence. Personal loyalties, local rivalries, issues such as immigration and temperance, loyalty, and simple inertia determined whether the decay of the Whigs in a given city, county, or state was swift or prolonged. The fallout from Kansas-Nebraska and the perturbing influence of immigration and nativism in party realignment are analyzed in Cole, *Whig Party in the South*, pp. 277–308; Michael F. Holt, *The Rise and Fall of the Whig Party: Jacksonian Politics and the Onset of the Civil War* (New York: Oxford, 1999), pp. 765–878; and Potter, *Impending Crisis*, pp. 225–265. Potter's tally of the contenders in 1854 (p. 249) includes Democrats; Rum Democrats; Hard-Shell, Soft-Shell, and Half-Shell Democrats; Whigs; Silver Gray Whigs; Free-Soilers; Republicans; People's Party men; anti-Nebraskans; Fusionists; Know-Nothings; Know-Somethings; Temperance men; and Hindoos. On the Know-Nothings (when asked about the movement, they replied, "I know nothing"), see Ray Allen Billington, *The Protestant Crusade, 1800–1860: A Study of the Origins of American Nativism* (New York: Macmillan, 1938); W. Darrell Overdyke, *The Know-Nothing Party in the South* (Baton Rouge: Louisiana State University, 1950); and John Higham, "Another Look at Nativism," *Catholic Historical Review* 44 (1958): 147–158.

63. Gienapp, *Origins of the Republican Party*, pp. 69–237 ("the key-note," "privileged class," and "The Republican Party is sounding," pp. 191–192).

64. Ibid., pp. 286–295. The moderate resolutions passed in Bloomington were drafted by a committee that Orville H. Browning assembled to ensure a hearing for all points of view *except* radical abolitionism. The convention's leaders included such disparate figures

as the die-hard Whig Abraham Lincoln, the machine Democrat John Wentworth, the German spokesman Francis Hoffman, the Know-Nothing stalwart Joseph Gillespie, and the fiery Free-Soiler Owen Lovejoy. What they shared was antipathy to Stephen Douglas and a hunger for victory.

65. Weed boasted of being in total control. He later wrote (in the third person), "It has been said that 'his failure to nominate Mr. Seward in 1856 was a great disappointment to Mr. Weed.' Nothing could be farther from the truth. Had Mr. Weed consented to the plan, Mr. Seward would have been nominated." Seward deferred to Weed's judgment, but would have preferred to bow out in advance rather than be nominated just so that Weed could keep the New York delegation in reserve. See Barnes, ed., *Memoir of Thurlow Weed*, pp. 244–245.

66. Gienapp, *Origins of the Republican Party*, pp. 305–346 ("those twin relics," p. 335; "no past political sins," p. 324; "spontaneous instinct," p. 317). The platform's planks regarding slavery would have been milder but for a rousing address made by Cassius Marcellus Clay of Kentucky at a preconvention caucus in Pittsburgh. On his remarkable life, see David L. Smiley, *Lion of White Hall: The Life of Cassius M. Clay* (Gloucester, Mass.: Peter Smith, 1969). Clay was born to a family wealthy in land and slaves, attended Transylvania and Yale, and entered politics as a fervent Whig in the mold of his distant cousin Henry Clay. After the Panic of 1837 convinced him that slavery was uneconomical as well as immoral, Clay dared to publish an abolitionist journal in his native slave state. His rewards were mobs, brawls, duels, and the nickname "Fighting Christian." Clay then scandalized northern friends by volunteering for Mexico (where he was taken prisoner), but he restored his reputation as an abolitionist by promoting free black communities in the Kentucky hills. In 1856 he served on the Republican National Committee, and in 1860 he finished second in the balloting for vice-presidential nominee. Clay then served as U.S. minister to Russia for seven years. Alas, an eccentric, protracted old age sullied all those achievements. He had been a handsome, clean-shaven youth, but now he let his white hair and beard grow wild. His sexual mischief caused the mother of his ten children to sue for divorce. At age eighty-four he married, then divorced, a fifteen-year-old servant. His political gyrations alienated old friends, and his paranoia led him to kill an alleged trespasser on his White Hall estate. He died in 1903, aged ninety-two.

67. Gara, *Presidency of Pierce*, pp. 157–184 ("the standard by which," p. 168); Gienapp, *Origins of the Republican Party*, pp. 375–411; Potter, *Impending Crisis*, pp. 259–265 ("election of Frémont," p. 262). Richard J. Carwardine, *Evangelicals and Politics in Antebellum America* (New Haven, Conn.: Yale University, 1993), pp. 235–278 ("truth and falsehood" and "Think that God's eye," p. 269). "There is no such person" in Paul F. Boller, *Presidential Campaigns* (New York: Oxford University, 1984), p. 98; "We Fremonters" in Stefan Lorant, *The Glorious Burden: The American Presidency* (New York: Harper and Row, 1968), p. 225. In addition to Toombs such notables as James M. Mason, John Slidell, Jefferson Davis, Andrew P. Butler, and Henry A. Wise voiced their hopes or fears that a Republican presidency would provoke southern secession.

68. "The election has ended," in Myers, ed., *Children of Pride*, pp. 263–264 (Letter of November 17, 1856); Gienapp, *Origins of the Republican Party*, pp. 413–448 ("If months," p. 448); "with no pretense" are Potter's own words in *Impending Crisis*, p. 265; "have not got" cited in Gara, *Presidency of Pierce*, p. 177.

69. One reason casual acquaintances thought Buchanan charming was his habit

of cocking his head in conversation, which gave the impression of deep concentration. In fact, he had poor vision in one eye and was just trying to see who was talking. Buchanan was so anal-retentive that he kept receipts and records of every pittance he spent every day of his life. But personal quirks aside, even an experienced diplomat was not the person the situation required. Diplomacy presupposes the parties are sovereign (not part of a union) and given to compromise (but neither side was after 1854). Everything said or done by Buchanan just made matters worse. For that reason, the mildly apologetic biography by the Pennsylvanian historian Philip S. Klein, *President James Buchanan* (University Park: Pennsylvania State University, 1962), is the closest anything comes to revisionism. See the assessments in Michael J. Birkner, ed., *James Buchanan and the Political Crisis of the 1850s* (Selinsgrove, Pa.: Susquehanna University, 1996). Buchanan's address of March 4, 1857, in *I Do Solemnly Swear: The Inaugural Addresses of the Presidents of the United States* (Philadelphia, Pa.: Chelsea House, 2001), pp. 135–143 ("This is, happily," p. 138). By the way, the official record of the Dred Scott decision refers to *Scott v. Sanford* (instead of Sandford) owing to a clerical error.

70. See Don E. Fehrenbacher, *The Dred Scott Case* (New York: Oxford University, 2001, orig. 1978), especially the passages on reactions to the decision, pp. 417–448 ("atrocious" and "detestation," p. 417); Curtis's minority opinion, pp. 403–414; and speculations as to the Court's motives, pp. 552–561 ("cupidity," "evil passions," and "Northern insult," pp. 557–558); Potter, *Impending Crisis*, pp. 267–296 ("five slave holders," p. 284); and Guelzo, *Crisis of the American Republic*, pp. 58–67 ("inferior class" and "no rights," pp. 64).

71. Elbert B. Smith, *The Presidency of James Buchanan* (Lawrence: University of Kansas, 1975), pp. 31–46; Potter, *Impending Crisis*, pp. 297–327 ("Mr. President," p. 316; "as much a slave state," p. 320). Robert J. Walker had deep roots in both Pennsylvania and Mississippi, and was known to be a patriotic Unionist. As a senator he had managed Polk's presidential campaign in 1844, then served Polk as secretary of the treasury. Lately he had built up a respected law practice in Washington City and even lent Buchanan his town house during the campaign of 1856. Privately Walker had turned against slavery, but he meant to honor the true opinions of Kansans as the president asked him to do. When Buchanan instead endorsed Lecompton, Walker considered it a betrayal and resigned. He would support the Union wholeheartedly in the Civil War. See James Patrick Shenton, *Robert John Walker: A Politician from Jackson to Lincoln* (New York: Columbia University, 1961).

72. Ralph H. Bowen, ed., *A Frontier Family in Minnesota: Letters of Theodore and Sophie Bost 1851–1920* (Minneapolis: University of Minnesota, 1981). This extraordinary collection fulfills the editor's promise: "These letters tell us a great deal about this psychic dimension of pioneering. Beyond physical obstacles and economic adversities, they provide what must be a virtually unique first-person account of how it actually felt to be an immigrant, an exile, a dweller on the forest's edge" (p. xii); "You ask how," p. 222; "the Indians," p. 224.

73. In 1820 the expedition led by Governor Lewis Cass, including Schoolcraft, boated across Lake Superior, struck inland to the Mississippi, and poled north. But they stopped at Lake Cass, just a day or two short of the river's source. Twelve years later Schoolcraft and the missionary William Boutwell pressed on to glory. They wanted a fancy classical name, but the only Latin words Boutwell could recall were *veritas* ("truth") and *caput* ("head"), so Schoolcraft dropped their first and last syllables, then

merged the words into Itasca. His journal and other documents are collected in Philip P. Mason, ed., *Expedition to Lake Itasca: The Discovery of the Source of the Mississippi* (East Lansing: Michigan State University, 1958). Minnesota's French heritage is honored at Voyageurs National Park, in the state's northeastern corner. Minnesota's Indian heritage is honored at Pipestone National Monument in the state's southwestern corner. It preserves a quarry sacred to Indians from the Appalachians to the Rockies because its red rock was carved into peace pipes. Minnesota's Viking heritage, based on the "Kensington stone" hoax, is preserved in the name of the Twin Cities' professional football team.

74. Theodore C. Blegen, *Minnesota: A History of the State* (Minneapolis: University of Minnesota, 1963), pp. 107–163 ("active and stirring," p. 135); Roger G. Kennedy, *Men on the Moving Frontier* (Palo Alto, Calif.: American West, 1969), pp. 39–73 ("lithe, athletic" and "heavy, phlegmatic," p. 39; "The frontier," p. 57). During his years with the Sioux Sibley took Red Blanket Woman as his wife, but separated from her when he was called back to live among white men. He then returned to his roots by marrying a Yankee woman who bore him nine children. As for the name Minnesota, nobody knows who first coined it. All that is known is that Mississippi was already taken; Protestant opinion was offended by Saint Croix or Saint Peter (the French name for the Minnesota River); and Cannon, Rum, and Crow-Wing were nonstarters. Some members of Congress plumped for Washington or Jackson. But Sibley carried the day for mellifluous Minnesota. See George R. Stewart, *Names on the Land: A Historical Account of Place-Naming in the United States* (Boston, Mass.: Houghton Mifflin, 1967), pp. 257–260.

75. Blegen, *Minnesota: A History*, pp. 159–210 ("peopled with," p. 164; "the real parties," p. 170). A transplanted Ohioan boosted Minnesota in another effective way. He sang: "The Gopher girls are cunning / The Gopher girls are shy / I'll marry me a Gopher girl / Or a bachelor I will die" (p. 202). On Ireland's extraordinary career in the Northwest and in the United States as a whole, see Marvin Richard O'Connell, *John Ireland and the American Catholic Church* (St. Paul: Minnesota Historical Society, 1988); and James H. Moynihan, *The Life of Archbishop John Ireland* (New York: Arno, 1976, orig. 1953).

76. Blegen, *Minnesota: A History*, pp. 213–230.

77. Sophie Bost wrote to her in-laws in November 1862 that she might understand how one could hesitate to take sides in the Civil War if not that most of the Union army was Republican whereas the rabble Irish were all pro-slavery and pro-peace at any price. "I can't view the matter in this way, seeing, as I have since 1850, the progress—religious, if not political—that the people of the North have made on the slavery issue. As was to be expected, it has been impossible to keep politics from getting entangled with this progress, which culminated in the election of Lincoln, an election which signified that slavery was *sectional*, but liberty *national* . . . and that slavery was a cancer that must be arrested by all means and destroyed if possible": Bowen, ed., *Frontier Family in Minnesota*, p. 219.

78. Chester M. Oehler, *The Great Sioux Uprising* (New York: Oxford University, 1959); Kennedy, *Men on the Moving Frontier*, pp. 61–73 ("as sure as there is a God," p. 63). The uprising began after some braves murdered five settlers. Red Cloud, knowing the whites were bound to retaliate anyway, used the occasion to make himself war chief. Ramsey appealed to the White House, but the most Lincoln could do was dispatch General John Pope, who seconded the governor's call for extermination. As hideous as the uprising was, Blegen puts it in perspective by noting that the number of whites killed by

the Sioux was less than those killed in New York's draft riots the following year. Sibley, his public career at an end, retired to Saint Paul, where he lived until 1891, lending his name to businesses and promoting the Minnesota Historical Society he had helped to found. Ramsey bestrode the state as governor and U.S. senator until the reformer Ignatius Donnelly ousted him in 1875. It was said that Ramsey used "hope, fear, avarice, ambition, personal obligations, money, whiskey, oysters, patronage, contracts, champagne, loans, the promise of favors, jealousy, personal prejudice, envy—everything that could be tortured into a motive" to promote his wheeling and dealing in business and politics. Ramsey's reply to such charges was, "Well, I can't see what all the fuss is about" (*Men on the Moving Frontier*, pp. 68–71). His face graven and his muttonchops graying, he continued to preside from his eclectic mansion in Saint Paul until 1903.

79. On the grasshopper plagues, see the discussion and maps in William Watts Folwell, *The History of Minnesota*, 4 vols. (St. Paul: Minnesota Historical Society, 1921–1930), 3: 93–111. On post–Civil War agriculture, see Merrill E.. Jarchow, *The Earth Brought Forth: A History of Minnesota Agriculture to 1885* (St. Paul: Minnesota Historical Society, 1949), pp. 80–261.

80. Blegen, *Minnesota: A History*, pp. 287–383; Leonard H. Bridges, *Iron Millionaire: The Life of Charlemagne Tower* (Philadelphia: University of Pennsylvania, 1952).

81. See Jennifer A. Delton, *Making Minnesota Liberal: Civil Rights and the Transformation of the Democratic Party* (Minneapolis: University of Minnesota, 2002). The farmer-labor alliance emerged in a context of rural hostility to eastern monopolies, banks, and railroads. But tensions among rural and urban workers, and among ethnic and religious groups, were overcome only when the Wilson administration's wartime Commission of Public Safety targeted labor unions, war dissenters, and "seditious" Germans and Irish. In 1918 the League endorsed Charles A. Lindbergh, Sr., as its gubernatorial candidate in the Republican primary. His strong showing encouraged the founding, in 1921, of the Farmer-Labor Party (FLP) that consigned the Democrats to third-party status in the interwar years. The FLP foundered in 1938, when its factions quarreled over communism and Harold Stassen, a Republican, co-opted much of its relief program. That led to a merger with the Democrats in 1944 and the emergence of today's Democratic Farmer-Labor Party under the energetic leadership of the young graduate student Hubert Humphrey.

82. The details and mood of those months are nicely described by Robert Sobel, *Machines and Morality: The 1850s* (New York: Thomas Y. Crowell, 1973), pp. 224–250 ("What can be the end," p. 232). An excellent full-length account is James L. Huston, *The Panic of 1857 and the Coming of the Civil War* (Baton Rouge: Louisiana State University, 1987). Huston argues the economic depression exacerbated sectional tensions when northerners cried out for the high tariffs and internal improvements so despised in the South.

83. See Kathryn Teresa Long, *The Revival of 1857–1858: Interpreting an American Religious Awakening* (New York: Oxford University, 1998), pp. 3–92 ("every nook and cranny," p. 13; "melancholy days," p. 51); and Timothy L. Smith, *Revivalism and Social Reform: American Protestantism on the Eve of the Civil War* (New York: Abingdon, 1957). Greeley's *Tribune* and the competing *New York Herald* decided that tough Wall Street financiers praying and sobbing were news. The journalistic obsession peaked during Lent in 1858, culminating in an "extra" edition of the *Tribune* on Good Friday. As for celebrity converts, none caused more excitement than Orville "Awful" Gardner, a bare-knuckle

boxing champion who had done time in Sing-Sing for breaking the jaw of a visiting busi-nessman. George Templeton Strong, the snide Wall Street lawyer, recorded a couplet: "Ye Saints rejoice, give cheerful thanks / For Awful Gardner's joined your ranks" (p. 42).

84. The pastor of New York's Church of the Puritans, George B. Cheever, believed that the revival was God's way of forcing Americans finally to face up to slavery. "Where the Church does not apply God's Word against sin," he cried, "there both the conscience toward God and the spirit of liberty are debauched and wasted, and the nation ripens for destruction." Cheever and Lewis Tappan urged the newly prayerful businessmen to sup-port an antislavery resolution at the American Tract Society in May 1858. Opponents defeated the resolution in defense of the "tranquillity and peace" of the nation, but some at least were moved to concede that slavery was sinful (Long, *Revival of 1857–1858*, pp. 112–115). The southern journals are quoted in Sobel, *Machines and Morality*, p. 248; Bar-rett quoted in Foner, *Free Soil, Free Labor, Free Men*, p. 313.

85. Smith, *Presidency of Buchanan*, pp. 65–68; Hirshson, *Lion of the Lord*, pp. 152–183 ("Money is my God," p. 161).

86. On the maneuvers preceding the Illinois conventions of 1858, see Johannsen, *Stephen Douglas*, pp. 570–572; and Allen C. Guelzo, *Abraham Lincoln, Redeemer President* (Grand Rapids, Mich.: Eerdmans, 1999), pp. 209–217. "'A House Divided': Speech at Springfield, Illinois," in Roy Prentice Basler, ed., *The Collected Works of Abraham Lincoln*, 9 vols. (New Brunswick, N.J.: Rutgers University, 1953–1955), 2: 461–466.

87. I do not claim even partial familiarity with the vast literature on Lincoln, pre-ferring to put my trust (I believe it is warranted) in Guelzo, *Redeemer President*, and the newer works cited in subsequent notes. That said, a broad range of scholarly interpreta-tion of Lincoln the man and the statesman can be harvested from Allan Nevins, *The Emergence of Lincoln*, 2 vols. (New York: Scribner, 1950); Benjamin P. Thomas, *Abraham Lincoln* (New York: Knopf, 1952); Don E. Fehrenbacher, *Prelude to Greatness: Lincoln in the 1850s* (Stanford, Calif.: Stanford University, 1962); Stephen B. Oates, *With Malice toward None: The Life of Abraham Lincoln* (New York: Harper and Row, 1977); G. S. Boritt, *Lincoln and the Economics of the American Dream* (Memphis, Tenn.: Memphis State University, 1978); G. S. Boritt, ed., *The Lincoln Enigma: The Changing Faces of an American Icon* (New York: Oxford University, 2001); Oscar Handlin and Lillian Hand-lin, *Abraham Lincoln and the Union* (Boston, Mass.: Little, Brown, 1980); Mark E. Neely, Jr., *The Last Best Hope of Earth: Abraham Lincoln and the Promise of America* (Cambridge, Mass.: Harvard University, 1993); and David H. Donald, *Lincoln* (New York: Simon and Schuster, 1995).

88. Guelzo, *Redeemer President*, pp. 26–142 ("a policy Match," p. 100; "handbook on infidelity" and "doctrine of necessity," p. 117); "cold, calculating" from "Address before the Young Men's Lyceum of Springfield" in Basler, ed., *Collected Works*, 1: 115. On Lin-coln's days as a toddler and his later connections to his birth state, see Lowell H. Harri-son, *Lincoln of Kentucky* (Lexington: University of Kentucky, 2000). Lincoln is rightly said not to have paid much attention to slavery until the 1850s, but he always despised it—especially after his second journey to New Orleans in 1831 when he witnessed the mistreatment of slaves from Kentucky sold down the river. Herndon recalled long de-bates during which Lincoln refused to concede an inch of his philosophy, yet Lincoln's ambition, said Herndon, was a "little engine that never knew rest"—Emanuel Hertz, ed., *The Hidden Lincoln, from the Letters and Papers of William H. Herndon* (New York:

Viking, 1938), p. 124. Lincoln's humor, which he said served to restore his spirits as a shot of whiskey served other men, is celebrated in P. M. Zall, ed., *Abe Lincoln Laughing: Humorous Anecdotes from Original Sources by and about Abraham Lincoln* (Knoxville: University of Tennessee, 1995).

89. One welcome trend in recent Lincoln scholarship is an emphasis on his ideas and rhetoric: that is, an insistence that we study and take seriously Lincoln's speeches and writings rather than just troll for his motives. I am convinced that civic religion and rhetoric are keys to an understanding of Lincoln the statesman. See John Channing Briggs, *Lincoln's Speeches Reconsidered* (Baltimore, Md.: Johns Hopkins University, 2005); Michael Lind, *What Lincoln Believed: The Values and Convictions of America's Greatest President* (New York: Doubleday, 2004); Joseph R. Fornieri, *Abraham Lincoln's Political Faith* (De Kalb: Northern Illinois University, 2003); Stewart Winger, *Lincoln, Religion, and Romantic Cultural Politics* (De Kalb: Northern Illinois University, 2003); and Douglas L. Wilson, *Lincoln's Sword: The Presidency and the Power of Words* (New York: Knopf, 2006).

90. "Speech at Peoria" in Basler, *Collected Works*, 2: 249–278; see the extended analysis with biblical references in Fornieri, *Lincoln's Political Faith*, pp. 104–132.

91. "Biblical republicanism" is the phrase coined in Fornieri, *Lincoln's Political Faith*, pp. 35–69. Lincoln as the embodiment of the American creed or dream is the theme in Jim Cullen, *The American Dream: A Short History of an Idea That Shaped a Nation* (New York: Oxford University, 2003), pp. 74–102. Lincoln as a Romantic nationalist just as steeped in millenarianism as the revivalist Christians of the era is argued in Winger, *Lincoln, Religion, and Romantic Cultural Politics*. "The great democrat" as a label for Lincoln is argued in Lind, *What Lincoln Believed*, pp. 20–27, 265–314. The provenance and argument of Lincoln's speech on discoveries and inventions are analyzed in depth in Briggs, *Lincoln's Speeches*, pp. 184–220.

92. On the imponderable but fascinating interplay of these two giants of the American civic religion, see Daniel Mark Epstein, *Lincoln and Whitman: Parallel Lives in Civil War Washington* (New York: Ballantine, 2004); quotations, pp. 8–9, 26–27.

93. On the Lincoln-Douglas debates, see Johannsen, *Stephen Douglas*, pp. 614–679 ("destiny which Providence" and "This is a young," pp. 671–672; "a war of sections" and "erecting a despotism," p. 642); Potter, *Impending Crisis*, pp. 328–355 ("The real issue" and "That is the issue," p. 339); Guelzo, *Redeemer President*, pp. 209–227; and Harry V. Jaffa, *Crisis of the House Divided: An Interpretation of the Issues in the Lincoln-Douglas Debate*, rev. ed. (Chicago, Ill.: University of Chicago, 1982). Jaffa, pp. 405–409, relates how Lincoln used Douglas's expansionism to expose the fraudulence of his popular sovereignty. Lincoln asked him at Freeport whether he favored all territorial annexations regardless of the impact on issues of slavery. Douglas predictably gave his usual recitation about America's destiny to expand from the equator to the pole. Would Judge Douglas then permit Mexicans to exercise sovereignty under the American flag? Well, no: no self-government for colored people. Jaffa concludes with the exaggerated, but nonetheless cutting, observation that Douglas's vision amounted to *Lebensraum* and racial supremacy. As for Lincoln, he never used his most cutting logic suggesting that black slavery threatened white liberty, perhaps because it was too verbose or abstract for the stump. But he did record how his argument ran: "If A. can prove, however conclusively, that he may, of right, enslave B.—why may not B. snatch the same argument, and prove

equally, that he may enslave A.?" If it were a matter of color, then you yourself might be enslaved by the first man you meet with a fairer skin; if intelligence, by the first man you meet with a superior intellect; if interest, then by the first man claiming an interest in enslaving you (Basler, *Works of Lincoln*, II: 222–223).

94. Legend of the trail junction recounted in Meinig, *Transcontinental America*, pp. 69–70; theories of the name assessed by Stewart, *Names on the Land*, pp. 153–55.

95. Samuel N. Dicken and Emily F. Dicken, *The Making of Oregon: A Study in Historical Geography* (Portland: Oregon Historical Society, 1979), is an outstanding resource.

96. Robert J. Loewenberg, *Equality on the Oregon Frontier: Jason Lee and the Methodist Mission, 1834–1843* (Seattle: University of Washington, 1976), pp. 195–241 ("unprincipled white men" and "not seem," p. 235); Malcolm Clark, Jr., *Eden Seekers: The Settlement of Oregon, 1818–1862* (Boston, Mass.: Houghton Mifflin, 1981), pp. 157–160 ("except in just and lawful wars," p. 159). H. H. Bancroft's *History of Oregon* appeared as Vols. 29 and 30 of his *Works* (1886–1888). A "solemn conclave of democracy" was the characterization of Dorothy O. Johansen and Charles M. Gates, *Empire of the Columbia*, 2nd ed. (New York: Harper and Row, 1967), p. 189.

97. On the 125 Indian tribes numbering perhaps 100,000 before the arrival of Europeans, see Dicken and Dicken, *Making of Oregon*, pp. 35–47. They were primitive hunters and gatherers with no agriculture or domestic beasts except dogs, but the river tribes excelled at fishing. The Cayuse directed their anger at whites for obvious reasons, but in fact they were already being pushed out of the Willamette by other tribes when the settlement of Oregon began. Clark, in *Eden Seekers*, pp. 196–219, describes the Cayuse war that dragged on until 1849. The Cayuse hoped for help from the feisty Nez Percé, but white commissioners persuaded them to remain neutral. After delay and confusion due to Oregon's having no money or government, the First Oregon Rifles formed under Colonel Cornelius Gilliam. A forty-nine-year-old native of Florida and a Baptist preacher, he had fought Seminoles, Black Hawks, and Mormons (in Missouri). Gilliam's first sally was routed. But the Cayuse, assured by their shaman that the whites could not harm them, risked a pitched battle. The disciplined volleys and charges of Gilliam's men routed the braves. William Henry Gray, in *A History of Oregon, 1792–1849, Drawn from Personal Observation and Authentic Information* (Portland, Ore.: Harris and Holman, 1870), quoted depositions allegedly proving that Catholic missionaries incited the war. The open letter signed by fifteen "young ladies," by contrast, was genuine. They solemnly pledged "to evince, on all suitable occasions, our detestation and contempt for any and all young men, who *can*, but *will not*, take up arms and march at once to the seat of war, to punish the Indians, who have not only murdered our friends, but have grossly insulted our sex. We never can, and never will, bestow our confidence upon a man who has neither patriotism nor courage enough to defend his country and the girls;—such a one would never have sufficient sense of obligations to defend and protect a *wife*. Do not be uneasy about your claims and your rights in the valley; while you are defending the rights of your country, she is watching yours" (pp. 573–574). Gray concludes by lamenting that since those days "the Papal superstition has increased among the Indians, thus rendering them more hopelessly depraved, and consigning them and their descendants to unending superstition and ignorance, or to utter oblivion as a race, to be superseded by an enlightened Christian, American people" (p. 624).

98. Ronald B. Lansing, *Nimrod: Courts, Claims, and Killing on the Oregon Frontier* (Pullman: Washington State University, 2005), is a fascinating account of the truth behind the legend of Nimrod O'Kelly ("all laws heretofore passed," p. 45). He accidentally killed a belligerent squatter in May 1852 when his shotgun discharged during a fight. He then turned himself in. But another murder had occurred nearby, convincing the jury that O'Kelly was lying. Still, his age made him sympathetic and his tall tales were dramatic (he claimed to have fought with Jackson at New Orleans). Twice his death sentence was commuted at the last minute. Meanwhile, his civil suits dragged on. At age seventy-five Nimrod even traveled to Washington City to press his land claims, but he died without vindication in 1864. His children finally won court approval of the "O'Kelly Patent" in 1881, thirty-five years after Nimrod staked it out and seventeen years after his death. The claim, worth about $400 in the 1850s, was then valued at $32,000.

99. On territorial politics and statehood, see Robert W. Johannsen, *Frontier Politics and the Sectional Conflict: The Pacific Northwest on the Eve of the Civil War* (Seattle: University of Washington, 1955); James E. Hendrickson, *Joe Lane of Oregon: Machine Politics and the Sectional Crisis, 1849–1861* (New Haven, Conn.: Yale University, 1967); Henry L. Simms, "The Controversy over the Admission of the State of Oregon," *Mississippi Valley History Review* 32 (1945): 255–274; and Clark, *Eden Seekers*, pp. 268–296 ("The true policy," p. 290). Lane was elected one of Oregon's first senators, then ran for vice president on the southern ticket in 1860. But he remained grudgingly Unionist after secession even when his son, a cadet at West Point, was commissioned in the Confederate army. Lane retired to his remote farm near Roseburg in Oregon, where he lived until 1881.

100. Oregon's tribes resisted confinement on reservations, but like Indians elsewhere failed to coordinate their action. The Modoc War in the Klamath (1872), the Nez Percé War in the Blue Mountains (1877), and the Bannock War in the southeastern desert (1878) were thus isolated and brief affairs. By 1883 the Indians had become so invisible (as one scholar observed) that whites could not even be bothered to hate them.

101. Jewel Lansing, *Portland: People, Politics, and Power, 1851–2001* (Corvallis: Oregon State University, 2003). On railroad construction, see Dicken and Dicken, *Making of Oregon*, pp. 105–133.

102. George MacDonald Fraser, *Flashman and the Angel of the Lord* (New York: Knopf, 1995), p. 206. In a back note (p. 388) Fraser elaborates: "There are two words to describe John Brown's appearance: grim and formidable." Even allowing for the stiffness common in nineteenth-century daguerreotypes, he says "the long Anglo-Saxon head, prominent nose and ears, wide mouth set like a trap, stern certainty of expression, and above all, the level, implacable eyes" call to mind at once the words "Ironside, Yankee, Puritan, and Covenanter." The question of Brown's mental balance has been addressed by all his biographers, but perhaps answered best by Allen C. Guelzo, "Terrorist or Madman? A Review of *John Brown: The Legend Revisited*, by Merrill D. Peterson," *Claremont Review of Books* (Fall 2003). Guelzo ("gigantic figure" is his phrase) concludes that liberal secular societies habitually dismiss religious or ideological zeal as symptomatic of some pathology, and that this renders them all the more vulnerable.

103. Potter, *Impending Crisis*, pp. 356–376 ("peculiar," p. 372; "a bundle" and "a man of clear head," p. 375; "then one-fourth of the people," p. 374). The "Secret Six" were Gerrit Smith, George L. Stearns, Franklin B. Sanborn, Theodore Parker, Samuel Gridley

Howe, and Thomas Wentworth Higginson (the only one who thought Brown would cease and desist after Forbes's betrayal). After the raid at Harpers Ferry all but Higginson denied involvement, claimed ignorance of Brown's real intentions, fled, or went mad (Smith). Potter points out that the absence of insurrections during the Civil War belies the widespread prewar belief that slaves were eager for violence.

104. This counterfactual situation is posed in Smith, *Presidency of Buchanan*, pp. 91–99.

105. Potter, *Impending Crisis*, pp. 376–384 ("make the utmost," p. 376; "I see a book," p. 377; "Never permit," p. 383; "The whole abolition crusade" in a letter of November 7, 1859, *Children of Pride*, pp. 527–528. Jones's full text is worth citing as a revelation of southerners' perceptions after Brown's raid: "Some of the papers friendly to the South hope that the South will be forbearing and magnanimous! Against the miserable lives of these men who have plotted arson, robbery, murder, and treason over a vast portion of our country, who may weigh millions in property, millions of lives, the virtue, the order, the peace and happiness of our people, the majesty of the laws, the sacredness of religion, our Constitution and our Union?"

106. See Albert D. Kirwan, *John J. Crittenden: The Struggle for the Union* (Lexington: University of Kentucky, 1962), pp. 336–365; and Thomas H. O'Connor, *Lords of the Loom: The Cotton Whigs and the Coming of the Civil War* (New York: Scribner, 1968), pp. 114–167. The meeting in Boston was called by Levi Lincoln, the seventy-seven-year-old "cotton Whig" from Worcester, who had served Massachusetts as a governor, senator, representative, and mayor. He rallied textile manufacturers and others who deplored the prospect of secession and war disrupting the cotton trade. Capitalists in New York and Philadelphia shared this concern. Indeed, O'Connor notes that not a single sector of northern business expected to profit from sectional strife. So much for the Marxist theory of an "irrepressible conflict" between economic systems.

107. The best source on the Democrats' split may still be Roy F. Nichols, *The Disruption of American Democracy* (New York: Macmillan, 1948); on the conventions of 1860, see pp. 288–322. See also Avery O. Craven, *The Growth of Southern Nationalism, 1848–1861* (Baton Rouge: Louisiana State University, 1953), pp. 323–334; William J. Cooper, Jr., *Liberty and Slavery: Southern Politics to 1860* (New York: Knopf, 1983), pp. 248–285; and Johannsen, *Stephen Douglas*, pp. 749–807 ("Secession from the Democratic," p. 772). Both tickets made a pretense of balance. Douglas's running mate was Herschel V. Johnson of Georgia, a southerner devoted to Union; and Breckinridge's was Joe Lane, who founded the state of Oregon and sympathized with the South. Neither choice made any impact at all.

108. The rollicking mood of the Republicans and the city of Chicago, whose population of 108,000 temporarily increased by more than one-third during the convention, is captured in Theodore J. Karamanski, *Rally 'Round the Flag: Chicago and the Civil War* (Chicago, Ill.: Nelson-Hall, 1993), pp. 1–31. Would Lincoln have been nominated in Indianapolis, the party's initial choice for host city? There is no telling, but Republicans in Illinois can be credited with hammering the Wigwam together at their private expense in order to land the convention.

109. Potter, *Impending Crisis*, pp. 405–429 ("An Anti-Slavery man," p. 420; "gravest of crimes," p. 423; "oceans of money," p. 425); Summers, *Plundering Generation*, pp. 266–280 ("King of the Lobby," p. 267; "bought and paid for" and "is it wise," p. 270; "We

won't pay it," p. 272). "Lincoln ain't here" is from Henry Clay Whitney, *Lincoln the Citizen* (New York: Baker and Whitney, 1907), p. 289, but may be apocryphal. For biographies of the dueling managers see Willard L. King, *Lincoln's Manager, David Davis* (Cambridge, Mass.: Harvard University, 1960); and Glyndon G. Van Deusen, *Thurlow Weed, Wizard of the Lobby* (Boston, Mass.: Little, Brown, 1947). On the exposé by the Covode Committee of Democrats' corruption, see Smith, *Presidency of Buchanan*, pp. 99–104, 121.

110. Potter, *Impending Crisis*, pp. 430–447 ("old Mumbo Jumbo," "Who's afraid," and "too much of good sense," pp. 431–432; "Mr. Lincoln is the next," p. 441). Potter anticipates my theme by observing about the Republicans (p. 433): "Tactically it was perhaps shrewd, if not wise, to pretend that there was no serious danger. Yet tactics did not require them to deceive themselves with their own pretense." On Douglas's futile southern tour, see Lionel Crocker, "The Campaign of Stephen A. Douglas in the South, 1860," in John Jeffrey Auer, ed., *Antislavery and Disunion, 1858–1861: Studies in the Rhetoric of Compromise and Conflict* (New York: Harper and Row, 1963), pp. 262–278; and Johannsen, *Stephen Douglas*, pp. 786–803.

111. Let me take this opportunity to plug one of the most useful resources on the Internet: "Dave Leip's Atlas of U.S. Elections" at http: www.uselectionatlas.org.

112. Michael Davis, *The Image of Lincoln in the South* (Knoxville: University of Tennessee, 1971), pp. 3–40 ("free love," p. 14; "This is just," pp. 16–17); Mitchell Snay, *Gospel of Disunion: Religion and Separatism in the Antebellum South* (New York: Cambridge University, 1993), pp. 151–175; Simpson, *A Good Southerner*, pp. 210–236; Dwight Dumond, ed., *Southern Editorials on Secession* (New York: Century, 1931); Charles G. Sellers, "The Tragic Southerner," in Charles R. Crowe, ed., *The Age of Civil War and Reconstruction, 1830–1900* (Homewood, Ill.: Dorsey, 1975), pp. 81–108; "the Japanese" quoted in Leonard L. Richards, *The Slave Power: The Free North and Southern Domination, 1780–1860* (Baton Rouge: Louisiana State University, 2000), p. 214; "only this" from the address at the Cooper Institute in Basler, *Works of Lincoln*, III: 547–548.

113. Potter, *Impending Crisis*, pp. 448–499; Smith, *Presidency of Buchanan*, pp. 129–138 ("having invested," p. 130). Cooper, in *Liberty and Slavery*, pp. 248–185, calls William Lowndes Yancey of Alabama the "orator of secession." Yancey cried to every packed hall, "It is the right, expressed in the Declaration of Independence, to do this thing, whenever the government under which we live becomes oppressive, and erect a new government which may promise to preserve our liberties" (p. 269). Carpenter, in *The South as a Conscious Minority*, quotes a letter from G. W. Johnson to Jefferson Davis: "There is no incompatibility between the right of secession by a State and the right of revolution by the people. The one is a civil right founded upon the Constitution; the other is a natural right resting upon the Law of God" (p. 196).

114. Excellent monographs have explored the complex motives of poor white secessionists, including fear of domination by corrupt northern business; fear of the Negroes' emancipation; and fear of losing their own moral, manly, egalitarian self-image. Ironically, some planters risked disunion for fear that a Republican regime might encourage poor whites to challenge their paternalist rule. See Steven A. Channing, *Crisis of Fear: Secession in South Carolina* (New York: Norton, 1970); William L. Barney, *The Secessionist Impulse: Alabama and Mississippi in 1860* (Princeton, N.J.: Princeton University, 1974); J. Mills Thornton III, *Politics and Power in a Slave Society: Alabama, 1800–1860* (Baton

Rouge: Louisiana State University, 1978); Michael P. Johnson, *Toward a Patriarchal Republic: The Secession of Georgia* (Baton Rouge: Louisiana State University, 1977); J. William Harris, *Plain Folk and Gentry in a Slave Society* (Middletown, Conn.: Wesleyan University, 1985). The importance of honor and moral outrage was argued anew in Bertram Wyatt-Brown, *Yankee Saints and Southern Sinners* (Baton Rouge: Louisiana State University, 1985). Barney, *The Secessionist Impulse*, pp. 153–230 ("Submit to be governed," p. 228); "Lincoln's triumph" in Smith, *Presidency of Buchanan*, p. 136.

115. "We propose to do," in Thornton, *Politics and Power in a Slave Society*, p. 454; "Our cause is holy" (a quotation from Charles D. Fontaine), in Barney, *The Secessionist Impulse*, p. 316; "I do not apprehend" from a letter of October 18, 1860, by Jones, Jr., in Myers, ed., *Children of Pride*, pp. 621–625. A quotation from Reinhold Niebuhr, though anachronistic, is apt at this juncture: "The pretensions of virtue are as offensive to God as the pretensions of power. One has the uneasy feeling that America as both a powerful nation and as a 'virtuous' one is involved in ironic perils which compound the experiences of Babylon and Israel": *The Irony of American History* (New York: Scribner, 1952), p. 160. Abraham Lincoln would soon reach the same conclusion.

Chapter 7

1. After much deliberation (read: wasted time) I have settled on the term Civil War to describe the conflict of 1861–1865. At first Confederates spoke of the "War of Northern Aggression" and Yankees of the "War of the Rebellion." The U.S. government itself muddled reality by adopting measures (such as the naval blockade) valid only in international conflicts. In 1863 Lincoln's Gettysburg address made the phrase "a great civil war" official, but needless to say southern loyalists rejected that term. Their cause might be lost, but it was real: the Confederate States of America had not been a fiction. Later historians entertained neutral terms such as the War between the States and War for the Union. But the term Civil War may be justified by the words and deeds of southern leaders themselves. In 1861 Virginia's secessionist Governor John Letcher accused Lincoln's cabinet of having "chosen to inaugurate civil war"; cited in John S. D. Eisenhower, *Agent of Destiny: The Life and Times of General Winfield Scott* (New York: Free Press, 1997), p. 370. In 1865 the most fiery southern partisan, Nathan Bedford Forrest, warned his troops, "Civil war, such as you have just passed through, naturally engenders feelings of animosity, hatred, and revenge." He bade them accept the fact that "a separate and independent confederacy had failed"; cited in Jay Winik, *April 1865: The Month That Saved America* (New York: HarperCollins, 2001), pp. 321–322. After the war southern state governments admitted the point (albeit under duress) by repealing their secession ordinances and swearing allegiance to an indissoluble Union. Had the outcome differed, histories might describe the War of Southern Independence. But it turned out to have been the Civil War all along.

2. Interesting global perspectives on the Civil War include David M. Potter, "Civil War," in C. Vann Woodward, ed., *The Comparative Approach to American History* (New York: Basic Books, 1968), pp. 135–345; Charles Bright and Michael Geyer, "Where in the World Is America?" in Thomas Bender, ed., *Rethinking American History in a Global Age* (Berkeley: University of California, 2002), pp. 63–99; Robert Wiebe, "Framing U.S. History," in Thomas Bender, ed., *Rethinking American History in a Global Age* (Berkeley: University of California, 2002), pp. 236–249; Michael Adas, "From Settler Colony to

Global Hegemon: Integrating the Exceptionalist Narrative of the American Experience into World History," *American Historical Review* 106, no. 5 (December 2001): 1692–1720.

3. I am not alone in this dismal interpretation. See David Mayers, *Wars and Peace: The Future Americans Envisioned 1861–1991* (New York: St. Martin's, 1998), pp. 5–22 ("knew the American people," p. 9); Edward L. Ayers, *In the Presence of Mine Enemies: War in the Heart of America 1859–1863* (New York: Norton, 2003), esp. pp. 143–149, 172–187, 390–415.

4. Elbert B. Smith, *The Presidency of James Buchanan* (Lawrence: University of Kansas, pp. 143–451 ("All our troubles," p. 144–145; "intemperate," p. 148; "overt," p. 149). The tortuous annual message is worth reading in full: see John Bassett Moore, ed., *The Works of James Buchanan*, 12 vols. (Philadelphia, Pa.: Lippincott, 1908–1911), 11: 7–43. Buchanan was ill served by a cabinet suited to Democratic politics as usual in the 1850s. Secretary of State Lewis Cass of Michigan was too old and sick to deal with the crisis. He resigned, returned, then resigned again. Attorney General Jeremiah Black, from Pennsylvania, was a stodgy legalist who sympathized with the South. Secretary of the Interior Jacob Thompson of Mississippi quit in protest over Buchanan's failed effort to supply Fort Sumter. Secretary of the Treasury Howell Cobb played the role of an obstructionist while awaiting the secession of his native Georgia. Secretary of War John B. Floyd of Virginia not only opposed the defense of Fort Sumter but was suspected of facilitating southerners' purchase of arms in the North while embezzling $840,000 on behalf of an in-law. Every member of the cabinet resigned, changed jobs, or both during "secession winter" except Secretary of the Navy Isaac Toucey, Jr. But Toucey was a Connecticut doughface well aware of his state's economic ties to the South. He moved not one ship in response to southern secession, thereby earning himself an ex post facto congressional censure; see William M. Fowler, *Under Two Flags: The American Navy in the Civil War* (Annapolis, Md.: Naval Institute, 2001), p. 34. See also Philip G. Auchampaugh, *James Buchanan and His Cabinet on the Eve of Secession* (Lancaster, Pa.: Lancaster, 1926).

5. On the convention in Montgomery, see Emory M. Thomas, *The Confederate Nation 1861–1865* (New York: Harper and Row, 1979), pp. 37–66 ("mania for unanimity," p. 44); Frank E. Vandiver, *Their Tattered Flags: The Epic of the Confederacy* (College Station: Texas A&M University, 1987, orig. 1970), pp. 18–32 ("great truth," pp. 24–25); Charles Robert Lee, *The Confederate Constitutions* (Chapel Hill: University of North Carolina, 1963), pp. 60–122.

6. On the failure of compromise efforts, see Smith, *Presidency of Buchanan*, pp. 152–165; David M. Potter, *The Impending Crisis 1848–1861* (New York: Harper and Row, 1976), pp. 514–554 ("The tug," p. 526); David M. Potter, *Lincoln and His Party in the Secession Crisis* (Baton Rouge: Louisiana State University, 1995, orig. 1942), pp. 112–187; Daniel W. Crofts, *Reluctant Confederates: Upper South Unionists in the Secession Crisis* (Chapel Hill: University of North Carolina, 1983), pp. 130–194. A newspaper in Milwaukee expressed the often overlooked furor over the threat secession posed to the Mississippi River: "There can be no doubt that any forcible obstruction of the Mississippi would at once lead to a war between the West and the South." The *Chicago Tribune* even raised the prospect of "blotting Louisiana off the map"; see Howard Cecil Perkins, ed., *Northern Editorials on Secession*, 2 vols. (New York: Appleton- Century, 1942), II: 545, 558). Potter notes the irony in the fact that this same session of Congress passed without controversy bills to admit Kansas as a free state and organize the Colorado, Nevada, and Dakota territories without

any provisions regarding slavery. The dying Stephen Douglas took sour satisfaction in this vindication of "popular sovereignty." Hence, all the territorial issues that shredded American politics in the 1850s were swept off the table (or under the rug) by the time of the Civil War.

7. On sentiment in the border states during "secession winter," see Ellis Merton Coulter, *The Civil War and Readjustment in Kentucky* (Chapel Hill: University of North Carolina, 1926); William A. Link, *Roots of Secession: Slavery and Politics in Antebellum Virginia* (Chapel Hill: University of North Carolina, 2003), esp. pp. 1–10, 213–232; Mary E. R. Campbell, *The Attitude of Tennesseeans toward the Union, 1847–1861* (New York: Vantage, 1961); Joseph Carlyle Sitterson, *The Secession Movement in North Carolina* (Chapel Hill: University of North Carolina, 1939); James M. Woods, *Rebellion and Realignment: Arkansas's Road to Secession* (Fayetteville: University of Arkansas, 1987); the general accounts in Albert D. Kirwan, *John J. Crittenden: The Struggle for the Union* (Lexington: University of Kentucky, 1962); Ralph A. Wooster, *The Secession Conventions of the South* (Princeton, N.J.: Princeton University, 1962); Potter, *Lincoln and His Party*, pp. 280–314; and Crofts, *Reluctant Confederates*, pp. 164–194 ("to live and die," p. 330). Crofts's statistics prove that the votes of former Whigs combined with those of Democrats in non-slaveholding counties explain the "Unionist counteroffensive" from January to March.

8. Crofts, *Reluctant Confederates*, pp. 215–253 ("forbearing and patient" and "wise and winning," p. 241); Potter, *Lincoln and His Party*, pp. 315–335 ("firm policy," p. 329).

9. *Charles Francis Adams (Jr.), 1815–1915: An Autobiography* (Boston, Mass.: Houghton Mifflin, 1916), p. 82. Lincoln's decision to grow a beard during the weeks before his inauguration remains mysterious. Perhaps, as legend has it, he got the idea from a little girl's letter, but that leaves the question of why he took up the suggestion. Beards began to come into fashion when British officers in the Crimean War copied the style of their Russian adversaries and Americans copied the British. Did Lincoln think whiskers would endow him with military gravitas? Did he intuit the value of trading the image of rail-splitter for that of a patriarch, Father Abraham? His biographer Richard Carwardine, in *Lincoln: A Life of Purpose and Power* (New York: Knopf, 2006), suspects that more than vanity was involved. It was as if Lincoln's change of face prefigured the changing face of the nation.

10. The only scholarly biography is Don W. Wilson, *Governor Charles Robinson of Kansas* (Lawrence: University of Kansas, 1975). See also the sketch in Homer E. Socolofsky, *Kansas Governors* (Lawrence: University of Kansas, 1990), pp. 81–85. After his term Robinson quit the Republicans to become a political maverick and eventually ended up in the Greenback Party. He never stopped serving his adopted state; he was a university regent, president of the state historical society, and a progressive reformer. Nor were Robinson's conflicts of interest stemming from land speculation unusual or even illegal. Literally every army officer posted to Fort Leavenworth and Fort Riley was a partner in town development schemes. See Tony R. Mullis, *Peacekeeping on the Plains: Army Operations in Bleeding Kansas* (Columbia: University of Missouri, 2004), pp. 119–152.

11. On population, see James R. Shortridge, *Peopling the Plains: Who Settled Where in Frontier Kansas* (Lawrence: University of Kansas, 1995), pp. 4–13, 76–82.

12. Again, the only scholarly biography is a local one. See Wendell Holmes Stephenson, *The Political Career of James H. Lane* (Topeka: Kansas Historical Publications, 1930). On Lincoln's visit, see David Dary, *True Tales of Old-Time Kansas*, rev. ed. (Law-

rence: University of Kansas, 1984), pp. 242–251. Lincoln arrived on November 30, 1859, and spent a week in Kansas. Everyone commented on his ungainly form: "The knees stood up like the hind joints of a Kansas grasshopper's legs" (p. 245). His folksy manner also made people wonder what passed for eloquence among Illinoisans. "But in ten or fifteen minutes," a reporter confessed, "I was unconsciously and irresistibly drawn by the clearness and closeness of his argument. Link by link it was forged and welded, like a blacksmith's chain" (p. 247). From the steps of Leavenworth's Planters Hotel Lincoln held 10,000 people captive despite bitter weather. A wry reference to the presidency also provided the setting for some Lincolnesque humor. When a man hurled some official papers into a flagging stove, a politico joked, "Mr. Lincoln, when you become president will you sanction the burning of government reports by cold men in Kansas Territory?" Abe replied, "Not only will I not sanction it, but I will cause legal action to be brought against the offenders" (p. 249). Though jocular, the remark spoke volumes in a land where Jayhawks and Ruffians committed cold-blooded murder without incurring even the least "legal action."

13. G. Raymond Gaeddert, *The Birth of Kansas* (Philadelphia, Pa.: Porcupine, 1974, orig., 1940), pp. 7–160 ("On all public questions," p. 130); Robert W. Richmond, *Kansas: A Land of Contrasts*, 3rd ed. (Wheeling, Ill.: Forum, 1989), pp. 66–104 ("hungry children," p. 86); and Albert Castel, *A Frontier State at War: Kansas, 1861–1865* (Ithaca, N.Y.: Cornell University, 1958). During the drought of 1860–1862 the original "Buffalo Bill," William Mathewson, saved many lives by shooting buffalo and carting the meat to homesteads. For Lane's career as a wartime liberator of blacks, see Richard Sheridan, "From Slavery in Missouri to Freedom in Kansas: The Influx of Black Fugitives and Contrabands into Kansas, 1854–1865," in Rita Napier, ed., *Kansas and the West: New Perspectives* (Lawrence: University of Kansas, 2003), pp. 157–180 ("I'll do it," p. 162).

14. More will be said of Indians on the Great Plains; for the fate of tribes in Kansas, see William E. Unrau and H. Craig Miner, *The End of Indian Kansas: A Study of Cultural Revolution, 1854–1871* (Lawrence: University of Kansas, 1978); and Joseph B. Herring, "The Chippewa and Munsee Indians: Acculturation and Survival in Kansas, 1850s–1870s," in Napier, ed., *Kansas and the West*, pp. 76–90. Herring tells a moving story of how two tribes avoided removal or extinction through the gradual embrace of farming and ranching, American law, Christianity, and intermarriage. Contrast that with the notion of assimilation expressed by a delegate to the Wyandotte convention. It was generally accepted, he said, that any Injun able to chew a tobacco plug and swig a half pint of whiskey was as good as a white man (Gaeddert, *Birth of Kansas*, p. 57).

15. See C. Robert Haywood, *Trails South: The Wagon-Road Economy in the Dodge City–Panhandle Region* (Norman: University of Oklahoma, 1986), pp. 150–189; and *Victorian West: Class and Culture in Kansas Cattle Towns* (Lawrence: University of Kansas, 1991), esp. pp. 90–111. The man who tamed Abilene was not James Butler "Wild Bill" Hickok. His reputation was crafted by a fanciful piece in *Harper's Monthly* and fixed in stone by his death in Deadwood in 1876. The real hero was Thomas James Smith, a square-faced, bushy-mustached lawman who forced cowboys to check their guns and gave backbone to tremulous citizens. He was killed chasing outlaws after just seven months. See Dary, *True Tales*, pp. 102–113. Haywood argues that we should save our sympathy, not for Luke Short or Doc Holliday, but for people like Philander Gillette Reynolds. A tireless entrepreneur who ran a vast network of stagecoach and wagon lines

on the trails, Reynolds delivered the U.S. mail and shipped the materials to build dozens of towns. But the railroads his own transport helped to construct swallowed his business along with the businesses of the cowboys, bartenders, and prostitutes. In *Victorian West*, an elegant cultural history, Haywood shatters the myth of the "lawless" cattle town. Rather, as the editor Nick Klaine of Dodge City wrote at the time, "Our people are as a class from the thickly populated East; men of moderate means drawn hither by the hope of acquiring a competence, united together by one common bond of sympathy, dependent upon one another for support and encouragement. Some of our rudest habitations are graced by faces and forms that would do honor to the most elegant home or social gathering" (p. 15).

16. Lying 500 miles from salt water (the Gulf of Mexico), Kansas has a severely continental climate. Temperatures from 121 degrees Fahrenheit to minus forty degrees have been recorded, and the state is known for its cyclones, blizzards, hailstorms, lightning, drought, and flooding along the Arkansas and Republican rivers and their tributaries. Statehood celebrations in January 1861 were distinctly muted by the twenty-four inches of wind-whipped snow blanketing eastern Kansas.

17. On the Mennonite "Volga Germans" from southern Russia, the provenance of "Turkey Red" wheat, and the "Exodusters," see Norman E. Saul, "Myth and History: Turkey Red Wheat and the 'Kansas Miracle,'" pp. 140–155; Randall B. Woods, "Integration, Exclusion, or Segregation? The 'Color Line' in Kansas, 1878–1900," pp. 155–171, in Paul K. Stuewe, ed., *Kansas Revisited: Historical Images and Perspectives* (Lawrence: University of Kansas, 1990); and Nell Irvin Painter, *Exodusters: Black Migration to Kansas after Reconstruction* (New York: Knopf, 1977). The Exodusters met some resistance. "Is Kansas a grand asylum for decrepit and destitute blacks?" asked a newspaper in Leavenworth; see Craig Miner, *Kansas: The History of the Sunflower State, 1854–2000* (Lawrence: University of Kansas, 2002), p. 152. But whereas poor, uneducated blacks never had equality, no Jim Crow laws or organized bigotry emerged until the heyday of the Ku Klux Klan in the 1920s. On homesteading and agriculture, see H. Craig Miner, *West of Wichita: Settling the High Plains of Kansas, 1865–1890* (Lawrence: University of Kansas, 1986); John Ise, *Sod and Stubble: The Story of a Kansas Homestead* (Lincoln: University of Nebraska, 1968); and George E. Hasse and Robin Higham, eds., *Rise of the Wheat State: A History of Kansas Agriculture, 1861–1986* (Manhattan, Kan.: Sunflower University, 1987). Imagine pushing a wheelbarrow loaded with buffalo dung half a mile over windy prairie while the sun beat on your bonnet and ankle-length dress. "It was comical," wrote an editor in west Kansas, "to see how gingerly our wives handled these chips at first. They commenced by picking them up between two sticks, or with a poker. Soon they used a rag, and then a corner of their apron. Finally, growing hardened, a wash after handling them was sufficient. And now? It is out of the bread, into the chips and back again—and not even a dust of the hands" (Richmond, *Land of Contrasts*, p. 150).

18. A good summary of nineteenth-century reform movements is in Miner, *Kansas*, pp. 143–189. On the centrality of the temperance movement, see Robert Smith Bader, *Prohibition in Kansas: A History* (Lawrence: University of Kansas, 1986).

19. William Allen White, from *Emporia Gazette* (August 15, 1896); Carl Becker, from *Essays in American History Dedicated to Frederick Jackson Turner*, 1910, excerpted in Stuewe, ed., *Kansas Revisited*, pp. 9–27 ("to thank God," p. 25; "To venture," p. 21).

20. This is a good time to recommend an excellent Web site containing, among

many other rich sources, all the presidential inaugural addresses: http://www.bartleby. com/124. On what Lincoln said, did not say, and in some cases could not say in the address, see Phillip S. Paludan, *The Presidency of Abraham Lincoln* (Lawrence: University of Kansas, 1994), pp. 52–58; on cabinet appointments, pp. 21–45.

21. The day before Lincoln's inauguration, Scott had warned Seward of the stakes at Fort Sumter. A war to force the seceded states back into the Union, he said, would drag on for years; involve an enormous waste of lives; and eventuate in vast devastation, $250 million in public debt, and a postwar military regime led by "a Protector or an Emperor." In fact, he underestimated the damage. But his Army perspective caused him to overestimate what it would take to relieve Fort Sumter. The Navy soon proved its ability to force passage even through heavy coastal fortifications. See Eisenhower, *Winfield Scott*, pp. 358–368.

22. No historical narrative except Europe's descent into World War I in July 1914 has been parsed in more detail than the run-up to the bombardment of Fort Sumter. The following works address the key question of whether Lincoln wanted war and was just trying to maneuver the rebels into firing the first shot: William and Bruce Catton, *Two Roads to Sumter* (New York: McGraw-Hill, 1963); William A. Swanberg, *First Blood: The Story of Fort Sumter* (New York: Scribner, 1957); Richard N. Current, *Lincoln and the First Shot* (Philadelphia, Pa.: Lippincott, 1963); summaries in Paludan, *Presidency of Lincoln*, pp. 49–67 ("Wanted" and "the man is not equal," p. 61); Potter, *Impending Crisis*, pp. 555–583; and Allen C. Guelzo, *The Crisis of the American Republic: A History of the Civil War and Reconstruction Era* (New York: St. Martin's, 1995), pp. 91–99.

23. See Norman B. Ferris, *Desperate Diplomacy: William H. Seward's Foreign Policy, 1861* (Knoxville: University of Tennessee, 1976), pp. 3–32; Lord Lyons quoted from Thomas Wodehouse Legh, 2nd Baron Newton, *Lord Lyons: A Record of British Diplomacy*, 2 vols. (London: Arnold, 1913), I: 30–33. Also: Crofts, *Reluctant Confederates*, pp. 297–301; Phillip Shaw Paludan, *"A People's Contest": The Union and Civil War 1861–1865* (New York: Harper and Row, 1988), pp. 32–41. Some great historians were baffled by Seward's belligerence; see, for example, Potter, *Lincoln and His Party*, pp. 367–371; Kenneth M. Stampp, *And the War Came: The North and the Secession Crisis, 1860–1861* (Baton Rouge: Louisiana State University, 1950), pp. 271–272; and Allan Nevins, *The War for the Union*, 4 vols. (New York: Scribner, 1959–1971), I: 72–74. Finally, it has been suggested that Seward did not take secession as seriously as the Radical Republicans, and that his secret contacts with secessionists bordered on disloyalty. In fact, Seward took secession and the prospect of war so seriously that he alone strategized as if the Confederacy were already a dangerous foreign power within a delicate transatlantic balance of power. To save any chance of repairing the Union, by peaceful or warlike means, it was imperative to "isolate the battlefield" by deterring foreign support for Jefferson Davis's regime or foreign resistance to a Union blockade.

24. Perhaps it was ex post facto rationalization, but Lincoln later made a rare boast: "The plan succeeded; they attacked Sumter—it fell, and thus did more service than it otherwise could." From the diary of Senator Orville Browning, cited in Paludan, *Presidency of Lincoln*, p. 66; "a state for a fort," p. 64.

25. Vandiver, *Their Tattered Flags*, pp. 35–44; William J. Cooper, Jr., *Jefferson Davis, American* (New York: Knopf, 2000), pp. 329–340 ("thousands of lives," p. 341; "many reasons," p. 339); James M. McPherson, *Battle Cry of Freedom: The Civil War Era* (New

York: Ballantine, 1988), pp. 264–275. The inaugural address of February 18, 1861, is in Lynda Lasswell Crist and Mary Seaton Dix, eds., *The Papers of Jefferson Davis*, 11 vols. (Baton Rouge: Louisiana State University, 1971–), 7: 46–50. Davis wrote to John A. Campbell on April 6: "We have waited hopefully for the withdrawal of garrisons which irritate the people of these states and threaten the respective localities, and which can serve no purpose to the United States unless it be to injure us. So far from desiring to use force for the reduction of Fort Sumter we have avoided any measure to produce discomfort or to exhibit discourtesy, until recently when we were informed that the idea of evacuation had been abandoned." (*Papers of Jefferson Davis*, 7: 92).

26. The politician Pleasant A. Stovall of Georgia (born 1857) later wrote that Robert Toombs burst into the climactic cabinet meeting to warn that firing on Fort Sumter would "strike a hornet's nest. . . . It is unnecessary; it puts us in the wrong; it is fatal." Vandiver, in *Tattered Flags*, p. 43, uses the alleged speech for dramatic effect; but Cass Canfield, in *The Iron Will of Jefferson Davis* (New York: Harcourt Brace Jovanovich, 1978), pp. 64–65, notes that although Toombs was a rival of Davis, it is hardly credible that the Fire-Eeater pleaded for peace. McPherson, *Battle Cry* ("oozing out," p. 272), is especially good on the pressures weighing on Davis. On April 10 Davis received a telegram from Louis T. Wigwall stating, "No one now doubts that Lincoln intends war" and urging a "bold stroke" sure to bring Virginia into the Confederacy. See Louise Wigfall Wright, *A Southern Girl in 1861: The Wartime Memories of a Confederate Senator's Daughter* (New York: Doubleday, Page, 1905), pp. 36–37. Meanwhile, the Union naval relief expedition to Fort Pickens came off swimmingly. Lieutenant David Dixon Porter retained the *Powhatan* even after receiving orders to send it back to the fleet bound for Sumter, because his original orders (trumped up by Seward) had the president's signature and their retraction had only the navy secretary's signature. Porter's *Incidents and Anecdotes of the Civil War* (1885), reprinted in Brian M. Thomsen, ed., *Blue and Gray at Sea: Naval Memoirs of the Civil War* (New York: Doherty, 2003), pp. 97–113, contains a colorful account of the bold action that reinforced Fort Pickens in the teeth of Braxton Bragg's superior forces. The Union held the fort and thus neutralized Pensacola for the entire war. Porter's tale begins with the memory of how gloomy Washington City became during "secession winter" as "the beautiful Southern women who formed the bright galaxy of stars in Washington society" departed for home and the festive bachelor Buchanan relinquished the White House to the dour Mary Lincoln (p. 101).

27. A stanza from "The Conflict of Convictions" in *Battle Pieces: The Civil War Poems of Herman Melville* (Edison, N.J.: Castle, 2000, orig. 1866), p. 15.

28. William A. Link, *Roots of Secession: Slavery and Politics in Antebellum Virginia* (Chapel Hill: University of North Carolina, 2003), pp. 232–244; Elizabeth R. Varon, *We Mean to Be Counted: White Women and Politics in Antebellum Virginia* (Chapel Hill: University of North Carolina, 1998), pp. 137–168 ("The effect," p. 162; "Mothers, wives," p. 163); "tyrant" and "Liberty or death" in Crofts, *Reluctant Confederates*, p. 351.

29. The "battle of Baltimore" is said to have yielded the war's first casualties—four dead soldiers, thirty-six wounded soldiers, and six dead civilians—but the encounter might better be styled the last in the long series spawned by antebellum mob violence. See Benjamin W. Bacon, "Ben Butler Severs the Gordian Knot," in *Sinews of War: How Technology, Industry, and Transportation Won the Civil War* (Novato, Calif.: Presidio, 1997), pp. 1–14; Mark W. Summers, "'Freedom and Law Must Die Ere They Sever':

The North and the Coming of the Civil War," in Gabor S. Boritt, ed., *Why the Civil War Came* (New York: Oxford University, 1996), pp. 177–200 ("We can fix," p. 180); Doris Kearns Goodwin, *Team of Rivals: The Political Genius of Abraham Lincoln* (New York: Simon and Schuster, 2005), pp. 351–357 ("nigger thieves," p. 352; "the very existence," p. 355). On Dahlgren, with whom Lincoln formed a close friendship, see Robert V. Bruce, *Lincoln and the Tools of War* (Urbana: University of Illinois, 1989), pp. 3–21. Marylanders stayed in the Union, but that did not stop them from singing the lilting anthem of protest, "Maryland, My Maryland," for the duration.

30. Leonidas Polk deserves the blame insofar as he kept Jefferson Davis in the dark about his intentions. For his part, Davis issued less than explicit instructions. See Steven E. Woodworth, *No Band of Brothers: Problems in the Rebel High Command* (Columbia: University of Missouri, 1999), pp. 12–18.

31. On the border states generally, see Paludan, *Presidency of Lincoln*, pp. 69–88 ("To lose Kentucky" and "for any purpose," p. 83). On Missouri, see Christopher Phillips, *Damned Yankee: The Life of General Nathaniel Lyon* (Columbia: University of Missouri, 1990); William E. Parrish, *Turbulent Partnership: Missouri and the Union, 1861–1865* (Columbia: University of Missouri, 1963); and William L. Shea and Earl J. Hess, *Pea Ridge: A Civil War Campaign in the West* (Chapel Hill: University of North Carolina, 1992).

32. On Scott's strategy, see Eisenhower, *Agent of Destiny*, pp. 381–390. American strategists, no less than Europeans of that era, were spellbound by the ghost of Napoléon conjured in the writings of Antoine Henri Jomini. But Professor Denis Hart Mahan of West Point (father of the navalist Alfred Thayer Mahan) argued that swift, decisive Napoleonic campaigns required large, highly professional armies and small theaters of operation (hence Napoléon's debacle in Russia). Neither condition prevailed in North America. Mahan therefore warned against efforts to deliver quick knockout blows in favor of occupying enemy territory, digging in, and forcing the enemy to make costly attacks. West Point's engineering ethos reinforced that stress on mass over maneuver. See the excellent summary in Guelzo, *The Crisis of the American Republic*, pp. 104–111.

33. The otherwise obscure General Barnard Bee coined the nickname when he bade his faltering South Carolinians to take heart from the doughty Virginians: "There stands Jackson like a stone wall." It proved ironic as well as iconic because "Stonewall" Jackson's reputation would be made, not by dogged defensive stands, but by audacious offensives such as his Shenandoah Valley campaign.

34. Bruce Catton, in *The Coming Fury* (New York: Washington Square, 1961), pp. 440–464, describes the first battle of Bull Run. He concludes that the "notion the Confederate army could have walked into Washington within twenty-four hours will hardly bear analysis" (pp. 469–470). Another Civil War legend that doesn't hold up is the amazing prescience revealed in C. Vann Woodward, ed., *Mary Chesnut's Civil War* (New Haven, Conn.: Yale University, 1981); "lulls us" and "will wake every inch," p. 111. The truth is that Mary Boykin Miller Chesnut reinvented herself after Dixie's demise like a real-life Scarlett O'Hara. She was indeed the wife of a prominent Confederate politician, an eyewitness to many events, and an ardent secessionist. But the 800-page diary she claimed to have written during the war was in fact written (or expanded and rewritten) in the 1880s and carefully tailored to the prejudices of the (mostly Yankee) booksellers' market. It is easy to seem prophetic when you know all the outcomes in advance. Woodward might have deceived himself about Chesnut for didactic purposes, and the

filmmaker Ken Burns for dramatic purposes, but Kenneth Lynn put *l'affaire Chesnut* to rest in *The Air-Line to Seattle: Studies in Literary and Historical Writing about America* (Chicago, Ill.: University of Chicago, 1983), p. 59: "She wrote a novel about the South during the Civil War and called it a diary." Ergo, I shall not be quoting again from Mrs. Chesnut.

35. Every major campaign, battle, and general in the Civil War has generated a virtual library. Readers seeking a comprehensive military history will be more than satisfied by Shelby Foote, *The Civil War: A Narrative*, 3 vols. (New York: Random House, 1958–1974); Bruce Catton, *The Centennial History of the Civil War*, 3 vols. (Garden City, N.Y.: Doubleday, 1961–1965); James M. McPherson, *Battle Cry of Freedom: The Civil War Era* (New York: Oxford University, 1988); Russell F. Weigley, *A Great Civil War: A Military and Political History* (Bloomington: Indiana University, 2000); Richard E. Berringer et al., *Why the South Lost the Civil War* (Athens: University of Georgia, 1986); and Herman Hattaway and Archer Jones, *How the North Won: A Military History of the Civil War* (Urbana: University of Illinois, 1983).

36. William Best Hesseltine, *Lincoln and the War Governors* (New York: Knopf, 1948); W. B. Yearns, *The Confederate Congress* (Athens: University of Georgia, 1960), pp. 3–16; Emory M. Thomas, *The Confederate Nation, 1861–1865* (New York: Harper and Row, 1979), pp. 98–144; Weigley, *A Great Civil War*, pp. 67–72. Memminger had served as chairman of the ways and means committee in the South Carolina legislature, but like most southern partisans he despised what little he knew about innovative methods of public finance. However, Weigley notes in Memminger's defense that the South's antitax culture and relative lack of specie left the treasury little choice besides policies that were inflationary and undermined confidence.

37. No one sang praises to Josiah Gorgas better than Gorgas himself. "I have succeeded beyond my utmost expectations," he wrote in 1864; see Frank E. Vandiver, *The Civil War Diary of Josiah Gorgas* (Tuscaloosa: University of Alabama, 1995), pp. 90–91. The extraordinary exertions and frustrations of Stephen Mallory are brilliantly described by Raimondo Luraghi in *A History of the Confederate Navy*, trans. Paolo E. Coletta (Annapolis, Md.: Naval Institute, 1996), esp. pp. 1–88 ("We are thus back," p. 20); and Joseph T. Durkin, *Stephen Mallory: Confederate Navy Chief* (Chapel Hill: University of North Carolina, 1954).

38. Among war contractors the du Ponts were a sterling exception from beginning to end. Their Eleutherian Mills sold the government almost 10 million tons of gunpowder, but they held the price below twenty-eight cents per pound even as the Confederates paid more than a dollar per pound to blockade runners. Henry du Pont, a graduate of West Point, also had a distinguished record in uniform, and Lammot du Pont performed valuable service importing saltpeter from England. See Marc Duke, *The Du Ponts: Portrait of a Dynasty* (New York: Dutton, 1976), pp. 156–176. On Union finance, see Bray Hammond, *Sovereignty and an Empty Purse: Banks and Politics in the Civil War* (Princeton, N.J.: Princeton University, 1970), pp. 1–70. "The age of shoddy" quoted in George Brown Tindall, *America: A Narrative History*, 2 vols. (New York: Norton, 1988), I: 676. "I don't know anything," Francis B. Carpenter, *Six Months at the White House with Abraham Lincoln* (New York: Hurd and Houghton, 1867), p. 252, quoted in Guelzo, *Crisis of the American Republic*, p. 163.

39. As the title suggests, Edward Hagerman, in *The American Civil War and the Origins of Modern Warfare: Ideas, Organization, and Field Command* (Bloomington: Indi-

ana University, 1988), stresses the advances made. See also Stansbury Hayden, *Military Ballooning during the Early Civil War* (Baltimore, Md.: Johns Hopkins University, 2000, orig. 1940); and Ira W. Rutkow, *Bleeding Blue and Gray: Civil War Surgery and the Evolution of American Medicine* (New York: Random House, 2005). Bruce, in *Lincoln and the Tools of War*, pp. 22–144, stresses the inertial resistance to new technology. He recounts how Lincoln, after observing a "wonderful new repeating battery of rifled guns, shooting fifty balls a minute," said to General McClellan, "I have a notion to go out with you and stand or fall with the battle" (p. 122). But such were Ripley's delays that Gatling had to wait until 1864 just to get serious consideration of his invention. By then Lincoln had ceased to hope for a technological breakthrough. Besides, Gatling was a Copperhead! Not until 1866 did Captain Stephen St. Vincent Benét (grandfather of the poet who wrote *John Brown's Body*) declare the Gatling gun a success and order some for the Army.

40. On the Union's naval mobilization in 1861, see David G. Surdam, *Northern Naval Superiority and the Economics of the American Civil War* (Columbia: University of South Carolina, 2001); Fowler, *Under Two Flags*, pp. 39–78; and Virgil Carrington Jones, *The Civil War at Sea*, Vol. 1, *The Blockaders, January 1861–March 1862* (New York: Holt, Rinehart, and Winston, 1960). Bache's importance is highlighted in Robert V. Bruce, *The Launching of Modern American Science, 1846–1876* (New York: Knopf, 1987); "is the image" quoted in Bruce Catton, *The American Heritage Picture History of the Civil War* (New York: Doubleday, 1960), p. 179; "Son, dat ain't" in Willie Lee Rose, *Rehearsal for Reconstruction: The Port Royal Experiment* (Indianapolis, Ind.: Bobbs-Merrill, 1964), p. 12.

41. Ferris, *Desperate Diplomacy*, pp. 97–125 (arrests of British consuls); 194–207 (*Trent* affair), "monstrous," p. 117; Hay's poem, p. 207. Brian Jenkins, *Britain and the War for the Union*, 2 vols. (Montreal: McGill-Queen's University, 1974), I: 47–102 (British neutrality), I: 181–204 (Mason and Slidell). H. C. Allen, "Civil War, Reconstruction, and Great Britain," pp. 3–96, and John A. Williams, "Canada and the Civil War," pp. 257–298, in Harold Hymans, ed., *The Shot Heard Round the World: The Impact Abroad of the Civil War* (New York, 1969).

42. Senator Wade said he was proud to be called a "radical" because so were Jesus, Martin Luther, and George Washington. See Hans L. Trefousse, *Benjamin Franklin Wade: Radical Republican from Ohio* (New York: Twayne, 1963); and Bruce Tap, *Over Lincoln's Shoulder: The Committee on the Conduct of the War* (Lawrence: University of Kansas, 1998). One of the first acts of his committee of "Jacobins" was to investigate the botched assault on Ball's Bluff in October. The Confederates, hoping to cut Union access to Washington, D.C., via the Potomac, had fortified posts on the river. Under pressure, McClellan ordered General Charles P. Stone to reconnoiter a rebel post at Ball's Bluff thirty miles upstream from the capital. Stone in turn gave the task to Colonel Edward D. Baker, who was a former senator and a good friend of Lincoln. Baker's clumsy assault cost him his life and half his 1,700-man brigade. Wade, however, made Stone the scapegoat in part because Stone had ordered the return of slaves seeking refuge with federal troops. Over the course of the war the joint committee repeatedly second-guessed Lincoln and his generals, issued "orders" they were usually pleased to ignore, and made detailed reports on the Union war effort that proved invaluable to the enemy. Robert E. Lee said that the committee's prints were worth two divisions of soldiers to him (Perret, *Lincoln's War*, pp. 324–333).

43. "McClellan is to me" in Warren W. Hassler, Jr., *General George B. McClellan: Shield of the Union* (Baton Rouge: Louisiana State University, 1957), p. xv; "I can do it all" in Tyler Dennett, ed., *Lincoln and the Civil War in the Diaries and Letters of John Hay* (Westport, Conn.: Negro Universities, 1939), p. 33. For years the standard biography was the critical work by Stephen W. Sears, *George B. McClellan: The Young Napoleon* (New York: Ticknor and Fields, 1988), and the primary source was Stephen W. Sears, ed., *The Civil War Papers of George B. McClellan: Selected Correspondence, 1861–1865* (New York: Ticknor and Fields, 1989). I rely mostly on the evenhanded account by Ethan S. Rafuse, *McClellan's War: The Failure of Moderation in the Struggle for the Union* (Bloomington: Indiana University, 2005); "vain and unstable" (Kenneth P. Williams), "possessed" (Sears), and "messianic complex" (James M. McPherson), quoted on p. 1; "in considerable degree," p. 157.

44. Rafuse, *McClellan's War*, pp. 8–158; "day of adjustment" and "dodge the nigger" from letters written in November 1861 in Sears, ed., *Civil War Papers*, pp. 128–132; "In considering the policy" in Roy P. Basler, ed., *The Collected Works of Abraham Lincoln*, 9 vols. (New Brunswick, N.J.: Rutgers University, 1953–1955), 5: 48–49).

45. "Nobody believes" in Allan Nevins and Milton Halsey Thomas, eds., *The Diary of George Templeton Strong*, 4 vols. (New York: Macmillan, 1952), 3: 256; "General, what shall," in McPherson, *Battle Cry of Freedom*, p. 368.

46. Why is Stanton still relegated to a supporting role at best in accounts not only of the Civil War, but of American history overall? Why have no major biographies appeared since (the admittedly excellent) Benjamin P. Thomas and Harold M. Hyman, *Stanton: The Life and Times of Lincoln's Secretary of War* (New York: Knopf, 1962)? A major magazine recently polled historians as to the 100 most influential persons in U.S. history. I was evidently the only one who suggested the man who made the federal government capable of mobilizing the nation for total war.

47. Hattaway and Jones, in *How the North Won*, argue that the decisive Henry and Donelson campaign put the South on the strategic defensive for the duration. The best sources on Grant include (obviously) *The Personal Memoirs of Ulysses S. Grant*, 2 vols. (New York: Charles H. Webster, 1885–1886); and the following biographies: Geoffrey Perret, *Ulysses S. Grant: Soldier and President* (New York: Random House, 1997); Brooks Simpson, *Ulysses S. Grant: Triumph over Adversity, 1822–1865* (New York: Houghton Mifflin, 2000); Jean Edward Smith, *Grant* (New York: Simon and Schuster, 2001). On Grant's first major campaign, see Franklin Cooling, *Forts Henry and Donelson: The Key to the Confederate Heartland* (Knoxville: University of Tennessee, 1987). On Foote's indispensable role, see Spencer C. Tucker, *Andrew Foote: Civil War Admiral on Western Waters* (Annapolis, Md.: Naval Institute, 2000), esp. pp. 116–197 ("a work of almost insuperable," p. 118); and Rowena Reed, *Combined Operations in the Civil War* (Annapolis, Md.: Naval Institute, 1978). Confederate batteries defeated the initial gunboat assault on Fort Donelson, destroying two ironclads and wounding Foote. But the "floating batteries" returned to support Grant's ground attack. The rebel commander, Simon Bolivar Buckner, was a buddy of Grant's at West Point. He expected more generous terms.

48. Melville, a stanza from "Donelson" in *Battle Pieces*, p. 37. The stern weather to which Melville referred was a godsend to Grant. Bitter storms—what we now call a "wintry mix"—caused the Cumberland River to flood Donelson's lower ramparts and inhibited a bold breakout attempt by the Confederate garrison.

49. For a revisionist account of the ironclad battle, see Robert W. Day, *How the Merrimac Won: The Strategic Story of the CSS Virginia* (New York: Crowell, 1957). Commander S. D. Greene, "In the *Monitor* Turret" (1885), reprinted in Thomsen, ed., *Blue and Gray at Sea*, pp. 408–416, provides a gripping memoir of life inside the first ironclads and the battle's almost unbearable tension as the vessels repeatedly fired their shells at point-blank range, then attempted to ram each other. The unheralded hero, according to Greene, was the *Monitor*'s skipper, Lieutenant John L. Worden. Although suffering the effects of illness and imprisonment in the South, he fought with skill and courage until a Confederate shell found the only vulnerable point: the peephole in the pilot's house. His powder burns were so bad that Worden's face seemed to be bleeding from every pore. Greene concluded: "Probably no ship was ever devised which was so uncomfortable for her crew, and certainly no sailor ever led a more disagreeable life than we did on the James River, suffocated with heat and bad air if we remained below, and a target for sharpshooters if we came on deck" (p. 101).

50. Sears, *McClellan: The Young Napoleon*, pp. 163–182; Rafuse, *McClellan's War*, pp. 175–207; Bacon, *Sinews of War*, pp. 50–63 ("the leap of a giant," p. 54). Johnston's decision to withdraw from Manassas to the Rappahannock was sound but heartbreaking. The hungry rebels had to destroy the largest meat-curing plant in Virginia lest it fall into the Yankees' hands. They marched south with their nostrils tantalized by the scent of burning bacon.

51. Perrot, *Ulysses S. Grant*, pp. 183–199 ("devil's own day" and "lick 'em tomorrow," p. 197; "complete victory," p. 198). James Lee McDonough, *Shiloh, in Hell before Night* (Knoxville: University of Tennessee, 1977); "a decisive victory," p. 81. See also Thomas L. Connelly, *Army of the Heartland: The Army of Tennessee, 1861–1862* (Baton Rouge: Louisiana State University, 1967), pp. 145–175; Bruce Catton, *Grant Moves South* (Boston, Mass.: Little, Brown, 1960). Charles Roland, in *Albert Sidney Johnson: Soldier of Three Republics* (Austin: University of Texas, 1964), attributes northerners' fear of Johnston to his unparalleled combat experience in the Texan revolution, the Mexican War, and the Mormon campaign, which he commanded; and to his imposing frame, charismatic personality, and offensive élan. War journalism matured at Shiloh because by then all major formations were accompanied by reporters who shared soldiers' misery and peril. A correspondent for the *Chicago Tribune* lost his head to a cannonball just two yards from General Grant. As in Mexico, reporters were also agents and pawns in the conflicts among generals. At first, Grant was scorched in the press for his poor defensive deployments; then he was hailed for a great victory. At first, Johnston was mourned and Beauregard was praised for valor; then both were damned for failing to order a climactic attack at dusk on the first day. On Shiloh and the Confederate conscription act, see Vandiver, *Their Tattered Flags*, pp. 118–132.

52. Melville, a stanza from "Shiloh: A Requiem" in *Battle Pieces*, p. 63.

53. Chester G. Hearn, *Admiral David Glasgow Farragut: The Civil War Years* (Annapolis, Md.: Naval Institute, 1998); Fowler, *Under Two Flags*, pp. 94–127. Farragut's 7,500 heavy mortar shells did little damage, owing to poor fuses; he had expected this. For the apt image "greatest fireworks," credit McPherson, *Battle Cry of Freedom*, p. 420; "buried his face" and "a peace" cited in Cooper, *Jefferson Davis, American*, p. 388. Admiral George Dewey, who served with Farragut during the Civil War, wrote in 1913 in his autobiography (excerpted in Thomsen, ed., *Blue and Gray at Sea*, p. 19): "Farragut has

always been my ideal of the naval officer, urbane, decisive, indomitable. Whenever I have been in a difficult situation, or in the midst of such a confusion of details that the simple and right thing to do seemed hazy, I have often asked myself, 'What would Farragut do?'"

54. The Seven Days battles from June 25 to July 1 included Beaver Dam Creek, Gaines' Mill, Savage Station, Frayser's Farm, White Oak Swamp, and Malvern Hill. The last was especially bloody, and unnecessary, since McClellan was already committed to a withdrawal. The Confederacy lost 20,141 men that week; the Union lost 15,855. See Stephen W. Sears, *To the Gates of Richmond: The Peninsula Campaign* (New York: Ticknor and Fields, 1992); Kevin Dougherty, *The Peninsula Campaign of 1862: A Military Analysis* (Jackson: University of Mississippi, 2005); and the articles in Gary W. Gallagher, ed., *The Richmond Campaign of 1862: The Peninsula and the Seven Days* (Chapel Hill: University of North Carolina, 2000); "the confounded Chickahominy" and "the shot that struck me" in Richard M. Ketchum, ed., *The American Heritage History of the Civil War* (New York: Doubleday, 1960), pp. 147, 149. The volume of literature on Lee is probably second only to that on Lincoln. Good discussions of Lee's character and subsequent status as "marble man" include Alan Nolan, *Lee Considered: General Robert E. Lee and the Civil War History* (Chapel Hill: University of North Carolina, 1991); Emory M. Thomas, *Robert E. Lee: A Biography* (New York: Norton, 1995); Thomas L. Connolly, *The Marble Man: Robert E. Lee and His Image in American Society* (New York: Knopf, 1977); Brian Holden Reid, *Robert E. Lee: Icon for a Nation* (London: Weidenfeld and Nicolson, 2005); and the older but excellent Stephen W. Sears, "Getting Right with Robert E. Lee," in Sears, ed., *The Best of the American Heritage: The Civil War* (Boston, Mass.: Houghton Mifflin, 1991), pp. 40–53.

55. On the "second American revolution," see Heather Cox Richardson, *The Greatest Nation on Earth: Republican Economic Policies during the Civil War* (Cambridge, Mass.: Harvard University, 1997); Leonard Curry, *Blueprint for Modern America: Non-Military Legislation of the First Civil War Congress* (Nashville, Tenn.: Vanderbilt University, 1968); Paludan, "*A People's Contest*," pp. 105–150; Richard F. Bensel, *Yankee Leviathan: The Origins of Central State Authority in America, 1859–1877* (New York: Cambridge University, 1990); Paludan, *Presidency of Lincoln*, pp. 108–118; and Hammond, *Sovereignty and the Empty Purse*, pp. 165–326, 261–282 ("in accordance with," pp. 273–274).

56. "My paramount object" in Roy C. Basler, ed., *The Collected Works of Abraham Lincoln*, 9 vols. (New Brunswick, N.J.: Rutgers University, 1952–1955), 5: 388; Lucas E. Morel, *Lincoln's Sacred Effort: Defining Religion's Role in American Self-Government* (New York: Lexington, 2000), pp. 178–179, assesses Lincoln's choice of words in the letter to Greeley. An interesting shred of evidence on Lincoln's parsing of pragmatism and principle dates from 1855, when his friend in Illinois Joshua Speed suggested that he align himself with the Know-Nothings. Lincoln recoiled. If he was on record as abhorring the oppression of Negroes, he asked, how could he be affiliated with those wanting to treat Catholic and foreign whites as if they were slaves? Should the Know-Nothings triumph, said Lincoln, "I shall prefer emigrating to some country where they make *no pretense* of loving liberty—to Russia, for instance, where despotism can be taken pure, and without the base alloy of hypocrisy" (Basler, *Works of Lincoln*, 2: 323.)

57. Allen C. Guelzo, *Abraham Lincoln: Redeemer President* (Grand Rapids, Mich.: Eerdmans, 1999), pp. 311–351 ("in the relaxed," p. 312; "suppose God," p. 325; "we had about," p. 335; "convinced that the war," p. 336; "God's purpose," pp. 326–327).

58. Guelzo, *Crisis of the American Republic*, pp. 126–139; Paludan, *Presidency of Lincoln*, pp. 137–154 ("so many persons," p. 146); McPherson, *Battle Cry of Freedom*, pp. 490–510 ("We won't fight," p. 493; "As laborers," p. 495; "military necessity," p. 504).

59. On Britain's successful quest for alternative sources of cotton, see Sven Beckert, "Emancipation and Empire: Reconstructing the Worldwide Web of Cotton Production in the Age of the American Civil War," *American Historical Review* 109 (2004): 1405–1438. India's cotton crop rose from 226.5 million pounds in the 1850s to 384.4 million in the 1860s, and eventually reached 920.1 million by the 1910s. The American South never recovered its hold on world markets.

60. See R. J. M. Blackett, *Divided Hearts: Britain and the American Civil War* (Baton Rouge: Louisiana State University, 2001); Frank L. Owsley, *King Cotton Diplomacy: Foreign Relations of the Confederate States of America*, 2nd ed. (Chicago, Ill.: University of Chicago, 1959), passim ("North and South must choose," p. 297); Alfred Grant, *The American Civil War and the British Press* (Jefferson, N.C.: MacFarland, 2000), esp. pp 121–134 on the *Times*; Jenkins, *Britain and the War for the Union*, 1: 205–234 ("temper of Congress," p. 222), 2: 85–182; Allen, "Civil War, Reconstruction, and Great Britain," in Hyman, ed., *Heard Round the World*, pp. 3–96; Ferris, *Desperate Diplomacy*, pp. 171–179 ("of no value," p. 178); McPherson, *Battle Cry of Freedom*, pp. 546–157 ("more hopeful," p. 555). Jenkins, 2: 1, cites an author whose British press clippings in 1862 led him to conclude for the *American Presbyterian Review*: "Our democracy is disliked by their aristocracy; our manufactures rival theirs; our commerce threatens at many points to supplant theirs. We are in dangerous proximity to some of their best colonies." Hence fear of America's future power shaped "the words and policy of many of England's greatest and best men." The French, by contrast, quickly lost interest in the Civil War and turned their attention to their own absurd venture in Mexico. By 1863 even the ardent socialist Louis-Auguste Blanqui thought that the "speeches of Father Lincoln, all steeped in Protestant hypocrisy," evoked ennui; and the procurator of Rouen, a textile town, said that American affairs "no longer have the power to impassion anyone" (David H. Pinkney, "France and the Civil War," in Hyman, ed., *Heard Round the World*, pp. 99–144; quotations, p. 105).

61. Rafuse, *McClellan's War*, pp. 300–333. Stephen W. Sears, *Landscape Turned Red: The Battle of Antietam* (New York: Ticknor and Fields, 1983); on Lee's lost orders, which Sears judges far less important than legend suggests, pp. 112–119; "a savage continual," p. 241; "Yes, Aunty," p. 297. Joseph L. Harsh, *Taken at the Flood: Robert E. Lee and Confederate Strategy in the Maryland Campaign of 1862* (Kent, Ohio: Kent State University, 1999). Have you ever been curious about just what Lee's lost orders said? Here they are (note Lee's pique over the Marylanders' failure to welcome him as a liberator):

SPECIAL ORDERS No. 191. HDQRS. ARMY OF NORTHERN VIRGINIA,
September 9, 1862.

    I. The citizens of Fredericktown being unwilling, while overrun by members of this army, to open their stores, in order to give them confidence, and to secure to officers and men purchasing supplies for benefit of this command, all officers and men of this army are strictly prohibited from visiting Fredericktown except on business, in which case they will bear evidence of this in writing from division commanders. The provost-marshal in Fredericktown will see that his guard rigidly enforces this order.

II. *Major Taylor will proceed to Leesburg, Va., and arrange for transportation of the sick and those unable to walk to Winchester, securing the transportation of the country for this purpose. The route between this and Culpeper Court-House east of the mountains being unsafe will no longer be traveled. Those on the way to this army already across the river will move up promptly; all others will proceed to Winchester collectively and under command of officers, at which point, being the general depot of this army, its movements will be known and instructions given by commanding officer regulating further movements.*

III. *The army will resume its march tomorrow, taking the Hagerstown road. General Jackson's command will form the advance, and, after passing Middletown, with such portion as he may select, take the route toward Sharpsburg, cross the Potomac at the most convenient point, and by Friday morning take possession of the Baltimore and Ohio Railroad, capture such of them as may be at Martinsburg, and intercept such as may attempt to escape from Harpers Ferry.*

IV. *General Longstreet's command will pursue the main road as far as Boonsborough, where it will halt, with reserve, supply, and baggage trains of the army.*

V. *General McLaws, with his own division and that of General R. H. Anderson, will follow General Longstreet. On reaching Middletown he will take the route to Harper's Ferry, and by Friday morning possess himself of the Maryland Heights and endeavor to capture the enemy at Harpers Ferry and vicinity.*

VI. *General Walker, with his division, after accomplishing the object in which he is now engaged, will cross the Potomac at Cheek's Ford, ascend its right bank to Lovettsville, take possession of Loudoun Heights, if practicable, by Friday morning, Keys' Ford on his left, and the road between the end of the mountain and the Potomac on his right. He will, as far as practicable, co-operate with Generals McLaws and Jackson, and intercept retreat of the enemy.*

VII. *General D. H. Hill's division will form the rear guard of the army, pursuing the road taken by the main body. The reserve artillery, ordnance, and supply trains, &c., will precede General Hill.*

VIII. *General Stuart will detach a squadron of cavalry to accompany the commands of Generals Longstreet, Jackson, and McLaws, and, with the main body of the cavalry, will cover the route of the army, bringing up all stragglers that may have been left behind.*

IX. *The commands of Generals Jackson, McLaws, and Walker, after accomplishing the objects for which they have been detached, will join the main body of the army at Boonsborough or Hagerstown.*

X. *Each regiment on the march will habitually carry its axes in the regimental ordnance wagons, for use of the men at their encampments, to procure wood, &c.*
By command of General R. E. Lee:

> *R. H. CHILTON, Assistant Adjutant-General.*

62. Hans L. Trefousse, *"First among Equals": Abraham Lincoln's Reputation during His Administration* (New York: Fordham University, 2005), pp. 16–64 ("a man of fixed principles," p. 36; "President Lincoln never," p. 38; "The Administration belongs," "the beginning of the end," and "the grandest proclamation," p. 50). "God has decided" in Sears, *Landscape Turned Red*, p. 518. "I think the constitution" in Basler, *Works of Lincoln*, 5: 529–537; "lookers-on" in Owsley, *King Cotton Diplomacy*, p. 351; "the character of the war" and "the Constitution as it is" in McPherson, *Battle Cry of Freedom*, pp. 559–560.

"Constitutional common sense" is the judgment of Herman Belz, in *Lincoln and the Constitution: The Dictatorship Question Reconsidered* (Fort Wayne, Ind.: Louis Warren Lincoln Library, 1984), p. 24. On the president's conception and drafting of the proclamation, see Allen C. Guelzo, *Lincoln's Emancipation Proclamation: The End of Slavery in America* (New York: Simon and Schuster, 2004).

63. Robert P. Broadwater, *The Battle of Perryville, 1862: Culmination of the Failed Kentucky Campaign* (Jefferson, N.C.: McFarland, 2005). "Their hearts are evidently" in Ketchum, ed., *American Heritage History*, p. 212. Bragg blamed Kentuckians' cupidity for their failure to put their arms where their mouths were. In fact, most of the Kentuckians who sympathized with the South had already enlisted or were waiting for proof that the Confederates were in Kentucky to stay.

64. Basler, *Works of Lincoln*, 6: 408. When Lincoln assumed office, there was one vacancy on the high court. The death of John McLean on April 4, 1861, created another, and the resignation of John Archibald Campbell of Alabama on April 20 created a third. After Congress reconvened the following winter, Lincoln nominated Noah Swayne of Ohio, Samuel Miller of Iowa, and David Davis of Illinois.

65. Paludan, in *"A People's Contest,"* pp. 91–102, makes the standard case that Republicans suffered a setback in the elections of 1862. McPherson, in *Battle Cry of Freedom*, pp. 560–562, stresses how modest their losses were. Paludan attributes the drop in Republican turnout to disaffection with abolition. McPherson notes that hundreds of thousands of likely Union supporters were absent on military service. In 1864 Lincoln would take pains to ensure soldiers a chance to vote. The cabinet crisis has been described in detail in Paludan, *Presidency of Lincoln*, pp. 170–181; and Goodwin, *Team of Rivals*, pp. 491–495. Having kept Chase and Seward on board, Lincoln joked, "I can ride on now, I've got a pumpkin in each end of my bag" (p. 494).

66. On the war economy of the Confederacy, see Thomas, *Confederate Nation*, pp. 190–214 ("bread or blood," p. 205); Vandiver, *Their Tattered Flags*, pp. 230–347 ("bread or peace," p. 236); Richard Cecil Todd, *Confederate Finance* (Athens: University of Georgia, 1954), pp. 110–198; Richard D. Goff, *Confederate Supply* (Durham, N.C.: Duke University, 1969); Douglas B. Ball, *Financial Failure and Confederate Defeat* (Urbana: University of Illinois, 1991); Robert C. Black III, *The Railroads of the Confederacy* (Chapel Hill: University of North Carolina, 1952); John E. Clark, Jr., *Railroads in the Civil War: The Impact of Management of Victory and Defeat* (Baton Rouge: Louisiana State University, 2001), pp. 26–87; Paul A. C. Koistenen, *Beating Plowshares into Swords: The Political Economy of American Warfare, 1606–1865* (Lawrence: University of Kansas, 1996), pp. 197–264 (on horses, pp. 224–226), and Paul W. Gates, *Agriculture and the Civil War* (New York: Knopf, 1965), pp. 3–108. Gates (p. 34) quotes an editorial in the *Mobile News* that sums up the popular mood: "Alas, poor human frailty! Excited by reports of sudden fortunes realised by small means, urged on by greed and love of gain, eminent physicians, distinguished lawyers and politicians, liberal and hospitable planters, high minded merchants, lowered the high standards of Southern chivalry, out-Jewed Jews, out-traded the sharp Yankee trader, and descended from that exalted position known on earth—Southern gentlemen—to become nothing better than common hucksters."

67. For a colorful swashbuckler's memoir, see Thomas E. Taylor, *Running the Blockade: A Personal Narrative* (Annapolis, Md.: Naval Institute, 1995, orig. 1896). Planning a trip to Bermuda? Be sure to ride the bus out to Saint George's at the island's northeastern

tip and visit the Confederate museum celebrating blockade runners. Estimates on successful voyages are in Owsley, *King Cotton Diplomacy*, p. 285; and Stephen R. Wise, *Lifeline of the Confederacy: Blockade Running during the Civil War* (Columbia: University of South Carolina, 1988), p. 221. Mark Thornton and Robert B. Ekelund, Jr., in *Tariffs, Blockades, and Inflation: The Economics of the Civil War* (Wilmington, Del.: Scholarly Resources, 2004), pp. 29–58, offer an econometric analysis of the "Rhett Butler effect," concluding that governmental efforts to regulate blockade running actually created more incentive to cheat. With freight rates in Nassau running as high as £80 in gold per ton, it was folly for smugglers to load up on bulky war matériel to sell at fixed prices to the Confederacy. David G. Surdam, in *Northern Naval Superiority and the Economics of the American Civil War* (Columbia: University of South Carolina, 2001), makes the counterintuitive macroeconomic case that loss of exports was more damaging to the South than loss of imports. Luraghi, in *History of the Confederate Navy*, pp. 345–349, argues that the Confederacy's navy succeeded because its task was never to defeat the blockade, but rather to defend the South from amphibious attack so that the land armies had a chance to prevail.

68. Albert B. Moore, *Conscription and Conflict in the Confederacy* (New York: Macmillan, 1924).

69. On the Confederacy's nation building, see Gary W. Gallagher, *The Confederate War* (Cambridge, Mass.: Harvard University, 1997); George C. Rable, *The Confederate Republic: A Revolution against Politics* (Chapel Hill: University of North Carolina, 1994); Thomas B. Alexander and Richard Beringer, *The Anatomy of the Confederate Congress* (Nashville, Tenn.: Vanderbilt University, 1972); Drew Gilpin Faust, *The Creation of Confederate Nationalism: Ideology and Identity in the Civil War South* (Baton Rouge: Louisiana State University, 1988). On the president of the Confederacy, see Cooper, *Jefferson Davis, American*; William C. Davis, *Jefferson Davis* (New York: HarperCollins, 1991); and Brian R. Dirck, *Lincoln and Davis: Imagining America, 1809–1865* (Lawrence: University of Kansas, 2001), esp. pp. 163–200. Dirck sums up: "It is a matter of the passage in Romans 8:31, familiar alike to Northerner and Southerner in the trying times of war: 'If God be for us, who can be against us?' Lincoln focused most of his energies on the 'if.' Davis did not" (p. 200). Gladstone's speech of October 1863 in Newcastle is cited in Jenkins, *Britain and the War for the Union*, 2: 172; "Liberty is always" in *Papers of Jefferson Davis*, 7: 412–419.

70. Prejudice against Jews was in general far less pronounced in the South than in the North, and southern Jews loyally served the Confederacy in business, government, and the army. See Robert N. Rosen, *The Jewish Confederates* (Columbia: University of South Carolina, 2000). On the remarkable Benjamin, see Robert Douthat Meade, *Judah P. Benjamin: Confederate Statesman* (Baton Rouge: Louisiana State University, 2001, orig. 1943), pp. 139–271. He was born on the island of Nevis to parents of Sephardic extraction; dropped out of Yale but studied law in New Orleans; and became a renowned lawyer and Whig politician. Indeed, his rise paralleled that of Benjamin Disraeli to a surprising degree, until everything went off the rails in 1861. Benjamin opposed secession, but resigned from the U.S. Senate out of loyalty to Jefferson Davis. His first appointment as the Confederacy's attorney general wasted his talents. By the time Davis promoted him to secretary of state, it was too late to salvage the Confederacy's diplomatic position. By the time Davis made Benjamin secretary of war, it was too late to salvage the war econ-

omy. Benjamin also opposed moving the capital to Richmond, opposed the cotton embargo, was among the first to consider arming the slaves, and even proposed emancipation in hopes of winning belated recognition from Britain. After the war he went into exile and built a third successful career as a British barrister.

71. See William W. Freehling, *The South versus the South: How Anti-Confederate Southerners Shaped the Course of the Civil War* (New York: Oxford University, 2001); Rable, *The Confederate Republic: A Revolution against Politics*; Jon L. Wakelyn, *Confederates against the Confederacy: Essays on Leadership and Loyalty* (Westport, Conn.: Praeger, 2002), "a crazy quilt," p. 215; and Richard E. Beringer et al., *The Elements of Confederate Defeat: Nationalism, War Aims, and Religion* (Athens: University of Georgia, 1988), pp. 23–31; "enthusiastic welcome" in Cooper, *Jefferson Davis, American*, p. 461.

72. The classic by Bell Irvin Wiley, *The Life of Johnny Reb: The Common Soldier of the Confederacy* (Baton Rouge: Louisiana State University, 1943), relies almost exclusively on letters interpreted in the cynical mode. Since American soldiers in World War II said that they fought just for their buddies and just to get the war over so they could go home, Wiley suspected a similar lack of ideological zeal among Civil War soldiers. But James M. McPherson, *What They Fought For 1861–1865* (Baton Rouge: Louisiana State University, 1994); and Reid Mitchell, *Civil War Soldiers* (New York: Viking, 1988), assembled impressive evidence that many soldiers on both sides understood the political, moral, and indeed religious values at stake and were prepared to risk death for them. James I. Robertson, Jr., in *Soldiers Blue and Gray* (Columbia: University of South Carolina, 1998), rose to the defense of his mentor, Wiley, and alighted on middle ground. "Like soldiers of all ages, Civil War troops were human: they rarely spoke in praise of things. They were quick to criticize; and because they were, their writings tend to be more negative and sarcastic than one might expect" (p. ix), yet "this nation's common people can and will show that they value some ideals more than they value their lives" (p. 228). The wonderful quotation from the southerner taken prisoner is cited by Foote, *Civil War*, 1: 65, but blame me for the phonetic drawl.

73. Drew Gilpin Faust, *Mothers of Invention: Women of the Slaveholding South in the American Civil War* (Chapel Hill: University of North Carolina, 1996), pp. 30–195 ("thinned out, p. 31; "My work," p. 134; "I fear the blacks" and "I am afraid," pp. 59–60; "War has hardened," pp. 190–191); Drew Gilpin Faust, "'Without Pilot or Compass': Elite Women and Religion in the Civil War South," in Randall M. Miller et al., eds., *Religion and the American Civil War* (New York: Oxford University, 1998), pp. 250–260; "It may well" in Drew Gilpin Faust, "Altars of Sacrifice: Confederate Women and the Narratives of War," *Journal of American History* 76 (1990): 1228; Mary Elizabeth Massey, *Bonnet Brigades: Women in the Civil War* (New York: Knopf, 1966), pp. 197–241 ("poor country women," p. 215); George Rable, *Civil Wars: Women and the Crisis of Southern Nationalism* (Urbana: University of Illinois, 1991), esp. pp. 91–220 ("blue devils," p. 166); Glenna Matthew, "'Little Women' Who Helped Make This Great War," in Gabor S. Boritt, ed., *Why the Civil War Came* (New York: Oxford, 1996), pp. 31–49; Guelzo, *Crisis of the American Republic*, pp. 306–318 ("have brought an everlasting," p. 314); "To the women" in Dirck, *Lincoln and Davis*, p. 240; "the strange part played" in Harry S. Stout, *Upon the Altar of the Nation: A Moral History of the American Civil War* (New York: Viking, 2006), p. 108; "I am sure" in Wiley, *Life of Johnny Reb*, p. 46.

74. Mark Weitz, *More Damning Than Slaughter: Desertion in the Confederate Army*

(Lincoln: University of Nebraska, 2005); Anne Sarah Rubin, *A Shattered Nation: The Rise and Fall of the Confederacy 1861–1868* (Chapel Hill: University of North Carolina, 2005), pp. 68–79.

75. Susan-Mary Grant, *North over South: Northern Nationalism and American Identity in the Antebellum Era* (Lawrence: University of Kansas, 2000), pp. 1–18, 111–172 ("We are wafted," p. 151; "the Union would exist," p. 131); Melinda Lawson, *Patriot Fires: Forging a New American Nationalism in the Civil War North* (Lawrence: University of Kansas, 2002), pp. 1–13, 129–159 ("the ideas of Massachusetts," p. 138); "the roundabout route" in Charles C. Coffin, *Abraham Lincoln* (New York: Harper, 1893), p. 341. For northern culture in general, see George Frederickson, *The Inner Civil War: Northern Intellectuals and the Crisis of the Union* (New York: Harper and Row, 1965).

76. On the Union's war mobilization, see Koistenen, *Beating Plowshares into Swords*, pp. 102–177; Gates, *Agriculture in the Civil War*, pp. 129–250 (data on exports, p. 227); John E. Clark, Jr., *Railroads in the Civil War: The Impact of Management on Victory and Defeat* (Baton Rouge: Louisiana State University, 2001), pp. 26–73, 160–212 ("pugnaciously efficient," p. 63); Thomas Weber, *The Northern Railroads in the Civil War, 1861–1865* (New York: King's Crown, 1952); George E. Turner, *Victory Rode the Rails: The Strategic Place of the Railroads in the Civil War* (Westport, Conn.: Greenwood, 1953); Francis A. Lord, *Lincoln's Railroad Man: Herman Haupt* (Teaneck, N.J.: Fairleigh Dickinson University, 1969); Russell F. Weigley, *Quartermaster General of the Union Army: A Biography of M. C. Meigs* (New York: Columbia University, 1959); and Paludan, *"A People's Contest,"* pp. 151–197. Devotion to a common cause did not prevent Haupt and Governor Andrew from feuding over financial damages incurred when a railroad tunnel that Haupt was hired to dig in Massachusetts ran into political opposition there. In September 1863 Andrew "engineered" a military commission for Haupt so he would not be at liberty to appear in court and press charges. Haupt angrily refused to be "kicked upstairs," whereupon Stanton fired him.

77. Thornton and Ekelund, *Tariffs, Blockades, and Inflation*, pp. 59–80; Lawson, *Patriot Fires*, pp. 40–64 ("Patriotism and Profit" and "Make the U.S.," pp. 54–55). The standard work on Cooke is Ellis P. Oberholtzer, *Jay Cooke: Financier of the Civil War* (Philadelphia: G. W. Jacobs, 1907). On labor, see David Montgomery, *Beyond Equality: Labor and the Radical Republicans, 1862–1872* (New York: Knopf, 1967).

78. On charity generally, see Robert H. Bremner, *The Public Good: Philanthropy and Welfare in the Civil War Era* (New York: Knopf, 1980), pp. xi–xviii, 35–110. Frederick Law Olmsted, a director of the U.S. Sanitary Commission, complained that his colleagues couldn't be bothered to study their tasks with care, "but *get along somehow and guess it will do.* Damn them" (p. 42). Thus he gave early voice to the sentiment "Close enough for government work." On Yankee women, see esp. Nina Silber, *Daughters of the Union: Northern Women Fight the Civil War* (Cambridge, Mass.: Harvard University, 2005). Silber challenges the notion that the Civil War advanced women's liberation, arguing that the war and abolition pushed feminism into the background, women faced constraints even during the war, and after the war most women returned to their normal "separate sphere." She finds it significant that one of the few novels of the time dealing with northern women at war, Louisa May Alcott's *Little Women* (1867), taught modesty and reserve, not self-assertion. A contemporary source tending to reinforce that conclusion is Linus P. Brockett and Mary C. Vaughn, *Woman's Work in the Civil War: A Record of Patriotism, Heroism, and Patience* (Philadelphia, Pa.: Zeigler, McCurdy, 1867). But

contrast Jeanie Attie, *Patriotic Toil: Northern Women and the American Civil War* (Ithaca, N.Y.: Cornell University, 1998). See also Lawson, *Patriot Fires*, pp. 14–39, 148–156; Massey, *Women in the Civil War*, pp. 43–86 ("My God, man," p. 49); and Nina Brown Baker, *Cyclone in Calico: The Story of Mary Ann Bickerdyke* (Boston, Mass.: Little, Brown, 1952). The number of women who "contributed" to men's morale through prostitution increased tenfold in Washington, D.C., and also increased around other concentrations of troops. Some abolitionists conflated cause and carnality. Massey quotes a friend of Lucretia Mott enticing another friend to visit inasmuch as there were "plenty of colored brothers around, if that's what you want" (p. 272).

79. For summaries of the historiography, see Phillip Shaw Paludan, "Religion and the American Civil War," pp. 21–42; George M. Frederickson, "The Coming of the Lord: The Northern Protestant Clergy and the Civil War Crisis," pp. 110–130; and Reid Mitchell, "Christian Soldiers? Perfecting the Confederacy," pp. 297–312, in Randall M. Miller, Harry S. Stout, and Charles Reagan Wilson, eds., *Religion and the American Civil War* (New York: Oxford University, 1998). See also Bell Irvin Wiley, *The Life of Billy Yank: The Common Soldier of the Union* (Baton Rouge: Louisiana State University, 1952), pp. 262–274; Steven E. Woodworth, *While God Is Marching On: The Religious World of Civil War Soldiers* (Lawrence: University of Kansas, 2001); Gardiner H. Shattuck, Jr., *A Shield and a Hiding Place: The Religious Life of the Civil War Armies* (Macon, Ga.: Mercer University, 1987); "the great theological" and "a thin simple" in Morel, *Lincoln's Sacred Effort*, pp. 184–185. Gerald Linderman, in *The Experience of Combat in the American Civil War* (New York: Free Press, 1987), esp. pp 156–168, 240–265, provides a realistic assessment of soldiers' faith and disillusionment. Soldiers invariably wanted to believe that their ordeal had a higher meaning and that the reward of faith was divine protection (no atheists in foxholes). But they "naturally distrusted the efficacy of prayer when they found that the most devout Christian was as liable to be shot as the most hardened sinner, and that a deck of cards would stop a bullet as effectively as a prayer book" (p. 159).

80. James H. Moorhead, *American Apocalypse: Yankee Protestants and the Civil War, 1860–1869* (New Haven, Conn.: Yale University, 1978); Guelzo, *Crisis of the American Republic*, pp. 318–327 ("feeble pretences," p. 323; "final and high" and "world-Republic," pp. 324–325); "we shall no more" in Daniel T. Rodgers, *Contested Truths: Keywords in American Politics since Independence* (New York: Basic Books, 1987), p. 137.

81. The most thorough compilation of documentation showing that Union soldiers were indeed highly motivated by their desire to abolish slavery is in Chandra Manning, *What This Cruel War Was Over: Soldiers, Slavery, and The Civil War* (New York: Knopf, 2007). See also Reid Mitchell, "The Perseverance of the Soldiers," in Gabor S. Borritt, ed., *Why the Confederacy Lost* (New York: Oxford University, 1992), pp.111–132 ("Never in a war," p. 132). The tension between biblical and civil religion during the Civil War and its resolution mostly in favor of the latter have been probed in depth by Harry S. Stout, *Upon the Altar of the Nation: A Moral History of the American Civil War* (New York: Viking, 2006); "in a word" and "most idolatrous," pp. 373–374. But see also the splendid summary in Paludan, *"A People's Contest*," pp. 339–374 ("cutting men down" and "holy war," pp. 359–360; "a pitiful God," et seq, pp. 363–364); "There is no God," in Guelzo, *Crisis of the American Republic*, p. 326. On the decline of the Sabbath, see Alexis McCrossen, *Holy Day, Holiday: The American Sunday* (Ithaca, N.Y.: Cornell University, 2000), pp. 46–50 ("there is no Sunday," p. 47). The lyrics to no fewer than 277 of the 600 entries in the Episcopal hymnal of 1940 were composed during the nineteenth century. Of

course, many of those were of British origin, including Sabine Baring-Gould's "Onward Christian Soldiers" (1864), but Baptist and Methodist hymnals would doubtless yield an even higher percentage of nineteenth-century American composers and compositions.

82. "I have lost" cited in Elizabeth Fox-Genovese, "Days of Judgment, Days of Wrath: The Civil War and the Religious Imagination of Women Writers," and "These United States" cited in Paludan, "Religion," both in Miller et al., eds., *Religion and the American Civil War*, pp. 240, 28.

83. Charles H. Ambler and Festus P. Summers, *West Virginia: The Mountain State* (Englewood Cliffs, N.J.: Prentice-Hall, 1958), pp. 3–119; John Alexander Williams, *West Virginia: A History* (Morgantown: West Virginia University, 2001), pp. 3–37 ("You stand," p. 30). West Virginia's average altitude of 1,500 feet makes it the loftiest state east of the Mississippi.

84. Wheeling enjoyed a boom in the 1820s and 1830s as the jumping-off point for settlers taking the National Road to Ohio, but its continued growth required a bridge over the Ohio River. No federal or state support was forthcoming, so private capital completed the span in 1849. Jealous Pennsylvanians sued, arguing that the bridge inhibited the passage of steamboats with high smokestacks. The case dragged on until 1852, when the Supreme Court favored Wheeling's contention that boats should be rigged with adjustable stacks. In 1854 Pittsburgh laughed last when a storm wrecked the bridge. Wheeling then placed its hopes in the Baltimore and Ohio railroad, which built through the town when (ironically) the Pennsylvania railroad persuaded its home state to deny the Baltimore and Ohio rights to Pittsburgh. But Wheeling lost out again when the Baltimore and Ohio bridged the Ohio River several miles south of town, then turned its spur line to Parkersburg into the main westbound trunk. It is an odd feeling to drive through Wheeling on route I–70. If traffic is light, you will "cross West Virginia" in about fifteen minutes.

85. Asbury, the tireless circuit rider who founded America's largest nineteenth-century denomination, met one of Jackson's grandmothers on the frontier in 1788. He became the family's spiritual mentor and delighted in the piety shown by Tom. But Asbury also noted how much land the extended Jackson family accumulated and how the highland elites would lord it over the poor as much as the lowland planters did. See Elmer T. Clark et al., eds., *The Journal and Letters of Francis Asbury*, 3 vols. (Nashville, Tenn.: Abingdon, 1958) 3: 576–577. Thomas Jackson surely deserved his appointment to the U.S. Military Academy and his professorship at the Virginia Military Institute, but family connections won him those billets.

86. John W. Shafer, *A Clash of Loyalties: A Border County in the Civil War* (Morgantown: West Virginia University, 2003), pp. 1–16. Shafer begins with an excellent summary of the traditional historiography that argued for relative homogeneity in the northwestern counties and growing disaffection toward the rest of Virginia, and revisionist historiography that stresses complexity. See also Daniel W. Crofts, *Reluctant Confederates: Upper South Unionism in the Secession Crisis* (Chapel Hill: University of North Carolina, 1989), pp. 133–153; Ambler and Summers, *The Mountain State*, pp. 150–194; and Williams, *West Virginia*, pp. 37–55 ("died, as he was born, a Virginian," p. 38), with its persuasive discussion of up-country elites. His quotation is from Mary Ann Jackson, *Memoirs of Stonewall Jackson by His Widow* (Louisville, Ky.: Prentice, 1985), p. 14.

87. On West Virginia's statehood, see Waitman T. Willey, *The Formation of West Virginia* (Wheeling, W. Va.: News Publishing, 1901); Charles H. Ambler, *Francis H. Pierpont, Union War Governor of Virginia and Father of West Virginia* (Chapel Hill: University of North Carolina, 1937); and George E. Moore, *A Banner in the Hills: West Virginia's Statehood* (New York: Appleton-Century-Crofts, 1963). "Talk about" cited in Crofts, *Reluctant Confederates*, p. 163. The "reorganized" Virginia delegation to Congress was credentialed by the Unionist senator Andrew Johnson of Tennessee, which had also seceded.

88. On the devastation visited on Jefferson County, Harpers Ferry, and the Baltimore and Ohio, see Festus P. Summer, *The Baltimore and Ohio in the Civil War* (New York, 1939); and Chester G. Hearn, *Six Years of Hell: Harpers Ferry during the Civil War* (Baton Rouge: Louisiana State University, 1996). Colonel "Stonewall" Jackson wrote of Joseph Johnston's instructions to demolish the railroad and armory in June 1861: "It was a sad work, but I had my orders" (*Six Years*, p. 75). In their stockholders' report for the year, railroad officials fixed its losses at forty-two locomotives and tenders, 386 cars, twenty-three bridges totalling 4,713 feet of span, and 36½ miles of track.

89. Benjamin Kelley, a general from western Virginia, was fifty-four years old in 1861. He had worked as a freight agent for the Baltimore and Ohio. There is evidence that he and President Lincoln encouraged the Wheeling convention to claim the eastern panhandle counties in order to help secure the vital railroad. Jefferson County was not strongly Unionist before the war. But its rump electorate voted almost unanimously to join West Virginia if only in hopes of getting better protection against Confederate raids. Just a week after the poll, Lee's army appeared in the Shenandoah Valley and the county braced for another invasion.

90. Ambler and Summers, *Mountain State*, pp. 229–263 ("have been true," p. 242); Williams, *West Virginia*, pp. 56–94 ("The Secessionists," p. 82). After West Virginia became a state, Pierpont moved his titular Virginia regime to Alexandria, safely under the guns of Washington City's defenses. When Union bayonets installed him in Richmond in 1865, Pierpont devoted himself to conciliating secessionists in hopes of truly "reconstructing" his beloved state. He pleaded for restoration of civil rights for former rebels, approved every pardon issued by President Andrew Johnson, and tried to bring former Confederates into the government. All this infuriated the Radical Republicans. In 1868 they prevailed on the Union general John M. Schofield to replace Pierpont with a military governor. The Virginians themselves lodged another barb in his breast. Their reconstituted assembly repealed the acts of Pierpont's wartime regime that facilitated West Virginia's defection. Federal courts ignored the gesture, but West Virginians have not forgotten that in the eyes of Virginians, their forefathers were secessionist traitors.

91. Most illustrious were the Rangers of Captain Jesse C. McNeill, who escaped from captivity in Missouri to fight for his home in Hardy County. As late as February 1865 the Rangers rode to Cumberland, Maryland, far behind enemy lines, and captured two generals. The Union general Philip Sheridan said that McNeill's Rangers were "the most dangerous of all the bushwhackers," and the Confederate "gray ghost" John Mosby said that the capture of the generals "surpasses anything I have done." On combat and politics in wartime West Virginia, see Richard Orr Curry, *A House Divided: A Study of Statehood Politics and the Copperhead Movement in West Virginia* (Pittsburgh, Pa.: University of

Pittsburgh, 1964), esp. pp. 141–147; Shafer, *Clash of Loyalties*, pp. 1–16; Crofts, *Reluctant Confederates*, pp. 133–153; and Ambler and Summers, *Mountain State*, pp. 207–128 (quotations above, p. 228).

92. Shafer, *Clash of Loyalties*, pp. 149–174 ("rebel ruffians," p. 163). Another measure designed to unite the fractured population was the choice of Charleston as the permanent capital in 1869. That gave residents of the large Kanawha valley a stake in a state heretofore associated too closely with Wheeling and Whigs.

93. Altina L. Waller, *Feud: Hatfields, McCoys, and Social Change in Appalachia, 1860–1900* (Chapel Hill: University of North Carolina, 1988). Journalists naturally spun Jonathan Hatfield's intercourse with Rose Anne McCoy as a Romeo and Juliet story. In fact, the lovers quickly drifted into the arms of others, and the clansmen were more interested in razorbacks and revenge. Their sporadic gunfights culminated in 1888, when the Hatfield brothers burned down a McCoy cabin, killing two children and seriously injuring their mother.

94. Williams, *West Virginia*, pp. 95–129 ("I had not the remotest," p. 120).

95. Ambler and Summers, *Mountain State*, pp. 444–465. During the strikes of 1920–1921, miners reverted to Hatfield-McCoy tactics by sniping at bosses and scabs, then formed paramilitary units stiffened by veterans of the Great War. The governor declared "a state of war, insurrection, and riot" (p. 456) along a twenty-five-mile front line skirting Blair Mountain. Mother Jones, now ninety-one, arrived bearing a letter from President Harding that promised a federal investigation of working conditions, but failed to stop the violence. In the end, the state tried 543 ringleaders, some on charges of treason, in the same courthouse where John Brown was condemned sixty-three years before.

96. Jefferson Davis is frequently faulted for dispersing his regiments in deference to state governors who insisted that the entire Confederacy be defended. That is not quite fair, in my judgment, since Davis abandoned that forlorn hope within twelve to eighteen months. What a strategy of "wearing out the Yankees" really required was a hedgehog defense of a logistic network whose nodes (after New Orleans fell) were Richmond, Atlanta, Columbus (Georgia), Chattanooga, Mobile, and Vicksburg. Davis's sins of omission *might* well have have contributed to the loss of Vicksburg and Chattanooga, but by 1863 he did not have enough army to go around.

97. "With all my other trials I have to contend against," grumbled Grant, "is added that of speculators whose patriotism is measured by dollars and cents." While Grant's farm boys slaved with picks and shovels to dig canals across horseshoes on the river, the oily merchants traded with the enemy for cotton produced by actual slaves and competed with the army for precious railroad capacity. In a rare fit of bad judgment and frustration, Grant singled out Jews for "violating every regulation" and ordered their expulsion. Lincoln and Halleck hastily countermanded the order, and Grant himself admitted that he deserved censure. See Perret, *Grant: Soldier and President*, pp. 236–238.

98. The nickname came from a correspondent's dispatch that read "Fighting—Joe Hooker, etc." "Fighting" was supposed to be a headline, but copy editors dropped the dash in error or from a sense of drama. Myth has it that Hooker zoned a bawdy district in Washington, D.C., and invited prostitutes to follow his army into the field, hence they came to be known as "hookers." But Norman Ellsworth Eliason, in *Tarheel Talk: An Historical Study of the English Language in North Carolina to 1860* (Chapel Hill: University

of North Carolina, 1956), documented use of the term "hooker" in 1845; and according to John Russell Bartlett, in *Dictionary of Americanisms: A Glossary of Words and Phrases Usually Regarded as Peculiar to the United States* (New York: Johnson Reprint, 1968), a hooker was defined as a strumpet in 1859. No general's name is needed to account for a "hooker's" being someone who "hooks" men as if they were fish taking bait. "May God" cited by T. Harry Williams, *Lincoln and His Generals* (New York: Vintage, 1952), p. 232.

99. Stephen W. Sears, *Chancellorsville* (Boston, Mass.: Houghton Mifflin, 1996); for a brief, stirring account, see Robert K. Krick, "Lee's Greatest Victory," in Sears, ed., *Civil War: Best of American Heritage*, pp. 123–146.

100. Michael B. Ballard, *Vicksburg: The Campaign That Opened the Mississippi* (Chapel Hill: University of North Carolina, 2004); Perret, *Grant: Soldier and President*, pp. 244–264; "Wan, hollow-eyed" in Peter Franklin Walker, *Vicksburg: A People at War* (Chapel Hill: University of North Carolina, 1960), p. 161; "all save the spirit" in Ketchum, ed., *American Heritage History*, p. 323.

101. Cooper, *Jefferson Davis, American*, pp. 434–437 ("to feed upon," p. 353); Archer Jones, *Confederate Strategy from Shiloh to Vicksburg* (Baton Rouge: Louisiana State University, 1961), pp. 206–114; Woodworth, *No Band of Brothers*, pp. 51–69.

102. Chamberlain's unit deserves every bit of acclaim it has received. But the men from Maine could not have made their brave stand at all if the regiments from New York and Pennsylvania stationed below them to the north had not also withstood ferocious attacks. A. M. Judson, in *History of the Eighty-Third Regiment Pennsylvania Volunteers* (Erie, Pa.: B. F. H. Lynn, 1865), gives a breathtaking account, all the more credible given that Captain Judson fought in almost every major battle in the East; recorded impressions while they were fresh; and filled his memoirs with criticism, sarcasm, and wit, not Romantic bravado. On day two at Gettysburg Judson observed a Union soldier crawl out from behind a rock to save enemy wounded, only to be shot by rebels who didn't realize what he was about. "A more sublime instance of courage and humanity was perhaps never before exhibited upon the battlefield" (p. 68). Regarding the forty-six men of his own unit killed that afternoon, Judson wrote: "These brave men have the glory of having laid down their lives on the soil of old Pennsylvania, in protecting her hearthstones from the tread of the invader, and in one of the most sanguinary battles which the history of this or any other war has recorded" (p. 69).

103. On Lee's secret plan for the third day, see Tom Carhart, *Lost Triumph: Lee's Real Plan at Gettysburg and Why It Failed* (New York: Putnam, 2005); "Come on, you Wolverines," p. 220. The battle site called East Cavalry Field is a remote, rarely attended portion of Gettysburg National Military Park, but it may have been as much a breakwater to the Confederacy's "high tide" as Little Round Top or Seminary Ridge. For the whole campaign, see Stephen W. Sears, *Gettysburg* (Boston, Mass.: Houghton Mifflin, 2003); and the electric historical novel by Michael Shaara, *The Killer Angels* (New York: Ballantine, 1975). For the pivotal fight on the second day, see Glenn W. LaFantasie, *Twilight at Little Round Top: July 2, 1863—The Tide Turns at Gettysburg* (Hoboken, N.J.: Wiley, 2005). For some more new information regarding Pennsylvania Germans' role in the campaign, African-Americans captured and sold back into slavery, and townswomen who opened their homes to the wounded, see Margaret S. Creighton, *The Colors of Courage: Gettysburg's Forgotten History—Immigrants, Women, and African Americans in the Civil War's Defining Battle* (New York: Basic Books, 2005).

104. Melville, a stanza from "Gettysburg: The Check" in *Battle Pieces*, p. 85.

105. "Grant is my man" in Williams, *Lincoln and His Generals*, p. 272. "A bird" cited in Joseph H. Parks, *General Edmund Kirby Smith, C.S.A.* (Baton Rouge: Louisiana State University, 1954), p. 427.

106. Colonel Shaw was white, but the troops (including two of Frederick Douglass's sons) were all black. Their stirring story is told in the Academy Award–winning film *Glory* (1989), directed by Edward Zwick and starring Matthew Broderick, Denzel Washington, and Morgan Freeman. The Confederates held the fort, only to abandon Battery Wagner on September 6. For the whole story of black belligerents in the Civil War, see Dudley Taylor Cornish, *The Sable Arm: Black Troops in the Union Army, 1861–1865* (New York: Norton, 1966). For a colorful eyewitness account of the diplomatic "Battle of the Rams," those two ironclads under construction in Liverpool, see *The Education of Henry Adams* (Cambridge, Mass.: Riverside, 1961, orig. 1918), pp. 167–179. Young Henry accompanied his father, Charles Francis Adams, during the latter's seven-year mission in London.

107. See Steven E. Woodworth, *Six Armies in Tennessee: The Chickamauga and Chattanooga Campaigns* (Lincoln: University of Nebraska, 1998); and Larry J. Daniel, *Days of Glory: The Army of the Cumberland, 1861–1865* (Baton Rouge: Louisiana State University, 2004). Thomas earned his epithet by leading the rugged resistance that allowed Rosecrans's army to break contact at Chickamauga. Grant got most of the credit for the triumph at Chattanooga, even though his battle plans either failed or were not executed. The foggy charge up Lookout Mountain on November 24 memorialized by journalists' purple prose was more a hike than a battle: the mountain had little military utility and was lightly defended. Grant's plan the next day called for Sherman to assault the Confederate left, over what turned out to be awful terrain. So the angry Ohio Valley men of the Army of the Cumberland really deserve credit for executing a "Pickett's charge" of their own and pulling it off. Chattanooga is a gorgeous, funky, historic town well worth a few days' visit.

108. On the equation of liberty and equality—the real constitutional revolution Lincoln wrought—see Herman Belz, *Emancipation and Equal Rights: Politics and Constitutionalism in the Civil War Era* (New York: Norton, 1978). No texts except biblical passages have inspired commentary so out of proportion to their length as the Gettysburg Address. David Herbert Donald, in *Lincoln* (New York: Simon and Schuster, 1995), is little moved by the words of a man he imagined a consummate pragmatist without much vision because of "the essential passivity of his nature" (p. 14). By contrast, John Diggins, in *On Hallowed Ground: Abraham Lincoln and the Foundations of American History* (New Haven, Conn.: Yale University, 2000), argues that Lincoln imbued the war with ideological content and elevated America's definition by cleansing the Constitution. Thus the new birth of freedom was indeed new, but the idea of freedom was old. Garry Wills, in *Lincoln at Gettysburg: Words That Made American History* (New York: Doubleday, 1992); and Pauline Maier, *American Scripture: Making the Declaration of Independence* (New York: Knopf, 1997), argue instead that Lincoln ignored Jefferson's original intent limiting liberty to white males. Perhaps that is so, but if it is, then Lincoln imbibed the Lockean spirit of natural rights more fully than Jefferson. Slavery denied Negroes' right to life, liberty, and property; denied them government by consent of the governed; and denied them the right to rebel against tyranny. I thank David Carlyon for sharing his

essay "Broadway to Gettysburg: Did Edwin Forrest Influence Abraham Lincoln?" which examines earlier speeches that might have inspired Lincoln's themes, cadence, and phrases. Lincoln's use of rhetoric to alter, rather than merely to manipulate or obfuscate, political realities is persuasively analyzed by Douglas L. Wilson, *Lincoln's Sword: The Presidency and the Power of Words* (New York: Knopf, 2006), and Gabor Boritt, *The Gettysburg Gospel: The Speech That Nobody Knows* (New York: Simon and Schuster, 2006). On hateful propaganda against Lincoln, see Michael Davis, *The Image of Lincoln in the South* (Knoxville: University of Tennessee, 1971), pp. 62–85, "No President," p. 98; *Royal Ape* was the title of a wartime production by the playwright William Russell Smith; "Rarely was man" is from Emerson's essay "Lincoln" cited in Diggins, *On Hallowed Ground*, pp. 37–40.

109. Goodwin, *Team of Rivals*, pp. 614–616 ("Why, here," "blushed," and "warmest campaign," p. 615). Perret, *Lincoln's War*, pp. 349–360; Perret, *Grant: Soldier and President*, pp. 287–298. Grant not only resented the "show business" entailed by being commander in chief but was of two minds about the promotion. He had hoped for the Pacific command, both to escape war's brutality and to return to a city he had hankered for ever since his brief posting there in 1852: San Francisco.

110. This mystery is untangled beautifully by Menahem Blondheim, "'Public Sentiment Is Everything': The Union's Public Communications Strategy and the Bogus Proclamation of 1864," *Journal of American History* 89 (2002): 869–899. "Public sentiment" in Basler, *Works of Lincoln*, 2: 553. Bates, *Lincoln in the Telegraph Office*, pp. 228–243, quotes the War Department telegraphy staff to the effect that Villard had no hard feelings over his arrest because "some choice scraps of news later found their way to the office of the syndicate, which supplied material for new 'scoops,' and had a soothing influence generally" (243). The editors in New York were released and back in business by May 22, thanks to the intervention, via Seward, of Thurlow Weed.

111. Frank L. Klement, *Dark Lanterns: Secret Political Societies, Conspiracies, and Treason Trials in the Civil War* (Baton Rouge: Louisiana State University, 1984), pp. 1–63; Paludan, *Presidency of Lincoln*, pp. 217–227. The Illinois Union League alone claimed 140,000 members. They urged Lincoln to visit his home state in advance of the elections of fall 1863. Instead he sent a letter that gained nationwide publicity. "You say you will not fight to free negroes. Some of them seem willing to fight for you; but no matter. Fight you then, exclusively to save the Union. I issued the proclamation on purpose to aid you in saving the Union" (Basler, *Works of Lincoln*, 6: 406–410). Nevins, in *War for the Union*, 3: 155, recognized that the elections of 1863 reversed the antiwar trend of the previous year and foreshadowed Lincoln's triumph the following year. On soldiers' hatred of copperheads, see Trefousse, *"First among Equals,"* pp. 85–116; and Wiley, *Life of Billy Yank*, pp. 286–288.

112. Klement, *Dark Lanterns*, pp. 64–186 ("to overthrow," p. 84), provides copious evidence of Republicans' exaggeration and exploitation of alleged plots. The self-styled chief of the Knights of the Golden Circle, George Washington Leigh Bickley, was an itinerant fraud whose mostly imaginary society dissolved on his arrest in July 1863. Phineas T. Wright, an obscure, fast-talking Democratic hack, conjured the Order of American Knights but had to watch it dissolve at its little convention of February 1864. Sanderson, the man who "exposed" them, was bucking for promotion to brigadier but died before Lincoln's landslide, which his report helped to propel. H. H. Dodd, leader of

Indiana's Sons of Liberty, was a disciple of Vallandigham genuinely fearful for civil rights, but he also knew his organization was more "pretense than a reality" (p. 108). Plots hatched or supported by the Confederate government were more palpable, as demonstrated by Thomas Fleming, "The Northwest Conspiracy," in Robert Cowley, ed., *What Ifs of American History: Eminent Historians Imagine What Might Have Been* (New York: Putnam, 2003), pp. 103–125; and Cooper, *Jefferson Davis, American*, pp. 496–498. Camp Douglas is often compared to Andersonville, the Confederate prisoner-of-war camp where so many Union men starved to death. But Camp Douglas, located on Chicago's South Side, was not cruel, just disease-ridden. One out of eight prisoners died. See George Levy, *To Die in Chicago: Confederate Prisoners at Camp Douglas 1862–1865* (Evanston, Ill.: Evanston, 1994). The Saint Albans raiders based in Quebec netted $208,000 and killed one citizen. The Royal Canadian Mounted Police arrested the lot, but a judge deemed their crime "an act of war" and released them. See Robin Winks, *Canada and the United States: The Civil War Years* (Baltimore, Md.: Johns Hopkins University, 1960); and Charles M. Wilson, "The Hit and Run Raid," in Sears, ed., *Civil War: Best of American Heritage*, pp. 174–185.

113. This laudatory assessment of Lincoln's record on civil liberties is based on Belz, *Lincoln and the Constitution*; Mark E. Neely, Jr., *The Fate of Liberty: Abraham Lincoln and Civil Liberties* (New York: Oxford University, 1991); Harold M. Hyman, *A More Perfect Union: The Impact of the Civil War and Reconstruction on the Constitution* (New York: Knopf, 1973); Daniel P. Franklin, *Extraordinary Measures: The Exercise of Prerogative Powers in the United States* (Pittsburgh, Pa.: University of Pittsburgh, 1991); Don E. Fehrenbacher, *Lincoln in Text and Context: Collected Essays* (Stanford, Calif.: Stanford University, 1987), pp. 113–142; Guelzo, *Crisis of the American Republic*, pp. 167–171; and Forrest McDonald, *The American Presidency: An Intellectual History* (Lawrence: University of Kansas, 1994), pp. 398–402. Critical assessments of presidential war powers include Wilfred E. Binkley, *The Powers of the President: Problems of American Democracy* (New York: Russell and Russell, 1973, orig. 1937); Clinton L. Rossiter, *Constitutional Dictatorship: Crisis Government in the Modern Democracies* (Princeton, N.J.: Princeton University, 1948); Edward S. Corwin, *The President: Office and Powers, 1798–1957*, 4th ed. (New York: New York University, 1957); and Arthur M. Schlesinger, Jr., *The Imperial Presidency* (Boston, Mass.: Houghton Mifflin, 1973).

114. Paludan, *Presidency of Lincoln*, pp. 259–274 ("I have determined," p. 267); "Upon the progress" in Basler, *Works of Lincoln*, 8: 332.

115. "Their campaigns combine" cited in Cass Canfield, *The Iron Will of Jefferson Davis* (New York: Harcourt, Brace, Jovanovich, 1978), p. ix. A good summary (among many) of the campaign is Noah A. Trudeau, *Bloody Roads South: The Wilderness to Cold Harbor, May–June 1864* (Boston, Mass.: Little, Brown, 1989). Captain Judson, whom we met at Gettysburg, was in the thick of things again. As his Pennsylvanians crept toward Wilderness Tavern, they glimpsed advancing rebels: "The moment they plunged into the woods they were lost to the sight. Here were two great armies forming line of battle for a desperate struggle, within half a mile of each other, scarcely a movement of either of which could be observed by the other. . . . We were about to charge an invisible foe, or, to use a common phrase, were about to go it blind. . . . On we went, o'er briar, o'er brake, o'er logs and o'er bogs, through the underbrush and overhanging limbs, for about three-quarters of a mile, yelling all the while like so many demons. . . . We had by this time got

into such a snarl that no man could find his own company or regiment. . . . It was now the Johnnies' turn to come the game of pull-the-link-horn over us, and right well did they improve the opportunity. Every man saw the danger, and without waiting for orders to fall back, broke for the rear on the double quick. The rebels, in their turn, commenced yelling and sending minnies [minié balls] after us, killing and wounding many of our men. . . . This day was a disastrous one for the Eighty-Third. . . . Our whole loss, in killed, wounded and missing, at the Wilderness and at Laurel Hill amounted to over three hundred" (Judson, *History of the Eighty-Third*, pp. 94–96).

116. Melville, a stanza from "The Armies of the Wilderness" in *Battle Pieces*, p. 97.

117. Foote, *Civil War*, 3: 186–191, on the dramatic scene at the Chancellorsville crossroad; Noah A. Trudeau, *The Last Citadel: Petersburg, Virginia, June 1864–April 1865* (Boston, Mass.: Little, Brown, 1991); "Many a man" in Mark D. Howe, ed., *Touched with Fire: Civil War Letters and Diary of Oliver Wendell Holmes, Jr., 1861–1864* (Cambridge, Mass.: Harvard University, 1946), p. 149. For Holmes's accounts of the campaign that wrenched his philosophy of life, see Louis Menand, *The Metaphysical Club: A Story of Ideas in America* (New York: Farrar, Straus and Giroux, 2001), pp. 49–69; "More desperate fighting," p. 53.

118. Albert Castel, *Decision in the West: The Atlanta Campaign of 1864* (Lawrence: University of Kansas, 1992); John F. Marszalek, *Sherman: A Soldier's Passion for Order* (New York: Free Press, 1993); Craig Symonds, *Joseph E. Johnston: A Civil War Biography* (New York: Norton, 1992). Davis had feuded repeatedly with Johnston, whom he considered arrogant and overrated. For his part, Johnston did not want the battered command Bragg left behind but did his duty very well. Soldiers in the Army of Tennessee were incredulous, angry, and deflated after learning of Johnston's dismissal. Hood's arm was shot up at Gettysburg, and his leg was amputated at Chickamauga..

119. Paludan, *Presidency of Lincoln*, pp. 275–285; "it seems exceedingly" in Basler, *Works of Lincoln*, 7: 514. Do "damn the torpedoes" and "Atlanta is ours" require citations? McPherson, in *Battle Cry of Freedom*, pp. 761, 774, does not think so (and neither do I, really). On Americans' descent during the war from limited, gentlemanly combat to ruthless destruction of the sort they prefer to attribute to foreign enemies, see Charles Royster, *The Destructive War: William Tecumseh Sherman, Stonewall Jackson, and the Americans* (New York: Random House, 1991).

120. Bates, *Lincoln in the Telegraph Office*, pp. 267–286 ("unusually weary," p. 278).

121. Mark Twain, *Roughing It* (New York: Penguin, 1987, orig. 1872), pp. 177–179. When Orion Clemens, who campaigned for Lincoln in Missouri, was named territorial secretary for Nevada, he hired his brother Sam as a clerk. The stagecoach limited passengers to twenty-five pounds of baggage. That of the Clemenses included clothes, Orion's Colt six-shooter, Sam's Smith and Wesson seven-shooter, tobacco, some coins, and a Webster's dictionary "weighing one thousand pounds." Greeley is cited in John B. Reid and Ronald M. James, *Uncovering Nevada's Past: A Primary Source History of the Silver State* (Reno: University of Nevada, 2004), p. 3; Jeanne Elizabeth Wier cited in Phillip I. Earl, *This Was Nevada* (Reno: Nevada Historical Society, 1986), p. x; William D. Rowley cited in Wilbur S. Shepperson, ed., *East of Eden, West of Zion: Essays on Nevada* (Reno: University of Nevada, 1989), p. viii.

122. On Native Americans, see James W. Hulse, *The Silver State: Nevada's Heritage Reinterpreted*, 3rd ed. (Reno: University of Nevada, 2004), pp. 18–32; on the war, see Ferol Egan, *Sand in a Whirlwind: The Paiute Indian War of 1860* (New York: Doubleday,

1972). As usual, a wise old chief—in this case, Numaga—counseled against war but was overruled by young braves. Especially tragic was the subsequent life of Sarah Winnemucca, granddaughter of chief Truckee. Educated by an army couple and at a Catholic school in San José, she spent her life pleading on behalf of the decimated, impoverished Paiutes only to be vilified by white Nevadans. She caught tuberculosis and died in Montana at age forty-seven. See Gae Whitney Canfield, *Sarah Winnemucca of the Northern Paiutes* (Norman: University of Oklahoma, 1983). The name Winnemucca, familiar to motorists on Interstate 80, was derived from the tale of a Paiute with only one moccasin, or *mau-cau*, hence *One-a-mau-cau*. It was the custom for young men in love to keep one foot bare as a sign that they were unavailable.

123. Ronald James, *The Roar and the Silence: A History of Virginia City and the Comstock Lode* (Reno: University of Nevada, 1998), is a thorough study rich in statistics. Best of all, James's scholarship leaves most of the legends about silver lodes intact. "No amount of historical research and source criticism," he writes, "is likely to overcome the persuasive nature of an oral tradition that assumed the role of local history" (p. 20). See also Russell R. Elliott and William D. Rowley, *History of Nevada*, rev. ed. (Lincoln: University of Nebraska, 1987), pp. 61–68 ("loud-mouthed," p. 63); Russell R. Elliott, "Nevada's Mining Heritage," in Shepperson, ed., *East of Eden, West of Zion*, pp. 41–55; Robert Laxalt, *Nevada: A Bicentennial History*, rev. ed. (Reno: University of Nevada, 1991), pp. 30–39; Barbara Land and Myrick Land, *A Short History of Reno* (Reno: University of Nevada, 1995), pp. 19–31 ("Get the facts," p. 31; "I have seen more rascality," p. 29); and the contemporary reports of a correspondent for *Harper's Monthly* in J. Ross Browne, *A Peep at Washoe and Washoe Revisited* (Balboa Island, Calif.: Paisano, 1959). Comstock himself sold out for just $11,000, squandered it, and was killed (perhaps by his own hand) in Montana in 1870. "There was *nothing* in the shape of a mining claim that was not salable," wrote Mark Twain. "We received presents of 'feet' every day. If we needed a hundred dollars or so, we sold some; if not, we hoarded it away, satisfied that it would ultimately be worth a thousand dollars a foot. I had a trunk about half full of 'stock.' When a claim made a stir in the market and went up to a high figure, I searched through my pile to see if I had any of its stock—and generally found it" (*Roughing It*, p. 315).

124. Nye's remarkable speech is included in Reid and James, *Uncovering Nevada's Past*, pp. 62–66. Nye considered gambling the worst of "all the seductive vices extant." He has been spinning in his grave for seventy-five years.

125. On statehood, see Elliott and Rowley, *History of Nevada.*, pp. 69–89; Hulse, *Silver State*, pp. 74–90; Laxalt, *A Bicentennnial History*, pp. 40–54; a lively memoir by William Morris Stewart, *The Reminiscences of Senator William M. Stewart* (New York: Neal, 1908); and Russell R. Elliott, *Servant of Power: A Political Biography of Senator William M. Stewart* (Reno: University of Nevada, 1983). The makeup of Nevada's convention was unusual in that only eight of the thirty-nine delegates were lawyers. The name Nevada was as unpopular as Washoe, since "snowy" was bad advertising, and almost the entire Sierra Nevada range lay in California. Humboldt and Esmeralda were suggested, but the convention decided to stick with what was familiar. The terms established by Congress were also unusually rigorous, including a specific recognition of federal supremacy, abolition of slavery, federal ownership of all public land, equal taxation for absentee owners of land in Nevada, religious toleration, and no taxation of federal property.

The "paramount allegiance" clause was modeled on one adopted by Maryland earlier in the year. For the text, see Reid and James, *Uncovering Nevada's Past*, pp. 31–32.

126. Elliot and Rowley, *History of Nevada*, pp. 90–151; Hulse, *The Silver State*, pp. 114–132. Even though the haul was several hundred miles shorter, the railroad charged Nevadans rates 25 percent higher than it charged Californians because the latter had the option of importing by sea. Congressman Rollin Daggett protested in 1881 that the Central Pacific had gouged $30 million out of his state over ten years, but the railroad executives knew how to stay on good terms with Senator Stewart. On Nevada's extraordinarily high percentage of foreign-born residents see Wilbur S. Shepperson, *Restless Strangers: Nevada's Immigrants and Their Interpreters* (Reno: University of Nevada, 1970); and James, *The Roar and the Silence*, pp. 91–118, 143–166. On the cattle barons such as John Sparks and Lewis "Old Broadhorns" Bradley, who both became governors (and were models for the Cartwrights on the television show *Ponderosa*), see Laxalt, *A Bicentennial History*, pp. 14–24.

127. According to Hulse, in *The Silver State*, pp. 162–179, the rancher Jim Butler named Tonopah (Paiute for "place of little water"), but his wife Belle, who was raised in mining camps, made the silver strike. They staked claims, assayed, and made fortunes with the help of a newly arrived lawyer from New York, Tasker L. Oddie. Oddie went on to a long, if undistinguished, career in state politics. See Loren Briggs Chan, *Sagebrush Statesman: Tasker L. Oddie of Nevada* (Reno: University of Nevada, 1973). On Newlands, see William D. Rowley, "Reno and the Desert of Buried Hopes," in Shepperson, ed., *East of Eden, West of Zion*, pp. 119–133.

128. Laxalt, *A Bicentennial History*, pp. 89–112; Land, *Short History of Reno*, pp. 115–119. Perhaps because their state offered little besides the hope of a lucky strike and naughty fun, the carnival trade came naturally to Nevadans. Even their most famous engineer got into the act. George Washington Gale Ferris was born in Galesburg, Illinois, and grew up in Carson Valley. He earned a degree from Rensselaer Polytechnic Institute, and built trestles and tunnels for West Virginia railroads. In 1893 the construction chief for Chicago's Columbian Exhibition asked him to design something spectacular, like the Eiffel Tower. Recalling the gigantic waterwheels on the Carson River, he bestowed on the world's children an amusement park ride: the Ferris wheel (Earl, *This Was Nevada*, pp. 96–99).

129. Chan, *Sagebrush Statesman*, pp. 135–143; Laxalt, *A Bicentennial History*, pp. 89–120 ("by corporate cats," p. 114); Elliot and Rowley, *History of Nevada*, pp. 273–374; Hulse, *The Silver State*, pp. 203–222. On the history of gambling and the casino industry in Nevada, see David G. Schwartz, *Roll the Bones: The History of Gambling* (New York: Gotham, 2006), pp. 351–421. The Canadian's words were spoken to me during my stop for the night on a cross-country trip. The fretting native was William A. Douglass, "Musings of a Native Son," in Shepperson, ed., *East of Eden, West of Zion*, pp. 95–110.

130. Guelzo, *Redeemer President*, pp. 397–430 ("This war," p. 428; "a King's cure," p. 400); Goodwin, *Team of Rivals*, pp. 676–681; Paludan, *Presidency of Lincoln*, pp. 297–302; and the comprehensive treatment by Michael Vorenberg, *Final Freedom: The Civil War, the Abolition of Slavery, and the Thirteenth Amendment* (New York: Cambridge University, 2001).

131. John Baptiste Calkin set Longfellow's "Christmas Bells" to music in 1872, trimmed its seven stanzas to five, and gave the carol the title "I Heard the Bells on

Christmas Day." The war inspired its lugubrious tone, but also the memory of Longfellow's wife, who died in 1861 when her dress caught fire while she melted sealing wax. The hopeful last verse ("God is not dead . . . The Wrong shall fail") clearly refers to the Union's impending victory.

132. Wiley Sword, *Embrace an Angry Wind: The Confederacy's Last Hurrah: Spring Mill, Franklin, and Nashville* (New York: HarperCollins, 1992); Burke Davis, *Sherman's March* (New York: Random House, 1980); John G. Barrett, *Sherman's March through the Carolinas* (Chapel Hill: University of North Carolina, 1956). For the investigations into the Union's alleged "worst atrocity," see Marion Brunson Lucas, *Sherman and the Burning of Columbia* (College Station: Texas A&M University, 1976).

133. See Ronald C. White, Jr., *Lincoln's Greatest Speech: The Second Inaugural* (New York: Simon and Schuster, 2001). The added interpretation regarding Lincoln's possible political aims in the speech is my own perilous contribution. But it is not meant to detract from the profound moral theology, which may be summed up as follows. If humanity is a fallen race and death is the penalty for sin, then what is a holy God to do with us? Perfect justice would entail effacement of the human race, as in Noah's time. Perfect mercy would entail indulgence of every depravity until the human race effaces itself. The only way humanity could be preserved was if God's justice obliged souls to suffer the consequences of earthly sin while his mercy offered salvation through a surrogate sacrifice. If that was God's formula on the cosmic scale, then a fortiori it must apply on the merely temporal scale. How else to explain what the new American Israel was enduring? On Lincoln's earlier remarks about the war as the wages of sin, see Morel, *Lincoln's Sacred Effort*, esp. pp. 192–193. A man whose thoughts really progressed was Frederick Douglass. He repeatedly doubted Lincoln's commitment to abolition and full civil rights, then warmed to the president at their first meeting in 1863, then embraced him in 1865. Lincoln made Douglass the first African-American guest at the White House when he ushered him into the postinaugural reception. See David W. Blight, *Frederick Douglass' Civil War: Keeping Faith in Jubilee* (Baton Rouge: Louisiana State University, 1989).

134. Jay Winik, *April 1865: The Month That Saved America* (New York: HarperCollins, 2001), pp. 73–202, 301–350 ("as Christian men," p. 166; "This will have," p. 189; "to pacifying the country," p. 195; "You have been," pp. 321–322). "Let 'em up easy" in Ketchum, ed., *American Heritage Civil War*, p. 592. The War Department's telegraph office got the long-awaited news on April 3: " We took Richmond at 8:15 this morning. . . . The city is on fire in two places. . . . G. Weitzel, Brig-Gen'l Comd'g" (Bates, *Lincoln at the Telegraph Office*, p. 360). For a change, Lincoln himself was not in the office; he was already in Richmond. It was Palm Sunday and the crowds cheered as if he were the Nazarene prophet entering Jerusalem. A black man cried, "I know that I am free, for I have seen Father Abraham and felt him" (Trefousse, *"First among Equals": Lincoln's Reputation*, p. 132).

135. Melville, a stanza from "The Martyr" in *Battle Pieces*, p. 142.

136. Bacon, *Sinews of War*, pp. 227–237.

137. Goodwin, *Team of Rivals*, pp. 745–747.

Chapter 8

1. Claudia Golden and Frank Lewis, "The Economic Costs of the American Civil War: Estimates and Implications," *Journal of Economic History* 35 (1975): 299–326; Wesley

C. Mitchell, "The Greenbacks and the Cost of the Civil War," in Ralph Andreano, ed., *The Economic Impact of the Civil War* (Cambridge, Mass.: Schenkman, 1962), pp. 66–78. But see also the dissent in Peter Temin, "Reply to Golden and Lewis," *Journal of Economic History* 38 (1978): 493. Temin argues that Golden and Lewis exaggerated the indirect costs.

2. Mark Thornton and Robert B. Ekelund, Jr., *Tariffs, Blockades, and Inflation: The Economics of the Civil War* (Wilmington, Del.: Scholarly Resources, 2004), pp. 81–103 ("New Deal," p. 99, has been traced to an editor in Raleigh who touted the advantages of a return to the Union); "The Economic Impact of the Civil War," in Robert W. Fogel and Stanley Engerman, eds., *The Reinterpretation of American Economic History* (New York: Harper and Row, 1971), pp. 371–376. During the 1930s, Charles and Mary Beard popularized the notion that the Civil War ushered in an era of hothouse industrial capitalism. The apparent role of World War II in pulling the United States out of the Great Depression encouraged Richard Hofstadter and others to make this the conventional wisdom. Thomas Cochran first challenged the consensus with "Did the Civil War Retard Industrialization?" *Mississippi Valley Historical Review* 48 (1961): 197–210. Since then increasingly sophisticated econometric studies have pretty much demolished the idea that the war was either a positive good in the long run or a necessary evil in the short run. See especially Roger L. Ransom, "Fact and Counterfact: The 'Second American Revolution' Revisited," *Civil War History* 45 (1999): 28–60. Richard Franklin Bensel, in *Yankee Leviathan: The Origins of Central State Authority in America, 1859–1877* (New York: Cambridge University, 1990), isolates the winners and losers resulting from the war.

3. On labor in the Civil War, see David Montgomery, *Beyond Equality: Labor and the Radical Republicans* (Urbana: University of Illinois, 1981); and Grace Palladino, *Another Civil War: Capital and the State in the Anthracite Regions of Pennsylvania, 1840–1868* (Urbana: University of Illinois, 1991). On the women's movement in the Civil War and Reconstruction, see Ellen Carol Du Bois, *Feminism and Suffrage: The Emergence of an Independent Women's Movement in America* (Ithaca, N.Y.: Cornell University, 1978); Jeanne Boydston, Mary Kelley, and Anne Margolis, *The Limits of Sisterhood: The Beecher Sisters on Women's Rights and Woman's Sphere* (Chapel Hill: University of North Carolina, 1988); Norma Basch, *Framing American Divorce: From the Revolutionary Generation to the Victorians* (Berkeley: University of California, 1999), esp. pp. 68–94 ("sin against nature," p. 69); and the classic by Elizabeth Cady Stanton, Susan Brownell Anthony, and Matilda Joslyn Gage, *History of Woman Suffrage*, 6 vols. (New York: Arno, 1969, orig. 1881–1922).

4. Angie Debo, *A History of the Indians of the United States* (Norman: University of Oklahoma, 1970), pp. 168–214; Francis Paul Prucha, *The Great Father: The United States Government and the American Indians*, 2 vols. (Lincoln: University of Nebraska, 1984), I: 411–436; Philip Weeks, *Farewell, My Nation: The American Indian and the United States in the Nineteenth Century*, 2nd ed. (Wheeling, Ill.: Harlan Davidson, 2001), pp. 89–122 ("south of Kansas," p. 93). On the Army in the West during and after the Civil War, see Robert Utley, *Frontiersmen in Blue: The U.S. Army and the Indian, 1848–1865* and *Frontier Regulars: The U.S. Army and the Indian, 1866–1890* (New York: Macmillan, 1967 and 1973); Robert Wooster, *The Military and United States Indian Policy, 1865–1903* (New Haven, Conn.: Yale University, 1988); Michael L. Tate, *The Frontier Army in the Settlement of the West* (Norman: University of Oklahoma, 1999).

5. I cannot recommend highly enough the rich, synthetic article by Pekka

Hämäläinen, "The Rise and Fall of Plains Indian Horse Cultures," *Journal of American History* 90 (2003): 833–862.

6. On the Texans' invasion of New Mexico, see Alvin M. Josephy, *The Civil War in the American West* (New York: 1991). On Sand Creek and its consequences, see Stan Hoig, *The Sand Creek Massacre* (Norman: University of Oklahoma, 1961); Prucha, *Great Father*, pp. 415–461; Weeks, *Farewell, My Nation*, pp. 112–120 ("to take captive" and "right and honorable," pp. 115–116; "Did we cease," p. 156); Debo, *History of the Indians*, pp. 184–200. For a fascinating firsthand account of Sand Creek and the mad Chivington, who equated "muscular Christianity" with genocide, see David Fridtjof Halaas and Andrew E. Masich, *Halfbreed: The Remarkable True Story of George Bent—Caught between the World of the Indian and the White Man* (New York: Da Capo, 2004). George was the son of William, builder of Bent's Fort; and Mis-stan-sta (Owl Woman), a Cheyenne princess. Having failed to mediate peace and having barely survived the massacre, Bent had to choose racial sides. He rode with the Cheyenne Dog Soldiers on their raids of revenge.

7. Debo, *History of the Indians*, pp. 201–214; Prucha, *Great Father*, pp. 479–533; Weeks, *Farewell, My Nation*, pp. 163–182 ("so long as the buffalo," p. 182); Ray Allen Billington, *Westward Expansion: A History of the American Frontier*, 5th ed. (New York: Macmillan, 1982), pp. 555–610. On the trade in buffalo hides, tongues, horns, heads, and meat, see David Dary, *Entrepreneurs of the Old West* (New York: Knopf, 1986), pp. 216–225.

8. Eric Foner, *Forever Free: The Story of Emancipation and Reconstruction* (New York: Knopf, 2005), pp. 41–67 ("negro paradise," p. 61; "Verily the work," p. 67); Allen C. Guelzo, *The Crisis of the American Republic: A History of the Civil War and Reconstruction Era* (New York: St. Martin's, 1995), pp. 353–357, 379–382; Willie Lee Rose, *Rehearsal for Reconstruction: The Port Royal Experiment* (Indianapolis, Ind.: Bobbs-Merrill, 1964); Patricia C. Click, *Time Full of Trial: The Roanoke Island Freedmen's Colony, 1862–1867* (Chapel Hill: University of North Carolina, 2001); "studied outrage" quoted by Claudine L. Ferrell, *Reconstruction* (Westport, Conn.: Greenwood, 2003), p. 10. The narrative and interpretation offered in this section draw most heavily on Foner, whose *Forever Free* is a synthetic masterpiece based on a lifetime of research as the era's foremost scholar. See also Foner, *Nothing but Freedom: Emancipation and Its Legacy* (Baton Rouge: Louisiana State University, 1983); *Reconstruction: America's Unfinished Revolution, 1863–1877* (New York: Harper and Row, 1988); and *Freedom's Lawmakers: A Directory of Black Officeholders during Reconstruction* (New York: Oxford University, 1993).

9. Hans Trefousse, *Andrew Johnson: A Biography* (New York: Norton, 1989); Albert Castel, *The Presidency of Andrew Johnson* (Lawrence: University of Kansas, 1979), pp. 1–16; David Warren Bowen, *Andrew Johnson and the Negro* (Knoxville: University of Tennessee, 1989).

10. Allan Nevins, "The Emergence of Modern America, 1865–1878," offers a chilling summary of Dixie in defeat in Mark C. Carnes and Arthur M. Schlesinger, Jr., eds., *A History of American Life* (New York: Scribner, 1996), pp. 721–732. On political and social organization by blacks during and after the Civil War, see Steven Hahn, *A Nation under Our Feet: Black Political Struggles in the Rural South from Slavery to the Great Migration* (Cambridge, Mass.: Harvard University, 2003). The former slave Henry Turner's "If I cannot" quoted in Foner, *Reconstruction, America's Unfinished Revolution*, p. 78.

11. Foner, in *Forever Free*, pp. 41–43, tells the poignant story of Robert Smalls, the enslaved pilot of the CSS *Planter*, who smuggled aboard his family and other African-Americans, slipped out of Charleston harbor, and surrendered the gunboat to the Union blockade. It was May 1862—well before the Emancipation Proclamation—but to Smalls this was a "freedom war" from the onset. He was granted a commission in the U.S. Navy; bought land near Beaufort; served five terms in Congress during Reconstruction; and held various government posts until 1913, when President Woodrow Wilson "cleansed" the federal bureaucracy of Republican (read: Negro) appointees. See Edward A. Miller, Jr., *Gullah Statesman: Robert Smalls from Slavery to Congress, 1839–1915* (Columbia: University of South Carolina, 1995).

12. Albert Castel, in *The Presidency of Andrew Johnson* (Lawrence: University of Kansas, 1979), pp. 213–230, summarizes the historiography to the 1970s. Representative works include Henry Wilson, *History of the Rise and Fall of the Slave Power in America* (Boston, Mass.: Houghton Mifflin, 1877), which damned Johnson's pro-southern bias; James Ford Rhodes, *History of the United States from the Compromise of 1850 to the Final Restoration of Home Rule at the South in 1877* (New York: Macmillan, 1906), which damned Radical Republicans for attempting to impose military rule and Negro suffrage but blamed Johnson's ineptness for giving them the chance; David Miller DeWitt, *The Impeachment and Trial of President Andrew Johnson* (New York: Macmillan, 1903), which praised Johnson's lonely, principled stand; Robert Watson Winston, *Andrew Johnson, Plebeian and Patriot* (New York: Henry Holt, 1928), and Lloyd Paul Striker, *Andrew Johnson: A Study in Courage* (New York: Macmillan, 1929), whose titles speak for themselves; Howard K. Beale, *The Critical Year: A Study of Andrew Johnson and Reconstruction* (New York: Harcourt, Brace, 1930), which praised Johnson's "Jacksonian" resistance to capitalism; Eric L. McKitrick, *Andrew Johnson and Reconstruction* (Chicago, Ill.: University of Chicago, 1960), which debunked Beale's economic determinism, restored the plight of the Negro to center stage, and showed how moderate rather than radical Republicans were the key players; Kenneth Stampp, *The Era of Reconstruction* (New York: Knopf, 1965), which lambasted Johnson with regard to civil rights; David Donald, *The Politics of Reconstruction, 1863–1867* (Baton Rouge: Louisiana State University, 1965), which traced the behavior of all the antagonists to their local constituencies; and Hans L. Trefousse, *The Radical Republicans: Lincoln's Vanguard for Racial Justice* (New York: Knopf, 1969), which came full circle by depicting the era as a tragic opportunity for racial justice lost through Johnson's obstruction. Finally, after the civil rights movement of the 1960s and 1970s bequeathed its own legacy of partial disappointments, Eric Foner crafted his authoritative accounts, derived from the complex realities of the 1860s and 1870s.

13. The collective biography of prominent carpetbaggers by Richard Nelson Current, *Those Terrible Carpetbaggers* (New York: Oxford University, 1988), challenged the myth of the venal, ignorant Yankee come south to bully and plunder; "only a smear word," p. 423. See also Foner, *Reconstruction, America's Unfinished Revolution*, pp. 294–295. Northern soldiers' fascination with the Negroes they encountered on their campaigns is well documented in Bell Irvin Wiley, *The Life of Billy Yank: The Common Soldier of the Union* (Baton Rouge: Louisiana State University, 1952), pp. 109–123. Various soldiers expressed their delight with the Negroes' spirituals, home cooking, humor, and willing help to the lost or wounded. That said, a Pennsylvanian wrote to his wife, "I

won't be unfaithful to you with a Negro wench . . . though it is the case with many soldiers" (p. 117).

14. As the mythical image of carpetbaggers was exploded by Current, so that of the scalawags was debunked by Carl N. Degler, *The Other South: Southern Dissenters in the Nineteenth Century* (New York: Harper and Row, 1974), pp. 191–263 ("the local leper," p. 191). On the amazing rise of the "Yeoman Farmer Home from the War," see Robert F. Durden, *The Dukes of Durham, 1865–1929* (Durham, N.C.: Duke University, 1975). On Holden, who as governor during Reconstruction routed the Klan only to suffer impeachment in 1871, see William C. Harris, *William Woods Holden: Firebrand of North Carolina Politics* (Baton Rouge: Louisiana State University, 1987).

15. On the political landscape after the war, see Michael J. Quill, *Prelude to the Radicals: The North and Reconstruction during 1865* (Washington, D.C.: University Press of America, 1980); and McKitrick, *Andrew Johnson and Reconstruction*, pp. 43–84. On economic disputes, see Howard K. Beale, "The Tariff and Reconstruction," and Stanley Cohen, "Northeastern Business and Radical Reconstruction: A Reexamination," in Andreano, ed., *Economic Impact of the American Civil War*, pp. 106–145; "being consumed," p. 115; "Why should this," pp. 133–134. On Treasury policy, see Paul Studenski and Herman E. Kroos, *Financial History of the United States*, 2nd ed. (New York: McGraw-Hill, 1963), pp. 161–191.

16. Castel, *Presidency of Andrew Johnson*, pp. 17–54 ("plebeians and mechanics" and "would vote with," p. 29; "If you could extend," pp. 44–45); "the hope of a 'white man's government'" from Frederic Bancroft, ed., *Speeches, Correspondence, and Public Papers of Carl Schurz*, 6 vols. (New York: Putnam, 1912–1913), 2: 256, cited in Guelzo, *Crisis of the American Republic*, p. 385; Foner, *Forever Free*, pp. 76–100 ("Are not our rights," p. 78). Under Johnson's decree the freed Adams and Eves were even expelled from the "negro paradise" founded by Grant on Davis Bend. Jefferson Davis himself spent two years in prison, the longest incarceration of any former rebel, until Johnson's pardon sent him home to find that even his former rivals and critics now considered him an icon of the lost cause. A thrilling historical novel depicting what might have been is Thomas Fleming, *The Secret Trial of Robert E. Lee* (New York: Tom Doherty, 2006). Fleming's panel of union generals appointed to try Lee (and by extension the entire Confederate planter elite) concludes the whole war was a mistake but the American people cannot bear to face that. "It will take a hundred years before the country can face the truth" (p. 327). In other words, pretense prevailed.

17. On the Blairs and the clashing political ambitions after the death of Lincoln, see Castel, *Presidency of Andrew Johnson*, pp. 37–42; and LaWanda C. F. Cox, *Politics, Principle, and Prejudice, 1865–1866: Dilemma of Reconstruction America* (New York: Free Press of Glencoe, 1963), pp. 1–49. On Seward's policy toward the French, see Glyndon G. Van Deusen, *William Henry Seward* (New York: Oxford University, 1967), pp. 422–493; and Thomas A. Bailey, *A Diplomatic History of the American People*, 8th ed. (New York: Appleton-Century-Crofts, 1969), pp. 348–359. Seward's instructions to Schofield as recorded in the latter's memoirs, *Forty-Six Years in the Army* (New York: Century, 1897), p. 385—"I want you to get your legs under Napoleon's mahogany and tell him he must get out of Mexico"—are often quoted for their color, but they leave the wrong impression. No American ultimatum or threat of war forced the emperor's hand. A host of domestic and foreign complications (including the Austro-Prussian war looming in 1866) com-

bined to make Napoléon III lose interest in Mexico. Seward was counting on that. Maximilian, the Austrian Hapsburg archduke (or "archdupe" as the newspapers quipped), truly believed he was called to give his fellow Catholics in Mexico a wise and good government. Proclaiming himself a Mexican citizen for life, he refused to depart with his French bodyguards and went before a firing squad on June 19, 1867. His beautiful twenty-seven-year-old wife, Carlotta, went mad; she lived in an insane asylum until 1927.

18. The constitutional implications of the Fourteenth Amendment were immense at the time and increased still more as future Supreme Court justices massaged the interpretations of words such as "person," "state," and "due process." For instance, corporations became persons but women did not, until another amendment in 1919. For authoritative discussions in laymen's terms, see William E. Nelson, *The Fourteenth Amendment: From Political Principle to Judicial Doctrine* (Cambridge, Mass.: Harvard University, 1988); Hermann Belz, *Abraham Lincoln, Constitutionalism, and Equal Rights in the Civil War Era* (New York: Fordham University, 1998), pp. 170–246; Hermann Belz, *Emancipation and Equal Rights: Politics and Constitutionalism in the Civil War Era* (New York: Norton, 1978), pp. 75–140; David A. J. Richards, *Conscience and the Constitution: History, Theory, and Law of the Reconstruction Amendments* (Princeton, N.J.: Princeton University, 1993), pp. 108–148; and Foner, *Reconstruction: America's Unfinished Revolution*, pp. 356–411. Stevens's "because I live" quoted in Foner, *Forever Free*, p. 118. In spring 1866 the House of Representatives also passed a law enfranchising Negroes in the District of Columbia (even though 99.5 percent of its white residents had defeated a referendum to that effect). The bill failed in the Senate, but provided an occasion for a delegation led by Frederick Douglass to lobby the president on behalf of voting rights for blacks. After they left, Johnson allegedly vented: "Those damned sons of bitches thought they had me in a trap! I know that damned Douglass; he's just like any nigger, and he would sooner cut a white man's throat than not." Everyone quotes this to prove Johnson's racism, and it certainly jibes with two well-known facts: as a self-styled frontier plebeian, Johnson had an unbridled tongue; and although he was a stern abolitionist he could be very irritating. But since the only source for the quotation was a secondhand account by a reporter for the *New York World*, it might have been fiction; see Hans L. Trefousse, *Impeachment of a President: Andrew Johnson, the Blacks, and Reconstruction* (New York: Fordham University, 1999), p. 15.

19. See James G. Hollandsworth, *An Absolute Massacre: The New Orleans Race Riot of July 30, 1866* (Baton Rouge: Louisiana State University, 2001). A merchant in the city's antebellum free black community testified before Congress as follows. "I was standing on the west side of the Mechanics' Institute when they began to shoot. I saw policemen shooting poor laboring men, men with their tin buckets in their hands and even old men walking with sticks. They tramped upon them and mashed their heads with their boots after they were down. I was in the Institute and heard Doctor Dostie say, 'Keep quiet. We have here the emblem of the United States. They cannot fire upon us when we have this emblem.' But they came in and fired upon us, although he took up the flag. I saw people fall like flies. When the policemen broke into the hall they did not respect the United States flag but cried, 'Damn that dirty rag.' . . . There is not one inch of ground in Louisiana that is loyal to the United States. They teach their children to be disloyal in the school-books they use," in Dorothy Sterling, ed., *The Trouble*

They Seen: The Story of Reconstruction in the Words of African Americans (New York: Da Capo, 1994), p. 97.

20. On the riots in New Orleans, Memphis, and elsewhere, see George C. Rable, *But There Was No Peace: The Role of Violence in the Politics of Reconstruction* (Athens: University of Georgia, 1984), pp. 43–58. On the politics of 1866, see Brooks D. Simpson, *The Reconstruction Presidents* (Lawrence: University of Kansas, 1998), pp. 67–110 ("He is obstinate," p. 110); Castel, *Presidency of Andrew Johnson*, pp. 55–98 ("swing around the circle," p. 90; "drunken tailor," p. 62). Cox was a nominal Republican elected governor on Lincoln's Unionist ticket. He was the sort Johnson needed to win over.

21. "I simply don't" in the appendix of Willa Cather, *O Pioneers!* (New York: Oxford University, 1999), p. 172; "a place to come" cited in Donald R. Hickey, *Nebraska Moments: Glimpses of Nebraska's Past* (Lincoln: University of Nebraska, 1992), p. ix; "A lot of good people" cited in James C. Olson and Ronald C. Naugle, *History of Nebraska*, 3rd ed. (Lincoln: University of Nebraska, 1997), p. 397. See also Alvin Saunders Johnson, *Pioneer's Progress: An Autobiography* (New York: Viking, 1952); and Frederick C. Luebke, "Time, Place, and Culture in Nebraska History," in James H. Madison, ed., *Heartland: Comparative Histories of Midwestern States* (Bloomington: University of Indiana, 1988), pp. 226–245. The estimate of migrants crossing the Nebraska country is from Merrill J. Mattes, *Platte River Road Narratives* (Urbana: University of Illinois, 1988). Driving the central route cross-country? Shun Interstate 80 in favor of U.S. 20. It's just as fast (with almost no traffic) and a spooky journey into the past.

22. Olson and Naugle, *History of Nebraska*, pp. 67–130 ("The great mass," p. 99); Hickey, *Nebraska Moments*, pp. 44–51 ("plied, begged," p. 45). "The Big Muddy" is from the journal (1857) of Erastus F. Beadle, *Ham, Eggs, and Corn Cakes: A Nebraska Territory Diary* (Lincoln: University of Nebraska, 2001), p. 32. Beadle returned to Cooperstown, New York, to become the most prolific publisher of frontier dime novels. During the Civil War some 3,300 Nebraskans enlisted to fight under Grant at Donelson and Shiloh. On the demise of Nebraska's Native Americans, see David Wishart, *Unspeakable Sadness: The Dispossession of the Nebraska Indians* (Lincoln: University of Nebraska, 1994); and Catherine Price, *The Oglala People, 1841–1879: A Political History* (Lincoln: University of Nebraska, 1996). No political correctness is required to appreciate the greatness of Red Cloud. He went on the warpath with reluctance and only for shrewd political aims rather than suicidal vengeance. The defeat his braves dealt Lieutenant Colonel William J. Fetterman's eighty troopers in December 1867 was the worst suffered by the U.S. Army prior to Little Bighorn. It won the closure of the Bozeman Trail. In 1870 Red Cloud went east to talk President Grant into founding a new supply post for the Sioux, and made a speech that won the hearts of New Yorkers. He later waged a seven-year struggle to have a crooked Indian agent removed, and he succeeded again. He even argued for peace before the climactic massacre of the Sioux at Wounded Knee in 1890. Red Cloud died a convert to Catholicism in 1909.

23. Hickey, *Nebraska Moments*, pp. 52–57 ("the worst copperhead" and "Every vote," p. 53; "The boldest," p. 57); Olson and Naugle, *History of Nebraska*, pp. 131–155. Lincoln was literally no more than a sod hut in the sun. It did not even lie on a river. The *Omaha Republican* predicted that "Nobody will ever go to Lincoln who does not go to legislature, the lunatic asylum, the penitentiary, or some of the state institutions" (p. 149). On

the role of Nebraska's statehood in the struggle over impeachment, see Trefousse, *Impeachment of a President*, pp. 63–64.

24. Robert W. Cherny, *Populism, Progressivism, and the Transformation of Nebraska Politics, 1885–1915* (Lincoln: University of Nebraska, 1981); Hickey, *Nebraska Moments*, pp. 125–132 ("Hast ever been," p. 125); Olson and Naugle, *History of Nebraska*, pp. 169–277.

25. The success of Arbor Day inspired the legislature to name Nebraska the Tree Planters' State, a motto that lasted until 1945, when the Cornhusker State was substituted. Would the University of Nebraska's "Big Red" football team be so fearsome if named the Tree Planters?

26. On the Depression-era novelties, see Olson and Naugle, *History of Nebraska*, pp. 304–348. On Arbor Day, the Nebraska National Forest, the new Capitol, and Boys Town, see Hickey, *Nebraska Moments*, pp. 140–158, 197–212; and Frederick C. Luebke, *Nebraska: An Illustrated History*, 2nd ed. (Lincoln: University of Nebraska, 2005), pp. 254–258. In addition to William Jennings Bryan, Ted Sorensen, Johnny Carson, and Red Cloud, Nebraska has cause to claim (by birth, residence, or association) Chief Standing Bear, General John "Black Jack" Pershing, the financier and presidential candidate Charles G. Dawes, "Buffalo Bill" Cody (whose Wild West show began in Nebraska), the novelist Willa Cather, the actor Henry Fonda, the dancer Fred Astaire, the talk show host Dick Cavett, and at least three baseball stars: Grover Cleveland Alexander, "Wahoo Sam" Crawford (who holds the lifetime record for triples), and "Bullet Bob" Gibson.

27. Patrick Lloyd Hatcher, *The Suicide of an Elite: American Internationalists and Vietnam* (Stanford, Calif.: Stanford University, 1990).

28. Foner, *Forever Free*, pp. 107–149 ("special favors" and "change the whole structure," p. 149; "You never saw a people," p. 132; Castel, *Presidency of Andrew Johnson*, pp. 117–137 ("The turning point," p. 137); "Republicans gave the ballot" quoted in George Brown Tindall, *America: A Narrative History*, 2nd ed., 2 vols. (New York: Norton, 1988), 1: 714; David Herbert Donald, Jean Harvey Baker, and Michael F. Holt, *The Civil War and Reconstruction* (New York: Norton, 2001), pp. 561–566.

29. For the rich and exciting history of the United States' acquisition of Alaska, see Walter A. McDougall, *Let the Sea Make a Noise: A History of the North Pacific from Magellan to MacArthur* (New York: HarperCollins, 2004, orig. 1994), pp. 112–172, 277–308; Howard I. Kushner, *Conflict on the Northwest Coast: American-Russian Rivalry in the Pacific Northwest, 1790–1867* (Westport, Conn.: Greenwood, 1975), pp. 106–158; Ernest N. Paolino, *The Foundations of American Empire: William Henry Seward and U.S. Foreign Policy* (Ithaca, N.Y.: Cornell University, 1973), pp. 106–107; Van Deusen, *William Henry Seward*, pp. 535–544; Castel, *Presidency of Andrew Johnson*, pp. 120–122; Bailey, *Diplomatic History of the American People*, pp. 360–372. The Russian American Company had long since ceased to profit from its trade in the pelts of seals and sea otters, and supplying and defending remote Alaska was far beyond the means of the czarist navy and merchant marine. It was spared capture during the Crimean War (1854–1856) only by an Anglo-Russian gentlemen's agreement to declare their American territories off-limits to combat. Czar Alexander II decided in 1859 to sell the colony and use the capital to build railroads at home, but the Civil War in America postponed the transaction. During the war Russia sided with the Union because it faced its own rebellion among the Poles; hence the postwar window of goodwill: see

Albert A. Woldman, *Lincoln and the Russians* (Cleveland, Ohio: World, 1952). The whole episode nonetheless left Seward frustrated. He believed the purchase of Alaska would cause Canada's Pacific coast to fall into the United States' hands as well; but it didn't. Parliament hastily passed the British North America Act of 1867, which granted Canada self-governing dominion status, created the province of British Columbia, and pledged to build the Canadian Pacific Railroad.

30. Trefousse, *Impeachment of a President*, pp. 48–145; Castel, *Presidency of Andrew Johnson*, pp. 139–178 ("blacks in the South" and "an incendiary document," pp. 152–154); Simpson, *Reconstruction Presidents*, pp. 107–130.

31. Trefousse, *Impeachement of a President*, pp. 146–79 ("the elect of an assassin," p. 154); "hopes of free institutions" (from *Congressional Globe*) quoted in Donald et al., *Civil War and Reconstruction*, p. 572. Butler literally waved the blood-stained nightshirt of a carpetbagger pulled from his bed and beaten by Klansmen. The "seven tall men" or "seven recusants" or "seven martyrs" as they were variously praised, scorned, or mourned included Trumbull (Illinois), Fessenden (Maine), Joseph S. Fowler (Tennessee), James W. Grimes (Iowa), John Henderson (Missouri), Edmund Ross (Kansas), and Peter Van Winkle (West Virginia). Van Winkle, because of the alphabetical polling of states, cast the decisive vote that redeemed the president. Contrary to a myth propagated in John F. Kennedy's *Profiles in Courage* and other books, none of the seven was especially vilified by his party or constituents on account of this vote: see Ralph Roske, "The Seven Martyrs?" *American Historical Review* 64 (1959): 323–330. Moreover, the heroic or treasonous stance (depending on your perspective) of the seven was not all that decisive. Other moderate Republicans stood ready to make the risky vote to acquit if necessary to prevent the riskier prospect that impeachment would pass (Trefousse, *Impeachment of a President*, pp. 165–169). The votes on the other charges were identical, and Chase gaveled the trial to a close on May 29.

32. A sharecropper in Alabama recounted his contretemps with visiting Klansmen on the eve of the election of 1868. "They asked me who I was going to vote for: Grant or Seymour. I claimed to them, as they had overpowered me, that I would vote for Seymour. I did that to get off. They wanted to know my politics and I answered, 'What is politics, sir?'—very ignorant like. They had another man out at the same time and they whipped him tremendous for talking politics. When the election came off I didn't vote. I was afraid to. I thought if I couldn't vote the Republican ticket I would not vote at all" (Sterling, ed., *The Trouble They Seen*, p. 381).

33. On the Fifteenth Amendment, see William Gillette, *The Right to Vote: Politics and the Passage of the Fifteenth Amendment* (Baltimore, Md.: Johns Hopkins University, 1969). On Grant, see Geoffrey Perret, *Ulysses S. Grant: Soldier and President* (New York: Random House, 1997), pp. 375–403 ("Ulys, do you want," p. 378; "Let us have peace," p. 379); and Jean Edward Smith, *Grant* (New York: Simon and Schuster, 2001); "on the assured success" cited in Donald et al., *Civil War and Reconstruction*, p. 574. On the little-known but rather important presidential campaign of 1868, see Charles Hubert Coleman, *The Election of 1868: The Democratic Effort to Regain Control* (New York: Columbia University, 1933). Though one could not yet call it a "southern strategy," Seymour's campaign made the first feint toward rebuilding a national Democratic majority on the strength of a coalition among immigrants, labor, and southern segregationists.

34. On the major federal agencies attempting to reshape the South, see James E. Sefton, *The United States Army and Reconstruction, 1865–1877* (Baton Rouge: Louisiana

State University, 1967); and George R. Bentley, *A History of the Freedmen's Bureau* (Philadelphia: University of Pennsylvania, 1955). The Freedmen's Bureau was able to accomplish as much as it did thanks to volunteers and funds provided by northern benevolent associations. When the bureau was terminated in 1869, it had reached the peak of its activity. Some 9,500 teachers, over half from New England and three-quarters female, staffed primary schools for Negro children. The American Missionary Association and American Freedmen's Union Commission carefully screened volunteers to ensure that they possessed "fervent piety" and "a genuine spirit of love for God and man" (Bentley, *Freedmen's Bureau*, p. 133), albeit Roman Catholics were excluded. See also Robert H. Bremmer, *The Public Good: Philanthropy and Welfare in the Civil War Era* (New York: Knopf, 1980), pp. 113–143.

35. Rable, *But There Was No Peace*, pp. 91–121. Romantic notions of honorable Johnny Rebs and vigilantes patrolling lawless Texas after the Civil War will not survive a reading of Barry A. Crouch and Donaly E. Brice, *Cullen Montgomery Baker: Reconstruction Desperado* (Baton Rouge: Louisiana State University, 1997). On the Klan, see Allen Trelease, *White Terror: The Ku Klux Klan Conspiracy and Southern Reconstruction* (Baton Rouge: Louisiana State University, 1971); and Elaine Frantz Parsons, "Midnight Rangers: Costume and Performance in the Reconstruction-Era Ku Klux Klan, *Journal of American History* 92 (2005): 811–836. The latter is a lurid, insightful account of Klan violence as performance art. Of course, only a few southern whites participated in or condoned the terror, but everyone knew of it and few stood up against it. After all, it served the damned Yankees and scalawags right if they could not police the country they presumed to rule. William Faulkner plumbed the psychological depths of racial violence in his novel *Light in August* (1932), in which a light-skinned orphan who believes himself black is accused of cohabiting with and killing a white woman. The sheriff's deputy coolly hunts him down and then dissolves into rage at the moment of capture. He castrates "the nigger" before killing him.

36. For overviews of the whole process, see Richard H. Abbott, *The Republican Party and the South, 1855–1877: The First Southern Strategy* (Chapel Hill: University of North Carolina, 1986), and *For Free Press and Equal Rights: Republican Newspapers in the Reconstruction South* (Athens: University of Georgia, 2004); William Gillette, *Retreat from Reconstruction, 1869–1879* (Baton Rouge: Louisiana State University, 1979); and Michael Perman, *The Road to Redemption: Southern Politics, 1869–1879* (Chapel Hill: University of North Carolina, 1984). One who rode with the South Carolina Red Shirts was Dr. Simon Baruch, a Jewish German immigrant whose son, Bernard Baruch, became a prominent financier in New York and a confidant of Franklin D. Roosevelt. Simon Baruch remained a rebel to the last, much to the family's embarrassment. See Patricia Spain Ward, *Simon Baruch: Rebel in the Ranks of Medicine, 1840–1921* (Tuscaloosa: University of Alabama, 1994).

37. Richard E. Welch, Jr., *George Frisbie Hoar and the Half-Breed Republicans* (Cambridge, Mass.: Harvard University, 1971), "needs of the Freedmen," p. 17; "Send down Freedom," p. 26. David M. Jordan, *Roscoe Conkling: Voice in the Senate* (Ithaca, N.Y.: Cornell University, 1971). The issues in the upper South are poignantly described in Ben H. Severance, *Tennessee's Radical Army: The State Guard and Its Role in Reconstruction, 1867–1869* (Knoxville: University of Tennessee, 2005). Even a well-disciplined and mostly white state militia in a state with large numbers of Unionists and a Radical governor (William G. "Parson" Brownlow) was unable to establish a lasting Republican

Party or equality for African-Americans. What chance, then, did less favored Radical regimes stand elsewhere in the South?

38. On the critical former slave states *not* subject to Congressional Reconstruction, see the articles in Richard O. Current, ed., *Radicalism, Racism, and Party Alignment: The Border States during Reconstruction* (Baltimore, Md.: Johns Hopkins University, 1969), esp. Current's introduction, pp. xiii–xxvi; Ross A. Webb, "Kentucky: 'Pariah among the Elect,'" pp. 105–145; Jacqueline Balk and Ari Hoogenboom, "The Origins of Border State Liberal Republicanism," pp. 220–244; and William Gillette, "Anatomy of a Failure: Federal Enforcement of the Right to Vote in the Border States during Reconstruction," pp. 265–304. Richard Paul Fuke, *Imperfect Equality: African Americans and the Confines of Racial Attitudes in Post-Emancipation Maryland* (New York: Fordham University, 1997); and Lowell Hayes Harrison and James C. Klotter, *A New History of Kentucky* (Lexington: University of Kentucky, 1997), chaps. 15–16, provide strong case studies of border states.

39. "*Alabama* claims" was the short phrase used for the Anglo-American dispute over all the various damage claims left from the Civil War. The British granted a big point of principle in the Treaty of Washington of 1871, which established nations' responsibility to show "due diligence" to forestall construction of warships for use against third parties with whom they are at peace. It also set up a five-country commission (Brazil, Italy, and Switzerland in addition to the principals) to arbitrate the specific claims. But Sumner disrupted its irenic deliberations with a bombastic speech insisting that Britain pay up to $8 billion in indirect costs for helping the Confederates prolong the war. By May 1872 threats of war crossed the Atlantic. That obliged Secretary of State Hamilton Fish and the U.S. arbitrator Charles Francis Adams to avert war and save the tribunal by carrying a brief against their own country. They declared the tribunal incompetent to rule on indirect claims and promised to deny them if granted competence. Their foreign colleagues took kindly to that; hence the award in September 1872 of $15.5 million, as compared with $1.9 million granted to Britain the following year for claims against the Union blockade.

40. In addition to Georgia and Texas, Greeley won handily in the border states of Missouri, Kentucky, Tennessee, and Maryland, and nearly carried West Virginia and Virginia. The sixty-six electoral votes he earned would not, however, be cast for Greeley. He both wore himself out and lost his wife during the campaign, and he died before the electoral college convened in 1873. Most of Greeley's electors cast their ballots, not for his running mate B. Gratz Brown, but for Thomas A. Hendricks, the Democratic governor-elect of Indiana. Hendricks rode that to the Democrats' vice presidential nomination four years later.

41. The majority opinion (five to four) in *Slaughterhouse* was written by Justice Samuel Freeman Miller, a Kentucky-born graduate of Transylvania College who practiced both medicine and law. An early organizer of the Republican Party, he was rewarded when Lincoln appointed him to the Court in 1862. Since his legal philosophy tilted in favor of states' rights, Miller has often been blamed for helping the South renege on civil rights for African-Americans. That was the opposite of his intent. Miller had left his native state for Iowa in 1850 out of disgust with slavery, and after the war he tried to protect freedmen's civil rights. But he firmly opposed judicial activism lest the Supreme Court become a "perpetual censor" over every state and local law in the land. See

Michael A. Ross, *Justice of Shattered Dreams: Samuel Freeman Miller and the Supreme Court during the Civil War Era* (Baton Rouge: Louisiana State University, 2003).

42. On the Supreme Court's role in Reconstruction, contrast Belz, *Emancipation and Equal Rights*, pp. 108–140; Harold M. Hyman, *A More Perfect Union: The Impact of the Civil War and Reconstruction on the Constitution* (Boston, Mass.: Houghton Mifflin, 1973), pp. 414–553. Belz argues that the postwar amendments amounted to a major, progressive revision of constitutional law despite the restrictive court rulings. Hyman argues the postwar legal framework amounted to no more than a modest adjustment of federal-state relations and blacks' civil rights and that by 1900 "the Fourteenth and Fifteenth Amendments had become almost irrelevant for Negroes" (p. 542). However, Hyman and William M. Wiecek, in *Equal Justice under Law: Constitutional Development, 1835–1875* (New York: Harper and Row, 1982), modify that extreme judgment.

43. On the Grant administration, see Simpson, *Reconstruction Presidents*, pp. 133–196 ("the spirit of hatred," pp. 178–179; "It could not have saved," p. 181; "our people are tired out," p. 177; "like an exploding shell" and "The whole public," p. 186); Heather Cox Richardson, *The Death of Reconstruction: Race, Labor, and Politics in the Post–Civil War North, 1865–1901* (Cambridge, Mass.: Harvard University, 2001), pp. 6–82, 122–155. On the restoration of Democratic majorities within the southern states themselves, see Michael Perman, *The Road to Redemption: Southern Politics, 1869–1879* (Chapel Hill: University of North Carolina, 1984); and Foner, *Forever Free*, pp. 189–213. Adelbert Ames, a West Point graduate from Maine, had fought courageously in the Civil War before undertaking his "Mission with a large M" in Mississippi. He testified bitterly before Congress about the state elections of 1875, and then repaired to Tewksbury, Massachusetts, to make millions as a textile manufacturer, land speculator, and inventor. In the Spanish American War, Ames, who was then sixty-three, served as brigadier at Santiago de Cuba. Then he began a third or fourth life as a social lion close to the Rockefellers, a golf aficionado, and a pillar of his Episcopal parish. "The days are many," he told a southern historian, "before Christ's Sermon on the Mount will be our practical religion. Mississippi like other states has a weary task before it." He died in 1933 at age ninety-seven, the oldest surviving Civil War general (Current, *Those Terrible Carpetbaggers*, pp. 412–416).

44. On the candidates, platforms, and election returns, see Polakoff, *Politics of Inertia*, pp. 210–311 ("Money and intimidation," p. 216); Allan Peskin, "Was There a Compromise of 1877?" *Journal of American History* 60 (1973): 63–75; Foner, *Forever Free*, pp. 189–213 ("honor bound," p. 197); Donald et al., *Civil War and Reconstruction*, pp. 629–641. On the violence in South Carolina, see Rable, *But There Was No Peace*, pp. 163–185; William Peirce Randel, *Centennial: American Life in 1876* (Philadelphia, Pa.: Chilton, 1969), pp. 240–257. The earlier story of a simple, bilateral deal between Hayes's managers and southern Democrats as told in C. Vann Woodward, *Reunion and Reaction: The Compromise of 1877 and the End of Reconstruction* (Boston, Mass.: Little, Brown, 1951), has been corrected. Many maneuvers coincided, as one can imagine. One bargain, however, is well established. Five Republicans from Ohio led by Senator John Sherman and Congressman James Garfield palavered with three Democrats from Louisiana at Wormley House in Washington on February 26, 1877. The Ohioans assured them that Hayes would carry out Grant's decision to evacuate troops and permit "home rule" in return for a promise that blacks would enjoy civil rights and education. Congress remained in session, permitting Hayes to be sworn in privately on March 3 (a day early because of the

Sabbath) and publicly on March 5; "the permanent pacification" in *I Do Solemnly Swear: The Inaugural Addresses of the Presidents of the United States, 1789–2001* (Philadelphia, Pa.: Chelsea House, 2001), p. 171. Hayes was not cynical—he believed that blacks would fare worse so long as Union soldiers remained to anger white southerners—but he was naive to think that the Republicans would fare better by appealing to the heirs of white southern Whigs rather than to blacks.

45. Phyllis Flanders Dorset, *The New Eldorado: The Story of Colorado's Gold and Silver Rushes* (New York: Macmillan, 1970), pp. 24–114 ("Here lies," p. 26; "Unparalleled," p. 77); Carl Abbott, Stephen J .Leonard, and Thomas J. Noel, *Colorado: A History of the Centennial State*, 4th ed. (Boulder: University of Colorado, 2005), pp. 43–59 ("We are bound," p. 47; "because we wished," pp. 48–49; Carl Ubbelohde, Maxine Benson, and Duane A. Smith, *A Colorado History: Revised Centennial Edition* (Boulder, Col.: Pruett, 1976), pp. 58–100 ("Pikes Peak or Bust" and "Busted by God," p. 58). On the origins of Denver, see Stephen J. Leonard and Thomas J. Noel, *Denver: Mining Camp to Metropolis* (Boulder: University of Colorado, 1990), pp. 3–31; and Gunter Barth, *Instant Cities: Urbanization and the Rise of Denver and San Francisco* (New York: Oxford University, 1975), pp. 128–181. On the naming of Colorado, see George R. Stewart, *Names on the Land* (Boston, Mass.: Houghton Mifflin, 1967), pp. 301–303 (Senator Gwin of California objected to the final bill because he loved the word Colorado and wanted to bestow it on the land later called Arizona). The tale of Ada Lamont was a tragedy of Shakespearean proportions. When the pastor she had recently married disappeared from their wagon train in October 1858, it was assumed that he had run off with a tart, who was also missing. The bitter Ada became a whore, then Denver's premier madam for a decade, until one day her husband's Bible was found on the prairie next to a skull containing a bullet. He had not been unfaithful, just bushwhacked. Ada then became a drunk, wandered off to the mining camps, and was found eventually dead of starvation (Dorset, *The New Eldorado*, pp. 48–49).

46. Ore near the surface was sufficiently oxidized so that exposure to mercury bled out the pure metal. But the deeper "refractory" rock consisted of carbonates so hard to process that 50 to 90 percent of the ore was wasted. In the 1860s all manner of methods were tried to make smelting economical, but all failed (quipped an expert) owing to "the scientific men without practice and the practical men without science, the honest men without capacity and the smart men without honesty" (Abbott et al., *History of the Centennial State*, p. 68).

47. Ubbelohde, et al., *Colorado History*, pp. 101–57 ("reverence for," p. 152); Dorset, *New Eldorado*, pp. 115–65 ("the brilliant feat," p. 162); Abbott, et al., *History of the Centennial State*, pp. 59–83; Duane A. Smith, *The Birth of Colorado: A Civil War Perspective* (Norman: University of Oklahoma, 1989). Another energetic but less strident campaign won women the vote in 1894, but very few ran for or were elected to state office. Female activists were more inclined to campaign against prostitution and alcohol.

48. Rodman Wilson Paul, *Mining Frontiers of the Old West 1848–1880* (Albuquerque: University of New Mexico, 2001, orig. 1963), pp 114–134; Dorset, *New Eldorado*, pp. 193–292. The "gold standard" on western mining in general is Clark C. Spence, *Mining Engineering and the American West: The Lace-Boot Brigade, 1849–1933* (New Haven, Conn.: Yale University, 1970).

49. "There the great braggart" is from Isabella Bird, *A Lady's Life in the Rocky*

Mountains (1873), cited in Leonard and Noel, *Denver: Mining Camp*, p. 43. On the agricultural growth of the state, see the fascinating geographical work by William Wyckoff, *Creating Colorado: The Making of a Western American Landscape 1860–1940* (New Haven, Conn.: Yale University, 1999), pp. 101–153 ("Greeley is located," p. 127). On Eaton's career, see Jane E. Norris and Lee G. Norris, *Written in Water: The Life of Benjamin Harrison Eaton* (Athens: Ohio University, 1990), pp. 108–224 ("thirty years ago," p. 224).

50. Ubbelohde et al., *Colorado History*, pp. 181–190 ("Utes Must Go!" p. 186). Why does everyone condemn the Franciscan missions in California for trying to keep Indians in agricultural communities and consider treating Protestants' and Progressives' efforts to do the same as tragic failures? Meeker took the Ute out of their holy and beautiful mountains and plunked them into schoolhouse chairs or onto the seats of farm vehicles to work fields enclosed by fences. Then he decided that soldiers were needed to force the Indians to receive his help. No wonder the Indians tried to play hooky. Chef Ouray wisely counseled nonviolence, but he could not control every young brave, and it took just one incident (albeit the "Meeker massacre" was extreme) to convince whites all over again that the only good Injun was a dead one. Even after the Ute removal, however, Colorado's western slope remained the wild west. Its most colorful characters were indubitably the Bassett women. Elizabeth and her husband emigrated from Little Rock, Arkansas, in the late 1870s just in time to witness the departure of the Ute. Feminists in deed, not just in theory, Elizabeth and her daughters punched cattle, toted pistols, faced down outlaws, and stood up for ranchers' rights. Ann, the "queen of the cattle rustlers," lived to 1956 and her sister Josie to 1964. See Grace McClure, *The Bassett Women* (Athens: Ohio University, 1985).

51. Jefferson Randolph "Soapy" Smith was one of the Georgians who migrated to Colorado in search of riches. A sleight-of-hand artist, he earned his nickname by wrapping a bar of soap in a fifty-dollar bill, then offering to sell "a bar" to dupes for just five dollars. He eventually commanded a small army of bunco artists working the gold towns with a panoply of swindles and scams. In 1898 Smith followed the Klondike gold rush "north to Alaska" and was a law unto himself in Skagway until vengeful dupes gunned him down. See McDougall, *Let the Sea Make a Noise*, pp. 411–412. On Denver's growth and politics, see Leonard and Noel, *Denver: Mining Camp*, pp. 116–179 ("The time will come," p. 140). On Colorado as a "businessman's state" and the incoherence of the various reform movements, see Abbott et al., *History of the Centennial State*, pp. 110–134, 240–257.

52. Patricia Limerick, *The Legacy of Conquest: The Unbroken Past of the American West* (New York: Norton, 1987), p. 143.

53. This brief summary is an inadequate effort to distill the essence of C. Vann Woodward, *The Strange Career of Jim Crow* (New York: Oxford University, 1955); J. Morgan Kousser, *The Shaping of Southern Politics: Suffrage Restriction and the Establishment of the One-Party South, 1880–1910* (New Haven, Conn.: Yale University, 1974); Jane Dailey, *Before Jim Crow: The Politics of Race in Post-Emancipation Virginia* (Chapel Hill: University of North Carolina, 2000); and Degler, *The Other South*, pp. 191–371. "The long controversy" cited in Foner, *Forever Free*, p. 199; "It is the religious duty" cited in Mark Wahlgren Summers, "Party Games: The Art of Stealing Elections in the Late-Nineteenth-Century United States," *Journal of American History* 88 (2001): 424. Robert

Francis Engs, *Freedom's First Generation: Black Hampton, Virginia, 1861–1890* (Philadelphia: University of Pennsylvania, 1979), and *Educating the Disenfranchised: Samuel Chapman Armstrong and Hampton Institute, 1839–1893* (Knoxville: University of Tennessee, 1999), are excellent local studies. No one punctures the myth of southern solidity post-Reconstruction like the feisty Irishman William "Little Billy" Mahone. A civil engineer from Virginia, Mahone rose to the rank of brigadier general in the Confederate army and led the charge that repaired the breach in the Battle of the Crater in 1864. After the war he took Lee's advice and devoted his life to repairing the southern economy as a railroad executive. In 1877, the year Reconstruction is said to have ended, Mahone assumed leadership of the Readjuster Party. In 1881 he led it to victory and was elected to the Senate, where he sat with the Republicans and dispensed patronage to black and white supporters. By the time his term ended, the Democrats' racial violence and fraud had ousted his party and pared back suffrage to prevent it from making a comeback.

54. Robert L. Ransom and Richard Sutch, *One Kind of Freedom: The Economic Consequences of Emancipation*, 2nd ed.(New York: Cambridge University, 2001); Charles F. Oubre, *Forty Acres and a Mule: The Freedmen's Bureau and Black Landownership* (Baton Rouge: Louisiana State University, 1978); James L. Roark, *Masters without Slaves: Southern Planters in the Civil War and Reconstruction* (New York: Norton, 1977); Stanley Lebergott, *The Americans: An Economic Record* (New York: Norton, 1984), pp. 249–267; R. Douglas Hurt, *American Agriculture: A Brief History* (Ames: Iowa State University, 1994), pp. 165–170, 222–228 ("all God's dangers," p. 226). "Education amounts" cited in Foner, *Reconstruction: America's Unfinished Revolution*, p. 369; "The South must take care" from Tourgée, *A Fool's Errand* (Cambridge, Mass.: Harvard University, 1961), p. 169. Tourgée left North Carolina for Chautauqua County, New York, where he ended a lifelong debate with himself by deciding that his name was pronounced "tour-ZHEE." After initial success he flopped as a writer and lecturer, flopped as a politician, and then in 1891 founded the National Citizens' Rights Association, which also flopped. In 1896 he assisted the plaintiff's counsel in *Plessy v. Ferguson*, but failed in that effort to overturn Louisiana's segregation of railroad cars. President McKinley at last rewarded Tourgée with the consulate in the French port of Bordeaux, where he died in 1905. See Otto H. Olsen, *Carpetbagger's Crusade: The Life of Albion Winegar Tourgée* (Baltimore, Md.: Johns Hopkins University, 1965).

55. W. E. B. Du Bois, *The Souls of Black Folk*, ed. David W. Blight and Robert Gooding-Williams (Boston, Mass.: Bedford, 1990, orig. 1903), pp. 145–146.

56. Paul Harvey, *Freedom's Coming: Religious Culture and the Shaping of the South from the Civil War through the Civil Rights Era* (Chapel Hill: University of North Carolina, 2005), pp. 1–46 ("Disguise it as we may," p. 27; "please the white folks" and "rag of contempt," p. 9; "rebel church," p. 13 ; "the gospel according" and "used to come 'round," p. 37; "divided the human race," p. 42. Also: Charles Reagan Wilson, *Baptized in Blood: The Religion of the Lost Cause, 1865–1920* (Athens: University of Georgia, 1980); Sydney E. Ahlstrom, *A Religious History of the American People* (New Haven, Conn.: Yale University, 1972), pp. 670–697, 715–729 ("shall be whirled aloft," p. 684); Samuel S. Hill, "Religion and the Results of the Civil War," in Randall H. Miller, Harry S. Stout, and Charles Reagan Wilson, eds., *Religion and the American Civil War* (New York: Oxford University, 1998), pp. 360–382. H. Shelton Smith, *In His Image, But . . . : Racism in Southern Religion* (Durham, N.C.: Duke University, 1972), demonstrates how southern

white supremacists often took cues from northern theologians. On Catholics and freed-men, see Thomas Spalding, *Martin John Spalding: American Churchman* (Washington, D.C.: Catholic University, 1973), pp. 130–205 ("a golden opportunity," pp. 199–200); and Stephen J. Ochs, *Desegregating the Altar: The Josephites and the Struggle for Black Priests, 1871–1960* (Baton Rouge: Louisiana State University, 1990). On Dixon, see Anthony Slide, *American Racist: The Life and Films of Thomas Dixon* (Lexington: University of Kentucky, 2004); and Cathy Boeckmann, *A Question of Character: Scientific Racism and the Genres of American Fiction, 1892–1912* (Tuscaloosa: University of Alabama, 2000). The conflation of politics, religion, and civic religion in the postbellum South is wonderfully illustrated by the tombstone of a Confederate veteran in Tennessee: "Belonged to the Ku Klux Klan, a deacon in the Baptist Church, and a Master Mason for forty years" (Harvey, *Freedom's Coming*, p. 38).

57. Du Bois, *Souls of Black Folk*, pp. 37–61, 117–147 ("the shadow," p, 40; "The prob-lem," pp. 45, 61); "a powerful theological" in Harvey, *Freedom's Coming*, p. 8. In Du Bois's title, the plural "souls" alludes to the dilemma of trying to be true to one's race and one's country in a society that doesn't permit this. "It is a peculiar sensation, this double con-sciousness, this sense of always looking at one's self through the eyes of others. . . . One ever feels his two-ness—an American, a Negro; two souls, two thoughts, two unrecon-ciled strivings; two warring ideals in one dark body, whose dogged strength alone keeps it from being torn asunder" (p. 38). Was Du Bois aware that his "two-ness" delivered the coup de grâce to Ralph Waldo Emerson's pretentious "supernal one-ness"?

58. Edmund Wilson, *Patriotic Gore: Studies in the Literature of the American Civil War* (New York: Oxford University, 1962); Kenneth S. Lynn, "Patriotic Gore," in *Visions of America: Eleven Literary Historical Essays* (Westport, Conn.: Greenwood, 1973), pp. 167–175; Daniel Aaron, *The Unwritten War: American Writers and the Civil War* (New York: Knopf, 1973), esp. pp. xiii–xix, 327–340 ("laid upon our literature," p. xv; "almost ignobly safe," p. 106; "the real war," p. 327; "We are accustomed," pp. 337–338); "that spiritual censorship," a quotation from Thomas Beer, p. xvii); Marcus Cunliffe, *The Literature of the United States* (New York: Penguin, 1970), pp. 181–223 ("The art of depict-ing," p. 197); Michael Kammen, *A Season of Youth: The American Revolution and the His-torical Imagination* (New York: Knopf, 1978), pp. 256–259 ("incredible, self-serving," p. 259). One Civil War novel deserving of praise is John W. De Forest's *Miss Ravenel's Conversion from Secession to Loyalty* (1867). De Forest, a native of Connecticut, was a veteran of combat and the Freedmen's Bureau in South Carolina. His grasp of southern sensibilities anticipated that of Tourgée; his battle scenes anticipated those of Crane.

59. Anne Rose, *Victorian America and the Civil War* (Cambridge, Mass.: Harvard University, 1992), pp. 17–66; Gary Dorrien, *The Making of American Liberal Theology: Imagining Progressive Religion 1805–1900* (Louisville, Ky.: Westminster John Knox, 2001), pp. 248–334; Ahlstrom, *Religious History of the American People*, pp. 763–804; Bruce Kuklick, *Churchmen and Philosophers from Jonathan Edwards to John Dewey* (New Haven, Conn.: Yale University, 1985), pp. 191–229; Edwin Scott Gaustad, *A Religious History of America*, rev. ed. (San Francisco, Calif.: Harper and Row, 1990), pp. 164–214; Henry Holt, *Garrulities of an Octogenarian Editor* (Boston, Mass.: Houghton Mifflin, 1923), p. 46, cited by Allan Nevins, "Emergence of Modern America," in Carnes and Schlesinger, eds., *History of American Life*, p. 785.

60. Spencer's *First Principles* was first published in the United States in 1864. A

sense of the range of reaction to evolution and social Darwinism can be had from these representative works. Charles Hodge in *What Is Darwinism?* (New York: Scribner, 1874), condemned evolution as utterly incompatible with Christian belief, but he was one of the last voices of the old commonsense school of theology dating back to Witherspoon and the Scottish Enlightenment. William Torrey Harris spoke for the new generation. He did more than anyone else to popularize Hegelian philosophy in the United States, founded the nation's first philosophical journal, and assumed Horace Mann's mantle as the nation's most influential educational reformer while serving as superintendent of the public schools in Saint Louis. In 1875 Harris warned against the danger that modern science can drift into a pantheism—the equating of God and creation—which holds that God is a becoming, not a being, and that man is capable of infinite progress and union with God by his own activity: "Pantheism, or God the Universe," reprinted in William H. Goetzmann, ed., *The American Hegelians: An Intellectual Episode in the History of Western America* (New York: Knopf, 1973), pp. 215–222. John Fiske, a philosopher who was brought to Harvard by Eliot, popularized Spencer in *Outlines of Cosmic Philosophy*, 2 vols. (Boston, Mass.: Houghton-Mifflin, 1874), and sought to reconcile science and faith by arguing that human religion and law also advanced through evolutionary processes culminating in ethical monotheism and Anglo-Saxon constitutionalism. Finally, Josiah Strong reached the mass middle class in *Our Country: Its Possible Future and Its Present Crisis* (New York: American Home Missionary Society, 1886), by invoking evolution to justify the social gospel and manifest destiny. James Turner, *Without God, without Creed: The Origins of Unbelief in America* (Baltimore, Md.: Johns Hopkins University, 1985), elegantly recounts the whole process.

61. The wonderful prizewinning account by Debby Applegate, *The Most Famous Man in America: The Biography of Henry Ward Beecher* (New York: Doubleday, 2007), appeared while this book was in press. I was gratified to discover that Applegate's judgments and findings regarding the sex scandal (pp. 391–455 in her book) do not contradict my own. See also Robert Shapley, *Free Love and Heavenly Sinners: The Story of the Great Henry Ward Beecher Scandal* (New York: Knopf, 1954), pp. 3–33 ("Orthodoxy is my doxy," p. 22); Dorrien, *Imagining Progressive Religion*, pp. 198–204 ("We must not stop," p. 199). Shapley's is a popular account stuffed with zesty quotations, illustrations, and outlandish characters. Dorrien's purpose is to tease out the congruence, often ironic, between Beecher's private behavior and public theology. Richard Wightman Fox, *Trials of Intimacy: Love and Loss in the Beecher-Tilton Scandal* (Chicago, Ill.: University of Chicago, 1999), is a masterpiece of historical research and imagination. Fox begins with the ending and works backward in time. As he shows, this is just what the principal characters did as they attempted to understand the trajectories of their own lives, justify their behavior, and manipulate the perceptions of others.

62. Shapley, *Free Love*, pp. 34–73; Dorrien, *Imagining Progressive Religion*, pp. 204–213, 234–239 ("I propose," pp. 207–208; "my friends," p. 211; "Theodore, I am," p. 239); Fox, *Trials of Intimacy*, pp. 133–177, 301–307; "A freedman's bureau" quoted in Chester F. Dunham, *The Attitude of the Northern Clergy toward the South, 1860–1865* (Toledo, Ohio: Gray, 1942), p. 234. Fox argues quite persuasively that Beecher could not have fathered the baby if Elizabeth's affair ended, as all seem to agree, no later than February 10, 1870. That in turn casts serious doubt on the rumors of an abortion. By contrast, Elizabeth's increasingly distraught psychological state makes a miscarriage highly likely.

63. Shapley, *Free Love*, pp. 77–164 ("Sabbath harlequinades," p. 21; "Wherever I find," p. 125; "forty thousand," p. 161); Dorrien, *Imagining Progressive Religion*, pp. 237–243 ("forty of his," p. 237; "the grave of love," p. 241); Fox, *Trials of Intimacy*, pp. 151–168. For more on Woodhull, see Barbara Goldsmith, *Other Powers: The Age of Suffrage, Spiritualism, and Scandalous Victoria Woodhull* (New York: Knopf, 1998).

64. Shapley, *Free Love*, pp. 185–273 ("holocaust," p. 209); Dorrien, *Imagining Progressive Religion*, pp. 244–252 ("convict its fondest," p. 246; "watching, caring," a quotation from Beecher's mentor William Ellery Channing, p. 252); "A God without wrath" from Niebuhr, *The Kingdom of God in America* (New York: Harper and Row, 1937), p. 193). In 1878 Elizabeth Tilton changed her story again, in a published letter confessing the affair. No one believed her. The most Beecher ever did that hinted at a confession was to endorse a Democrat, Grover Cleveland, for president in 1884. When Cleveland's illegitimate child came to light and Republicans accused him of being a philanderer, it recalled to Beecher "the gloomy night of my own suffering" (Shapley, *Free Love*, p. 262). Tilton, ill-used by all parties, emigrated to France in 1883 and died there in 1907, having outlived the others. The truth about the scandal is simply unknowable. As Fox brilliantly demonstrates, the historian is confronted not with evidence either sufficient or insufficient to render a verdict, but rather with stories concocted by the Tiltons and Beecher about themselves and the others' stories. That makes nine stories in all, and all of them changed over time in light of changes in the stories put out by others (*Trials of Intimacy*, pp. 1–9). Thus, Fox shows how Beecher's final retort to Elizabeth's final reversal was to put out a new story to the effect that Elizabeth had been crazy all along. It reminded him "of a story of a Negro waiter who was asked if it was the second bell for breakfast that had been rung. 'No, sah; it's not the second bell, it's de second ringing ob de fust bell.' In this maliciously chosen analogy, Elizabeth Tilton had no speech at all; her bell just kept ringing" (p. 41). Fox concludes with the wry observation that twentieth-century historians eager to damn Victorian hypocrisy expressed a certainty about Beecher's guilt (and even Elizabeth's alleged abortion) that his contemporaries never possessed. Historians also tell stories about others' stories.

65. Santayana, "Materialism and Idealism in America," a lecture of 1918 at Bryn Mawr College, reprinted in Douglas L. Wilson, ed., *The Genteel Tradition: Nine Essays by George Santayana* (Lincoln: University of Nebraska, 1998), pp. 117–130 (quotation, p. 129); Sherman quoted in Nevins, "Emergence of Modern America," in Carnes and Schlesinger, eds., *History of American Life*, p. 733. "Systematizing" is the central post–Civil War theme in Alan I. Marcus and Howard P. Segal, *Technology in America: A Brief History* (New York: Harcourt Brace Jovanovich, 1989), pp. 133–254. William Appleman Williams, in *The Contours of American History* (New York: Franklin Watts, 1973), pp. 300–305, 325–330, also stresses the ubiquitous concept of system in post–Civil War American culture and economics, albeit he has in mind the Liberal system of "laissez nous faire." Of course, the corporate systematization traced by the business historian Alfred O. Chandler developed within the context of laissez-faire culture, which in turn seemed validated by the theory of natural selection by survival of the fittest. Wilfred McClay, *The Masterless: Self and Society in Modern America* (Chapel Hill: University of North Carolina, 1994), elegantly summarizes the same process, calling it "consolidation."

66. McDougall, *Let the Sea Make a Noise*, pp. 293–297, 317–322 ("the grandest," p. 319). Politics was the business of Leland Stanford, a farmer's son from New York

by way of Wisconsin. He became governor of California in 1862. Finance was the business of Collis P. Huntington, a farmer's son from Connecticut. Construction was the business of Charles Crocker, a farmer's son from New York by way of Indiana. Bookkeeping was the business of Mark Hopkins, a merchant's son from New York by way of Indiana. All four were forty-niners. All four ended up giving their names to great institutions in California: the Leland Stanford, Jr., University; the Huntington Library, the Crocker Bank, and the Mark Hopkins Hotel. The Chinese sojourners who hacked the railroad over the High Sierra under the most dangerous and uncomfortable conditions deserve every bit of praise they regularly receive nowadays. But if not for the top tier of management; the federal subsidies; the integrated efforts of iron mills in Pittsburgh, boiler plants in Philadelphia, shipping lines in New York; battalions of engineers, railroaders, chemists, loggers, carpenters, and masons; and the sailors and stevedores who shipped the massive bulk of materials needed around Cape Horn to California—if not for them, those coolies would never have been in California at all.

67. Alfred D. Chandler, *The Visible Hand: The Managerial Revolution in American Business* (Cambridge, Mass.: Harvard University, 1977), pp. 122–144; Lebergott, *The Americans: An Economic Record*, pp. 122–144; 277–196 ("were generally made," p. 287); John W. Oliver, *History of American Technology* (New York: Ronald, 1956), pp. 415–432; John F. Stover, *American Railroads*, 2nd ed. (Chicago, Ill.: University of Chicago, 1997).

68. Robert Luther Thompson, *Wiring a Continent: The History of the Telegraphic Industry in the United States, 1832–1866* (Princeton, N.J.: Princeton University, 1947); John B. Dwyer, *To Wire the World: Perry M. Collins and the North Pacific Telegraph Expedition* (Westport, Conn.: Praeger, 2001); Bern Dibner, *The Atlantic Cable* (Norwalk, Conn.: Burndy, 1959).

69. Peter Temin, *Iron and Steel in Nineteenth-Century America: An Economic Inquiry* (Cambridge, Mass.: MIT, 1964), pp. 169–193; Oliver, *History of American Technology*, pp. 297–361 ("This country has thrown," p. 319). For relief from New Deal bashing of captains of industry like Carnegie, see Harold C. Livesay, *Andrew Carnegie and the Rise of Big Business* (Boston, Mass.: Little, Brown, 1975).

70. Harold F. Williamson and Arnold R. Daum, *The American Petroleum Industry*, 2 vols. (Evanston, Ill.: Northwestern University, 1959), Vol. 1: *The Age of Illumination, 1859–1899*; Daniel Yergin, *The Prize: The Epic Quest for Oil, Money, and Power* (New York: Simon and Schuster, 1991), pp. 19–113 ("the beginning," p. 35); Ron Chernow, *Titan: The Life of John D. Rockefeller, Sr.* (New York: Random House, 1998); Lebergott, *The Americans: An Economic Record*, pp. 322–336; Oliver, *History of American Technology*, pp. 331–345 ("every wheel," p. 343).

71. William Appleman Williams, *The Contours of American History* (New York: Franklin Watts, 1973), pp. 300–330; Chandler, *The Visible Hand*, pp. 145–314 ("scientific management," p. 275). Americans spoke of the Panic of 1873 as if the failure of Jay Cooke's financial firm had been the sole cause of the crash. Cooke had overextended himself promoting the Northern Pacific Railroad and hoped to float extensive bonds in Europe. When instead European markets tumbled, he was ruined. Nor was the result a mere panic. It was a long period of sluggish growth punctuated by recessions that lasted until 1896. Europeans called it their "great depression." Finally, the American responses to the downturn—including cartelization, concentration, labor strife, high tariffs, and

growing government involvement in the economy—were not distinctive, but reflected what was occurring in Germany, France, Italy, and to a lesser extent Britain. On the long-term consequences of 1873 and its aftermath, see Hans Rosenberg, "Grosse Depression und Bismarckzeit," translated as "Political and Social Consequences of the Great Depression of 1873–1896 in Central Europe," in James J. Sheehan, ed., *Imperial Germany* (New York: Franklin Watts, 1976), pp. 39–60.

72. On the rise of amateur and professional sports in America, see Steven A. Riess, ed., *Major Problems in American Sport History* (Boston, Mass.: Houghton Mifflin, 1997), pp. 82–168. On the nationalization of baseball, see George B. Kirsch, *Baseball in Blue and Gray: The National Pastime during the Civil War* (Princeton, N.J.: Princeton University, 2003); and Eric M. Leifer, *Making the Majors: The Transformation of Team Sports in America* (Cambridge, Mass.: Harvard University, 1995), pp. 1–78. The White Stockings were the forerunners of the Chicago Cubs, not the White Sox founded in 1901 for the new American League. The first official recognition of the importance of baseball in American life was President Andrew Johnson's reception at the White House for players from the Brooklyn Athletics and the Washington Nationals, plus the eminent sportswriter Henry Chadwick, in August 1865.

73. Nevins, "Emergence of Modern America," pp. 733–756 ("far less raw" and "chromo civilization," p. 782). Mark Twain and Charles Dudley Warner, *The Gilded Age, A Tale of Today*, ed. Louis J. Budd (New York: Penguin, 2001); "Author Mark Twain," p. xi. Matthew Josephson, *The Robber Barons: The Great American Capitalists, 1861–1901* (New York: Harcourt, Brace, 1934); "the fearful sabotage," p. 453. Edward Chase Kirkland, in *Industry Comes of Age: Business, Labor, and Public Policy, 1860–1897* (New York: Holt, Rinehart, and Winston, 1961), first punctured the myths about the gilded age. Burton W. Folsom, Jr., *The Myth of the Robber Barons*, 3rd ed. (Young America's Foundation, 1996), is a "post–Reagan revolution" defense of the "great American capitalists" who hastened industrial growth for all. On the unreadiness of the courts and culture to handle the issues of the urban, industrial age, see Morton Keller, *Affairs of State: Public Life in Late Nineteenth Century America* (Cambridge, Mass.: Harvard University, 1977). Richard White, in "Information, Markets, and Corruption: Transcontinental Railroads in the Gilded Age," *Journal of American History* 90 (2003): 19–43, argues that in the case of Pacific railroads the corruption was anything but creative. Not until 1883 was a second transcontinental road completed (the Northern Pacific), and then only after bankruptcy and recapitalization. The theme of "creative corruption" is introduced in Walter A. McDougall, *Freedom Just Around the Corner: A New American History 1585–1828* (New York: HarperCollins, 2004), pp. 1–16.

74. William L. Riordan, *Plunkitt of Tammany Hall: A Series of Very Plain Talks on Very Practical Politics* (New York: Dutton, 1963, orig. 1905); "There's an honest graft," p. 3; "This civil service law," p. 11; "They were mornin' glories," p. 20; "The Democratic Party," p. 88; "I have made a careful study," p. 41; "I see a vision," p. 89. Riordan, a reporter for the *New York Evening Post*, interviewed the forty-year veteran of Tammany Hall at Graziano's shoeshine stand in the old county courthouse off Foley Square. Plunkitt, who wore a top hat and had a handlebar mustache, referred to the spot as his office, although he could have afforded a whole downtown office building. Riordan's little book won a rave from the *Review of Reviews*: "One who masters the philosophy of these charming discourses will have mastered the whole secret of New York metropolitan politics—Tammany's secret" (ix).

75. Arthur Mann's introduction to Riordan, *Plunkitt of Tammany Hall*, pp. vii–xxii ("I'm in favor," p. xvi); Ralph G. Martin, *The Bosses* (New York: Putnam, 1964), pp. 15–37 ("The typical businessman" and "All your high principles," p. 21). Rufus E. Shapley, *Solid for Mulhooly: A Political Satire* (New York: Arno, 1970, orig. 1889), with illustrations by Thomas Nast, purports to be the biography of an Irish bog-trotter raised in a trough with two pigs and a bottle ("His Paddy-gree"). He confesses that his youth was "spent in a condition of poverty and squalor not apparently conducive to exceptional mental growth, but which is, nevertheless, as experience has demonstrated, especially calculated to develop a genius for leadership in American politics" (p. 20). A friendly judge makes Mike Mulhooly a citizen after just two years in New York (the rule was five), and he "did not neglect to vote at the election immediately following his naturalization. Indeed . . . so great was his fear that his vote might not be properly counted in his own Election District, that he took the precaution to deposit another constitutional expression of his will in an adjoining District; and, to still further protect his newly-acquired rights of citizenship, he repeated this precaution against fraud in two other Districts more remote from his home" (p. 26). Employed as a barkeep, Mike opines, "Let me mix a nation's cock-tails and I care not who makes its laws" (p. 28). Climbing the ladder at Tammany, he learns that "a Boss wields a power almost as absolute, while it lasts, as that of the Czar of Russia or the King of Zululand. The Leaders, the Ring, and the Boss combined, constitute the modern system of American politics which has been found to work so successfully in all large cities, especially in those which are fortunate enough to have secured a working majority of Leaders from Ireland" (p. 46).

76. The principal sociologist arguing the value of urban machines in this era was Robert Merton, the mentor of the famous Philadelphian social critic Digby Baltzell. Peter McCaffrey, *When Bosses Ruled Philadelphia: The Emergence of the Republican Machine, 1867–1933* (University Park: Pennsylvania State University, 1993), dissented, arguing that the machine took more than it gave. For good thumbnail accounts of cities across the country, see Martin, *The Bosses*, pp. 15–37 ("A boss might as well," p. 18).

77. When Tweed was jailed, a guard asked his name, age, address, citizenship, and occupation. To the last Tweed proudly replied, "Statesman." His amazing saga wasn't over. In December 1875, while under house arrest during a civil suit, Tweed escaped. Despite his portliness and celebrity, he somehow escaped detection for six months before turning up in Cuba, where he donned a sailor's suit and booked passage to Spain. The alarm reached Spanish authorities on the transatlantic cable, but Tweed still might have slipped ashore had soldiers not spotted him as the funny fat man in those cartoons. The U.S. Navy bore him home, where he died of heart failure in 1878 at age fifty-five. Tweed wrote his own epitaph: "New York City politics was always dishonest. . . . This population is too hopelessly split up into races and factions to govern except by the bribery of patronage, or by corruption." See Kenneth D. Ackerman, *Boss Tweed: The Rise and Fall of the Corrupt Pol Who Conceived the Soul of Modern New York* (New York: Carroll and Graf, 2005); Sven Beckert, *The Monied Metropolis: New York City and the Consolidation of the American Bourgeosie, 1850–1896* (New York: Cambridge University, 2001), pp. 172–195; Edwin G. Burrows and Mike Wallace, *Gotham: A History of New York City to 1898* (New York: Oxford University, 1999), pp. 1002–1019; Samuel Augustus Pleasants, *Fernando Wood of New York* (New York: Columbia University, 1948); Seymour J. Mandelbaum, *Boss Tweed's New York* (New York: Wiley, 1965), pp. 7–86 ("something for everyone," p. 70;

"wisest and best," p. 81; "There is not," p. 85); Harold Mehling, *The Scandalous Scamps* (New York: Henry Holt, 1956), pp. 87–107 ("New York City politics," p. 88); David Loth, *Public Plunder: A History of Graft in America* (Westport, Conn.: Greenwood, 1966, orig. 1938), pp. 192–255 ("No jury," p. 201; "A statesman," p. 203).

78. Martin, *The Bosses*, pp. 37–61 (goo-goos, p. 40; "decent machine," p. 44).

79. As usual, Mark Wahlgren Summers, *The Era of Good Stealings* (New York: Oxford University, 1993), is the most wry and scholarly exposé of an era's foibles; on electoral fraud see Summers, "Party Games: The Art of Stealing Elections" ("Why, dang it all," p. 426). On the gold conspiracy, see Kenneth D. Ackerman, *The Gold Ring: Jim Fisk, Jay Gould, and Black Friday, 1869* (New York: Dodd, Mead, 1988). On scandals in the Grant administration, see Perret, *Grant, Soldier and President*, pp. 388–446; and Shelley Ross, *Fall from Grace: Sex, Scandal, and Corruption in American Politics from 1702 to the Present* (New York: Ballantine, 1988), pp. 100–111. Adams's anathema "sanctity of fiduciary relations" is from Charles Francis Adams, Jr., and Henry Adams, *Chapters of Erie* (Ithaca, N.Y.: Great Seal, 1956, orig. 1871), p. 8. For more of Henry Adams's high dudgeon on corruption in the allegedly "Gilded Age," check out his *Democracy, An American Novel* (New York: Holt, 1882).

80. In February 1865 a delegation of civic leaders including the editor of the *Chicago Tribune* visited the War Department to argue for a reduction of the city's enlistment quota. Workers were so scarce in the booming city that they begrudged every one sent to the army. Stanton promised future relief in exchange for an immediate draft of 6,000 men, but they still protested. At that point President Lincoln, who had listened in silence, gave them a stern talking-to: "Gentlemen, after Boston, Chicago had been the chief instrument in bringing this war on the country. The Northwest had opposed the South as the Northeast has opposed the South. It is you are largely responsible for making blood blow as it has. You called for war until we had it. You called for emancipation and I have given it to you. Whatever you asked you have had. Now you come here begging to be let off from the call for men. . . . You ought to be ashamed of yourselves. I have a right to expect better things of you. Go home and raise your 6,000 extra men": Ida M. Tarbell, *The Life of Abraham Lincoln*, 2 vols. (New York: Harper, 1909), 2: 149.

81. Donald L .Miller, *City of the Century: The Epic of Chicago and the Making of America* (New York: Simon and Schuster, 1996), pp. 89–175; William Cronon, *Nature's Metropolis: Chicago and the Great West* (New York: Norton, 1991), pp. 97–262 ("The cities have not made," p. 97); Theodore J. Karamanski, *Rally 'Round the Flag: Chicago and the Civil War* (Chicago, Ill.: Nelson-Hall, 1993), pp. 159–184 ("bovine city," p. 170).

82. Timothy B. Spears, *Chicago Dreaming: Midwesterners and the City, 1871–1919* (Chicago, Ill.: University of Chicago, 2005, pp. 3–23 ("shock city," p. 5); "wickedest city" in Karamanski, *Rally 'Round the Flag*, p. xii. Spears includes a quotation from the German sociologist Max Weber. He toured Chicago in 1904, but his observations might have been made in any era: "As far as one can see from the clock tower of the firm Armour and Son—nothing but cattle lowing, bleating endless filth—in all directions—for the town goes on for miles and miles until it loses itself in the vastness of the suburbs." The frenetic economic activity and the violence—labor, criminal, and ethnic—caused him to liken the bustling, brawling city to "a human being with his skin removed, and in which all the physiological process can be seen going on" (p. 4).

83. On politics, see Robin L. Einhorn, *Property Rules: Political Economy in Chicago, 1833–1872* (Chicago, Ill.: University of Chicago, 1991), pp. 104–230. On corruption, see Herbert Asbury, *Gangs of Chicago: An Informal History of the Chicago Underworld* (New York: Knopf, 1940), pp. 67–94; James L. Merriner, *Grafters and Goo-Goos: Corruption and Reform in Chicago, 1833–2003* (Carbondale: Southern Illinois University, 2004), pp. 10–35; and the classic by Bessie Louise Pierce, *A History of Chicago*, 3 vols. (New York: Knopf, 1940), 2: 246–353 ("what posterity must think," p. 302).

84. Karen Sawislak, *Smoldering City: Chicagoans and the Great Fire, 1871–1874* (Chicago, Ill.: University of Chicago, 1995), pp. 1–67 ("There has never been," p. 27); Lisa Krissoff Boehm, *Popular Culture and the Enduring Myth of Chicago, 1871–1968* (New York: Routledge, 2004), pp. 1–26 ("Politicians have worshiped," pp. 18–19; "Do you wonder," p. 26); Ross Miller, in *American Apocalypse: The Great Fire and the Myth of Chicago* (Chicago, Ill.: University of Chicago, 1990), p. 93, compares the Civil War and the Chicago fire as media events. By an extraordinary coincidence a huge forest fire the very same day burned to a crisp the lumber town of Peshtigo north of Green Bay, Wisconsin, killing more than 2,000 people. But Peshtigo was a mere colony whereas Chicago was the Midwest's imperial metropolis; property mattered more than people; and the Chicago fire, being man-made, seemed more fraught with meaning than a natural disaster.

85. Sawislak, *Smoldering City*, pp. 68–259 ("never taught," p. 116; "wise, prudent," p. 127; "knavish combinations," p. 137; "Practice economy," p 197; "to have been esteemed," p. 244); Merriner, *Grafters and Goo-Goos*, pp. 36–60; Asbury, *Gangs of Chicago*, pp. 95–154.

86. Francis Paul Prucha, *American Indian Policy in Crisis: Christian Reformers and the Indian, 1865–1900* (Norman: University of Oklahoma, 1976); Prucha, *The Great Father*, pp. 485–606; Weeks, *Farewell, My Nation*, pp. 166–202 ("Gentlemen, your advice," p. 167; "no longer a pleasure," p. 170); Debo, *History of the Indians*, pp. 233–266 ("I tell no lies," p. 238); William H. Armstrong, *Warrior in Two Camps: Ely S. Parker, Union General and Seneca Chief* (Syracuse, N.Y.: Syracuse University, 1978); Paul Andrew Hutton, *Phil Sheridan and His Army* (Lincoln: University of Nebraska, 1985); Shirley A. Leckie, *Elizabeth Bacon Custer and the Making of a Myth* (Norman: University of Oklahoma, 1993); John S. Gray, *Custer's Last Campaign: Mitch Boyer and the Little Big Horn Reconstructed* (Lincoln: University of Nebraska, 1991); Gary Clayton Anderson, *Sitting Bull and the Paradox of Lakota Nationhood* (New York: HarperCollins, 1996). A curious footnote to Little Bighorn is that "Long Hair," as Sitting Bull called Custer, had decided to clip off his long yellow curls just before the campaign. Shades of the biblical Samson?

87. Weeks, *Farewell, My Nation*, pp. 205–241 ("dumping ground" and "a thoroughly competent," p. 208; "Kill the Indian," p. 231); "Wish well to the Indians" quoted by Prucha, *The Great Father*, p. 599. On the "law and order" movement for Native Americans, see William Thomas Hagan, *Indian Police and Judges: Experiments in Acculturation and Control* (New Haven, Conn.: Yale University, 1966). On the Indian school movement, see Richard Henry Pratt, *Battlefield and Classroom: Four Decades with the American Indians*, ed. Robert M. Utley (New Haven, Conn.: Yale University, 1964). Concentration was not quite complete in 1877: the Apache renegades led by Geronimo remained at large until 1886.

88. William Peirce Randall, *Centennial: American Life in 1776* (Philadelphia, Pa.: Chilton, 1969), pp. 186–191, 283–306 ("The masses," p. 187).

89. On the exhibits, in addition to Randall, see Alan I. Marcus and Howard P. Segal, *Technology in America: A Brief History* (New York: Harcourt, Brace, Jovanovich, 1989), pp. 135–140, and Nevins, "Emergence of Modern America," pp. 802–806. The most wonderful exhibits from the Centennial Exhibition of 1876 used to be on permanent display in the Arts and Industries Building of the Smithsonian Institution on Washington's Mall. Alas, the building is now "closed for renovation," suggesting that nineteenth-century technology has lost its appeal for the public or the Smithsonian's bosses.

90. Randall, *Centennial*, pp. 186–191, 283–306 ("The president's absence," p. 301; "entirely free," from Custer's *My Life on the Plains*, p. 146; "the pretentious national Exhibition" is Nevins's summary remark about numerous digs made by European visitors at what they deemed the vulgar patriotism of the exhibition: "Emergence of Modern America," p. 806. James Butler Hickok wandered the Great Plains serving intermittently as an army scout and then a U.S. marshal. Legend obscures much of his life, but not the facts of his death. The thirty-nine-year-old gunslinger was so eager to get into a crowded poker game that he broke his own rule about sitting with his back to the door. A drifter, Jack McCall, could thus creep up like Booth in Ford's Theater and discharge a revolver in the back of his victim's head. Hickok slumped forward, revealing that his hand contained both black aces and both black eights, known ever after as the "dead man's hand." The only point of dispute is whether his fifth card was the queen or the jack of diamonds. Of course, white men would never have been in Deadwood had they honored the solemn treaties their government made with the Sioux. See Joseph G. Rosa, *They Called Him Wild Bill: The Life and Adventures of James Butler Hickok* (Norman: University of Oklahoma, 1964). As for the Northfield raid, the James gang targeted that bank because it was owned by the family of the Union general Adelbert Ames. The robbery was foiled by a clerk who died rather than open the safe, then by feisty townspeople who killed three bandits and formed a posse that captured the infamous Younger brothers. Jesse and Frank James, who had plundered with Quantrill's raiders during the Civil War, fled from Minnesota to Tennessee, where they went straight for a while before getting back into the robbery game. It earned Jesse a bullet in 1882.

91. Randall, *Centennial*, pp. 59–66 ("Ah, this is interesting," p. 61; "Size is not grandeur," pp. 65–66; "Henceforth, America," pp. 360–361).

Chapter 9

1. Mark Twain, *Following the Equator: A Journal around the World* (Hartford, Conn.: American, 1897), pp. 353–354. I tracked this magnificent snippet of satire to its source in order to grasp the context of Twain's observations. But I first encountered it in the pages of a delightful science book. John M. Marzluff and Tony Angell, *In the Company of Crows and Ravens* (New Haven, Conn.: Yale University, 2005), not only describes corvid biology and behavior, but speculates on the ways crows and people have influenced each other's social evolution (they quote Twain on pp. 36–37).

2. Anthony Trollope, *North America* (New York: Knopf, 1951, orig. 1862); "All idea of truth," p. 503; "I do not know," p. 507; "The Irishman," pp. 527–528.

3. William Hepworth Dixon, *New America*, 2 vols. (London: Hurst and Blackett, 1867), "While Americans are busy," I: 355; "Brothers and sisters," II: 210–211; "Well, says

the judge," II: 358–359. The judge granted that his party's platform condemned polygamy, but preferred it be extirpated through moral rather than physical force. Referring to Samuel Bowles, the editor of a Radical Republican journal published in Springfield, Massachusetts, he said, "If you can persuade Brigham to lie down with Bowles, I am willing to see it."

4. See Sir George Campbell, MP, *White and Black: The Outcome of a Visit to the United States* (London, 1879); Disraeli quoted from his novel *Sybil, or The Two Nations* (1845), p. 149.

5. Michael P. Federici, professor of political philosophy at Mercyhurst College in Erie, Pennsylvania, first apprised me of Brownson's relevance to the themes of this volume. I am also indebted to his insightful introduction in Michael P. Federici, ed., *Orestes A. Brownson: Works of Political Philosophy* (Wilmington, Del.: ISI, 2007). After Brownson's death his son compiled his amazing output of essays, reviews, rebuttals, critiques, novels, memoirs, autobiography, and treatise on American constitutionalism in Henry F. Brownson, ed., *The Works of Orestes A. Brownson*, 20 vols. (Detroit, Mich.: Thorndike House, 1884). "His predominant passion" is from Isaac Hecker's reminiscence in *Catholic World* (1887) cited in Arthur M. Schlesinger, Jr., *A Pilgrim's Progress: Orestes A. Brownson* (Boston, Mass.: Little, Brown, 1966, orig. 1939); p. 276.

6. For reasons that will soon be apparent Brownson was virtually ignored by American historians and philosophers for more than a century after his death. The sole major exception was Arthur Schlesinger, Jr., who was encouraged to study Brownson in the 1930s by his father and by his mentor Perry Miller. In recent decades, however, Brownson studies have proliferated. The works consulted for this chapter (plus the pages describing his youth) include Schlesinger, *Pilgrim's Progress*, pp. 3–28; R. A. Herrera, *Orestes Brownson: Sign of Contradiction* (Wilmington, Del.: ISI, 1999), pp. 1–12; Gregory S. Butler, *In Search of the American Spirit: The Political Thought of Orestes Brownson* (Carbondale: Southern Illinois University, 1992), pp. 1–43; and Patrick W. Carey, *Orestes A. Brownson: American Religious Weathervane* (Grand Rapids, Mich.: Eerdmans, 2004), pp. x–xx, 1–29. See also Americo D. Lapati, *Orestes A. Brownson* (New York: Twayne, 1965); Leonard Gilhooly, *Contradiction and Dilemma: Orestes Brownson and the American Idea* (New York: Fordham University, 1972); and Peter Augustine Lawler, "Orestes Brownson and the Truth about America," *First Things* (December 2002): 23–28.

7. "New Views on Religion, Society, and the Church" in *Works of Brownson* 4: 1–56 ("Man will be sacred," p. 48); Butler, *In Search of the American Spirit*, pp. 44–77; Carey, *American Religious Weathervane*, pp. 30–55; Herrera, *Sign of Contradiction*, pp. 13–24 ("robustness, manliness," p. 34); Schlesinger, *Pilgrim's Progress*, pp. 29–65 ("We know no party," p. 64).

8. "Our Indian Policy" and "A Discourse on Lying" from *Boston Quarterly Review* (April 1839 and April 1840) do not appear in Brownson's collected works but are posted online at www.orestesbrownson.com ("those not usually christened" and subsequent quotations, pp. 3–4). On this period in Brownson's life, see Carey, *American Religious Weathervane*, pp. 55–96; Herrera, *Sign of Contradiction*, pp. ix–xx, 25–45 ("cultivated bluster," p. xii); Schlesinger, *Pilgrim's Progress*, pp. 66–88 ("universal fraud," p. 67).

9. "Democracy and Liberty" and "Origin and Ground of Government" (*Democratic Review*, 1843) in *Works of Brownson* 15: 258–404. On these years, see Herrera, *Sign of Contradiction*, pp. 47–58; Schlesinger, *Pilgrim's Progress*, pp. 88–149.

10. "To pretend, as some do" cited in Lawler, "Orestes Brownson and the Truth," p. 25.

11. See "Bushnell's Discourses" (1849–1851) in *Works of Brownson* 7: 1–116.

12. Brownson's autobiographical account is "The Convert" in *Works of Brownson* 5: 1–199 ("ignorant, degraded," pp. 157–158). The interpretation here is based on Carey, *American Religious Weathervane*, pp. 97–153; Herrera, *Sign of Contradiction*, pp. 59–71; Schlesinger, *Pilgrim's Progress*, pp. 150–184; and Butler, *In Search of the American Spirit*, pp. 78–115. These authors tend largely to agree that Brownson came to a cool decision based on his logical thought process and his abiding desire to reconcile spiritual truth with material justice. Even though Brownson himself wrote of finding in God the father he never knew as a child, historians have resisted the temptation to account for his "aberration" by some trendy psychological explanation. So far as Anglicans were concerned, Brownson did not mince words. He once referred to the Oxford Tractarians as "priggish dons attracted to Catholicity by aesthetic and intellectual, not religious, motives" (Herrera, *Sign of Contradiction*, p. xiii).

13. "Catholicity Necessary to Sustain Popular Liberty" (October 1845), in *Works of Brownson* 10: 1–16. See also Carey, *American Religious Weathervane*, pp. 154–163 ("revolutionized," p. 157); Herrera, *Sign of Contradiction*, pp. 72–86; "When Brownson embraced" in Schlesinger, *Pilgrim's Progress*, pp. 182–183. On Hecker, see Joseph F. Gower and Richard M. Leliaert, eds., *The Brownson-Hecker Correspondence* (South Bend, Ind.: University of Notre Dame, 1979); and "An Improbable Friendship" in Herrera, *Sign of Contradiction*, pp. 153–66.

14. In his "Fable for Critics" James Russell Lowell meanly charged, "He shifts quite about, then proceeds to expound / That 'tis merely the earth, not himself, that turns round, / And wishes it clearly impressed on your mind / That the weathercock rules and not follows the wind. . . . / He offers the true faith to drink in a sieve,—/ When it reaches your lips, there's naught left to believe / But a few silly- (syllo-, I mean) -gisms that squat 'em / Like tadpoles, o'erjoyed with the mud at the bottom" (cited in Schlesinger, *A Pilgrim's Progress*, p. 278).

15. Carey, *American Religious Weathervane*, pp. 163–192 ("uncalled-for," p. 182); Herrera, *Sign of Contradiction*, pp. 87–99; ("tenure with Catholicism," p. xii); "Slavery and the Mexican War" (July 1847) in *Works of Brownson* 16: 25–59.

16. On Brownson's physical aging, see Schlesinger, *Pilgrim's Progress*, pp. 187–188. "A Christian Foundation for Politics" is the title coined in Butler, *In Search of the American Spirit*, p. 116. The ensuing discussion adumbrates Butler's excellent chapter on Brownson's reconsideration of faith and politics after his conversion.

17. Butler, *In Search of the American Spirit*, pp. 116–160. The original sources for Butler's quotations are as follows: "represents the irrational" from "Transcendentalism," *Works of Brownson* 6:23; "You cannot go on" and "pretended, peaceful" from "Channing on Social Reform, *Works of Brownson* 10: 204–205; "the Messiah" and "Satan is my hero" from *The Spirit-Rapper* in *Works of Brownson* 9: 1–234 (quotations on pp. 41–43, 69); "cruel and despotic" from "The Fugitive Slave Law," *Works of Brownson* 17: 37.

18. Butler, *In Search of the American Spirit*, pp. 160–162; "the true founders" from "Saint-Bonnet," *Works of Brownson* 14: 232–233; "is but a forgetting" from "Newman," *Works of Brownson* 3:136. The Broadway Tabernacle lectures of 1856 appeared as "The

Church and the Republic; Or, The Church Necessary to the Republic and the Republic Compatible with the Church" (July 1856) in *Works of Brownson* 12: 1–32 ("without the presence" and "stand on a basis," pp. 13–14).

19. "Mission of America" (October 1856) in *Works of Brownson* 11: 551–584 ("This manifest destiny," pp. 567–568). "Protestantism is outgrown," Brownson boldly asserted. Either atheism and self-worship in forms such as Transcendentalism would inherit the future, or else Catholicity would.

20. On Brownson during the Civil War, see Herrera, *Sign of Contradiction*, pp. 101–137 ("thick-headed," p. 109); Carey, *American Religious Weathervane*, pp. 193–281; "Slavery and the Church" (October 1862) and "Catholics and the Anti-Draft Riots" (October 1863) in *Works of Brownson* 17: 317–352, 412–447 ("sneer not," p. 446).

21. Orestes Brownson, *The American Republic: Its Constitution, Tendencies, and Destiny* (New York: P. O'Shea, 1866), pp. 1–25, 192–217 ("the wild theories," p. 3; "Next after religion," p. 19; "whatever its terms," p. 195). For excellent adumbrations of *The American Republic* see Federici's introduction to *Brownson: Works of Political Philosophy*; Herrera, *Sign of Contradiction*, pp. 139–165; and Butler, *In Search of the American Spirit*, pp. 163–192.

22. Brownson, *The American Republic*, pp. 218–276.

23. Ibid., pp. 348–358 ("The most marked," p. 348; "scorns all geographical," p. 351).

24. Ibid., pp. 358–391 ("humanity, superior," p. 355; "favorite guise," p. 362).

25. Ibid., pp. 392–439 ("always be progressive," p. 368; "The effect of this mission," pp. 428–429; "The American people," p. 439).

26. Good samples of Brownson's post–Civil War works include "Beecherism and Its Tendencies" from *Catholic World* (January 1871), "Religion and Science" (April 1874), and "The Papacy and the Republic" (January 1873) in *Works of Brownson* 3: 460–484, 3: 519–536, and 13: 326–350 respectively. Brownson, chides, "Mr. Thomas K. Beecher, who is more frank and outspoken than his cunninger, more cautious, and more timid brother" plainly stated that "all churches are equally good or equally bad, and the best church for a man is that in which he feels most at his ease." Yet he goes on to praise the Catholic church for "her exclusiveness or denial of the pretentions of all other churches. . . . This Beecher can swallow any number of contradictions without making a wry face; for he seems to hold that whatever *seems* to a man to be true is true for him." Henry Ward Beecher is not so forthright. His language "is singularly indefinite"; his statements "an india-rubber band" (460–461). Tracing Beecherism back to the Transcendentalists, Brownson judges it "essentially unintellectual, illogical, and irrational," even more so than the Protestantism of the reformers themselves (463). Hence Beecherism "jumps astride every popular movement, or what appears to it likely to be a popular movement, of the day [including] liberty of divorce, and virtually for polygamy and concubinage or free love, and free religion, while it retains enough of its original Calvinist spirit to require the state to take charge of our private morals, and determine by statute what we may or may not eat, drink, or wear, when we may go to bed or get up. . . . It substitutes change for stability, passion for reason, opinion for faith, desire for hope, philanthropy for charity, fanaticism for piety, humanity for God, and, in the end, demonism for humanity, since man, as he renounces God, inevitably comes under the power of Satan" (479). Thus did the Calvinist spirit degenerate into mere public opinion and seek to legislate morality "under the same pretence" that Calvin did in Geneva and the Puritans in the New England colonies (484). But Beecherism thrived because (unlike

the blundering Unitarianism) it continued to style itself evangelical. Whence we are left to conclude that in American religion honesty is never the best policy.

27. Lawler, "Orestes Brownson and the Truth," pp. 25–28; and Butler, *In Search of the American Spirit*, pp. 193–218, offer pithy and penetrating summations of the pale shadow Brownson cast on American letters. William James is brilliantly described in Louis Menand, *The Metaphysical Club: A Story of Ideas in America* (New York: Farrar, Straus, and Giroux, 2001), pp. 337–375 ("Truth *happens*," p. 353; "if the hypothesis," pp. 355–356).

28. Henry Adams, *The Degradation of the Democratic Dogma* (New York: Peter Smith, 1949, orig. 1919), p. 85.

29. Douglass to Theophilus Gould Steward, Anacostia (July 27, 1886), in William L. Andrews, ed., *The Oxford Frederick Douglass Reader* (New York: Oxford University, 1996), pp. 312–314.

INDEX